W9-AYZ-622

RADAR TECHNOLOGY

RADAR TECHNOLOGY

ELI BROOKNER

CONSULTING SCIENTIST
RAYTHEON COMPANY
WAYLAND MASSACHUSETTS

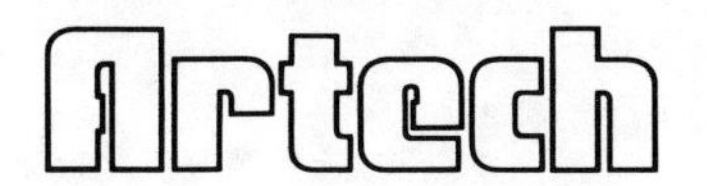

Artech

Jane Carey

Mark Walsh

Maura Zeman

ARTECH HOUSE, INC.
610 Washington Street
Dedham, Massachusetts 02026

Printed and bound in the United States of America

Library of Congress Catalog Card Number: 77-13055
International Standard Book Number: 0-89006-021-5

To Ethel, who for 21 years has inspired, encouraged, slept with, fed, argued, and in all ways lived with me. Hand in hand, belly to belly, and back to back we go through life together (although a little apart) — to this person, I humbly dedicate this book.

Acknowledgments

I am indebted to a large number of persons regarding the preparation of this book. First, I would like to thank William Kirby (of Raytheon) for suggesting that the lecture series be in book form, and for his encouragement and general help at the beginning of this undertaking. Also, I would like to thank RCA, for videotaping the first series under the guidance of K. Palm; Raytheon, for videotaping the second and third series under the guidance of I.M. Holliday; the Boston IEEE, for allowing all three evening radar series to be videotaped (the audio tracks of these videotapes were a help to the authors in preparing their chapters for the book); Ernest E. Witschi of the Boston IEEE Office, for administering the business end of the evening radar lecture series; Raytheon management, for supporting the radar series (especially Dr. Joseph F. Shea, Senior Vice-President); and W.B. Farnsworth, Z.J. Poznanski, and the Raytheon management, for giving financial support. I appreciate the help of Dr. H.H. Behling of RCA in editing Dr. W.C. Curtis' chapter after Dr. Curtis passed away; Eliot Cohen of NRL for his extensive updating of Chapter 19; the expert help of Neal Vitale of Artech House in putting the book together; and P.J. Collins of Artech House for her book design and paste up work.

Others who aided through discussions or comments on the text are David Laighton, C.F. Bacher, D.J. Hoft, Dr. J.D. Harmer, Dr. M. Michelson, G. Works, R.A. Handy, W.C. Brown, M. Bressel, J.F. Skowron, F.E. Daum, Dr. T.W.R. East, J.M. Howell, Jr., M. Kaye, V.G. Hansen, Dr. H. Groginsky, R.H. Cantwell, L. Woo, P.C. Barr, Dr. J.P. Sage, J.D. Collins, J.J. Kovaly, T.E. Parker, W. Feist, A.J. Jagodnik, B.C. Shifrin, R.W. Bierig, John S. George, J.T. Haley, Dr. J. Osepchuk, Dr. P. Miles, Dr. William M. Hall, Dr. M.D. Grossi, A.H. Katz, and D.B. Odom (Raytheon); Dr. R.C.M. Li, Dr. R.C. Williamson, Dr. R.W. Ralston, Ernest Stern, Dr. M.B. Schulz and Dr. G.N. Tsandoulas (Lincoln Laboratory); Dr. Earl E. Swartzlander, Jr., Dr. R. Davidheiser, R.A. Allen, and D. Claxton (TRW); Dr. T.W. Bristol (Hughes); D.T. Bell, Jr., W.R. Shreve, and Dr. L.T. Claiborne (Texas Instruments); Max Yoder (Office of Naval Research); Dr. M. Caulton (RCA); Dr. D.J. LaCombe (General Electric); Dr. M.F. Tompsett, C.H. Séquin, R.H. Walden, and G.H. Robertson (Bell Laboratories); Drs. H.H. Zappe and G.C. Feth (IBM); R.C. Heimiller and Dr. P.L. Jackson (ERIM); Dr. L.J. Porcello (Scientific Associates, Inc.); Dr. H.G. Kosmahl (NASA); H. Jury (Hughes); W.E. Poole and R. Fry (MSC); E. Cohen (NRL); D.J. Chapman (Niagra Mohawk Power Corp.); Dr. W. Weinstock (RCA); J.J. Whelehan, Jr. (AIL); W.S. Jones (Westinghouse); W.E. Brown, Jr. (JPL); and Dr. R.W. Tarmy (Harris ESD).

Joanne Portsch and Esther Stevelman of the Raytheon Equipment Development Laboratory library located and provided copies of the many references used by me in preparing this book.

The following individuals and companies helped provide the many radar photographs and radar parameters in Chapter 1 – J.E. Reed (Bendix); Dr. J.T. Nessmith (RCA); Dr. W.L. Rubin (Sperry Gyroscope); M.A. Johnson (General Electric); F.H. Shorkley (Western Electric); F.C. Williams (Hughes); J.S. Grischy (ITT Gilfillan); B.M. Brown (Selenia); D.C. Betts (Westinghouse); M. Blanchet (Thomson-CSF); C.M.R. Merrill (Marconi); Hollandse Signaalapparaten B.V.; Plessey Radar; and J. Kassabian, H.A. Beltis, W.T. Comisky, H.I. Lipson, V. Karentz, E. DuBois, and K. Canfield (Raytheon).

Sterling Wong (Raytheon) helped provide most of the art for my material. Thanks are due to the secretaries at Raytheon who transcribed the tapes from the lecture series and typed the authors' sections – Nettie George, Natalie Fowle, Paula Anderson, Mildred Sawyer, L.L. DiMaria, Marcia Olsen, Rose Z. Mooradian, B.R. Piper, Marge G. Wright, E.A. Antonucci, Anne Moniz, Ruth Duncanson, and P. MacMurray.

Finally, my greatest indebtedness and appreciation is due my family, my wife Ethel and our sons Larry and Richard. The major part of the book was prepared during the time normally devoted to my family; without their understanding, patience, and tolerance, this book would not have been possible. Larry made many valuable suggestions for improving Chapter 1. Ethel was a constant inspiration and gave continual encouragement during the long and sometimes arduous preparation of the manuscript.

July 1977

Eli Brookner
Lexington, Massachusetts

Acknowledgments

Introduction

Radar Technology is designed to fill the needs of engineers with varying backgrounds – those just entering the field of radar; the engineer working in a special area of radar (such as the transmitter or receiver) who desires to become familiar with other facets of radar; the experienced system engineer who wants to update his or her knowledge of the recent developments in digital logic, digital memory (ROMs, PROMs, EAROMs, EEROMs, EBAMs, bubble memories, and Josephson Junction memories), SAW devices (as pulse compression lines, resonator filters, oscillators, Fourier transform systems, delay lines, bandpass filters, buffers, burst waveform processors, and Doppler filter banks), acousto-electronic devices, CTDs, CCDs, BBDs (these three as pulse compression devices, bandpass filters, buffers, Fourier transform devices, and pulse Doppler filters), microprocessors, microcomputers, dome antennas, solid-state technology, tube technology, and fiber optics technology; the experienced system engineer who wants to obtain more detailed information of other aspects of radar, such as laser technology and theory; the specialist who desires to learn more about the field, such as the signal processing engineer who desires to learn more about the step transform algorithm for linear FM pulse compression; the design engineer who wants a convenient and extensive set of design curves (Chapter 1), or detailed simple detection calculation curves and procedures (Chapters 2 and 3 and the Appendix); and the engineer or manager who desires the parameters of the many radars in use today, an extensive table of which is provided in Chapter 1 along with photos for most of these radars with captions that contain additional information. The material is presented in a lucid form so that it can be followed easily.

The book is divided into six parts. Part 1 covers Radar System Fundamentals; Part 2, Signal Waveform Design and Processing; Part 3, Propagation Effects; Part 4, Synthetic Aperture Radar Techniques; Part 5, Radar Systems and Components; Part 6, Special Topics, such as Laser Radars, Tracking and Smoothing, the Kalman Filter, Pulsed Radar Transmitter Spectrum Control, and Fiber-Optics. The contents of the various parts and chapters of the book are described in detail in the following paragraphs.

Part 1 – Radar System Fundamentals

Introduction (Brookner)

Brief history of the development of radar given.

Chapter 1 – Fundamentals of Radar Design (Brookner)

Fundamental Radar System Tradeoffs – Frequency Selection; Type of Scanning (Mechanical, Electronic, or Hybrid; 2-D or 3-D Scanner; V-Beam or Stacked Beam); Waveform; Polarization; Signal Processing (MTI or pulse Doppler); and Transmitter (tube or solid-state). Included is a table summarizing parameters of 96 radars – 2-dimensional surveillance and search radars, 3-dimensional radars (stacked and V-beam), heightfinders, electromechanical steering systems, all electronic phase-phase steering radars, frequency scanning radars, hybrid scanners, and tracking systems; American, Russian, French, Dutch, and English. Illustrations are presented for nearly all these radars. Also given are state-of-the-art of low noise receivers and power TRAPATTs, IMPATTs, and FETs; and extensive and convenient environment design curves (sea and land clutter, rain clutter, atmospheric attenuation, and sky temperature).

Chapter 2 – Detection and Measurement (Barton)

The most important factors for detection and measurement are covered in a convenient, clear, and concise summary. Topics included for detection are ideal detector, video integration loss, target fluctuation loss (all Swerling models), collapsing loss, diversity gain, equations for determining number of diversity samples, beam shape loss, and false alarm relationships. Covered for target estimation are ideal, monopulse, search, conical scan, and offset axis estimations; angle and range noise errors; glint, target scintillation, and multipath errors.

Chapter 3 – Radar Cumulative Probability of Detection (Brookner)

The mistake of overestimating radar power by 14 dB and standoff jammer power by 8 dB by using single-pulse detection criteria when cumulative probability of detection criteria should be used is pointed out, with the differences between these two types of detection elaborated upon in clear physical terms. Simple curves and equations (including the cumulative-probability-of-detection range equation) are given for easily evaluating the cumulative probability of detection. Also pointed out and explained is the possibility of underestimating the required single-pulse detection signal-to-noise ratio by 4 dB or more if synchronized conditions are assumed when, in fact, unsynchronized conditions should be used.

Chapter 4 – Philosophy of Radar Design (Barton)

Barton documents for the first time real-world aspects of radar design – a must for every radar system designer. The fundamental track and search radar equations for noise and mainlobe- or sidelobe-jamming-limited conditions are given. The basic clutter-limited radar equations for discrete-point, extended-surface, and extended-volume clutter are given.

Chapter 5 – Some Radar Design Problems and Considerations (Shrader)

Pitfalls that the radar supplier must avoid are presented (such as designing by committee, extrapolating from failure, and not knowing what the customer wants). Real cases, such as the negative dB noise figure spec, are given as examples.

Chapter 6 – Pitfalls of Radar System Development (Fowler)

Pitfalls which the customer must avoid are presented. Fowler's six laws affecting DOD system acquisition are presented, together with real, illustrative examples. For example, Fowler's fifth law – any war game, system analysis, or study which can't be explained on the back of an envelope is not just worthless, it's probably dangerous (the computer being the enemy of reason).

Part 2 – Signal Waveform Design and Processing

Chapter 7 – Waveform Selection and Processing (Sinsky)

Given for the first time is a simple description of the classic ambiguity function, knowledge of which is essential to the design of large time-bandwidth waveforms. Instead of couching the description in mathematical terms, its physical meaning is explained.

Chapter 8 – Large Time-Bandwidth Radar Signals (Cook)

Discussed are the advantages and disadvantages of large time-bandwidth waveforms. Several sample large time-bandwidth product waveforms are presented. Introduced is the matched filter concept.

Chapter 9 — Fast Fourier Transform (Sheats)

Sheats gives a very lucid description of the Fast Fourier Transform algorithm. For the first time, a simple and clear explanation of the clever pipeline hardware implementation of the FFT is given.

Chapter 10 — Signal Processing Linear Frequency Modulated Signals (Purdy)

Several analog and digital state-of-the-art techniques for processing the most used large time-bandwidth waveform — the linear chirp waveform — are presented.

Chapter 11 — Radar Digital Processing (Perry/Martinson)

A novel and efficient technique for processing large time-bandwidth product linear chirp waveforms is described. Its application to strip map synthetic aperture processing is presented.

Chapter 12 — Surface Acoustic Wave Devices for Radar Systems (Worley)

The state-of-the-art of the relatively new Surface Acoustic Wave (SAW) devices is reviewed.

Chapter 13 — A 1 GHz, 2,000,000:1 Pulse Compression Radar (Haggarty/Meehan/O'Leary)

A 2,000,000:1 linear chirp waveform generator and processor is described. Included are techniques for correcting target velocity time compression.

Present and Future Trends in Signal Waveform Design and Processing Technology and Techniques (Brookner)

Covered are the latest results on new Winograd Fourier Transform Algorithm (WFTA), a super fast Fourier transform algorithm; memory technology, such as RAM, ROM, PROM, EAROM, EEROM, EPROM (all of which are defined), Charge-Coupled Devices (CCDs), Electron-Beam-Addressed Memories (EBAMs), and Josephson Junction memories; logic devices, such as CCD butterflies on a chip and GHz GaAs MESFET and TED technology; high speed A/D's; microprocessors (μP) and microcomputers (μC) (a clear description of these being given with their differences defined); new SAW device applications, such as continuously variable delay lines, MTI processors (coherent and incoherent), delay line crystal oscillators, bandpass filters, variable linear chirp waveform generators and processors, Fourier transform analyzers, Circulating Memory Filters (CMFs), high-Q resonator oscillators and filters, and burst waveform processors; acoustoelectronic devices, such as convolvers, memory correlators, integrating convolvers, burst waveform processors (these devices have potential of processing 100 ms, 100 MHz signals with arbitrary shapes), and CCD/memory-correlator high speed buffers; and analog Charge-Transfer Devices (CTDs), CCDs, and Bucket Brigade Devices (BBDs) are explained, and their technologies and applications (such as for matched filter processors, bandpass filters, and pulse Doppler processors) are discussed.

Part 3 — Propagation Effects

Chapter 14 — Pulse-Distortion and Faraday-Rotation Ionospheric Limitations (Brookner)

Definitive treatment of the effects of the ionosphere on pulse distortion, including easy-to-use curves and equations. A simple equation for determining Faraday rotation is given.

Chapter 15 — Effects of the Atmosphere on Laser Radars (Brookner)

Simple to understand exposition of the complex subject of the effects of atmospheric turbulence on propagation of optical radar signals. Many of the results presented apply to microwave signal propagation. Comparisons are made of propagation through fog and rain at optical and microwave frequencies and pulse distortion through clear and inclement weather is covered.

Part 4 — Synthetic Aperture Radar Techniques

Chapter 16 — Synthetic Aperture Fundamentals (Curtis)
Chapter 17 — High Resolution Radar Fundamentals (Synthetic Aperture & Pulse Compression) (Kovaly)
Chapter 18 — Synthetic Aperture Radar Spotlight Mapper (Brookner)

To better understand a subject, it should be viewed from different aspects. This treatment is given synthetic aperture radars in Chapters 16, 17, and 18, by Curtis, Kovaly, and Brookner, respectively. Chapter 16 introduces strip synthetic aperture mapping from the point of view of Doppler processing as well as long synthetic apertures; Chapter 17 relates it to the improved resolution obtained in range with a linear FM waveform through pulse compression; and Chapter 18 considers the subject of spotlight mapping from the viewpoint of Doppler processing and relates the results to those of Chapters 16 and 17 for focused and unfocused strip mappers, and to standard antenna principles. Chapter 18 covers Doppler-beam-sharpening and partially-focused processing, simple real-time digital processors, system errors, and T-space.

Present and Future Trends in Synthetic Aperture Radar Systems and Techniques (Brookner)

Recent developments in Synthetic Aperture Radar (SAR) systems and techniques are presented. Covered are a synthetic interferometer radar for topographic mapping; a harmonic synthetic aperture radar for the detection of metal objects in foliage (the METRRA system); the Apollo Lunar Sounder Radar System; the hologram matrix radar (which gives ice depth with a CW waveform); ERIM simultaneous X- and L-Band dual-polarized SAR; the SEASAT satellite SAR system; and the Venus SAR mapping system.

Part 5 — Radar Systems and Components

Introduction (Brookner)

A brief history of the development of microwave tubes (magnetrons, amplitrons, CFAs, TWTs, and klystrons) is given. Covered are the instances when one is to be used in preference to another.

Chapter 19 — L-Band Solid-State Technology (Dodson)

Some of the latest developments in L-band solid-state modules and radars are presented.

Chapter 20 — Airborne Solid-State Radar Technology (Harwell)

Described are the MERA and RASSR X-band solid-state phased-array airborne radar systems, together with the latest developments in this field.

Chapter 21 — Antenna Arrays (Knittel)

Introduced are the basics of array fundamentals, grating lobes, mutual coupling, types of radiating elements, blindness phenomena and their relationship to end-fire grating lobes, element impedance matching, wide angle impedance matching, radiating element design procedures, and waveguide simulators. Numerous example array antennas are given.

Chapter 22 — Linear-Beam Tubes (Staprans)

Covered are the state-of-the-art of cavity-type high-power linear-beam tubes (such as klystrons, TWTs, and twystrons), advantages and disadvantages of each of these types, and their AM and PM modulations and noise characteristics.

Chapter 23 — Cross Field Microwave Tubes (Smith)

Noise characteristics and techniques of tuning (frequency and mechanical), priming, and locking are covered.

Present and Future Trends in Radar Systems and Components (Brookner)

Latest developments and future trends in UHF and L-, S-, and X-band solid-state devices, module technology, and reliability are summarized. Also covered are distributed- and lumped-element and monolithic technology.

Technologies are described and performances are reviewed for new dome antennas, continuous-aperture scanning via series ferrite scanning technique, and adaptive arrays (fully-adaptive, partially-adaptive, and sidelobe cancellor systems).

Covered are newest technology in very highly efficient TWTs, CFAs, and klystrons; long life tubes and cathodes (M-type cathodes and Thin-Film Field Emission Cathodes [TFFECs]); high gain (30 dB) cathode-driven CFAs; electronically-tuned pulse magnetrons (using multipactor-cavity electrodes); extended interaction oscillators (EIOs); extended interaction amplifiers (EIAs); millimeter wave gyrotrons (cyclotron resonance masers); and relativistic-electron beam tubes.

Part 6 – Special Topics

Chapter 24 – Laser Radars (Jelalian)

Excellent coverage of fundamentals of laser theory and technology for radar applications. Covered are laser equation (with its similarities and differences to microwave equation), laser noise terms, laser radar detection and estimation, laser transmitters and detectors, and several laser sample systems.

Chapter 25 – Tracking and Smoothing (Morrison)

Complicated subject of tracking and smoothing is presented in simple terms, along with an interesting insight into the development of this field starting with the work of Gauss. Introduced are the process equation, observation relation, transition matrix and equation, concept of state variables, polynomial smoothing, least square smoothing, minimum variance smoothing, and Kalman filtering.

Chapter 26 – The Kalman Filter (Sheats)

The Kalman Filter is related to the classical Weiner filter.

Chapter 27 – Pulse Radar Transmitter Spectrum Control (Weil)

Discussed is the important subject of spectrum control, a topic that must be dealt with in all radars built.

Part 6 – Present and Future Trends (Brookner)

The use of lasers for extremely high precision distance measurements (1 part in 10^7) is presented, and the state-of-the-art of fiber-optic communication technology is summarized. Some recent results on pulsed radar out-of-band radiation control are given.

Appendix – How to Look Like a Genius in Detection Without Really Trying (Brookner)

Simple cookbook procedures for determining the detection performance of a radar system are presented. The procedures are given in the form of simple flow diagrams and easy-to-use worksheets which are illustrated extensively with example calculations. The procedures apply to a wide variety of target models – Marcum nonfluctuating, all Swerling, Wienstock, Rice, log-normal, and chi-square. Included are the beam shape factor loss (for one- and two-coordinate scanning), CFAR loss, amplitude quantization loss, RF/IF filter mismatch loss, collapsing loss, target fluctuation model loss, and video integration loss.

Radar Technology is an outgrowth of four radar lecture series. Three were evening Boston IEEE Radar Lecture Series that the editor organized and chaired – "Modern Radar Theory (Fall/Winter 1972), "Modern Radar Technology" (Fall/Winter 1973), and "Modern Radar Techniques, Components and Systems" (Fall/Winter 1976). The evening series were very well received – nearly 1,000 registered totally for the three. Moreover, the video tapes of the first series have been shown at:

1. Syracuse, New York, by the IEEE chapter there (150 attended).
2. San Diego, by the local IEEE chapter.
3. Los Angeles, by the University of Southern California on their industrial TV circuit. They made the course available to the Aerospace Corporation, ITT, Hughes Aircraft Company, and North American Rockwell, among others (250 registered for the viewing).
4. Kwajalein in Marshall Islands (by RCA and Raytheon at two different times, with Lincoln Laboratory personnel attending).
5. Eastern Test Range in Florida, by RCA.
6. Morristown, New Jersey, by RCA.
7. Raytheon facilities in Wayland, Sudbury, and Waltham, Massachusetts.
8. Georgia Institute of Technology (they have made copies of the New York IEEE tapes and show periodically to new personnel).
9. United Aircraft, Nordon Division in Connecticut, by the local IEEE chapter there.
10. Motorola in Scotsdale, Arizona.
11. ITT.
12. Teledyne.
13. Massachusetts Institute of Technology.

Requests for the tape were received from overseas (India and Italy). The fourth lecture series was held as part of the Northeast Electronics Research and Engineering Meeting (NEREM) held in October 1974. The series was entitled "Radar Systems and Components," and the editor organized and chaired one of the four sessions. The attendees to the series included engineers of all backgrounds – students, starting engineers, specialists, very experienced engineers, mechanical engineers, and reliability engineers.

Based on the initial and continuing interest expressed in these four series and in the video tapes (for the evening series, the first two are in the possession of the IEEE, and the third is at Raytheon), it was felt that a book covering the material of the various lecture series would be of value to engineers with similarly varied backgrounds throughout the world. *Radar Technology* has been prepared with this goal in mind.

The information from the 1972 and 1973 lecture series has been revised extensively and updated to include the most recent technology developments. Material not covered in the lecture series has been added as has a Present and Future Trends section at the end of each of the book's six parts. These new portions were included by the editor to cover important systems, technologies, and techniques not covered, and to give the state-of-the-art as of 1977 and future trends. Also, historical information has been included in some of the introductory sections to the six parts.

Author List

D.K. Barton
Raytheon Company
Bedford, Massachusetts

E. Brookner
Raytheon Company
Wayland, Massachusetts

C.E. Cook
MITRE Corporation
Bedford, Massachusetts

W.C. Curtis *(deceased)*
RCA
Burlington, Massachusetts

B.C. Dodson, Jr.
Naval Research Laboratory
Washington, DC

C.A. Fowler
MITRE Corporation
Bedford, Massachusetts
(Formerly with Raytheon Company)

R.D. Haggarty
MITRE Corporation
Bedford, Massachusetts

T.E. Harwell
Texas Instruments Inc.
Dallas, Texas

A.V. Jelalian
Raytheon Company
Sudbury, Massachusetts

G.H. Knittel
Lincoln Laboratory
Lexington, Massachusetts

J.J. Kovaly
Raytheon Company
Bedford, Massachusetts

L.W. Martinson
RCA
Moorestown, New Jersey

S.J. Meehan
MITRE Corporation
Bedford, Massachusetts

N. Morrison
Knight-Ridder Newspapers
Miami, Florida
(Formerly with Bell Laboratories)

G.C. O'Leary
Lincoln Laboratory
Lexington, Massachusetts
(Formerly with MITRE Corporation)

R.P. Perry
RCA
Moorestown, New Jersey

R.J. Purdy
Lincoln Laboratory
Lexington, Massachusetts

L. Sheats
Raytheon Company
Wayland, Massachusetts

W.W. Shrader
Raytheon Company
Wayland, Massachusetts

A.I. Sinsky
Bendix Communications Division
Baltimore, Maryland

W.A. Smith
Raytheon Company
Waltham, Massachusetts

A. Staprans
Varian Associates
Palo Alto, California

T.A. Weil
Raytheon Company
Wayland, Massachusetts

J.C. Worley
Motorola Inc.
Fort Lauderdale, Florida
(Formerly with Sperry Rand Research Center)

Table of Contents

Part 1
Radar System Fundamentals

Some of the milestones in the development of modern radar are given below:

1886-1888 Heinrich Hertz demonstrates the generation, reception, and scattering of electromagnetic waves.

1903-1904 Christian Hulsmeyer develops and patents a primitive form of collision avoidance radar for ships.

1922 M.G. Marconi suggests (in his acceptance speech for the IRE Medal of Honor) an angle-only radar for ship collision avoidance.

1925 First short pulse echoes from the ionosphere are observed on cathode ray tube by G. Breit and M. Tuve of Johns Hopkins University.

1934 First photo of short pulse echo from aircraft made by R.M. Page of the Naval Research Lab.

1935 First demonstration of short pulse range measurements of aircraft targets, by British and Germans.

1937 First operational radar built – the Chain Home in Britain, designed by Sir Robert Watson-Watt.

1938 Signal Corps SCR-268 becomes first operational anti-aircraft fire control radar; 3100 sets eventually produced. Range, $>$ 100 nmi; frequency, 200 MHz.

1938 First operational shipboard radar, the XAF, aboard the battleship *USS New York,* has a range of 12 nmi for surface ships and 85 nmi for aircraft.

1941 (Dec.) By this date, 100 SCR-270/271 Signal Corps early warning radars have been produced. One of these radars, located in Honolulu, detects Japanese invasion of Pearl Harbor. The returns, however, are misinterpreted as friendly aircraft.

During the 1930s, parallel efforts on the development of radar were being carried out in France, USSR, and Japan. During the war years, the efforts in the United States, with aid from the British, greatly outstripped those of Germany and Japan and played a major role in the allied successes.

The progress made in radar since these early days is illustrated by the sample radars presented in Chapter 1. The parameters for 96 radars are given in Table 1 of that chapter.

Chapter 1 discusses the fundamental radar system trade-offs (frequency selection, type of scanner, polarization, waveform selection, type of signal processing, and type of transmitter). Also given is the state-of-the-art of low noise receivers and of IMPATT, GaAs FET, and TRAPATT solid state power devices.

In Chapter 2, Barton summarizes the many important factors for determining a radar's detection and measurement-accuracy capabilitites that have emerged over the years. Chapter 3 points out that, if a single-pulse detection criterion is used when a cumulative-probability-of-detection criterion should be used, the mistake can be made of overestimating the radar transmitter power by 14 dB and the standoff jammer by 8 dB. Chapter 4 by Barton discusses the real world aspects of radar design never before documented. Finally, Chapters 5 and 6 present some of the many other pitfalls in radar design that must be avoided – Shrader (Chapter 5) from the supplier's point of view and Fowler (Chapter 6) from the customer's.

Chapter 1 Brookner

Fundamentals of Radar Design

The purpose of this introductory chapter is to provide and explain some of the thinking involved in the design of a radar.

First and foremost consideration must be given to the overall requirements — the problem to be solved must be well defined — then the following factors can be considered:

1) *Frequency selection*
2) *Mechanical vs. all-electronic vs. hybrid scanning system*
3) *Choice of polarization*
4) *Radar waveforms to be used*
5) *Type of processing to be used — incoherent, MTI, or pulse Doppler*
6) *Tube vs. solid-state transmitter*

The trade-offs relating to the above factors are discussed in the following sections. It is hoped that with the brief introduction that follows the reader is able to delve more deeply into the subject by reading the ensuing chapters and more advanced texts and literature on the subject.

Selection of Radar Frequency

A primary factor influencing the selection of radar frequency is the radar's function — is it required to search only, track only, or to do both? If search only, a low frequency usually is preferred, such as UHF. This preference results from the sensitivity of a radar being dependent on its power-aperture product, as indicated by the following radar equation

$$R^4 = \frac{\sigma T_F}{4 \pi \Omega \, SNR \, k \, T_S \, L_T} P_{AV} A_E \qquad (1)$$

where:

A_E = effective receiver aperture area (square meters)

k = Boltzmann's constant = 1.38 x 10^{-23} joules/Kelvin

L_T = total system losses, power ratio $\geqslant 1$

P_{AV} = average transmitter power (watts)

R = radar slant range (meters)

SNR = signal-to-noise ratio per pulse, power ratio

T_F = radar frame time (seconds)

T_S = system temperature (Kelvin)

Ω = angular region to be searched out (steradians)

σ = target cross section (square meters)

The above search form of the radar equation does not include frequency. Hence, the maximum search range is explicitly independent of radar frequency. However, large aperture areas are more feasible at low frequencies. Thus, indirectly, the search range is dependent on radar frequency. The BMEWS radar is an example of a low frequency (UHF) search radar (Figure 1). (Note — the parameters for most of the radars illustrated in this chapter, plus some others, are given in Table 1.)

If the radar is to be used for tracking, a higher frequency is preferred. The main concern in tracking is simultaneously achieving enough signal-to-noise ratio per pulse for detection and the de-

Figure 1 **Ballistic Missile Early Warning System (BMEWS) parabolic — torus antennas of UHF AN/FPS-50 search radars used to detect ICBMs as they rise over the horizon at Clear, Alaska; see Table 1a. BMEWs AN/FPS-49 track radar in dome; see Figure 10. (Brookner)**

sired angular accuracy. The form of the radar equation which gives the signal-to-noise ratio per pulse is

$$SNR = \frac{\sigma}{4 \pi R^4 k \, T_S \, B} \; \frac{P_{PK} A_E^2}{\lambda^2} \qquad (2)$$

where:

B = signal bandwidth (hertz)

P_{PK} = peak transmitted power (watts)

λ = radar wavelength (meters)

Equation (2) indicates that, for a given aperture size, the signal-to-noise ratio per pulse is inversely dependent on λ^2. Hence, a smaller wavelength (or, equivalently, higher frequency) provides a larger signal-to-noise ratio per pulse. More importantly, however, the angle accuracy depends on λ^4. This dependence follows from the basic angular accuracy equation given by (see Chapter 2)

$$\sigma_\theta^2 = \frac{\theta_3^2}{2 k_M^2 \, SNR \, N} \qquad (3a)$$

where:

k_M = antenna angle accuracy factor, equal to 2.1 for a tapered-Taylor weighted circular aperture having 40 dB sidelobes

θ_3 = the antenna 3 dB beamwidth (in radians) = $k_0 \lambda / D$ (3b)

D = the antenna aperture diameter (meters)

k_0 = weighting parameter relating λ/D to θ_3, equal to 1.310 for 40 dB tapered-Taylor weighted circular aperture (with $\bar{n} = 4$ [1])

(text continues on page 31)

[a] Instrumented range.

[b] Selenia G-14 dual beam antenna; used on Argos-10, ASR-803, AASR-804, ARSR-805, ATCR-2T and ATCR-4T.

[c] Estimated by the author using other available parameters and information.

[d] Marconi S2021 L-band transmitter; 1305-1365 MHz to order; 2 μs std, 1.5-5 μs on order.

[e] Marconi S2011 L-band transmitter; 1305-1365 MHz to order; 2.5-5 μs on order.

[f] Marconi S2010 S-band transmitter

[g] Marconi S2012 S-band transmitter

[h] Marconi S7100 digital double-canceller, quadrature-phase-detection, MTI signal processor; CR = 40 dB; max. rg. for MTI ≥ 120 nmi; up to 6 prfs; VB 100 times that of mean prf when 6 prfs used; random prf selection possible for pulse interference or inpulse spoof jamming rejection.

[i] Array power capability.

[j] See glossary at end of chapter for definitions

[k] U.S. AN/XYZ designation, where

X = A = airborne
= C = air transportable (inactivated)
= F = fixed, static
= G = ground
= M = mobile, installed on vehicle dedicated to transporting radar
= S = shipborne
= T = ground, transportable
= U = general utility (two or more platforms: airborne, shipboard, and ground)
Y = P = radar
Z = G = fire control or search light directing
= N = navigational aids (landing, altimeters, beacons, etc.)
= Q = special, or combination of purposes
= S = detecting and/or range and angle measurement system
= Y = surveillance (search, detect, multiple target tracking) and control (both fire control and air control)

[l] At feed horn in search mode (does not include space fed lens losses); 11 dB for track

[m] As well as coherent MTI

[n] ≥ 40 Kft for 45° ≥ El ≥ 0.7° ($10 m^2$ fluctuating target; 80% probability of detection)

Footnotes for *Table 1*

Radar	AASR-804 (48)	AN/FPS-8 (-88, PA version) [j]	AN/FPS-14	AN/FPS-18
Type	2-D airport surveillance radar	2-D surveillance; used at commercial airports and military bases.	2-D SAGE air defense; DEW line gap filler.	2-D SAGE air defense; DEW line gap filler.
Manufacturer	Raytheon Canada	GE	Bendix	Bendix
Location	Ground, static	Ground, static (MPS-11, mobile version)	Ground, static	Ground, static
Frequency Band	L	L	S	S
Frequency (MHz)	1250 - 1350	1280 - 1380	2700 - 2900	2700 - 2900
Power, Peak (MW)	0.5	1	0.45	1
, Av (kW)		1.1	0.54	1.2
Antenna Size (m) (or Type)	12.8 x 6.7 [b]			
Antenna Gain (dB)	36 (LB); 34.5 (UB)			
Beam Shape	CSC^2	CSC^2		
Type Scanner	Mechanical	Mechanical		
No. of Beams	1	1		
Beamwidth, El (deg)	4 (CSC^2 to >40°)	30		
, Az (deg)	1.25	2.5		
Angular Coverage, El	≥ 70 Kft for 17° ≥ EL ≥ 1.5° (10 m^2) [n]		0 to 20 Kft (2.5 m^2)	0 to 20 Kft (2.5 m^2)
Az (deg)	360	360		
Scan Rate, El		-		
Az	6 rpm	0 - 10 rpm		
PRF	360 pps (av)	360 pps	1200	1200
PRF Stagger	Quadruple			
Pulse Width, Transmit (μs)	2.0	3	1	1
, Compressed (μs)	NA		NA	NA
Pulse Compression Ratio	NA		NA	NA
Type Tube			Fixed Tuned Magnetron	Klystron
Noise Figure (dB)	4	9		
Range	150 to 200 nmi [a] 150 nmi (10 m^2, 80%, 0° EL)		48 nmi	48 nmi
MTI	I.F. = 33-36 dB [j]		23 dB SCV [j]	30 dB SCV
Polarization			Vert/Circ.	Vert/Circ.
Frequency Agility			No	No
ECCM Features				
Misc. Comments	VB = 1200 kts; freq. diplexed; DMTI			
Fig. No.	(See 47)	37	34	34

Table 1a 2-D Surveillance and Search Radars

Radar	AN/FPS-20	AN/FPS-24	AN/FPS-50	AN/FPS-66
Type	2-D long-range air surveillance	2-D surveillance	BMEWS search radar	ATC 2-D surveillance radar
Manufacturer	Bendix	GE	GE	Bendix
Location	Ground, static	Ground, static	Ground, static (Thule, Greenland; Clear, Alaska)	Ground, static
Frequency Band	L	VHF	UHF	L
Frequency (MHz)	1250-1350	214-236	425	1250-1350
Power, Peak (MW)	2.5	5 per channel	5	2
, Av (kW)	5	25	300	4.32
Antenna Size (m) (or Type)			122 x 50	
Antenna Gain (dB)				35
Beam Shape		CSC^2	Pencil	
Type Scanner			Mechanical (organ-pipe)	Mechanical
No. of Beams			2	1
Beamwidth, El (deg)		30	1.0 [c]	
, Az (deg)		2.9	0.4 [c]	1.3
Angular Coverage, El	0 to 100 Kft ($2.5 m^2$)		-	0.2° - 45°
Az(deg)		-	-	360
Scan Rate, El		-	-	-
Az	3.3, 5, 6.6, or 10 rpm	0.25 rpm CW or CCW; 5 rpm CW		5 rpm
PRF	360	278	27	360 pps (av)
PRF Stagger				
Pulse Width, Transmit (μs)	6	6,18	2000	6
, Compressed (μs)	-	Used with 18 μs pulse	NA	
Pulse Compression Ratio	NA		NA	
Type Tube	Klystron			
Noise Figure (dB)	9	4.5		8
Range	220 nmi			200 nmi
MTI	30 dB SCV			SCV = 25 dB
Polarization	Horiz.			H, V, C [j]
Frequency Agility				
ECCM Features		Highly sophisticated		
Misc. Comments	Dual channel system			VB=80 knots; Pol. diplexing
Fig. No.	35	41	1	

Table 1a 2-D Surveillance and Search Radars

Radar	AN/FPS-88 (PA version of −8)	AN/FPS-107	AN/MPS-11	AN/SPS-10
Type	2-D surveillance	2-D surveillance radar	2-D surveillance; mobile version of AN/FPS-8	2-D surface search and navigation radar(30)
Manufacturer	GE	Westinghouse	GE	Sylvania
Location	Ground, static	Ground, static	Ground, mobile	Shipborne
Frequency Band	L	L	L	C(30)
Frequency (MHz)	1280 - 1380	1250 - 1350	1280 - 1380	
Power, Peak (MW)	1	10	1	0.3(30)
, Av (kW)	1.1	15	1.1	0.05(30)
Antenna Size (m) (or Type)				3m wide
Antenna Gain (dB)				
Beam Shape	CSC^2		CSC^2	
Type Scanner	Mechanical	Mechanical	Mechanical	Mechanical
No. of Beams	1	1	1	1
Beamwidth, El (deg)	58		30	
, Az (deg)	1.3		2.5	1.5
Angular Coverage, El		100 Kft, 40°		
Az (deg)	360	360	360	
Scan Rate, El	-		-	
Az	5, 10 rpm		0-10 rpm	
PRF		244 pps		
PRF Stagger		None		
Pulse Width, Transmit (μs)	3	6	3	
, Compressed (μs)		None		
Pulse Compression Ratio		None		
Type Tube				
Noise Figure (dB)	≤2.5 (PA)		9	
Range		260 nmi (1 m^2)		
MTI		3 PC		
Polarization	Circular (13)	V, C		
Frequency Agility		None		
ECCM Features	Yes			
Misc. Comments	Dual channel operation(13)			
Fig. No.	(see 37)	45	(see 37)	39

Table 1a 2-D Surveillance and Search Radars

Radar	AN/SPS-29, -37, -43	AN/SPS-49	AN/SPS-58A	AN/TPN-24
Type	2-D early warning surveillance	Long range 2-D air search radar	2-D surveillance radar	2-D airport surveillance radar (ASR) of AN/TPN-19 system
Manufacturer	Westinghouse	Raytheon	Westinghouse	Raytheon
Location	Shipborne	Shipborne	Shipborne	Ground, transportable
Frequency Band	P	L	L	S
Frequency (MHz)				2,700 to 2,900
Power, Peak (MW)			10	0.45
, Av (kW)				0.47
Antenna Size (m) (or Type)	(Bedspring)	7.3 by 2.4 (32)		4.3 wide, 2.4 high
Antenna Gain (dB)				33.6
Beam Shape	Fan beam			Cosecant squared in elevation, beamswitch varies low elev. pattern
Type Scanner	Mechanical			Mechanical
No. of Beams	1			1
Beamwdith, El (deg)				-
Az (deg)				1.73
Angular Coverage, El				To 30° (to 53 Kft on $2m^2$)
, Az (deg)				
Scan Rate, El		-		-
Az		6 rpm (32)		15 rpm
PRF				1050 (average)
PRF Stagger				16:20:17:22
Pulse Width, Transmit (μs)				1
, Compressed (μs)				NA
Pulse Compression Ratio				NA
Type Tube				Magnetron
Noise Figure (dB)				2.5 (PA)
Range				60 nmi (for $1m^2$, 30 Kft)
MTI			Recursive Elliptic 4 pole filter	Coherent and non-coherent
Polarization	Horizontal	H (32)	V	Vert. or Cir.
Frequency Agility				Dual freq. div.
ECCM Features		Yes		Dickie fix; freq. dup.; FTC
Misc. Comments		Integral IFF feed (32)		Rain rej. (ICR) = 20 dB [j] Sub clutter vis. = 25 dB
Fig. No.	(see 38)	40	46	12

Table 1a 2-D Surveillance and Search Radars

Radar	AN/TPS-35	AN/UPS-1, -1A, -1B -1C	Argos-10	ARSR-1, 2[33]
Type	2-D air surveillance radar for AF AN/TSQ-47	2-D tactical air surveillance/ air defense or ATC; surface search	2-D early warning search radar	Enroute 2-D air surveillance radar
Manufacturer	RCA	RCA	Selenia (GE Xtrm)	Raytheon
Location	Ground, transportable	Ground, mobile	Ground, static	Ground, static
Frequency Band		L	L	L
Frequency (MHz)		1250 - 1350		1280 - 1350
Power, Peak (MW) , Av (kW)		1 1	> 1 [13]	4 2.9
Antenna Size (m) (or Type)			12.8 x 6.7 [b]	14.3 x 6.7
Antenna Gain (dB)		27.5 | 28.5	36 (LB); 34.4 (UB) [c]	34
Beam Shape		$CSC^2\ \theta$ (-1A,-1C) | fan beam (-1, -1B)	CSC^2	Fan
Type Scanner	Modified	Mechanical	Mechanical	Mechanical
No. of Beams		1	2 in elevation	1
Beamwidth, El (deg) , Az (deg)	AN/UPS-1	11 | 10 3.7 | 3.5	4 (CSC^2 to > 40°) 1.25 [c]	 1.2
Angular Coverage, El	(See AN/UPS-1)	CSC^2 to 40° | 10°	to >40° [c]	0.2° - 45°
Az (deg)		360	360	360
Scan Rate, El Az		- 0-15 rpm		- 6 rpm
PRF				360 pps (av)
PRF Stagger			Multiple	Yes
Pulse Width, Transmit (μs)		(2500 ft resol.)		2
, Compressed (μs)			Yes	-
Pulse Compression Ratio				NA
Type Tube		QK-358 magnetron		
Noise Figure (dB)		3 dB (PA)		4
Range		> 160 nmi ($1m^2$)	250 nmi	200 nmi
MTI		SCV = 25 dB	3-pulse digital	SCV = 27 dB
Polarization				H, C
Frequency Agility			Adaptive or Random	
ECCM Features		Classified	May incorporate SB	
Misc. Comments		Used in USMC AN/TSQ-18 and AF AN/TSQ-47	Clutter map, MCC, RCM	VB = 1150 knots; no diplexing
Fig. No.	36	36	48	42

Table 1a **2-D Surveillance and Search Radars**

Radar	ARSR-3	ARSR-805[48]	ASR-6[33]	ASR-7[33]
Type	Enroute 2-D air surveillance radar	Enroute 2-D air surveillance radar	ATC 2-D airport surveillance radar	ATC 2-D airport surveillance radar
Manufacturer	Westinghouse	Raytheon Canada	Texas Inst.	Texas Inst.
Location	Ground, static	Ground, static	Ground, static	Ground, static
Frequency Band	L	L	S	S
Frequency (MHz)	1250 - 1350	1250 - 1350	2700 - 2900	2700 - 2900
Power, Peak (MW)	5	5	0.4	0.425
, Av (kW)	3.3		0.4	0.425
Antenna Size (m) (or Type)		12.8 x 6.7 [b]		
Antenna Gain (dB)	34.5 (LB), 33.5 (UB)[33]	36 (LB), 34.5 (UB)	34	34
Beam Shape	Fan	CSC^2	Fan	Fan
Type Scanner	Mechanical	Mechanical	Mechanical	Mechanical
No. of Beams	2 beam in elevation	1	1	1
Beamwidth, El (deg)		4° (CSC^2 to >40°)		
, Az (deg)	1.2	1.25	1.4	1.4
Angular Coverage, El	60 Kft, to 40°	(Better than for AASR-804)	0.2° - 30°	0.2° - 30°
, Az (deg)	360	360	360	360
Scan Rate, El	-	-	-	-
Az	5 rpm	6 rpm	15 rpm	13 rpm
PRF	310 - 364 pps (av)	360 pps (av)	700 to 1200	713, 950, 1050, 1120, 1173, 1200 pps
PFR Stagger	8 prf's	Quadruple		
Pulse Width, Transmit (μs)	2	2.0	0.833	0.833
, Compressed (μs)	NA	NA		
Pulse Compression Ratio	NA	NA		
Type Tube		Amplitron		
Noise Figure (dB)		4	4	4.75
Range	>200 nmi ($2m^2$)	200 nmi [a]	60 nmi	60 nmi
MTI	3 PC [j]	I.F. = 33 to 36 dB	SCV = 25 dB	SCV = 25 dB
Polarization	L, C		V, C	V, C
Frequency Agility	39 freq			
ECCM Features				
Misc. Comments	$\dot{V}B$ = 1200 knots; Pol. diplexing	VB = 1200 kts, freq diplexed; DMTI	VB = 1250 knots; no diplexing	VB = 2000 knots; no diplexing
Fig. No.	43	(See 47)		

Table 1a 2-D Surveillance and Search Radars

Radar	ASR-8(33)	ASR-803(48)	ASR-808(48)	ATCR-2T (with G-14 antenna)
Type	ATC 2-D airport surveillance radar	2-D airport surveillance radar	2-D airport surveillance radar	Enroute 2-D air surveillance radar
Manufacturer		Raytheon Canada	Raytheon Canada	Selenia
Location	Ground, static	Ground, static	Ground, static	Ground, static
Frequency Band	S	L	S	L
Frequency (MHz)	2700 - 2900	1250 - 1350	2700 - 2900	1250 - 1350
Power, Peak (MW)	1	0.5	0.5	2
, Av (kW)	0.618	0.88	0.96	2.5 [c]
Antenna Size (m) (or Type)		12.8 x 6.7 [b]	5.5 x 2.74	12.8 x 6.7 [b]
Antenna Gain (dB)	33.5 (LB), 32.5 (UB)	36 (LB); 34.4 (UB)	34 (LB), 31.5 (UB)	36.5 (LB); 34.4 (UB)
Beam Shape	Fan	CSC^2	CSC^2	CSC^2
Type Scanner	Mechanical	Mechanical	Mechanical	Mechanical
No. of Beams	1	2 in elevation	1	2 in elevation
Beamwidth, El (deg)		4° (CSC^2 to >40°)	5° (CSC^2)	4° (CSC^2 to >40°)
, Az (deg)	1.4	1.25	1.4	1.25
Angular Coverage, El	0.2° - 30°	≥ 40 Kft for 30° ≥EL ≥ 4°	≥ 30 Kft for 30° ≥ EL ≥ 4°	80 Kft
Az (deg)	360	360	360	
Scan Rate, El	-	-	-	-
Az	12.5 rpm	12 rpm	15 rpm	6 rpm
PRF	1030 pps (av)	800 pps (av)	960 pps (av)	300 - 800(13)
PRF Stagger		Quadruple	Quadruple	Sextuple or pseudo-sextuple(13)
Pulse Width, Transmit (μs)	0.6	1.1	1.1	2.8
, Compressed (μs)		NA	NA	NA
Pulse Compression Ratio		NA	NA	NA
Type Tube				Mag. M5051 or M5052(13)
Noise Figure (dB)	4	4	4	2.5
Range	60 nmi	80 nmi [a]; 138 nmi max.	60 nmi [a]; 77 nmi max.	180 nmi ($2m^2$)
MTI	SCV = 28 dB	I.F. = 33 - 36 dB	8 bit dig. MTI	Digital double canceller(13)
Polarization	V, C	H, C	V, C	H, C
Frequency Agility				
ECCM Features				
Misc. Comments	VB = 800 knots; freq diplexing	VB = 1200 kts; freq diplexed; DMTI	Freq diplexed Log/FTC; VI	CM, Log/FTC(13) [j]
Fig. No.		47		49 (with G-7 antenna)

Table 1a 2-D Surveillance and Search Radars

Radar	ATCR-3T	ATCR-4T (with G-7 antenna)	LP 23 (civilian); TRS 2050 (military)	M-33	Nike Ajax ACQ
Type	2-D airport surveillance radar; GCA surv or approach radar	2-D airport surveillance radar; GCA surv or approach radar	2-D long surveillance air defense and ATC radar		Acquisition
Manufacturer	Selenia	Selenia	Thomson - CSF		Western Electric
Location	Ground, static	Ground, static	Ground, static		Ground, mobile
Frequency Band	S	L	L		S
Frequency (MHz)	2700 - 2900	1250 - 1350	1250 - 1350[13]		
Power, Peak (MW)	0.5	0.5	2.2		0.5/1.0
, Av (kW)	0.450[13]	0.500[13]			630/1300
Antenna Size (m) (or Type)			13 x 9		(Pill Box)
Antenna Gain (dB)	34.0	32.5 (LB), 31 (UB)	36 (LB); 35.5 (UB)		
Beam Shape	Modified CSC^2 [13]	Modified CSC^2 [13]	CSC^2 to 40°		
Type Scanner	Mechanical	Mechanical	Mechanical		Mechanical
No. of Beams	1	2 in elevation[13]	2 in elevation		1
Beamwidth, El (deg)	5[13]	7.5 ±1 [13]			
, Az (deg)	1.5	1.3	1.2		
Angular Coverage, El	35 Kft	50 Kft			
Az (deg)	360				
Scan Rate, El	-	-	-		-
Az	15 rpm	12 rpm	3 or 6 rpm		10,20,30 rpm
PRF	1000 pps (typ)	1000 (typ)	250 pps (4 μs); 375 pps (3 μs)		1000
PRF Stagger			Yes		
Pulse Width, Transmit (μs)	1	1	4 or 3		1.3
, Compressed (μs)	NA	NA			
Pulse Compression Ratio	NA	NA			
Type Tube	Mag. RK 5586[13]	Mag. 5J26	Mag. (10 Khr life)		
Noise Figure (dB)	3[13]		3 (PA)		
Range	80 nmi ($2m^2$)	100 nmi ($2m^2$)	185 nmi ($2m^2$)	20/60/90 nmi [a]	60/120/200 nmi [a]
MTI	Digital[13]	Digital double canceller[13]	SCV ≥ 25 dB; digital		
Polarization	H, C	H, C	C		
Frequency Agility					
ECCM Features			Pulsed interf rej ckts		
Misc. Comments		Can use G-14 antenna	Dual or triple freq div av		
Fig. No.	50	51 (with G-7 antenna)	54		4

Table 1a 2-D Surveillance and Search Radars

Radar	Model 3900	Improved Nike Hercules Mobile HIPAR	S631
Type	2-D commercial marine surface radar	2-D acquisition for Hercules surface-to-air missile system	Search: L and S band back-to-back L: High angle cov | S: Low angle cov
Manufacturer	Raytheon	GE	Marconi
Location	Shipborne	Ground, static or mobile	Ground, static
Frequency Band	X	L	L and S (back-to-back)
Frequency (MHz)	9375 ±50		1250 - 1310 (standard) | 2900 - 3000 (standard)
Power, Peak (MW)			0.8[d] | 2.3[e] | 1.2[f] | 2.25[g]
, Av (kW)			1.5 | 3.0 | 1.5 | 3.3
Antenna Size (m) (or Type)	1.2 (end fed-slotted array)		14 x 4.5
Antenna Gain (dB)			37 (Par), 34 (CSC^2) | 45 (Par), 39 (CSC^2)
Beam Shape	Fan	Mobile system switchable from CSC^2 to fan beam (13)	CSC^2 or parabolic reflector in vertical direction
Type Scanner	Mechanical	Mechanical	Mechanical
No. of Beams	1		2 (back-to-back)
Beamwidth, El (deg)	22		3.4 (Par), to 35 (CSC^2) | 1.5 (Par), to 35 (CSC^2)
, Az (deg)	2		1.24 (Par), 1.24 (CSC^2) | 0.55 (Par), 0.55 (CSC^2)
Angular Coverage, El	22°		0-35° (CSC^2)
Az (deg)	360		360
Scan Rate, El	-		
Az	30 rpm		3 and 6 or 4 and 8 rev/min
PRF	3000 pps (SP), 1500 pps (LP)		220 - 850 | 220 - 750 | 270 - 750 | 200 - 600
PRF Stagger	None		Staggered
Pulse Width, Transmit (μs)	0.1, 0.67		1.5-5 | 2.5-5 | 2-5 | 2.5-5
, Compressed (μs)	NA		NA
Pulse Compression Ratio	NA		NA
Type Tube			Magnetron
Noise Figure (dB)	11		2.8 | 3.8 (PA)
Range	To 32 nmi [a]		
MTI	None	Unique MTI(13)	MTI | MTI
Polarization			
Frequency Agility			
ECCM Features	None	Yes(13)	
Misc. Comments			Squintless linear feed
Fig. No.	44	55	22

Table 1a 2-D Surveillance and Search Radars

Radar	S650 (S1050 ant)	S650H (S1055 ant)	S670 (S1070 ant)	S654	WW-2 AN/CPS-1 long range ground radar (12), (45)
Type	2-D enroute or terminal area surveillance			2-D enroute, terminal area or approach surv for civil and military	2-D early warning and surveillance of enemy a/c; control of firendly a/c
Manufacturer	Marconi			Marconi	GE
Location	Ground, static			Ground, static	Ground, mobile
Frequency Band	UHF			L	S
Frequency (MHz)	582-592, 592-602, or 602-610			1250 - 1310 (std)	~ 3,200
Power, Peak (MW)	0.50			2.3[e] / 0.800[d]	
, Av (kW)	0.8			3.0 / 1.5	
Antenna Size (m) (or Type)	3.97 x 16	4.32 x 16	3.97 x 20.6	5.78 x 9.75	7.6 x 2.44 / 7.6 x 1.52
Antenna Gain (dB)	31	30	32	33.5 (LB), 31 (UB)	
Beam Shape	Singly curved parabola	Shaped for high el cov	Singly curved parabola	Fan	Parabolic shape in vert dir / CSC^2
Type Scanner	Mechanical			Mechanical	Mechanical
No. of Beams	1			2 in elevation	2 (back-to-back)
Beamwidth, El (deg)					3 / 30
, Az (deg)	2.1	2.1	1.7	1.7	0.9 / 0.9
Angular Coverage, El					30 Kft
Az (deg)	360			360	360
Scan Rate, El				-	-
Az	5 and 10 rpm or 7.5 and 15 rpm		7.5 rpm	5 and 10 or 7.5 and 15 rpm	1, 2 or 4 rpm
PRF	250-400 pps (4 μs) or 400-480 pps (3 μs)			220-750 pps / 220-850 pps	
PRF Stagger	Yes			Yes	
Pulse Width, Transmit (μs)	4 or 3			2.5,3,4,5[e] / 2 std[d]	
, Compressed (μs)	NA			NA	
Pulse Compression Ratio	NA			NA	
Type Tube	Klystron (fed by TWT)			Mag (AFC; 500 kHz/s)	
Noise Figure (dB)	<4.0			2.8 (PA)	
Range					200 nmi (WW-2 heavy bomber)
MTI	CR = 40 dB; max rg ≥ 120 nmi[h]			CR = 40 dB; rg ≥ 120 nmi[h]	
Polarization	H			C or C, L	V / H
Frequency Agility					
ECCM Features	Random prf selection possible			Random prf selection	
Misc. Comments	480-650 pps possible with reduced power and 3 μs			CFAR, single or dual mode	Antenna mount wt. 12 klbs
Fig. No.	53 (for S650; others similar)			52	56

Table 1a **2-D Surveillance and Search Radars**

Radar	AN/CPS-6B	AN/FPS-7	AN/TPS-27	AN/TPS-34
Type	3-D WW-2 V beam radar	3-D stacked beam surveillance radar	3-D stacked beam surveillance radar	3-D V beam tactical early-warning radar for forward air def
Manufacturer	GE	GE	Westinghouse	Sperry Gyroscope
Location	Ground, transportable	Ground, static	Ground, transportable	Ground, transportable (by helicoptor)
Frequency Band	S	L	S	L
Frequency (MHz)		1250 - 1350	2700 - 3900	
Power, Peak (MW)	0.9/xtm; 2 (EW)	10	2.5	5
, Av (kW)		14.5	4.5	
Antenna Size (m) (or Type)	7.6 x 3.0; 9.8 x 3.0 (45°)			9.8 m
Antenna Gain (dB)				
Beam Shape	2 fan beams; one vert, one at 45° tilt	7 stacked beams, CSC^2	Stacked beams in el	2 CSC^2 fan beams; one vert, one at 45°
Type Scanner	Mechanical	Mechanical	Mechanical	Mechanical
No. of Beams	2 (see Figure 62 caption)	7 in el		2
Beamwidth, El (deg)		18		
, Az (deg)	1 (vert beam)	1.4		
Angular Coverage, El	0-24°; 40 Kft	18°; ~ 150 Kft (13)		
Az (deg)	360	360	360	
Scan Rate, El		-		
Az	0 - 15 rpm	3.3, 5, 6.6, 10 rpm		
PRF	300 or 600 pps	244	300	
PRF Stagger			-	
Pulse Width, Transmit (μs)	1,2 (300 pps); 1 (600 pps)	6	6	
, Compressed (μs)			None	
Pulse Compression Ratio			None	
Type Tube		Klystron(13)		
Noise Figure (dB)		9.5		
Range	150 nmi (V-beam) (EW) 240 nmi (S-beam) (EW) 120 nmi (Height finding)	270 nmi (13)	155 nmi ($1m^2$)	250 nmi ($0.5m^2$)
MTI	Yes	Yes(13)	2 pulse canc	
Polarization			V	V, 45° L
Frequency Agility			16 freqs	
ECCM Features		Yes(13)		
Misc. Comments		Dual channel xtrm; 10 preset freqs		8 hrs to assemble(13)
Fig. No.	62	57	59	63

Table 1b **3-D Radars, Stacked or V-Beam**

Radar	AN/TPS-43	Russian Barlock (P-50) (14), (13)	TH.D.1955
Type	3-D stacked-beam surveillance radar	3-D stacked beam GCI search and early warning radar	3-D stacked beam air defense-surveillance radar
Manufacturer	Westinghouse	Russian	Thomson-CSF
Location	Ground, air transportable	Ground, transportable	Ground, static
Frequency Band	S	S	S
Frequency (MHz)	2900 - 3100	2695 - 3125	
Power, Peak (MW)	4	1 (per beam); diff freq per beam	20
, Av (kW)	6.7		20
Antenna Size (m) (or Type)		2 lg truncated par with clipped corners	16 x 6
Antenna Gain (dB)	76 2-way for lowest beam		46
Beam Shape		6 beams in el	CSC^2 energy dist
Type Scanner	Mechanical	Mechanical	Mechanical
No. of Beams	6 stacked in elev[13]	6 in el	12 in el
Beamwidth, El (deg)			
, Az (deg)	1.1 [13]	0.7	
Angular Coverage, El	10 Kft, 20°		0°-20° (CSC^2)
Az (deg)		360	360
Scan Rate, El		-	-
Az	6 rpm [13]	6 (EW), 12 rpm (GCI)	6 rpm
PRF	250 (av)	375	
PRF Stagger	6 prf's		
Pulse Width, Transmit (μs)	6.5	1.8 - 3.1	
, Compressed (μs)			Yes
Pulse Compression Ratio			Yes
Type Tube	Linear beam twystron[13]		Klystron (4-5 K hrs)
Noise Figure (dB)	5		
Range	260 nmi ($5m^2$)	175 nmi	>215 nmi for fighter a/c at any flyable alt
MTI	Dig 4 pulse canc		
Polarization	Vert		At operators discretion
Frequency Agility	16 freq		Yes
ECCM Features	JATS, CPACS, SB[13][j]		Many
Misc. Comments	MTBF:185 hr; MTTR:1 hr		Height acc 2-4 mr at lg rg
Fig. No.	58	60	61

Table 1b 3-D Radars, Stacked or V-Beam

Radar	AN/FPS-4 (tower mounted)	AN/FPS-6, 89	AN/FPS-89, 6	AN/MPS-8 (trailer mounted)
Type	Height finder	Height finder	Height finder	Height finder
Manufacturer	RCA	GE	GE	RCA
Location	Ground, static	Ground, static	Ground, static	Ground, mobile
Frequency Band		S	S	
Frequency (MHz)		2700 - 2900	2700 - 2900	
Power, Peak (MW)		4.5	4.5	
, Av (kW)		3.6	3.6	
Antenna Size (m) (or Type)		8 (height)		
Antenna Gain (dB)		38.5	38.5	
Beam Shape		Pencil fan beam	Pencil fan beam	
Type Scanner		Mechanical	Mechanical	
No. of Beams	Same as	1	1	Same as
Beamwidth, El (deg)	AN/TPS-10D	0.9	0.9	AN/TPS-10D
, Az (deg)	except	3.2	3.2	except trailer
Angular Coverage, El	tower-mounted			mounted
Az (deg)	and	360	360	and
Scan Rate, El	nontransportable	20 or 30 nods/min	20 or 30 nods/min	hence mobile
Az		180° in 4 s	180° in 4 s	
PRF		300 - 405 pps	300 - 405 pps	
PRF Stagger				
Pulse Width, Transmit (μs)		2	2	
, Compressed (μs)				
Pulse Compression Ratio				
Type Tube				
Noise Figure (dB)				
Range				
MTI				
Polarization				
Frequency Agility				
ECCM Features				
Misc. Comments			Improved AN/FPS-6	
Fig. No.	See Fig. 64	65	See Fig. 65 for FPS-6	See Fig. 64

Table 1c Height Finders (Including 3-D Height Finders)

Radar	AN/MPS-14	AN/SPS-8	AN/SPS-8A	AN/SPS-8B	AN/TPS-10D (tripod mounted)
Type	Height finder		Shipboard height finder and 3-D surveillance radar		Height finder
Manufacturer	GE		GE		RCA
Location	Ground, mobile		Shipboard		Ground, transportable
Frequency Band			S		X
Frequency (MHz)			3430 - 3570		9230 - 9404
Power, Peak (MW)		0.7 - 2.0		1.3	0.30
, Av (kW)		1		1.8	
Antenna Size (m) (or Type)					
Antenna Gain (dB)		37.5		41	42
Beam Shape		Fan pencil beam		Pencil beam	Fan pencil beam
Type Scanner	Same as		Mechanical		Mechanical
No. of Beams	AN/FPS-6		1		1
Beamwidth, El (deg)	except mobile.	1.1		1.2	0.75
, Az (deg)		3.5		1.5	2.05
Angular Coverage, El	6-trucks,		0° to 36° (over 12° az sect)		-2° to +23°, -5 to +60 Kft
Az (deg)	3-trailer		360		360
Scan Rate, El	system	300, 600 or 1200 scans/min		360, 720 or 970	30 and 60 osc/min
Az	designed for		1,2,3,5,10 rpm		1/3 (scan) and 3(slew) rpm
PRF		500 or 1000 pps		450 or 700 pps	539 rpm
PRF Stagger	quick transport				
Pulse Width, Transmit (μs)	to strategic	1 or 2		2	0.55 (0-60 nmi), 2 (0-120 nmi)
, Compressed (μs)	or tactical sites.				
Pulse Compression Ratio					
Type Tube					Mag RK 6002/QK 221
Noise Figure (dB)					12
Range					120 nmi
MTI					
Polarization					
Frequency Agility					
ECCM Features					
Misc. Comments			Robinson Scanner		STC, FTC, IAGC, DBB [j]
Fig. No.	See Fig. 65		70		64

Table 1c Height Finders (Including 3-D Height Finders)

Radar	Russian Cake Series (14)	S613	S669	TRS 2205 (VOLEX III)
Type	Height finder	Height finder	Height finder	3-D surveillance radar; height finder
Manufacturer	Russia	Marconi	Marconi	Thomson-CSF
Location	Ground	Ground, transportable, static, or mobile	Ground, static	Ground, static
Frequency Band	S	C	S	S
Frequency (MHz)		5300 – 5340 or 5480 – 5520	2900–3000 (std) / 2740–2900 or 2900–3100	2900 – 3100
Power, Peak (MW)		1	1.2[f] / 2.25[g]	1
, Av (kW)		1.5	1.5 / 3.3	2
Antenna Size (m) (or Type)	(Peel shaped par ref)	4.27 x 1.3	12.2 x 2.13	6.8 x 3.4 (low beam) / 3.4 x 2.3 (high beam)
Antenna Gain (dB)	~36	39.5	40.5	40 / 37
Beam Shape	Fan pencil beam	Fan pencil beam	Fan pencil beam	Pencil
Type Scanner	Mechanical	Mechanical	Mechanical	Robinson
No. of Beams	1	1	1	2 (low & high)
Beamwidth, El (deg)	~1.5	0.9	0.6	1 / 2
, Az (deg)	~3.5	3.0	3.7	2 / 2
Angular Coverage, El		-5° to +55°		-0.5 to 7 / +5 to +20
Az (deg)	360	360	360	360
Scan Rate, El	30–40 nods/min	45°/sec	3° to 30° per 3 s cycle	720 scans/min
Az		18.5 rpm; 180° in 2 s	16 rpm; 179° in 4 s	3 rpm
PRF		300 pps (std)	270–750 pps / 200–600 pps	500 pps
PRF Stagger				
Pulse Width, Transmit (μs)		5 (std) (order dn to 2)	2–5 / 2.5–5	4
, Compressed (μs)			NA	
Pulse Compression Ratio			NA	
Type Tube		Long anode block mag	Magnetron	2 Magnetrons
Noise Figure (dB)		4.5 (PA)	3.8	<5 (PA)
Range	> 100 nmi 160–190 nmi for Sponge Cake			155 nmi ($2m^2$) / 110 nmi ($2m^2$)
MTI				
Polarization		C		H / C
Frequency Agility				
ECCM Features			Burnthrough,	Freq div avail; CFAR
Misc. Comments			±1500 ft at 150 nmi	Δh = ±1 Kft at 100 nmi
Fig. No.	66	67	68	71

Table 1c Height Finders (Including 3-D Height Finders)

Radar	AN/CPN-4	AN/FPN-16	AN/FPN-62 (Normal PAR)	AN/MPQ-4
Type	Precision approach plus acquisition radars	Precision approach radar	Precision approach radar	Mortar-location radar
Manufacturer		ITT Gilfillan	Raytheon	GE
Location	Ground, air transportable	Ground, static	Ground, static X	Ground, mobile
Frequency Band				Ku
Frequency (MHz)			9000 - 9160	16,100
Power, Peak (MW)			0.045 (min)	0.08
, Av (kW)			0.0267	0.15
Antenna Size (m) (or Type)				
Antenna Gain (dB)				
Beam Shape			Vert fan / Horiz fan	Pencil
Type Scanner			Eagle	Foster dual scanner
No. of Beams	PAR	Parameters	2	2 in el 2° apart
Beamwidth, El (deg)	parameters	similar to	2 / 0.55	0.8
, Az (deg)	similar to	those of	0.85 / 3.5	1
Angular Coverage, El	those of	AN/FPN-62	2° / 7°	
Az (deg)	AN/FPN-62		20 / 3.5	25°
Scan Rate, El				-
Az			2 looks/s	167 scans/s
PRF			3300 pps (av)	8600
PRF Stagger			Ratios 10:12:16:14:10:4	
Pulse Width, Transmit (μs)			0.18 ±0.02	0.25
, Compressed (μs)				
Pulse Compression Ratio				
Type Tube			QKH1811 coaxial mag	
Noise Figure (dB)			4.1 (with PA), 9.0 (without)	12
Range	≥ 8 nmi	≥ 8 nmi	≥ 15 nmi ($1 m^2$)	10,000 m (max)(13) 170 m (min)(13)
MTI			Double; SCV ≥ 23 dB	
Polarization				
Frequency Agility				
ECCM Features				
Misc. Comments	Predecessor to AN/FPN-62	Predecessor to AN/FPN-62	No blind vel bet 20 & 600 Kt	acc 50 m at 10 Km(13)
Fig. No.	See Fig. 74	See Fig. 74	74	73

Table 1d Electromechanically Steered Systems

Radar	AN/MPN-11, -13	Fan Song B[14]		Fan Song E[14]	
Type	Precision approach plus acquisition radars	SA-2 Guideline missile fire control radar and TTR (TWS)		SA-2 Guideline missile fire control radar and TTR (TWS)	
Manufacturer		Russia		Russia	
Location	Ground, mobile	Ground, transportable		Ground, transportable; one on cruiser	
Frequency Band		S		C	
Frequency (MHz)		2965 - 2990 3025-3050		4910 - 4990 and 5010-5090	
Power, Peak (MW)		0.60		1.5	
, Av (kW)					
Antenna Size (m) (or Type)					
Antenna Gain (dB)					
Beam Shape		Vert fan	Horiz fan	Vert fan	Horiz fan
Type Scanner		Lewis		Lewis	
No. of Beams		2		2	
Beamwidth, El (deg)	PAR	10	2	7.5	1.5
, Az (deg)	parameters	2	10	1.5	7.5
Angular Coverage, El	similar to	10	10	7.5	7.5
Az (deg)	those of	10	10	7.5	7.5
Scan Rate, El Az	AN/FPN-62				
PRF				900-1020 pps search 1740-2070 pps track	
PRF Stagger					
Pulse Width, Transmit (μs)				0.4-1.2, 0.2-0.9	
, Compressed (μs)					
Pulse Compression Ratio					
Type Tube					
Noise Figure (dB)					
Range	10 nmi	32-64 nmi unambig. rg.		40-80 nmi (unambig. rg.)	
MTI					
Polarization					
Frequency Agility					
ECCM Features				AD, LORO	
Misc. Comments	Predecessors to AN/FPN-62				
Fig. No.	See Fig. 74	72a		72b	

Table 1d Electromechanically Steered Systems

Radar	ANTARES Height-finding radar	AR-3D	AN/SPS-32	AN/SPS-33
Type	Auto height finding or low-el long-rg surv	3-D air-defense radar	2-D surveillance (used with AN/SPS-33)	Tracking radar (used with AN/SPS-32)
Manufacturer	Thomson-CSF	Plessey	Hughes	Hughes
Location	Ground, static	Ground, mobile or static	Shipborne: USS Long Beach and Enterprise	Shipborne: USS Long Beach and Enterprise
Frequency Band	S	S		
Frequency (MHz)				
Power, Peak (MW)	1/mag	1.11		
, Av (kW)	1/mag	10		
Antenna Size (m) (or Type)	9 x 8 (cyl-par)	4.9 high by 7.1 wide	6 high by 12.2 wide(13)	7.6 high by 6 wide(13)
Antenna Gain (dB)	45	41.5		
Beam Shape	Pencil	Pencil	Vert fan beam(18)	Pencil
Type Scanner	Az mech; dig ϕ sh. in el	Freq in el; mech in az	Freq in az(18)	Freq in el; ϕ in az(18)
No. of Beams	3 in el (one/mag)	1		
Beamwidth, El (deg)		2		
, Az (deg)		1		
Angular Coverage, El	$-2°$ to $+35°$; ≥100 Kft (80%)	30°; to 210 Kft	To near zenith(18)	To near zenith(18)
Az (deg)	360	360	360 (4 faces)	360 (4 faces)
Scan Rate, El		-		
Az	6 rpm	6 rpm		
PRF		250 pps		
PRF Stagger				
Pulse Width, Transmit (μs)		36		
, Compressed (μs)		0.1		
Pulse Compression Ratio		130:1 (lower beam)		
Type Tube	3 mag (one/beam)	Grid mod klystron(13), (34)		
Noise Figure (dB)	3.5 (PA)			
Range	195 nmi	300 nmi (≤ 150 Kft, $15 m^2$, 90%, Swerling 3)	Sev hundred nmi(18)	
MTI		I.F. = 27 dB; to 75 nmi		
Polarization	C	H		
Frequency Agility				
ECCM Features		Az & el of jammer displayed		
Misc. Comments	ϕ comparison amp monopulse	200 MHz signal BW		100's targets tracked automatically(18)
Fig. No.	82	79	85	85

Table 1e Electronically Scanned Systems (Excluding Phase-Phase Scanning)

Radar	AN/SPS-39	AN/SPS-48, 48(A), 48C	AN/SPS-52, -52B	AN/TPS-32, -64
Type	3-D air surveillance radar	3-D air surveillance and weapons designation radar	3-D air surveillance and weapon support	3-D surveillance or assault radar with GCI operation
Manufacturer	Hughes	ITT Gilfillan	Hughes	ITT Gilfillan
Location	Shipborne	Shipborne: CG, DLG, DDG, CV and CLCC	Shipborne: Destroyers, frigates, cruisers, carriers	Ground, transportable
Frequency Band		Probably S(13)	Probably S(13)	S(13)
Frequency (MHz)				2905 to 3080(13)
Power, Peak (MW), Av (kW)				2.2, 0.665, 0.06
Antenna Size (m) (or Type)	Parabolic cylinder reflector (18)	Rect slotted array	Planar array	
Antenna Gain (dB)				41
Beam Shape	Pencil(18)	Pencil	Pencil	Fan pencil
Type Scanner	Freq in el; mech in az	Freq in el; mech in az	Freq in el; mech in az	Freq in el; mech in az
No. of Beams	1	Multiple	1	Multiple
Beamwidth, El (deg)				0.84
, Az (deg)				2.15
Angular Coverage, El	100° (with 10% BW)(18)		To 90°	20°, 100 Kft
Az (deg)	360 (18)	360	360	360
Scan Rate, El	Few ms/scan(18)			
Az				6 rpm
PRF				265 to 917 pps(13)
PRF Stagger				
Pulse Width, Transmit (μs)				1500 ft res; 30
, Compressed (μs)				
Pulse Compression Ratio				
Type Tube				
Noise Figure (dB)				
Range	100 to 160 nmi(13)			200 nmi ($1 m^2$, 90%, Mach 3) 300 nmi (max) [a]
MTI		in -48(A)	-52 analog; -52B DMTI	Yes
Polarization				H
Frequency Agility				
ECCM Features		Extensive		
Misc. Comments	Serpentine feed(18)	ADT in -48C [j]		Grd version of SPS-48(30)
Fig. No.	75	76	78	77

Table 1e Electronically Scanned Systems (Excluding Phase-Phase Scanning)

Radar	AN/TPS-59	AN/TPS-64	Matador 3-D TRS 2210[13]	Signaal 3-D Multi-target Tracking Radar (MTTR)[35]	
Type	Long range surveillance for US Marine Corps Tactical Air Operations	3-D surveillance or assault radar with GCI operation	3-D mobile air defense or gap filler with automatic data remoting	Air search and tracking radar search	track
Manufacturer	GE	ITT Gilfillan	Thomson-CSF	Hollandse Signaalapparaten (HSA)	
Location	Ground, transportable	Ground, transportable	Ground, extremely mobile (<1 hr inst)	Shipborne: Guided weapon frigate	
Frequency Band	L		S		
Frequency (MHz)	1215 - 1400				
Power, Peak (MW) , Av (kW)	0.0349 6.28		0.6/mag 2/mag		
Antenna Size (m) (or Type)	9.1 x 4.9 (14.5 dBsm)		(Cylindro-parabolic ref)	~6[13] (2 par dishes)	~6[13] (2 planar arrays)
Antenna Gain (dB)	38.9		40		
Beam Shape	Pencil			pencil	pencil
Type Scanner	Phase in el; mech in az		Az mech; dig ø sh in el	Elect-mech in el; mech in az	Freq and mech[35]
No. of Beams	1		3 in el (one/mag)		
Beamwidth, El (deg)	1.6	Parameters	1.9		
, Az (deg)	3.2	same as	1.5		
Angular Coverage, El	0° to 19°, 100 Kft	for AN/TPS-32	24°	To 15° nominally; ~CSC^2	
Az (deg)	360		360		
Scan Rate, El	-				
Az	6 and 12 rpm		5 rpm		40 meas/min
PRF			500 pps		
PRF Stagger					
Pulse Width, Transmit (μs)			5.8		
, Compressed (μs)	0.4 (min)				
Pulse Compression Ratio	1, 64, 128, 256 (avail)				
Type Tube	Solid state		3 mags (one/beam)		
Noise Figure (dB)	(540 K sys temp)		3.5 (PA)		
Range	3 to 300 nmi 240 nmi (1m^2, 70%)		130 nmi (fighter, 80%)		
MTI	Grd & weather sup:55 & 37 dB		SCV = 30 dB		
Polarization			C		
Frequency Agility					
ECCM Features			Modern		
Misc. Comments	MTBF = 1400 hr; MTTR = 40 min		El monopulse	Tracks up to 100 tgs simultaneously; see text.	
Fig. No.	90	See Fig. 77	81	80	

Table 1c Electronically Scanned Systems (Excluding Phase-Phase Scanning)

Radar	AN/APQ-140	AN/FPS-85	AN/FPS-108 (COBRA DANE)(32),(36)	AN/GPN-22
Type	Multifunction: search, track, ground mapping, terrain following & avoidance	Multifunction Array for search and track of missiles and satellites	Multifunction Array for search and track of missiles and satellites	High-performance PAR (HI-PAR). Derived from AN/TPN-25
Manufacturer	Raytheon	Bendix	Raytheon	Raytheon
Location	Airborne	Ground (Eglin, AF Base Florida)	Ground, static (Shemya, Alaska)	Ground
Frequency Band	Ku	UHF	L	X
Frequency (MHz)		442 ± 5	1215-1250 (NB); 1175-1375 (WB)	9000 to 9200
Power, Peak (MW)		32	15.4(32),(36)	0.3 (0.012 with CFA off)
, Av (kW)		160	920(32), (36)	1 (0.04 with CFA off)
Antenna Size (m) (or Type)	(Reflect array)(32)	26.9 x 26.9 (trans); 58 (Rec)	29	4.7 x 4.7 (offset hyperbola)
Antenna Gain (dB)				43.5 dB (directivity)
Beam Shape	$CSC^2\theta$ fan beam and pencil beams(32)	Pencil beam	Pencil	Pencil
Type Scanner	Phase-phase(32)	Phase-phase	Phase-phase	Phase-phase (limited)
No. of Beams				1
Beamwidth, El (deg)		1.4 (trans); 0.80 (Rec)		0.75
, Az (deg)		1.4 (trans); 0.80 (Rec)		1.45
Angular Coverage, El	±70°(32)	0°-105°	±60° (NB); ±22° (WB)	8° (-1° to +7°)
Az (deg)	±70 (32)	120	±60 (NB); ±22 (WB)	20 (-10 to +10)
Scan Rate, El				22 meas/s on 6 tgs
Az				
PRF	1000 beams/s av(32)	20 frames/s		3500 (av)
PRF Stagger				3:2 (search pulse to pulse)
Pulse Width, Transmit (μs)		1,5,10,16,25,32,64,125, 128,250	150-1500; 2000;1000;1000	2 0.5 (search); 1 (track)
, Compressed (μs)			0.2-0.2; 1; 0.005; 0.04	~0.01[c] (track)
Pulse Compression Ratio		1600,250,25,128,12.8,64, 32,16	750-7500; 2000; 2 x 10^5; 25,000	100:1 (track)
Type Tube			96 TWTs	TWT/CFA (final)
Noise Figure (dB)			(PA)	4.3 (PA) [l]
Range		Satellite ranges	1000 nmi (-20 dBsm, 1000 μs, SNR = 16.5 dB, 60 pps)	20 nmi (1m^2, clear) 14 nmi (1m^2, 2 in/hr)
MTI		Digital		SCV = 24 dB; dig; 3 pulse
Polarization	V,H,RC,LC(32)		V	C (20 dB ICR)
Frequency Agility	Pulse-to-pulse(32)	Pulse-to-pulse, 51 frequencies		
ECCM Features				Dicke fix; non-coh dig MTI[m]
Misc. Comments	4 horn monopulse(32)	9 beam rec cluster	Tracks >100 objects	TWS; duel freq. div.
Fig. No.	19	16	18	87

Table 1f Phased Arrays (Phased-Phased Steered Systems)

Radar	AN/SPY-1[20]	AN/TPN-25	Missile site radar (MSR)
Type	Multifunction phase array for AEGIS surface to air missile defense	Precision approach radar (PAR) of AN/TPN-19 system	Multifunction: Short-range track and missile guidance
Manufacturer	RCA	Raytheon	Raytheon
Location	Shipborne	Ground, highly mobile for tactical support	Ground, static
Frequency Band	S	X	S[13], [31], [32]
Frequency (MHz)		9000 to 9200	
Power, Peak (MW)		0.3 (0.012 with CFA off)	>1[32]
, Av (kW)		1 (0.04 with CFA off)	>100[32]
Antenna Size (m) (or Type)	3.67 x 3.84 (~hexagon)	2.8 x 3.6 (offset hyperbola)	~4m[13],[32] (space fed)
Antenna Gain (dB)		43.5 (directivity)	
Beam Shape	Pencil	Pencil	Pencil
Type Scanner	Phase-phase	Phase-phase (limited)	Phase-phase
No. of Beams		1	
Beamwidth, El (deg)	~1.8 [c]	0.75	
, Az (deg)	~1.8 [c]	1.3	
Angular Coverage, El	Hemispherical (four faces)	15° (−1° to +14°)	0–90° (4 faces)[32]
Az (deg)		20 (−10 to +10)	360 (4 faces)[32]
Scan Rate, El / Az		22 meas/s on 6 tgs	
PRF		3500 (av)	
PRF Stagger		3:2 (search pulse to pulse)	
Pulse Width, Transmit (μs)		2 0.5 (search); 1 (track)	
, Compressed (μs)		~0.01 ns [c] (track)	
Pulse Compression Ratio		100:1 (track)	
Type Tube		TWT/CFA (final)	
Noise Figure (dB)		4.3 (PA) [l]	
Range		20 nmi ($1 m^2$, clear) 15 nmi ($1 m^2$, 2 in/hr)	>600 nmi[13]
MTI	Yes	SCV = 24 dB; dig; 3 pulse	
Polarization	V	C (20 dB ICR)	
Frequency Agility		Dual freq diversity	
ECCM Features		Dicke fix, non-coh MTI [m]	
Misc. Comments		TWS; dual freq div	Monopulse feed[32]
Fig. No.	83	86	13

Table 1f Phased Arrays (Phased-Phased Steered Systems)

Radar	Perimeter acquisition radar (PAR)	Patriot (formerly SAM-D)
Type	Multifunction: Long-range search and track of missiles and satellites	Multifunction: Search and track a/c and missiles air defense
Manufacturer	GE	Raytheon
Location	Ground (North Dakota)	Ground, mobile
Frequency Band	UHF(31), (13)	C
Frequency (MHz)	442(13), (31)	
Power, Peak (MW)	7.3(13)	>0.1[i] (32)
, Av (kW)		>10[i] (32)
Antenna Size (m) (or Type)	~30 m octagon(13)	
Antenna Gain (dB)		
Beam Shape	Pencil	Pencil
Type Scanner	Phase-phase	Space fed(32), (50) (Separate rec. & trans. feeds.)
No. of Beams		
Beamwidth, El (deg)		
, Az (deg)	~±60 (13)	
Angular Coverage, El		
Az (deg)		
Scan Rate, El		
Az		
PRF		
PRF Stagger		
Pulse Width, Transmit (μs)		
, Compressed (μs)		
Pulse Compression Ratio		
Type Tube	128 TWTs	
Noise Figure (dB)		
Range	to ~ 4000 Km(13)	
MTI		
Polarization		
Frequency Agility		
ECCM Features		
Misc. Comments		5000 Ferrite phase shifters(32), (50)
Fig. No.	15	14

Table 1f Phased Arrays (Phased-Phased Steered Systems)

Radar	AN/FPS-16	BEMEWS AN/FPS-49 or -49A	BEMEWS AN/FPS-92	AN/MPS-25
Type	Tracker	Missile and satellite tracker; search over limited vol	Missile and satellite tracker; search over limited vol	Tracker
Manufacturer	RCA	RCA	RCA	RCA
Location	Ground, static (also on instrumentation ships)	Ground, static. (Thule, Greenland; Flyingdales Moor, UK)	Ground, static (Clear, Alaska)	Ground, mobile
Frequency Band	C	UHF	UHF	
Frequency (MHz)	5450 - 5825	425 ± 5%	425 ± 5%	
Power, Peak (MW)	1	5.0	5.0	
, Av (kW)	1			
Antenna Size (m) (or Type)	3.66 or 4.88	25.6	25.6	
Antenna Gain (dB)	44.5 or 47	38.0 dB min at mid-band	38.0 dB min at mid-band	
Beam Shape	Pencil beam	Pencil (5-horn monopulse)	Pencil (5-horn monopulse)	
Type Scanner	Mechanical	Mechanical	Mechanical	
No. of Beams	1	1	1	
Beamwidth, El (deg)	1.1 or 0.8	2	2	Mobile version
, Az (deg)	1.1 or 0.8	2	2	of
Angular Coverage, El	-10° to 190°	180°	180°	AN/FPS-16
Az (deg)	360			
Scan Rate, El	25°/s	10°/s (13)		
Az	45°/s			
PRF	142 to 1364 (12 values)	27	27	
PRF Stagger				
Pulse Width, Transmit (μs)	0.25, 0.5, 1.0	2000	2000	
, Compressed (μs)		2000	1 (non-linear FM)	
Pulse Compression Ratio		1	2000	
Type Tube	Magnetron			
Noise Figure (dB)	11 (max)	3 (PA)	3 (PA)	
Range	150 nmi ($1 m^2$)	~5000 Km (13)	~5000 Km (13)	
MTI				
Polarization	L (C with mod kit)	Trans:RC; Rec: H & V		
Frequency Agility				
ECCM Features				
Misc. Comments		500 Hz track dop res	925 Hz track dop res	
Fig. No.	2	10	10	See Fig. 2 for FPS-16

Table 1g Trackers

Radar	AN/SPG-55B (Mod 8)		M33 and Nike-Ajax TTR	Nike Ajax MTR (Target Mode)	Nike Ajax MTR (Beacon Mode)
Type	Tracker	Illuminator for Terrier missile homing	Target tracking radar	Missile tracker	Missile tracker
Manufacturer	Sperry Gyroscope		Western Electric	Western Electric	Western Electric
Location	Shipborne		Ground, mobile	Ground, mobile	Ground, mobile
Frequency Band	C	X	X	X	X
Frequency (MHz)					
Power, Peak (MW)	1	0.005	0.250	0.140	0.25
, Av (kW)		5	62.5	62.5	140
Antenna Size (m) (or Type)			Metal-slat lens	Metal-slat lens	Metal-slat lens
Antenna Gain (dB)					
Beam Shape	Pencil				
Type Scanner	Mechanical		Mechanical	Mechanical	Mechanical
No. of Beams	1		1	1	1
Beamwidth, El (deg)	1.6	0.8(T), 3.0(R)			
, Az (deg)	1.6	0.8(T), 3.0(R)			
Angular Coverage, El					
Az (deg)					
Scan Rate, El					
Az					
PRF	427		1000 pps	1000 pps	4000 pps
PRF Stagger					
Pulse Width, Transmit (μs)	12,13,0.1	CW	0.25	0.25	0.25
, Compressed (μs)	0.1,1,0.1	-			
Pulse Compression Ratio	127, 13, 1	-			
Type Tube					
Noise Figure (dB)					
Range	300,000 yds[a]		100 nmi[a]	50 nmi[a]	50 nmi[a]
MTI					
Polarization	L, V				
Frequency Agility	Pulse-to -pulse				
ECCM Features					
Misc. Comments	Acq & track poss at X-band			Capable of beacon track	Capable of skin track
Fig. No.	7		3,4	4	4

Table 1g Trackers

Radar	Nike Hercules TTR or MTR (Target Mode)	Nike Hercules MTR (Beacon Mode)	Nike Hercules MTR (Ajax Mode)
Type	Target or Missile tracker	Missile tracker	Missile tracker
Manufacturer	Western Electric	Western Electric	Western Electric
Location	Ground, mobile	Ground, mobile	Ground, mobile
Frequency Band	X	X	X
Frequency (MHz)			
Power, Peak (MW)	0.25	0.2	0.140
, Av (kW)	31.3	100	140
Antenna Size (m) (or Type)			
Antenna Gain (dB)			
Beam Shape			
Type Scanner	Mechanical	Mechanical	
No. of Beams	1	1	
Beamwidth, El (deg)			
, Az (deg)			
Angular Coverage, El			
Az (deg)			
Scan Rate, El			
Az			
PRF	500	2000	4000
PRF Stagger			
Pulse Width, Transmit (μs)	0.25	0.25	0.25
, Compressed (μs)			
Pulse Compression Ratio			
Type Tube			
Noise Figure (dB)			
Range			
MTI			
Polarization			
Frequency Agility			
ECCM Features			
Misc. Comments			
Fig. No.	9	9	9

Table 1g Trackers

N = number of pulses integrated

σ_θ = 1-σ angle accuracy (radians)

Substituting Equation (2) in Equation (3a) and using Equation (3b) yields the tracking accuracy form of the radar equation

$$\sigma_\theta^2 = \frac{k_0^2 \pi^2 R^4 k T_S B}{2 k_M^2 N \sigma} \quad \frac{\lambda^4}{P_{PK} A_E^3} \tag{4}$$

Note that, whereas the search range is dependent on the power-aperture product, $P_{AV} A_E$, SNR is dependent on the product of the power and squared-aperture, $P_{AV} A_E^2$; the angular accuracy radar is dependent on one over the product of the power and cubed-aperture, $1/P_{AV} A_E^3$. The AN/FPS-16, M-33 TTR*, Nike-Ajax TTR, Hawk AN/MPQ-39, TARTAR AN/SPG-51, Terrier AN/SPG-55, NATO Sea Sparrow director/illuminator, and Nike-Hercules high-power TTR are all examples of high frequency (S-band or higher) tracking radars (Figures 2 through 9).

Not all tracking radars operate at high frequencies. The BMEWS AN/FPS-49 tracking radar works at 425 MHz (Figure 10). In this case, the long range requirements naturally led to the selection of a large aperture and long wavelength. A large aperture is needed to meet the sensitivity requirements for long range, as there usually is a limit on the transmitter power that can be provided economically. The state-of-the-art of transmitter power tubes was a major factor in the selection of UHF, instead of L-band, for the FPS-49 – high power L-band tubes were not available at the time it was built (around 1958). Large apertures in turn require long wavelengths; the large aperture offsets the loss of angle accuracy that accompanies use of long wavelengths.

*See glossary at end of chapter.

Figure 3 **Target Tracking Radar (TTR) of M-33 Anti-Aircraft Fire Control System, controls firing of 90mm and 120mm guns. Nike-Ajax TTR (see Figure 4) and MTR have same profiles as M-33 TTR; see Table 1g. (Photo courtesy of Western Electric.)**

Figure 2 **AN/FPS-16 C-Band Instrumentation Tracking Radar;** see Table 1g. (The mobile trailer-mounted version is designated as AN/MPS-25; an Air Force version is designated as AN/FPQ-13.) Some models incorporate laser and IR tracking as adjuncts. (Photo courtesy of RCA.)

Figure 4 **At right is TTR of Nike-Ajax Surface-to-Air Missile System (Nike-Ajax MTR has same profile as TTR); at left is Nike-Ajax Acquisition Radar (ACQ) which has same profile as M-33 ACQ. See Tables 1a and 1g. (Photo courtesy of Western Electric.)**

Figure 5 Hawk AN/MPQ-39 X-Band CW Illuminator, acquires targets designated by MPQ-34 or -35; see Figures 11 and 21. Separate antennas are used for transmit and receive enabling simultaneous transmission and reception. (Photo courtesy of Raytheon Company.)

Figure 6 AN/SPG-51 C-Band Track-Illuminator for TARTAR Missile System. (Photo courtesy of Raytheon Company.)

Figure 7 Terrier Missile System AN/SPG-55 Track and Guidance Radar, uses C-band Monopulse Tracker with integral X-band CW Illuminator; see Table 1g. For the AN/SPG-55B Mod 8, acquisition in severe clutter is possible using the X-band CW signal. (Photo courtesy of Sperry Gyroscope.)

Figure 8 NATO Sea Sparrow CW Director/Illuminator. (Photo courtesy of Raytheon Company.)

Figure 9 Nike-Hercules Surface-to-Air and Surface-to-Surface Missile System TTR using Polarization-Twist Cassegrainian Reflector (MTR has same profile). Nike-Hercules replaced Nike-Ajax in the late 1950s and early 1960s; see Table 1g. (Photo courtesy of Western Electric.)

Figure 10 **BMEWS AN/FPS-49 Search/Track UHF Monopulse Radar; see Table 1g. (Photo courtesy of RCA.)**

Figure 11 **Hawk AN/MPQ-35 Pulse Acquisition Radar. (Photo courtesy of Raytheon Company.)**

In some instances the combined search and track function must be met. For this case a number of alternatives are available. One could build a system using a search radar and one or more tracking radars. The NATO Sea Sparrow point defense system uses an X-band radar (Figure 8) for tracking and typically relies on an L-band radar for search and acquisition; the old Nike-Ajax system relies on an S-band radar for search and acquisition and an X-band radar for tracking (Figure 4). Similarly, the Hawk system relies on an X-band tracking radar (Figure 5) and lower frequency radar for high elevation angle search and acquisition (Figure 11); the TPN-19 airport surveillance radar uses an S-band radar for search and acquisition (Figure 12) and an X-band radar for precision tracking.

A second alternative is to build a radar (often at a compromise frequency) that can search and track simultaneously. A multifunction phased-array radar or a track-while-scan radar can be used [1,2]. Examples of the former are the Missile Site Defense Radar (MSR) and Patriot (formerly SAM-D) radar (Figures 13 and 14). The MSR is an S-band multifunction phased-array radar capable of both search and track of a large number of reentry vehicles (RVs). The Patriot radar is a C-band air-defense multifunction phased-array radar capable of detecting and tracking many aircraft and missiles simultaneously. Examples of track-while-scan radars are given in the next section.

If one radar is to do both track and search functions the best "compromise" frequency is not always S- or C-band. For deep-space search and track requirements, a VHF or UHF system is desirable in most instances, as large apertures for long range detection are the driving function. Three examples of UHF multifunction phased-array radars are the Perimeter Acquisition Radar (PAR), AN/FPS-85 space radar, and SLBM early warning PAVE PAWS radar (Figures 15, 16, and 17).

If severe space object identification requirements are imposed for deep-space search and track radars, a higher frequency than UHF

Figure 12 **ASR AN/TPN-24 of AN/TPN-19 Air Traffic Control System. Stacked horns illuminate parabolic reflector to shape vertical fan beam without resorting to extra antenna height; see Table 1a. (Photo courtesy of Raytheon Company.)**

Figure 13 Missile Site Radar (MSR), a multifunction phased array short range radar used in Safeguard anti-ballistic missile (ABM) system; see Table 1f. Located at Spartan launch sites and used to steer Spartan interceptors as well as track the enemy missiles. (Photo courtesy Raytheon Company.)

Figure 14 Multifunction Phased Array Radar of PATRIOT (Formerly SAM-D) Tactical Air-Defense System; designed as future replacement for Hawk and Nike-Hercules Systems; see Table 1f. (Photo courtesy of Raytheon Company.)

Figure 15 Perimeter Acquisition Radar (PAR), a long range multifunction phased array radar used in Safeguard ABM system to detect and accurately track ICBM missiles. Hands off to MSR [31] ; see Figure 13 and Table 1f. (Photo courtesy of General Electric.)

Figure 16 AN/FPS-85 Long Range Phased-Array Multifunction UHF Radar. At left is 5,184 element transmitter, at right is thinned array (4,660 active elements, 19,500 passive crossed dipoles) receiver. For non-raid conditions, 30% of time is spent on space surveillance, 50% on space tracking for NORAD, and 20% on SLBM early warning surveillance [13] ; see Table 1f. (Photo courtesy of Bendix Corporation.)

Figure 17 Multifunction UHF Phased-Array Radar AN/FPS-115 (PAVE PAWS) for Early Warning and Attack Characterization of SLBMS; secondary mission is support of Air Force Spacetrack Program. Each of the two faces has 1792 active elements and 885 dummy elements over a 72.5 ft. diameter. Power output per face is about 145 kW av., 585 kW peak. Range, 3000 nmi for 10 m^2 target. The active array area can be increased to 102 ft. (Brookner)

Figure 18 AN/FPS-108 (Cobra Dane) Long Range Multifunction Phased-Array Radar. Designed for three missions — intelligence gathering on Soviet missile systems undergoing test firings, spacetrack, and ICBM early warning. Thinned array of 15,360 active, 34,769 passive elements [36] ; See Table 1f. (Brookner)

may be needed. If target imaging is required, very wideband transmitter signals (200 MHz or greater) may be used and frequencies of L-band or higher are required. An example is the L-band Cobra Dane Radar located at Shemya, Alaska (Figure 18). Ionosphere (see Chapter 14), hardware, and frequency-allocation limitations prevent one from obtaining such bandwidths at UHF and below.

There are applications where the "compromise" frequency has to be very high, such as in the airborne multifunction-array radar. This type of radar has to do search, track, terrain avoidance, navigation updates, and synthetic aperture spotlight mapping (for target identification, navigation updating, and damage assessment; see Chapter 18, Figure 1, and Chapter 20, Figure 3 for illustration). The aperture size for this type of radar must be small. Consequently, high frequency is needed to achieve the required angular accuracy for the tracking function; typically, X- or Ku-band is chosen. Fortunately, the target cross sections and the ranges at which they must be detected and tracked are such that detection and tracking can be achieved with a reasonable amount of transmitter power for the small apertures used. Examples of airborne multifunction-array radars are the AN/APQ-140 (Figure 19) and RASSR (see Chapter 20).

The requirement of detection of low flying targets (100 feet) imposes high frequencies (like X-band) for ground- or ship-based surface search radars. Alternately, a high siting of the radar could be used (Figure 20).

Another reason for selecting a high carrier frequency is to achieve a very narrow elevation beamwidth in order to reduce the degrading effects of ground or sea multipath when tracking low flying targets. A system which uses this approach is the Netherlands' HSA (Hollandse Signaalappa Flycatcher) radar, which can switch between X-band and Ka-band; the Ka-band is used for very low elevation angles when weather permits.

A low frequency search radar is often used in conjunction with a high frequency search radar in order to cover high and and low elevation angles simultaneously. One example is the Hawk system, which uses a low frequency radar (Figure 11) for high elevation angles and a high frequency (X-band; Figure 21) radar for horizon search.

Sometimes two widely spaced frequencies instead of one are selected for a search radar to provide increased invulnerability to jamming. The Marconi S631 surveillance radar is such an example (Figure 22). It uses two back-to-back antennas with one at L-band, the other at S-band (using either parabolic or cosec^2 reflectors).

Other trade-off factors that enter into the selection of radar frequency are the dependence on frequency of the receiver noise figure (Figures 23 and 24), sea and land clutter [3,4] (Figures 25 and 26), rain clutter (Figure 27), and atmosphere [5] and rain [6] attenuation (Figures 28 and 29). These figures indicate that clutter, atmospheric attenuation, and receiver noise figure tend to increase with higher frequency.

When lowering the operating frequency to UHF or VHF and siting at a northerly latitude, aurora clutter [7,8] becomes a factor. This clutter increases with decreasing carrier frequency. The maximum aurora volumetric scattering at UHF is about 10^{-8}m^2/m^3 (seen by the Prince Albert radar in Saskatchewan, Canada) [7]. Sky cosmic noise also becomes a problem at UHF and below (Figure 30) [9].

For a ground-based radar, the O_2 and H_2O absorption lines [5] (Figure 28) make millimeter waves unattractive except for a covert or short range radar. An example of the latter is the HSA "Flycatcher" which uses Ka-band to remove ground multipath. For spaceborne radar applications, where atmospheric attenuation and radar clutter are not problems, millimeter wave radars are very attractive.

Figure 19 **AN/APQ-140 Airborne Multifunction Phased-Array; see Table 1f. (Photo courtesy of Raytheon Company.)**

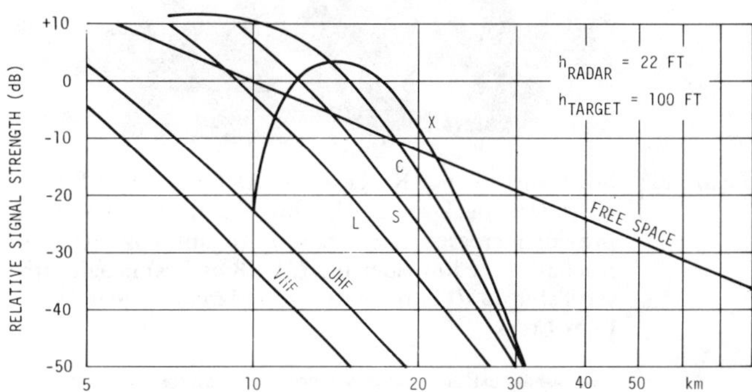

Figure 20 **Comparison of Signal Received for Free Space Propagation to Signal Received in Presence of Perfectly Reflecting Earth, isotropic antenna and target assumed. Signal return normalized to that obtained at a range of 10 km for free space propagation. h_{RADAR} and h_{TARGET} represent heights of radar and target, respectively. (From D.K. Barton with permission [29].)**

Figure 21 **X-Band AN/MPQ-34 CW Low Elevation Angle Acquisition Radar for the Hawk System, which provides azimuth, Doppler velocity, and range data (through FM techniques) [30]. (Photo courtesy of Raytheon Company.)**

Figure 22 **S631 Surveillance Radar for Air Defense Radar Systems. Uses back-to-back L- and S-band antennas which provide increased invulnerability to jamming. Antennas have very low sidelobes (28 dB or less) in azimuth; see Table 1a. (Photo courtesy of Marconi Radar Systems Ltd.)**

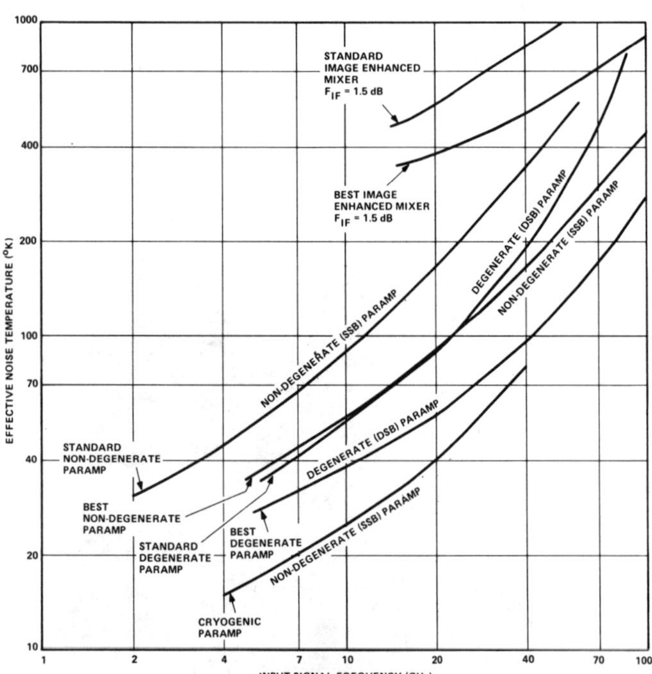

Figure 24 **State of the Art of Cooled (Cryogenic) and Uncooled Parametric Amplifiers. The non-degenerate single sideband (SSB) amplifiers are used for radars whereas the degenerate double sideband (DSB) amplifiers are used for radiometers. The best amplifier performances indicated are obtained using selected circulated and varactor components. (Figure courtesy of J. Whelehan of AIL.)**

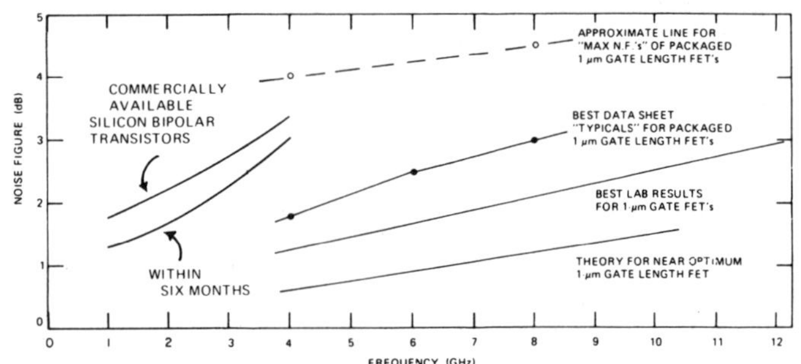

Figure 23 **State of the Art of Bipolar Transistors and FET Low Noise Devices [37, 38]**

Low carrier frequencies are sometimes selected because they can provide unambiguous target Doppler velocity measurements. Assume that pulse Doppler processing is used – N echo pulses resulting from N radar transmissions T_{PP} seconds apart are processed coherently. This processing Fourier transforms these N echo pulses to obtain the frequency of the echo signal. Assuming only approaching targets, the maximum Doppler frequency that can be measured unambiguously is equal to the radar PRF (pulse repetition frequency), f_R, which in turn is equal to one over the radar pulse-to-pulse period, T_{PP}. The corresponding maximum target Doppler velocity that can be measured unambiguously, $v_{D,U}$, is*

$$v_{D,U} = \frac{\lambda f_R}{2} \tag{5}$$

Thus, to increase $v_{D,U}$, the carrier frequency should be lowered.

For example, assume target range is to be measured unambiguously out to 100 nmi. A (100 nmi) (12.355 μs/nmi) = 1235 μs pulse-to-pulse period is required, resulting in an 809 pulses/second radar pulse repetition rate (PRR). Assume a radar frequency of 140 MHz (VHF), λ = 7.03 ft, and $v_{D,U}$ = 2844 ft/s (1685 nmi/h). If targets traveling up to Mach 2 (~2000 ft/s) must be observed, they can be seen unambiguously in Doppler velocity and in slant range.

If a radar carrier frequency of 10,000 MHz (X-band) were selected for the example above, $v_{D,U}$ would equal ~40 ft/s (23.6 nmi/h). For this carrier frequency a Mach 2 target Doppler velocity would not be measured unambiguously. The maximum Doppler velocity measured would be in the fiftieth Doppler velocity ambiguous interval (i.e., folded over fifty times in Doppler).

Not only is the target Doppler velocity ambiguous at higher carrier frequencies but the Doppler-velocity frequency interval over which no clutter is present becomes smaller. For Sea State 3 conditions, the sea clutter 3 dB Doppler velocity spectrum width is about 5 ft/s. One has to be at least ±2 (5 ft/s) away from the ambiguous velocity lines to be free of sea clutter. Thus, for the X-band carrier frequency, only 20 ft/s of the 40 ft/s unambiguous region is not clutter free; about half of the Doppler velocity frequencies are free of clutter. For the VHF carrier frequency, only 20 ft/s of the 2844 ft/s (0.7%) unambiguous region is not free of clutter. The situation at X-band would not be as bad as indicated if the maximum radar range were less. For example, if the range were 25 nmi, the clutter free region would be 87.5% of the Doppler frequencies.

If clutter is not a problem and the target Doppler velocity must be determined very accurately, a higher carrier frequency may be preferred, the Doppler velocity accuracy being proportional to the radar carrier frequency assuming that Doppler ambiguities are no problem or can be resolved. This fact is seen readily by calculating

*If receding targets must also be dealt with, Equation (5) becomes $V_{D,U} = \lambda f_R/4$. Note that the unambiguous velocity interval is the same in both cases.

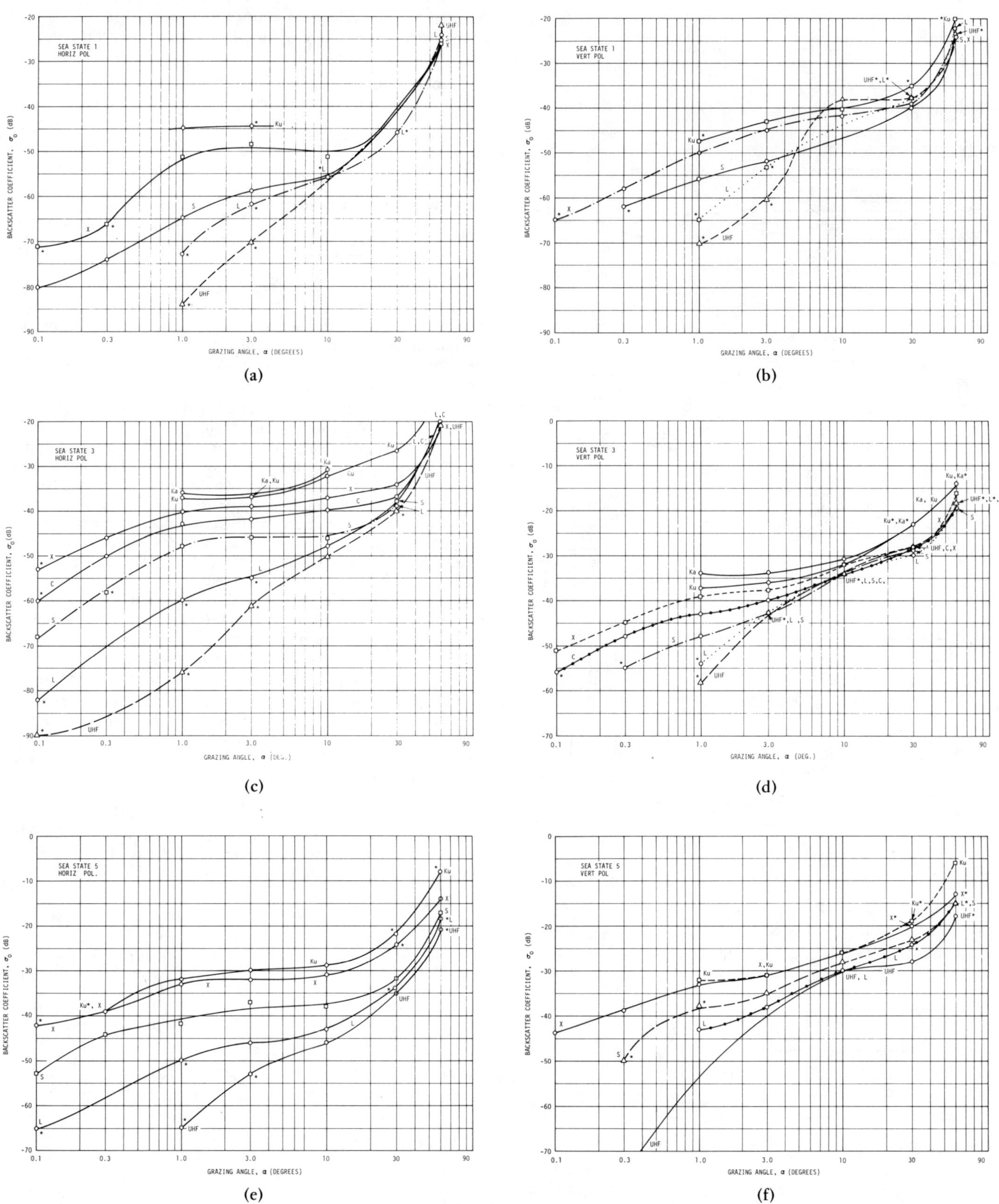

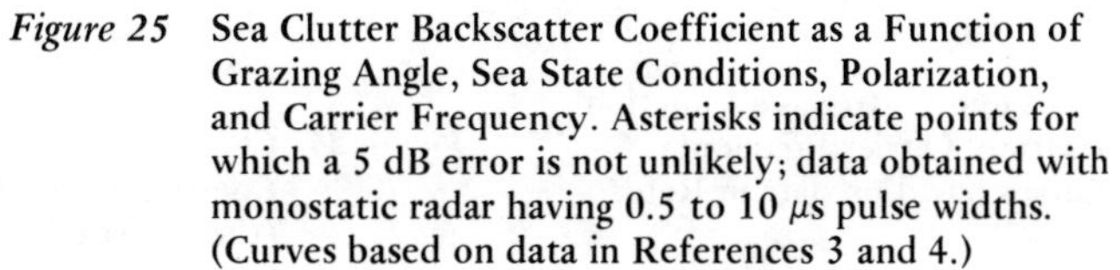

Figure 25 **Sea Clutter Backscatter Coefficient as a Function of Grazing Angle, Sea State Conditions, Polarization, and Carrier Frequency. Asterisks indicate points for which a 5 dB error is not unlikely; data obtained with monostatic radar having 0.5 to 10 μs pulse widths. (Curves based on data in References 3 and 4.)**

the radar Doppler velocity resolution and noting that the radar Doppler velocity accuracy is typically about one-tenth the Doppler velocity resolution.

The resolution of the Doppler frequency measurement, Δf_D, is approximately one over the integration time

$$\Delta f_D = \frac{1}{NT_{PP}} = \frac{f_R}{N} \tag{6}$$

The Doppler velocity resolution thus becomes

$$\Delta v_D = \frac{\lambda \Delta f_D}{2} = \frac{\lambda f_R}{2N} = \frac{v_{D,U}}{N} \tag{7}$$

Hence the Doppler velocity resolution and, in turn, accuracy improve with increasing frequency. For the above VHF and X-band examples it follows that the Doppler velocity resolutions, for the radar are 284 ft/s (169 nmi/h) and 3.98 ft/s (2.36 nmi/h), respectively, if N = 10.

Increasing the carrier frequency can, as indicated before, result in an undesirable ambiguous velocity problem; see Equation (5). If velocity ambiguities occur, they can be removed by using multiple PRFs.* Doppler velocity ambiguities do not always present a problem. An example is satellite-target two-dimensional imaging (slant range and Doppler or, equivalently, cross range) where only the relative velocities of scatterers on the target need be measured unambiguously. These relative velocities are small compared to the satellite velocity. Hence, a higher carrier frequency can be used, which results in ambiguous satellite velocity measurements but unambiguous relative velocity measurements of scattering points on the satellite. The ALCOR radar is at C-band to achieve two-dimensional satellite-target imaging.

Sometimes one specific requirement defines the radar frequency. Such was the case for the Wake Measurement Radar (WMR), the first ever put on board a reentry vehicle for wake measurements (Figures 31, 32, and 33) [10, 11]. The radar obtained wake returns by looking out the rear of the RV (Figure 33). Its frequency was made as close as possible to a ground-based C-band Nike TTR to facilitate comparison of the data from the two radars.

Mechanical vs. All-Electronic vs. Hybrid-Scanning Systems

The choice of antenna scanner depends on the system requirements — maximum number of targets to be detected at any one time, number of targets to be tracked, speed and maneuverability

*or/and multiple carriers.

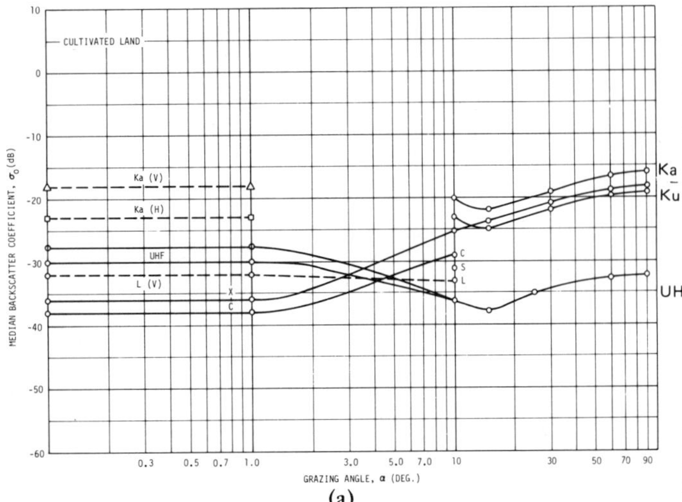

(a)

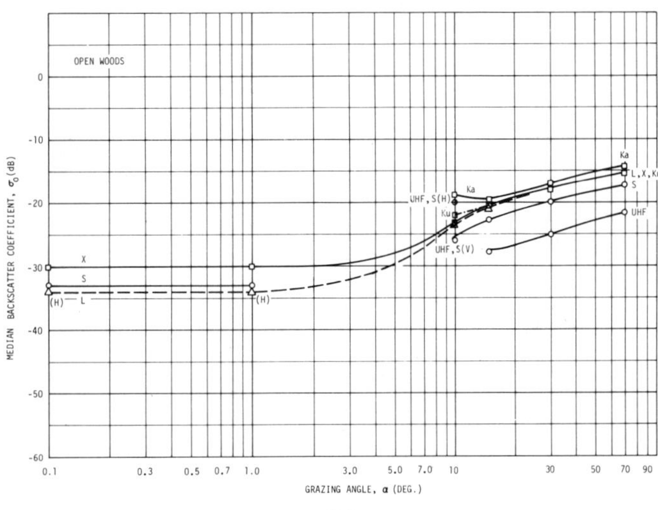

(b)

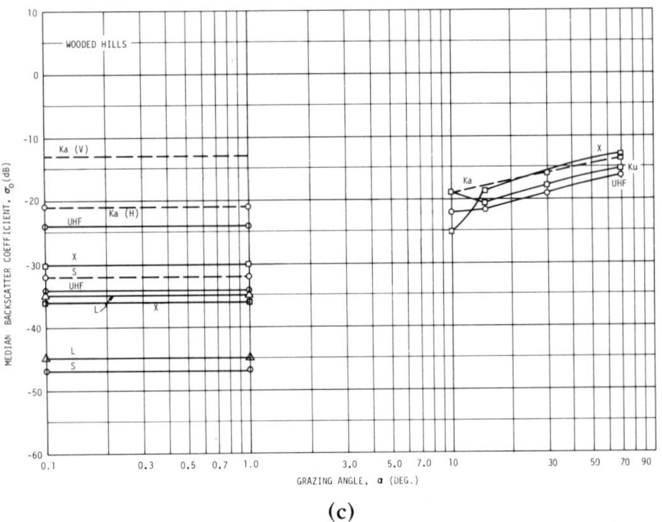

(c)

Figure 26 Backscatter Coefficient of Land Clutter as a Function of Grazing Angle, Polarization, Carrier Frequency, and Type of Terrain. (Curves based on data in Reference 4.)

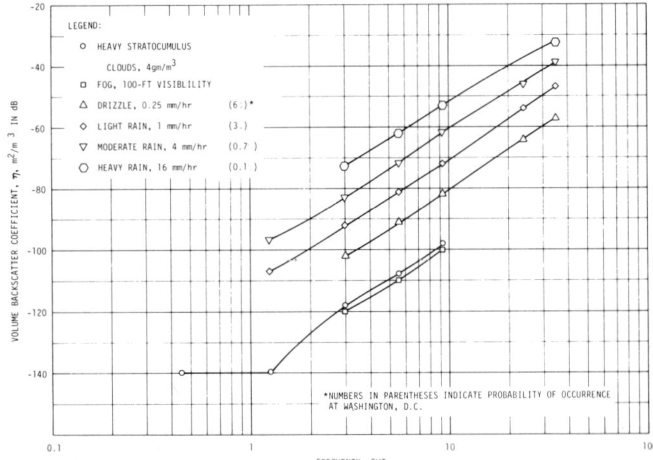

Figure 27 Cloud, Fog, and Precipitation Volume Backscatter Coefficient as a Function of Carrier Frequency. (Curves based on data in Reference 4.)

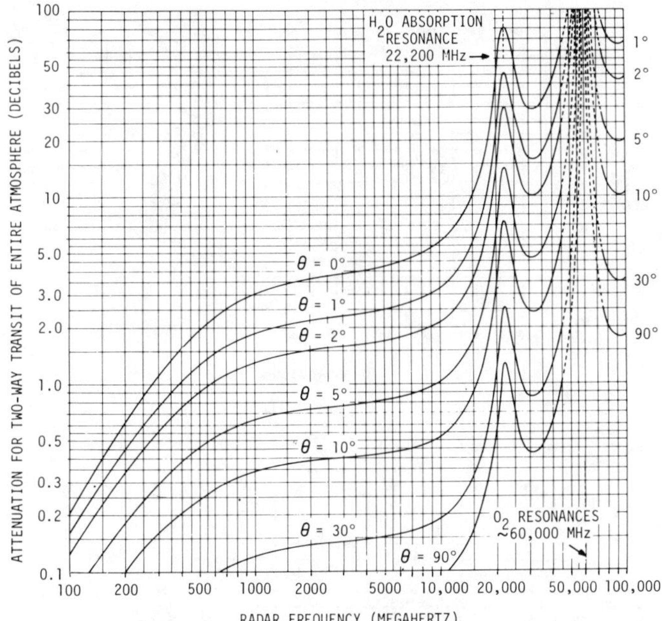

Figure 28a **Radar Attenuation Due to Atmospheric Absorption for Propagation Through Entire Troposphere as a Function of Carrier Frequency for Various Elevation Angles. Ionospheric loss, which may be significant below 400 MHz, and lens loss, which is significant for grazing angles θ below 5°, are not included. (From Blake with permission [5].)**

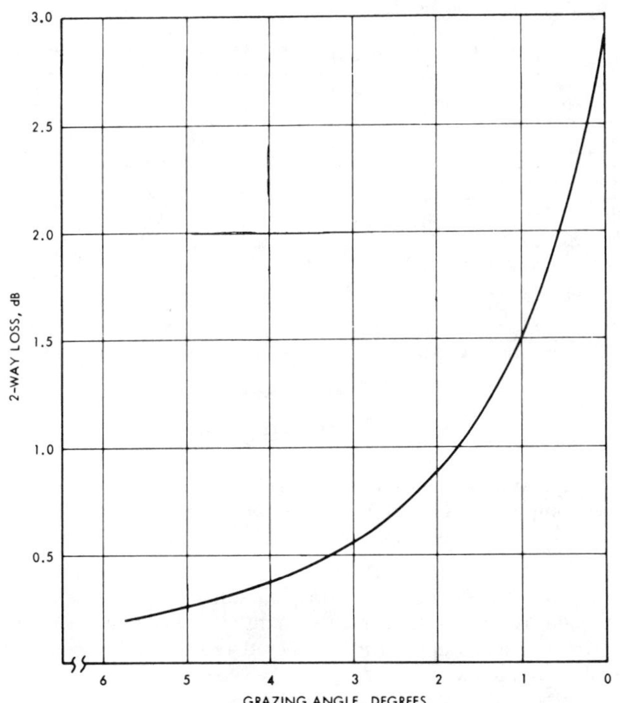

Figure 28b **Atmospheric Lens Loss (arising from the atmosphere acting as a divergent lens for grazing angles below 5°). (From Weil with permission [42].)**

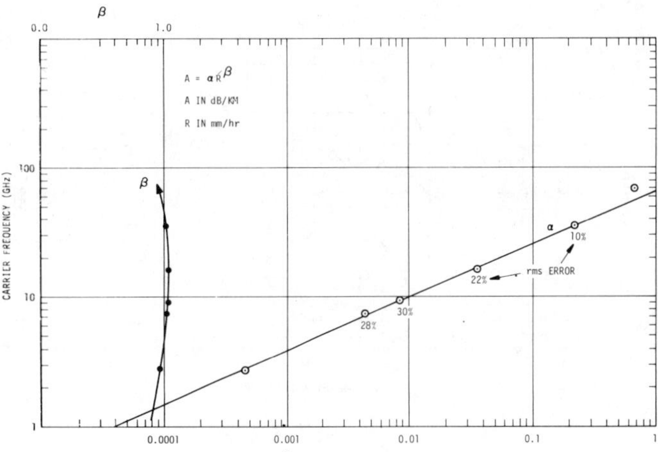

Figure 29 **One-way Rain Attenuation Coefficient, A, vs. Frequency. (From Hall with permission [44]; curves based originally on data published by Crain [6].)**

of the targets, and elevation and azimuth angle over which the targets can occur. The scanner ranges from the simplest and least costly mechanical scanners which are used when the system requirements are not severe to the most complex and most costly electronic scanners which are needed for the most demanding environment. In between are compromise hybrid and electromechanical scanning systems. Finally, there are also those systems which require no scanning. All these systems are discussed in the following paragraph with examples given of each.

As indicated, the simplest and least expensive type of scanning is usually achieved mechanically. If 360° of azimuth coverage over a large elevation angle (about 20°) is required, a vertically oriented fan beam rotated in azimuth may be used. Examples of such radars are the AN/MPQ-35, AN/TPN-24, AN/MPQ-34, AN/FPS-14, AN/FPS-20, AN/UPS-1C, AN/FPS-8, AN/SPS-29, AN/SPS-37, AN/SPS-43, AN/SPS-10, AN/SPS-49, AN/FPS-24, ARSR-2, ARSR-3, Model 3900, AN/FPS-107, AN/SPS-58A, ASR-803, Selenia's Argos-10, ATCR-2T, ATCR-3T, and ATCR-4T, Marconi's S654 and S650, Thomson-CSF's LP-23, and improved Nike Hercules HIPAR (Figures 11, 12, 21, and 34 to 55).

Typically the fan beam has a cosec^2 pattern in elevation in order to make most efficient use of the radar energy [12]. Some radars have used a cosec^2 pattern back-to-back with a narrow beam pattern (obtained with a simple parabolic dish), the latter providing long range coverage. This arrangement is the case of the WW-2 Long Range Ground Radar (Figure 56) [12]. (The Marconi S631 of Figure 22 is also capable of such a set-up.) Recently, air traffic control radars and one early warning radar have incorporated a second receive-only fan beam which is pointed upward so as not to see the ground or low elevation angle returns. The system transmits from a standard fan beam but initially receives (for the first 20 nmi, for example) from the high angle looking beam so as not to see close-in ground clutter, birds, or insects. Beyond 20 nmi, the signal is received from the fan beam used to transmit the signal. The ARSR-3, ASR-803, Marconi S654, Thomson-CSF LP23, Selenia ATCR-2T and -4T, and Argos -10 are such radars (Figures 43, 47 to 49, 51, 52, and 54).

For improved tracking capability, elevation angle information is also desired and a stacked beam system can be used. Examples are the AN/FPS-7, AN/TPS-43, AN/TPS-27, Russian Barlock, and Thomson-CSF TH.D. 1955 radar (Figures 57 to 61). Receivers are becoming cheaper with the advent of microwave integrated circuitry and LSI circuitry. Thus, the stack beam system can be a very attractive one.

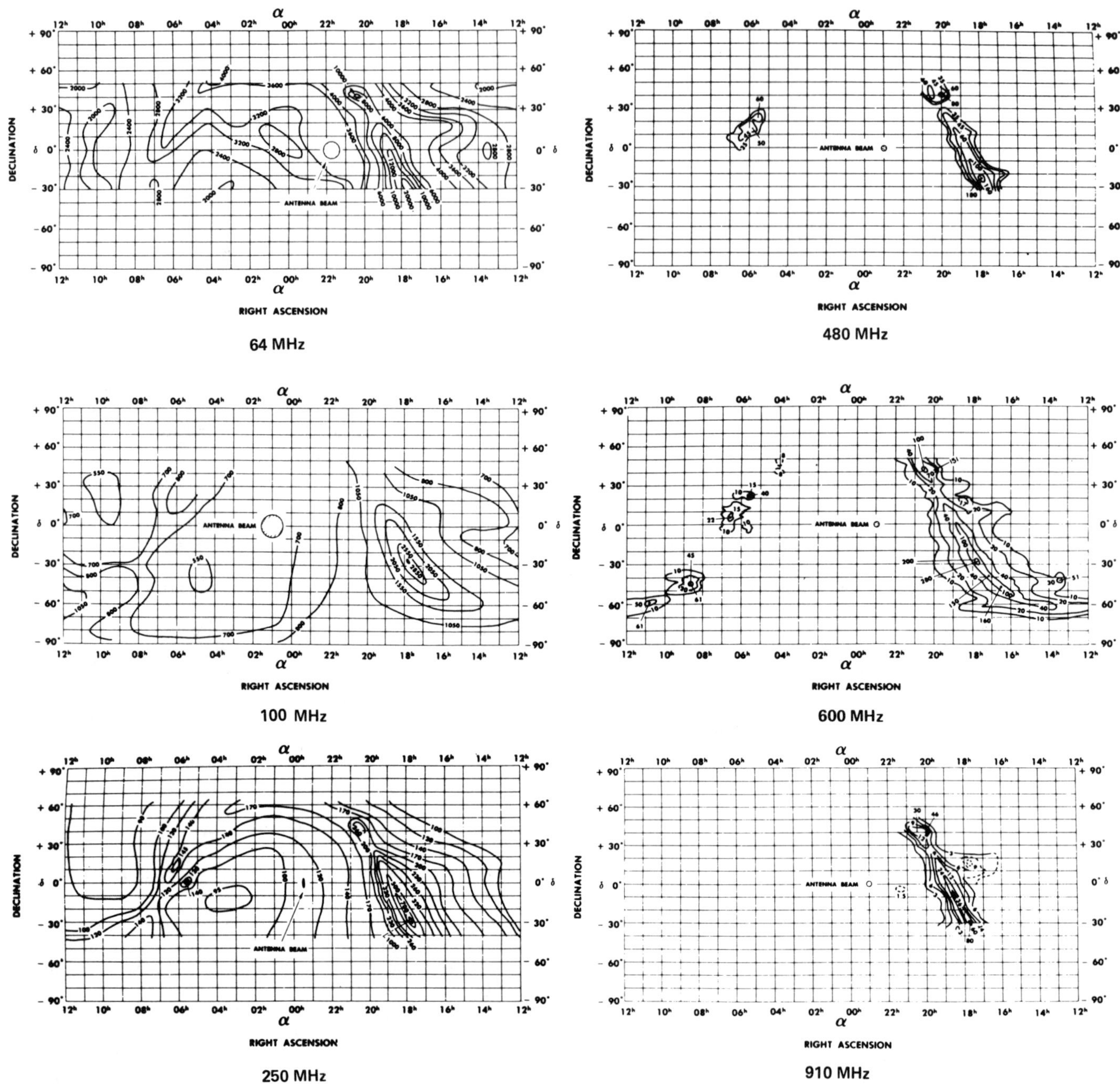

Figure 30 **Sky Cosmic Noise Temperature for Different Carrier Frequencies (Curves compiled by Ko [9].)**

Alternately, to obtain elevation angle information, a V-beam scanning system can be used. For this system two fan beams are scanned simultaneously. One of the beams is oriented vertically while the other is tilted at an angle from the vertical (such as 45°). The original V-band system is the MIT WW II AN/CPS-6B (Figure 62) [12] ; another example is the AN/TPS-34 (Figure 63).

Both the stack beam and V-beam systems provide target azimuth, elevation, and range information, and thus are referred to as 3-D radars. The simple fan beam systems (such as those of Figures 34 to 55) are called 2-D radars, as they provide only target azimuth and range data. (For flat ground sitings and careful elevation lobing pattern calibration, the latter arrangements can provide coarse target altitude data [41] .) Because the 3-D radars provide elevation, azimuth, and range information, targets can be tracked while the radar is doing search; such systems are called track-while-scan radars. For many applications, 2-D radars also can be used as track-while-scan radars.

The stacked beam system, although more complex than the V-beam system, provides better performance. The V-beam radar has a blip association problem for high target densities. The stack beam radar is less vulnerable to jamming because of its narrow beamwidth in both elevation and azimuth. The V-beam radar requires twice as much power as a simple 2-D search radar while the stacked beam radar requires substantially less power than its 2-D counterpart.

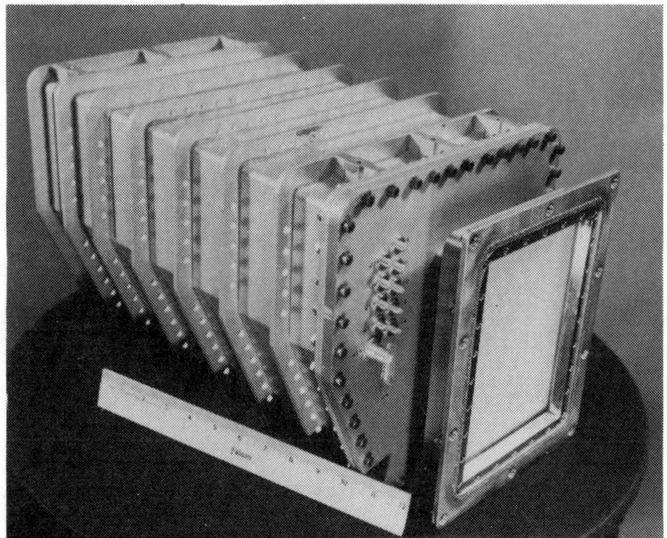

Figure 31 **Pressurized Transmitter Package of Wake Measurement Radar (WMR), only reentry measurement instrumentation radar placed on board a reentry vehicle [10, 11]. Seen at right foreground is the fused silica quartz antenna window. The receiver front end with the first IF amplifier is also contained in this package. (Photo courtesy of Raytheon Company.)**

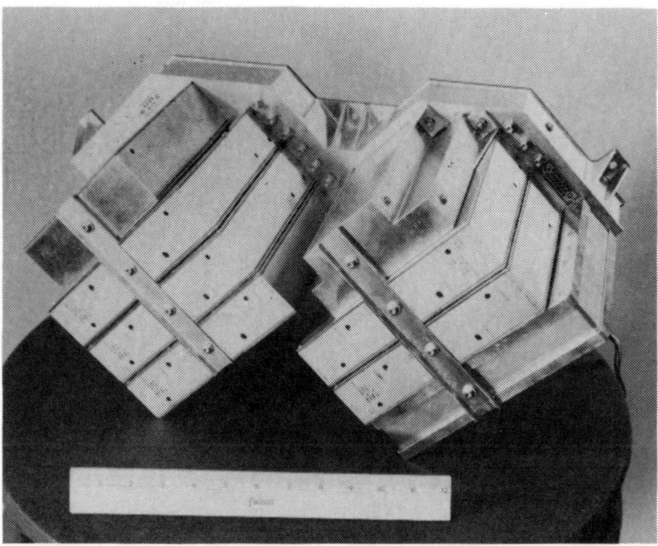

Figure 32 **WMR Ten Channel Receiver Processor, Digital Control Circuitry, Radar Clock, and Telemetry Conditioning Circuitry (Photo courtesy of Raytheon Company.)**

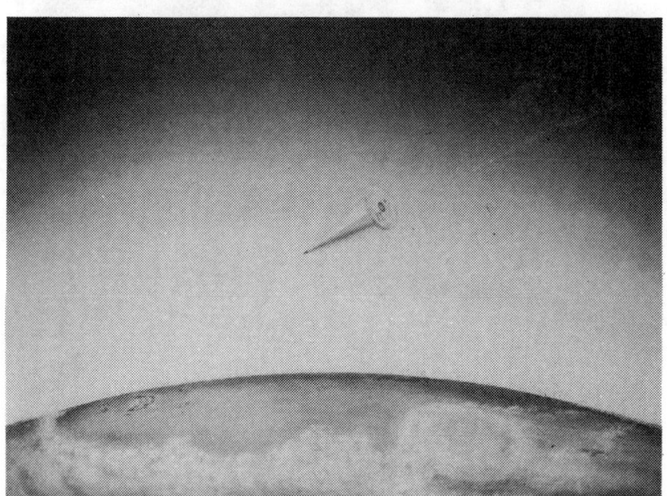

Figure 33 **Pictorial of WMR Looking out Rear of Reentry Vehicle (Photo courtesy of Raytheon Company.)**

Figure 34 **Medium Range AN/FPS-14 Gap Filler Search Radar for SAGE Air Defense System; see Table 1a. (Photo courtesy of Bendix Corporation.)**

An alternate simple way to obtain target elevation angle or, equivalently, height information, is to use a height finder radar in conjunction with a 2-D one. For this type of system a beam (usually slightly wider in azimuth) is scanned mechanically in elevation to provide height information on the target. Examples of height finders are the AN/TPS-10D, AN/FPS-6, Russian Cake Series, and Marconi S613 and S669 (Figures 64 to 68).

Typically, the height finder is directed to obtain the elevation angle of a target located by a 2-D search radar. Some height finders, such as the Marconi S669, can provide 3-D volumetric coverage (over 360° in azimuth) in case of failure of the 2-D radar. This radar also has a jammer burn-through mode.

The above height finders nod their main reflector up and down to scan a beam vertically. Such scanning of the beam can also be achieved by moving the antenna feed up and down, i.e., by electromechanical scanning. Most practical feeds of this type convert a small circular mechanical rotation of the feed to a large, rapid, saw-tooth elevation scanning of the beam. One such scanner is the Robinson used on the WW-2 SCI height-finding radar (Figure 69) [12]. A more recent height finder employing such a scanner is the S-band AN/SPS-8 (Figure 70). Thomson-CSF uses two Robinson scanners on their S-band TRS 2205 (Volex III) 3-D ground surveillance radar to scan two pencil beams in elevation (one from 0.5° to 7°, the other from 5° to 20°) while obtaining 360° coverage by mechanical rotation (3 RPM) in azimuth (Figure 71). An elevation scan speed of 720 scans per minute is achieved. Ordinarily this radar is used in conjunction with a 2-D radar but it can be used by itself as a 3-D radar. The Thomson-CSF DRBI 10 3-D Naval S-band surveillance radar and PICADOR mobile 3-D air defense S-band radar use a Robinson elevation scanner with mechanical azimuth scanning but only have one beam [13].

(text continues on page 51)

Figure 35 Long Range AN/FPS-20 Air Surveillance Radar; see Table 1a. It is one of a family of long range dual-channel search and GCI radars; others are AN/FPS-20A, -20B, -64, -65, -66, -67, -100, AN/GPS-4 and AN/MPS-7 [13]. (Photo courtesy of Bendix Corporation.)

Figure 36 AN/UPS-1C Transportable Search Radar; see Table 1a. (Photo courtesy of RCA.)

Figure 37 Medium Power Aircraft Search and Early Warning AN/FPS-8 Radar; see Table 1a. (Mobile version is MPS-11; parametric amplifier receiver version is AN/FPS-88.) (Photo courtesy of General Electric.)

Figure 38 AN/SPS-43 Long Range Search Radar. Its predecessors are the AN/SPS-37 and -29, all three are in service; see Table 1a. (Photo courtesy of Westinghouse Electric Corporation.)

Figure 39 US Navy AN/SPS-10 Surface Search Radar; see Table 1a. (Sylvania)

Figure 40 New US Navy Long Range Air Search AN/SPS-49 Radar; see Table 1a. (Photo courtesy of Raytheon Company.)

Figure 42 ARSR-2 Enroute 2-D Air Surveillance Radar; see Table 1a. (Photo courtesy of Raytheon Company.)

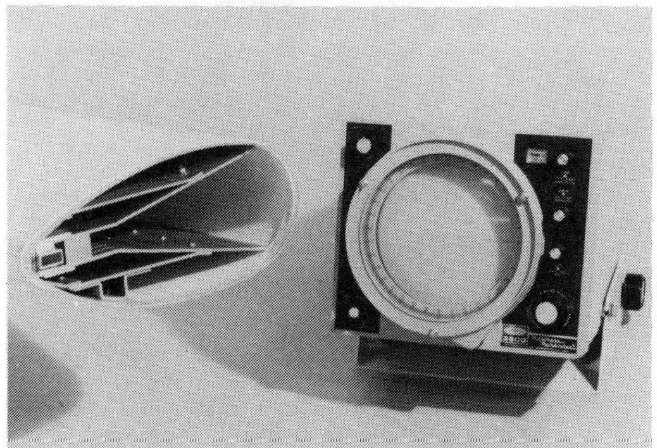

Figure 44 2-D Commercial Small-Boat Model 3900 Navigation Radar; see Table 1a. (Photo courtesy of Raytheon Company.)

Figure 41 AN/FPS-24 2-D Search Radar; see Table 1a. (Photo courtesy of General Electric.)

Figure 43 New ARSR-3 Enroute 2-D Air Surveillance Radar; see Table 1a. High elevation angle receive-only beam used in conjunction with normal transmit/receiver beam; see text. (Photo courtesy of Westinghouse Electric Corporation.)

Figure 45 AN/FPS-107 2-D Surveillance Radar; see Table 1a. (Photo courtesy of Westinghouse Electric Corporation.)

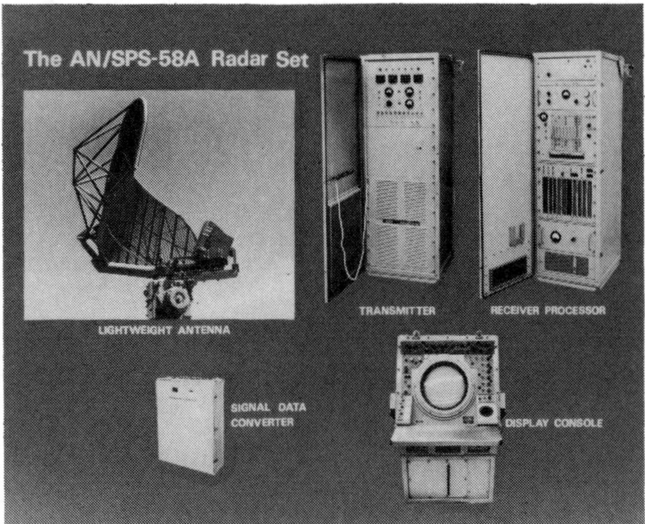

Figure 46 Shipboard AN/SPS-58A 2-D Surveillance Radar; see Table 1a. (Photo courtesy of Westinghouse Electric. Corporation.)

Figure 47 ASR-803 2-D Airport Surveillance Radar; see Table 1a. Selenia G-14 antenna used, as on ARGOS-10 (Figure 48) and on some versions of ATCR-2T and -4T (Figures 49 and 51). (Photo courtesy of Raytheon Canada Ltd.)

Figure 48 ARGOS-10 2-D Early Warning Radar for Operation Against High-Performance Aircraft and Air-to-Surface Missiles, which uses sophisticated ECCM circuitry and dual beam antennas (see Figures 43 and 47 and text); see Table 1a. (Photo courtesy of Selenia.)

Figure 49 ATCR-2T L-Band Enroute 2-D Air Surveillance Radar at Vienna Site. G-7 antenna used (as with ATCR-4T of Figure 51) here instead of G-14 (of Figures 47 and 48). G-7 has dual elevation beams, as do G-14 and ARSR-3; see Table 1a. Built either with single channel or dual channels for frequency diversity or to have a hot standby second transmitter. (Photo courtesy of Selenia.)

Figure 50 ATCR-3T S-Band 2-D Airport Surveillance Radar in Geneva; see Table 1a. Single and dual channel systems built. (Photo courtesy of Selenia.)

Figure 51 ATCR-4T L-Band 2-D Airport Approach Radar and GCA Surveillance Radar in Hong Kong; see Table 1a. Single and dual channel systems built; has clutter map. (Photo courtesy of Selenia.)

Figure 52 S654 L-Band 2-D Enroute, Terminal Area, or Approach Surveillance Radar for civil and military use; see Table 1a. Single and dual mode systems built. First blind velocity can be one hundred times the mean PRF. Uses dual beam antenna, such as that for ARSR-3 and ARGOS-10; see text. (Photo courtesy of Marconi Radar Systems Ltd.)

Figure 53 S650 2-D Enroute or Terminal Area Surveillance Radar at Heathrow Airport; see Table 1a. (Photo courtesy of Marconi Radar Systems Ltd.)

Figure 54 TRS 2050 Long Range L-Band Surveillance Radar for Air Defense and Air Traffic Control (civilian version designated as LP23); see Table 1a. Dual and triple channel frequency diversity available, uses novel DMTI in which the phase variations are thresholded to detect moving targets. This technique eliminates the velocity response characteristic of conventional MTI. Second-time-around returns eliminated by digital video correlation and video blanking. (Photo published with permission of Thomson-CSF.)

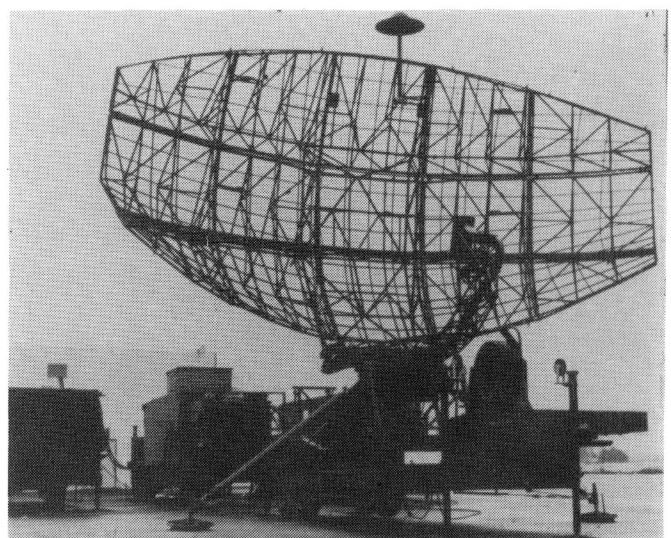

Figure 55 Mobile L-Band HIPAR (High Power Acquisition Radar) of Improved Nike-Hercules System deployed in 1960s for defense against high performance aircraft, air-to-ground missiles, and tactical ballistic missiles. Used also for latest Nike family system configuration – the Nike-Hercules SAMCAP fielded in late 1960s and early 1970s; see Table 1a. (Photo courtesy of Western Electric.)

Figure 56 S-Band WW-2 Long-Range Radar using two back-to-back 25 ft cylindrical reflectors – one for low elevation long range (8 ft parabolic dish), the other for high elevation (5 ft, csc^2 beam); see Table 1a. (Photo from Volume 1 of *MIT Radiation Laboratory Series.*)

Figure 57 L-Band AN/FPS-7 Vertically Stacked-Beam 3-D Surveillance System, with a standby transmitter (fed to dummy load) which permits maintenance without downtime; see Table 1b. (Photo courtesy of General Electric Company.)

Figure 58 S-Band AN/TPS-43 Stacked-Beam 3-D Surveillance Radar; see Table 1b. Altitude accuracy +305m at 100 nmi [13]. (Photo courtesy of Westinghouse Electric Corporation.)

Figure 59 S-Band AN/TPS-27 Stacked-Beam 3-D Tactical Surveillance Radar, predecessor to AN/TPS-43; see Table 1b. (Photo courtesy of Westinghouse Electric Corporation.)

Figure 60 S-Band Russian Barlock Stacked-Beam 3-D Radar; see Table 1b. The six beams operate in different frequency bands; 2695-2715, 2715-2750, 2815-2835, 2900-2990, 2990-3025, and 3080-3125 MHz. Entire van rotates [14]. (Photo courtesy of *Aviation Week and Space Technology.*)

Figure 61 TH.D. 1955 S-Band Long-Range 3-D Stacked-Beam Radar, one of the most powerful air defense surveillance radars; see Table 1b. Selected by NATO for use with NADGE system. (Photo published with permission of Thomson-CSF.)

Figure 62 First 3-D Radar, the WW-2 AN/CPS-6B V-beam radar; see Table 1b. Five transmitters having five different carrier frequencies are used, the vertical beam being formed by three beams in elevation and the 45° beam by two. (Photo from Volume 1 of *MIT Radiation Laboratory Series.*)

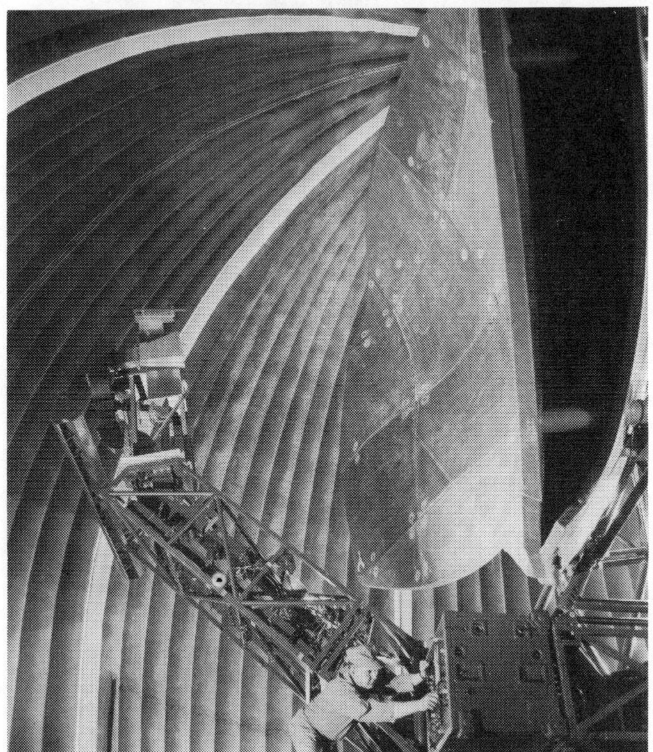

Figure 63 US Marine Corps L-Band AN/TPS-34 3-D V-Beam Mobile Radar used for beachheads and forward positions; see Table 1b. Believed to be first tactical V-beam system. (Photo courtesy of Sperry Gyroscope.)

Figure 64 X-Band AN/TPS-10D Nodding Height-Finder Mounted on Tripod; see Table 1c. The AN/MPS-8 and AN/FPS-4 trailer and tower-mounted versions are functionally identical; 450 units built between 1948 and 1955. (Photo courtesy of RCA.)

Figure 66 Russian Cake Series S-Band Truck-Mounted Height Finder; see Table 1c. Sponge Cake version has range accuracy of ±1 nmi and height accuracy of 500 m, both at range of 100 nmi [14]. (Photo courtesy of *Aviation Week and Space Technology*.)

Figure 65 US Army S-Band AN/FPS-6 Nodding Height-Finder, successor to AN/TPS-10 series (mobile version is AN/MPS-14; improved version is AN/FPS-89); see Table 1c. 450 units built between 1953 and 1960. (Photo courtesy of General Electric.)

Figure 67 C-Band S613 Air Defense Height-Finder, with simple nodding motion or computer controlled for automatic operation with height extraction for up to 22 heights per minute; see Table 1c. (Photo courtesy of Marconi Radar Systems Ltd.)

Figure 68 S-Band S669 Static Air Defense Height-Finder with the following modes – single shot (one nod per target), automatic searchlight (continuous nodding at target bearing), manual searchlight, burn through, volumetric scan, and sector volumetric scan; see Table 1c. (Photo courtesy of Marconi Radar Systems Ltd.)

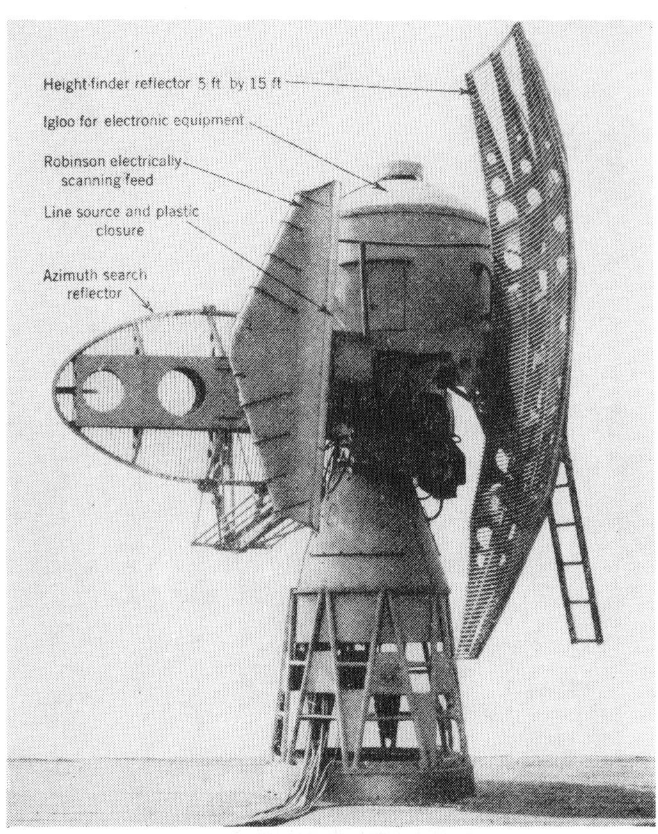

Figure 69 S-Band (3 GHz) WW-2 SCI Height Finder using Robinson Scanner, with a 5 x 15 ft grating reflector and 8 x 2 x 1 ft rolled trapezoidal Robinson feed. Beamwidth 3.5° in azimuth, 1.2° in elevation; 10 scans per second of 10.5° elevation sector [15]. (Photo from Volume 1 of *MIT Radiation Laboratory Series.*)

Figure 70 Shipborne S-Band 3-D Surveillance and Height-Finder AN/SPS-8 Radar using Robinson scanner; see Table 1d. (Photo courtesy of General Electric.)

Figure 71 3-D TRS 2205 (VOLEX III) Radar using two reflectors fed by separate Robinson scanners; see Table 1d. (Photo published with permission of Thomson-CSF.)

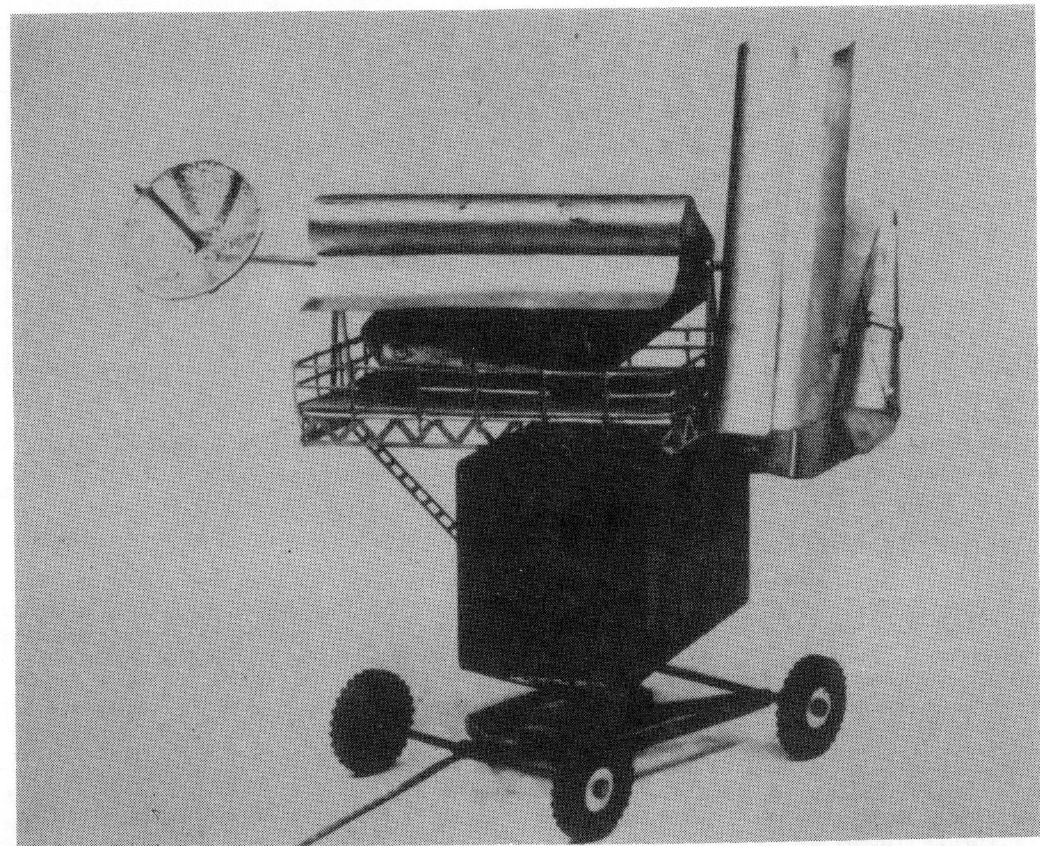

Figure 72a **Soviet S-Band Fan Song B Tracking and Guidance Radar for SA-2 Guideline Missile. It has two trough-like Lewis electromechanical scanners mounted at right angles. Separate frequency bands are used for each scanner; see Table 1d. The small parabolic dish is used to send UHF guidance pulses to the missile. (Photo courtesy *Aviation Week and Space Technology*.)**

Figure 72b **Newer Soviet C-Band Fan Song E Missile Fire Control Radar for SA-2 Guideline Surface-to-Air Missile. It also uses Lewis scanner horizontally and vertically oriented trough-shaped antennas; see Table 1d. Two parabolic dishes above horizontally-oriented trough-antenna provide LORO countering DECM. When enemy uses DECM on Lewis scanner to shift target angle, radiation from Lewis scanners is halted and initiated from wide beam parabolic dishes with reception occurring on Lewis scanners. This tactic nullifies enemy knowledge of being locked onto as intended SAM victim; see Reference 14. (Photo courtesy of *Aviation Week and Space Technology.*)**

Some systems scan two fan beams, one in elevation and the other in azimuth to obtain 3-D information on the target. This approach is used by the Russians on the Fan Song radars [14] (Figure 72) – two electromechanical scanners (called Lewis scanners [15]) are used, one for elevation scanning, the other for azimuth scanning. The Lewis scanner, like the Robinson scanner, uses the mechanical rotation of a feed in a circle to achieve either elevation or azimuth scanning (depending on the orientation of the scanner). Whereas the Thomson-CSF TRS-2205 provides 3-D coverage over a 360° by 20° region, the Fan Song S-band radar (Figure 72a) provides 3-D coverage over only a 10° by 10° sector, as the fan beams are 10° by 2° and scanned 10° (see Table 1). The Fan Song E (Figure 72b) is at C-band and scans two fan beams 7.5° by 1.5° in perpendicular directions (see Table 1).

Electromechanical scanning has been used to obtain rapid horizon coverage with a pencil beam, as with the AN/MPQ-4 Mortar locating radar which obtains rapid horizon coverage over a limited azimuth (25°) by the use of a Foster dual-scanner [15] electromechanical feed (Figure 73). This feed uses the circular rotation of a cone inside a conical shell to sequentially scan 25° in azimuth with a pencil beam at two different elevation angles. Detection of a mortar shell at both elevation angles allows backtracking to the mortar location. In contrast to the Lewis and Robinson scanners, the Foster scanner has a greater vulnerability to jamming because the beam scan angle is dependent on the radar carrier frequency.

Another electromechanical scanner is the Eagle scanner used during World War II on the airborne AN/APQ-7 high-resolution navigation and bombing radar [12]. The antenna consists of a horizontally-oriented waveguide which feeds a set of dipole radiators. The energy is fed down one end of the waveguide; the width of the waveguide is varied to scan the beam 30° off broadside. By feeding the energy from the other end, the beam scans 30° in the opposite direction [12]. The Eagle scanner was used during the later phases of the war on the MPN-1 Precision Acquisition Radar (PAR) Ground Control Approach (GCA) system. For this system, two Eagle scanners scanned two fan beams, one in azimuth, the other in elevation. This system has been upgraded through the years (CPN-4, MPN-5, MPN-11 through MPN-15 and FPN-16). Its most recent version is the solid state (except for the transmitter power tube and thyratron) Normal PAR (AN/FPN-62) which has just been developed (Figure 74).

Another type of electromechanical scanner uses a linear array of feeds into which energy is sequentially switched mechanically by means of rotary switch – the organ-pipe scanner [15-17]. Instead of mechanical or electromechanical steering in elevation, electronic steering can be used in elevation with mechanical steering in azimuth to obtain 3-D coverage. Radars of this type are the AN/SPS-39 (Figure 75), AN/SPS-48 (Figure 76), AN/TPS-32 (Figure 77), AN/SPS-52 (Figure 78), and Plessey AR-3D (Figure 79) [2, 16, 18]. For these systems, varying the radar frequency scans a pencil beam in elevation. Frequency scanning is less expensive than phase scanning; however, it has the disadvantage of a pre-programmed frequency vs. elevation scan being required. As a result, the system is vulnerable to jamming, the radar frequency being known for each elevation scan angle.

An interesting system that employs two back-to-back search antennas along with two back-to-back tracking antennas on the same mount is the Hollandse Signaalapparaten (HSA) 3-D Multi-Target Tracking Radar (3-D MTTR) (Figure 80) [19]. The search antennas use electromechanical switching between five feeds to obtain elevation scanning up to 15°; the mechanical rotation of the mount provides azimuth scanning. The tracking antennas are made up of slotted waveguides which permit frequency scanning of a beam. The waveguides are placed at an angle so that frequency scanning provides vertical scanning of the beam when the mount rotates. Interleaved with this vertical scanning is the scanning of a beam horizontally across the target. A fifth multi-element antenna on the mount provides high angle coverage.

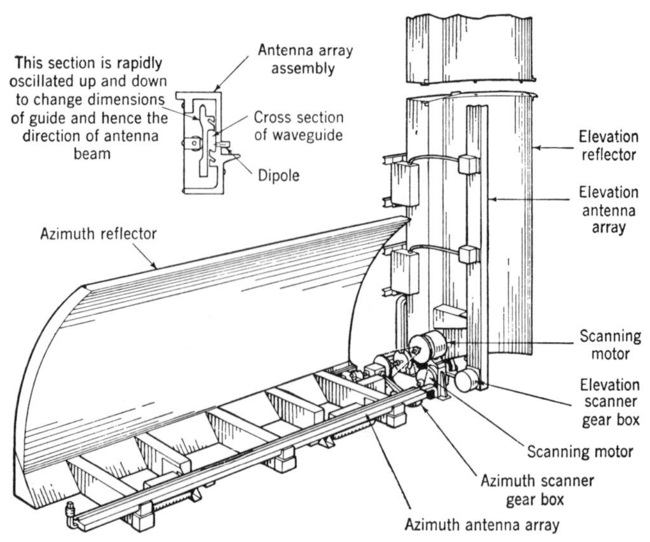

Figure 74a **Two Eagle scanners used to scan electromechanically vertically and horizontally oriented fan beams in azimuth and elevation, respectively. (Drawing from Volume 2 of *MIT Radiation Laboratory Series* [41].)**

Figure 74b **X-Band Precision Approach Radar AN/FPN-62 (also called Normal PAR) which uses two Eagle scanners to scan two fan beams – one horizontally, the other vertically. This radar is a direct descendant of the WW-2 AN/MPN-1, which was followed by the AN/CPN-4, AN/MPN-5, -11, -13, -15 and AN/FPN-16 radars; see Table 1d. The latest version is solid state throughout except for the power amplifier tube and the transmitter thyratron. (Photo courtesy of Raytheon Company.)**

Figure 73 **Ku-Band AN/MPQ-4 Mortar-Location Radar using Foster scanner; see Table 1d and text. (Photo courtesy of General Electric.)**

Figure 75 **US Navy AN/SPS-39 Frequency Scanned 3-D Radar; see Table 1e. Energy management is used to radiate more energy at low elevation angles for long range targets by using nonlinear frequency scan. PRF is lower for lower elevation angles, longer range targets [18]. (Photo courtesy of Hughes Aircraft Company.)**

Figure 76 **US Navy AN/SPS-48 3-D Long Range Surveillance Frequency Scanning Radar; see Table 1e. It is fed by serpentine delay line and uses flexible power management [18]. (Photo courtesy of ITT Gilfillan.)**

The Thomson-CSF Matador 3-D TRS 2210 radar (Figure 81) combines stacked beams (three in elevation) with phase scanning in elevation. This mobile radar is similar to the Thomson-CSF ANTARES height-finder radar (Figure 82) used for air defense and civil air traffic control.

For the most severe and demanding target environments, 3-D coverage is obtained by the use of a multifunction phased-array radar which permits simultaneous search and track [1, 2]. Because of its increased complexity (with respect to both hardware and software) and cost, this approach is warranted only when the mechanical, electromechanical, and hybrid approaches discussed cannot do the job. Such a situation occurs when the radar is required to track a large number of greatly accelerating targets. Examples of such radars are the MSR and Patriot radars (Figures 13 and 14). It is difficult to build a mechanically scanning radar that would provide the high data rates (around 10 samples per second) required for the large number of targets that must be tracked simultaneously. Mechanical track-while-scan radars are limited to sampling rates of about one sample per second. The MSR also has the requirement that it be hardened against nuclear attacks, an end which is not achieved easily with mechanically rotating antennas.

Other examples of electronically scanned arrays which use phase-steering in both elevation and azimuth are the AEGIS (Figure 83) [20], Cobra Dane (Figure 18), AN/FPS-85 (Figure 16), and PAVE PAWS (Figure 17). Finally, there is the Dome antenna which provides slightly greater than hemispherical coverage. This antenna uses a planar array having phase-phase steering. To provide greater than hemispherical coverage, this planar array is placed under a dome lens (Figure 84) (see the "Present and Future Trends" section of Part 5).

A compromise to the much more expensive phase scanning in two dimensions is the use of frequency scanning in one dimension and phase scanning in the other. An example of a system that uses frequency/phase steering is the AN/SPS-33 of Figure 85 [18]. For this tracking radar, a pencil beam is steered in elevation by frequency scanning and in azimuth by phase scanning.

The long-range detection and designation radar used in conjunction with the AN/SPS-33 is the AN/SPS-32 which frequency scans a fan beam in azimuth (Figure 85). This antenna is used on the nuclear carrier *USS Enterprise* and nuclear cruiser *USS Long Beach.* To obtain 360° of coverage, antennas are mounted on four sides of the ship's square superstructure.

Another compromise is to use an antenna which electronically scans over a limited field of view and to shift the electronic coverage to other regions by mechanical steering. A system of this type is the PAR AN/TPN-25 of the TPN-19 (Figure 86). Such a limited scan antenna can be rotated mechanically to the center of the 20° (azimuth) by 15° (elevation) region to be covered electronically [21, 22]. The fixed base version of this radar is the PAR AN/GPN-22 (Figure 87).

The simplest type of scanning is a fixed beam system – one that requires no scanning of the antenna beam – such as the WMR radar. It uses a single beam pointed out the rear of the reentry vehicle (see Figures 31-33) [10, 11].

For further details on other scanners as well as the ones described, the reader is referred to the references cited, especially 1, 2, 12, 15, 16, and 18.

Polarization

Proper selection of polarization can help to reduce radar clutter. At S-band and above, rain clutter can be significant. If a right-hand circularly (RHC) polarized signal is transmitted, a left-hand circularly (LHC) polarized echo signal is received from spherically

Figure 77 S-Band AN/TPS-32 3-D Frequency Scanned Radar used in Marine Tactical Data System; see Table 1e. It is a land based derivative of the AN/SPS-48. As with the AN/SPS-39 and -48, energy management is used in elevation; AN/TPS-64 has same basic operational and technical parameters. (Photo courtesy of ITT Gilfillan.)

Figure 78 US Navy AN/SPS-52 3-D Frequency Scanned Radar; see Table 1e. (Photo courtesy of Hughes Aircraft Company.)

Figure 79 S-Band AR-3D 3-D Frequency Scanned Radar; see Table 1e. Uses linear array feed in front of parabolic cylinder reflector. Provides 1° of scan per 10 MHz of carrier frequency change; energy management in elevation is used. Signal is swept 200 MHz on transmit and separated into 20 MHz channels on reception. An acoustic surface line compresses signal to 0.1 μs in each channel for fine elevation angle measurement (0.15°). (Photo courtesy of Plessey Radar.)

Figure 80 The Signaal 3-D Multi-Target Tracking Radar (MTTR) which uses both electromechanical and frequency scanning; see Table 1e and text. Capable of tracking up to 100 aircraft [19]. (Photo courtesy of Hollandse Signaalapparaten.)

shaped rain drops. Aircraft targets, being complex in shape, scatter both RHC and LHC polarized signals. Thus, by transmitting a RHC polarized signal and receiving only the RHC polarized echoes, most rain clutter can be eliminated (see Figure 88).

The proper choice of polarization also can help to reduce the magnitude of land and sea clutter. Generally, vertical polarization has a larger sea clutter cross section than does horizontal polarization; compare Figures 25a, 25c, and 25e with Figures 25b, 25d, and 25f. However, if very high resolution is used by the radar, a horizontally polarized clutter return has a spikier appearance than does a vertically polarized one (see Figure 89) [23]. Spiky clutter is a problem for an automatic detection system – it generates a large number of false alarms.

Certain targets have a larger cross section for one polarization than they do for the other. In some cases, the illumination of the target with multiple polarizations (i.e., the use of polarization diversity) is desirable because it increases the signal detectability. However, the expense of such a tactic has prohibited this approach from becoming practical to date. Multiple polarizations also can be used for target identification and for greater invulnerability to jamming (i.e., for Electronic Counter-Counter Measures (ECCM)). Circular polarization is desirable in order to avoid the Faraday rotation loss obtained with linear polarized low frequency radars (at L-band or below) when propagating through the ionosphere (see Chapter 14).

Choice of Waveform and Signal Processing (Incoherent, MTI, or Pulse Doppler)

If clutter is not a problem and no Doppler velocity measurements are needed, a simple high peak-power pulse waveform can be used by the radar. Such pulses can be generated easily with a low cost magnetron transmitter. Typically a 1 MW peak power 1 μs pulse is transmitted, producing a 500 ft resolution. Long range applications, such as space surveillance, require more energy than that supplied by a 1 MW peak power 1 μs pulse. A possible solution is to use a longer pulse having the same peak power. If the 500 ft resolution must be maintained for tracking accuracy, the longer pulse width (1 ms, for instance) must be phase-coded for the resolution to be maintained. This coding can be achieved by using a chirp waveform (i.e., by linearly changing the carrier frequency of the pulse by 1 MHz during the time of the 1 ms pulse; see Chapters 7 and 8). If such a pulse had the same 1MW peak power, it would have one thousand times the energy of the simple 1 μs pulse, yet it would have the same 500 ft range resolution.

If clutter is a problem, coherent Moving Target Indicator (MTI) or pulse Doppler processing can be used. Clutter resulting from radar returns from the ground or sea is essentially at zero Doppler velocity. Hence, if a notch filter could be placed at zero Doppler frequency (with the width of this filter equal to the width of the clutter spectrum), the clutter would be eliminated. MTI processing, in effect, puts a notch filter or rejection band at zero Doppler velocity and multiples of the radar pulse repetition frequency. A two-pulse MTI canceller simply involves subtracting the echo of the present transmission from the preceding return. If the scatterer is not moving, (such as with a clutter patch), the phase and amplitudes of the two pulses are the same and so cancel. Pulse Doppler processing involves, as indicated in the first section of this chapter, the coherent processing of a large number of pulses, N (typically eight or more). These pulses are Fourier analyzed to obtain the return signal spectrum. Generally, the signal echoes having zero Doppler are interpreted as clutter and ignored; the echoes having non-zero Doppler are interpreted as indicating a moving target. (Non-zero Doppler clutter is discussed shortly.) Furthermore, the Doppler velocity of the target can be determined from the Doppler frequency of the signal, assuming that the radar PRF is high enough so as not to produce ambiguous velocities. (If such ambiguities arise, they can be eliminated by observing the target with bursts of N pulses having different PRFs.)

It is apparent that pulse Doppler processing has the advantages over MTI both of providing target Doppler velocity information along with clutter rejection, and of providing greater detection sensitivity. This second point follows from the fact that the N signal echo pulses are integrated coherently. Consequently, a detection sensitivity improvement of $10 \log_{10} N$ is obtained relative to the sensitivity achieved with a single pulse. If N = 10, pulse Doppler processing provides a detection sensitivity improvement of 10 dB or, equivalently, a detection range increase of 2.5 dB or 78%. With MTI processing, the system detection sensitivity is, on the average, the same as that achieved with a single pulse.

Pulse Doppler processing also has the advantage that it can suppress non-zero Doppler clutter. Such clutter is extended in range in contrast to targets of interest which usually occupy only a few range cells. To prevent range-extended rain clutter from ringing false alarms, the detection threshold of the range cell under observation is set to be a fixed amount larger than the average energy in any of the twenty or so range cells before or after it. Thus, range cells including rain and no target do not exceed the threshold; if the target is strong enough it will be detected even in the range cells with rain. The range cells with no rain have performance which is essentially undegraded [48]. This type of processing is called Constant False Alarm Rate (CFAR) processing.

An MTI system can eliminate rain clutter; however, a modification is required – a second MTI circuit in cascade with the zero Doppler clutter MTI must be used. The notch for the second MTI is placed at the mean frequency of the rain clutter by using an adaptive circuit which measures the mean rain clutter frequency.

A pulse Doppler radar also can be operated to prevent strong zero Doppler clutter from producing false alarms in the Doppler filters at non-zero Doppler frequencies. Ordinarily, a strong zero Doppler point clutter produces strong outputs at the non-zero Doppler filters because those filters do not have infinite rejection to zero Doppler returns. These outputs can be rejected by adaptively adjusting the thresholds at the non-zero Doppler filters to be a fixed amount larger than the expected zero-Doppler output level. Because the outputs of the zero-Doppler filter are measured as a function of time, the thresholds needed at the output of the non-zero Doppler filters as a function of time (range) can be selected adaptively.

The methods of pulse Doppler CFAR and adaptive threshold processing described in the above two paragraphs give pulse Doppler a significant advantage over MTI for automatic detection systems (i.e., systems attempting to detect targets without the intervention of an operator – a major problem to be dealt with in a modern radar system). For MTI arrangements, a hard limiter often is used in the receiver prior to the MTI circuit in order to achieve the CFAR performance. This approach has two disadvantages. First, it reduces the clutter suppression capability of the MTI circuit (typically by 10 to 30 dB, depending on the type of MTI circuit) [16]. Second, when strong clutter is present, targets that would be larger than the clutter after MTI processing if no limiter were used are no longer detected. On PPI scopes, such strong clutter produces what are called *black holes.* Pulse Doppler processing, with CFAR, does not have this disadvantage.

Recently, Lincoln Laboratory developed a technique which uses the cascade of an MTI and pulse Doppler processor to achieve clutter rejection. This combination, called a Moving Target Detector (MTD), is an approximation to an optimum clutter rejection filter [24,25]. A disc memory is used in the system to store stationary clutter returns observed from scan to scan. This memory allows the removal of such returns by the use of adaptive thresholds in range, Doppler, and angle. Also, this system requires that a number of scan returns be observed before a target is declared present and track on it is initiated. Thus, track initiation is used to help eliminate false clutter returns. Multiple PRFs are used to detect weak targets that ordinarily would be masked by rain

Figure 81 Matador 3-D TRS2210 Mobile Radar which uses phase scanning in elevation of three stacked-beams; see Table 1e and text. Three carrier frequencies are radiated simultaneously and, because of the feed dispersive character, they generate the three beams in elevation. (Photo published with permission of Thomson-CSF.)

Figure 82 ANTARES (Antenna Tracking Altitude, Azimuth, and Range by Elevation Scan) S-Band Height-Finder; see Table 1e. When height information is not needed, it does low elevation angle search; it can handle 100 targets/min. It is similar to Matador 3-D TRS2210 radar using elevation phase scanning of three stacked elevation beams. (Photo published with permission of Thomson-CSF.)

Figure 83 US Navy AEGIS AN/SPY-1 Multifunction Phased-Array; see Table 1f and Reference 20. (Photo courtesy of RCA.)

Figure 85 AN/SPS-32 and -33 Frequency Scan Acquisition Radar and Frequency-Phase Tracking Radar aboard the *USS Long Beach*, nuclear powered missile cruiser; see Table 1e. First operational use of billboard antennas. (Photo courtesy of Hughes Aircraft Company.)

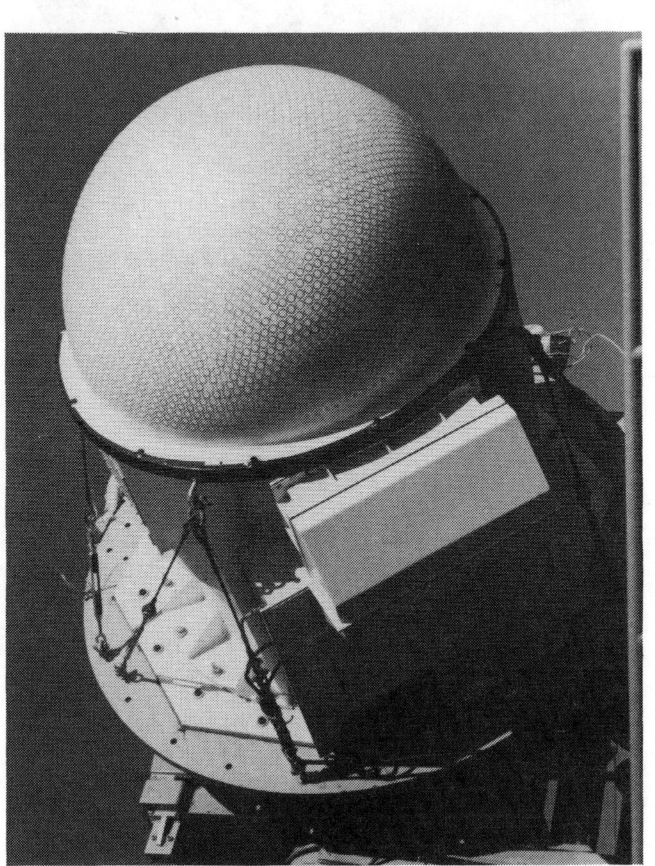

Figure 84 Demonstration Model of New Dome Antenna Multifunction Phased Array which provides greater than 2π steradian coverage from one antenna. (Photo courtesy of Sperry Gyroscope.)

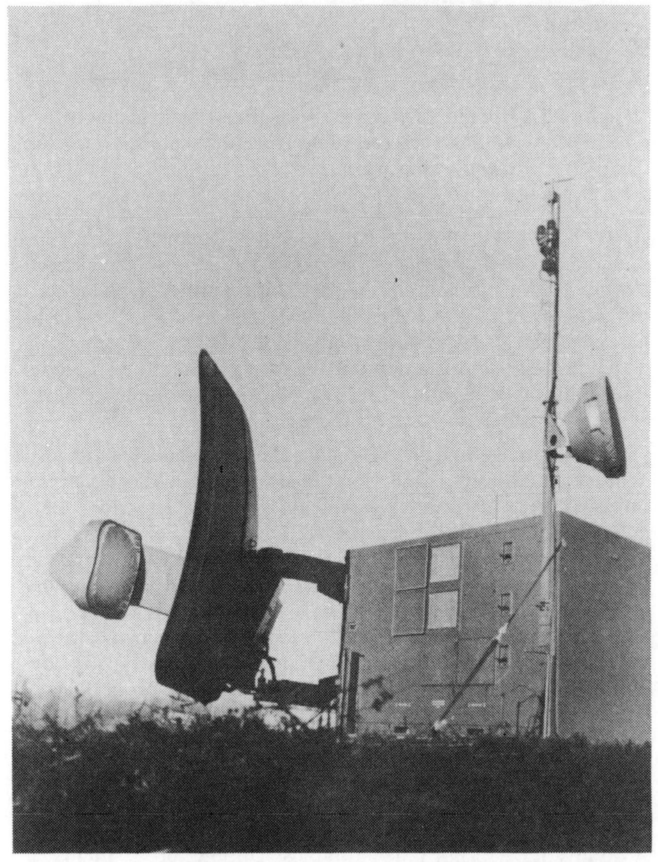

Figure 86 X-Band AN/TPN-25 Precision Approach Radar (PAR) of AN/TPN-19 System. Uses limited phase-phase scanning over 15° x 20° sector; see Table 1f and text. (Photo courtesy of Raytheon Company.)

Figure 87 **X-Band AN/GPN-22 PAR, fixed base version of AN/TPN-25; see Table 1f. (Photo courtesy of Raytheon Company.)**

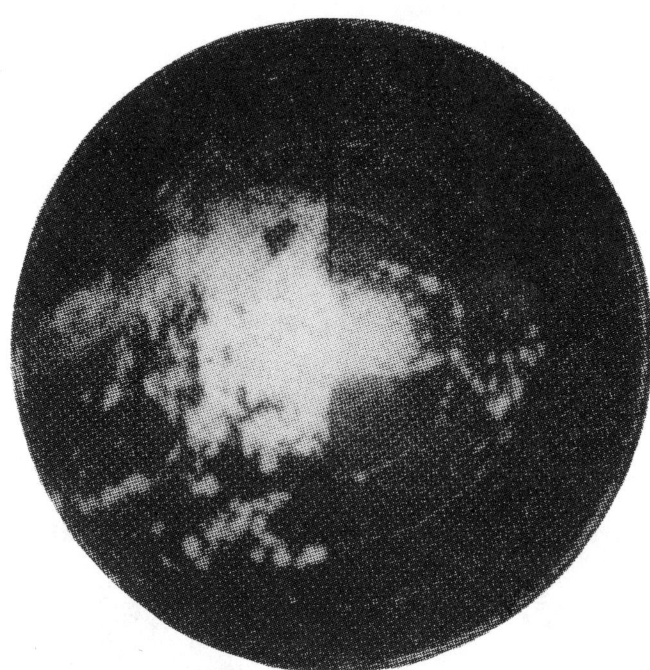

Figure 88 **PPI indicating reduction of very heavy rain clutter by use of circular polarization on a modified AN/CPS-5 L-Band Radar. Left half of display has linear polarization, right half has circular; no MTI used. (Photo courtesy of Warren White of Airborne Instruments Laboratory; see Reference 46.)**

clutter having the same ambiguous Doppler frequency. The MTD scheme has the advantage of being able to detect and track zero Doppler aircraft targets (in the zero Doppler filter) while ignoring bird and insect targets (which occur in the same filter). At zero Doppler, aircraft targets are at broadside and hence have much larger cross sections than do birds or insects. Therefore, a higher threshold can be used in a zero Doppler filter to eliminate bird echoes but still detect aircraft targets.

One disadvantage to using pulse Doppler processing instead of MTI is that a longer search scan time is required. Whereas a two-pulse MTI requires only $2T_{PP}$ seconds for target detection, the pulse Doppler processor requires NT_{PP} seconds. For N = 10, it takes five times longer to scan out the search volume with pulse Doppler processing than it does with MTI. This problem can be serious for a 3-D track-while-scan radar which scans a single pencil beam by phase scanning in elevation and by mechanical scanning in azimuth. If 10 s were required with two-pulse MTI processing, 50 s would be required with pulse Doppler processing. A scan-to-scan period of 50 s may be too large to maintain track on many targets of interest. If it were essential to use pulse Doppler processing because of its improved detection sensitivity and performance in clutter, a stacked beam system could be used to realize a 10 s scan-to-scan period, using five beams in elevation.

A disadvantage of both MTI and pulse Doppler processing is their vulnerability to jamming. For both techniques more than one pulse is required to detect a signal imbedded in clutter. A responsive jammer could measure the frequency of the first transmitted pulse and then center the jammer frequency to spot jam the following pulses. MTI has an advantage here in that fewer pulses are used per observation of the target. Assume a two-pulse MTI canceller – if frequency agility is used between pairs of MTI pulses and the enemy attempts to use a responsive spot jammer, signals not in clutter can be detected without loss of sensitivity.

Sometimes it is possible to eliminate the need for MTI or pulse Doppler processing by reducing the radar system resolution cell size (by narrowing the antenna beam and pulse widths) or by lowering the radar carrier frequency.

If the clutter is not too much stronger than the target return (only about 3 dB), a technique other than MTI or pulse Doppler processing is available – frequency agility – which involves transmitting a number of pulses at frequencies sufficiently far apart to decorrelate the clutter returns. By video integrating the returns, clutter suppression is achieved. The application of this technique is illustrated in Figure 7, Chapter 23; this approach has an inherent jamming invulnerability. Frequency agility also has the advantage of reducing target angle error measurements (called the angle glint error [16]; see Chapter 2).

A CW waveform (i.e., a long constant frequency signal) can be used for clutter rejection. It provides target Doppler velocity unambiguously but has the disadvantage of not giving target range information. By modulating the carrier frequency, range information can be obtained. For example, alternately transmitting up- and down-ramped carriers provides range and Doppler velocity information on the target. The Hawk AN/MPQ-39 (Figure 5) and the NATO Sea Sparrow director/illuminator (Figure 8) are examples of CW radars. Because CW radars must receive the weak target echo signal while still transmitting, means must be provided for preventing the strong transmitted signal from saturating the weak one being received. Often separate antennas for receive and transmit (see Figures 5 and 8) and special circuitry for nulling out the transmitted signal seen in the receiver (the feed-through nulling technique) are used [16,26,27].

The aspects of waveform selection and signal processing are discussed in greater detail in Chapters 7 and 8; see also Chapters 10, 11, and 13.

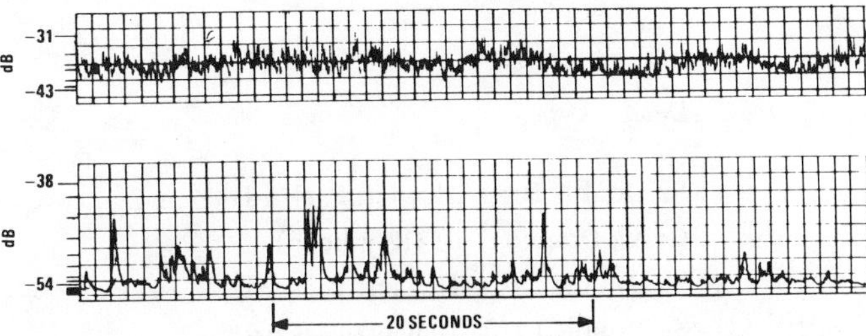

Figure 89 A-Scope High Resolution Sea Clutter Returns showing that although horizontal polarization has a lower average backscatter return than vertical polarization, it is very spiky. Upper trace backscatter coefficient is for vertical polarization; lower trace is for horizontal polarization. (From Long [23].)

Tube vs. Solid State Transmitter

It is difficult to give a general answer regarding whether one should select a tube instead of a solid state transmitter, or vice versa. The state-of-the-art of solid state transmitter components has advanced rapidly, and they are becoming more and more competitive with tube transmitters.

At UHF, a large phased array radar for space applications can be built more economically using a solid state transmitter rather than a tube transmitter. An example of such a UHF system is the PAVE PAWS solid state radar (Figure 17). For this system, a distributed solid state transmitter is used (i.e., each active antenna radiating element is fed by a transistor power amplifier called a solid state module). But it is not always desirable to use a solid state transmitter for UHF. If short range and low duty cycle operation is required, a tube system may be preferred, solid state devices being most effective when high duty cycles, long pulses, and low peak power are used.

The advances in bipolar transistors, IMPATTs, TRAPATTs, and FETs are such that solid-state soon may be more economical or desirable than tube systems for frequencies at L-band or above; see Tables 2 through 6 and Reference 28. The Naval Electronics Laboratory presently is developing a transmit-receive module for an airborne X-band array.

A distributed solid state transmitter is made up of a large number of modules, typically on the order of a few thousand for a large phased array system. For a tube system, the transmitter usually consists of only a few tubes, typically 1 to 10. The old AN/FPS-85 is an exception – it uses a distributed tube transmitter of 5184 tetrodes (one for each radiating element) instead of solid state modules, as solid state equipment was not available when the radar was built and rebuilt (see Chapter 6) from 1965 to 1968. The L-band Cobra Dane which has 96 TWTs is another exception [36, 50]. The failure of a few devices in a solid state transmitter is less catastrophic than is the failure of a few tubes in a tube transmitter; hence, systems using solid state transmitters tend to be more attractive.

A factor in the favor of solid-state distributed arrays made up of transmit/receive modules is that they permit the phase shifting to be executed before power amplification so that the inherent loss (which can be on the order of 1.6 dB 2-way) is not experienced. The advances in transistor and FET receiver amplifiers (see Figure 23) are such that very low noise figure receivers can be incorporated in the modules thereby making the resulting components more competitive with paramp receivers used in tube systems. Commercially available now (from Nippon Electric Company [NEC]) is an uncooled GaAs FET 60 dB gain amplifier which provides 125 K guaranteed temperature over the band from 3.7 to 4.2 GHz with a measured temperature of $<$ 95 K at center frequency [49].

Figure 90 AN/TPS-59 Solid State L-Band US Marine Radar using phase scanning in elevation of 54 rows of radiating elements; see Table 1e. Can detect up to 500 targets at 6 rpm rate [47]. (Photo courtesy of General Electric.)

Solid state devices also can be combined in parallel into a single port which feeds the horn that illuminates a dish antenna. Used in this way, the solid state transmitter replaces a tube transmitter and generally is referred to as a *solid state bottle.*

Other examples of solid state radars are the RASSR (see Chapter 20) and AN/TPS-59 (Figure 90). Those interested in seeing examples of other radars are referred to References 1, 2, 12, 13, 15-19, and 34. (See also the "Present and Future Trends" section of Part 5.)

Single Package

Frequency Band	Frequency GHz	Peak Power, P_{PK} (W)	Efficiency (%)	Duty Cycle, d (%)	Device Type
X	10	30	8-11	10[1,2]	Si, Double Drift (Hewlett Packard)[3]
X	10	22	15	30[1,2]	GaAs, Double Drift (Raytheon)
Ku	14	26	8-9	10[1,2]	Si, Double Drift (Hewlett Packard)
Ku	16	20	8-9	10[1,2]	Si, Double Drift (Hewlett Packard)[4]
Ka	35	0.75	12	CW	GaAs, Single Drift (Raytheon)
Ka	40	0.7	8	CW	Si, Double Drift (TRW)
V	60	0.5	8.5	CW	Si, Double Drift (TRW)

Multiple Packages (S_I Devices)

Frequency Band	Peak Power (W)	DC to RF Efficiency (%)
X	500[1]	5-6
Ku	200[1]	5-6

[1] Pulse Width – 100 ns to 10 μs depending on cooling
[2] Can go to 100% duty by holding P_{PK} d constant
[3] HP 50820710
[4] HP 50820716

Table 2 **State-of-the-Art IMPATT Sources [43]**

Frequency (GHz)	Power Output (W)	Power Gain (dB)	Efficiency* (%)	V_D (V)	Manufacturer
4	3.0	6	47	12	Fujitsu
6	2.4	5	34	12	Fujitsu
8	1.9	4	24	12	Fujitsu
8	2.5	4	25	8	TI
9	1.0	4	16	8.5	RCA
15	0.45	5	12.5	8.5	RCA
18	0.23	4.5	5.4	8.5	RCA
22	0.14	4.8	9	8.5	RCA

*This figure is Power Added Efficiency, $\eta_{PA} = \dfrac{P_{OUT} - P_{IN}}{P_{DC}}$

Table 3 **State-of-the-Art CW Power GaAs FET Devices (Courtesy Pitzalis [39])**

	PEAK POWER OUTPUT (WATTS)	PULSE WIDTH (μSEC)	DUTY CYCLE (%)	1 dB BANDWIDTH (%)	POWER GAIN (dB)	EFFICIENCY (%)	SOURCE
TRAPATTS	240*	0.2	.1	8.3	4.8	29	HUGHES
	195	NARROW	–	NARROW	6.3	20	R.C.A.
	98	0.2	.1	8.3	6	30	HUGHES
	82	10	?	8	4.9	15	R.C.A.
	55	50	1	5	5	24	HUGHES/NRL
	45	50	.001	10	7	21	SPERRY
	3	C.W.	100	OSCILLATOR	–	20	B.T.L.
IMPATTS	21 (3 DIODES)	C.W.	100	NARROW	LOCKED OSCILLATOR LOCKING POWER = 3.5W	12-13	B.T.L.
	12.1	C.W.	100	OSCILLATOR	–	21	LINCOLN LABS
	3.4	C.W.	100	OSCILLATOR	–	37	LINCOLN LABS
SI BIPOLAR TRANSISTORS	5-8	C.W.	100	10-15	4-5	30	M.S.C., R.C.A.
	1 (LINEAR)	C.W.	100	BROADBAND	10	30	R.C.A., M.S.C.
GaAs FETs	4	C.W.	100	NARROW	6	43.5	FUJITSU
	1	C.W.	100	NARROW	6	24	R.C.A.
	1**	C.W.	100	OCTAVE (2-4GHz)	9	35	N.R.L./R.C.A. DEVICE
	0.26	C.W.	100	NARROW	9.6	68	R.C.A.

*Two devices—hybrid combined
**Two devices—balanced stage

Table 4 State-of-the-Art Microwave Power Devices, 3-4 GHz (Courtesy Cohen [40])

		PEAK POWER OUTPUT (WATTS)	PULSE WIDTH (μ SEC)	DUTY CYCLE (%)	BANDWIDTH (%)	POWER GAIN (dB)	EFFICIENCY (%)	SOURCE
TRAPATTS		27	1	.1	OSCILLATOR	–	42.5	HUGHES
		10	0.25	.1	10	6	12	SPERRY
		6	C.W.	100	OSCILLATOR	–	24	HUGHES
		5	C.W.	100	5.5	5	16.4	HUGHES
IMPATTS	Si DOUBLE DRIFT	16	0.8	25	OSCILLATOR	–	12.3	H.P.
	GaAs HI-LO	12.8	0.8	25	OSCILLATOR	–	25	VARIAN
	GaAs S.B. READ	8	C.W.	100	OSCILLATOR	–	35	RAYTHEON
	GaAs FLAT PROFILE	5	C.W.	100	7	25 (4 STAGES)	10	TYPICAL
	GaAs S.B. READ	4.5	C.W.	100	6.5	4.5	22	RAYTHEON
	GaAs HI-LO	2.9	C.W.	100	OSCILLATOR	–	18.8	VARIAN
Si BIPOLAR TRANSISTORS	4 DEVICES	3	C.W.	100	5	6.0	?	T.I.
	1 DEVICE	1	C.W.			4.5	20	T.I.
GaAs FETs		3.6	C.W.	100	NARROW	4.0	23	T.I.
		2.9	C.W.	100	NARROW	4.0	25	T.I.
		2.2	C.W.	100	NARROW	3.2	21.6	FUJITSU
		1.3	C.W.	100	NARROW	4.2	30	R.C.A.
		0.7	C.W.	100	NARROW	6.0	46	T.I.

Table 5 State-of-the-Art Microwave Power Devices, 8-10 GHz (Courtesy Cohen [40])

MATERIAL	DEVICE TYPE	FREQUENCY (GHz)	PEAK POWER OUTPUT (WATTS)	PULSE WIDTH (μ SEC)	BANDWIDTH (%)	POWER GAIN (dB)	EFFICIENCY (%)	SOURCE
SILICON	DOUBLE DRIFT	16.5	11.0	0.8	OSCILLATOR	–	14	H.P.
	DOUBLE DRIFT	29–39	1.2	C.W.	OSCILLATOR	–	10	HUGHES
	DOUBLE DRIFT	39	11.0	0.35	OSCILLATOR	–	10	HUGHES
	DOUBLE DRIFT	55	1.6	C.W.	OSCILLATOR	–	11.5	FUJITSU
	SINGLE DRIFT	60	1.0	C.W.	UP TO 6 GHz	9 (2 STAGES)	2	HUGHES
	DOUBLE DRIFT	66	0.7	C.W.	OSCILLATOR	–	6	HUGHES
	DOUBLE DRIFT	92	0.2	C.W.	OSCILLATOR	–	3.9	HUGHES
	SINGLE DRIFT	94	0.1	C.W.	2 GHz	4	2	HUGHES
	DOUBLE DRIFT	140	0.72	0.3	OSCILLATOR	–	4	HUGHES
	DOUBLE DRIFT	170	0.03	C.W.	OSCILLATOR	–	2	HUGHES
	SINGLE DRIFT	185	0.08	C.W.	OSCILLATOR	–	2.3	NIPPON T&T
	DOUBLE DRIFT	205	0.09	0.05	OSCILLATOR	–	0.5	HUGHES
	SINGLE DRIFT	285	0.008	C.W.	OSCILLATOR	–	0.35	NIPPON T&T
GALLIUM ARSENIDE	S.B. READ	13.7	3.2	C.W.	OSCILLATOR	–	24	RAYTHEON
	DOUBLE DRIFT	21	1.2	C.W.	OSCILLATOR	–	15.6	HITACHI
	SINGLE DRIFT	34.8	0.7	C.W.	OSCILLATOR	–	12.4	R.C.A.
	S.D. ION IMP.	37.5	0.5	C.W.	OSCILLATOR	–	9.6	LINCOLN LABS.
	S.B.	51	0.2	C.W.	OSCILLATOR	–	11.0	HITACHI
	S.B.	53	0.4	C.W.	OSCILLATOR	–	5.2	HITACHI

Table 6 State-of-the-Art IMPATT Devices, above 12 GHz (Courtesy Cohen [40])

Glossary

AD = angle deception
ADT = automatic detection and tracking
AFC = automatic frequency control
ARSR = air route surveillance radar
ASR = airport surveillance radar
ATC = air traffic control
BMEWS = Ballistic Early Warning System
BW = bandwidth
C = circular
CCW = counterclockwise
CG = missile cruisers
CLCC = amphibious command ships
CM = clutter mapper
CPACS = coded pulse anti-clutter system
CR = cancellation ratio
CV = carriers
CW = clockwise
DBB = detector balanced bias
DDG = missile destroyers
DEMC = deceptive electronic countermeasures
DLG = missile frigates
DMTI = digital MTI
EW = early warning
FTC = fast time control (helps combat CW jamming)
GCA = ground control approach
GCI = ground control intercept
H = horizontal polarization
HIPAR = High Power Acquisition Radar
IAGC = instantaneous AGC (Automatic Gain Control)
I.F. = MTI improvement factor
JATS = jamming analysis and transmission selection
L = linear
LB = lower beam
LC = left circular polarization
LORO = lobe-on-receive-only mode (used to counter Deception Electronic Countermeasures [DECM])
LP = long pulse
MCC = moving clutter cancellation
MTBF = mean time before failure
MTI = moving target indicator
MTR = missile tracking radar
MTTR = mean time to repair
NB = narrow bandwidth
PA = parametric amplifier
PAR = precision approach radar
PC = pulse canceller
R = receive
RC = right circular polarization
RCM = rain contour mapping
SB = sidelobe blanking
SCV = subclutter visibility
SP = short pulse
STC = slow time control
T = track
TTR = target tracking radar
TWS = track while scan
UB = upper beam
USMC = US Marine Corps
V = vertical polarization
VB = blind velocity
VI = video integration
WB = wide bandwidth

References

[1] Brookner, E. (ed.): *Practical Phased-Array Systems, Microwave Journal* Intensive Course, Dedham, Massachusetts, 1975.

[2] Kahrilas, P.J.: *Electronic Scanning Radar Systems (ESRS) Design Handbook,* Artech House, Dedham, Massachusetts, 1976.

[3] Nathanson, F.E.: "Monostatic Sea Reflectivity," in *Models for the Multi-Mission Radar Study,* Technology Service Corp. Report TSC-PD-A0994-10, 23 September 1975.

[4] Nathanson, F.E.: *Radar Design Principles, Signal Processing, and the Environment,* McGraw-Hill, New York, 1969. York, 1969.

[5] Blake, L.V.: "A Guide to Basic Pulse-Radar Maximum-Range Calculation, Part-1 Equations, Definitions, and Range Calculation," *NRL Report 6930,* Naval Research Laboratory, Washington, DC, 23 December 1969.

[6] Crane, R.K.: "Propagation Phenomena Affecting Satellite Communication Systems Operating in the Centimeter and Millimeter Wavelength Bands," *Proceedings of the IEEE, Vol. 59,* 1 February 1971, pp. 173-188.

[7] Jaye, W.E.; Chesnut, W.G.; and Craig, B.: "Analysis of Auroral Data from the Prince Albert Radar Laboratory," *SRI Report 7465,* Stanford Research Institute, Menlo Park, California, September 1969.

[8] *Radar Propagation in the Arctic* (ARARG-CP-97), AGARD Specialists Meeting of the Electromagnetic Wave Propagation Panel, held at the Max-Planck-Institute, Germany, 13-17 September 1971.

[9] Ko, H.C.: "The Distribution of Cosmic Radio Background Radiation," *Proceedings of the IRE, Vol. 46,* January 1958, pp. 208-215.

[10] Brookner, E.; and Doskocil, A.C.: "Reentry Vehicle Borne Wake Measurement Radar," (AD 394-083L), Volume 1, *Fourteenth Annual Tri-Service Radar Symposium Proceedings,* Fort Monmouth, New Jersey, 4-7 June 1968, pp. 366-399.

[11] Brookner, E.; Bartlett, C.J.; Edwards, R.; and Tarver, R.: "Flight Measurements for a Reentry Vehicle Borne Radar," Volume 1, *Seventeenth Annual Tri-Service Radar Symposium Proceedings,* Fort Monmouth, New Jersey, 25-27 May 1971, pp. 235-261.

[12] Ridenour, L.N.: *Radar Systems Engineering* Volume 1 in MIT Radiation Laboratories Series, McGraw-Hill Co., New York, 1947.

[13] Pretty, R.T. (ed.): *1976 Jane's Weapon Systems,* Jane's Yearbooks, Paulton House, London, England.

[14] Miller, B.: "Soviet Radar Expertise Expands," *Aviation Week,* 15 February 1971, pp. 14-16; "Soviet Radars Disclose Clues to Doctrine," *Aviation Week,* 22 February 1971, pp. 42-50; "The Growing Threat-4, Soviets Closing Gap in Avionics, Computer," *Aviation Week,* 25 October 1971, pp. 40-46.

[15] Hansen, R.C. (ed.): *Microwave Scanning Antennas* (Volume 1 – "Apertures"), Academic Press, New York, 1964.

[16] Skolnik, M.I.: *Radar Handbook,* McGraw-Hill Co., New York, 1970.

[17] Skolnik, M.I.: *Introduction to Radar Systems,* McGraw-Hill Co., New York, 1962.

[18] Oliner, A.A.; and Knittel, G.H.: *Phased Array Antennas,* Artech House, Dedham, Massachusetts, 1972.

[19] *Record of the IEEE 1975 International Radar Conference,* Arlington, Virginia, 21-23 April 1975.

[20] Patton, W.T.: "Compact, Constrained Feed Phased Array for AN/SPY-1," Lecture No. 8 of *Microwave Journal* Intensive Course; E. Brookner (ed.): *Practical Phased-Array Systems, Microwave Journal* Intensive Course, Dedham, Massachusetts, 1975.

[21] Mailloux, R.J.: "Limited Scan Arrays – Parts 1 and 2," Lectures 9 and 10 of *Microwave Journal* Intensive Course; E. Brookner (ed.): *Practical Phased-Array Systems, Microwave Journal* Intensive Course, Dedham, 1975.

[22] Ward, H.R.; Fowler, C.A.; and Lipson, H.I.: "GCA Radars: Their History and State of Development," *Proceedings of the IEEE, Vol. 62,* June 1974, pp. 705-716.

[23] Long, M.W.: "On a Two-Scatterer Theory of Sea Echo," *IEEE Transactions on Antennas and Propagation, Vol. AP-22,* September 1974, pp. 667-672.

[24] Meuhe, C.E.: "Digital Signal Processor for Air Traffic Control Radars," *1974 NEREM Record, Radar Systems and Components,* Boston, Massachusetts, 28-31 October 1974.

[25] Muehe, C.E.: "Digital Signal Processor for Air Traffic Paper No. 25-3, *IEEE ELECTRO/76 Professional Program,* Boston, Massachusetts, 11-14 May 1976.

[26] O'Hara, F.J.; and Moore, G.M.: "A High Performance CW Receiver Using Feed-thru Nulling," *Microwave Journal, Vol. 6,* September 1965, pp. 63-71.

[27] Harmer, J.D.; and O'Hara, W.S.: "Some Advances in CW Radar Techniques," *Proceedings of the National Conference Military Electronics,* Washington, DC, June 1961, pp. 311-323.

[28] Cohen, E.: "TRAPATTs and IMPATTs: Current Status and Future Impact on Military Systems," *1975 IEEE EASCON Record,* 1975, p. 130-A.

[29] Barton, D.K.: *Internal Raytheon Memorandum;* see also, Barton, D.K.: "Radar Multipath," *Microwave Journal, 1976 Microwave Engineers' Handbook,* pp. 35-41.

[30] Barton, D.K.: "Real-World Radar Technology," *IEEE International Radar Conference,* Arlington, Virginia, 1975, pp. 1-22.

[31] Lowenhar, H.: "Strategic Defense, ABM Radars: Myth vs. Reality," *Space/Aeronautics, Vol. 52,* November 1969, pp. 56-64.

[32] *Electronic Progress, Vol. 16,* Raytheon Company, Winter 1974.

[33] Shrader, W.: "Radar Technology Applied to Air Traffic Control," *IEEE Transactions on Communications, Vol. COM-21,* May 1973, pp. 591-605.

[34] Hill, R.T.: "Trends in European Radar Technology," Paper No. 25-1, *IEEE ELECTRO/76 Professional Program,* Boston, Massachusetts, 11-14 May 1976.

[35] Van Rinkhuyzen, H.G.: "Three Dimensional Multi-Target Tracking Radar, 3D-MTTR," *IEEE International Radar Conference,* Arlington, Virginia, 1975, pp. 37-40.

[36] Filer, E.; and Hartt, J.: "Cobra Dane Wideband Pulse Compression System," Paper No. 61, 1976 *IEEE EASCON,* Washington, DC, 26-29 September 1976.

[37] Barrera, J.S.: "GaAs Field-Effect Transistors," *Microwave Journal, Vol. 19,* February 1976, pp. 28-31.

[38] Brand, F.: "Current Status of Microwave Technology," *Microwave Journal, Vol. 19,* May 1976, pp. 18-21.

[39] Pitzalis, O.: "Bipolar Transistor and FET Devices: Present and Future," *Boston IEEE Lecture Series "Modern Radar Techniques, Components and Systems,"* 28 October 1976.

[40] Cohen, E.D.: "TRAPATTs and IMPATTs: State-of-Art and Applications," *Boston IEEE Lecture Series "Modern Radar Techniques, Components and Systems,"* 18 November 1976.

[41] Hall, J.S.: *Radar Aids to Navigation,* Volume 2 of MIT Radiation Laboratory Series, McGraw-Hill, New York, 1947.

[42] Weil, T.A.: "Atmospheric Lens Effect; Another Loss for the Radar Range Equation," *IEEE Transactions on Aerospace and Electronic Systems, Vol. AES-9,* January 1973, pp. 51-54.

[43] Jerinic, G.; and Bierig, R.W.: *Private Communication,* Raytheon Company, 1976.

[44] Hall, W.M.: *Internal Raytheon Memorandum.*

[45] Cady, W.M.; Karelitz, M.B.; and Turner, L.A.: *Radar Scanners and Radomes,* Volume 26 of MIT Radiation Laboratory Series, McGraw-Hill, New York, 1948.

[46] White, W.D.: "Circular Radar Cuts Radar Clutter," *Electronics, Vol. 27,* March 1964, pp. 158-160.

[47] Klass, P.J.: "Marines to Test New Surveillance Radar," *Aviation Week and Space Technology, Vol. 105,* 6 December 1976, pp. 56-58.

[48] Brookner, E.: "How to Look Like A Genius in Detection Without Really Trying," *1974 IEEE NEREM,* Boston, Massachusetts, pp. 37-64; see also Appendix of this book.

[49] *Microwave Journal, Vol. 19,* December 1976, pp. 56-57.

[50] Torrero, E.A.: "Military and Aerospace," *IEEE Spectrum, Vol. 14,* January 1977, pp. 74-78.

Chapter 2 Barton

Detection and Measurement

The subject of radar detection has been discussed extensively in the literature [1-6]. In this section, the most important points involved in determining the signal-to-noise ratio required to obtain given detection and false-alarm probabilities are summarized, considering target effects and processing losses. The second half of this chapter covers the subject of radar measurement, and constitutes a summary of material in the Barton and Ward *Handbook of Radar Measurement* [7].

Detectability Factor in the Radar Equation

The most basic form of the radar equation gives the ratio of receiving single-pulse energy, E_1, to noise power density, N_o, at the receiver input

$$\frac{E_1}{N_o} = \frac{P_T \tau G_T G_R \lambda^2 \sigma}{(4\pi)^3 k T_I R^4 L_1} \qquad (1)$$

where P_T is peak transmitted power, τ is pulse width, G_T is transmitting antenna gain, G_R is receiving antenna gain, λ is wavelength, σ is target cross section, k is Boltzmann's constant, T_I is effective receiver input temperature, R is range, and L_1 is a system loss factor. Most of the terms in the equation are clearly defined, but since loss factors are being dealt with, it should be noted that L_1 is essentially the excess attenuation of RF energy reaching the receiver, relative to the theoretical free-space propagation. It typically includes microwave losses in transmission lines and RF components, in the atmosphere, and any discrepancy between values of P_T, G_T, and G_R used in the equation and those actually achieved in the radar. Loss factors in signal processing and those which result from scanning and signal fluctuation about its mean value are discussed later. Equation (1) merely is typical of many such equations which can be written to describe the ratios of signal to noise, to jamming, and to clutter or nonrandom interference of any sort.

In calculating the maximum range of a radar, the energy ratio is set equal to the *detectability factor,* D_X, required for given probabilites of detection and false alarm, and the equation is solved for the resulting range, R_M

$$E_1/N_o = D_X$$

$$R_M{}^4 = \frac{P_T \tau G_T G_R \lambda^2 \sigma}{(4\pi)^3 k T_I L_1 D_X} \qquad (2)$$

How D_X depends on the detection and false-alarm probabilities, target type, and various radar factors is discussed here. First, it should be noted that CW radars and other coherent systems may use coherent integration of signals over an observation time, t_o; in such a case, the transmitted energy, $P_{AV} t_o$, replaces $P_T \tau$ in Equation (1) and E_1/N_o represents the *single-sample energy ratio.*

The radar equation is a simple algebraic expression (at least when certain propagation factors are omitted, as is the case here); the fact that incorrect range calculations are so common is the result not of arithmetic error but of incorrect assessment of the detectability factor for the case at hand. The purpose of the following discussion is to provide a reasonable basis for estimating that factor without recourse to automatic computers or voluminous books of tables and graphs. The available computer programs, tables, and graphs are extremely accurate for a limited number of cases, but the simple procedures to be outlined are adequately accurate for a much more general set of conditions actually encountered in system engineering.

The terms "detectability factor" and "visibility factor" are used interchangeably in much of the literature, although the latter was used originally in reference to visual detection on a cathode-ray-tube display. The current IEEE definitions are:

> *Detectability Factor* (in radar) – In pulsed radar, the ratio of single-pulse signal energy to noise power per unit bandwidth that provides stated probabilities of detection and false alarm, measured in the IF amplifier and using an IF filter matched to the single pulse, followed by optimum video integration; in CW radar, the ratio of single-look signal energy to noise power per unit bandwidth, using a filter matched to the time on target.

> *Visibility Factor* (in radar) – In pulsed radar, the ratio of single-pulse signal energy to noise power per unit bandwidth that provides stated probabilities of detection and false alarm on a display, measured in the IF portion of the receiver under conditions of optimum bandwidth and viewing environment; in CW radar, the ratio of single-look signal energy to noise power per unit bandwidth using a filter matched to the time on target. The equivalent term for radar using automatic detection is detectability factor; for operation in a clutter environment, a clutter visibility factor is defined.

Clearly, the value of detectability factor depends on the required probabilities of detection and false alarm, the type of target fluctuation, and the number of pulses (or samples) used in the video integration process. Also, since the basic definition is based on use of a matched IF filter and optimum video integration, further "loss" factors are needed to describe the effects of practical receiver and processor design. These factors are discussed later. First, though, the basic signal and noise statistics which determine the detectability factor are considered.

Signal and Noise Statistics

Any signal, however small, would be detectable if there were no noise at the receiver input or generated in the receiver. The usual "thermal" noise can be described statistically as in Figure 1. In the IF stages of the receiver, prior to envelope detection, the noise voltage has a Gaussian distribution with zero mean and a variance equal to noise power, N (when arbitrarily referred to a one ohm load resistance). The detector at the receiver output rectifies the IF waveform, and averages the output to recover the envelope of the IF wave; for pure noise, this envelope has a Rayleigh distribution, conventionally plotted for positive voltages only as shown in the figure. If the IF wave is considered as the sum of in-phase and quadrature components I and Q (each having a Gaussian distribution with variance N/2), the detected envelope represents the sum of these components. Its mean value is $1.25\sqrt{N}$, and it extends upwards with decreasing probability to values several times as large. If a square-law detector is used, the output of which is pro-

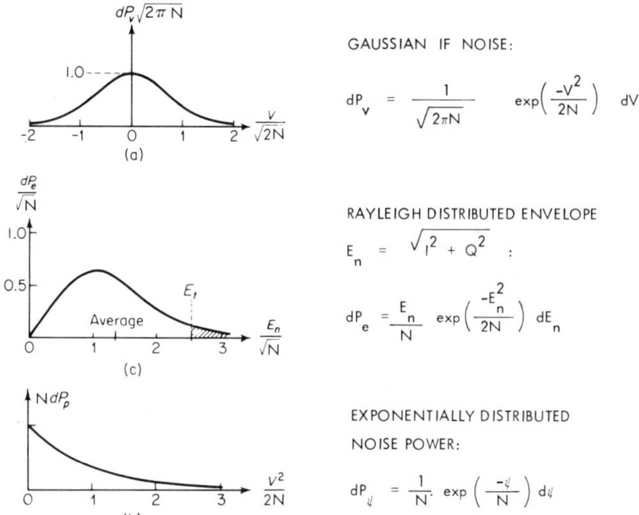

Figure 1 Amplitude Distributions of Random Noise [2]

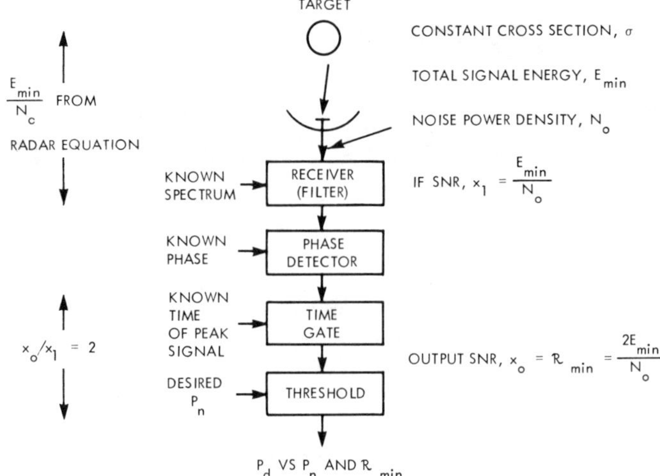

Figure 2 Ideal (Minimum Energy) Detection Process [12]

portional to $I^2 + Q^2$, its output voltage distribution is the same as the IF power distribution – an exponential function with mean value N as shown in the figure. For convenience, this discussion focuses on the Rayleigh-distributed voltage envelope, even though most radar detectors approximate the square-law process.

Often in radar the detection threshold, E_T, is placed at the output of the receiver, passing only signals which exceed that threshold. This arrangement would be used for a simple detector operating on a single pulse, or as the first step in what is called the *binary integration process.* To minimize false alarms from noise, this threshold is set at a level several times $\sqrt{N}$, and the area above E_T on the distribution (shown shaded in Figure 1) represents the probability of false alarm, P_N. To determine the resulting detection probability, P_D, when a signal is introduced, a particular system implementation must be considered. The ideal implementation, requiring the minimum signal energy to achieve a given P_D, is shown in Figure 2; this situation serves as a reference against which to compare practical systems, so that losses in detectability can be assigned to each departure from the ideal process. In the figure, the energy considered is scattered from the target and captured by the receiving antenna, where it competes with environmental and receiver noise of density $N_o = kT_I$ (W/Hz). The ideal process assumes exact knowledge of the signal spectrum, RF phase, and time of arrival of the signal. The receiver filter is then matched to the signal spectrum, producing a single sample in which the total signal energy, E_{MIN}, is coherently integrated to produce, at a time after reception of the entire signal, an IF output signal-to-noise power ratio $x_1 = E_{MIN}/N_o$ (a property of the matched filter). Since the signal phase is known, an in-phase reference voltage is used in the phase detector to recover the filtered signal envelope while rejecting the Q component of noise. Then, at the known time of occurrence of the peak of the filtered signal, the output, having a power SNR $x_o = \mathcal{R}_o = 2E_{MIN}/N_o$, is applied to a threshold to determine if the target is present. Curves are available [1 (p. 437), 3 (p. 297), and 6 (p. 114)] which indicate how P_D varies with $\mathcal{R}_o$ for a threshold set to give a particular value of P_N using this ideal process. These curves are merely plots of the error function for suitably selected arguments.

Single-Sample Detection with Practical Receiver

Since the usual purpose of the practical radar is to detect the presence of unknown targets in some volume in space, the ideal process must be modified in several respects (Figure 3). The practical receiver is matched only approximately to the single received pulse or energy sample, providing an IF SNR (power) ratio

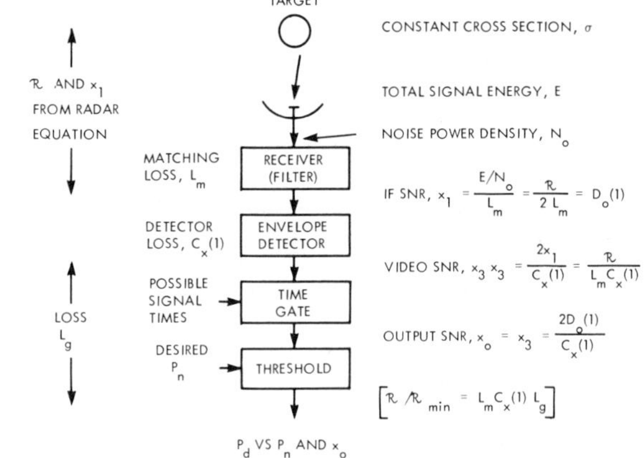

Figure 3 Single-Sample Detection with Mismatched Filter [12]

$$x_1 = \frac{E/N_o}{L_M} = \frac{\mathcal{R}}{2L_M} \tag{3}$$

It is this SNR which must be set equal to the detectability factor $D_o(1)$ to obtain the specified performance in the single-sample case. The subscript o denotes the steady-target case (as opposed to Swerling Case 1, 2, 3, or 4), while the argument (1) indicates that the detection is based on one sample. The matching loss, L_M, describes the reduction in power SNR relative to what would have been obtained in a matched filter.

A second compromise is made in use of an envelope detector to recover the signal, the RF phase of which is unknown. Where the ideal process (Figure 2) achieved an output video SNR $x_o = \mathcal{R}_{MIN} = 2x_1$, the envelope detector gives an effective output SNR

$$x_o = \frac{2x_1}{C_X(1)} = \frac{\mathcal{R}}{L_M C_X(1)} \tag{4}$$

where $C_X(1)$ is the "detector loss" for the single-sample process, describing the loss of information which was contained in the knowledge of RF phase, for the single-sample process. Finally, since signals must be sought over an appreciable interval in time delay, a number of successive noise samples must be passed by the gate to the threshold, giving many independent opportuni-

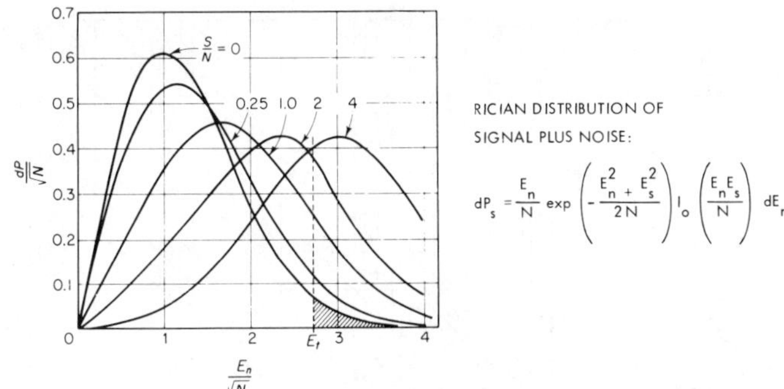

Figure 4 Amplitude Distributions of Signal Plus Noise [2]

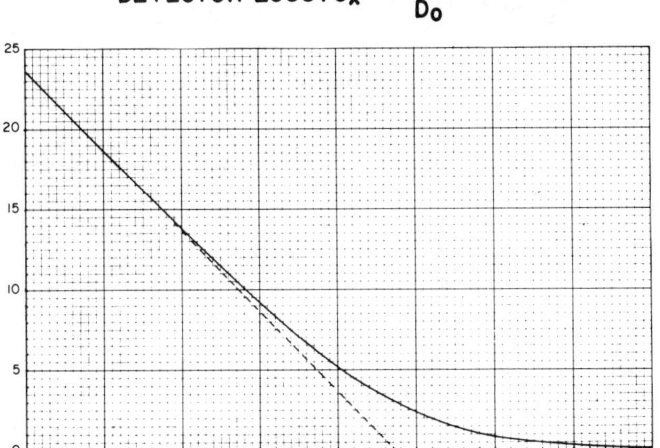

Figure 5 Detector Loss vs. IF SNR [12]

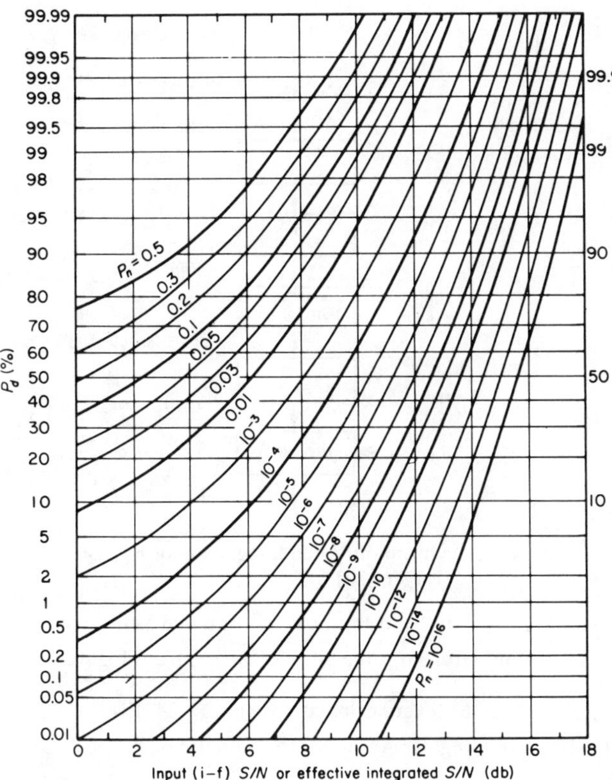

Figure 6 Detectability Factor vs. Detection Probability [2]

ties for a false alarm and hence requiring a higher threshold voltage to maintain the same probability of an output alarm in the same search time. The increase in signal energy required to obtain P_D with this higher threshold (lower P_N) can be denoted by L_G, the *gating interval loss.* The entire increase in required signal energy, as compared to the ideal process, is given by the product of the three loss factors:

$$\mathcal{R}/\mathcal{R}_{MIN} = L_M C_X(1) L_G \tag{5}$$

The calculation of $D_o(1)$ originally was carried through by S.O. Rice [8] ; the distributions for a sinusoidal signal plus noise (Figure 4) are now known as Rician distributions.

Having established the threshold E_T, based on the Rayleigh distribution of Figure 1 and a given P_N, the detection probability is found by integrating the Rician distribution function dP_S from E_T to ∞. For a given P_D, the required single-pulse SNR $D_o(1)$ is higher than for the known-phase process by the detector loss factor, C_X, plotted in Figure 5. Values of $D_o(1)$ vs. P_D are shown in Figure 6 [1 (p. 437), 3 (p. 308), and 6 (p. 115)]. It turns out that the factor C_X for single and multiple-pulse noncoherent detection can be approximated as

$$C_X = \frac{D_o + 2.3}{D_o} \tag{6}$$

where D_o is the IF SNR at the input to the envelope detector. The concept of detector loss is very useful in both detection and measurement calculations, and it tends to emphasize the fact that the loss is directly attributable to the failure of the envelope detector to recover all the information which would have been used in a detection process with known signal phase.

Detection Based on Video Integration

Most radar detection procedures, both in pulsed and CW radar systems, use the integrated sum of many pulses or samples as the input to the threshold device (Figure 7). The single-sample IF (power) SNR $x_1 = (E_1/N_o)/L_M$ is calculated from the radar equation and IF matching loss as explained previously, and set equal to $D_o(n)$ for the given P_D and P_N. The video SNR after envelope detection is $x_3 = 2x_1/C_X(n)$, and this figure is increased by the factor n in the matched video integrator to obtain the output to the threshold

$$x_o = \frac{\mathcal{R}}{L_M C_X(n)} = \frac{2nD_o(n)}{C_X(n)} \tag{7}$$

The three loss factors which describe the increase in energy relative to the ideal case are the same as in Figure 3, except that the n-pulse detector loss replaces $C_X(1)$. Since $D_o(n)$ is less than $D_o(1)$, the n-pulse loss from Equation (6) is larger than the single-pulse loss.

In n-pulse video integration, more signal energy is required but the single-pulse energy is less than in the one-pulse detection process (Figure 3) by the *integration gain*, $G_I(n)$. The increase in total signal energy for n pulses is often described by the *integration loss*, $L_I(n)$, (plotted in Figure 8).

$$L_I(n) \equiv \frac{nD_o(n)}{D_o(1)} = \frac{C_X(n)}{C_X(1)} \tag{8}$$

Thus, if $D_o(1)$ is found from Figure 6 and given values of P_D and P_N, $D_o(n)$ can be found from Figure 8 and a given n

$$D_o(1) = G_I(n)D_o(n) = \frac{nD_o(n)}{L_I(n)} \tag{9}$$

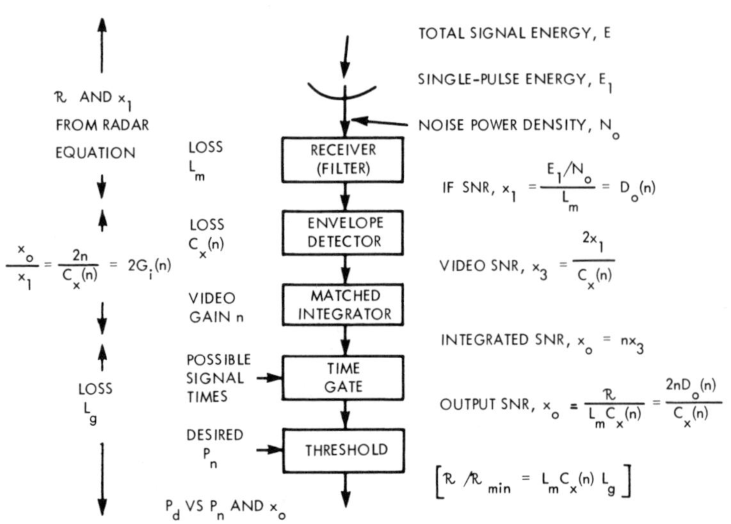

Figure 7 **Optimum Video Integration of n Pulses on Steady Target [12]**

The other losses L_M and L_G must be evaluated as in previous cases. As an example, for $P_D = 0.5$ and $P_N = 10^{-6}$, from Figure 6,

$D_o(1) = 11.2$ dB (exact calculations give 11.24 dB).
If a 10-pulse integration is used, from Figure 8

$L_I(10) = 2.4$ dB (exact calculations give 2.41 dB).

Hence, the integration gain (which would have been 10 dB for coherent integration in a matched IF filter) is

$$G_I(10) = \frac{n}{L_I(n)}\text{, or 7.6 dB}$$

In past literature, attempts have been made to present a single curve to represent integration loss as a function of n. The concept of detector loss makes it clear that a family of curves, such as that in Figure 8, is necessary to describe the changes in L_I as a function of actual SNR. Note the L_I is not identical to C_X, but rather it is the increase in C_X resulting from using an IF SNR $D_o(n)$ rather than $D_o(1)$.

Collapsing Loss

In his classic work on detection theory [9], Marcum discusses the effect of "collapsing" radar data, as occurs when information from a three-coordinate scanning process is collapsed onto a two-dimensional display. Each resolution cell on the display then combines (integrates) receiver noise from several different regions in the scan, only one of which, at a given time, contains a signal. Obviously there is a loss in effective SNR in such a process, but it is not as severe as would be computed by simply adding the several noise components into a combined noise power (or power density). Instead, a collapsing ratio, ρ, is defined as

$$\rho = \frac{m+n}{n} = 1 + \frac{m}{n} \tag{10}$$

where n is the number of signal-plus-noise samples integrated and m is the number of additional samples of noise alone (integrated along with the n samples). For example, if 10 stacked beams in elevation are passed through separate receivers and envelope detectors before being summed in a single video integrator (or display), the collapsing ratio, ρ, is equal to 10. If square-law detectors are used, the result is the same as if the signal energy were divided equally among the 10 channels, giving 10 times the number of pulses integrated but reducing each pulse SNR by 10 dB. For that case, the integration loss would be $L_I(\rho n)$. The collapsing loss can be found as the factor by which this loss figure exceeds $L_I(n)$

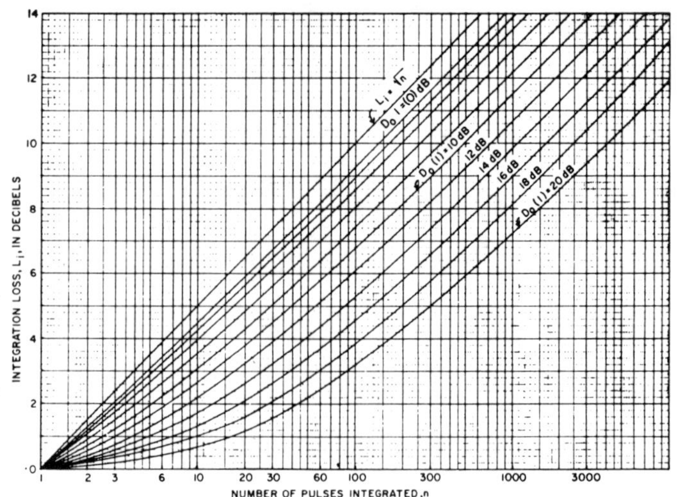

Figure 8 **Integration Loss vs. Number of Pulses Integrated [12]**

$$L_C = \frac{C_X(\rho n)}{C_X(n)} = \frac{L_I(\rho n)}{L_I(n)} \tag{11}$$

If a linear detector is used, it has been shown [10] that the collapsing loss is higher than for a square-law detector.

False-Alarm Relationships

There exists some confusion in the literature as to the relationships between false-alarm probability, false-alarm time, and "false-alarm number." A false alarm is an unwanted threshold crossing resulting from noise (or other interference) when the target is absent. The threshold setting depends on the false-alarm probability, the fraction of detection decisions that result from false alarms. In general, the user of the system is concerned primarily with the false-alarm time (a measure of how long the radar operates without a false alarm) or the false-alarm rate (the average number of false alarms per unit time).

For an ungated receiver without video integration, the false-alarm rate γ is simply

$$\gamma = BP_N \tag{12}$$

where B is the IF bandwidth. If the output is gated to reach the threshold circuit for an interval t_G during each repetition period t_P

$$\gamma = \frac{Bt_G P_N}{t_P} = \frac{\eta P_N}{t_P} \tag{13}$$

where $\eta = Bt_G$ is the number of range cells included in the gate. From this expression it is apparent that information on the range (time delay) of potential targets permits t_G to be reduced and P_N to be increased without increasing γ (see Figure 2). The average time between false alarms is simply

$$\overline{t_{FA}} = \frac{1}{\gamma} = \frac{t_P}{Bt_G P_N} = \frac{t_P}{\eta P_N} \tag{14}$$

for the system without integration, or

$$\overline{t_{FA}} = \frac{n}{\gamma} = \frac{nt_P}{Bt_G P_N} = \frac{nt_P}{\eta P_N} \tag{15}$$

when n-pulse integration is used and when an alarm extending over the n-pulse interval is counted as a single event.

The confusion arises because Marcum, in the absence of any established usage, defined as *false-alarm time* the interval during which the probability of there being no false alarms is equal to 0.5. It follows from the Poisson distribution that

$$\text{Marcum's } t_{FA} = 0.69\ \overline{t_{FA}} \tag{16}$$

Marcum further defined two "false alarm numbers"

$$n_{FA} = \frac{t_{FA}\eta}{t_P} = \frac{0.69\ n}{P_N} \tag{17}$$

$$n'_{FA} = \frac{t_{FA}\eta}{nt_P} = \frac{0.69}{P_N} \tag{18}$$

where the subscript FA has been added to distinguish these numbers from the numbers of pulses integrated. Equation (17) is the number of opportunities for a detection decision during t_{FA}, while Equation (18) is the number of independent opportunities for such a decision. Unfortunately, use of the false-alarm number has been perpetuated in today's literature.

A further confusion arises in considering collapsing loss, which depends upon whether the number of opportunities for decision remains constant or decreases with increasing ρ. Obviously, if collapsing results from such steps as decreasing video bandwidth, the number of opportunities for decision is reduced and P_N may be permitted to increase, partially offsetting the effect of collapsing and thereby minimizing L_C. A discussion of this effect is included in Reference 6.

Target Fluctuation Effects

The previous discussions have considered a steady target signal in a noise background. Yet, most real radar targets are not steady, but rather fluctuating signals, introducing a further statistical uncertainty in the detection process. The basic work in this area was done by Swerling [11] at RAND Corporation. Four different fluctuating models were defined as below:

Amplitude Distribution	Swerling Case Number: *Slow Fluctuation*	Swerling Case Number: *Fast Fluctuation*
$dP = \frac{1}{\overline{\sigma}} \exp \frac{-\sigma}{\overline{\sigma}}\ d\sigma$	Case 1	Case 2
$dP = \frac{4\sigma}{\overline{\sigma}^2} \exp \frac{-2\sigma}{\overline{\sigma}}\ d\sigma$	Case 3	Case 4

The amplitude distributions are Rayleigh (for Cases 1 and 2) and the square root of a chi-squared distribution with four degrees of freedom (for Cases 3 and 4). (See glossary at end of Chapter 3, third definition.) Slow fluctuation (Cases 1 and 3) presumes constant cross section during the n-pulse integration process, with independent values observed during each scan or detection attempt. Fast fluctuation (Cases 2 and 4) presumes independence from pulse to pulse during the integration. Physically, Cases 1 and 2 correspond to aircraft and other complex targets observed at microwave frequencies, while Cases 3 and 4 result when one dominant scatterer is modulated by an assortment of smaller scatterers nearby. Whether the fluctuation is fast or slow depends on the dynamics of the problem and the repetition period of the radar, but use of frequency agility is one way of ensuring fast fluctuation.

Target fluctuation changes the Rician distribution of signal plus noise to a Rayleigh form (Cases 1 and 2) or a chi-squared form (Cases 3 and 4). For single-pulse detection, a simple equation describes the Rayleigh case

$$D_1(1) = D_2(1) = \frac{\ln P_N}{\ln P_D} - 1 \tag{19}$$

where the subscript denotes the Swerling target case. For n-pulse integration, reference can be made to families of curves [3, 4], or approximate values based on plotted curves (Figure 9) can be used, for fluctuation loss L_F

$$L_{F1} = \frac{D_1(n)}{D_o(n)} \approx \frac{D_1(1)}{D_o(1)} \quad \text{(for Case 1 target)} \tag{20}$$

$$L_{F3} = \frac{D_3(n)}{D_o(n)} \approx \frac{D_3(1)}{D_o(1)} \quad \text{(for Case 3 target)} \tag{21}$$

It turns out that the fluctuation loss for these cases is almost independent of the number of pulses integrated and only slightly dependent on P_N, as shown in Figure 9. Furthermore, for rapidly fluctuating targets,

$$L_{F2} = \frac{D_2(n)}{D_o(n)} \approx [L_{F1}(1)]^{1/n} \quad \text{(for Case 2 target)} \tag{22}$$

$$L_{F4} = \frac{D_4(n)}{D_o(n)} \approx [L_{F3}(1)]^{1/n} \quad \text{(for Case 4 target)} \tag{23}$$

Thus, in decibels, the loss for fast fluctuation is 1/n times that for slow fluctuation. The empirical relationships, originally described in Reference 12, also permit us to estimate losses for partially correlated targets, where n_E independent cross-section samples are observed during an n-pulse integration ($n_E \leqslant n$)

$$L_F(n_E) \equiv \frac{D_E(n, n_E)}{D_o(n)} \approx [L_{F1}(1)]^{1/n_E} \tag{24}$$

The detectability factor, $D_E(n, n_E)$, for any arbitrary case of a partially correlated Rayleigh target can then be found from data in Figures 6, 8, and 9.

$$D_E(n, n_E) = \frac{D_o(1) L_I(n) L_F(n_E)}{n} \tag{25}$$

The benefits of frequency or time diversity, relative to the slowly fluctuating Rayleigh target, are described by a diversity gain

$$G_D(n_E) \equiv \frac{D_E(n, 1)}{D_E(n, n_E)} \approx \frac{L_F(1)}{L_F(n_E)} \tag{26}$$

From Figure 9 it is obvious that both the fluctuation loss and the potential for diversity gain are greatest when high P_D is required.

In order to determine the number of diversity samples available in a given observation time, t_o, the corelation intervals of the target must be known in time and frequency [7 (p. 172)]. By definition, for Cases 2 and 4, n_E = n as a result of target dynamics. For longer correlation times, t_C,

$$n_E = 1 + (t_o/t_C) \tag{27}$$

where

$$t_C = \frac{\lambda}{2\omega_A L_X} \tag{28}$$

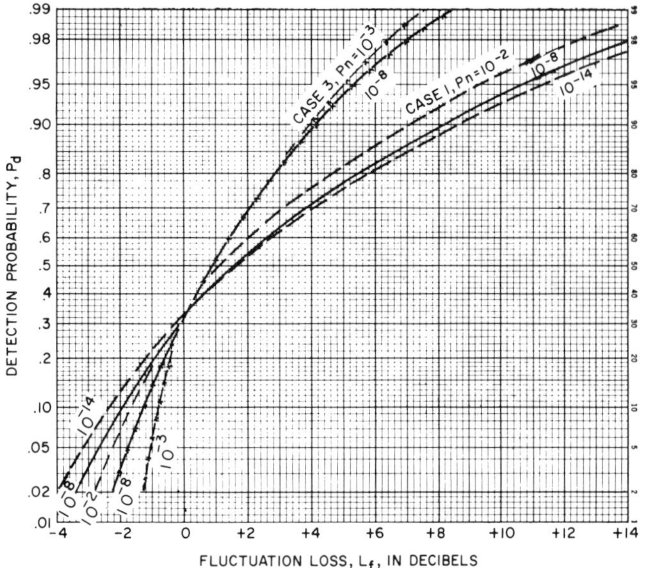

Figure 9 **Fluctuation Loss vs. Detection Probability**

for a target of span L_X rotating at a rate ω_A with respect to the line of sight. In frequency, where diversity samples are taken uniformly over a band Δf,

$$n_E = 1 + (\Delta f/f_C) \tag{29}$$

where

$$f_C = \frac{c}{2L_R} \tag{30}$$

for a target with scatterers distributed radially over a length L_R. As noted in Reference 7, effective values of L_X and L_R are used for nonuniform scatterer distributions.

The Swerling cases correspond to the chi-squared distribution with 2 and 4 degrees of freedom, while the steady target is the limiting case as the number of degrees of freedom approaches infinity. Many targets can be represented by chi-squared distributions with 2K degrees of freedom, where $K < 1$ gives a broad range of fluctuation and $K > 2$ gives reduced range. The general expression for fluctuation loss, relative to Case 1 (where K = 1), is

$$L_{FK} \approx [L_{F1}]^{1/K} \tag{31}$$

Other distributions used to describe targets include the log-normal

$$dP = \frac{1}{\sqrt{2\pi}\,\sigma_X \psi} \exp\left[-\frac{(\ln\psi - \ln\psi_M)^2}{2\sigma_X^2}\right] d\psi \tag{32}$$

Here, ψ_M is the median cross section, and σ_X is the standard deviation of the natural logarithm of the cross section ($\sigma_X = 0.23\,\sigma_Y$ where σ_Y is in decibels). Fluctuation loss can be very large for log-normal targets with $\sigma_X > 1$, especially at higher values of detection probability [13].

Combined Losses and Detectability Factor

In the radar equation (Equation (2)), the detectability factor, D_X, represents the single-pulse energy ratio, obtained with antenna gains G_T and G_R, which is needed to achieve given probabilities of detection and false alarm. In the preceding sections, the relationships have been established among P_D, P_N, received energy ratio E/N_o, matching loss L_M, integration loss L_I, collapsing loss L_C, and several detectability factors $D_1 \ldots D_E$ describing different target and diversity cases. A further step is necessary to determine D_X for a search radar or acquisition scanning radar, and that is to account for the variation in antenna gain over the observation time t_o. By convention, t_o is defined as the time during which the target lies within a rectangular cell of dimensions equal to those of the one-way half-power beamwidths of the antenna, centered on the beam axis. In a fan-beam, two-dimensional search radar, scanning in one coordinate (usually azimuth), the average energy received during a scan past the target is less than that which would be received with full gain in an inteval t_o by the *beamshape loss*, L_P, which is approximately equal to 1.33. If the radar exchanges many pulses with the target during this interval, and if the video integrator is matched to the envelope of the received pulse train, adequate detection performance can be obtained if the center pulse in the train is larger by the factor 1.33 than that which would be calculated for a uniform burst of n pulses [14]. If an unweighted integrator is used, the corresponding factor is 1.45 (or 1.6 dB), as given by Blake [15]. For a two-coordinate scan, the factor is L_P^2 if the signals received over the entire scan are integrated together. However, if the number of pulses per beamwidth (in the rapid-scan coordinate) or scan lines per beamwidth (in the slower-scan coordinate) is not large, or if successive pulses or scans are combined by processes other than integration, the beamshape loss becomes dependent on detection probability, target fluctuation, and pulse (or scan-line) density [16]. An example of how the products $D_o L_P$ and $D_1 L_P$ can vary with P_D, for an arbitrary two-coordinate scan pattern, is shown in Figure 10. For the steady target at $P_D = 0.5$, the loss is $L_P^2 \approx 3.2$ dB; for higher P_D, it increases to 5 dB or more. For Case 1 target fluctuation, the loss at $P_D = 0.5$ is $L_P^2 \approx 3$ dB, decreasing to 0 dB for $P_D \approx 0.92$ (and thereby compensating in part for the larger L_F). The effect of additional detection opportunities at the skirt of the beam is to provide time diversity in the detection process.

The combined effect of several loss factors is to produce a detectability factor

$$D_X = L_M L_P L_X D(n) \tag{33}$$

where L_M describes IF filter matching, L_P is beamshape loss, L_X is the combined effect of collapsing and other processor losses, and D(n) is the detectability factor for the steady target, D_o, or for the fluctuating target D_1, D_2, D_3, D_4, or D_E found in the literature of detection theory [3, 4, 5, 9, 11, 12, or 13]. Further discussion of factors L_M and L_X is given in Reference 6*.

Basic Measurement Processes

When a radar measurement is to be made, a series of three steps is involved:

1) The desired target is resolved from its background and from other objects in at least one radar coordinate – range delay, Doppler frequency or angle.
2) The target signals in adjacent or overlapping resolution cells are detected and compared.
3) Interpolation within the nearest resolution cell is performed, on the basis of the signal comparison.

If only coarse location is needed, the third step can be omitted and the center of the strongest cell, identified in the second step, can be taken as the target position, with a resulting rms error on the order of 30 percent of the cell width in each coordinate. If interpolation is applied, the theoretical minimum error is determined by the width of the cell (x_3 at its half-power points) and the SNR

$$\sigma_X = \frac{x_3}{k_X\sqrt{\mathcal{R}}} \tag{34}$$

**Editor's Note* – See Appendix for additional cookbook procedures for detection calculations. One procedure given is modification, formalization, and extension of method outlined in this chapter by Barton.

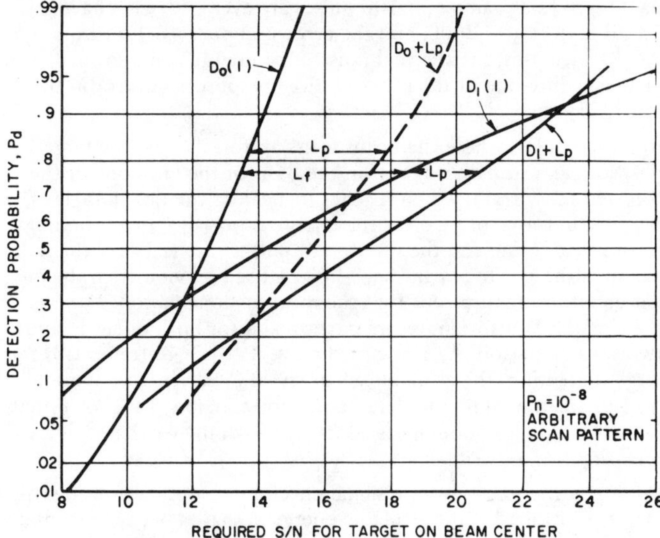

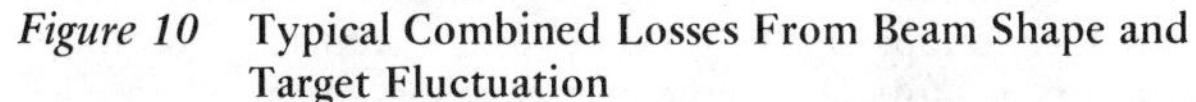

Figure 10 Typical Combined Losses From Beam Shape and Target Fluctuation

The product $k_X\sqrt{\mathcal{R}}$ is the *splitting ratio* or *interpolation ratio* of the measurement. The error slope k_X is a measure of the sensitivity of the measurement process; when error is expressed as in Equation (34) as a fraction of the cell width, k_X is a dimensionless quantity near unity (and perhaps as large as two, for angular measurement).

An illustration of the process for angular measurement is shown in Figure 11. Two receiving beams are generated, offset by $\pm\theta_K$ from the axis of measurement. In a monopulse radar, the signals from the two beams usually are compared by forming a sum and a difference signal at RF. In the vicinity of the axis, the ratio Δ/Σ is a linear measure of the target deviation, θ, from the axis

$$\frac{\theta}{\theta_3} \equiv \frac{1}{k_M}\frac{\Delta}{\Sigma} \tag{35}$$

$$k_M \equiv \left.\frac{\partial(\Delta/\Sigma)}{\partial(\theta/\theta_3)}\right|_{\theta=0} \tag{36}$$

The division of Δ by Σ, known as normalization of the error signal, makes the deviation reading independent of signal amplitude, which may vary slowly with range and rapidly with target fluctuation. The same process, using range gates or filters in place of antenna beams, can be used to measure signal delay (range) or frequency (Doppler shift).

The shape of the Δ pattern approximates the first derivative of the Σ pattern or the elementary pattern used to form the offset pair. Hence, the measurement sensitivity, k_X, which is the derivative of the Δ pattern, is the second derivative of the Σ pattern – a measure of its curvature or sharpness at the axis. Antenna patterns in angle, or signal ambiguity functions in delay and frequency, which have narrow, sharp peaks have greater sensitivity to target deviations from the axis and lower errors in the presence of noise. However, the sharpness of the response is subject to strict limits, imposed by the aperture width and wavelength (for angle), the signal bandwidth (for time delay), and the signal duration or time width (for frequency). Figure 12 shows how the normalized monopulse slope k_M varies with the Σ-channel sidelobe ratio, G_{SR}, almost independently of the type of illumination used. The increase in k_M with increasing taper reflects the fact that beamwidth, θ_3, increases faster than the rms error in measurement. The second half of the plot shows that the absolute sensitivity of measurement decreases with increasing taper. The slope parameter, K_R, is defined by Hannan [20] as the *difference slope ratio.*

$$K_R \equiv K/K_o \tag{37}$$

$$K \equiv \left.\frac{\partial(\Delta/\Sigma_o)}{\partial\theta}\right|_{\theta=0} \tag{38}$$

where K_o is the maximum possible value of K for the given aperture size, and Σ_o is the maximum possible Σ voltage gain for that aperture.

Expressions for Noise Angle Error

Considering only the thermal noise component of measurement error, Equation (34) can be particularized to the several different processes of angular measurement as shown below:

Ideal

$$(\sigma_\theta)_{MIN} = \frac{\lambda}{\mathcal{L}_o\sqrt{\mathcal{R}_o}} = \frac{1}{K_o\sqrt{\mathcal{R}_o}}$$

where $\mathcal{R}_o$ is $\mathcal{R}$ for uniform illumination.

Monopulse

$$\sigma_\theta = \frac{1}{K\sqrt{\mathcal{R}_o}} = \frac{\sqrt{\eta_A}}{K\sqrt{\mathcal{R}_M}}$$

$$\frac{\sigma_\theta}{\theta_3} = \frac{1}{k_M\sqrt{\mathcal{R}_M}} \quad \text{where } k_M \approx 1.5$$

where $\mathcal{R}_M$ equals $\mathcal{R}$ on beam axis.

Search

$$\frac{\sigma_\theta}{\theta_3} = \frac{\sqrt{L_P}}{k_P\sqrt{\mathcal{R}_M}} \approx \frac{0.7}{\sqrt{\mathcal{R}_M}}$$

where k_P equals slope factor for scanning radar.

Conical Scan

$$\frac{\sigma_\theta}{\theta_3} = \frac{\sqrt{2L_K}}{k_S\sqrt{\mathcal{R}_M}} \approx \frac{1}{\sqrt{\mathcal{R}_M}}$$

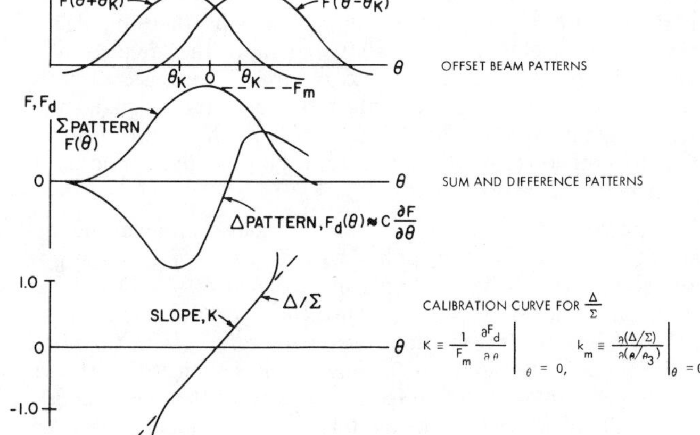

Figure 11 Basic Angular Measurement Process

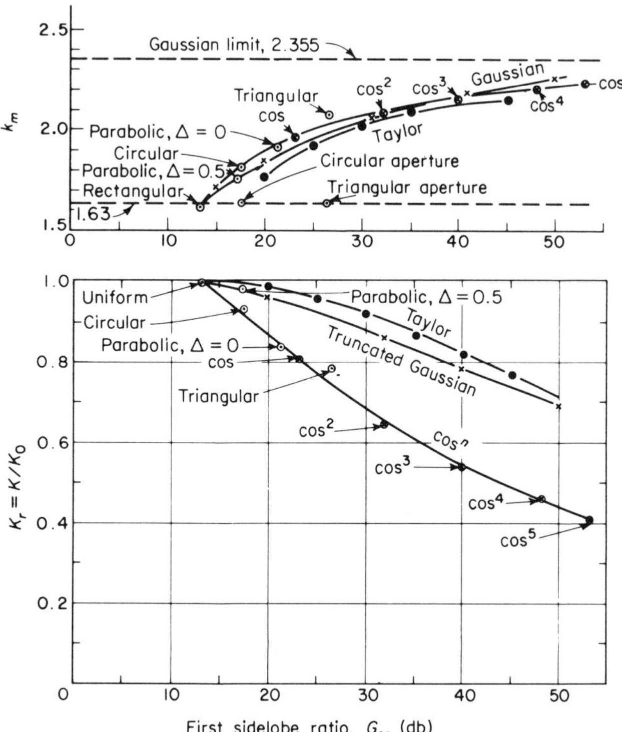

Figure 12 **Angle Difference Slopes vs. Sum-Channel Sidelobe Ratio [7]**

where k_S equals conical-scan error slope.

For off-axis track, add second term $\sigma_2 = \theta/\sqrt{\mathcal{R}}$

The minimum noise error, for a given aperture area A and wavelength, is obtained with a uniformly illuminated aperture for the monopulse sum pattern, collecting signal energy corresponding to $\mathcal{R}_o$, and a linear-odd difference illumination, which gives a slope $K_o = \lambda/\mathcal{L}_o$. Here, $\mathcal{L}_o$ is the rms aperture width as defined by Manasse or Skolnik

$$\mathcal{L}_o^2 = \frac{1}{A}\int_A (2\pi x)^2 \, dA \qquad (39)$$

in which x is the coordinate normal to the propagation path at the aperture, in the plane in which the angle is measured. For a rectangular aperture of width w, $\mathcal{L}_o = \pi w/\sqrt{3} = 1.81$ w. If this rms aperture width is, for example, 100 wavelengths, then the ideal measurement yields an error of 0.01 radian, further divided by the square root of the received energy ratio. Since this energy ratio is at least 40 (more likely 100) for a detectable signal, the error would be on the order of 0.001 radian. This sort of calculation gives the lower limit to error, regardless of the method used to perform the measurement.

If a tapered illumination is used in a monopulse antenna, the second equation listed above applies, with an actual on-axis energy ratio $\mathcal{R}_M = \eta_A \mathcal{R}_o$ (where the aperture efficiency $\eta_A < 1$) and an actual difference slope $K < K_o$. When the error is normalized to 3-dB sum beamwidth, the third expression results, with the normalized monopulse slope k_M (a dimensionless quantity) as shown in Figure 12. In order to reduce sidelobes to a tolerable level, a tapered illumination must be used, leading to an increase in beamwidth (which would have been $\theta_o = 0.88\ \lambda/w$ for uniform illumination), a decrease in $\mathcal{R}_M$, and an increase in error. This normalized form of the error expression is one of the most useful, because a reasonable estimate of error can be made knowing only the beamwidth and the energy ratio, without detailed knowledge of antenna design. For example, if 100 pulses are averaged, each having 10 dB SNR, $\mathcal{R}_M = 2000$, and the error is about $\theta_3/70$ for $k_M = 1.5$. As $\mathcal{R}_M$ increases, the thermal noise error may become so small as to be masked by one of the other numerous sources of error in the radar.

In search radar, where the beam scans uniformly past the target position, an estimate is made on the basis of the envelope of the received pulse train, which reaches its peak at the target angle. The expression for error involves the energy ratio $\mathcal{R}_M$ (corresponding to this peak value and the number of pulses received over the radar beamwidth) and the beamshape loss L_P used previously in the radar equation. It turns out that the ratio of slope k_P to $\sqrt{L_P}$ is approximately 1.4 for any beam pattern and for both one-way and two-way operation. As a rule of thumb, if $\mathcal{R}_M \geqslant 40$ (or 16 dB) for target detection, the available accuracy of a search radar is about $0.11\ \theta_3$ for the minimum detectable target. Below it is shown that the search radar seldom gives accuracy much higher than this level, on a single-scan basis.

Conical scan radars also use signal envelope demodulation to sense target position, but the available energy is shared between azimuth and elevation channels. Comparing the conical scan equation above with that for monopulse, the ratio $k_S/\sqrt{2L_K}$ replaces k_M, and this ratio is near unity (L_K is the crossover loss of the squinted beam, usually about 2 dB). Accordingly, the conical scan error is from 1.5 to 2 times greater than for monopulse, reflecting an information loss of 3.5 to 6 dB.

Finally, if the target is not tracked near the antenna axis, there is an additional noise error which is proportional to the off-axis angle, θ, and inversely proportional to the energy ratio, $\mathcal{R}$, actually received at that angle. This error can be attributed to noise introduced by the sum-channel receiver (in monopulse) in the normalization process, which converts the error voltage into an output angle reading. If the position of the target is measured as $\theta_3/2$ off the null axis, with the two-way gain 6 dB below its on-axis value, the error is more than twice its on-axis value because of the presence of both increased difference-channel noise and the second component caused by sum-channel noise.

Expressions for Range Noise Error

Range (or time delay of the signal) is measured by processes similar to those used in angle. In Figure 13(a), the difference channel is formed by differentiating the output of a matched filter. When the derivative passes through zero during the time the matched-filter output is above the detection threshold (indicating presence of a strong signal), the output time mark is generated to indicate a signal peak. The SNR out of the matched filter must be large to prevent gross errors caused by noise peaks not associated with a signal. In Figure 13(b), the video signal from a wideband receiver is matched approximately in Σ and Δ gates (envelope correlators), and the Δ/Σ ratio is used to control the tracking loop. This circuit performs successfully even if the single-pulse SNR is below unity, integration being performed in the servo filter.

The equations for time delay error are listed below:

Ideal

$$(\sigma_T)_{MIN} = \frac{1}{\beta_A\sqrt{\mathcal{R}}} = \frac{1}{K_o\sqrt{\mathcal{R}}} \approx \frac{\tau_o}{1.63\sqrt{\mathcal{R}}}$$

Mismatched Filter

$$\frac{\sigma_{T1}}{\tau_{3A}} = \frac{1}{K\tau_{3A}\sqrt{\mathcal{R}_1}} \approx \frac{1}{\sqrt{\mathcal{R}_1}}$$

Pulse Compression

$$\frac{\sigma_{T1}}{\tau_{3X}} = \frac{1}{K\tau_{3X}\sqrt{\mathcal{R}_1}} = \frac{1}{k_M\sqrt{2S/N}}$$

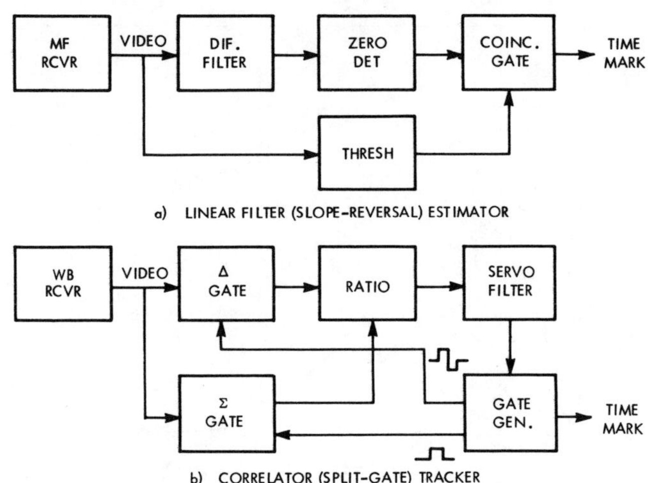

Figure 13 Range Measurement Processes

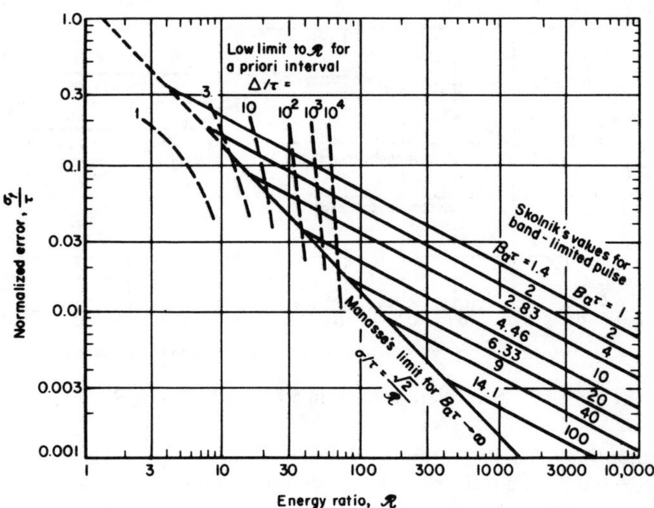

Figure 14 Measurement Accuracy vs. Energy Ratio for Rectangular Pulse [7]

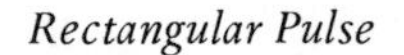

Rectangular Pulse

$$\frac{(\sigma_T)_{MIN}}{\sigma} = \frac{\sqrt{2}}{\mathcal{R}} \geqslant \frac{1}{\sqrt{2B\tau\mathcal{R}}}$$

These equations are similar to those for angle error, with rms signal bandwidth, β_A, replacing $\mathcal{L}_o/\lambda$, where

$$\beta_A{}^2 = \frac{1}{E} \int_{-\infty}^{\infty} [2\pi(f-f_o)]^2 A^2(f)\, df \tag{40}$$

and A(f) is the signal spectrum, centered at f_o, with total energy E. As before, $\mathcal{R} = 2E/N_o$ is the energy ratio at the receiver input. When normalized to the half-power width, τ_o, of the filtered pulse, a form is obtained with a constant 1.63 similar to the value of k_M for an ideal monopulse estimator. With a mismatched filter and an input pulse width τ_{3A}, the normalized slope $K\tau_{3A} \approx 1$ yields a single-pulse estimate dependent only on the energy ratio. Use of uniform-spectrum pulse compression with weighting on the received spectrum gives an expression for error (normalized to the compressed output pulse) which is exactly analogous to monopulse angular measurement. The curves for k_M as a function of sidelobe ratio (Figure 12) are directly applicable.

Ranging on a rectangular pulse is an interesting special case, since this signal has an infinite rms bandwidth. The expression derived by Manasse [17] shows that the error is inversely proportional to energy ratio (not to the square root of energy ratio, as might be expected). Skolnik [18] derived an approximation for the band-limited rectangular pulse in which the square-root relationship reappeared. Barton and Ward [7] show that Manasse's result is consistent with Skolnik's if an adaptive receiver-estimator is used in which bandwidth is adjusted in accordance with energy ratio to avoid multiple zero crossings at the output of the differentiator. The results are combined in Figure 14, indicating, for example, that an energy ratio of 300 permits the receiver bandwidth to be set at $B_A \approx 60/\tau$ for $\beta_A \approx 11/\tau$ and an accuracy of $0.005\ \tau$. Of course, if the transmitted signal were bandlimited to some smaller value, the Skolnik curves would apply at all energy ratios above the limit set by Manasse's expression. Thus, for a transmitter limit $B_A = 10/\tau$, the error would be proportional to $\sqrt{\mathcal{R}}$ for $\mathcal{R} > 40$, and would drop only to $0.012\ \tau$ at $\mathcal{R} = 300$.

Other Error Sources

Thermal noise is only one source of measurement error, and often it is of secondary importance in radar. Angle tracking is especially affected by target glint and scintillation errors, summarized below:

Glint $\quad \sigma_\theta = 0.35\ L_X/R$

$$\sigma_R = 0.35\ L_R$$

$$\sigma_F = 0.35\ (2L_X\omega_A/\lambda)\ \text{Hz}$$

All radars – monopulse, sequential, or conical scanning – respond to the wavefront which arrives at the antenna. When the target is a complex scatterer like an aircraft, rather than a single point scatterer, the wavefront contains local ripples which lead to glint error. This error can be approximated by a normal distribution with standard deviation equal to 0.35 times the physical spread of the scattering sources. The equations describe this spread as L_X across the beam (in azimuth or elevation) and L_R along the beam axis. A typical value for aircraft might be $L_X = L_R = 20$m, leading to position errors of 7m and corresponding angular errors of 0.35 mr at R = 20 km. Glint errors can be reduced by smoothing, but generally several seconds of averaging are required to obtain significant reduction. Components of error caused by glint must be added (in rss fashion) to thermal noise and other errors.

Scintillation arises from the same target scattering processes that cause glint, but the scintillation error is dependent on the response of the radar to varying signal amplitude. A null-tracking monopulse system has no scintillation error. A conical scan radar has a scintillation error dependent on the spectral density of signal modulation at the scan rate (typical errors being $\approx 0.02\ \theta_3$). A linearly scanning search beam is subject to scintillation error if the signal fluctuates during the passage of the beam over the target. Swerling [19] gave results for the case where fluctuation was fast enough to give pulse-to-pulse independence, and Barton and Ward [7] generalized the situation as shown in Figure 15. The worst error occurs when the correlation time of the signal, $t_C{}'$, is comparable to the time-on-target, $t_o = \theta_3/\omega$; in such a case, the error, σ_S, is approximately $0.12\ \theta_3$. The same curve describes the error caused by frequency scan over a frequency-sensitive target. From Equation (30), the correlation frequency is $f_C = c/2L_R$; when this figure is comparable to the frequency shift needed to scan the beam through one beamwidth, the error is large. It is this scintillation error which limits the accuracy of most search radars to $\approx 0.1\ \theta_3$, even for strong signals.

Atmospheric propagation also contributes to error, both with bias (or slowly varying) components and with random fluctuations. Magnitudes of both types of error are given in the literature [7]. Finally, multipath reflections from the earth's surface cause large errors in elevation data on targets near the horizon. Rough estimates of these errors, normalized to the elevation beamwidth, are shown in Figure 16. A tracking radar, unless constrained by heavy

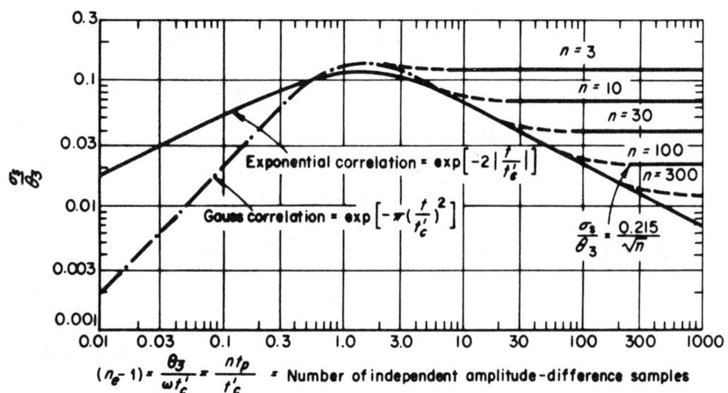

Figure 15 **Scintillation Error in a Linear Scanning Radar [7]**

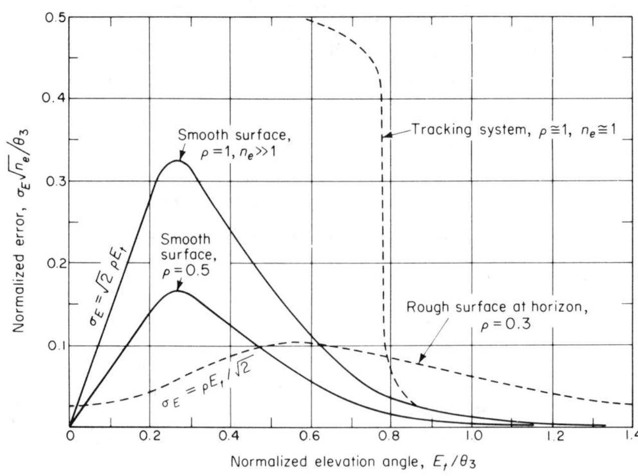

Figure 16 **Typical Multipath Error for Gaussian Beam [7]**

smoothing, loses elevation track when the target approaches within $\approx 0.8\ \theta_3$ of a reflective surface, locking on the horizon with periodic jumps to false tracking points above the target or below the horizon. For surfaces with lower reflection coefficients, or when long-term smoothing is effective in averaging over all phase angles of reflection, the error is less but still severe. Rough surfaces give smaller errors but these errors extend to angles up to $1.5\ \theta_3$. In general, it is impossible to obtain elevation data better than $\approx 0.1\ \theta_3$ for targets below $\approx \theta_3$ over any surface.

Total Measurement Error

A complete analysis of measurement error is made by estimating separately each of the many applicable components — thermal noise, clutter and other interference, target noise, propagation, and components due to radar circuits and components. These several components, being statistically independent of each other, can then be combined by rss addition

$$\sigma_T^2 = \sigma_1^2 + \sigma_2^2 + \ldots\ldots$$

Separate estimates are made for each measured coordinate. As indicated below, some of these sources are dependent upon target elevation, some on range, some on both, and some are independent of either.

Rss Combination (In Each Coordinate) of

Radar Instrumental Error
Noise/Interference Component } *Range*
Target Noise } *Dependent*
Refraction and Fluctuation *(Range-Elevation Dependent)*
Multipath *(Elevation Dependent)*

Separation by Spectral Characteristics

Bias vs. Variable Error
Detailed Spectral Analysis for Smoothing and Differentiation (Limitations of White-Noise Analysis)

The independence permits us to prepare a vertical coverage chart (similar to the conventional search coverage chart) with contours at various error levels.

For some applications, a spectral analysis is useful. This analysis can be a simple separation into *bias* and *variable* components, or a more refined analysis of spectral density from zero frequency to the upper limit of the data system. If target velocity is to be obtained by differentiation, or if smoothed position data are to be used, the spectral analysis is of great importance. It is especially important to avoid mathematical analyses based on random white noise (independent errors from sample to sample at the output), unless thermal noise is the only significant component. There is an inclination to assume that single-pulse estimates from 10,000 pulses can be averaged to obtain a 100-fold improvement in accuracy. The important factor is not the number of pulses or samples averaged, but the number of independent error samples obtained during the averaging interval. If only one or two such independent samples are obtained, the effect of smoothing is minimal, and an analysis based on white noise theory is extremely misleading.

A good error analysis can predict the actual performance within a factor of about two. Unless considerable care is taken, using material presented above and in the references, the estimated error can be off by a factor of 10, leading to unsatisfactory system design.

References

[1] Skolnik, M.I.: *Introduction to Radar Systems,* McGraw-Hill, New York, 1962.

[2] Barton, D.K.: *Radar System Analysis,* Artech House, Dedham, Massachusetts, 1976.

[3] DiFranco, J.V.; and Rubin, W.L.: *Radar Detection,* Prentice-Hall, Englewood Cliffs, New Jersey, 1968.

[4] Meyer, D.P.; and Mayer, H.A.: *Radar Target Detection,* Academic Press, New York, 1973.

[5] Blake, L.V.: "Prediction of Radar Range," Chapter 2 in *Radar Handbook* (M.I. Skolnik, ed.), McGraw-Hill, New York, 1970.

[6] Barton, D.K.: *The Radar Equation* (Vol. 2 in *Radars*), Artech House, Dedham, Massachusetts, 1974.

[7] Barton, D.K.; and Ward, H.R.: *Handbook of Radar Measurement,* Prentice-Hall, Englewood Cliffs, New Jersey, 1969.

[8] Rice, S.O.: "Mathematical Analysis of Random Noise," *BSTJ 23, No. 3,* July 1944 and *BSTJ 24, No. 1,* January 1945; reprinted, N. Wax, Dover Publications, 1954.

[9] Marcum, J.I.: "A Statistical Theory of Target Detection by Pulsed Radar," *RAND Report RM-754,* December 1947, with *Appendix RM-753,* July 1948; reprinted, *IRE Transactions on Information Theory, Vol. IT-6, No. 2,* April 1960.

[10] Trunk, G.V.: "Comparison of Collapsing Losses in Linear Square-Law Detectors," *Proceedings of the IEEE, Vol. 60, No. 6,* June 1972; reprinted in Reference 6.

[11] Swerling, P.: "Probability of Detection for Fluctuating Targets," *RAND Report RM-1217,* March 1954; reprinted *IRE Transactions on Information Theory, Vol. IT-6, No. 2,* April 1960.

[12] Barton, D.K.: "Simple Procedures for Radar Detection Calculation," *IEEE Transactions on Aerospace and Electronics Systems, Vol. AES-5, No. 5,* September 1969; reprinted in Reference 6.

[13] Heidbreder, G.R.; and Mitchell, R.L.: "Detection Probabilities for Log Normally Distributed Signals," *IEEE Transactions on Aerospace and Electronics Systems, Vol. AES-3, No. 1,* January 1967.

[14] Hall, W.M.: "Antenna Beam-Shape Factor in Scanning Radars," *IEEE Transactions on Aerospace and Electronics Systems, Vol. AES-4, No. 3,* May 1968; reprinted in Reference 6.

[15] Blake, L.V.: "The Effective Number of Pulses per Beamwidth for a Scanning Radar," *Proceedings of the IRE, Vol. 41, No. 6,* June 1953; reprinted in Reference 6.

[16] Hall, W.M.; and Barton, D.K.: "Antenna Pattern Loss Factor for Scanning Radars," *Proceedings of the IEEE, Vol. 53, No. 9,* September 1965; reprinted in Reference 6.

[17] Manasse, R.: "Range and Velocity Accuracy from Radar Measurements," *Lincoln Lab Report 312-26,* February 1955 (AD236236).

[18] Skolnik, M.I.: "Theoretical Accuracy of Radar Measurements," *IRE Transactions on Aeronautical and Navigational Electronics, Vol. ANE-7, No. 4,* December 1960.

[19] Swerling, P.: "Maximum Angular Accuracy of a Pulsed Search Radar," *Proceedings of the IRE, Vol. 44, No. 9,* September 1956.

[20] Hannan, P.W.: "Optimum Feeds for All Three Modes of a Monopulse Antenna," *IEEE Transactions on Antennas and Propagation, Vol. AP-9, No. 5,* September 1961; reprinted in *Monopulse Radar* (Volume 1 of *Radars*), Artech House, Dedham, Massachusetts, 1974.

Chapter 3 Brookner

Cumulative Probability of Detection

Introduction

If detection sensitivity is based on single-scan detection when in fact cumulative probability detection applies, the radar transmitter power may be overestimated by 14 dB, a stand-off jammer's power may be overestimated by 8 dB. An error of 4 dB can occur when synchronized conditions are assumed when nonsynchronized conditions in fact exist. These types of errors are elaborated upon in this chapter. No background in the area of cumulative probability of detection is assumed.

The results of Chapters 1 and 2 indicate that the radar range varies as the 4th root of the transmitter power; see Equations (1) in Chapter 1 and Equation (2) in Chapter 2. This range law is based on single-scan detection, i.e., on the observation of the returns from one scan of the radar. In this chapter, it is shown that when detection is based on the observations made on more than one scan of the radar, the detection range is dependent on a root of the transmitter power other than the 4th root – such as the 3.3th root of the transmitter power. Detection based on the accumulation of illuminations made on an approaching target prior to reaching a range R is called cumulative probability of detection, in contrast to single-scan detection on which the equations already mentioned are based.

Difference Between Single-Scan and Cumulative Detection

Consider the situation of a scanning radar placed at a site to be defended. Throughout it is assumed that the radar is a narrow beam scanning radar such that there is no ground multipathing present, i.e., there exist no paths between the radar and target involving a reflection from the ground which could cause lobing of the antenna pattern in elevation. Such is the case for a narrow beam electronic scanning array.

The defense radar must detect any approaching aircraft by some range R in order for the defense to react in time. When using single-scan detection, the target is placed at the range R in question (see Figure 1). (Nothing is said about how the target arrived at range R and no use is made of prior observations of the target.) A determination is made of the probability of detection based on a single radar scan observation.

How does target detection actually occur? For simplicity, let us assume only one pulse illuminates the target during one scan period of the radar. Figure 1 shows the scan volume with the antenna beam illuminating the volume occupied by the target. The time T_F for the radar to scan out the search volume once is given by

$$T_F = (\Omega_S/\Omega_B)T_P \qquad (1)$$

where:

Ω_S = search volume (steradians)

Ω_B = antenna beam size (steradians)

T_P = pulse-to-pulse period

To detect the presence of a target a threshold test is used. If the received echo pulse voltage at the receiver video output exceeds a threshold voltage, a target is said to be present (see bottom half of Figure 1). The target range is given by the time at which the threshold is exceeded. If the threshold is exceeded at a time τ_R seconds after the transmitter pulse is sent out, the target range is $R = \tau_R c/2$. It is possible for the receiver noise to exceed the threshold in the absence of the signal; such an event is referred to as a false alarm. If the range interval over which targets are being searched is from R_1 to R_2 and the width of the transmitted radar pulse is τ, the number of possible independent resolvable time instants at which a target echo signal can appear is $\eta = (R_2-R_1)/(c\tau/2)$. (Note that η has the same definition as that in Chapter 2. If no target is present, η represents the number of opportunities for a false target indication per pulse-to-pulse period.)

Although a single pulse is assumed to be transmitted per scan, the results are equally applicable to the case of the transmission of N pulses per scan with the echo returns from these pulses added coherently in the receiver (i.e., the returns are added in phase at the output of the receiver IF). These N coherently integrated pulses are envelope detected and then applied to the threshold test. The results obtained when incoherent integration is used per scan are similar to those obtained with coherent integration. (Later the results are extended to the case in which N pulses are incoherently integrated per scan.)

Because initially it is assumed that only one pulse illuminates the target per scan, one can talk interchangeably of single-scan and single-pulse detection.

Let us now consider the cumulative probability of detection criteria. In an actual physical situation the target does not appear at the range R out of the blue. Instead, an attacking aircraft probably approaches on a radial path from the radar horizon. As a result the radar has had opportunities to observe the target at ranges greater than R ($R + \Delta R$, $R + 2\Delta R$, $R + 3\Delta R$, . . . where ΔR is the distance travelled by the aircraft between scans), as depicted in Figure 2. In determining the cumulative probability of detection, one asks the question "What is the probability of detecting the target by range R taking into account the possibility that the target could have been detected at earlier ranges, $R + \Delta R$, $R + 2\Delta R$, $R + 3\Delta R$, . . . ?" It is thus apparent that, for the assumption of the same transmitted energy per pulse, the cumulative probability of detection by range R is larger than the single-pulse probability of detection at range R. Equivalently, less energy per pulse is needed to achieve a specified cumulative probability of detection, say 90%, by range R than to achieve a 90% probability of detection at range R based on single-scan detection.

Let us refer to the range as determined by the cumulative probability of detection as the cumulative-probability-of-detection range, and to the range determined by the single-scan (or equivalently, single-pulse) detection criteria as the single-scan (or equivalently, single-pulse) detection range.

Dependence of Single-Scan and Cumulative-Probability-of-Detection Ranges on Transmitter Power

Now consider the dependence of these two ranges on the radar transmitted power. Assume that a 90% probability of detection is specified as being desired. For the single-scan probability of detection situation, this figure is interpreted as meaning that a 90%

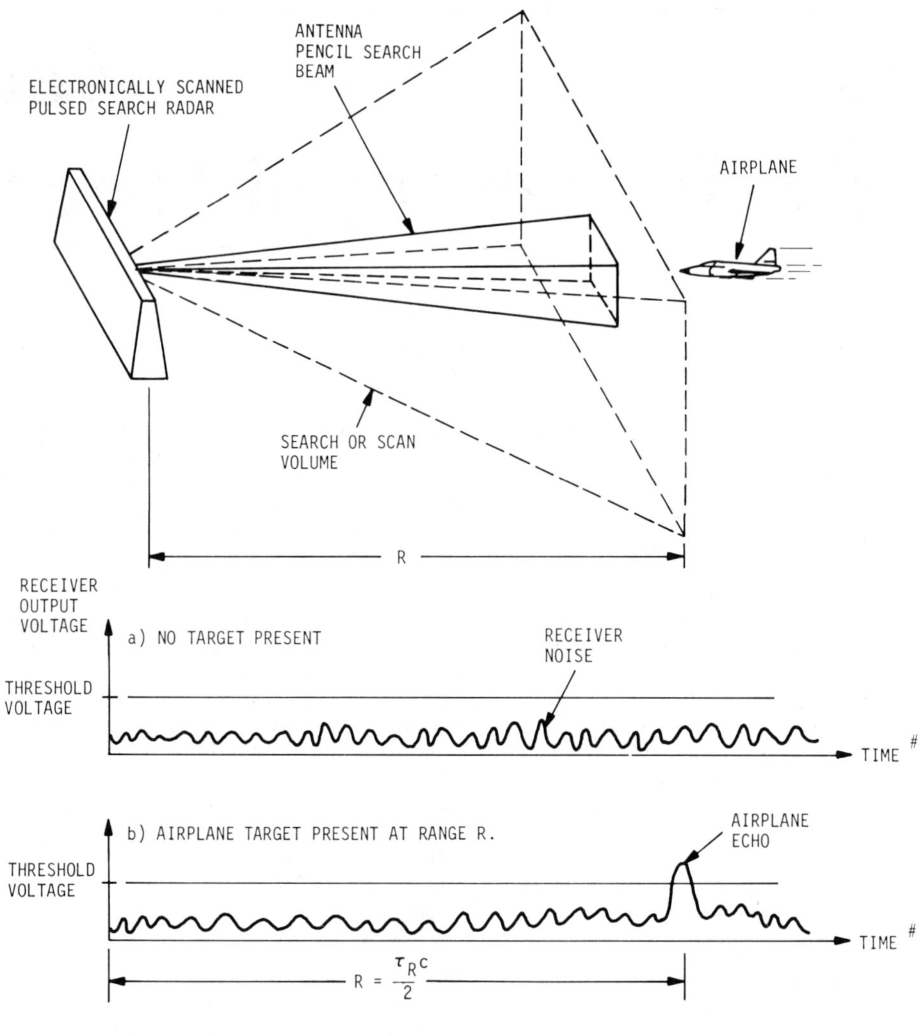

Figure 1 **Search Radar Single-Scan Probability of Detection. Assumption Made that One Pulse Illuminates the Target per Scan; thus, Single-Scan Probability of Detection Equals Single-Pulse Probability of Detection.**

probability of detection per scan (or per pulse) is required at range R. Assume that a signal-to-noise ratio of 20 dB per pulse is needed to achieve this performance and that a transmitted energy per pulse of E joules gives this signal-to-noise ratio per pulse when a 1 m^2 target located at range R is observed. The question raised is what increase in energy per pulse is needed to achieve this detection probability at twice the range R, that is, at range 2R. The classical single-pulse noise limited range equation given by Equation (2) of Chapter 1 and Equation (1) of Chapter 2 indicate that the signal-to-noise ratio per pulse is proportional to the energy transmitted per pulse (for Equation (2) of Chapter 1, $P_{PK}/B = P_{PK}\tau = E$, where τ is the width of the transmitted pulse, after pulse compression if a chirp waveform is used) and inversely proportional to R^4:

$$\text{SNR/pulse} \propto \frac{E}{R^4} \tag{2}$$

Hence, if the range is increased by a factor of two, the energy per pulse must be increased by a factor of 2^4 or 12 dB in order to still achieve a signal-to-noise ratio per pulse of 20 dB and, in turn, a probability of detection of 90%. Solving for R, one finds that the single-scan detection range is proportional to the fourth root of the energy per pulse:

$$R \propto [E/(\text{SNR/pulse})]^{1/4} \tag{3}$$

This relation is the classical fourth-root-range-law. It is also referred to as the R-to-the-fourth power law because the required E is proportional to R^4.

The cumulative-probability-of-detection range, R_C, is not proportional to the fourth root of the energy per pulse. Instead it is proportional to a root of the energy per pulse less than 4:

$$R_C \propto E^{1/S_1} \tag{4}$$

where typically $3 \leq S_1 \leq 4$.

Thus we have a range law other than the fourth root of E which one uses classically for the single-scan or single-pulse noise limited situation.

TARGET MODEL	POWER LAW, S_1	INCREASED ENERGY REQUIRED FOR FACTOR OF TWO INCREASE IN RANGE	
		ABSOLUTE	RELATIVE TO R^4 LAW
SWERLING I	3.50	10.5	-1.5
SWERLING III	3.55	10.7	-1.3
NONFLUCTUATING	3.74	11.2	-0.8
4th POWER LAW	4.0	12.0	0.0

Table 1 Cumulative-Probability-of-Detection Range Laws for Different Target Models. Assumptions Made that P_C = Cumulative Probability of Detection = 90% and n = False Alarm Number per Pulse = 10^6.

Table 1 gives values for the power law, S_1, for various target fluctuation models — Swerling I, Swerling III, and nonfluctuating.

The target amplitude fluctuations for an airplane result from the fact that the target consists of many scatterers, the returns from which add at the receiver constructively or destructively depending on the aspect angle of the target. A target consisting of many scatterers having about equal cross section gives rise to the Swerling I target model for the return signal amplitude. For this target model, the signal power has a chi-square distribution with two degrees of freedom. A target consisting of one dominant scatterer with many smaller scatterers gives rise to a Swerling III target model for the return signal amplitude. For this target model, the signal power has a chi-square distribution with four degrees of freedom. A nonfluctuating target is, as the name implies, a target whose amplitude does not fluctuate in time.

The results are given in Table 1 for a cumulative probability of detection of 90% and a false alarm number of 10^6. The false alarm number equals the number of independent range gate samples that must be observed in the absence of a signal before the noise samples falsely indicate the presence of one or more targets with a probability of 50%. This definition is the same as that used by Marcum [1] and throughout the literature since then, including the extensive detection curves of Meyer and Mayer [2]. For this reason this definition is used here in spite of the fact that it perhaps is not the best definition for false alarm number (see Chapter 2); also, it is easily related to the other definitions (see Glossary at end of chapter and Chapter 2). Table 1 indicates that, for a Swerling I type target, the power law is 3.5 instead of 4. As a result, to increase the range by a factor of 2, the energy per pulse must be increased by 10.5 dB instead of 12 dB as would have been predicted by the single-scan detection range law relationship — a saving of 1.5 dB.

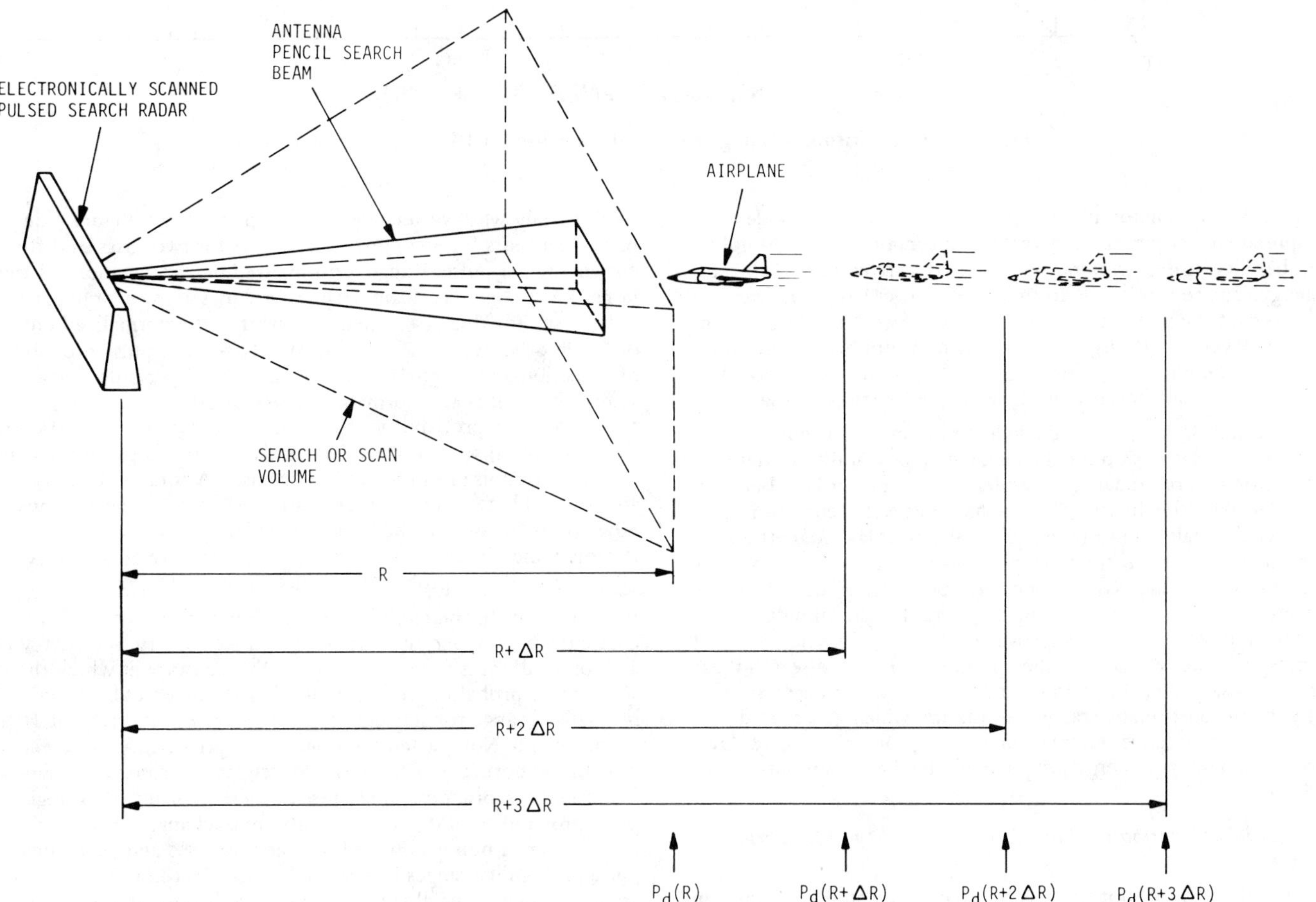

Figure 2 Search Radar Cumulative Probability of Detection

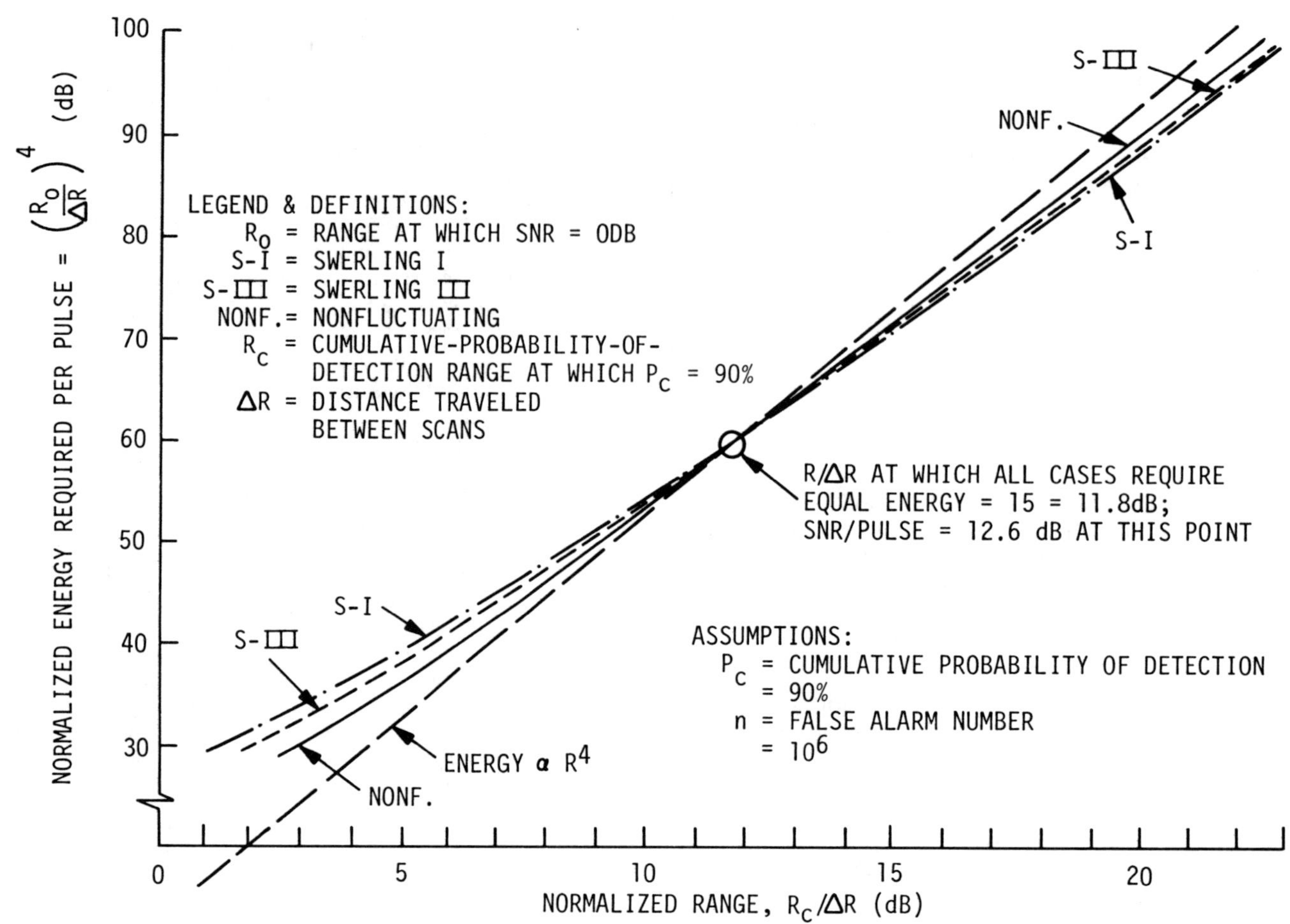

Figure 3 **Comparison of Energy Required per Pulse for Different Target Models**

Figure 3 shows pictorially how, for the various target models, the required energy per pulse increases as a function of the cumulative-probability-of-detection range. The results are compared to the energy required for the R to the fourth power law. This figure shows that the power law is better for the Swerling I target than for the Swerling III target model, which in turn has a better power law than the nonfluctuating target model; all the target models have a power law better than the R to the fourth power law.

Mallett and Brennan [3] have indicated a one-third-root-of-E power law relationship for the cumulative-probability-of-detection range for a search radar. Our results indicate power laws between 3 and 4. Why the disparity? There has been a tendency to misinterpret the Mallet and Brennan one-third-root law. Mallett and Brennan specify that the third-power law applies when R/R_C has the optimum value* (to be specified later). The tendency has been to interpret this law as applying in general for the cumulative-probability-of-detection range when actually it does not. As range varies, in order to maintain the optimum R/R_C for a search radar, it is necessary that the energy per pulse vary with range (specifically, be proportional to range), a situation which generally does not occur in practice. For the cumulative probability power laws specified here, the energy per pulse was held constant with range [4, 5].

Physical Explanation for Cumulative-Probability-of-Detection Power Laws

So far, the typical root law relations have been stated for the cumulative-probability-of-detection range without giving an indication of physically why we get these strange power laws. Figure 4 depicts the results for an approaching target for two cases of different energy per pulse. Range is normalized by the distance ΔR the target moves between scans. Figure 4a shows a low energy per pulse case. For this case, when the target is at a normalized range of 42 (in other words, R = 42(ΔR)), a 10% single pulse probability of detection and a signal-to-noise ratio of 7 dB per pulse are assumed. When it is at a normalized range of 30 (R = 30(ΔR)), the single-pulse probability of detection is assumed to be 50% and the signal-to-noise ratio 13 dB per pulse. As the target approaches the signal-to-noise ratio per pulse increases. A total of 13 pulses are received between normalized ranges 42 and 30, the 10% and 50% points. Also sketeched is the cumulative probability of detection, which increases more rapidly. At the normalized range of 30, the cumulative probability of detection is 99%.

What happens to the cumulative probability of detection if we increase the energy per pulse by 40 dB? If we increase the energy by 10^4, or 40 dB as is the case in Figure 4b, the range at which the single-pulse probability of detection is 10% increases by 10 dB. The normalized range for single-pulse probability of detection of 10% becomes 420. Now, when the single-pulse probability of detection is 50%, the normalized range is 300, ten times its previous value. The signal-to-noise ratios per pulse at these 10% and 50% single-pulse probability of detection points are unchanged. However, the number of pulses observed between the 10% and 50% single-pulse probability ranges is increased from 13 to 121. Thus, there are many more pulses here for making up the cumulative probability of detection. The cumulative probability of detection of 99% is reached 36 pulse observations after the normalized range

*The third power law also applies as long as $\Delta R/R_C$ is held constant.

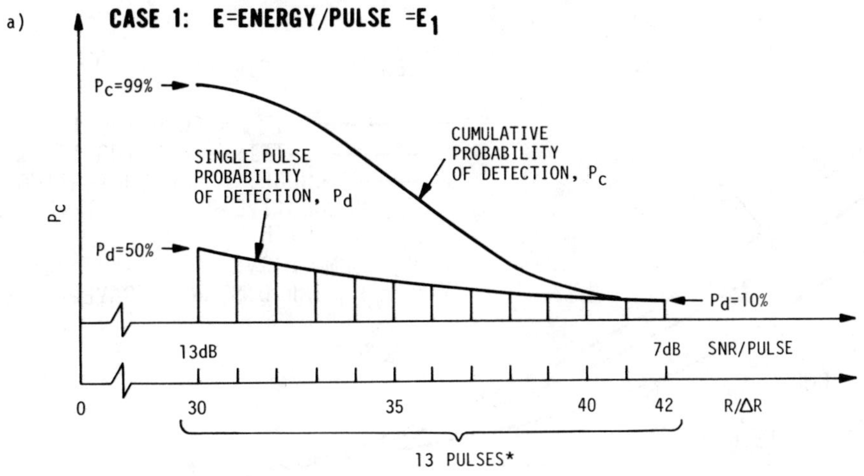

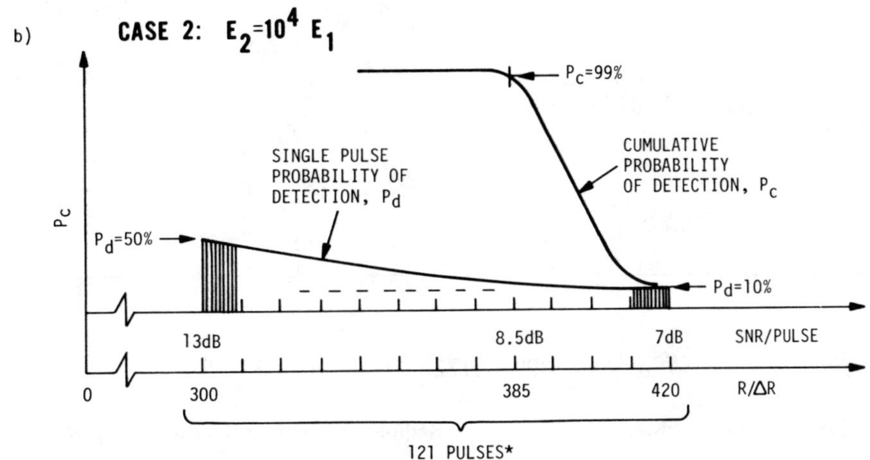

* IT BEING ASSUMED ONE PULSE ILLUMINATES THE TARGET PER SCAN.

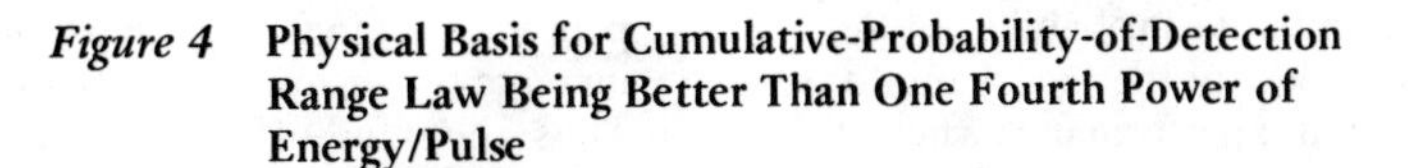

Figure 4 **Physical Basis for Cumulative-Probability-of-Detection Range Law Being Better Than One Fourth Power of Energy/Pulse**

420 is reached, i.e., at the normalized range 385. Thus instead of having a 99% cumulative probability of detection at the 50% single-pulse probability of detection point, it is reached at a longer range with a lower single-pulse probability of detection. It occurs at the 8.5 dB single-pulse signal-to-noise ratio (see Figure 4b). Thus where we detect the target with a 99% cumulative-probability-of-detection is, in effect, further out than the one-fourth root law would predict. If we had an R^4 relationship, the target would have been detected with a 99% cumulative probability of detection at 10 times the normalized range of 30, i.e., at $R/\Delta R = 300$. Instead, we are detecting at a normalized range of about 385. Thus, because we have many more observations of the target in the same interval of the single-pulse probability curve, we detect it sooner as we increase the power; therefore, we have a range relationship that is better than R^4. That fact is the key to the whole thing. It is that simple.*

*The values given for the two cases of Figure 4 were not worked out for any actual situation. They were selected to illustrate physically what is happening and the values given are close to the numbers found for the Swerling I target model.

SNR/Pulse Required for Specified Cumulative-Probability-of-Detection Range

Another interesting property pointed out by the examples of Figure 4 is that the single-pulse signal-to-noise ratio at which the cumulative probability of detection is 99% is 13 dB for the low energy example and 8.5 dB for the higher energy case. Thus, less received energy per pulse is needed to detect the target in the latter instance; this fact gives us further insight into the physical reasons why we have a range law less than R^4. The higher energy example has a larger $R/\Delta R$ and smaller signal-to-noise per pulse for a given cumulative detection probability.

Figure 5 shows how the required signal-to-noise ratio per pulse for a specified cumulative probability of detection, 90% in this example, decreases as the normalized range $R_C/\Delta R$ increases. The results are given for the three target fluctuating models listed in Table 1 — the Swerling I, Swerling III, and the nonfluctuating models. Note that at the normalized range of 13 dB, the required single-pulse signal-to-noise ratio is 12.6 dB for Swerling I (as well as the other models). If we go out a factor of 8 in range (9 dB) to $R_C/\Delta R \doteq 22$ dB, the signal-to-noise ratio required per pulse decreases — by about 5 dB, 4.3 dB, and 2.9 dB for the three models, respectively.

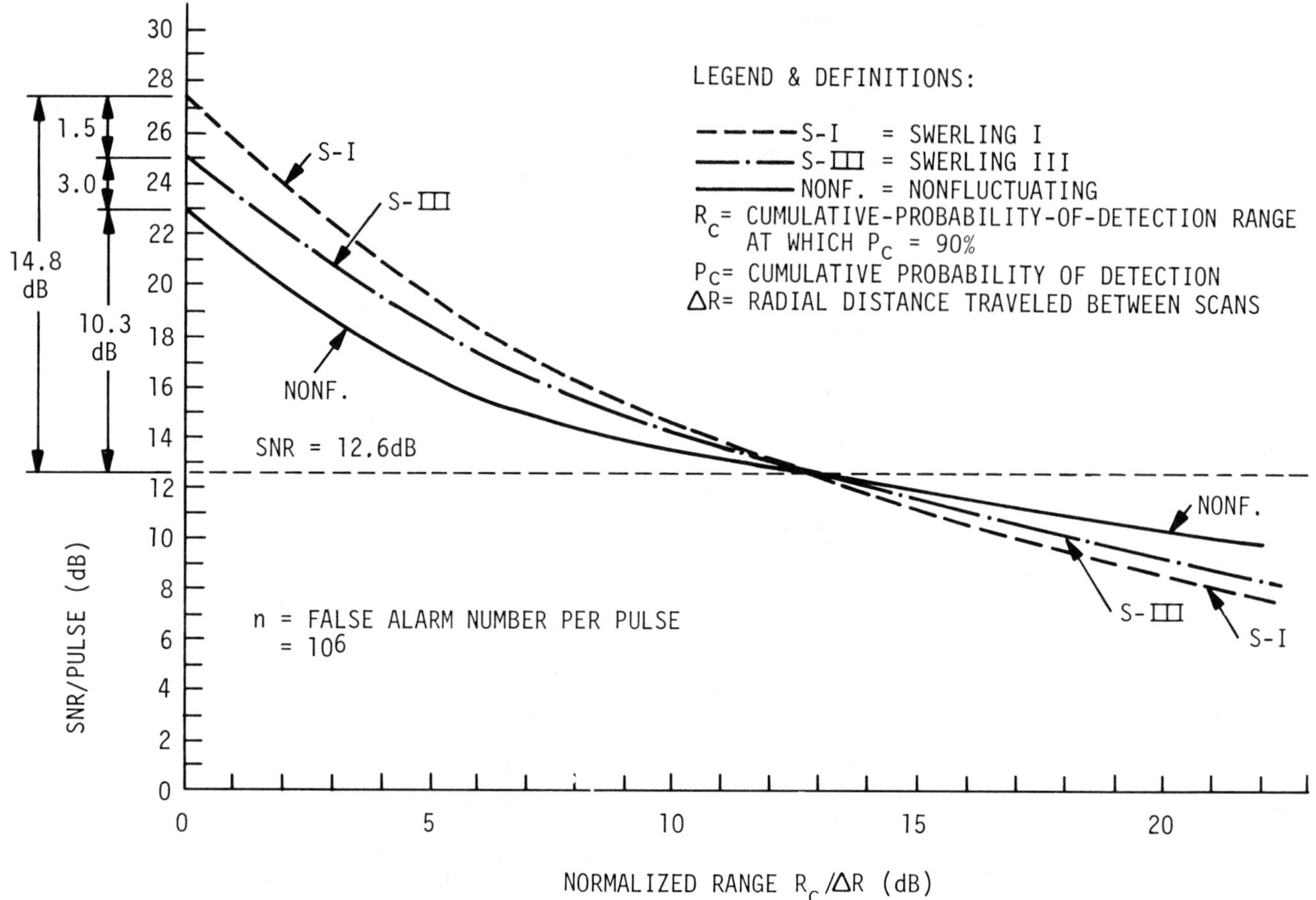

Figure 5 **Comparison of SNR per Pulse Requirements for Different Target Models**

As noted above, Figure 5 indicates that when $R_C/\Delta R = 13$ dB, all three target models require the same signal-to-noise ratio per pulse, 12.6 dB, to yield the same 90% cumulative probability of detection. This result is not surprising in view of the fact that all three target models give identical single-pulse performance when their single-pulse signal-to-noise ratios are about 11 dB; the single-pulse probabilities of detection are all about 38% for this level [2,6]. (Note that the same false alarm number is used for single-scan detection and cumulative-probability detection. This fact provides a fair basis for comparison because the false alarm rates are then the same for both detection types.)

Variation of Required Jammer Power with Cumulative-Probability-of-Detection Penetration Range

So far the detection problem has been discussed from the radar designer's point of view, covering how the detection range varies with radar transmitter power. We now reverse the situation and approach it from the jammer's standpoint — how does the aircraft penetration range vary with increased jammer power? A stand-off noise jammer is assumed (i.e., a jammer at a fixed range from the radar, which is radiating noise into the sidelobes of the search radar).

Assume initially that the stand-off jammer is designed to radiate enough power so that a cumulative probability of detection of 90% is realized at a normalized penetration range ($R_C/\Delta R$) of approximately 22 dB for a Swerling I target. It is desired to increase stand-off jammer power such that this range is decreased by a factor of 8 (9 dB) to a normalized range of 13 dB. As discussed previously, Figure 5 indicates that for the new normalized range of 13 dB, the radar needs a signal-to-noise ratio per pulse which is 5 dB higher than required at the normalized detection range of 22 dB. Hence, in order for the jammer to improve the aircraft's penetration range by a factor of 9 dB, the jammer power has to increase by an amount 5 dB less than predicted by the R^4 law. Thus, the stand-off jammer's power has to increase by 31 dB instead of 36 dB.

As more stand-off jammer power is used the penetration range decreases further to the low normalized range values of the curves of Figure 5. For the low normalized ranges, the signal-to-noise ratio per pulse curves are steeper. As a result, to go from a normalized range of 13 dB to one of 4 dB requires about 8.4 dB less power than specified by the R^4 power law, an increase of 27.6 dB being required rather than 36 dB.

The above jammer example was given using a 90% probability of detection for the defense radar. The offense usually designs to assure that a smaller probability of detection is achieved by the defense. For a lower cumulative probability of detection the power law increases. However, savings in required jammer transmitted power such as those given above are still obtained, except at slightly lower normalized ranges. (The reader can verify this fact using Figures 8 through 10 or Equation (6) all of which are discussed later.)

Is It More Effective to Increase PRF or the Energy Per Pulse in Order to Increase R_C

For the example of Figure 4a, ask the following question — "If we have a certain amount of average transmitter power available (which is not the maximum amount available in the radar or the maximum for which we can design), is it better to increase the

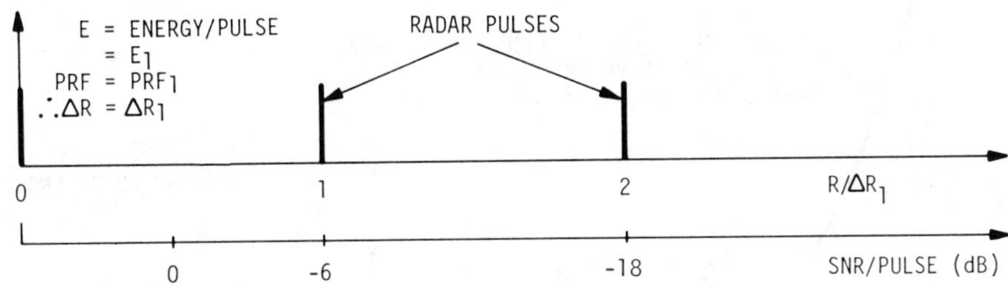

b) EFFECT OF INCREASING ENERGY/PULSE:

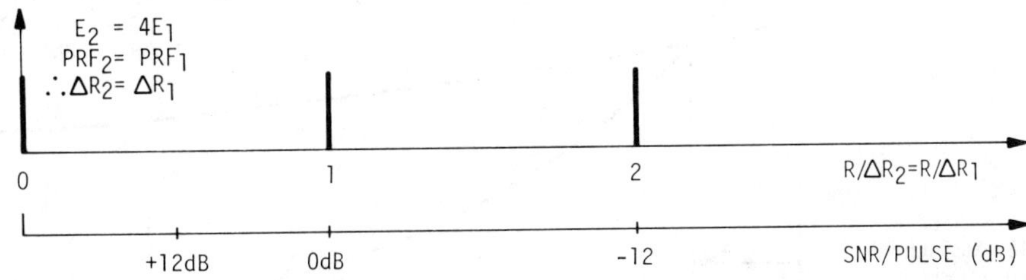

c) EFFECT OF INCREASING PRF:

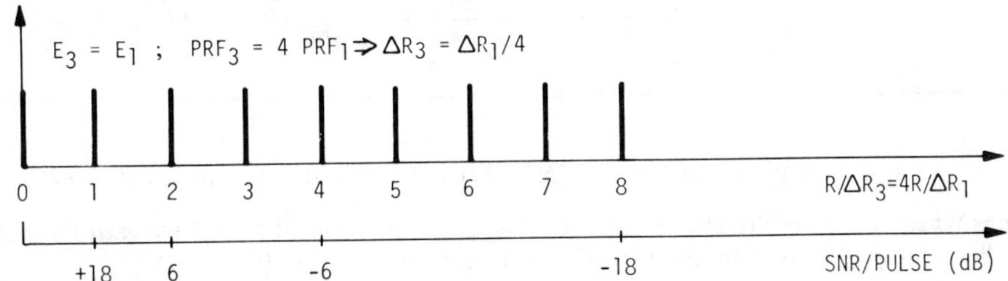

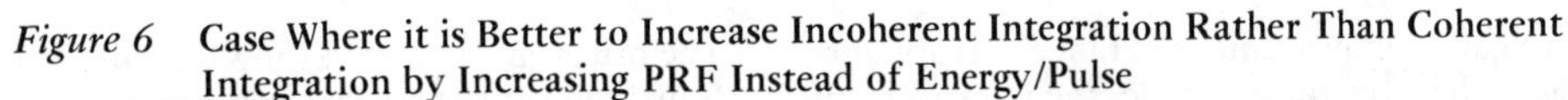

Figure 6 **Case Where it is Better to Increase Incoherent Integration Rather Than Coherent Integration by Increasing PRF Instead of Energy/Pulse**

average transmitted power by increasing the PRF or by increasing the energy per pulse, in order to increase the radar range based on cumulative probability of detection?" For the example indicated in Figure 4a, 13 pulses are received between the 10% to the 50% single-pulse probability of detection ranges. It turns out that if a large number of pulses go into the cumulative probability of detection calculation (and 13 pulses is a large number for the question posed), it is better to increase the energy per pulse rather than the PRF. The reason is that coherent integration is better than incoherent integration. Increasing the number of pulses means that, in effect, you have to incoherently integrate more pulses, the cumulative probability of detection process being equivalent to incoherent binary integration. On the other hand, if you increase the energy per pulse you have the same number of pulses being processed incoherently but you have more coherent integration, the increased energy being processed coherently in each received pulse. Thus, for the situation depicted in Figure 4a, it pays to increase the energy per pulse. But that is not always true.

Figure 6a indicates an example of the reverse, in a very low pulse repetition rate case. The first time the target comes into view over the horizon it is assumed to be at the normalized range $R/\Delta R_1 = 2$, i.e., at a true range $2 \times \Delta R_1$, in which ΔR_1 is the distance traveled by the target between scans. At this normalized range the signal-to-noise ratio per pulse is assumed to be -18 dB – a level which is too low for detection (only one pulse is being transmitted per scan period).

The next opportunity to see the target occurs ΔR_1 later at $R = \Delta R_1$ or, equivalently, at the normalized range $R/\Delta R_1 = 1$. At this range, the signal-to-noise ratio per pulse is -6 dB, 12 dB higher than on the previous observation but at a level still too low for target detection. The next chance for observation does not occur until the target is right on top of the radar at $R = 0$. Thus the incoming target is not detected by the radar for this case.

Let us examine what happens if the transmitted energy per pulse is increased by 6 dB. As indicated in Figure 6b, at $R = 2\Delta R_1$ the signal-to-noise ratio per pulse is -12 dB; at the range $R = \Delta R_1$ (the last opportunity to see the target), the signal-to-noise ratio per pulse is 0 dB, still too low for good detection.

Let's try increasing the PRF by a factor of 4 (6 dB), so that the average transmitted power is increased by 6 dB, the same amount that it was increased before except for the fact that the earlier change was effected by increasing the energy per pulse. Now there are more opportunities to see the target – four times as many, as illustrated in Figure 6c. The distance traveled by the target between scans becomes $\Delta R_3 = R_1/4$; the last opportunity to see the target occurs at a range 1/4th as large, i.e., at $R = \Delta R_1/4 = \Delta R_3$. At this range the signal-to-noise ratio per pulse is 24 dB greater than it was at the last opportunity to see the target for the original repetition rate of Figure 6a. The signal-to-noise ratio per pulse is now +18 dB. This figure provides a single-pulse probability of detection of about 80% for a Swerling I target model, for a false alarm number $n = 10^6$. Thus, whereas increasing the energy per pulse 6 dB does not permit target detection, increasing the PRF by the same amount does.

The example of Figure 6a therefore illustrates a case where it pays to increase the amount of incoherent integration rather than the

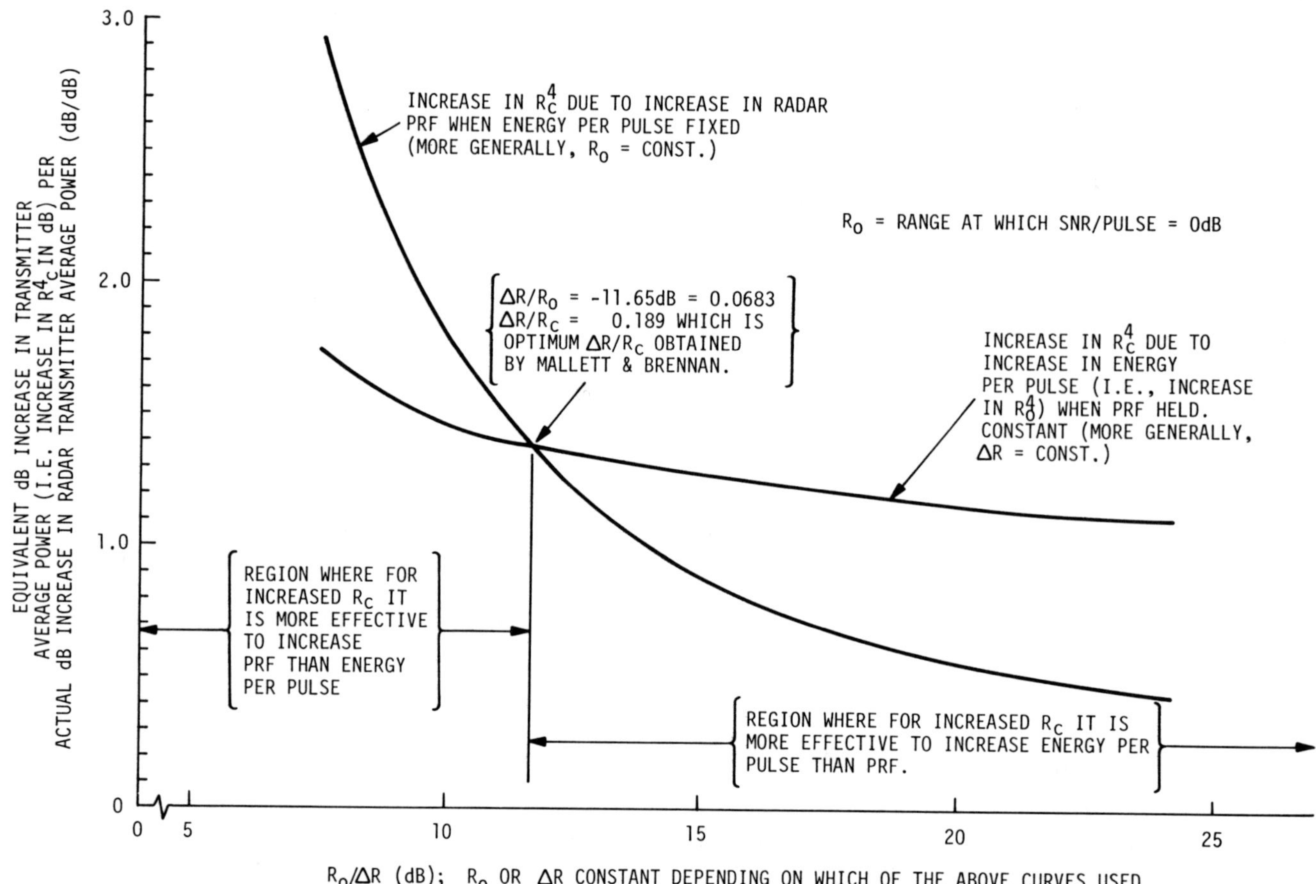

Figure 7 **Regions Where it is More Effective to Increase Energy per Pulse or PRF for Increased Range Based on Cumulative Probability of Detection; Swerling Case I Target Model, $P_C = 90\%$, $n = 10^6$**

coherent integration to get an improvement in range performance. As a side comment, note that, for this example, the improvement achieved was much better than that predicted by the 3.5 power law of Table 1. Range goes from zero to some finite number, so you are doing much better than an R^4 law – an infinite amount better!

In summary, what these two last examples (those of Figures 4a and 6a) tell us is that in some cases it pays to increase the PRF while for others it pays to increase the energy per pulse in order to increase the range of the system. If you have a very low PRF, it pays to increase the PRF; if you have a high PRF, it pays to increase the energy per pulse. There is some in-between PRF for which one should make neither changes and leave both these system parameters alone if possible. At this point one has the optimum PRF or, equivalently, the optimum value for $R_C/\Delta R$. The transmitter power is being most effectively used for detection at this point – minimum transmitter energy is required.

Optimum PRF or Energy Per Pulse for Maximum R_C

Figure 7 shows that such an optimum exists. Shown plotted are two curves. One is the decibel increase in range to the fourth power, i.e., $R_C{}^4$, per decibel increase in average transmitter power resulting from increasing the *energy per pulse,* the PRF being held fixed. The abscissa for this curve is the normalized parameter $R_0/\Delta R$ which in effect is the average transmitter power, because R_0 represents the range at which the signal-to-noise ratio equals unity or, equivalently, 0 dB, with ΔR being held constant and the PRF being fixed. When the classical fourth-root range law relation applies, the ordinate has the value of 1dB/dB. Thus, when the curve is above the 1 dB/dB level, the power law is better than that obtained classically, i.e., S_1 of Equation (4) is less than 4.

The other curve plotted shows the decibel increase in range to the fourth power per decibel increase in average transmitter power due to increasing *PRF,* with the energy per pulse held fixed. This curve is plotted again versus $R_0/\Delta R$, but now R_0 is held constant while ΔR is varying; in effect, the curve is plotted versus the radar PRF since

$$\Delta R = v_d T_F = v_d \frac{\Omega_S}{\Omega_B} T_P = v_d \frac{\Omega_S}{\Omega_B} \frac{1}{PRF} \tag{5}$$

where v_d equals target Doppler velocity and where use was made of Equation (1).

These curves verify the fact that, for low PRFs (i.e., for $R_0/\Delta R \leqslant 11.65$ dB), you get a larger increase in $R_C{}^4$ per decibel increase in average power when you increase the PRF than when you increase the energy per pulse. For example, at the low PRF of $R_0/\Delta R = 7$ dB, there is about a 3 dB increase in $R_C{}^4$ per decibel increase in power resulting from increasing the radar PRF as opposed to a 1.8 dB increase in $R_C{}^4$ per decibel increase in power resulting from increasing the energy per pulse.

At high PRFs (i.e., for $R_0/\Delta R \geqslant 11.65$ dB) the reverse is shown to be true. For example, at $R_0/\Delta R = 24$ dB, there is a 1.1 dB increase in $R_C{}^4$ per decibel increase in average power due to an increase in the energy per pulse, as opposed to an increase of 0.44 dB in $R_C{}^4$ per decibel increase in average power resulting from an increase in the PRF.

The two curves of Figure 7 cross over at the optimum PRF point, the point $R_0/\Delta R = 11.65$ dB.

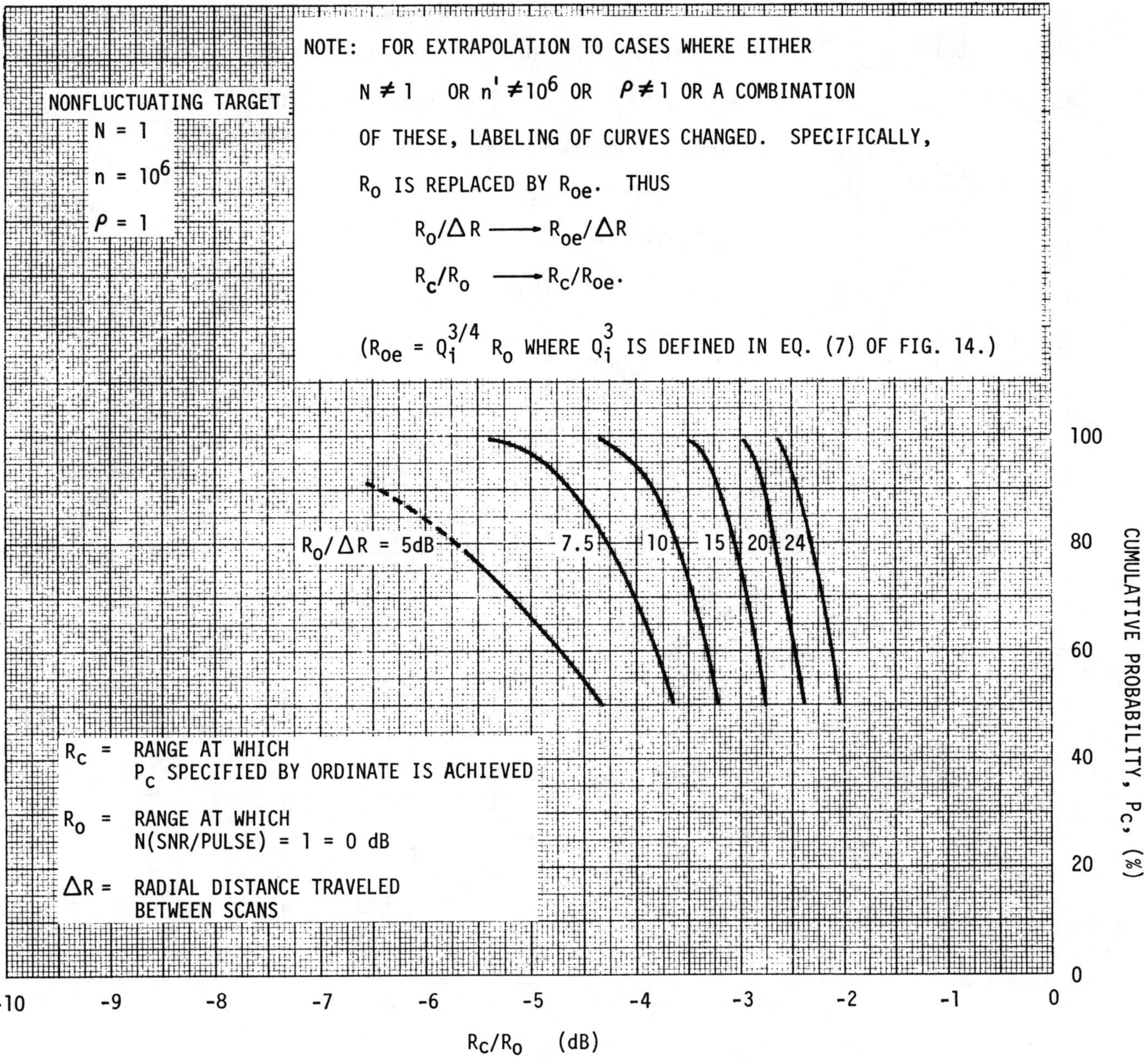

Figure 8 Universal Cumulative-Probability-of-Detection Curves: Nonfluctuating Target

Universal Cumulative-Probability-of-Detection Curves

Figures 8 to 10 show cumulative-probability-of-detection curves [7] plotted versus the normalized cumulative-probability-of-detection range parameter, R_C/R_0, for the Swerling I, III, and nonfluctuating target models. The curves represent a replot of the original Mallett and Brennan [3] curves in a more convenient form. These cumulative-probability-of-detection curves are given for different values of the PRF expressed in terms of the normalized range parameter, $R_0/\Delta R$. The horizontal spacing between successive curves gives the increase in the normalized range R_C/R_0, (and, in turn, R_C) for a specified increase in PRF, or, equivalently, an increase in average transmitter power for a fixed energy per pulse. Hence, the larger this spacing, the lower the power law, i.e., the smaller the root, S_1, of the transmitter average power to which range is proportional (the power being varied by varying the PRF). Because the horizontal spacing between the curves increases with increasing cumulative probability of detection (see Figures 8 to 10), it follows that the power law, S_1, decreases with increasing cumulative probability of detection. This fact is verified in Figure 11. (The extrapolation indicated by the note in Figures 8 through 10 is covered later in this chapter.)

Cumulative-Probability-of-Detection Range Equation

Just as one has the classical noise-limited radar range equation for the range based on single-pulse or single-scan detection, as given by Equation (1) of Chapter 1 and Equation (2) of Chapter 2, there exists also a range equation for range based on the cumulative probability of detection. This equation is called the cumulative-probability-of-detection range equation [4, 5]. The general form of this equation is given by

$$R_C' = R_C/\Delta R = R(P_C)/\Delta R = R_1'(P_C)_0 \left[\frac{R_0}{20\,\Delta R}\right]^{4/S_1} - 0.5 \qquad (6)$$

for

$$R_C/\Delta R \gg 1 \qquad (6a)$$

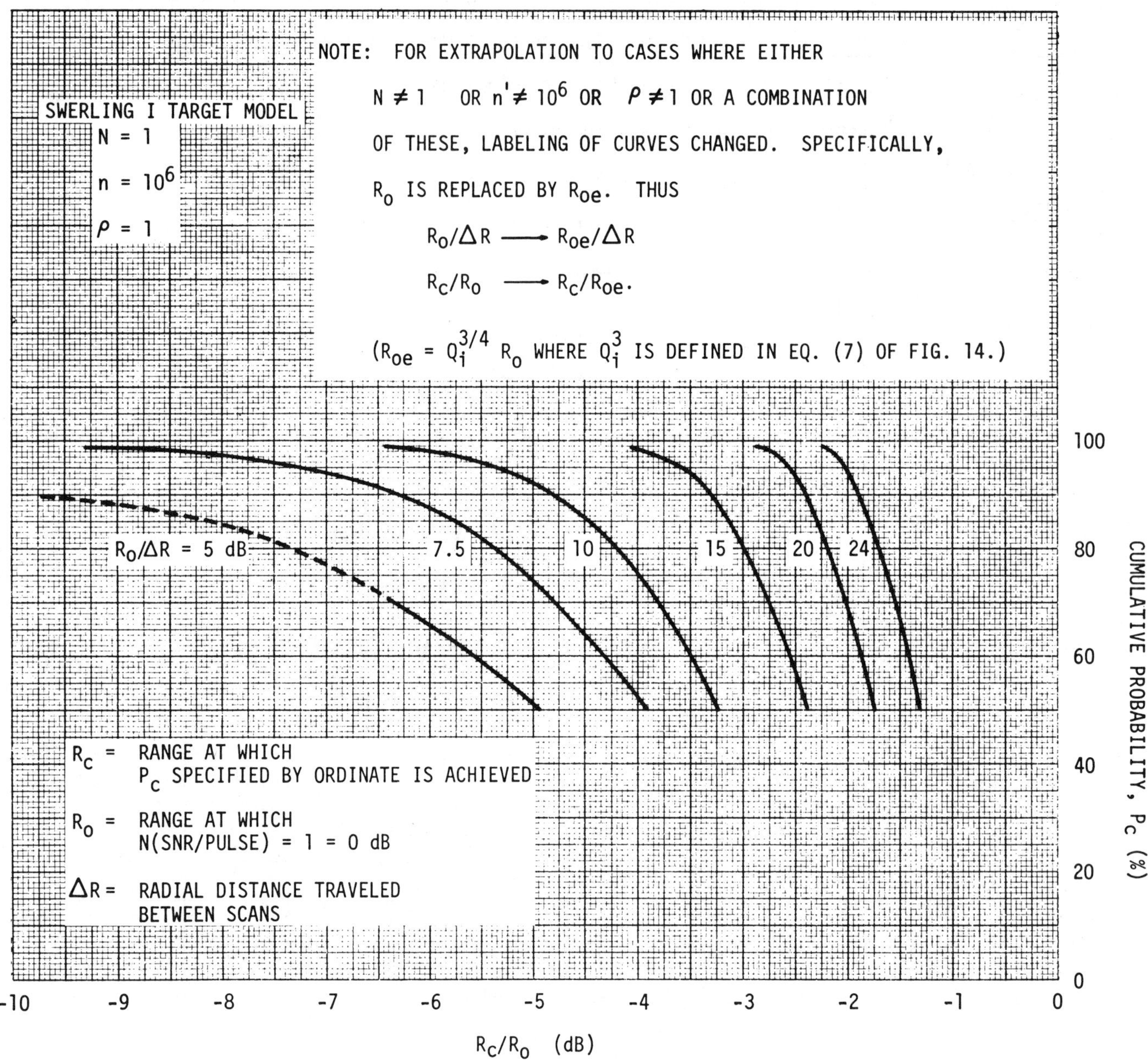

Figure 9 Universal Cumulative-Probability-of-Detection Curves: Swerling I Target Model

where:

$R(P_C) = R_C$ = cumulative-probability-of-detection range at which cumulative probability of detection = P_C

R_0 = range at which power SNR = 1 = 0 dB*

ΔR = radial distance traveled between scans.

$R_I'(P_C)_0$ = parameter obtained from Figure 12. Physically it is approximately equal to $R_C/\Delta R$ for $R_0/\Delta R = 20$. (The approximation is exact if the target is illuminated by the radar at the exact instant it reaches range R_C. Generally, the target is not illuminated at range R_C, but instead is illuminated last at ranges between R_C and $R_C + \Delta R$ just prior to reaching R_C. The definition of $R_I'(P_C)_0$ becomes clearer after the discussion of synchronized radar sampling for Figure 13.

*For one pulse, i.e., N = 1. For N = 1, R_0 = range at which N(SNR) = 1 = 0dB.

This equation permits the determination of the range at which a specified cumulative probability of detection is achieved. To use this equation, two parameters must be determined — the power law, S_1, and the constant, $R_I'(P_C)_0$. The parameters S_1 and $R_I'(P_C)_0$ are plotted in Figures 11 and 12 as a function of the cumulative probability of detection for the three target models specified before. Using these figures, the cumulative-probability-of-detection-range equation permits the determination of the cumulative-probability-of-detection range to an accuracy of about ±0.25 dB when $25 \leq R_0/\Delta R \leq 250$ for values of the cumulative probability of detection between 10% and 99% for the Swerling I target model, and between 50% and 99% for the Swerling III and non-fluctuating target models.

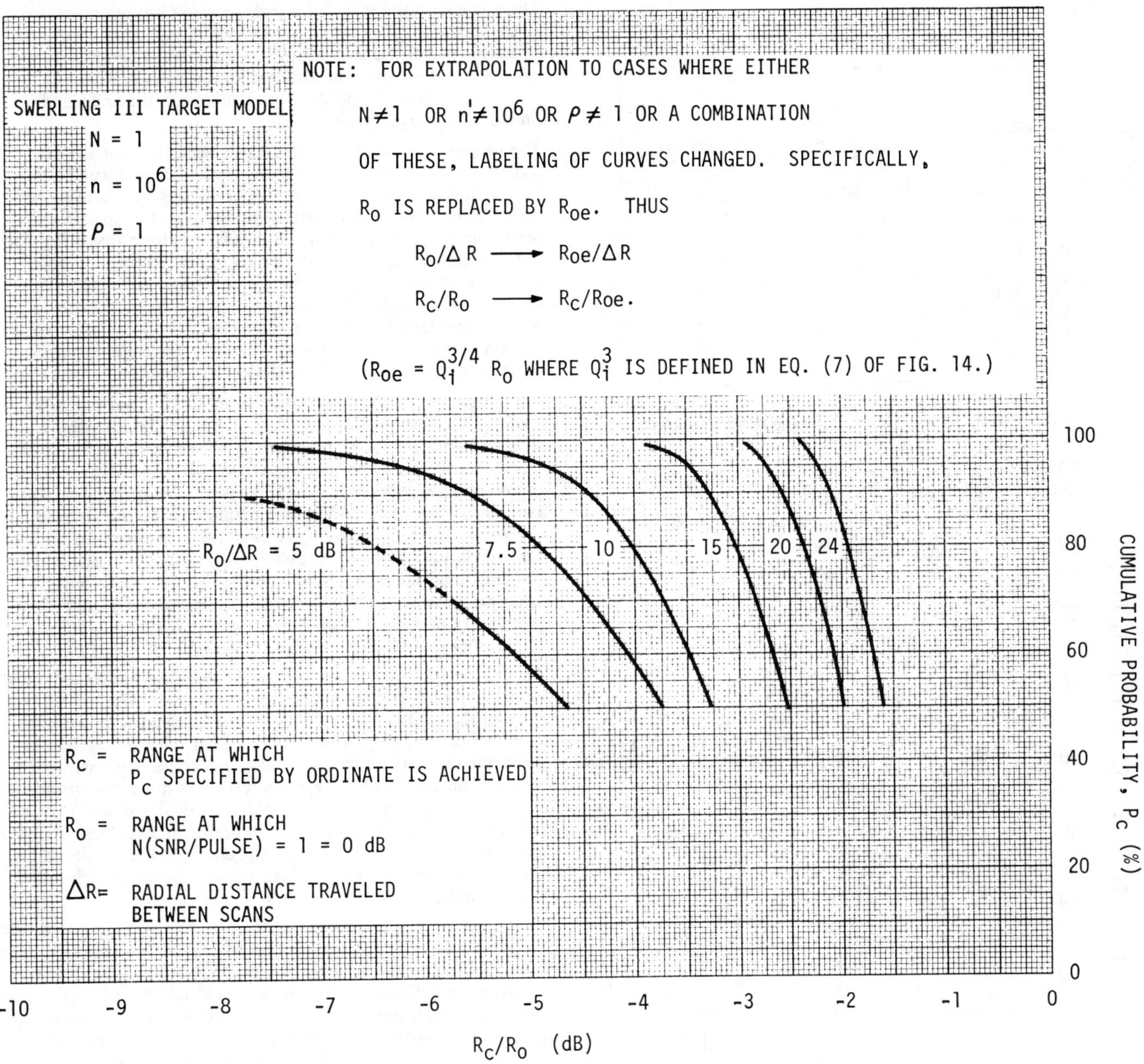

Figure 10 Universal Cumulative-Probability-of-Detection Curves: Swerling III Target Model

Effect of Nonsynchronization of Radar Sampling

Figure 13 compares the signal-to-noise ratio per pulse required for single-scan (single-pulse) detection to that required with cumulative probability of detection (the curves of Figure 5). The results are presented for 90% single-pulse detection and 90% cumulative probability of detection with the false alarm number equal to 10^6 for both cases. The single-pulse probability of detection curves shown give signal-to-noise ratios required per pulse that are not constant for all ranges — a result different from what might be expected at first. Why is there this apparent discrepancy?

To explain this phenomenon we return to the example of Figure 6. Recall that the target can only be observed at the ranges at which the radar illuminates the target. In between, the radar cannot observe the target. Assume that it is desired to detect an approaching target at range R with single-pulse probability P_D. Assume that the range R for this approaching target is between ranges at which the target is illuminated by the radar. If the target is to be detected at this in-between range with probability P_D, it must have already been detected with probability P_D at the last range that the target was illuminated. More energy therefore is needed to detect the target at R if it is an in-between range than if the radar illuminated the target at this range. Specifically, let the last range at which the target was illuminated be $R + \epsilon$, where $0 < \epsilon < \Delta R$. Then $(R + \epsilon)^4/R^4$ more energy is required to detect the target at the in-between range R if the radar sampling occurs at range $R + \epsilon$.

Assume that the power signal-to-noise ratio required for single-pulse detection with probability P_D is X_0 ($X_0 \doteq 21$ dB for a Swerling I target for P_D = 90% when the false alarm number is 10^6 — see asymptote of Figure 13). This signal-to-noise ratio is required for the return from the target illuminated at the last observable range $R + \epsilon$. Although no pulse is transmitted to illumin-

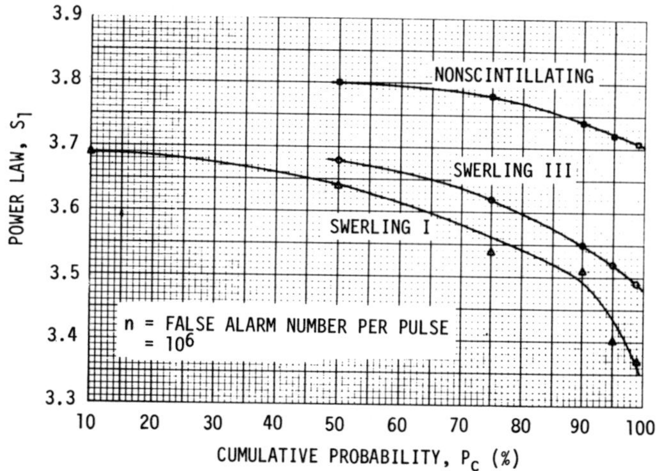

Figure 11 **Power Law vs. Cumulative Probability of Detection**

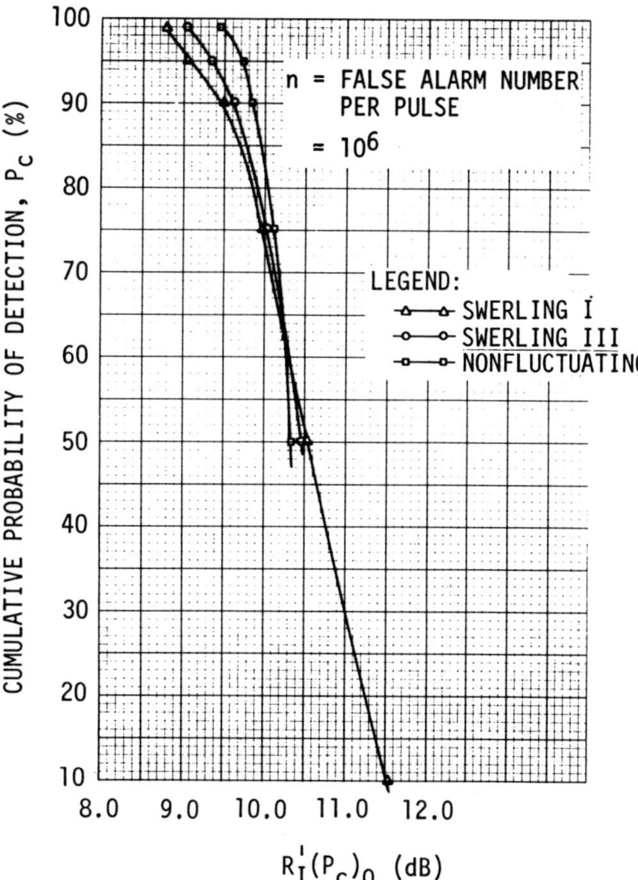

Figure 12 **Cumulative Probability of Detection, P_C, vs. $R_I'(P_C)_0$ ***

ate the target at range R, assume that one is. The power signal-to-noise ratio of this return is then $X_1 = X_0(R + \epsilon)^4/R^4$. The single-scan power signal-to-noise ratio given in Figure 13 is the average of X_1 over all possible values of ϵ between 0 and ΔR. The cumulative-probability-of-detection signal-to-noise ratios of Figure 13 are obtained using a similar average.

* Caution: $R_I'(P_C)_0$ in dB in Figure 12 but a power ratio in Equation 6, as are all other parameters in Equation 6.

The increase in the required signal-to-noise ratio per pulse is due to the radar sampling instants not being synchronized with the occurrence of the target at range R, but instead being arbitrarily synchronized to when the target is at range $R + \epsilon$, with ϵ having any value between 0 and ΔR with equal probability depending on when the target arrives.

The increase indicated in Figure 13 of the signal-to-noise ratio required per pulse for a specified single-scan probability of detection with decreasing normalized range $R_C/\Delta R$ means that one does not actually have a fourth root law even for single-scan probability of detection when one accounts for nonsynchronized radar range sampling. Considering the nonsynchronized radar sampling, the single-scan probability of detection performance depends on the radar sampling rate (PRF). Also, just as there is an optimum sampling rate for best cumulative-probability-of-detection performance, there is one for best single-scan-probability-of-detection performance (see Mallett and Brennan, Reference 3).

The range law values for S_1 specified in Table 1 and Figure 11 assume synchronized radar sampling, i.e., the radar sampling is such that it illuminates the target when it is at range R. However, the signal-to-noise ratios per pulse specified for cumulative probability of detection in Figures 5 and 13 assume nonsynchronized range sampling of the radar. Moreover, it follows from the increase shown in Figure 13 of the signal-to-noise ratio per pulse required for a specified single-scan probability of detection, that an appreciable part of the increase indicated in Figures 5 and 13 in the signal-to-noise ratio per pulse required for a specified cumulative probability of detection arises from the increase needed because of the nonsynchronized sampling of the radar.

The factor, -0.5, in the cumulative-probability-of-detection range equation (Equation (6)) results from the nonsynchronized sampling of the radar. Without the -0.5, the cumulative-probability-of-detection range equation would predict the range for the assumption of synchronized sampling and would apply without the need for the inequality of Equation (6a). With the factor included, the range for a nonsynchronous radar is calculated, and it is approximately one-half a scan-to-scan period, 0.5 (ΔR), smaller than for synchronized sampling when $R_C/\Delta R \gg 1$. The results of Figures 3 and 7 through 10 take into account the nonsynchronized radar sampling.

The curves of Figure 13 indicate the penalty that is imposed when specifying the required signal-to-noise ratio per pulse needed based on using the single-scan detection criteria when, in fact, the cumulative-probability-of-detection criteria should be used. For example, consider the Swerling I target model. A signal-to-noise ratio per pulse of about 21 dB is required for the single-scan detection criteria when $R/\Delta R$ = 23 dB; a signal-to-noise ratio per pulse of about 7 dB is required when using the cumulative probability of detection criteria, a difference of 14 dB.

The results of Figure 13 also indicate the error that results when the synchronized single-pulse signal-to-noise ratio is used rather than the (correct) unsynchronized signal-to-noise ratio. For example, for $R_C/\Delta R$ = 3 dB, the unsynchronized signal-to-noise ratio per pulse is 25.4, or 4.4 dB higher than for the synchronized case.

Extension to $N > 1$, $n' \neq 10^6$ and $\rho \neq 1$

The cumulative probability of detection results given up to this point apply for the case where the number of pulses on target per scan is one (N = 1), the false alarm number is 10^6 ($n = 10^6$), and no collapsing loss is present ($\rho = 1$). The extension to cases where $N > 1$, $n' \neq 10^6$ and $\rho \neq 1$ is given in Figures 14 and 15; see the glossary at the end of this chapter for definition of n′ and ρ and other symbols used in these figures (see also Chapter 2 by Barton, in which n and n′ are labeled n_{FA} and n_{FA}'). Figure 14 gives the

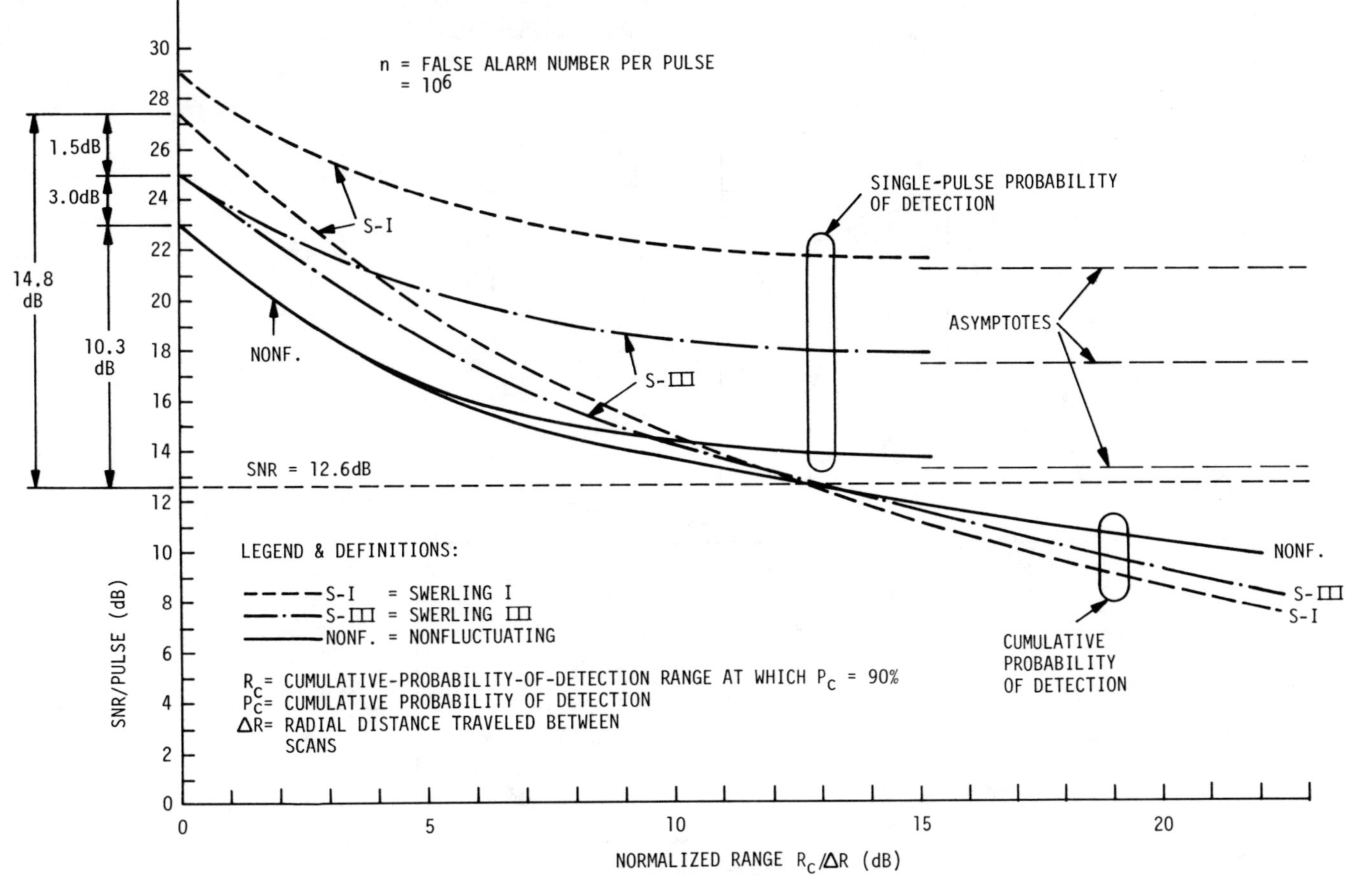

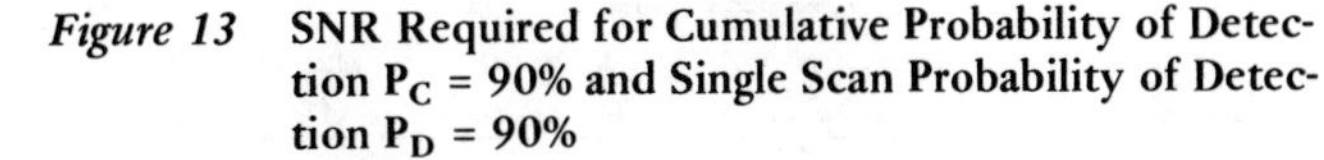

Figure 13 **SNR Required for Cumulative Probability of Detection P_C = 90% and Single Scan Probability of Detection P_D = 90%**

extension using the approximate cumulative-probability-of-detection range equation; Figure 15 uses the more precise universal cumulative-probability-of-detection curves of Figures 8 to 10. These extensions are given in terms of simple flow charts. Table 2 gives the accuracy of the procedures of Figures 14 and 15.

To help in using the flow charts of these two figures, seven examples are worked out in detail in Table 3 using both methods. The bottom row of Table 3 indicates that, for the examples included, the accuracy of the approximate procedure relative to the more precise one is 0.0 dB, except for one example in which it is 0.1 dB.

A convenient, easy to use worksheet is provided in Table 4, for both procedures. This worksheet indicates from which figure or equation each entry is obtained.

Figures 14 and 15 show how to find the cumulative-probability-of-detection range R_C given P_C; the more precise method of Figure 15 can be used to solve the reverse problem of finding P_C given R_C. This latter procedure is done by following the heavy flow lines in the opposite direction and replacing the multipliers marked by asterisks by divide operations. Figures 8 to 10 can also be used to solve for ΔR given P_C, R_C, and R_0. To apply Figures 8 to 10 to to the solution for R_0 given P_C, R_C, and ΔR requires a trial and error iteration. The method of Figure 14 cannot be used directly to solve for P_C given R_C because S_1 and $R_1'(P_C)_0$ are dependent on P_C, thereby necessitating an iterative procedure. The approximate procedure outlined by Barton in Chapter 2 can be used to solve for D (1, K) and D(ρN, K) of Equation (7) in Figure 14. However, evaluation of Q_i using Figures 16a and 16b is almost as accurate and more convenient. The procedure for finding D(ρN, K) from the Meyer plots [2] and similar curves is outlined in Reference 7 and Table 7 of the Appendix of this book.

Summary

1) The cumulative-probability-of-detection-range law is not a third or fourth root law, but rather some law generally between these two values. Even the single-pulse-probability-of-detection range law is generally a root law between these values.
2) An overestimate in the radar power of as much as 14 dB may occur if detection sensitivity is based on single-scan detection instead of cumulative detection.
3) An underestimate of the required single-pulse detection signal-to-noise ratio of 4 dB or more can occur if the synchronized conditions are assumed when unsynchronized conditions should be used.
4) 8 dB or less of stand-off jammer power may be needed to decrease the penetration range by a factor of ten when based on cumulative probability detection than when based on single-pulse detection.
5) A cumulative-probability-of-detection-range equation has been developed which permits the determination of the cumulative-probability-of-detection range without resorting to a computer computation. The accuracy of this equation is given in Table 2.
6) Easy to use universal curves are provided for obtaining the cumulative probability of detection.
7) Simple flow charts (Figures 14 and 15) and a worksheet (Table 4) are provided for using the cumulative-probability-of-detection range equation and universal cumulative-probability-of-detection curves.

PROBLEM SOLVED:

GIVEN: P_c, P_{fa}, N, $\mathcal{P}$, R_o, ΔR, TARGET MODEL (NONSCINTILLATING TARGET OR SWERLING CASE 1 OR CASE 3 TARGET MODELS)

FIND : R_c

P_c | TARGET MODEL | R_o | ΔR

FIG. 11 → S_1

FIG. 12 → $R_1'(P_c)_o$

CUMULATIVE- PROBABILITY RANGE EQUATION

$$R_c' = \frac{R_c}{\Delta R} = R_1'(P_c)_o\left(\frac{R_o}{20\Delta R}\right)^{4/S_1} - 0.5 \quad (6)$$

FOR $n = 10^6$, $\mathcal{P} = 1$, $N = 1$.

$R_o/\Delta R$

0.5

$R_{co}' \equiv R_c'$ FOR $n = 10^6$, $\mathcal{P} = 1$, $N = 1$

R_c'

R_c FOR $n' \neq 10^6$, $\mathcal{P} \neq 1$, $N \neq 1$,

I.E., FOR SPECIFIED, n', $\mathcal{P}$, N.

Q_i^{3/S_i}

CORRECTION FACTOR FOR $n \neq 10^6$, $\mathcal{P} \neq 1$, $N \neq 1$* GIVEN BY Q_i^{3/S_1} WHERE

$$Q_i^3 = \frac{D(1,K) \text{ FOR } n = 10^6}{ND(\mathcal{P}N,K) \text{ FOR } n' \ (\neq 10^6)} \quad (7)$$

WHERE D(1,K) AND D($\mathcal{P}$N,K) ABOVE ARE EVALUATED FOR P_d = 10% FOR K OF TARGET MODEL MEYER PLOTS (2), FEHLNER CURVES (8) OR SIMILAR DETECTION CURVES (9).

A MUCH MORE CONVENIENT PROCEDURE TO EVALUATE Q_i^3, BUT A SLIGHTLY LESS ACCURATE ONE, IS TO USE FIG. 16a OR b TO OBTAIN Q_i DIRECTLY. FOR NONFLUCTUATING AND SWERLING CASE 1 TARGETS $Q_i = Q_1$ OF FIG. 16a IS USED WHILE FOR SWERLING CASE 3 TARGETS $Q_i = Q_2$ OF FIG. 16b IS USED.

$\mathcal{P}$ | N | n'

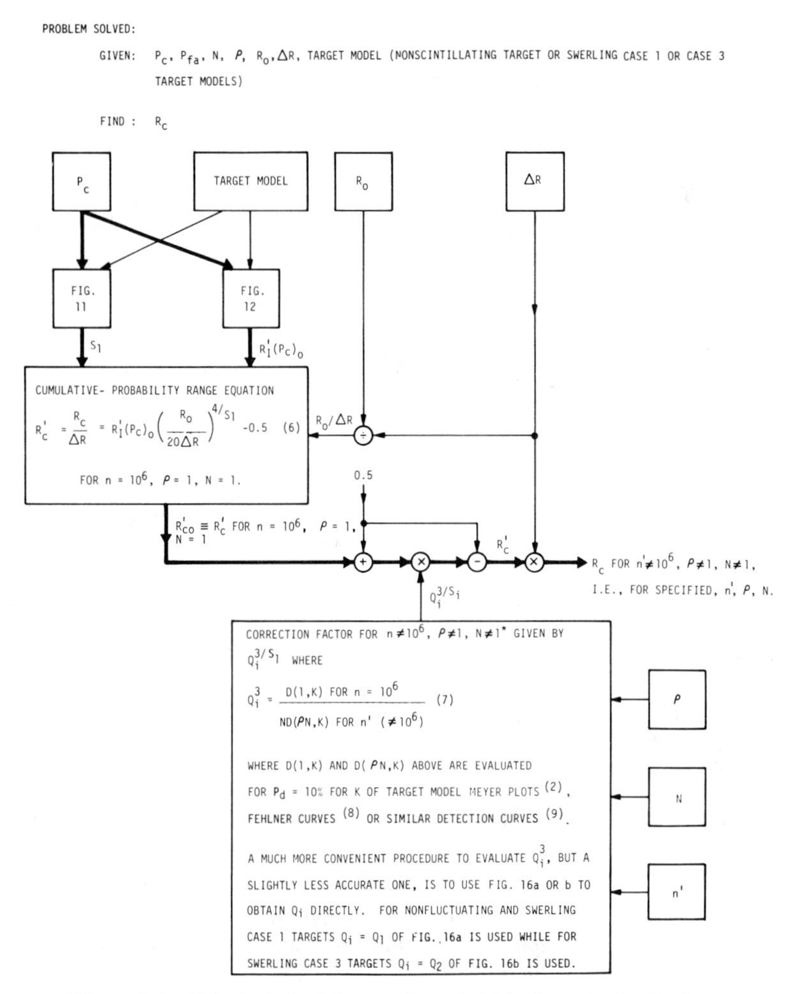

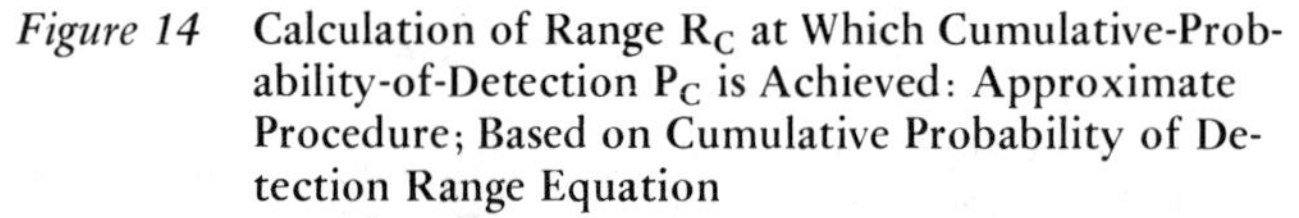

Figure 14 Calculation of Range R_C at Which Cumulative-Probability-of-Detection P_C is Achieved: Approximate Procedure; Based on Cumulative Probability of Detection Range Equation

PROBLEM SOLVED:

GIVEN : P_c, P_{fa}, N, $\mathcal{P}$, R_o, ΔR, TARGET MODEL (NONFLUCTUATING TARGET OR SWERLING CASE 1 OR 3 TARGET MODELS)

FIND : R_c

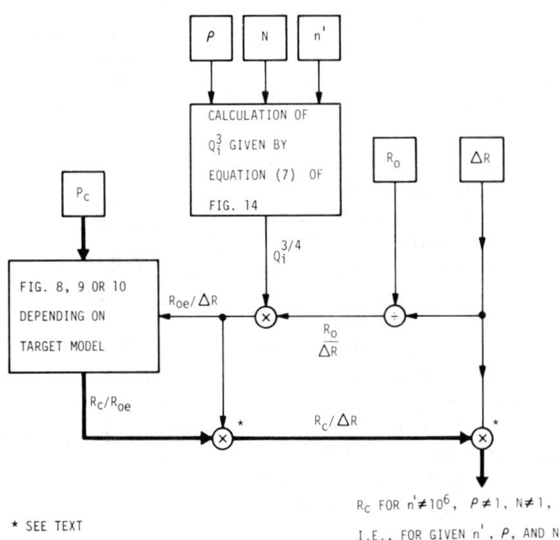

Figure 15 Calculation of Range R_C at Which Cumulative Probability of Detection P_C is Achieved (and Reverse Problem of Find P_C Given R_C): More Precise Procedure; Based on use of Universal Cumulative Probability of Detection Curves

P_d RANGE: 10%* ≤ P_C ≤ 99% FOR SWERLING CASE 1
50% ≤ P_C ≤ 99% FOR SWERLING CASE 3 AND NONFULUCTUATING MODEL

TARGET MODEL	NUMBER OF PULSES VIDEO INTEGRATED ρN	P_{fa}	n'	METHOD	RANGE OF $R_0/\Delta R$	ACCURACY
NONFLUCTU-ATING OR SWERLING CASES 1 OR 3	1	0.693×10^{-6}	10^6	APPROXI-MATE (FIG. 14)	$25^{\#} \leq R_0/\Delta R \leq 250$	$\leq \pm 0.25$ dB
				MORE PRECISE (FIG. 15)	$3^{\neq} \leq R_0/\Delta R \leq 250$	$\leq \pm 0.1$ dB
	1	$10^{-12} \leq P_{fa} \leq 10^{-5}$	$10^{12} \leq n' \leq 10^5$	APPROXI-MATE (FIG. 14)	$25^{\#} \leq R_0/\Delta R \leq 250$	$\leq \pm 0.35$ dB
				MORE PRECISE (FIG. 15)	$3^{\neq} \leq R_0/\Delta R \leq 250$	$\leq \pm 0.2$ dB
SWERLING CASES 1 AND 3	$1000 \geq \rho n \geq 1$	$10^{-12} \leq P_{fa} \leq 10^{-5}$	$10^{12} \leq n' \leq 10^5$	APPROXI-MATE (FIG. 14)	$25^{\#} \leq R_{0e}/\Delta R \leq 250$	$\leq \pm 0.45$ dB
				MORE PRECISE (FIG. 15)	$3^{\neq} \leq R_{0e}/\Delta R \leq 250$	$\leq \pm 0.3$ dB

#FOR 50% ≤ P_C ≤ 90% LOWER LIMIT FOR $R_0/\Delta R$ CAN BE DECREASED TO 10 WITH RESULTING ERROR 0.1 dB LARGER.
≠FOR $R_0/\Delta R$ = 5 dB = 3 ACCURACY VALUES ONLY APPLY FOR SOLID PORTION OF CURVES OF FIGS. 8 TO 10.

NOTE: FOR THE n' ≠ 10^6 OR N ≠ 1 OR $\rho \neq 1$ EXTRAPOLATED CASES, THE ACCURACY VALUES ARE BASED ON THE ASSUMPTION THAT Q_i^3 IS CALCULATED USING EQ. (7) OF FIG. 14. IF FIGS. 16a AND b ARE USED INSTEAD THE ACCURACY WILL BE BETWEEN ABOUT 0 AND 0.3 dB WORSE, THE ERROR BEING LARGEST WHEN $R_0/\Delta R$ IS LARGE AND N IS LARGE.

*BECOME 50% FOR PRECISE METHOD BECAUSE CURVES ONLY PLOTTED ABOVE P_C = 50%; SEE FIGURES 8 THROUGH 10.

Table 2 Accuracy of Determination of Cumulative-Probability-of-Detection Range, R_C

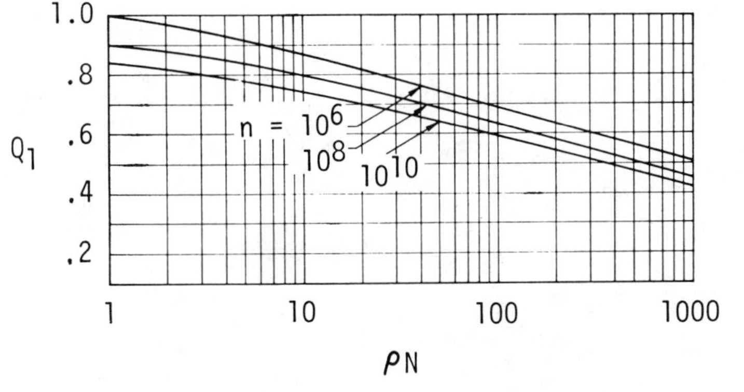

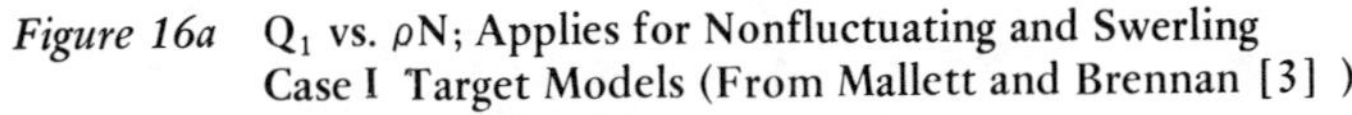

Figure 16a Q_1 vs. ρN; Applies for Nonfluctuating and Swerling Case I Target Models (From Mallett and Brennan [3])

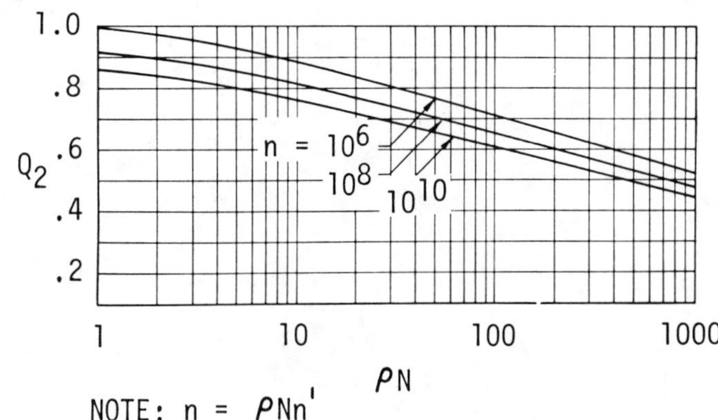

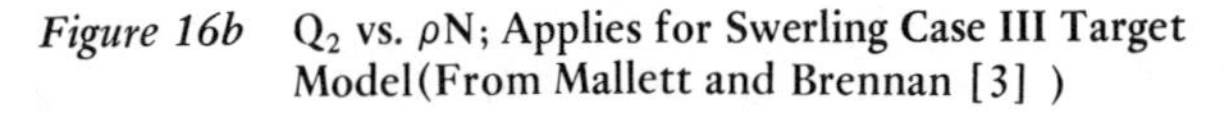

Figure 16b Q_2 vs. ρN; Applies for Swerling Case III Target Model(From Mallett and Brennan [3])

METHOD OF CALCULATION	PARAMETER	WHERE PARAMETER DERIVED FROM	NONFLUCTUATING TARGET	SWERLING FLUCTUATING TARGETS					
GIVEN	EXAMPLE NO.		1	2	3	4	5	6	7
	SWERLING CASE	GIVEN	0	1	1	3	1	1	3
	N		1	1	1	1	30	15	30
	ρ		1	1	1	1	1	2	1
	P_{fa}		6.93×10^{-7}	6.93×10^{-7}	3×10^{-11}	6.93×10^{-7}	6.93×10^{-7}	6.93×10^{-7}	6.93×10^{-7}
	n'		10^6	10^6	2.3×10^{10}	10^6	10^6	10^6	10^6
	$R_o/\Delta R$		30	30	30	30	30	30	30
	K	SEE NOTE BELOW #	∞	1	1	2	1	1	2
APPROXIMATE PROCEDURE (BASED ON CUMULATIVE-PROBABILITY RANGE EQUATION) FIG. 14.	S_1	FIG. 11	3.74	3.51	3.51	3.55	3.51	3.51	3.55
	$R'_I(P_c)_o$ (dB)	FIG. 12	9.83	9.5	9.5	9.62	9.5	9.5	9.62
	$R'_{co} \equiv R'_c$ FOR $n = 10^6$, $\rho = 1$, $N = 1$	EQ. 6 OF FIG. 14	14.34	13.65	13.65	13.97	13.65	13.65	13.97
	$D(1,K)^{(1)}$ (dB)	MEYER PLOTS (2)			7.15		7.15	7.15	7.6
	$ND(\rho N,K)^{(1)}$ (dB)	MEYER PLOTS (2)			9.8		11.42	11.42	12.02
	Q_i^3	EQ. 7 OF FIG. 14	1	1	0.543	1	0.374	0.374	0.361
	Q_i^{3/S_1}	$[Q_i^3]^{1/S_1}$	1	1	0.840	1	0.756	0.756	0.751
	APPROXIMATE $R_C/\Delta R$ FOR GIVEN n', ρ, AND N	$(R'_{co}+0.5)Q_i^{3/S_1}-0.5$	14.34	13.65	11.39(2)	13.97	10.19	10.19	10.37
MORE PRECISE PROCEDURE (BASED ON CUMULATIVE-PROBABILITY OF DETECTION CURVES) FIG. 15.	$Q_i^{3/4}$	$[Q_i^3]^{1/4}$	1	1	0.859	1	0.782	0.782	0.775
	$R_{oe}/\Delta R$	$(R_o/\Delta R)Q_i^{3/4}$	30	30	25.76	30	23.46	23.46	23.26
	R_C/R_{oe} (dB)	FIG. 8, 9 OR 10	-3.22	-3.33	-3.54	-3.32	-3.63	-3.63	-3.53
	MORE PRECISE $R_C/\Delta R$ FOR GIVEN n', ρ, AND N.	$\left(\frac{R_C}{R_{oe}}\right)\left(\frac{R_{oe}}{\Delta R}\right)$	14.29	13.93	11.40(2)	13.96	10.17	10.17	10.30
	ERROR IN $R_C/\Delta R$ (dB)	APPROXIMATE VS MORE PRECISE PROCEDURE	0.0	- 0.1	0.0	0.0	0.0	0.0	0.0

(1) $D(1,K)$ AND $D(\rho N,K)$ EVALUATED FOR P_d = 10%.

(2) EXACT CALCULATION INDICATES THAT R'_c SHOULD BE 0.2dB HIGHER IN WHICH CASE APPROXIMATE AND MORE PRECISE CALCULATIONS ARE WITHIN 0.2 dB OF CORRECT VALUES.

NOTE: =∞ , 1, AND 2 FOR SWERLING 0, I, AND III MODELS, RESPECTIVELY

Table 3 **Example Calculations: Find R_C for Specified P_C (= 90%)**

METHOD OF CALCULATION	PARAMETER	WHERE PARAMETER DERIVED FROM	SWERLING FLUCTUATING TARGET				
	EXAMPLE NO.		Ill. EX. *				
GIVEN	SWERLING CASE	GIVEN	3				
	N		30				
	ρ		1				
	P_{fa}		6.93×10^{-7}				
	n'		10^6				
	R_0		60				
	ΔR		2				
	$R_0/\Delta R$		30				
	K	SEE NOTE BELOW #	2				
APPROXIMATE PROCEDURE (BASED ON CUMULATIVE-PROBABILITY RANGE EQUATION) FIG. 14	S_1	FIG. 11	3.55				
	$R_I'(P_c)_o$(dB)	FIG. 12	9.62				
	$R_I'(P_c)_o$		9.16				
	$R_{co}' \equiv R_c'$ FOR $n = 10^6$, $\rho = 1$, $N = 1$	EQ. 6 OF FIG. 14	13.97				
	$D(1,K)$(dB)	MEYER PLOTS (2)	7.6				
	$ND(\rho N,K)$(dB)	MEYER PLOTS (2)	12.02				
	Q_i^3	EQ. 7 OF FIG. 14	0.361				
	$Q_i^{3/S1}$	$[Q_i^3]^{1/S1}$	0.751				
	$R_c/\Delta R$	$(R_{co}'+0.5)Q_i^{3/S1}-0.5$	10.37				
	APPROXIMATE R_c FOR GIVEN n', ρ, AND N	$(R_c/\Delta R)\Delta R$	20.73				
MORE PRECISE PROCEDURE (BASED ON CUMULATIVE PROBABILITY OF DETECTION CURVES) OF FIG. 15	$Q_i^{3/4}$	$[Q_i^3]^{1/4}$	0.775				
	$R_{oe}/\Delta R$	$(R_0/\Delta R)Q_i^{3/4}$	23.26				
	$R_{oe}/\Delta R$ (dB)		13.66				
	R_c/R_{oe} (dB)	FIG. 8, 9 OR 10	- 3.53				
	$R_c/\Delta R$ (dB)	R_c/R_{oe} (dB)+$R_{oe}/\Delta R$ (dB)	10.13				
	$R_c/\Delta R$		10.30				
	MORE PRECISE R_c FOR GIVEN n', ρ, AND N.	$\left(\frac{R_c}{\Delta R}\right)\Delta R$	20.60				
	ERROR IN $R_c/\Delta R$ (dB)	APPROXIMATE VS MORE PRECISE PROCEDURE	0.0dB				

* $P_c = 90\%$

SAME AS EXAMPLE NO. 7 OF TABLE 3.

\# NOTE: $K = \infty$, 1, AND 2 FOR SWERLING 0, I AND III MODELS, RESPECTIVELY.

Table 4 Cumulative-Probability-of-Detection Range, R_C, Work Sheet: Given P_C = ______, R_C = ?

Glossary

$D_0(N)$ — IF output power SNR/pulse required for case of nonfluctuating target (sometimes called Swerling Case 0 Target Model) to achieve probability of detection P_D for false alarm probability P_{FA} after analog video integration of N signal-plus-noise video samples. Additional assumptions — ideal matched IF receiver (i.e., L_m of Figure 7 of Chapter 2 equals zero) and no collapsing loss, i.e., $\rho = 1$. The results hold independently of whether a linear or square law envelope detector is used when obtaining $D_0(N)$ from the Meyer plots [2], Fehlner curves [8] or other such curves [14] (see Reference 7). Note that the definition for $D_0(N)$ is same as in Chapter 2 except that symbol N is used in place of n; see below. (The SNR/pulse is measured at the time at which the signal peaks at the output of the matched filter. However, the ratio of average signal power over an IF cycle to average noise power is specified rather than the peak power, occurring at the IF cycle peak, to average noise power. This definition is that used generally, [1-8, 10-15]. It is the one used in this book. It is, though, 3 dB lower than the SNR/pulse obtained from the curves in Reference 9.

$D_0(\rho N)$ — Definition same as $D_0(N)$ except collapsing loss is present, i.e., N signal-plus-noise video samples together with $m(=\rho N\text{-}N)$ noise alone video samples integrated. The results obtained from Meyer [2], Fehlner [8], DiFranco and Rubin [9] or Whalen [14] curves apply only for a square-law envelope detector when $\rho \neq 1$ [7, 16]. (Note that $D_0(\rho N) \neq D_0(N_1)$ for $N_1 = \rho N$. Without the parameter ρ no collapsing loss is assumed. It is assumed instead that N_1 signal-plus-noise samples are integrated.)

$D(N,K)$ — Definition same as for $D_0(N)$ except for fluctuating target, i.e., for chi-square target cross section model having K duo-degrees of freedom. For $K = \infty$, 1, and 2, one has Swerling Case 0, I, and III target models, respectively. Assumptions and restrictions given for $D_0(N)$ apply.

$D(\rho N,K)$ — Definition same as $D_0(\rho N)$ except for fluctuating chi-square target model having K duo-degrees of freedom. Assumptions and restrictions given for $D_0(\rho N)$ apply.

K — Number of duo-degrees of freedom for fluctuating chi-square target model; see definitions for $D(N,K)$ and $D(\rho N,K)$.

N — Number of signal-plus-noise video samples integrated per scan.

n — False alarm number for N=1, i.e., number of video samples in absence of signal observed before one or more false alarms occur with probability 50% [1]. Equal to $0.693/P_{FA}$ [1, 2, 8, 9].

n′ — False alarm number for ρN video integrated signal-plus-noise samples, specifically, number of video integrator output samples observed before one or more false alarms occur with probability 50% [1, 2, 8, 9].

$= 0.693/P_{FA}$

$= n/\rho N$ [1]

P_C — Cumulative probability of detection, that is, the probability of one or more single-scan detections by range R_C of an approaching target.

P_D — Probability of single-scan detection. (In general, based on integration of N signal-plus-noise video samples and $m = \rho N\text{-}N$ noise alone video samples.)

P_{FA} — False alarm probability at output of receiver after integration.

$= 0.693/n'$

$P_n = P_{FA}$

Q_1, Q_2, Q_i — See Equation (7) of Figure 14 and ensuing discussion.

R_C — Range at which P_C is achieved, equivalently, cumulative-probability-of-detection range.

R'_C — Normalized R_C, $R_C/\Delta R$

$R'_{CO} = R'_C$ for $n = 10^6$, $\rho = 1$, and $N = 1$; see Figure 14 and Tables 3 and 4.

$R'_I(P_C)_0$ — Coefficient for cumulative-probability-of-detection range equation; see Equation (6) and Figure 14. (Physically, $R'_I(P_C)_0$ is approximately equal to $R_C/\Delta R$ for $R_0/\Delta R = 20$; the approximation is exact if the target is illuminated at range R_C [4, 5]; see text.

R_0 — Range at which N(SNR/pulse) = 0 dB.

S_1 — Power law for cumulative-probability-of-detection range equation; see Equation (6) and Figure 14.

ΔR — Slant range distance traveled by approaching target between scans.

ρ — Collapsing ratio. $(N + m)/N$, where m = number of noise alone video samples integrated together with N signal-plus-noise video samples, [1, 2, 7, 10].

Note — See Appendix for cookbook procedures for calculating $D_o(\rho N)$ and $D(\rho N, K)$.

References

[1] Marcum, J.I.: "A Statistical Theory of Target Detection by Pulsed Radar," *RAND Corporation Reports RM-754* (1 December 1947) and *RM-753* (1 July 1948); also reprinted as part of a special monograph in the *IRE Transactions on Information Theory, Vol. IT-6, No. 2,* April 1960, and as AD-101287 and AD-101882.

[2] Meyer, D.P.; and Mayer, H.A.: *Radar Target Detection, Handbook of Theory and Practice,* Academic Press, 1973.

[3] Mallett, J.D.; and Brennan, L.E.: "Cumulative Probability of Detection for Targets Approaching a Uniformly Scanning Search Radar," *Proceedings of the IEEE, Vol. 51, No. 4,* April 1963, pp. 596-601; and "Correction," *Proceedings of the IEEE, Vol. 52, No. 6,* June 1964, pp. 708-709.

[4] Brookner, E.: "Cumulative Probability of Target Detection Relationships for Pulse Surveillance Radars," *Raytheon Report,* Raytheon Company, Wayland, Massachusetts, November 1972.

[5] Brookner, E.: "Cumulative Probability of Target Detection Relationships for Pulse Surveillance Radars," *Journal of The Institution of Engineers* (India), *Vol. 51, No. 10, Part 3,* May 1971, pp. 125-134; also Talk at Symposium on Radar Techniques and Systems, Sponsored by Radar Engineering Group, Telecommunication Engineers Division, Institution of Engineers, New Delhi, India, *Abstract,* p. 7, 1-3 May 1970.

[6] Barton, D.K.: *Radar Systems Analysis,* Artech House, Inc., Dedham, Massachusetts, 1976.

[7] Brookner, E.: "How to Look Like a Genius in Detection Without Really Trying," *1974 IEEE NEREM,* Boston, Massachusetts, pp. 37-64; see also Appendix of this book.

[8] Fehlner, L.F.: "Marcum's and Swerling's Data on Target Detection by a Pulsed Radar," *Report TG-451,* Applied Physics Laboratory, The John Hopkins University, Silver Spring, Maryland; also AD-602-121, July 1962. See also Fehlner, L.F.: "Supplement to Marcum's and Swerling's Data on Target Detection by a Pulsed Radar," *Report TG-451A,* Applied Physics Laboratory, The John Hopkins University, Silver Spring, Maryland, September 1964.

[9] DiFranco, J.V.; and Rubin, W.L.: *Radar Detection,* Prentice-Hall, Inc., Englewood Cliffs, N.J., 1968.

[10] Barton, D.K.: "Simple Procedures for Radar Detection Calculation," *IEEE Transactions on Aerospace and Electronic Systems, AES-5, No. 5,* September 1969, pp. 837-846.

[11] Heidbreder, G.R.; and Mitchell, R.L.: "Detection Probabilities for Log-Normally Distributed Signals," *IEEE Transactions on Aerospace and Electronic Systems, Vol. AES-3, No. 1,* 1967; also Aerospace Corporation, *Report TR-699 (9990) – 6,* April 1966.

[12] Skolnik, M.I.: *Introduction to Radar Systems,* McGraw-Hill, New York, 1962.

[13] Skolnik, M.I. (ed.): *Radar Handbook,* McGraw-Hill, New York, 1970.

[14] Whalen, A.D.: *Detection of Signals in Noise,* Academic Press, 1971.

[15] Hovanessian, S.A.: *Radar Detection and Tracking Systems,* Artech House Inc., Dedham, Massachusetts, 1973.

[16] Trunk, G.V.: "Comparison of the Collapsing Losses in Linear and Square-Law Detectors," *Proceedings of the IEEE, Vol. 60, No. 6,* June 1972, pp. 743-744.

Chapter 4 Barton

Philosophy of Radar Design

Introduction

In this chapter we discuss the process of radar system synthesis. Ideally, this synthesis starts with a well-defined problem, or a set of objectives to be accomplished in a real and known environment. There is a logical flow of system engineering tasks which translate the defined problem into a successful radar design. In most actual cases, however, the logical paths are subject to interference from various intangible or political factors which compromise the design process. These influences often lead to a line of development which, from the start, is doomed to technical or economic unfeasibility. Some guidelines are suggested to identify paths to a successful production design – one which leads to construction of more than two samples of a particular radar.

Ideal Process of Radar Synthesis

New radar designs may be produced by either of two methods – an old radar may be modified by continuous changes and evolve into a new design, or the new radar design may be synthesized from a statement of requirements. The second process is referred to as "ideal" in the sense that it should lead to the most economical design for the defined task. Figure 1 shows the steps in such a process, in which three types of inputs are needed – a set of objectives provided by the customer or user, the knowledge of radar system engineering provided by the literature and experience of the radar profession, and the funds for design and development, again provided by the customer or sponsor of the program. The desired result, labelled 4 , is the development of a successful radar.

Most of the material in this book and in the literature is concerned with the second type of input, which reflects the normal tasks of radar system engineers. The first step is the translation of a physical description of the environment into electromagnetic terms, which are needed to establish radar requirements. Theoretical constraints on the radar design then can be established from well-known precepts of detection and estimation theory combined with the radar equation and considerations of resolution and waveform design. At this point, the practical constraints, based on up-to-date knowledge of existing designs, must be considered. Estimates of RF losses, realistic compromises in receivers and signal processors, and noise components or errors introduced by practical components and construction techniques must be made to determine how closely the theoretical limits may be approached. Finally, with the major system characteristics defined, a series of economic and performance trade-off studies is needed to identify and select the circuit techniques and components most appropriate for the new radar.

Before discussing in more detail these several steps in radar synthesis, a few general remarks are in order, related primarily to the first and third types of input to the process – objectives and funding. Figure 1 is based on the assumptions that a practical, producible radar design is the desired output, and that such a design attracts the support and funding needed to bring it into existence, either as a replacement for older radars or as a new system. A "successful radar" design can be defined as one which is built in quantities greater than two, since many unhappy experiments lead to one or two models for engineering and environmental test but end up as museum curiosities. In most cases where a third radar is built, it serves as evidence that the design has passed the threshold of success, and that the equipment can be used in some field of operation.

There are other possible objectives, however, which lead to radar design and development programs. For example, a perpetual program of research, development, and simulation may be, in some organizations, preferable to the production of operational radar. Many more engineers can be kept employed in such a paperwork program, and it may be cheaper than placing a simple design into quantity production. This chapter explores the variations in synthesis steps which can lead to this result, intentionally or otherwise. Another possible objective is to find an application for a specific invention or technique which happens to have been developed. This process of having the solution and seeking the problem is more common than might be believed. In some cases, it can even lead to successful designs. Another possible objective is profit, which ideally should lead to a desire for a successful production design following the process of Figure 1. However, there are many nonlinearities and distortions in the system of financial rewards which may make it more profitable to produce a less effective design or to produce nothing but paper and software rather than radar equipment. Some of these alternate objectives are discussed later, after the ideal process has been reviewed. It is important to note that even the closest approaches to an ideal or rational synthesis process give only a small probability (perhaps ten percent) of producing a successful radar, because radar design is not an easy task. Historically, with the mixture of confused objectives and uncertain funding, the probability of success is nearer to one percent. Accordingly, the radar system engineer is fortunate if he can work on one successful design during his career. Let us see how this prospect can be improved.

Problem Definition

The initial input to the problem definition step is a statement of objectives for the radar. A distinction should be made between primary objectives and secondary or other objectives. In considering search and tracking (Table 1), the primary objective tends to determine the size and operating frequency of the system. Secondary objectives often are added to a performance specification in the hope that they can be met with little additional cost or complexity. Subsequently, consideration is given to what happens when these lesser objectives are allowed to control system design, or when both volume search and accurate tracking are stated as "primary." Normally, however, a search radar objective can be defined in terms of a range-elevation contour at which a given cumulative detection probability has been achieved on a given target, or within which a given single-scan probability of detection has been achieved at a given scan rate. The interface specification defines how much data is provided to assist the radar and its operators in performing their functions or in adapting to different targets or environments, as well as what radar outputs are needed, in what form, and what the radar's physical environment is like.

In specifying tracking radar performance, the range-elevation contours for different levels of accuracy in each measured coordinate, along with data rate and number of targets to be tracked, consti-

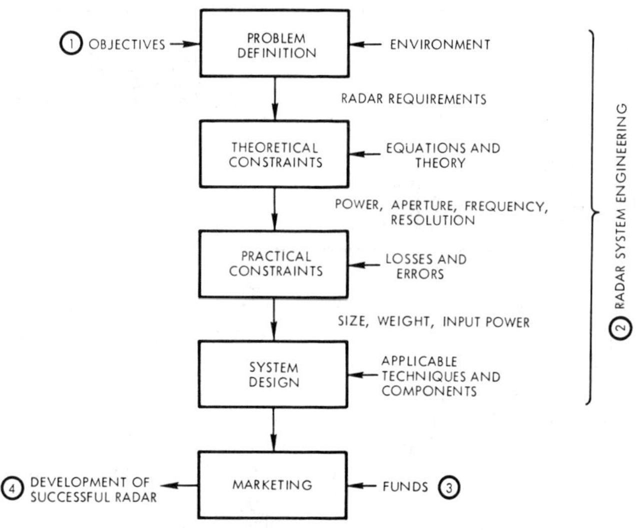

Figure 1 **Ideal Process of Radar Synthesis**

	Search Radar	Tracking Radar
Primary Objectives	Detection Probability in Given Volume of Space – 1) Cumulative Probability 2) Single-Scan Probability	Accuracy in a Given Volume of Space – 1) Angle Measurement 2) Range and Doppler Data Rate and Number of Targets Tracked
Secondary Objectives	Target Resolution Accuracy Data Rate	Target Resolution Acquisition Capability 1) Scan Volume 2) Probability vs. Time
Interfaces	Inputs Available From Other Portions of the Overall System; Outputs Required; Environmental Conditions	

Table 1 **Problem Definition – Objectives Determined by Operational Needs of Ultimate User**

tute the primary specifications. Secondary objectives are resolution requirements (between adjacent targets or between targets and unwanted scattering sources) and contours for acquisition in a given time interval. An important interface specification is the type and accuracy of designation data furnished to the tracker (e.g., two-dimensional search radar data to ±1° and ±0.5 mi). This information is needed to establish an acquisition scan procedure with adequate lock-on probability for the tracker. During the synthesis process it is necessary to ensure that a reasonable balance is obtained between the burden placed on the search radar or other designation source and that placed on the tracker for acquisition. Otherwise, the result may be a narrow-beam precision tracker the major task of which is searching a hemispherical volume for potential targets.

While only search and track objectives are shown in Table 1, similar lists can be established for other types of radar – navigation, weather observation, etc. The important thing is to identify which of several possible objectives are important enough to control the system design, and which are merely desirable features to be obtained as a by-product. One vital task of the radar system engineer is to help negotiate the priority of objectives to avoid imposing unnecessary burdens on the resulting equipment.

The second major input to the problem definition step is the description of the radar environment – not the ambient temperature

Constraint	Data Sources
Line of Sight	
Horizon Range	Equation (1)
Range-height-angle charts	Blake [1, 2]
Terrain Masking	Geodetic Survey Maps
*Attenuation**	
Clear Atmosphere	Blake [4]
Rain, Clouds, Snow	Gunn and East [5], Barton [3] Bean, Dutton, and Warner [7]
*Clutter**	
Weather	Gunn and East [5], Barton [6], Bean, Dutton, and Warner [7]
Land, Sea	Moore [8], Skolnik [9], Nathanson [10], Barton [6]
Refraction	
Tropospheric Bias	Barton [11]
Trophospheric Fluctuation	Barton and Ward [12], Chapter 15
Ionsophere	Barton [11], Chapter 14
Ducting	Bean and Dutton [13]
Surface Reflection	
Lobing (Fading)	Blake [2]
Multipath Error	Barton and Ward [12], Barton [14] Blake [2]
Diffraction	
Interference (ECM) Model	Arbitrary Selection

*see also Chapter 1

Table 2 **Environmental Constraints**

and vibration to which the equipment is exposed, but the electromagnetic environment which affects the transmission and reception of the signal. Table 2 lists the major environmental constraints and identifies the primary sources of related system engineering data. Microwave propagation phenomena are of considerable importance here; it should be noted that much of the applicable data is derived from the literature of radio communications. For search radar, a simple evaluation of line-of-sight range for low-altitude targets can be made using an effective earth radius, ka, in

$$R_H = \sqrt{2\,ka\,h_R} + \sqrt{2\,ka\,h_T} = 4130\,(\sqrt{h_R} + \sqrt{h_T}\,) \quad (1)$$

for k = 4/3, ka = 8.5 x 10^6 m, and values of R_H, h_R, and h_T in meters. Range-height-angle charts have been prepared by Blake [1, 2], and his computer programs permit charts to be prepared for the exponential reference atmosphere, using any convenient range and height scales. Unfortunately, for low-altitude targets, the dominant effects may be the occurrence of nonstandard refraction and terrain masking. While it is convenient to use the published charts and Equation (1) in system design, the engineer who can look out a window in an office may be reminded that the real world is not only spherical but roughened by hills, trees, and irregularities which are difficult to characterize mathematically. Statistical analyses of masked regions may be obtained, but the effects of diffraction in partially filling the holes at short range are almost impossible to model. Engineering of shipboard radars is easier than for land-based equipment, but only if the ships remain well offshore. The only realistic procedure for system engineers is to use the smooth sphere theory and ignore irregularities, and to keep in mind that the results are inapplicable to low-altitude targets.

Attenuation and clutter are defined somewhat better, provided the environmental specification is reasonably complete. Again, it is important to consider the possibility of variations from the expected values. For tracking radars, accurate estimates of refraction may be needed; for the standard atmosphere, these figures are available in several books. Deviations from standard atmosphere may be estimated by ray tracing if the actual refractivity profile is known. Again, there is a problem with low-altitude targets when conditions produce ducting; while the conditions which produce ducting are described in the literature, there is insufficient data on statistics of occurrence and magnitude of its effect. Refractive fluctuations, caused by tropospheric irregularities, are discussed in [12 (Appendix D)] ; ionospheric effects, in Chapters 14 and 15 by Brookner.

The subjects of surface reflection and diffraction and their effects on search coverage and tracking accuracy have received wide discussion in the literature. Again, the problems are well-defined for medium- and high-altitude targets, and ill-defined for low-altitude targets. In the absence of adequate experimental data, there are, at least, several models which are reasonable for different types of terrain or water surfaces. These models can be used to guide the system synthesis process and to evaluate alternative designs with respect to performance on low-altitude targets.

The third major input to problem definition is the description of the radar target. Table 3 lists the characteristics of different target models, and sources of related data. The basic choice is between a statistical description of the target (Swerling, chi-squared, or log-normal) and a deterministic model (a steady target, or one of the simple shapes for which mathematical formulas can be derived, or an arbitrary shape on which complete reflectivity patterns have been measured). If statistical models are used, the amplitude distribution is only part of the model; the frequency spectra and correlation intervals in time and radar frequency are also important.

For tracking or guidance applications, target glint models also are needed. Failure of the customer to provide target descriptions in the detail listed here can present the system engineer with a major problem. If agreement cannot be obtained on a reasonable model, there is the risk of overspecifying the radar and adding unnecessarily to its cost, or, alternatively, of providing a design which will fail to meet test requirements on the target which ultimately is used. One approach to this problem is to explore the sensitivity of the design to variation in target characteristics, and to favor the design options which are most tolerant of fluctuations. This choice should lead to a design which degrades gracefully as new types of targets are introduced after deployment.

Determination of Major Parameters

Once the problem has been defined, the general type and size of the needed radar can be determined. For either search or tracker acquisition, the minimum energy ratio required per search beam position can be determined according to procedures outlined in Chapters 2 and 3. For reliable tracking, the minimum energy ratio per target can be established without regard to accuracy, subject to later increase if this level proves insufficient to provide accurate data with the selected radar resolution. Before applying the radar equation, the time available for a search frame or a tracking sample must be established. This figure can be determined from the cumulative detection probability analysis or from the given data rate for search radar, and from the data rate and number of targets for tracking radar. The radar equations then can be used to scale the radar system.

Radar range equations appear in many forms, involving many radar parameters which are assumed as known. A set of equations particularly arranged for system synthesis are presented in Reference 15, and are summarized in Tables 4 and 5. These equations relate the major parameters of the radar to the performance and environmental requirements, constraints, and loss terms. The major parameters are the average power, aperture area, beamwidth, transmitting gain, sidelobe ratio, tunable bandwidth, signal bandwidth, and MTI (or Doppler) improvement factor. Performance and environmental requirements include range, detectability factor, solid angle and frame time for search observation, time for track, jammer noise power, jammer range, target size, unambiguous range of the radar, clutter reflectivity, and input noise temperature. Radar frequency is not used in the equations, although it is a factor in antenna and clutter descriptions. Thus, for search in a thermal noise environment, the requirements can be translated into a power-aperture product which is independent of waveform and frequency (except to the extent the input temperature may be frequency dependent). In surface clutter, the search requirement is for a given product of signal bandwidth and improvement factor, unless targets are elevated enough to place the clutter outside an achievable elevation beamwidth.

A preliminary choice of frequency band for the radar can be made by looking at sets of major characteristics which simultaneously meet the several environmental requirements. For example, if search in the clear environment requires a given power-aperture product, while stand-off jamming sets a frequency-dependent minimum product for $P_{AV}G_SB_J$ and weather clutter sets a frequency-dependent minimum for BI, it may turn out that only one or two frequency bands yield reasonable solutions. In each band, the waveform requirements for range and Doppler resolution are different, and the number or shape of the beams changes to provide volume coverage in the required time. Examples are given in the references as to how this multidimensional trade-off study may be carried out. Occasionally it turns out that one set of requirements can be met only at frequencies below, say, 2 GHz, while another set can only be met above 5 GHz. If this dilemna appears at the theoretical stage of the problem, without even considering the practical losses and errors, a change in input requirements clearly is necessary. No amount of juggling of radar characteristics can overcome the theoretical limitations.

Practical Constraints

A theoretical solution to the radar synthesis problem is necessary but not sufficient to ensure successful development. A large number of practical constraints must also be considered, as shown in Table 6. The loss factors already listed in the radar equations of Tables 4 and 5 impose one type of constraint – either simple RF transmission losses, varying with the frequency band, or more complex losses caused by target fluctuation or signal processing procedure, which are statistical in nature and which vary with detection probability. For example, if an attempt is made to obtain 90 percent detection probability on a single coherent signal sample, a heavy penalty is paid if the target fluctuates or the efficiency of signal processing varies as the signal moves from one range gate or Doppler filter to another. These losses affect not only frequency choice but choice of technique – raster scanning, range-Doppler matrix processing, scan-to-scan integration, or accumulation of detection probability. The size of the signal processor has al-

Target Characteristics	Data Sources
Radar Cross Section Patterns of Simple Shapes Average Cross Section vs. Aspect Angle Polarization Matrix	Kell and Ross [16] Weinstock [18]
Amplitude Distribution*	
Swerling Models	Blake [2], Skolnik [17], Barton [11]
Chi-Squared Models	Weinstock [18]
Log-Normal Models	Heidbreder and Mitchell [19]
Frequency Spectra and Correlation Intervals* Glint in Angle, Range and Doppler*	Barton and Ward [12] Dunn and Howard [20]

*see also Chapter 2

Table 3 **Target Models**

Noise Source	Tracking Requirement	Search Requirement
Receiver Noise	$P_{AV}G_TA_R = \dfrac{(4\pi)^2kT_IR_M^4DL_1L_2}{\sigma t_o}$	$P_{AV}A_R = \dfrac{4\pi\psi_SkT_IR_M^4DL_S}{\sigma t_S}$
Mainlobe Jamming (Self-Screening)	$P_{AV}G_TB_J = \dfrac{4\pi P_JG_JR_M^2DL_3}{\sigma t_o}$	$P_{AV}B_J = \dfrac{P_JG_J\psi_SR_M^2DL_4}{\sigma t_S}$
Sidelobe Jamming (Stand-Off)	$P_{AV}G_TG_SB_J = \dfrac{4\pi P_JG_JR_M^4DL_3}{R_J^2\sigma t_o}$	$P_{AV}G_SB_J = \dfrac{P_JG_J\psi_SR_M^4DL_4}{R_J^2\sigma t_S}$

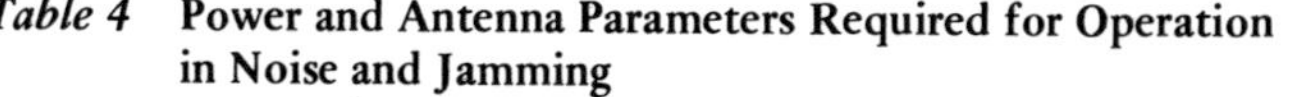

Table 4 **Power and Antenna Parameters Required for Operation in Noise and Jamming**

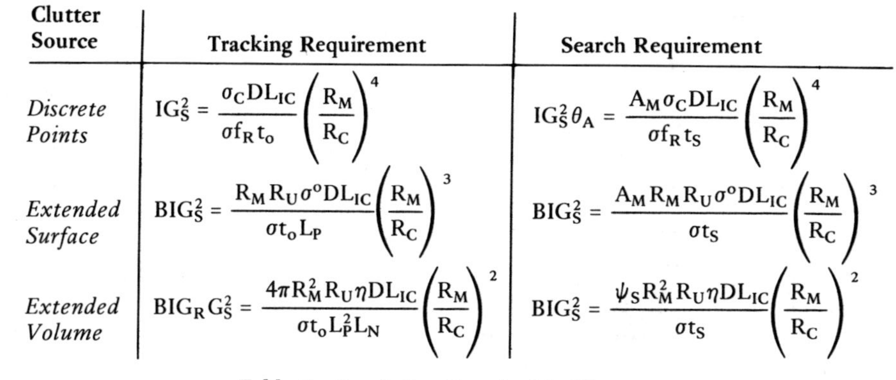

Clutter Source	Tracking Requirement	Search Requirement
Discrete Points	$IG_S^2 = \dfrac{\sigma_CDL_{IC}}{\sigma f_Rt_o}\left(\dfrac{R_M}{R_C}\right)^4$	$IG_S^2\theta_A = \dfrac{A_M\sigma_CDL_{IC}}{\sigma f_Rt_S}\left(\dfrac{R_M}{R_C}\right)^4$
Extended Surface	$BIG_S^2 = \dfrac{R_MR_U\sigma^oDL_{IC}}{\sigma t_oL_P}\left(\dfrac{R_M}{R_C}\right)^3$	$BIG_S^2 = \dfrac{A_MR_MR_U\sigma^oDL_{IC}}{\sigma t_S}\left(\dfrac{R_M}{R_C}\right)^3$
Extended Volume	$BIG_RG_S^2 = \dfrac{4\pi R_M^2R_U\eta DL_{IC}}{\sigma t_oL_P^2L_N}\left(\dfrac{R_M}{R_C}\right)^2$	$BIG_S^2 = \dfrac{\psi_SR_M^2R_U\eta DL_{IC}}{\sigma t_S}\left(\dfrac{R_M}{R_C}\right)^2$

Table 5 **Resolution Required in Clutter**

A_M = azimuth search sector
A_R = effective receiving aperture
B = signal processing bandwidth
B_J = width of jammer spectrum, assumed equal to agility bandwidth of radar
D = energy ratio required
f_R = pulse repetition rate
G_J = jammer antenna gain
G_R = radar receiving gain
G_S = mainlobe/sidelobe gain ratio
G_T = radar transmitting gain
I = Steinberg's MTI improvement factor
k = Boltzmann's constant
L_I = integration loss in noise
L_{IC} = integration loss in clutter
L_M = receiver matching loss
L_S = total search loss
L_1 = loss from transmitter to receiver input, in excess of free-space loss
L_2 = product of L_M, L_I, and collapsing losses
L_3 = portion of L_1L_2 applicable to jamming case (excluding losses common to signal and jamming)
L_4 = portion of L_S applicable to jamming case
P_{AV} = average transmitter power
P_J = total jammer power
R_C = range to clutter
R_J = range to jammer
R_M = maximum range of radar
R_U = unambiguous range
T_I = effective input noise temperature
t_o = observation time
t_S = total search time (frame time)
η = volume clutter reflectivity
θ_A = azimuth beamwidth (one-way, 3-dB)
σ = target cross section
σ^o = surface clutter reflectivity
ψ_S = solid angle searched, or equivalent full-range angle for csc^2 pattern'

List of Symbols for Tables 4 and 5

ways been one practical constraint, in that it is seldom desirable to build a very complex, multiple-channel processor to save a few tenths of a decibel in loss.

A primary concern in tracking radar is measurement error and how the allowable error is budgeted among the many possible contributors. The measurement sensitivity or error slope is an important property, since many errors scale inversely with this slope. However, the angle slope can be maximized only at the expense of high sidelobes in the difference pattern, and these sidelobes increase the errors from clutter, ECM, and multipath. The same considerations apply in the range-Doppler domain.

Weight and size trade-offs are most applicable to antennas and transmitters, yet these devices have not yet begun to yield to microminiaturization. These two subsystem areas also are affected most by availability of special components, since the computer and communications fields do not provide the right types of device for radar. The effect of all these factors on choice of frequency band is one of the major areas for system engineers to prove their mettle. Under favorable circumstances, the power-aperture-gain relationships in the radar equations, scaled up to account for practical losses and errors, permit a solution in at least one of the established radar bands. In a few cases, the clutter environment in

Major Parameters

RF Loss Budgets
Processing Loss Budgets
Error Budgets
Measurement Slopes vs. Sidelobes
Weight and Size Estimates
} As Functions of Frequency and Equipment Complexity

Reliability and Maintenance Considerations
Availability of Components (RF Tubes, etc.)

Select Frequency Band(s), From Which Follow
Average Power
Aperture Size
Beamwidth
Sidelobe Levels
Gain
Waveform(s)
Bandwidth
Agility
Coherent and Incoherent Integration Times
Signal Processing Procedure
Overall Size
Weight
Input Power

Table 6 **Determination of Major Parameters (Practical Constraints)**

this band is such that a waveform and processor can be designed to see the signal. Two or more different waveforms may be needed to solve different aspects of the problem; if the system engineer is particularly fortunate, there will be enough time to use these waveforms sequentially rather than having to build parallel radar channels throughout the system. There may even be so many options available that an attempt will be made to find the optimum choice through use of cost-minimizing equations. But such equations are seldom reliable or useful, because of the discontinuities introduced by available production tubes and similar components. The engineer and the customer should be wary of believing any optimization based on continuous cost curves. Obviously, there must be some basis for estimating the cost of a proposed system; but, unless it can be related closely to existing production equipment, the estimate must be considered inaccurate.

Selection of Applicable Techniques

The final step in system synthesis is to select and combine the available and applicable techniques and components which are to perform the functions defined in the previous steps. The word applicable should be emphasized. In each subsystem area, there appear a variety of devices and techniques competing for a place in the system, as indicated in Figure 2. It is most important to have our system requirements clearly in mind at this point so that the subsystem techniques which are applicable to solving the problem

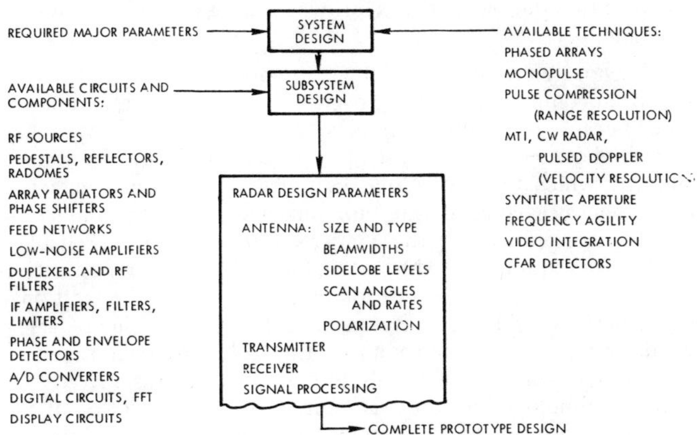

Figure 2 **Selection of Applicable Techniques**

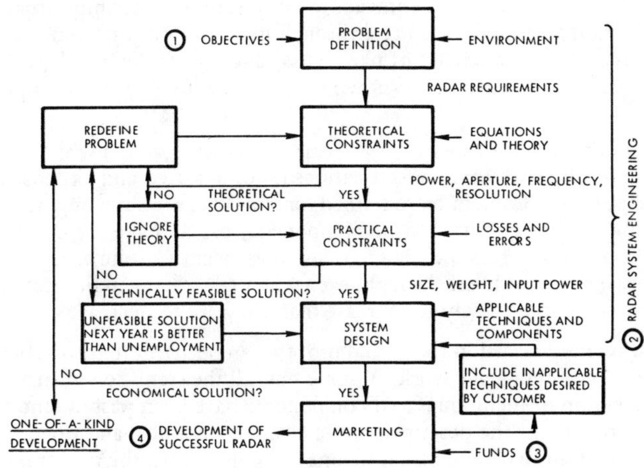

Figure 3 **Actual Process of Radar Synthesis**

can be chosen. The system engineer must write out the subsystem specifications and work with design specialists to select from the choices available. Ideally, the radar can be configured from existing devices using proven techniques, and the task is complete (at least until system test is needed). However, the real system design process invariably becomes more complicated than that described above; the steps and alternatives actually facing the system engineer, are explored below.

Actual Process of Radar Synthesis

The process shown in Figure 3 is the same as the ideal process of Figure 1 but with some added blocks and associated paths or loops which represent desirable and logical iterations to refine the stated objectives in view of costs or other consequences discovered in the system design process. For example, theory may show that two radar requirements are incompatible; redefinition of the problem, though, would avoid the conflict. If requirements are to search throughout the hemisphere around the radar site and to track each target with extreme accuracy at high data rate, a large power aperture product may be necessitated. Available power sources may then dictate a large antenna, which is only practical in the lower frequency bands. Yet the accuracy requirements may force the use of a very narrow antenna beam, or the presence of ionospheric refraction may dictate use of a higher frequency. This type of inconsistency arises frequently, and the technology of multi-function phased arrays may not always provide a solution at any cost, because no single frequency is suitable.

There are two choices available in the actual synthesis process. One is to separate the problem into two or more parts (e.g., separate search and tracking radars at different frequencies), each of which has a solution. Historically, this solution has been the workable option. But there is another approach which has been adopted in many cases. Marketing intelligence may indicate that the customer believes there is a single solution and therefore will insist on a proposal geared to his approach. The system engineer can believe his theory and analysis and try to prove the customer wrong, or the assumption can be made (even honestly) that a customer with an authorized program must have access to some new theory or data which permits the problem to be solved. If the sticky areas of analysis can be evaded in the proposal, further work under contract may (somehow) resolve the problem.

Another possibility is that the theoretical solution exists, but that practical problems rule out a successful development in the available time. Should the system engineer at this point call a halt and seek redefinition of the problem? Should the program wait until the necessary tubes or components become available? The conclusion, under the existing competitive system, may be that a delayed discovery of unfeasibility is preferable to termination of the project (or award of a contract to a less scrupulous competitor). Perhaps there will be a breakthrough in the meantime from the research labs, or an act of nature that overcomes the problem will occur (radars have been known to catch fire just before acceptance test).

Another point of difficulty is economics. If the design which complies fully with input specifications is complicated and expensive, should the problem be redefined, or should the system engineer rely on future reductions in component cost to make the design economically feasible? There is a strong incentive to proceed in hope of a cost breakthrough, even when experience indicates that costs are more likely to increase than they are to decrease.

A further hazard in the actual process can be attributed to efficient market intelligence work. Assume that all the steps down through selection of techniques and components have been passed, and the writing up of the description of a clean, economical and practical radar is all that remains. Someone has gone out to the customer's laboratory and talked to the persons who are expected to evaluate the proposal, and returns to report that they are enthusiastic about a specific circuit or technique. The customer expects this feature to be used in the radar – it had better be included even if it has no application to the problem as defined. This hazard exists in most radar developments because the government laboratory associated with a given procurement generally has its work concentrated in three or four areas. The lab may have made significant progress in these areas, but the next radar procurement in which they are involved may not need their results. The interests of the lab may lead them to attach their high technology to the radar development anyway, creating a problem for the system designer. Should the ideal process of selecting applicable techniques be held to firmly, or should the inapplicable new technology be incorporated somehow?

A Case History

Several years ago, a study contract was awarded to prepare the design of a radar, having very high clutter rejection, for detection of low flying aircraft over land. For about a year, specialists in waveform design, information theory, and optimum filter theory performed the study, leading to the definition of an optimum waveform and processor – 1 μs pulses repeated at intervals of approximately 1 ms and processed in an MTI canceller. It was impossible to go back to the customer, in the mid-1960s, to recommend that the World War II AN/TPS-1D be put back in production on the basis of optimum waveform and processor synthesis.

What actually happened was that someone discussed the problem with the customer who pointed out that the real interest was in solid-state radar antenna modules at X-band. The entire program was redirected toward RF module design, and the state of the art dictated low peak power with high duty cycle operation – the opposite of what was needed for clutter rejection. Needless to say, there was no radar designed or built under the program. The technique which was in vogue at that time just didn't match the problem, and it remains years away from application in radars of that class.

This case illustrates that a producible radar is not necessarily the goal of a radar system design effort. The user's need for a particular type of radar may be used as a vehicle for establishing and carrying out developments in technology areas related only slightly to that need. This procedure may not be wrong, and it may be the only way to obtain funding for development of advanced techniques. It does, however, create hazards for the radar system engineers in government and industry, whose synthesis and analysis techniques should make clear the discrepancy between the stated objective and the work being pursued. The diversion of such system resources to supporting unrelated developments also explains some of the failures of radar programs to reach production status.

System Engineering Ethics

The final stumbling block in the path to a successful radar is the need for funds at the point marked 3 in the synthesis process. There are many cases in which the synthesis takes so long in arriving at the right radar for the right price that the original requirement (or the funds to support it) have disappeared. Even in that ten percent of the cases where a rational process has been used, the requirements may have changed enough to prevent entering production. If the lists of AN nomenclature from AN/FPS-1 to AN/FPS-100 or more, the similar AN/SPS, and other lines of radar are checked looking for production designs, the yield is small. If those programs which stopped short of receiving AN numbers are included, the yield is not much over one percent.

What is the responsibility of the radar system engineer in correcting this situation? When a contract definition phase study arrives at one of the decision points in our diagram, and the technical factors indicate a need to redefine the requirements, should the engineer tell the customer that the requirements are wrong as so stated? That process is the "ideal,"but it doesn't work often. The real problems are not technical, and must be solved by a combination of system engineering, marketing, and management skills. The engineer may not be making the best contribution to a successful radar development by insisting on solutions entirely within a discipline. The final step in the synthesis process thus amounts to a stepping back and allowing other specialists to assume responsibility in their areas. The engineering material in this book and in the literature is applicable only to the technical side of the radar development problem.

Conclusions

The points discussed above can be summarized under five general remarks, as follows.

1 – Most radar developments are unsuccessful, at least if the measure of success is production of three or more units for use in the field. Engineers and their professional societies and employers cannot be proud of this situation, but it seems unlikely to change.

2 – Radar is only marginally feasible in most applications, if economic and operational factors are considered. We tend to believe that radar technology is so well advanced that it can solve any problem in its general area of application. Even with the most competent team of engineers, specialists and managers, it is rare and fortunate that a successful design will emerge, for reasons described earlier.

3 – Where a feasible approach does exist, it can be eliminated easily by pyramiding of requirements and inclusion of secondary objectives. Many potential successes have been converted to failures through overspecification and resulting complexity and cost, because of a failure to appreciate point 2 above. Through overconfidence, we add features and solutions to problems that come to our attention during synthesis but which were not part of the original objectives.

4 – Previously unsuccessful designs can now be produced with improved components. Having overreached in the past, and having failed because of resulting complexity, cost, size, weight, or other undesired consequences, we tend to proceed with newer system concepts which still stretch our resources. We overlook the fact that modern signal processing, antenna, and related techniques could be applied to solve these older system problems in reliable, light-weight equipment. The system block diagrams and operational capabilities might appear obsolete, but the cost and operational convenience would be so advantageous that the resulting radars would stay in production for a long time. This approach requires a reversal in organizational trends, in that our emphasis on "advanced technology" must be reoriented toward application of advanced components to old systems.

5 – The "rule of the third best" remains applicable in spite of thirty years of advances in technique since its enunciation by Watson-Watt. Before and during the Battle of Britain in World War II, he warned that the "best" design had to be rejected because it would never be achieved, and that the "second best" would be achieved too late to be deployed for use by the armed forces when they needed it. The third best would be adequate and available in time, and it was what won the Battle of Britain.

We can identify the best or the "optimum" design with even more precision today, using synthesis and analysis techniques described in this volume and elsewhere. We should find this optimum and use it as a reference, but should avoid attempts to build it. For example, the matched filter is a valuable concept in signal detection theory, as it permits us to compare other designs. If a circuit is claimed to give 3 dB improvement with respect to the matched filter, we can discount the proposal as unsound. On the other hand, the proposal of a matched filter should be considered evidence of incompetence in system engineering, since a more economical approach with small extra loss is always available.

The basic philosophy of system design outlined here involves a thorough understanding of radar theory, and a willingness to compromise both technical and nontechnical realities of the engineer's environment.

References

[1] Blake, L.V.: "Radio Ray (Radar) Range-Height-Angle Charts," *NRL Report 6650,* January 1968; see also *Microwave Journal, Vol. 4,* October 1968.

[2] Blake, L.V.: "A Guide to Basic Pulse-Radar Maximum-Range Calculations," *NRL Report 6930,* December 1969 (AD701321); see also Chapter 2 of *Radar Handbook* (M.I. Skolnik, ed.), McGraw-Hill, New York, 1970, and summary in Reference 3.

[3] Barton, D.K.: *The Radar Equation* (Vol. 2 of *Radars*), Artech House, Dedham, Massachusetts, 1974.

[4] Blake, L.V.: "Radar/Radio Tropospheric Absorption and Noise Temperature," *NRL Report 7461,* October 1972; see also Reference 3, pp. 207-209.

[5] Gunn, K.L.S.; and East, T.W.R.: "The Microwave Properties of Precipitation Particles," *Quarterly Journal of the Royal Meteorological Society, Vol. 80,* October 1954; reprinted in Reference 6.

[6] Barton, D.K.: *Radar Clutter* (Vol. 5 of *Radars*) Artech House, Dedham, Massachusetts, 1975.

[7] Bean, B.R.; Dutton, E.J.; and Warner, B.D.: "Weather Effects on Radar," Chapter 24 in *Radar Handbook* (M.I. Skolnik, ed.), McGraw-Hill, New York, 1970.

[8] Moore, R.K.: "Ground Echo," Chapter 25 in *Radar Handbook* (M.I. Skolnik, ed.), McGraw-Hill, New York, 1970.

[9] Skolnik, M.I.: "Sea Echo," Chapter 26 in *Radar Handbook* (M.I. Skolnik, ed.), McGraw-Hill, New York, 1970.

[10] Nathanson, F.E.: *Radar Design Principles,* McGraw-Hill, New York, 1969.

[11] Barton, D.K.: *Radar System Analysis,* Artech House, Dedham, Massachusetts, 1976.

[12] Barton, D.K.; and Ward, H.R.: *Handbook of Radar Measurement,* Prentice-Hall, Englewood Cliffs, New Jersey, 1969.

[13] Bean, B.R.; and Dutton, E.J.: *Radio Meteorology,* U.S. Printing Office, Washington, 1966.

[14] Barton, D.K.: *Radar Resolution and Multipath Effects* (Vol. 4 of *Radars*), Artech House, Dedham, Massachusetts, 1975.

[15] Barton, D.K.: "Radar Equations for Jamming and Clutter," *IEEE EASTCON Record,* 1967; reprinted in Reference 3.

[16] Kell, R.E.; and Ross, R.A.: "Radar Cross Section of Targets," Chapter 27 in *Radar Handbook* (M.I. Skolnik, ed.), McGraw-Hill, New York, 1970.

[17] Skolnik, M.I.: *Introduction to Radar Systems,* McGraw-Hill, New York, 1962.

[18] Weinstock, W.W.: "Radar Cross-Section Target Models," in *Modern Radar* (R.S. Berkowitz, ed.), John Wiley & Sons, New York, 1965.

[19] Heidbreder, G.R.; and Mitchell, R.L.: "Detection Probabilities for Log-Normally Distributed Signals," *IEEE Transactions on Aerospace and Electronic Systems, Vol. AES-3, No. 1,* January 1967.

[20] Dunn, J.H.; and Howard, D.D.: "Target Noise," Chapter 28 in *Radar Handbook* (M.I. Skolnik, ed.), McGraw-Hill, New York, 1970.

Chapter 5 Shrader

Some Radar Design Problems and Considerations

This chapter and the following one relate some of the pitfalls observed while working with the development of radar systems. This chapter presents the supplier's point of view, while Fowler's Chapter 6 gives the customer's point of view. This distinction is not a clear dividing line; many of our thoughts overlap.

Radar has been around for many years – a good estimate is 40 years, depending upon where you place the starting point. Some 30,000 radar systems have been built, and billions of dollars have been spent on radar. For example, in 1972 alone, over 900 million dollars were spent on radar; for that year, that figure was approximately 1/10 of one percent of the gross national product. There have been over 2,000 papers on radar published in the various IEEE publications and 23 tri-service radar symposiums; yet, when we reviewed this literature, there was a remarkable dearth of papers on the problems of developing radar systems. Thus a few examples are given.

One treatment related to this subject is Chapter 4 by David Barton. He presents the conflicting problems of building a successful radar, and of building a radar which would sell to the customer because of unnecessary gimmicks. Another coverage is a paper by Robert A. Frosch entitled "A New Look at Systems Engineering," published in the September 1969 *IEEE Spectrum*. This paper is recommended highly to anyone who is trying to develop a successful radar system.

The first three topics in this chapter have a common denominator; and that is – communicate with the customer, establish a good rapport, and stay on good terms. The fourth topic is the *modeling syndrome* – its warnings are not to believe your models too greatly and not to spend too much time with them. The fifth topic is *Beware The Committee!;* it is advice on what to do when someone brings in a committee to start guiding your radar program. Finally, there is a discussion of erroneous extrapolation from failures.

References are made to various companies with letters such as A, B, C, and D. These cases presented are based on fact. However, some of the facts have been altered in some of these cases, either to emphasize the point being made or to protect the guilty.

Communicate with a Prospective Customer (Know What the Customer Wants)

You must know what a customer wants before the RFP (Request For Proposal) is issued, because not getting the contract is the biggest pitfall of all. Usually, to get a contract, you must compete successfully against other companies; to compete successfully you must know what the customer wants. This information is never defined clearly in the RFP. The only way to know what a customer wants is to talk before the RFP is issued.

Case of the Pressurized Waveguide

An example is the case of the pressurized waveguide system. Four companies were competing for a production contract for what was basically a simple, low-priced radar system. Company A had built a prototype of the system; therefore, they felt that they knew exactly what the customer wanted. They thought that they were a shoe-in for the job, and didn't bother to go to the customer before the RFP was issued. The same was true of Company B. Companies C and D, however, had been talking to the customer. When the RFP came out, it contained what had been the original specifications for the radar – the peak power, the noise figure, antenna gain, and all the other parameters. But it contained one strange sentence – (paraphrased) if you could give the same performance without a pressurized waveguide, it would be preferred. This statement appeared to be an invitation to trade-off peak power versus noise figure – to deviate from the specification requirements. Companies C and D proposed systems that had unpressurized waveguide, incorporating a parametric amplifier. These systems had good noise figures and transmitter power could be reduced. Companies A and B, however, proposed systems that matched the numbers in the specification exactly. This procurement was in two steps, the first of which was qualification through the technical proposal. Companies A and B technically were disqualified. The reasons the customer gave for the disqualification were trivial reasons drummed up by the customer, just minor technical flaws in the proposal. Companies C and D went on to compete in price for the contract, and one of them eventually won the job with the lowest price. They, of course, had proposed what the customer wanted.

The Procurement Specification Syndrome (Propose What the Customer Wants)

When the customer issues a request for proposal, it's up to the companies to propose what the customer wants. An RFP and the specifications for a system do not complete the process. Often, they do not clearly say what the customer wants. If you have established a good rapport with the customer before the RFP is issued, it is probable that you know the intent in spite of what the words in the RFP might say.

The Case of the Negative Noise Figure

Once upon a time, a customer issued a request for a proposal for a radar system in which the system was required to have a -6 dB noise figure – which is theoretically impossible. This occurrence was in the early days of radar, when the definition of noise figure was not understood well. Company E decided to set the customer straight in their proposal, and they proposed a system with a +6 dB noise figure. Company E was disqualified from the competition because they did not propose what the customer had asked for. It is possible that the real reason for disqualification lay deeper than this technicality – but that reason was the one the customer gave. Other companies did propose a -6 dB noise figure assuming that they would straighten out the definition in the customer's mind later on. This example is one of not proposing what the customer wanted – at least what he requested – and thereby losing the contract.

Case of the Klystron Transmitter

Another example related to proposing what the customer wants was a response to an RFP for a shipborne radar system. In this case, the customer had indicated that he wanted a radar system in which the final power amplifier stage was a tetrode. The customer had had good experience with tetrodes and he felt that this radar system should have a tetrode. However, Company F, after an extensive investigation into the various microwave tubes available for this job, decided to propose a klystron system. When the customer read the proposal and listened to the technical presentations from the various competing companies, the case for the klystron was accepted. The facts presented for the klystron overwhelmingly supported its use in the particular system, and Company F won the contract. (Incidentally, in this particular instance, Company F backed up its prime proposal with one that used a tetrode transmitter – they covered all bets.)

I've just given two examples of not adhering to the customer's RFP. One resulted in the company losing the contract; the other won. Knowing when to deviate from an RFP and when to conform is the most critical decision in preparing technical proposals. Having established good communication with the customer prior to his issuance of the RFP makes solving this problem much easier.

The Equipment Specification Syndrome
(Provide What the Customer Needs)

Eventually, a company gets the contract to build equipment. With this contract comes specifications – you now have to provide what the customer needs. A definition of a loyal employee is one who doesn't do what the boss says to do, but does what the boss would say to do if the boss had all the facts. This situation is the one you're in as the company that has the contract. The specifications should not be followed blindly; rather, you should develop what is going to be a good system, and which is the system that the customer needs. One quote from Frosch – "When the customer is buying an equipment, he's doing so, so that he can fill a need that he had." Frosch is speaking from the viewpoint of the Department of Defense. He also states that ". . . the idea of a 'complete' specification is an absurdity. . ." A specification is simply a subset of points that hopefully describes equipment that can do the job that's desired. Occasionally, equipment can meet the subset of points and be totally inadequate for the job.

Case of the Radar without STC

Company G obtained a contract to build a high powered aircraft surveillance radar. The specification did not require sensitivity time control. (STC is a technique for reducing the sensitivity of the radar at close ranges.) This omission was probably an oversight on the customer's part but, whatever the reason, the specification did not require it. Anybody who has worked with high powered aircraft surveillance radars knows that they are capable of detecting birds and insects out to ranges of 30 to 50 miles. However, the specification did not require STC; the company built the system without STC. When the first prototype system was installed, lo and behold, you could not see an aircraft within 30 miles of the radar – because all the unwanted targets were saturating the display. The customer was not happy. There was a crash program to put STC in the radar but, as often happens with crash programs, the job was not done correctly. This problem was never fixed in the prototype system and it became a significant black mark against the system. It was a major reason why the system did not go into production.

The Modeling Syndrome

Preparing a model against which to test various systems, or to test alternate approaches to a solution of a given problem, can be very useful. Models can be particularly useful if employed to give an understanding of which factors are significant in solving a problem. But one must not blindly believe the result of a modeling exercise. The questions must be – is the result reasonable? What features made the result come out this way? Is the model comprehensive and accurate enough to guide further development?

Case of the Sidelobe Canceller Model

Sidelobe cancellers remove unwanted energy received through the main radar sidelobes from distant jammers. The basic method of cancelling this energy is to mount omnidirectional antennas near the radar antenna. The energy received by these auxiliary antennas is used to provide a bucking signal to cancel the unwanted jamming in the main channel. Company I had built a sidelobe canceller system that was about to undergo acceptance testing by the customer. The system was a two-loop system and the problem was to demonstrate how well it could cancel two jammers. The situation is such that, if one knows the exact location of the phase center of the main antenna, the jammers can be positioned where there is 6 dB better cancellation than that which occurs on the average. In fact, this company tried to put the jammers in this precise place, relative to the phase centers of the antennas in their model. However, the model was wrong – the phase center of the main antenna was not where it was thought to be and, furthermore, it moved as the main antenna rotated. The result was that the system didn't work at all. The problem was easily corrected by moving one jammer relative to the other. This action resulted in the average cancellation being 6 dB less than the ultimate that they had hoped to achieve, but at least the system now worked. The analysis in this case, as are too many analyses, was based on a model which was erroneous.

MTI Radar Models

The following is another example of the modeling syndrome. One of the unfathomable matters in radar design is that, despite the existence for years of good moving target indicator (MTI) radars, every time someone receives a contract to build a new MTI radar they start from ground zero. The first thing they do is to go to, for instance, Skolnik's *Radar Handbook* to find the reflectivity value of land clutter; then they start calculating. A few reasonable people easily could make a few simple measurements and set up a chart stating that, if you're going to build a radar with an x microsecond length pulse and a y degree beamwidth, it can detect aircraft in the severest land clutter if there is z dB of subclutter visibility. But so far, in the 20 years plus that good MTI radars have existed, nobody has made such a chart. The basic modelling calculations that attempt to predict where a radar will detect and track aircraft in land clutter are close to meaningless. I think that these analyses are bad because the basic data that goes into them are almost always wrong, inadequate, or both. These data include assumptions on the distribution of ground clutter, how stationary it is in the statistical sense, and how stationary the clutter distribution is as a function of range, and on the required probability of detection (P_d) on a single look. Does a radar with a one-second look interval need the same single look P_d as a radar with a ten-second look interval? Some persons think so. (See Chapter 3 in which Brookner points out the significant differences between single-look and cumulative-look detection.)

In the above case of MTI radars, it is unlikely that modelling can ever lead to correct answers. Careful measurements and correlation of data by some unbiased agency are needed; perhaps a government laboratory should take over the problem.

Beware the Committee

A committee can be a useful source of ideas, but it can also be irresponsible. Company K was developing a radar system which incorporated a sidelobe canceller. The customer felt that Company K did not have enough experience with this technique; therefore, the customer rounded up all the experts in the country to form a committee to advise Company K. The first member of the committee said, "Aha, the last time we built a sidelobe canceller we used such and such a technique; therefore, you must use that technique." The next member of the committee said, "Ah, but we used a different technique and you should definitely have that in your system." Another member of the committee said, "but if the jammer does this and that and the other thing, those first two techniques won't work and you have to have this third technique to cope with that." So, Company K was faced with combining all of these requirements into one system. The techniques were added one on top of the other. Company K ended up with a system design with many modes of operation, but which was too complex to be built and operate, and too expensive to sell.

Committees have many good ideas, but it is not their responsibility that the system then work or be cheap enough to sell. Thus, when you see a committee of experts coming your way, you immediately must do everything you can to get strong enough to fight them or, if you prefer, filter their ideas so that you can produce a useful system. In this particular example, as the system got more and more complex, the consultants started arguing that each other's techniques would not work. Finally, the customer fired the committee and Company K went on and built the system the way they thought best (and it worked).

Do Not Extrapolate from Failure

One has to be very careful that, if something goes wrong with a radar system or for some reason it fails to do what it is intended to do, the cause of failure is determined correctly. A typical problem that has been perpetuated over the years is related to MTI (moving target indication) radar systems; that is, if an MTI radar system is unable to track an aircraft through clutter, it is assumed that the MTI system does not have enough subclutter visibility. In fact, it was not recognized that it was the basic system which was not working. The customer who specified the unsuccessful system, when next writing a specification, will increase whatever subclutter visibility that was required the last time by 20 dB or so. In almost all cases of MTI radar systems that are unable to track aircraft targets through land clutter, there is a flaw in the basic system (that is, when the systems were developed, some basic precept of building an MTI radar system was ignored). The problem wasn't that x dB of subclutter visibility wasn't enough; in practice, the systems had virtually no capability of detecting aircraft in clutter, no matter how strong the signal-to-clutter ratio. The x dB (whether x is 20, 25, or 30) of subclutter visibility would have been adequate for most of the systems if, in fact, the systems had had the subclutter visibility that was claimed for them. The moral is – don't extrapolate from failures. Rather, specify new systems as extrapolations from previous successes.

Chapter 6

Fowler

Pitfalls of Radar System Development

The Defined Roles – Thresholds

In late 1968 the commander of the Israeli tank forces in the Six Day War visited the US and received a briefing on the Army's new electronic gunfire control system for tanks. The system included a digital computer, gyros, stabilized optics, and all of the very latest in US fire control technology. When the General expressed enthusiasm for this development, he was asked if he thought that his army could maintain this complex system in the field recognizing the difficult environment in which tanks operate. The general responded that the questioner had asked the wrong question. The proper question was would this sophisticated system provide reliable first round hit of small hard targets? If it would, then it could be maintained even if that maintenance required special crews, helicopter supplied spare parts, or whatever else. But if the new system only reduced the correction required to achieve second round hits, then he wouldn't consider adopting it because his gunners regularly got second round hits with their existing (and much simpler) fire control system. The general had clearly identified a performance *threshold;* a *step function* in the plot of equipment performance vs. real capability. It is fair to say that the US system design was not based on achieving this threshold but instead was geared to achieve the best that the technology would permit.

Although incremental improvements in performance frequently provide corresponding improvements in capability, there is in many cases this threshold effect. We should spend the bulk of research and development funds in seeking and then trying to achieve such thresholds, thereby minimizing our efforts in the flat and linear regions. For example, improving the performance of an MTI system on a radar by 10 dB is not worth very much if 20 dB are required to provide reliable coverage of targets in the regions of interest. And no degree of mechanical cleverness could have achieved the beam agility threshold of electronic scan.

The Proposed System – Feasibility

In the middle 1960's Company A proposed a modification to a non-coherent airborne fire control radar to provide a look-down capability to "see" aircraft against a ground-clutter background. Questions were raised as to the level of performance achievable in this system, so a feasibility test was required. "Feasibility" was established in a few tests run in a far western desert region and production was approved. When the first production unit of the system underwent routine testing in a different part of the country, where the clutter was stronger and there was automobile traffic (non-existent in the desert region), the system performance was found to be totally inadequate. As Barry Duggan said [1], "The wear on a hypothesis is proportional to the distance of extrapolation from the experimental results." Or as Gene Fubini said, sometimes a system whose feasibility has been established "won't stay feased."

Reliability – Theoretical and Actual

As the subject of reliability has developed, it has produced many fine and lasting benefits, particularly for electronics. However, because the results contain numbers, one tends to believe that they are precise. But these numbers are frequently unreliable because they are extreme extrapolations of measured component data or because they are based on averages of component performance in good and bad circuit designs. One thing is clear – where the designs are not solid or where some of the components are not "stable" (e.g., most new microwave power tubes), the actual *Mean Time Between Failures* is much lower than that calculated and the failures are unrelated to the λ's that determine the calculated value. This point has been particularly substantiated in radars.

Dr. John Allen and colleagues at Lincoln Labs studied airborne radars and their reliability. Figure 1, Allen's chart taken from his report [2], is a plot of the contractually specified MTBF's of different airborne radars as a function of the year the program was started. Also shown is the achievable MTBF on those radars for which data existed. This graph led Allen to speculate that, although the *specified* MTBF has steadily increased, the *real* MTBF always has turned out to be *10 hours.*

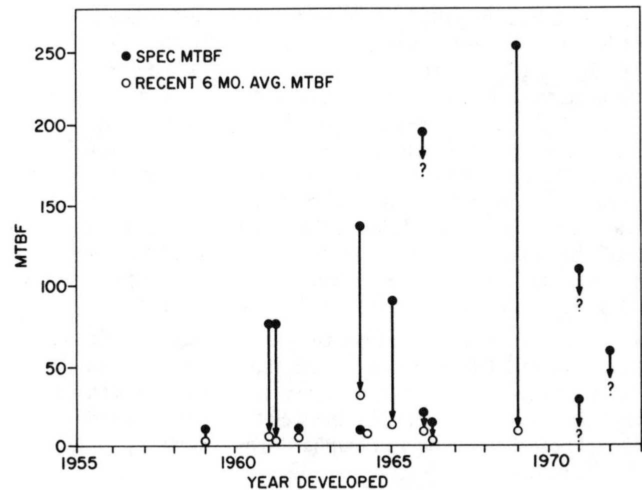

Figure 1 Allen's Chart

On the other hand, when designs are solid and the components are proven, actual MTBF exceeds the calculated value. The life of the electronics in many satellites has exceeded that predicted by a large margin. When radars have mature designs and components, they too will exceed the "specified" and "calculated" MTBF's. But to achieve such an end, a design pitfall pointed out by former Navy R&D Secretary, Robert Frosch [3], must be avoided – the urge to make everything equally reliable (or, rather, equally unreliable) like the one-horse shay. For example, in a recently designed radar, a highly reliable Zener diode in a small control circuit card in a transmitter had the usual derating factors for the assumed conditions. However, under unusual conditions a transient appeared and blew the diode; to make matters worse, the diode was located in an inaccessible spot in the transmitter. The next size diode would have lasted forever. There are two lessons to be learned. First – for most components, and particularly those in inaccessible places, play it very, very safe. Remember the story of the man who took Carter's Little Liver Pills for fifty years; when he died, they had to cut him open and take out his liver and beat it to death. Second – put those questionable parts where they can be easily replaced. Tube type TV's are still on the market; a probable reason is that almost anybody can find and replace a bad tube.

The Disappearing Rationale

Company B had the job of designing a universal signal processor for a series of radars. One of the elements of the processor was a velocity shaping MTI canceller for use against ground clutter and chaff. One of the radars was a C-band height finder which, like the other radars, had both a low and a high PRF mode. Calculations indicated good anti-chaff capability in the high PRF mode, but a very marginal capability in the low. After due consideration, it was decided to include in the C-band radar the low PRF mode in the canceller for the sake of commonality with the other signal processors; all were lower frequency radars in which performance at the low PRF mode was still significant. After the concepts were established and the basic design completed, the responsibility for the program was transferred to the production division of the company, as was customary in those days. Some years later the original designers saw a report indicating very poor anti-chaff performance of this radar. It was then learned that, as a cost reduction feature before the radar went into production, the high PRF mode was eliminated. The coordination between government, the radar's prime contractor, and the signal processing company was sufficient to eliminate the high PRF mode from the processor; but, since the rationale had been lost, the relatively useless low PRF mode canceller was faithfully produced for each radar.

Company C had the job of designing a combat surveillance radar for the Army. Stringent weight limits were set so that an individual soldier could carry the package on long hauls. This requirement necessitated extraordinary efforts in design, and the use of exotic materials and processes with substantial penalties in ruggedness and cost. When the equipment was delivered, a seasoned operation officer seeing it for the first time said it looked too flimsy for combat use. When informed that unusual measures were needed to get the weight down so that a soldier could carry it, he said that all equipment of that sort is carried in helicopters and jeeps these days. Sure enough, when the equipment went into the field, all extensive carries were by helicopter or vehicle.

These stories are examples of what is known in some circles as the Fourth Law of DOD operation, which states that the rationale on which the decision to develop a new system is based has largely disappeared or become inapplicable well before the system is deployed, because of changes in emphasis, roles, system parameters, and/or features.

Hardware vs. Software

A system designed by Company D many years ago has a mechanically directed antenna system; control of the pointing is by a set of foot pedals used by the operator. Its modern successor, a system designed by Company E, has an electronically scanned antenna the pointing of which is controlled by a general purpose digital computer. Its pointing sequences, unlike the cumbersome mechanical system, are essentially infinitely variable because of the flexibility of the digital computer. In a special test not too long ago, variation in the pointing sequence was required. In the mechanically scanned system, the operator had to push the foot pedals somewhat differently than normal. In the computer controlled system, the problem had to be referred to a programmer who took three days to prepare a new tape. A lot of those old analog systems had very *flexible inflexibility;* some of our new digital systems have very *inflexible flexibility.*

Many electrical hardware designers and equipment operators are not yet fully aware of the commanding position assumed by the software programmer. This statement is not to take sides in any real or imagined contest between hardware and software persons, but rather to point out the untenable position in which the operator is placed. With the old systems, the operator could push some buttons, turn some knobs, or even change some wires, to adapt to the unforeseen and unexpected situation. In today's modern "flexible system," that operator has to go to a programmer who frequently needs access to another computer to write a new program. As Bob Frosch said, "It is *almost* true that no military system is ever used for the precise purpose for which it is designed" [4]. Those in the radar business, like many others, have a lot to learn in designing systems so that a normal operator – in contrast to a skilled programmer – can interact with the computer in a natural and flexible way to adjust and control the system under changing environments and circumstances.

Testing

In the late sixties, the Defense Department came to realize that even extremely detailed paper designs and analyses were no proof of a workable and useful system. Extreme concurrency of production with development led to situations such as that which existed in a very complex multi-million dollar radar avionics system managed by Company F. Nearly 100 systems were rolling down the production line with a design which was known not to have a single sub-system which did not require significant redesign before it would work even at room temperature in a laboratory environment. This situation leads to solutions which cause the minimum impact to already produced hardware and thus are not normally optimum from performance, reliability, and maintainability standpoints. For many this demonstration was the conclusive proof of the fallaciousness of the approach so prevalent at that time – troubles encountered in the development model can be fixed up easily in the production model. The sensible comment that "if it's easy to fix, fix it; if it's not, we certainly don't want to go into production" began to be heard from more and more levels in the Defense Department. This reaction in turn led to the milestone (or, as some of the impatient service people liked to say, the "millstone") concept. Under this procedure, certain key accomplishments and demonstrations of performance (milestones) were established as a requirement before authorization to enter production would be given. This approach had the salutary effect of focusing government and company management attention on the capabilities and shortcomings of the R&D hardware rather than on efforts associated with tooling up the production lines.

But as the pendulum sensibly swung away from paper proof of design toward hardware verification, attention, interest, and dollars shifted to those who specified the tests and certified the suitability of each piece of hardware to enter production. Already one can sense the start of the clique. In fact, a reasonable prediction is that, within 5 years, there will be a society of testers and evaluators of electronic hardware which will hold annual symposia and publish their own journal. The danger, of course, is that, as the spotlight focuses on them, the persons who have to certify readiness for production start to get "antsy" as the decision date approaches, and feel the normal reluctance to sign off and certify that something is OK. Figure 2 shows for a system designed by Company G a plot of the hours of testing required before production authorization vs. time; the actual years indicate the recent, rapid movement of the pendulum. Also shown are the accomplishments of test hours by the contractor. Note that, as the contractor approached each contractually established test standard, the standard itself was increased. This plot demonstrates not only the

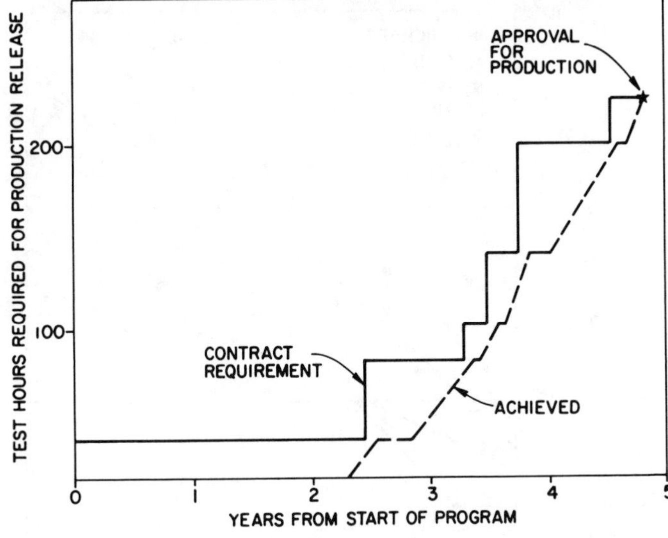

Figure 2 The "In" Decision Chart

"antsy-ness" problem but also the growing power and authority of the certifiers. It even suggests a name for the new organization – *The Society of Testers of Operational Performance* or *STOP.*

It is indeed a difficult job – but a vital one – to distinguish between the proving out of the usefulness and readiness for production of a system and testing for the sake of testing. It must be noted that very often the "key specifications" do not reflect even remotely the truly important parameters and features of a useful piece of hardware.

Computer Print-Out vs. Back-of-an-Envelope

Some years ago, one of the services presented its case for a new defensive missile fire control system that contained many new and expensive features. To prove the case, the computer analysis of an elaborate wargame scenario was used. This game compared the performances of the proposed and the existing systems against an attacking force of hundreds of bombers, air-launched missiles, fighter escorts, and stand-off jammers. The results showed an overwhelming superiority of the new system, which, in turn, assured its continuing development and production. But some engineers were curious to learn which of the new features was principally responsible for the outstanding success. Was it the longer detection range of the radar or was it the longer range of the missile? Was it the multishot capability of the system or . . .? None of these questions, however, was answerable from the published description of the war game. After a detailed examination of sub-subroutines in the elaborate program, it was discovered that the outcome of this entire war game was not significantly affected by any of these major features but was dominated completely by the antenna azimuth sidelobe levels ascribed to each system and the assumed characteristics and locations of the stand-off jammers. Very low sidelobes of the new system were assumed, whereas the very high measured sidelobes of the old system principally were due to the installation – a relatively minor investment could have brought the sidelobe levels of the old system close to those assumed for the new.

This example is but one of many illustrating the Fifth Law which states that, "Any war game, systems analysis, or study whose results can't easily be explained on the back of an envelope is not just worthless, it is probably *dangerous.*"

It has been said (by me) that the computer is the enemy of reason. Most everyone has heard the phrase "garbage in – garbage out" applied to computers. It is, however, much more important to avoid "garbage in – gospel out."

Use the Range Equation on All the Targets

As the science of radar has matured, the sophistication of calculations of radar performance has increased so that today's designer routinely calculates performance against selected types of scintillating targets with appropriate statistical models; includes all sources of noise temperature affecting receiver performance; and accurately assigns integration gains and losses of scanning systems, signal processors, etc. It is even relatively common to check performance degradation due to rain, ground clutter, sea clutter, and chaff. Even so, there occasionally occurs a situation in which there is a failure to use the range equation for all the "targets."

Companies H and I designed systems to detect moving personnel at ranges of two miles or less; frequent shielding by hedgerows and vegetation, however, introduced one way attenuations of 20 to 30 dB. The approach in each case was to modify an air search radar capable of detecting a one square meter aircraft at a range of about 60 miles to permit detection of slow moving (one to two miles per hour radial velocity) targets. This over-powered radar would then detect a one square meter man moving behind the attenuating vegetation or hedgerow. Table 1 lists the radar cross section of objects which, if moving radially one to two miles an hour, would be detected by the radar at various ranges. Note the capability of detecting a single man out to almost six miles behind a 20 dB screen and to two miles behind a 30 dB screen. However, also note that a few honeybees *above* the screen have the requisite cross section at the ranges of interest and a full grown pigeon practically draws sparks out of the signal processor [5]. Unfortunately, this simple direct view of the problem was obtained after the systems were built and tests showed them to be totally ineffective in the presence of birds and insects.

Another interesting example of the need to fully utilize the radar range equation is found in synthetic aperture radars. Early systems produced imagery which looked like poor photographs. It was widely assumed that if the resolution were increased, the imagery would become more and more optical in quality; for example, the image of an aircraft would have the outline shape of an aircraft. However, this clarity was not manifest; the fuselage and wing edges were not visible most of the time. Bill Caputi of AIL [6] created the model (Figure 3) of a C-121 Constellation aircraft. The patterns and arrow indicate directivity and the numbers indicate the radar cross sections in square feet for a 3 foot by 3 foot resolution patch. This model indicates the highly directional nature of the return from the fuselage and wing edges. When the

Range (miles)	σ Detectable (dBSM)	Targets of Interest: Original Design	Targets of Interest: Modified Design	Other Targets
64	0	Small Aircraft		
32	-12			
16	-24			A Pigeon
8	-36		Man behind 20 dB screen	A Sparrow
4	-48			Four Bumblebees
2	-60		Man behind 30 dB screen	
1	-72			Six Honey Bees

Table 1 L-Band Radar Performance

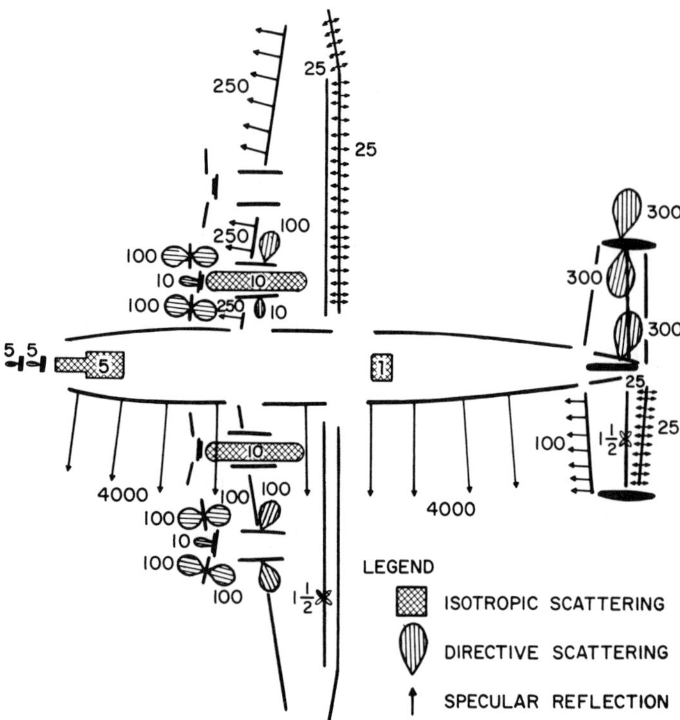

Figure 3 Caputi's Model

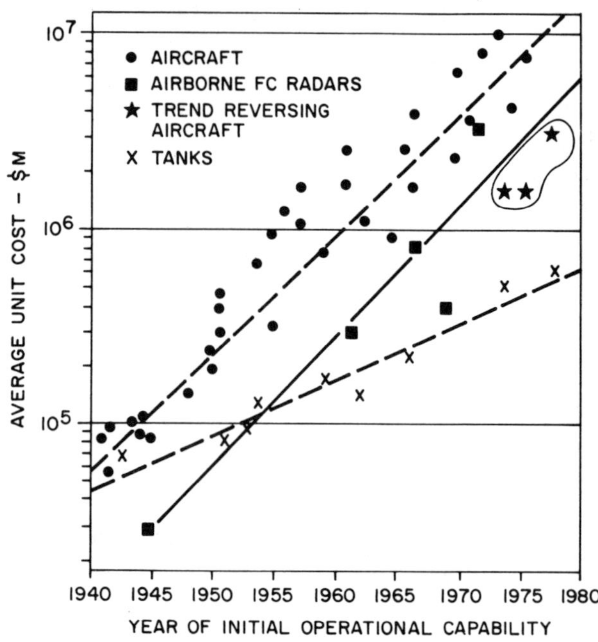

Figure 4 Augustine's Chart

radar beam is normal to these sections, cross sections ranging from 25 to several thousand square feet are obtained. However, a few degrees off normal, the cross sections drop down to the order of 0.001 square feet. This level compares with 0.1 to 1.0 square foot cross section for a resolution element of vegetation (for example, adjacent to a runway or parking area) and 0.03 square foot for the concrete apron or runway. Thus, the aircraft outline "shapes" obtained occasionally in earlier imagery resulted from returns from several non-aspect sensitive scatterers and from normal incidence to some aspect sensitive scatterers fortuitously connected by a "poor" resolution system. When the resolution is increased, the detectable scatterers become separated. Use of the radar range equation on each "target" permits an accurate calculation of the image to be obtained from any viewing aspect, thereby taking the "mystery" out of SLAR imagery.

Costs

The extreme seriousness of the continually increasing complexity and associated high cost of military hardware need not be discussed here, but some aspects of the problem do deserve comment. In 1971 Norman Augustine of LTV published [7] the chart shown in Figure 4. This chart shows the unit cost of each new model of aircraft and tank as a function of the year in which it was introduced. The cost shown is the average unit cost over the first production buy; the year shown is the one in which initial operational capability was achieved. The data beyond 1970 represents program projections. Note that the ordinate is logarithmic. The straight lines fitted to the data show a 3 dB per decade increase in the cost of tanks and a 6 dB increase for aircraft. In non-engineering language, the cost of tanks has doubled every ten years while the cost of aircraft has quadrupled.

Augustine has also noted [8] that US World War I aircraft nicely straddle the backward extension of the 6 dB per decade line and the Wright Brothers plane falls right on it. Further, a similar plot of commercial airlines, starting with the Ford Trimotor in the twenties, going through all the DC numbers, and including the superjets, shows the same slope [8]. Apparently, the aircraft industry is just a 6 dB per decade industry. This observation leads one to speculate that the airlines have the same kind of problem that the Defense Department has, and wonder if they are aware of it. (Incidentally, five percent per year inflation corresponds to about 2 dB per decade or a 60% increase every ten years; Ford automobiles, starting with the Model T and going through today's models, exhibit a 1.6 dB or 45% increase per decade. This figure would be very encouraging but for the fact that maintenance costs are not included.)

Mr. Augustine has pointed out the consequences of a forward extrapolation of the data in Figure 4 – if one assumes that the budget providing weapons for our nation's defense is held at today's level (based upon the current mood of Congress and the country, this view may well be optimistic, even including inflation), on the 18th of April in '36 (that is, the year 2036) our Air Force will be able to buy precisely one aircraft. Our armored forces will have been reduced to a single tank nine years earlier under the same assumptions. If this situation doesn't worry you, consider that an extrapolation of about 90 years produces an aircraft whose cost equals the US gross national product. But there's some evidence that the lesson is being learned. There are three trend-reversing aircraft indicated as stars on the chart – the A-10, the new close-air-support aircraft which is under development by Fairchild Hiller; the Northrup International Fighter, or F5E; and lightweight fighter formerly under competitive prototype development by Northrup and General Dynamics, Fort Worth. Also shown are the costs of a World War II airborne fire control radar system and five more recent systems [9]. The trend indicated is about 7 dB per decade. At this rate, the entire current annual defense procurement budget would be needed to buy one of these systems in the year 2033.

There are several factors that create or account for these high costs. First, the assumption of large production quantities and good learning curves can lead to a unit cost which disguises the inherent cost of the system. When more realistic budgets are recognized, the results are doubly painful. Most of the time, the government estimates its needs on the high side; the company marketing staff, with its necessary enthusiasm, extrapolates this figure upward in what is sometimes called a "Sputnik excursion." Although one would not quibble with the metabolic necessity of marketing organizations to calculate such an excursion, one could point out the equal importance of making a "McGovern excursion" before settling on a saleable *design,* let alone the sales forecast.

For this reason and others, a system frequently is designed that has an inherent cost that is too high by any measure. Some years ago, Company J had designed a radar, the projected unit-price of which (in a reasonable quantity) came out substantially higher than was saleable or warranted by the utility of the equipment. An intensive effort was made to find simplifications – portions of the hardware that could be substantially simplified or even omitted. Over a period of time, these ideas were generated and submitted to the pricing group. Each time the cost reduction was disappointingly small. When the cost decrease line was extrapolated to the point where nothing was left of the equipment, 40% of the cost remained (Figure 5). This Hysteresis effect (the Second Law) comes about because the estimator does not proportionally adjust the basic design or the "overhead" associated with the program when backing off incrementally from a point design and a program cost.

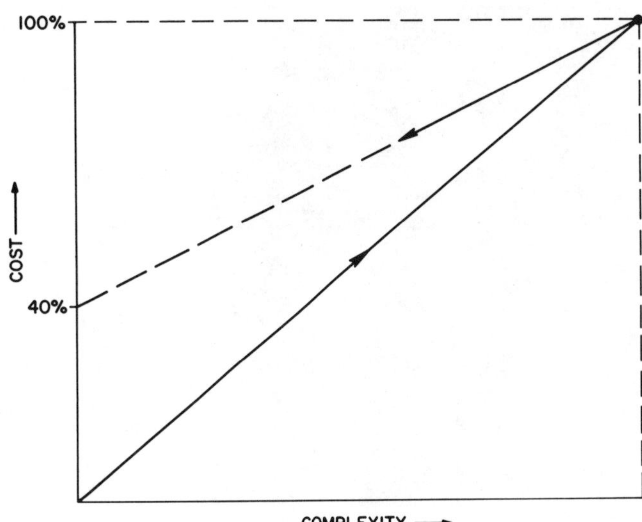

Figure 5 Fowler's Second (Hysteresis) Law

Another factor which insidiously increases the cost of a system is the tendency to design features which have a cost proportional to the overall cost of the system. For example, the Condor missile is a very expensive missile; its TV/control data link represents about one third of its cost. When a TV/control data link for the much cheaper Walleye missile was designed, (with nearly identical functions), a cost limit of about one third the total missile cost was established almost automatically – and was achieved.

Similarly when the application of sidelobe cancellers to several new radars was considered, a 2-loop canceller seemed appropriate for the one million dollar radar, a 6-loop canceller for the \$3M radar, and one with over 20 loops for the \$10M one.

The Third (or Ratio) Law states that the cost of a feature or capability in a system tends to be proportional to the cost of the entire system and is not related in any absolute way to the feature or capability being provided.

Of course it frequently makes sense to use more expensive features on more expensive systems. However, this fact does not explain why the mess tables on nuclear powered ships should cost more than those on conventionally powered ones. It's important to remember that the old adage "the whole equals the sum of the parts" applies to costs as well.

Unexplainable Radar Phenomena (URP's)

There are two unexplainable radar phenomena (or URP's) that deserve comment.

One phenomenon is the remarkable fact that every radar program is on an optimum schedule with respect to cost. In thousands of cases where the government has examined the cost associated with altering the program schedule with a contractor (or a prime contractor has examined with a sub-contractor), it has turned out that either speeding up or slowing down the program costs a substantial amount of money. In other words, *every single* radar program, regardless of technical, manpower, or component delivery problems, is always kept on the optimum schedule. Unbeknownst to most, there is a recorded case in which one radar company simultaneously (and successfully) negotiated increased costs for speeding up and for slowing down the building of the same hardware. The government, because of a shortage of funds, directed the contractor to slow down the rate of expenditure, stretch out the delivery of the hardware, and to submit a proposal to cover any costs. A few months later, because of a sudden urgent need for the hardware, the government directed the contractor to speed up deliveries. With the normal delays, both contract changes were negotiated at the same time. In any event, one of the unchallengeable (but unexplainable) successes of the radar industry is the one hundred percent achievement of optimum cost/schedule programs.

For reasons that no one – scientist or astrologer – has ever fully understood, radars, more so than radio stations, Elint systems, computers, or even oil refineries (if there still are such things), have a tendency to catch on fire. For various reasons, including superstition, this topic is not dwelt upon, but to illustrate the point, here are some before and after pictures. Figure 6 shows the first civil air traffic control radar installation at Queens College in Queens, New York; Figure 7 is a reproduction from the 28 May edition of the *Long Island Star;* and Figure 8 shows the hardware after the fire.

Figure 9 shows the FPS-85, the world's first large phased array radar, at Eglin Air Force Base in Florida. Figure 10 shows the famous fire in the mid-sixties.

To indicate that lessons indeed have been learned, Figure 11 shows an artist's conception of the Cobra Dane radar installation on Shemya in the Aleutians. Ignore the latest state-of-the-art phased array being designed and built by Company K; instead, focus on that large, contractually required, utilitarian underground water tank.

Figure 6 Queens College Radar Installation

Figure 7 Queens College Radar Installation on Fire (from *Long Island Star*)

Figure 8 Queens College Radar Installation (after fire)

For reference, Fowler's six laws affecting DOD system acquisition are listed below.

First Law – If the decision cycle time to initiate a program exceeds three years, the system becomes unstable and the decision is unlikely to be made at all.

> **Note** – Average Washington tour of military and civilian decision makers is about three years.

Second Law – (Hysteresis Law) – Once a system has been defined and costed, the cost reductions resulting from simplifications and deletions follow a trend which results in 40% of the cost remaining when none of the system remains.

Third Law (Ratio Law) – The cost of a feature or capability in a large system tends to be proportional to the cost of the entire system and not related in any absolute way to the feature or capability being provided.

Figure 9 FPS-85

Figure 10 FPS-85 (on fire)

Figure 11 Artist's Conception of Cobra Dane; note the large water tower (which is actually underground)

Note – More expensive mess tables are used in nuclear powered ships than in conventional ships.

Fourth Law – The rationale on which the decision to develop a new system is based has largely disappeared or become inapplicable well before the system is deployed, because of changes in emphasis, roles, system parameters, and/or features.

Fifth Law – Any war game, system analysis, or study whose results can't easily be explained on the back of an envelope is not just worthless, it's probably dangerous.

Note 1 – The computer is the enemy of reason.

Note 2 – The Services have learned to "out-analyze" Systems Analysis and, in the process, themselves.

Sixth Law – An "integrated system" development program (although sometimes necessary) hampers the development of both the vehicle and the sensors and weapons.

References

[1] *The Journal of Irreproducible Results, Vol. 16, No. 2,* p. 40, December 1967.

[2] Allen, J.L.; Lebou, I.L.; and Muehe, C.E.: *Report to DDR&E on Tactical Airborne Radar (U),* Fig. 9, p. 71, 11 May 1971. (Secret)

[3] Frosch, R.A.: "Remarks by the Assistant Secretary of the Navy," *IEEE Transactions on Aerospace and Electronic Systems,* p. 471.

[4] Frosch, R.A.: "A New Look at Systems Engineering," *IEEE Spectrum,* p. 27, September 1969.

[5] Pollon, G.E.: "Models of Clear Air Clutter for Radar Analysis," *16th Tri-Service Radar Symposium,* 1970. (Secret)

[6] Caputi, W.J.: (AIL) *Personal Correspondence*

[7] Augustine, N.R.: "An R&D Perspective of Land Warfare" (U), *Journal of Defense Research, Series B, Vol. 3B, No. 3,* Fig. 46, p. 272, Fall 1971. (Secret)

[8] Augustine, N.R.: *Personal Correspondence*

[9] Fubini, E.G.; et al.: *Report of the Defense Science Board Task Force on Avionics* (U), Fig. 15, p. 36, February 1973. (Secret)

Present and Future Trends in Radar Systems

There are emerging trends toward phased array systems, digital processing, and solid state transmitters. These trends are illustrated by the recent completion of the cost-effective Cobra Dane phased array, the awarding of a production contract for the construction of 10 AN/TPQ-37 artillery-locating phased array radars, the production of 12 AN/TPN-25 (Figure 86 in Chapter 1) limited-scanned phase-phase array radars, the announcement of the planned installation of the AN/SPY-1 (Figure 83 in Chapter 1) phase-phase array on board AEGIS ships, the development of the AN/TPS-59 (Figure 90 in Chapter 1) solid state radar using one dimensional phase scanning (which was recently selected by the Belgian Air Force for air defense), and the initiation of the construction of the PAVE PAWS (Figure 17 in Chapter 1) solid state phase-phase array radar. The Naval Air Development Center (NADC) has under development a solid state L-band phased array for aircraft applications in which it will eliminate the need for a rotodome antenna [1] ; see Figures 4 and 5 of the "Present and Future Trends" section of Part 5. The Naval Research Laboratory has under development the module for an X-band airborne phased array; see the "Present and Future Trends" section of Part 5. This tendency towards phased array systems does not mean that simple mechanical and electromechanical scanning systems are on the way out; far from it, the majority of radars still will use these simpler scanning systems and will for some time to come. The design of the World War II AN/MPN-1 precision approach radar which uses the Eagle Scanner is still being used in updated versions; the AN/FPN-62 (Figure 74 in Chapter 1) is a solid state version (except for the transmitter tubes) of the original system, using the same antenna design.

More interest is being shown in space-borne radars, with the space shuttle (to be operational in 1980) making such systems more economical. Consideration is being given to the use of space-borne radars for air traffic control, surveillance for the military, national resource monitoring, high resolution earth mapping (an extension of the NASA SEASAT radar), and a bistatic anti-collision system for coastal ships (the transmitter being on a geosynchronous satellite with a passive receiver on the ships) [2, 3] . The deployment of large low altitude reflectors for over-the-horizon air traffic control radar systems also has been suggested [2] .

A space-borne radar designed for observing other space objects against the sky background has the interesting potential for very low receiver noise temperatures ($\leqslant$ 30 K) when using cryogenically cooled receivers [4] . This action permits the use of low power transmitters.

The present and future trends in signal processors, antennas, solid state transmitter components, tubes, and laser systems are discussed in Parts 2 through 6.

References

[1] Davis, M.E.; Smith, J.K.; and Grove, C.E.: "L-Band T/R Module for Airborne Phased Array," *Microwave Journal, Vol. 20,* pp. 54-61, February 1977.

[2] Bekey, I.; and Mayer, H.: "1980-2000 Raising Our Sights for Advanced Space Systems," *Astronautics and Aeronautics, Vol. 14,* pp. 34-63, July/August 1976.

[3] Kurzhals, P.R.: "New Directions in Space Electronics," *Astronautics and Aeronautics, Vol. 15,* pp. 32-41, February 1977.

[4] Davidheiser, R.: "Is There a Josephson Junction in Your Future?" *Microwaves, Vol. 16,* pp. 50-55, March 1977.

Part 2
Signal Waveform Design and Processing

Chapter 7 by Sinsky gives the first simple description of the well-known ambiguity function, knowledge of which is essential in designing large time-bandwidth waveforms. Instead of typically couching the description of this function in mathematical terms, he presents it in terms of its physical meaning. Cook in Chapter 8 discusses the advantages and disadvantages of large time-bandwidth waveforms. Both Sinsky and Cook cover waveform design, give several examples of large time-bandwidth product waveforms, and introduce the matched filter concept.

The remaining chapters in this part deal primarily with the processing of large time-bandwidth waveforms. The implementation of the matched filter processor by the use of the Fast Fourier Transform (FFT) is described very lucidly in Chapter 9 by Sheats. He also gives the first simple and clear explanation of the clever pipeline hardware implementation of the FFT. Purdy in Chapter 10 presents several analog and digital state-of-the-art techniques for processing the most commonly used large time-bandwidth waveform, the linear chirp waveform. Purdy also reviews the state-of-the-art of digital memories, logic, and A/D converters. A novel and efficient technique for processing large time-bandwidth product linear chirped waveforms developed by Perry and Martinson is described in Chapter 11.

Chapter 12 by Worley reviews the state-of-the-art of the relatively new Surface Acoustic Wave (SAW) devices. Finally, in Chapter 13 Haggarty et al., describe a technique for generating and processing an extremely large time-bandwidth product (2,000,000:1) linear chirp waveform. The technique permits the generation of the signal from a very stable 5 MHz oscillator. The slope of the chirp waveform can be controlled (essentially continuously) so as to allow correction for echo signal waveform distortion due to target motion. The change in slope is made while still maintaining coherence relative to the master oscilloscope.

The "Present and Future Trends" section of Part 5 reviews the state-of-the-art (through mid-1977) of digital memories (RAMs, ROMs, PROMs, EAROMs, EEROMs, EPROMs, CCDs, bubbles, EBAMs, and Josephson Junctions); logic (e.g., DCCDs and Josephson Junctions); A/D converters; microprocessors (μPs) and microcomputers (μCs); CTDs, CCDs, and BBDs (as Fourier analyzers, bandpass filters, MTIs, buffers, pulse Doppler processors, and delay lines); SAW devices not covered in Chapter 12 (e.g., resonator filters and oscillators, delay line oscillators, bandpass filters, burst waveform processors, fixed and variable delay lines, Fourier transform processors, pulse compressors for arbitrary waveforms, generators of chirp waveforms of arbitrary slopes, and MTI processors); and acoustoelectronic devices (convolvers, matched filter processors, burst waveform processors, amplifiers, and buffers). A clear explanation of all of the above devices is given, as is a projection of these and other digital and analog technologies. The new FFT algorithm (WFTA) is also reviewed.

Chapter 7 Sinsky
Waveform Selection and Processing

This chapter is intended to familiarize the radar engineer with the general methods of waveform selection and the signal processing techniques that can be used with the various waveform types. The presentation is divided roughly into two parts. The first part is to familiarize the reader with the criteria for waveform selection and then introduce the necessary tools for analysis. The required mathematics are introduced only after a physical description of the problem is presented, in order to permit an intuitive understanding of the subject matter. The radar ambiguity function is introduced as the primary analytical tool for the analysis of waveform performance. An understanding of this simple function should give the radar engineer an insight into existing system performance and permit him to select the most suitable waveforms for future radar systems. The analytical techniques presented herein should enable the reader to better understand the existing literature, much of which assumes a familiarity with the subject matter.

The second part of the presentation describes several of the most commonly used waveform types. The performance of these waveforms is described in terms of the radar ambiguity function. Commonly used methods for generation and processing of these waveforms are presented and described.

This chapter was prepared for the practicing radar engineer who must understand the principles of waveform selection and processing but who is not necessarily a mathematical wizard. For this reason, the presentation begins with the fundamental physical concepts basic to all electrical engineers and it derives only sufficient mathematics to permit a better physical understanding of the problem. Tedious mathematics is relegated to several appendices.

Waveform Design Considerations

There is no one waveform that satisfies all requirements. If there were, then this subject would be trivial since someone would have determined this optimum waveform and someone else would have designed the necessary signal generation and processing scheme. The radar waveform selection process begins when the targets and their environment have been established and when the required target information has been chosen. The familiar example of an airport surveillance radar illustrates this point. The targets are aircraft the cross section, range, velocity, azimuth and elevation of which are known, within limits. The environment includes ground clutter, multipath, refraction, weather, and thermal noise. The required information includes each aircraft's range and azimuth to some desired degree of accuracy while minimizing the probability of false target indications resulting from the environment. The optimum transmitter waveform for this application must contain sufficient energy to achieve detection on the smallest aircraft at the longest range. It must have sufficient bandwidth to provide the necessary range accuracy and resolution, and it must have a time duration long enough to permit velocity discrimination between the moving targets and the ground clutter.

Figure 1 illustrates the type of problem which arises. The target of interest is at a range, R_o, moving at constant velocity, V_o. The environment is composed of other radar reflectors the ranges and velocities of which are (R_1, V_1), (R_2, V_2), . . . (R_N, V_N). The symbol V for velocity is defined throughout as a scalar quantity, the

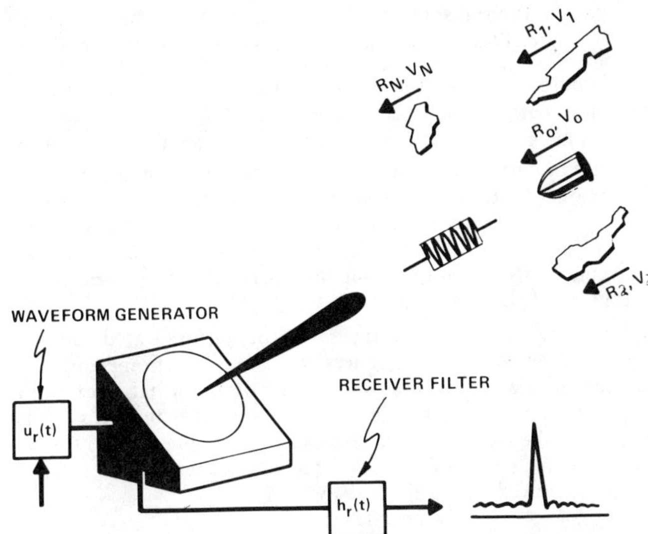

Figure 1 **Waveform Selection and Processing**

sign of which is positive when the object is moving toward the radar. The magnitude of V is equal to the rate of change of slant range. There is a frequency shift of the transmitted signal spectrum associated with the target motion; this shift is designated by the symbol f_d and is defined as positive for targets moving toward the radar. It further is assumed that the transmit signal spectrum is uniformly shifted by the frequency f_d due to relative target velocity V. This approximation is a good one when the ratio of signal spectrum extent to carrier frequency is relatively small. The effects of higher time derivatives of target motion such as acceleration are also assumed to be small over the time duration of the waveform. These latter two assumptions are valid for the large majority of radar problems; they result in considerable simplification of the mathematics. For this reason it makes sense to study the radar waveform design problem with these two constraints before attempting to consider the effects of wide band waveforms on accelerating targets. (For a discussion of the situation where the assumption is not true that the signal spectrum is uniformly shifted in frequency, see Chapter 13.)

Equation (1) defines the relationship between slant range, R, and the 2-way signal delay, τ; Equation (2) defines the relationship between target velocity, V, and the corresponding Doppler shift, f_d.

$$\tau = +2R/c \tag{1}$$

1 nanosecond is 0.492 foot

$$f_d = +\frac{2V}{c} f_o \tag{2}$$

1 hertz is 1 foot per second at f_o = 492 MHz

where:

- τ is the 2-way signal delay in seconds.
- R is the target slant range in feet (a scalar quantity).
- c is the velocity of light in feet per second (984×10^6).
- f_d is the spectrum frequency shift (Doppler shift) in hertz.
- V is the target velocity relative to radar in feet per second (a positive quantity for targets moving toward the radar). $V = -dR/dt$
- f_o is the transmitter signal carrier frequency in hertz. (this quantity is defined more precisely in the next section).

It is important to note the sign convention associated with the quantities defined above since some confusion in the literature results from careless handling. The confusion generally results from a failure to define f_d as a positive quantity for closing targets. Since target motion toward the radar always results in an upward shift in return frequency, the logical convention for the radar engineer is to choose the sign of f_d as positive for closing targets. This convention is adhered to throughout, particularly as regards the definition of the radar ambiguity function.

Referring again to Figure 1, once the ground rules stated above are clearly understood, the general waveform selection problem can be solved when the user of the radar decides what information is required on the target of interest and to what extent this information can be perturbed by the other objects appearing in the radar antenna's field of view. A list of the target parameters which can be measured with a single waveform are:

1) *Slant Range (R)*
2) *Velocity (V)*
3) *Cross Section*
4) *Orthogonal Angular Location*

The accuracy with which one can measure these quantities depends on the signal strength; but, equally important, accuracy depends on bandwidth, waveform length, and antenna beamwidth. This discussion is confined to the waveform properties most suitable for measuring R and V. Radar cross section measurement accuracy for targets in the clear is primarily a function of signal strength. Angular accuracy also depends mainly on signal strength, once the receiving antenna beamwidth has been established. The perturbing effects on cross section and angle measurements due to radar reflectors in the main lobe of the antenna beam but at different ranges and velocities can be estimated from the radar ambiguity function which is defined in the next section. This function describes the magnitude of interference resulting from objects at other ranges and velocities relative to the target of interest. Measurement interference due to objects at different angles depends on the antenna pattern characteristics as well as the conventional range-velocity ambiguity function. A general treatment of interference effects would, of course, include the radar's antenna characteristics as well as the waveform used. A four-parameter ambiguity function could then be derived to determine the perturbing effect of radar objects at other ranges, velocities, azimuth angles, and elevation angles. In order to obtain a clear understanding of one problem at a time, the antenna pattern effects are not included in this discussion. The problem addressed herein assumes that all targets and interfering objects are equally illuminated by the radar antenna. Antenna effects can be estimated by applying a weighting factor to the cross section of each interfering target proportional to the relative two-way antenna pattern. (For a discussion of the ambiguity function with antenna effects incorporated see Reference 16.)

The above discussion is intended to describe the nature of the problem and to establish the ground rules and limitations of the material to follow. A clear understanding of this relatively simple problem should suggest the necessary modifications required to add the additional variable factors not considered herein.

Now that the preliminaries are out of the way, a more specific set of waveform design considerations can be discussed.

Signal Energy

For a particular waveform the ultimate accuracy with which one can measure target range and velocity is proportional to the received signal energy. In a search radar the probability of detection increases with increased signal energy for a fixed false alarm rate. Thus, it would appear that better radar performance is attained at the expense of higher transmitted energy. Though this conclusion is accurate, it is misleading since one can very often obtain much better range and/or velocity accuracies with considerably less transmitted energy by increasing the waveform's bandwidth and/or time duration, respectively. On the other hand, a search radar can often improve its detection capabilities by decreasing waveform bandwidth, because the number of false alarms are reduced in proportion to signal bandwidth for a fixed probability of detection. It follows that judicious waveform design can reduce the required transmitter pulse energy and therefore the transmitter's average power output. This point is extremely important since, in general, transmitter cost can be reduced by many more dollars than can any signal generation and processing cost increase associated with a more sophisticated waveform. Equation (3) states the relationship between receive signal energy and a radar's detection performance.

$$S/N = L \cdot E/N_o \qquad (3)$$

where:

- S/N is the peak signal to mean noise power ratio of the signal plus noise envelope after it emerges from the signal processor. (See Figure 2a for an illustration.)
- L is a loss factor, equal to or less than unity, associated with the use of a non-matched filter; for the matched filter, the value of L is unity. (The value of L is defined in terms of the cross ambiguity function to be defined later.)
- E is the total energy of the signal return at a reference point at the input of the radar (in joules or watt seconds).
- N_o is the noise power density (in watts per hertz or joules) referred to the same point as for E; it includes all sources of thermal noise, such as sky noise, front end noise, and signal processor noise. (The value of N_o is the noise power in watts measured at the output of an ideal lossless 1 Hz rectangular bandpass filter.)

Figure 2a depicts the detection model. An input signal containing energy, E, embedded in white noise having a spectral power density of N_o is filtered and envelope detected. A threshold is selected to permit an acceptable probability of false alarm. Signals exceeding this threshold are declared as targets. The matched filter results in best detection performance since it maximizes the ratio of S/N which in turn results in the maximum probability of detection for a given false alarm rate.

Equation (4) describes the relationship between the standard deviation of the range or velocity measurement and the E/N_o ratio. Figure 2b describes the measurement process in terms of a range readout.

$$\sigma = \frac{K}{\sqrt{E/N_o}} \qquad (4)$$

where:

σ is the standard deviation of the measurement error about the true value of range or velocity.

K is a constant whose value is a function of the waveform parameters and the filter mismatch loss, L; the value of K has the dimensions of σ and is, of course, different for the range and velocity measurement errors. (Its value is defined more precisely in the next section.)

Measurement accuracy and detection capability are seen to be dependent on the ratio of E/N_o for a particular waveform when measurements are being made in the presence of thermal noise alone.

Range/Velocity Accuracy

The previous discussion indicated in a general way how detection performance and measurement accuracy were dependent on the E/N_o ratio for a waveform in a particular radar system. In this section, the value of K in Equation (4) is defined in terms of the particular waveform selected. The relationship between K and the waveform parameters should begin to shed some light on the waveform selection process. Equations (13) and (14) give the relationship between the range or velocity measurement error and the waveform parameters.

Before proceeding to these equations and their interpretation, it is necessary to establish a compatible set of symbols for describing the general class of waveforms to be discussed. Signals radiated into space are real signals; they can be displayed on an oscilloscope as real volts vs. time. The familiar complex notation is used to simplify the mathematical manipulations that must be performed but, in fact, all of the analysis could be done without introducing the symbol $j = \sqrt{-1}$. The complex notation is used in the analysis, but only after defining the symbols in terms of a real function. Equation (5) defines the general class of waveform with which we are dealing.

$$u_r(t) = a(t) \cos [2\pi f_o t + \varphi(t)] \tag{5}$$

where:

$u_r(t)$ is a real signal (in volts) which is a function of time, t.

$a(t)$ is an amplitude modulation factor having the dimension of volts; it is a real function of time.

f_o is the carrier frequency in hertz; for most waveforms it is simply the familiar radar operating center frequency. The value f_o is a constant independent of time; it is defined in a special way in Equation (13).

$\varphi(t)$ is the phase modulation about the carrier frequency; it is measured in radians and is a function of time.

The link between any real function of time, such as $u_r(t)$, and its complex representation is given by Equation (6).

$$u_r(t) = 1/2\ [u_c(t) + u_c^*(t)] \tag{6}$$

where:

$$u_c(t) = a(t)e^{j[2\pi f_o t + \varphi(t)]}$$
$$= a(t) \cos [2\pi f_o t + \varphi(t)] + j\, a(t) \sin [2\pi f_o t + \varphi(t)]$$

$$u_c^*(t) = a(t)e^{-j[2\pi f_o t + \varphi(t)]}$$
$$= a(t) \cos [2\pi f_o t + \varphi(t)] - j\, a(t) \sin [2\pi f_o t + \varphi(t)]$$

From the definition of $u_c(t)$ and its complex conjugate, $u_c^*(t)$, it is obvious that addition of these complex values and division by two results in the original real signal, $u_r(t)$. Simple logic indicates that, if given $u_c(t)$ or $u_c^*(t)$ and the definition above, one could re-establish the real signal, $u_r(t)$. This process, of course, is the game — manipulation of the complex quantity in the computations and subsequent interpretation of the result in light of Equation (6).

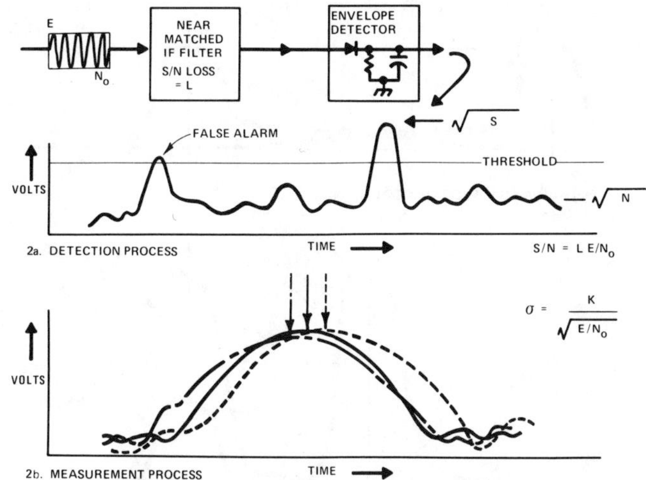

Figure 2 **Detection and Measurement Process**

One further simplification can be made in the notation. A complex modulation, defined by Equation (7), eliminates the necessity to manipulate the carrier frequency in all the calculations.

$$u(t) = a(t)e^{j\varphi(t)} \tag{7}$$

Note that $u_c(t) = u(t)e^{j2\pi f_o t}$

where:

$u(t)$ is, in general, a complex quantity having the dimension of volts; it is referred to as the "complex modulation." $u(t)$ is simply $u_c(t)$ with the carrier frequency removed.

Equations (5), (6), and (7) serve to define the signal waveforms in the "time domain." Some computations are simplified further if the Fourier transform of the time waveforms are used and some calculations are made in the frequency domain. Throughout this chapter the transform is symbolized by the capital letter; the time waveform, by the lower case equivalent. The transforms of $u_r(t)$, $u_c(t)$, and $u(t)$ are designated as follows:

$u_r(t) <=> U_r(f)$

$u_c(t) <=> U_c(f)$

$u(t) <=> U(f)$

The symbol $<=>$ means "transforms to." The Fourier transform of a function having the dimensions of volts has the dimension volts per hertz. Thus the Fourier transform is the voltage spectrum density of a voltage time waveform.

The Fourier transform is defined in the usual way by Equation (8). The inverse transform of U(f), defined in Equation (9), is equal to the original time function, u(t). The proof of Equation (9) can be found in most texts on the subject.

$$U(f) = \int_{-\infty}^{\infty} u(t)e^{-j2\pi ft}dt \tag{8}$$

$$u(t) = \int_{-\infty}^{\infty} U(f)e^{j2\pi ft}df \tag{9}$$

Equation (10) gives the relationship between the transform of the real signal, $u_r(t)$, and the transform of its complex representation, $u_c(t)$.

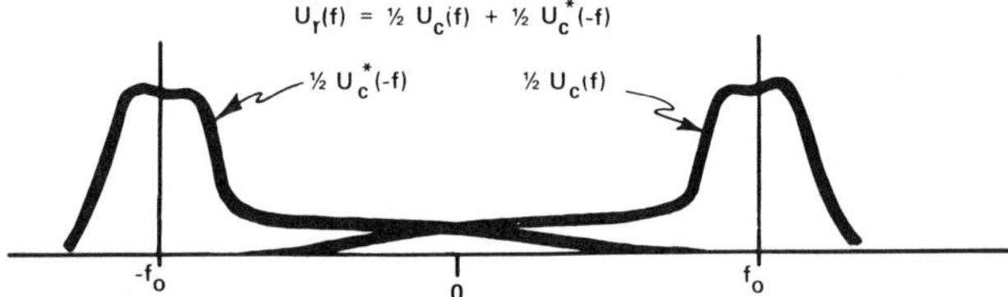

3a SPECTRUM MAGNITUDE OF REAL SIGNAL

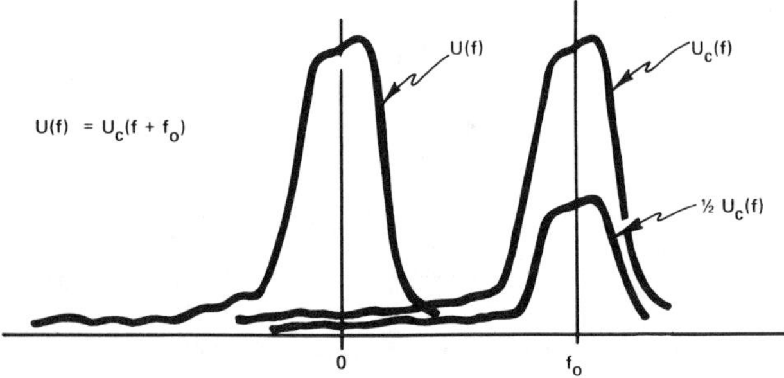

3b. SPECTRUM MAGNITUDE OF COMPLEX SIGNAL

Figure 3 **Fourier Transform Representation of Signals**

$$U_r(f) = ½\, U_c(f) + ½\, U_c^*(-f) \tag{10}$$

Figure 3a graphs the magnitude of the transform $U_r(f)$ as the sum of the two complex transforms on the right side of Equation (10). The magnitude of the transforms $U_c(f)$ and $U_c^*(-f)$ are seen to be mirror images. The $U^*(-f)$ is simply the complex conjugate of the frequency reversed U(f). It can be seen from the figure that, if the spectral content in $U_c(f)$ is quite small to the left of zero frequency, and consequently the spectral content of $U_c^*(-f)$ is small to the right of zero frequency, then the complex signal's transform divided by 2 is nearly equal to the positive half of the real signal's transform. Equation (11) states this small fractional bandwidth approximation.

$$U_r(f) \approx ½\, U_c(f) \quad \text{when} \quad 0 \leqslant f \leqslant \infty \tag{11}$$

Figure 3b graphs the transform of the complex modulation u(t) as it relates to the transform of the carrier frequency shifted complex signal, $u_c(t)$. The complex modulation's transform is simply a frequency shifted version of that of the complex signal, which in turn looks very much like the real signal's transform for small fractional bandwidth signals. Equation (12) relates the transform of the complex modulation u(t) to the transform of $u_c(t)$.

$$U(f) = U_c(f + f_o) \tag{12}$$

In plain language, the essential part of the radar waveform characteristics are contained in the complex modulation, u(t), and its Fourier transform, U(f). For the derivation minded individual, it should be noted that Equations (10), (11), and (12) follow directly from the definitions of $u_r(t)$, $u_c(t)$, and u(t) as stated in Equations (6) and (7). The above results follow from the definition of the Fourier transform and the use of the identities found in Appendix I.

Equations (13) and (14), the relationships between range and velocity measurement error and the waveform parameters, can now be comprehended. These equations are valid when the target is "in the clear" – it is perturbed by only white noise and is match filtered.

$$\sigma_\tau = \frac{1}{\beta_o \sqrt{2}\, \sqrt{E/N_o}} \tag{13}$$

where:

σ_τ is the standard deviation about the mean of the two-way time delay of the signal envelope (in seconds).

β_o is the effective signal bandwidth, having the dimensions of seconds^{-1}, and is defined as:

$$\beta_o^2 = \frac{(2\pi)^2 \int_{-\infty}^{\infty} f^2\, |U(f)|^2\, df}{\int_{-\infty}^{\infty} |U(f)|^2\, df}$$

The carrier frequency f_o is chosen so that

$$f_o = \frac{\int_0^{\infty} f\, |U_c(f)|^2\, df}{\int_0^{\infty} |U_c(f)|^2\, df}$$

The result then is

$$\int_{-\infty}^{\infty} f\, |U(f)|^2\, df = 0$$

when $U_c(f)$ is negligible for $f < 0$.

$$\sigma_{f_d} = \frac{1}{t_o \sqrt{2}\, \sqrt{E/N_o}} \tag{14}$$

where:

σ_{f_d} is the standard derivation about the mean of the measured target Doppler frequency shift; it has the dimensions of hertz.

t_o is the effective signal duration, having the dimensions of seconds, and is defined as

$$t_o^2 = \frac{(2\pi)^2 \int_{-\infty}^{\infty} t^2\, |u(t)|^2\, dt}{\int_{-\infty}^{\infty} |u(t)|^2\, dt}$$

The time origin for u(t) has been chosen so that

$$\int_{-\infty}^{\infty} t\, |u(t)|^2\, dt = 0$$

For the matched filter, the constant K in Equation (4) is $1/(\sqrt{2}\beta_o)$ regarding range accuracy and $1/(\sqrt{2}\, t_o)$ in the case of velocity. The values of β_o and t_o are measures of the bandwidth and time duration of the radar signal, respectively. In more precise language, β_o is the rms deviation of the power spectrum, $|U(f)|^2$, of the complex modulation function, u(t). The average value of the modulation power spectrum deviation is zero by choice of the carrier frequency definition given in Equation (13). Similarly, t_o is the rms time deviation of the signal power about its average value, as noted in Equation (14). In the next section, the quantities β_o and t_o are explained in terms of the ambiguity function. Qualitatively, it is noted that large β_o means wide modulation bandwidth, which in turn results in smaller time delay errors for a given signal energy. In the next section it is shown that large β_o results in narrow pulse widths after match filtering. Large values of t_o result from long time duration waveforms and produce narrow signal spectrums (which correspond to smaller Doppler estimation errors).

Figure 4 illustrates the use of Equations (13) and (14) for estimating the measurement error in time delay (range) and Doppler (velocity) for a signal having an idealized rectangular time envelope and a rectangular spectrum, such as would result from a large time-bandwidth product linear FM pulse.

Resolution, Clutter Rejection, and Ambiguous Responses

The E/N_o ratio of a received waveform improves detection performance and reduces the statistical errors in range and velocity as it increases. Also, increasing bandwidth and time duration of the waveform improves our ability to measure accurately range and velocity, respectively. Unfortunately, these three quantities –

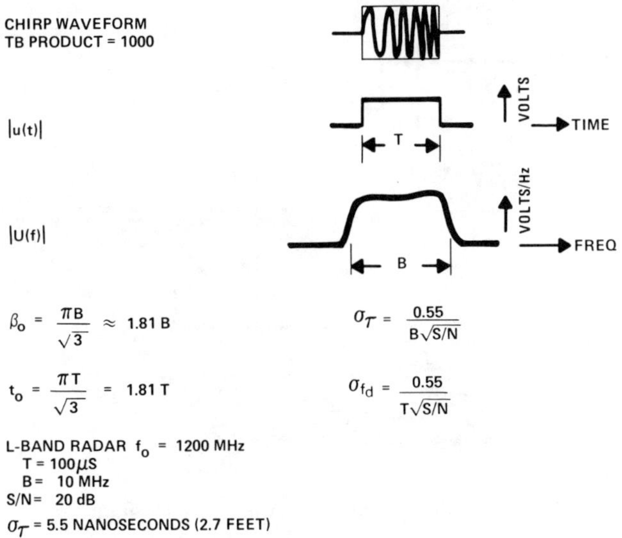

Figure 4 Range and Velocity Accuracy Example

namely E/N_o, β_o, and t_o — do not adequately describe the ability of a particular waveform to resolve closely spaced radar reflectors, nor do they indicate how susceptible the radar output is to clutter-like reflectors distantly located in range and/or velocity from an object of interest. Neither do these simple parameters indicate the possibility of some target, at a different range and Doppler, causing a similar response to that of a target in a region of interest. This information is essential to a complete understanding of the radar's performance in its environment. More important, these factors should be carefully considered before selecting the type of waveform and its processor. Figure 5 illustrates the three types of resolution problems mentioned above. The performance of the radar in the presence of interference can be determined by computing (or measuring) the output response (before envelope detection) of the receiving system to a single point target at all possible combinations of ranges and velocities. The response of the system to a combination of many reflectors simultaneously is then the linear superposition of each response taken one at a time. This procedure is, in fact, the method used to determine the usefulness of a particular waveform and its processing filter in the prescribed environment.

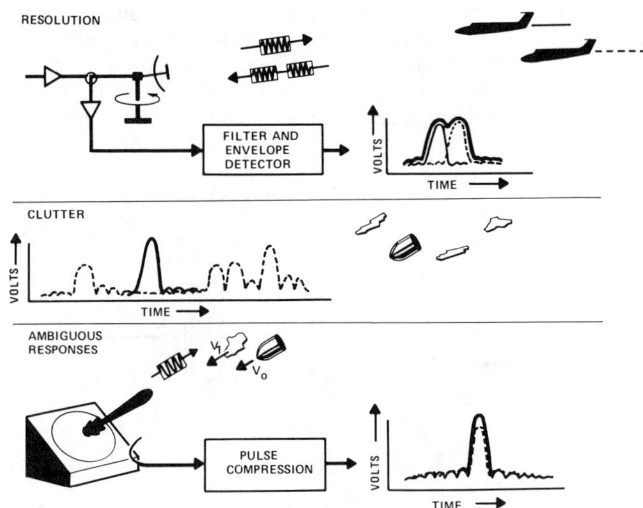

Figure 5 Types of Interference

The technique alluded to above was suggested by P.M. Woodward in his now famous monograph [1] published in the year 1953. Woodward defined a simple mathematical function, now referred to as the *Ambiguity Function*, which computes the output response of the matched filter to a range delayed, Doppler shifted signal. Since 1953 a great deal of effort has gone into the study of this simple function. References 2-11 listed at the end of this chapter are only some of the many in existence. The purpose of the next section is to define the ambiguity function and explain its usefulness for predicting waveform performance. The ambiguity function is derived rather than presented as a definition. This approach should give a better insight into its meaning and remove some of the confusion which might result from a frontal assault on the literature.

The Ambiguity Function

At this point in most literature, a formal statement of the ambiguity function is written in terms of a convolution integral. The formula is then treated as a definition and the writer proceeds to expound on the magical properties of the function. The physical significance of the function is explained variously in terms of the output of a matched filter or in terms of a correlation function. This treatment is adequate for those already partly familiar with the subject, but, for the uninitiated, some confusion generally results. To compound this confusion, the literature is not consistent in the definition of the ambiguity function. For all of these reasons, the ambiguity function is developed here only after first describing in words why the radar designer needs such a function and what the function provides. After this message is presented, the necessary mathematics are developed from a physical model. Once having developed the math and defined the symbols, we study the properties of the function. Some of the properties are intuitively evident from the model, some are not; but in any event their significance is apparent. The properties of the ambiguity function give considerable insight into the performance possibilities of various types of waveforms. A further intention of this chapter is to provide enough detail to permit the radar engineer to compute and graph his own waveform ambiguity functions.

Why is it needed?

As indicated in the previous section, some way is needed to describe quantitatively the ability of a waveform and its processing filter to resolve two or more radar reflectors at arbitrarily different ranges and velocities. A knowledge of this ability permits us to assess performance in a multitarget environment — the general radar situation.

What is it?

The radar ambiguity function is a formula which quantitatively describes the interference caused by a point target return located at a different range and velocity from a reference target of interest. The magnitude of this interference is normalized by assuming that the signal energy received at the radar's antenna from each point target is equal. When the radar receiver is not the idealized "matched filter" the function is appropriately referred to as the *Cross-Ambiguity Function.* The name *Ambiguity Function* is reserved for the matched filter situation.

The definition of our function can be understood more clearly by referring to Figure 6. Let us assume that we are viewing an A-scope output which displays target video as a function of time. A target located at range R_o and moving toward the radar at a velocity V_o is being tracked. The processing filter has its center frequency slightly increased from the transmitter's center frequency so that the moving reference target's Doppler shifted carrier corresponds to the processing filter's center frequency. Our eyes are fixed on an A-scope, noting only the output voltage in the reference time slit as the reference target response. This slit is located at τ_o, the time delay corresponding to our reference target's range.

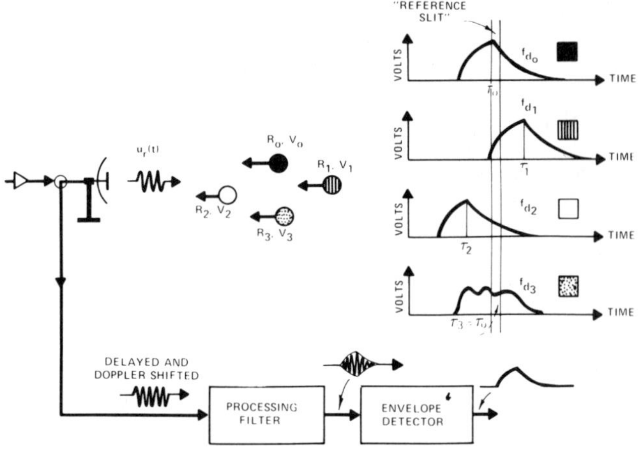

Figure 6 What is the Radar Ambiguity Function?

Next we assume that the slit location is fixed, as is the processing filter's shifted center frequency. We then observe the voltage in the slit for different targets, taken one at a time. The assumption is that only one target is present while the observations are made. We also assume that each target return has the same signal strength as our reference target. Figure 6 shows a succession of different target responses for targets at different ranges and velocities. The envelope detected response level observed in the slit is seen to depend not only on target range but on target velocity as well (as noted in the last A-scope reading in the figure). Even though one of the targets is at the same range as the reference target, its difference in velocity, and the resultant Doppler shift, cause its response to be reduced in the slit.

If we now make a three-dimensional graph having orthogonal coordinates labeled τ, f_d, and response volts, we can graph the results just obtained for the four targets in Figure 6. Figure 7 illustrates this graph for the four hypothetical targets. The logical extension of Figure 7 would be to graph a much larger number of response points above the τ, f_d plane. Figure 8 illustrates this extension. The graph was computed specifically for a 1 microsecond transmitted pulse of sine wave and a receiver consisting of a single pole

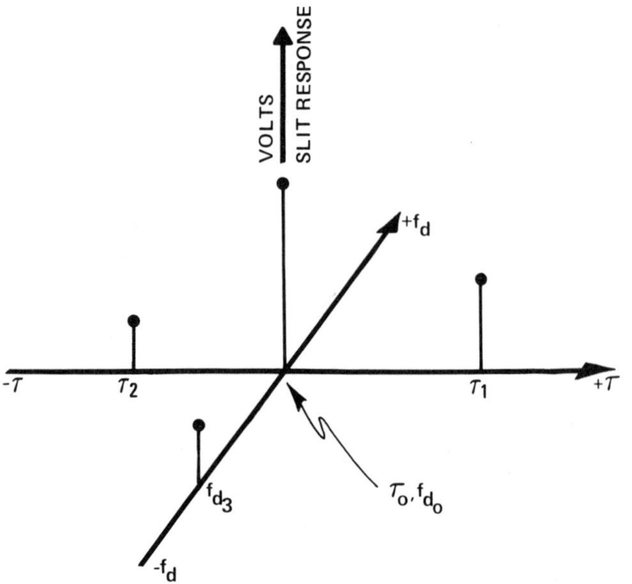

Figure 7 The Cross Ambiguity Diagram

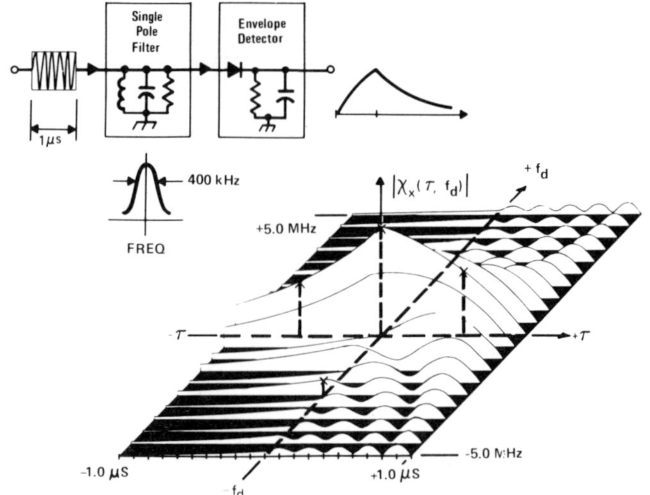

Figure 8 Computed Cross Ambiguity Diagram

bandpass filter having a 400 kHz bandwidth followed by an envelope detector. The graph in Figure 8 is seen to consist of a series of vertical slices – one for each Doppler shift. Each slice is the computed response of the receiver configuration to a Doppler shifted signal input. Figure 9 gives the reader a better feel for the continuous three-dimensional character of the surface; it was obtained by simply increasing the number of computed Doppler cuts.

Figures 7, 8, and 9 are referred to as *Cross Ambiguity Diagrams.* If the receiver filter had been ideally matched to the rectangular pulse, the resulting diagrams would have been referred to simply as the *Ambiguity Diagram.* The symbol $|X_x(\tau, f_d)|$ is used in Figure 8 to designate the voltage response; the square of this function is defined as the *Cross Ambiguity Function.* In the matched filter case, the subscript x is left off the symbol X and $|X(\tau, f_d)|^2$ signifies the *Ambiguity Function.*

The utility of the ambiguity diagram lies in the ability to quickly assess the interference level with which a target of interest must compete when it is in the vicinity of the other radar reflectors. Before proceeding to the derivation of the ambiguity function, several comments can be made, in light of an inspection of Figure 8, which are useful in our derivation. First, it is important to note that each of the constant Doppler cuts of the ambiguity diagram is simply the *time reversed* filter output response to that particular Doppler shifted version of the input signal. The reason for this time reversal is the way in which the ambiguity function is defined. Referring to Figure 6, targets farther away in range cause interference in our "slit" corresponding to the beginning of the time waveform; targets that occur before our reference target produce slit interference corresponding to the end section of the time waveform. The ambiguity function, then, can be computed

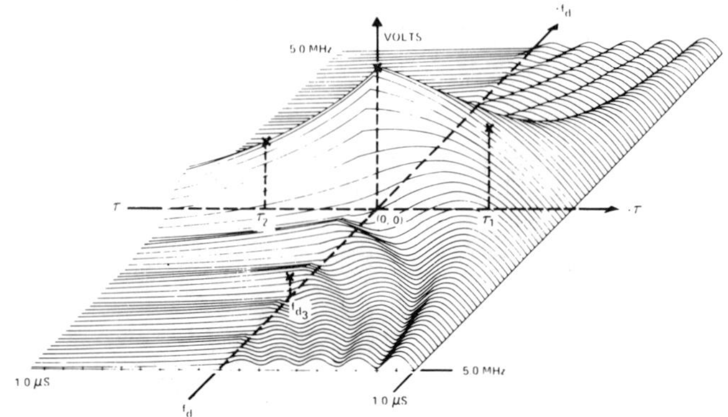

Figure 9 Cross Ambiguity Diagram, a Continuous Contour

simply by deriving the equation for the output response of any filter to an arbitrarily Doppler shifted time waveform and reversing the time axis of the response. In this manner, in fact, is precisely how we proceed. It should be pointed out here that some literature [2, 3] defines the ambiguity function as the filter output response and not as the interference response as is done here. This difference results in a time reversal of each Doppler slice of the diagram. As long as the reader notes this difference in definition, no confusion should result. The choice in this chapter is to define the ambiguity function in terms of an interference function, as Woodward [1] did in his original work. As defined here, the ambiguity function satisfies one of the criteria cited by Woodward – a target with relative delay τ and Doppler shift f_d is unresolved from an identical target at some reference delay and doppler (τ_o and f_{d_o}) when the value of the ambiguity function, $|X(\tau, f_d)|^2$, has the same value as $|X(\tau_o, f_{d_o})|^2$*. This comment can be true only, in general, when the time reversed filter response is implemented in the derivation to follow. Certain classes of waveforms which produce time symmetric responses result in the same ambiguity diagram regardless of the time reversal. This comment only tends to further confuse the uninformed reader who is not aware of the particular writer's definition; as mentioned earlier, there is also some confusion associated with the definition of positive Doppler. This pitfall is explained in the next section.

The Derivation

In the preceding section we presented a physical model and described the ambiguity or cross-ambiguity response in terms of that model. In this section, we follow through on the definition and derive the necessary mathematics to quantitatively compute the functions for any waveform-filter combination. Once having written the equations, we investigate some of their general properties. These characteristics reveal some interesting possibilities and constraints which are fundamental to all waveforms.

At this point, we could simply state the formula which relates the output response of any filter to an arbitrary input waveform. This formula could then be modified by adding a Doppler shift to the input waveform and time reversing the output response. The squared envelope of this result would be the desired function $|X_x(\tau, f_d)|^2$. However, we back up one step and derive the filter response to an arbitrary input waveform from basic principles and then perform the simple modification of this result to produce our ambiguity or cross-ambiguity function. This tactic results in a better insight into the nature of the ambiguity function; in addition, it demonstrates the simplicity of computing filter response functions. Many radar engineers are reluctant to perform these simple response calculations and more often than not relegate the job to the filter designer. This approach keeps the filter designer in work but does not allow the radar engineer to assess the many filter response possibilities available.

Figure 10 makes use of the principle of superposition to derive the filter output response. The formula expressing this output response is an integral referred to specifically as a convolution integral. The reason for the adjective "convolution" becomes apparent as we proceed. The solution as always, begins by stating the problem. Referring to Figure 10, we are asked to compute the real voltage waveform, $v_r(t)$, which emerges from a filter when the real input waveform is $u_r(t)$. The symbols for real and complex signals defined in Equations (5), (6), and (7) are used. The real waveform, $u_r(t)$, is an arbitrarily selected signal having any type of modulation and which may or may not include a carrier. First, we subdivide the input signal into a large number of very narrow sections. Each section has a width of Δx seconds and a height of $u_r(x_k)$ volts. Since the whole is equal to the sum of its parts, the entire input waveform, $u_r(t)$, can be represented by N discrete slices, as indicated in Figure 10. Next, we determine the filter output due

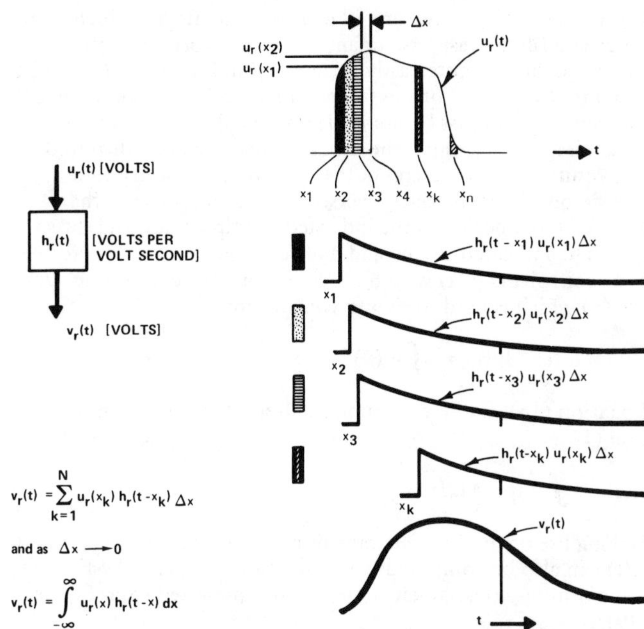

Figure 10 **Filter Response and the Convolution Integral**

to each of these N slices taken individually, then add the N output responses. This method can be used because our filter is a linear device and superposition therefore applies. The only problem is to determine the filter output response to a very narrow slice of voltage. The symbol $h_r(t)$ represents the response of the filter to a slice of infinitesimal width. In fact, to be precise, $h_r(t)$ is defined as the impulse response of the filter – it is the output waveform produced by an input signal having infinitesimal length. The impulse has the dimensions of volt-seconds. The impulse response, $h_r(t)$, therefore has the dimensions of volts per volt-second because it represents the filter's output voltage due to an input having a one volt-second input. The impulse response of any filter can be determined by one of the methods listed below:

1) Direct measurement of output response to a very narrow input pulse.
2) Direct computation from the circuit constants.
3) Computation of the Fourier transform of the measured frequency response, $H_r(f)$. Thus $h_r(t) \Longleftrightarrow H_r(f)$.
4) Same as 3) except that $H_r(f)$ is computed from the circuit constants.

Returning to Figure 10, each of the N input slices results in an output response which is the weighted and delayed impulse response of the filter. For example, the kth input slice has a width of Δx seconds, an amplitude of $u_r(x_k)$ volts, and it occurs at time x_k. Thus, to get the filter output due to this slice, the impulse response must be delayed by x_k seconds from the response due to the beginning of the input signal. Also, the response must be weighted by the relative strength of the input slice ($u_r(x_k)\,\Delta x$). The filter output to the kth input slice is thus $h_r(t-x_k)u_r(x_k)\,\Delta x$. Invoking the superposition theorem for linear circuits, the output of the filter due to all N input slices is the summation of the N outputs.

$$v_r(t) = \sum_{k=1}^{N} u_r(x_k)\,h_r(t-x_k)\,\Delta x$$

If the individual slice widths Δx get smaller and smaller, and N therefore approaches infinity, the result approaches a limit – Equation (15).

$$v_r(t) = \int_{-\infty}^{\infty} u_r(x)\,h_r(t-x)\,dx \tag{15}$$

*The peak of the ambiguity function is assumed to be at τ_o, f_{d_o}.

Equation (15) is the celebrated convolution integral which computes any filter's response to any input waveform. Equation (15) can be visualized more easily as depicted in Figure 11. In order to compute the output of the filter at a particular instant of time, t_1, the input signal must be multiplied by the delayed and time reversed impulse response; the product must then be integrated. The result is $v_r(t_1)$. To compute $v_r(t)$ for any arbitrary time t, one simply slides the time reversed impulse response to that value of t and performs the indicated multiplication and integration. The indicated sliding and multiplication process is referred to as convolving $u_r(t)$ with $h_r(t)$. Equation (16) defines the symbol (*) which is used to signify convolution.

$$v_r(t) = u_r(t) * h_r(t) = \int_{-\infty}^{\infty} u_r(x)\, h_r(t-x)\, dx \tag{16}$$

Equation (16) can be written in an alternate form, given in Equation (17).

$$v_r(t) = \int_{-\infty}^{\infty} u_r(t-x)\, h_r(x)\, dx \tag{17}$$

Making use of the Fourier transform relationships in Appendix I, $v_r(t)$ can also be computed from Equation (18), and the Fourier transform of $v_r(t)$, namely $V_r(f)$, can be computed from Equation (19).

$$v_r(t) = \int_{-\infty}^{\infty} U_r(f)\, H_r(f)\, e^{j2\pi ft}\, df \tag{18}$$

where:

$U_r(f) \Longleftrightarrow u_r(t)$

$H_r(f) \Longleftrightarrow h_r(t)$

$$V_r(f) = U_r(f)\, H_r(f) \tag{19}$$

Equation (19) states the familiar fact that the spectrum of the output signal is simply the product of the input signal spectrum and the filter's transfer function. If the fractional bandwidth of the waveform is small (so that the complex spectrum, $U_c(f)$, can be neglected for negative frequencies), as depicted in Figure 3a, and the complex transfer function, $H_c(f)$, is negligible for negative frequencies, Equations (20) and (21) follow from Equations (16) and (19).

$$v_c(t) \approx \int_{-\infty}^{\infty} u_c(x)\, h_c(t-x)\, dx$$

$$V_c(f) \approx U_c(f)\, H_c(f) \tag{20}$$

$$v(t) \approx \int_{-\infty}^{\infty} u(x)\, h(t-x)\, dx \tag{21}$$

$$V(f) \approx U(f)\, H(f)$$

In words, Equation (20) indicates that the output response of our filter expressed in complex notation can be closely approximated by the convolution of the input signal and the filter impulse response which are both represented in complex form. Equation (21) further states that the carrier frequency can be dropped from the equations; the complex modulation of the output signal thus can be computed from the convolution of the complex modulation of the input signal and the filter's impulse response. The proofs of Equations (20) and (21) are found in Appendix II. For the remainder of this chapter, the approximation signs are replaced by equal signs in Equations (20) and (21).

The ambiguity function can now be computed – which is what we originally set out to do. Since we are interested in the envelope of the filter output response to the input waveform, we work with Equation (21). We assume that the filter is not matched to the input wavefrom, u(t), which, in turn, leads us to the cross-ambiguity function. The ambiguity function is simply a special case of the cross-ambiguity function.

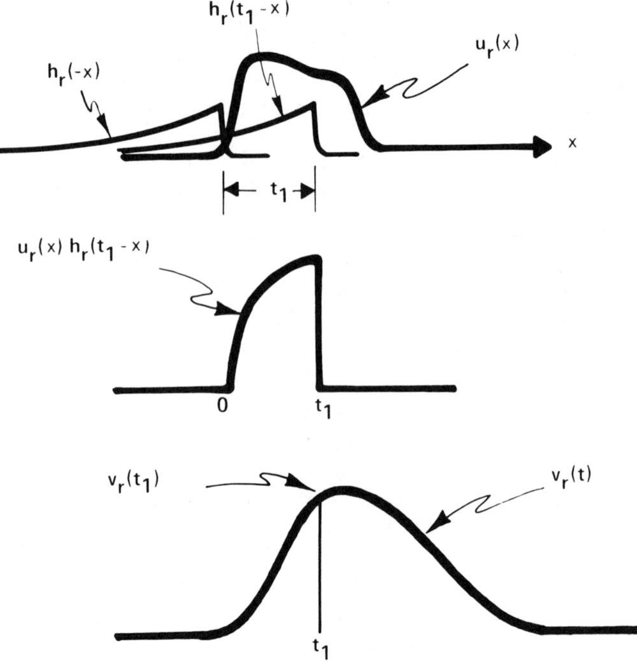

Figure 11 **Graphic Description of the Convolution Integral**

We begin by placing a Doppler shift on the input signal, u(x). The input thus becomes $u(x)\, e^{j2\pi f_d x}$ where x is the dummy time variable. The complex modulation on our filter output is given by Equation (22).

$$v(t, f_d) = \int_{-\infty}^{\infty} u(x)\, e^{j2\pi f_d x}\, h(t-x)\, dx \tag{22}$$

where:

$v(t, f_d)$ is the complex modulation out of the filter as a function of time for an input signal which is Doppler shifted by f_d hertz.

f_d is the Doppler shift of the signal spectrum. A positive value of f_d corresponds to increased frequency.

The final step in deriving our cross-ambiguity function requires the reversal of the time order of $v(t, f_d)$. To prevent any possible confusion between the time waveform which emerges from our filter and the ambiguity function, we change the name of the time variable from t to τ. We also rename the integral $X_x(\tau, f_d)$ since it no longer represents an output waveform but instead indicates an interference response. Equation (23) incorporates these changes.

$$X_x(\tau, f_d) = \int_{-\infty}^{\infty} u(x)\, h(-\tau-x)\, e^{j2\pi f_d x}\, dx \tag{23}$$

$|X_x(\tau, f_d)|^2$ is the cross-ambiguity function. It represents the magnitude of the interference produced by a target delayed in time by τ and Doppler shifted by f_d hertz from the reference target.

Let us assume that our filter is the matched filter for a waveform the complex modulation of which is $w(t)^*$. This assumption is true if $h(t) = w^*(-t)$. The substitution into Equation (23) results in the more familiar form of the cross-ambiguity relation given by Equation (24).

$$X_x(\tau, f_d) = \int_{-\infty}^{\infty} u(x)\, w^*(\tau + x)\, e^{j2\pi f_d x}\, dx \tag{24}$$

The ambiguity function is arrived at by allowing w(t) = u(t), resulting in the ambiguity relationship given by Equation (25).

*The matched filter has the property that it is the filter that provides the maximum possible output signal-to-noise ratio for the waveform, w(t), for which it is designed.

$$X(\tau, f_d) = \int_{-\infty}^{\infty} u(x)\, u^*(\tau + x)\, e^{j2\pi f_d x}\, dx \qquad (25)$$

It should be noted that $w^*(\tau + x)$ and $u^*(\tau + x)$ in Equations (24) and (25) have the dimensions of an impulse response which are volts per volt second. In order to keep the equations dimensionally correct this fact must be remembered. $X_x(\tau, f_d)$ has the dimensions of volts; it is referred to as the *Cross-Correlation Function.* Correspondingly, $X(\tau, f_d)$ is the *Correlation Function* consistent with Woodward's definition. The squared magnitudes of $X_x(\tau, f_d)$ and $X(\tau, f_d)$ are defined as the *Cross-Ambiguity* and the *Ambiguity* functions, respectively; the graphs of these squared functions are referred to as the *Cross-Ambiguity* and *Ambiguity* diagrams, respectively, in accordance with a proposed standard definition given in Reference 12. We have taken the liberty in this chapter of labeling the graphs of $|X_x(\tau_1, f_d)|$ and $|X(\tau_2, f_d)|$ as cross ambiguity and ambiguity functions, because the squared magnitude graphs do not show the sidelobe structure to the desired degree of detail.

Using the Fourier transform relations in Appendix I, Equations (24) and (25) can be rewritten as Equations (26) and (27).

$$X_x(\tau, f_d) = \int_{-\infty}^{\infty} U(f-f_d)\, W^*(f)\, e^{-j2\pi f\tau}\, df \qquad (26)$$

$$X(\tau, f_d) = \int_{-\infty}^{\infty} U(f-f_d)\, U^*(f)\, e^{-j2\pi f\tau}\, df \qquad (27)$$

Again, note that $W^*(f)$ and $U^*(f)$ are transfer functions and have the dimensions of volts per volt which, of course, makes them dimensionless. $U(f-f_d)$, on the other hand, is the spectrum of $u(t)\, e^{j2\pi f_d t}$ and has the dimensions of volts per hertz. The resulting integral, therefore, has the dimensions of volts.

Several final comments must be made before proceeding to the analysis of the ambiguity functions. Much of the existing literature has the sign preceding the Doppler term, f_d, reversed from that appearing here. Specifically, Equations (24) and (25) would each have a minus sign in the exponent of e, and a plus sign would precede f_d in each of Equations (26) and (27). It appears that this discrepancy derives from Woodward's original formulation in which he defined the symbol φ as Doppler shift but apparently intended $+\varphi$ to correspond to a downward shift in signal spectrum. A discussion of this discrepancy and a proposed standardization of the radar ambiguity function so it is consistent with Equations (25) and (27) appears in Reference 12. This reference further proposes that the graph of the ambiguity function be called the ambiguity diagram.

It is also conventional in the literature to assign the dimensions of energy to the ambiguity and cross-ambiguity functions. This convention results from the assignment of the same dimensions to the filter's impulse response as are assigned to the input signal. For example, in Equation (25) the quantity $u^*(\tau + x)$ would be assigned the same dimensions as are to $u(x)$. If both of these quantities are given the dimensions of volts, then the dimensions of $x(\tau, f_d)$ become volt^2-seconds – which have the dimensions of joules x ohms. The idea of a normalized 1 ohm input and output resistor adds nothing to the analysis; usually, it is ignored and the volt^2-seconds are simply equated to energy in joules. The unnormalized value of the ambiguity function at $\tau = 0$ and $f_d = 0$ is then said to be equal to 2E, where E is the energy in the real input waveform. (Note that $\int |u(t)|^2\, dt = 2\int |u_r(t)|^2\, dt = X(0, 0)$.)

In this chapter, the preference is to assign the dimensions of volts to $u(x)$, volts per volt-second to $u^*(\tau + x)$, and seconds to dx, all resulting in the dimensions of volts for $X(\tau, f_d)$. The mathematics still show that $X(0, 0) = \int u(x)\, u^*(x)\, dx$, but the dimensions are volts, not joules. It is hoped that the information is presented here in such a way for the reader to make an individual peace with the dimensions of the ambiguity function. In any event, the conclusions concerning relative waveform performance are, of course, independent of the choice of dimensions.

General Properties

This section summarizes several of the more important properties of the ambiguity and cross-ambiguity functions.

Peak Value of the Ambiguity Function

$$X(0, 0) = \int_{-\infty}^{\infty} |u(x)|^2\, dx = 1 = |X(0, 0)|^2 \qquad (28)$$

This equation follows directly from Equation (25); it says that the value of the correlation function at $\tau = 0$ and $f_d = 0$ is unity when the signal and its matched filter are normalized as indicated above. Most literature defines the ambiguity function in the normalized form, namely when $X(0, 0) = 1$.

Signal to Noise Loss – Peak Value of the Cross-Ambiguity Function

$$|X_x(\tau_o, f_{d_o})|^2 = L_{S/N} \qquad (29)$$

where:
$$X_x(\tau, f_d) = \int_{-\infty}^{\infty} u(x)\, w^*(\tau + x)\, e^{j2\pi f_d x}\, dx$$

and where:
$$\int_{-\infty}^{\infty} |u(x)|^2\, dx = \int_{-\infty}^{\infty} |w(x)|^2\, dx = 1$$

Equation (29) says that the signal to noise loss associated with use of the non-matched filter is computed as the peak value of the cross-ambiguity function located at (τ_o, f_{d_o}). Note that this value, in general, is not (0, 0). The value of L is less than unity, except for the matched filter case. The value of L is unity only when $w(x) = u(x)$. Figure 12 illustrates this result.

Constant Volume Constraint

$$\int_{-\infty}^{\infty}\int_{-\infty}^{\infty} |X_x(\tau, f_d)|^2\, d\tau\, df_d = \int_{-\infty}^{\infty} |u(x)|^2\, dx \int_{-\infty}^{\infty} |w(x)|^2\, dx \qquad (30)$$

$$= 1 \text{ when: } \int_{-\infty}^{\infty} |u(x)|^2\, dx = 1 \text{ and } \int_{-\infty}^{\infty} |w(x)|^2\, dx = 1$$

This result is probably the single most important one derived from a study of the cross-ambiguity function. It states that the volume under the normalized cross ambiguity surface is a constant independent of the type of waveform or filter employed. This conclusion is true provided that the normalization of waveform and filter is made as indicated above.

The logic goes as follows:

Assume that a filter is selected that matches the waveform $w(x)$, instead of $u(x)$. Normalize the filter so that its impulse response satisfies $\int |w(x)|^2\, dx = 1$. Now, for any input waveform, $u(x)$, having fixed signal energy (i.e., $\int |u(x)|^2\, dx = 1$), the volume is constant and equal to unity. When $u(x) = w(x)$ the same equality, of course, applies. Figure 13 illustrates the constant volume constraint for a large time bandwidth product signal.

Helstrom's Uncertainty Ellipse [5]

The curve resulting from the intersection of a plane parallel to the τ, f_d plane and the normalized ambiguity function surface is an ellipse the shape of which is given by Equation (31). This equation is a good approximation near $\tau = f_d = 0$. The ellipse computed when the plane is at a height of 0.75 is referred to as *Helstrom's Uncertainty Ellipse.* The equation can be derived from a series expansion of the ambiguity function.

$$|X(\tau, f_d)|^2 \approx 1 - \tau^2 \beta_o^2 + 2\pi f_d \alpha - f_d^2 t_o^2 \qquad (31)$$

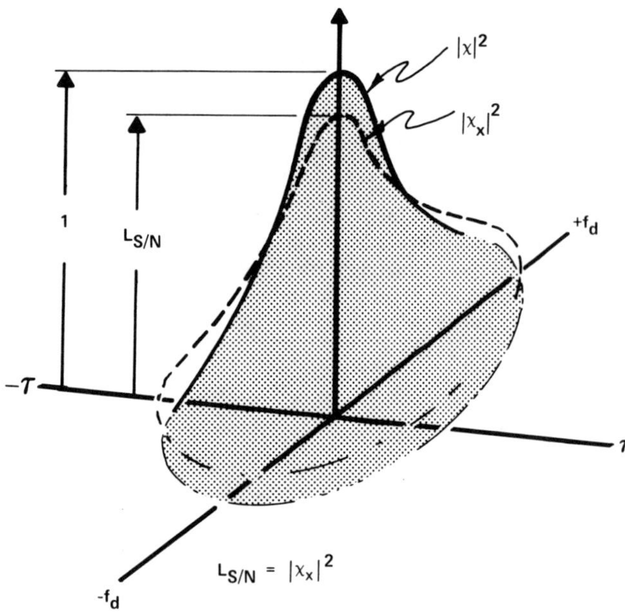

Figure 12 Signal to Noise Loss

Helstrom's ellipse is defined when $|X(\tau, f_d)|^2 = 3/4$

$$1/4 = \tau^2 \beta_o{}^2 - 2\pi f_d \alpha + f_d{}^2 t_o{}^2$$

All terms, except α, in Equation (31) have been previously defined. α is a measure of the delay-Doppler coupling of the waveform; it is given by Equation (32).

$$\alpha = 2\pi \int_{-\infty}^{\infty} t\varphi'(t)\, |u(t)|^2\, dt \Bigg/ \int_{-\infty}^{\infty} |u(t)|^2\, dt \tag{32}$$

where: $\varphi'(t)$ is the time derivative of the waveform's phase modulation, in radians per second. $\varphi(t)$ is defined by Equation (5).

Figure 14 illustrates the uncertainty ellipse.

The Radar Uncertainty Principle

$$t_o{}^2\beta_o{}^2 - \alpha^2 \geqslant \pi^2 \tag{33}$$

Equation (33) indicates that there is no limit on how small Helstrom's uncertainty ellipse can be. Stated differently, a waveform can simultaneously have as great a bandwidth and time duration as desired; thus, in turn, range and velocity resolutions of a waveform can be increased without limit. Also, the equation states that the range and velocity accuracy can be improved simultaneously without limit for a given ratio of E/N_o.

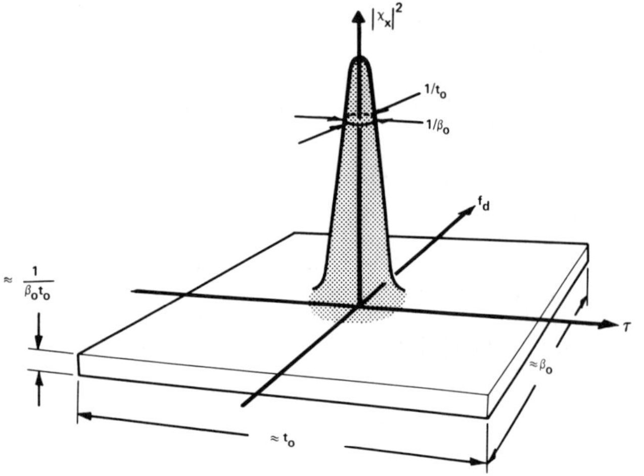

Figure 13 Constant Volume Theorem

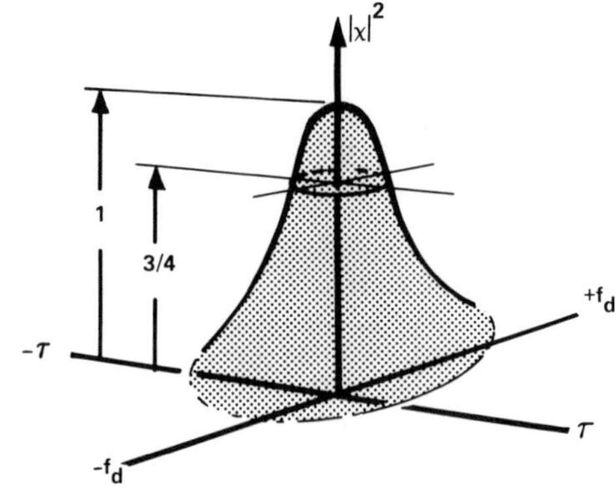

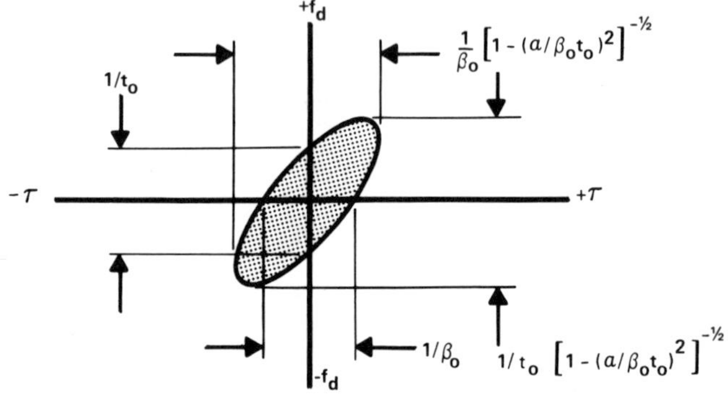

$$1/4 = \tau^2 \beta_o{}^2 - 2\tau f_d\, \alpha + f_d{}^2\, t_o{}^2$$

Figure 14 Helstrom's Uncertainty Ellipse

Typical Waveform Types and Their Uses

In this section several commonly used waveforms are compared in terms of their ambiguity functions. The three waveforms selected for detailed comparison are:

1) Simple 1 microsecond pulse of sine wave.
2) Linear FM pulse with 10 MHz sweep and 1 microsecond duration.
3) A 13 bit Barker phase coded pulse 1 microsecond duration.

All three pulses are chosen to have the same energy so that the differences are due only to the modulation. The effect of non-matched filtering is also demonstrated.

Simple Pulse of Sine Wave (Match Filtered)

Figure 15 graphs $|X(\tau, f_d)|$ (the square root of the ambiguity function) for a simple 1 microsecond pulse of sine wave. The graph was computed as a series of cuts parallel to the τ axis, extending from -5.0 MHz to +5.0 MHz. Each cut extends from -1.0 μs to +1.0 μs along the τ direction; it is separated by 0.5 MHz from the next cut. Beyond ±1.0 μs, the function is zero; beyond ±5.0 MHz, the function is continuous but diminished in amplitude. Also shown separately are a zero time cut along the Doppler axis and a zero Doppler cut along the time axis. The 3 dB widths are noted. This procedure is repeated for the Linear FM and Barker coded pulses.

The ambiguity function was computed by substituting into Equation (25) and performing the indicated integration. The procedure is summarized below:*

*Rect (x/T) = 1 for $-T/2 \leqslant x \leqslant T/2$
= 0 otherwise

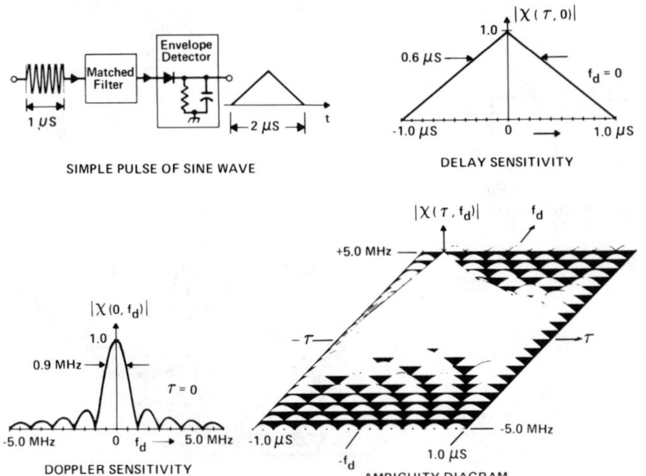

Figure 15 **Ambiguity Diagram for Simple Pulse**

1) $u(x) = \sqrt{1/T}\ \text{rect}\ (x/T)$; T = 1 μs (pulse length)

2) Substitute into Equation (25) using the proper limits of integration and solve for $|X(\tau, f_d)|^2$.

3) $$|X(\tau, f_d)|^2 = \left[\frac{\sin \pi f_d(T - |\tau|)}{\pi f_d(T - |\tau|)}\left(1 - \frac{|\tau|}{T}\right)\right]^2 \quad \text{for } -T \leqslant \tau \leqslant T$$

$$= 0 \text{ for } |\tau| > T$$

Figure 15 graphs $|X(\tau, f_d)|$ instead of its square; strictly speaking, as mentioned previously, it should not be called an ambiguity diagram. The choice was made to graph $|X(\tau, f_d)|$ because each cut on the τ axis becomes the voltage response (time reversed) that would emerge from the filter and thus appeals to the reader's intuition of what should be expected.

Simple Pulse of Sine Wave (Mismatched)

Figure 16 graphs $|X_x(\tau, f_d)|$ the square root of the cross ambiguity function for the simple 1 μs pulse. The filter selected is a single tuned bandpass filter having a 3 dB bandwidth of 0.4 MHz. The selected bandwidth results in the lowest signal to noise loss for the 1 μs pulse. This case was worked to illustrate the signal to noise loss principle and to show the effect of filter mismatch on interference sensitivity.

The cross ambiguity function was computed by substituting into Equation (23) and performing the necessary integration. The procedure is summarized below:

1) $u(x) = \sqrt{1/T}\ \text{rect}\ (x/T)$, T = 1 μs

2) $h(x) = \sqrt{2\pi B}\ e^{-\pi Bx}, \qquad x \geqslant 0,\ B = 0.4$ MHz

3) Substitute into Equation (23) using the proper limits of integration and solve for $|X_x(\tau, f_d)|^2$

4) $$|X_x(\tau, f_d)|^2 = \frac{2B}{\pi(B^2 + 4f_d^2)}\Big[1 + e^{2\pi B(\tau - T)} - 2e^{\pi B(\tau - T)} \cos(2\pi f_d(T - \tau))\Big]$$

$$\text{for } 0 < \tau \leqslant T$$

$$= \frac{2Be^{2\pi B\tau}}{\pi(B^2 + 4f_d^2)}\Big[1 + e^{-2\pi BT} - 2e^{-\pi BT} \cos(2\pi f_d T)\Big]$$

$$\text{for } -\infty < \tau \leqslant 0$$

Figure 17 compares the square root of the ambiguity function of the match filtered pulse with the corresponding cross-ambiguity

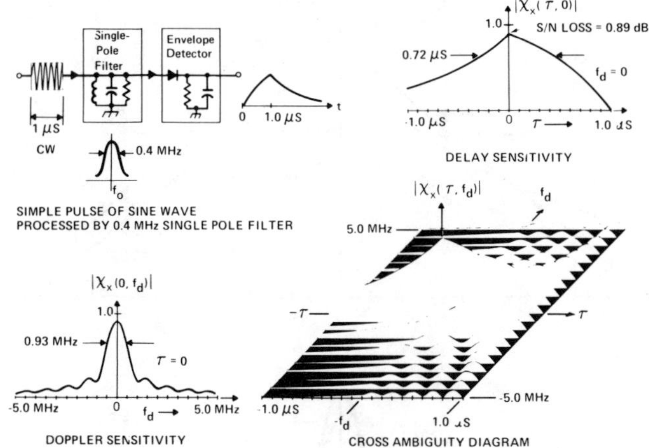

Figure 16 **Cross Ambiguity Diagram for Simple Pulse**

result for the mismatched pulse. The mismatched pulse experiences a 0.89 dB signal to noise loss and a pulse broadening at the 3 dB points of about 20%. The interference to targets closer to the radar than the reference target is considerably higher, as evident from the zero Doppler cut. This increase is due, of course, to the exponential ring down of the filter. The difference along the Doppler axis is in the absence of the sharp nulls at multiples of 1 MHz which occur for the matched case. The peak interference level is not affected appreciably by the mismatch.

[*Editor's Note* – For the above example a mismatched filter is used because it is simpler to implement than is the matched filter. Consideration has been also given in the literature to the use of a mismatched filter so as to obtain good clutter suppression. This arrangement is called the inverse filter. The inverse filter was derived for the case in which the clutter is assumed to have infinite extent in range, Rayleigh amplitude statistics, equal expected cross sections per range resolution cell, the same Doppler velocity for all range cells, and zero Doppler spread. A simpler rectangular pulse is assumed. Even for the assumption that the clutter Doppler velocity equals that of the target, it was found that the inverse filter provides infinite signal-to-clutter ratio in the absence of receiver noise. It is shown in Reference 17, though, that for the practical cases in which receiver thermal noise is present (the actual situation), no mismatched filter exists which can provide good clutter rejection for a simple rectangular pulse if the signal-to-clutter ratio at the output of the matched filter (matched for when only thermal noise is present) is much less than one.

Signal design such as the use of a wideband linear FM chirp waveform or coherently processed pulse burst waveform (or both) is needed to obtain good clutter suppression; see examples that follow and Chapter 8. The signal should be designed so that the area of ambiguity, A_ν, along the clutter Doppler line, ν, is small. Equivalently

$$\frac{\tau_3 N_o}{\gamma} \gg \int_{-\infty}^{\infty} |U(f)|^2\, |U(f - \nu)|^2\, df = A_\nu$$

where:

τ_3 is the 3 dB signal pulse width (after matched filter processing).

γ is the ratio of the average clutter cross section per range cell to the target cross section [17].

The above equation assumes that the energy in the signal is normalized to unity.]

Linear FM (Matched Filter)

Figure 18 graphs $|X(\tau, f_d)|$ for a rectangular pulse having linear frequency modulation. The ambiguity function was computed by

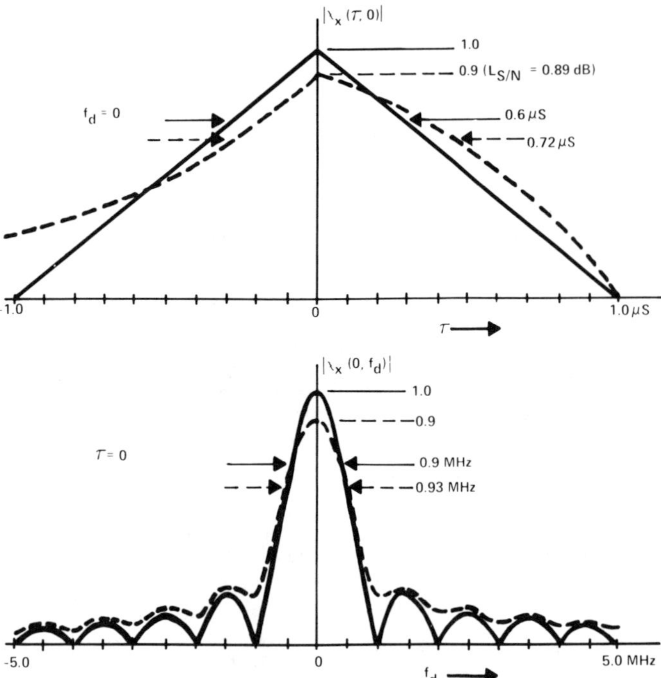

Figure 17 **Effect of Simple Pulse Filter Mismatch**

substituting into Equation (25) and integrating. The procedure is summarized below:

1) $u(x) = \sqrt{1/T}\ \text{rect}\ (x/T)\ e^{j\pi bt^2}$ $\quad T = 1\ \mu s$

$b = B/T = 10\ \text{MHz}/1\ \mu s$

2) Substitute into Equation (25) using the proper limits of integration and solve for $|X(\tau, f_d)|^2$

3) $$|X(\tau, f_d)|^2 = \left[\frac{\sin[\pi(f_d - b\tau)(T - |\tau|)]}{\pi(f_d - b\tau)(T - |\tau|)}\left(1 - \frac{|\tau|}{T}\right)\right]^2$$

$$\text{for } -T \leqslant \tau \leqslant T$$

$$= 0 \text{ for } |\tau| > T$$

The linear FM ambiguity function differs from that for the simple pulse in one important way, as can be noted by comparing the equation above with the one for the simple pulse — the variable f_d in the simple pulse equation is modified to $(f_d - b\tau)$ in the linear FM. This change results in a tilting of the ambiguity function along a diagonal the slope of which is b. Note that for positive b (corresponding to an increasing frequency FM signal) the ambiguity ridge runs toward $+\tau$ and $+f_d$. Thus, a target at farther range $(+\tau)$ but moving toward the radar $(+f_d)$ is unresolved from a stationary target at some closer range when increasing frequency chirp is used. The word ambiguity has a very real meaning for this particular waveform since real confusion can result from the interpretation of a target's range and velocity based on a single measurement.

Linear FM (Spectrum Weighting Filter)

Figure 19 graphs $|X_x(\tau, f_d)|$ for the same rectangular pulse of linear FM as above, but the receiving filter is designed to reduce the interference magnitude to -35 dB along the τ axis at $f_d = 0$ by weighting the frequency spectrum of the incoming linear FM signal to produce a spectrum which corresponds to a Taylor $\bar{n} = 6$, 35 dB sidelobe design. The procedure for filter design and computation of the resulting cross ambiguity function is outlined below:

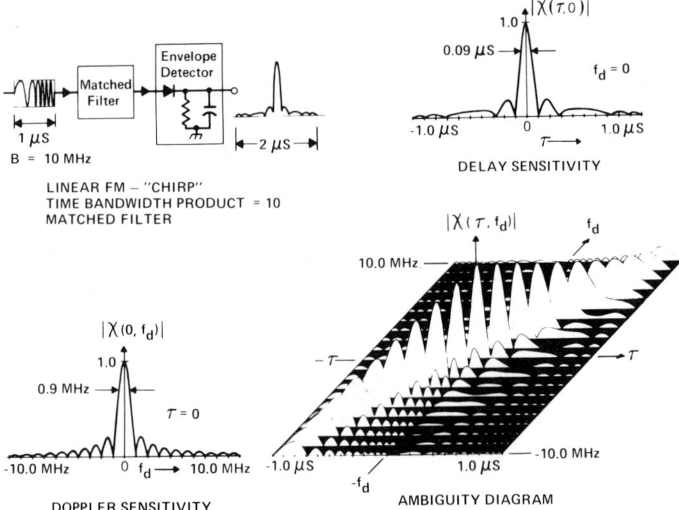

Figure 18 **Ambiguity Diagram for Pulse of Linear FM**

1) $u(x) = \sqrt{1/T}\ \text{rect}\ (x/T)\ e^{j\pi bt^2}$

$T = 1\ \mu s,\ B = 10\ \text{MHz}$

$b = B/T$

2) The Fourier transform of u(x), namely U(f), is computed

3) The filter transfer function is computer to be

$$H(f) = \frac{U^*(f)}{|U(f)|^2}\ T(f)$$

where: T(f) is the series of ideal Taylor weights extending from -B/2 to +B/2 MHz. The weights are for Taylor $\bar{n} = 6$, SL = 35 dB.

4) Equation (26) is used to evaluate the cross ambiguity function. H(f) above replaces W*(f), since the assumption is that H(f) is the matched filter for the hypothetical waveform the spectrum of which is W(f).

$$|X_x(\tau, f_d)|^2 = C_N \left|\int_{-B/2}^{B/2} U(f - f_d)\, H(f)\, e^{-j2\pi f\tau}\, df\right|^2$$

where:

$$C_N = \frac{1}{\int_{-2B}^{2B} |U(f)|^2\, df \int_{-B/2}^{B/2} |H(f)|^2\, df}$$

Dispersive Delay Network
Taylor Spectrum Weighting Filter
Envelope Detector
1 μS
B = 10 MHz
LINEAR FM – "CHIRP"
TIME BANDWIDTH PRODUCT = 10
MISMATCHED FILTER FOR TAYLOR $\bar{n}$ = 6, 35 dB TIME SIDELOBES
$|X_x(\tau, 0)|$
1.0
S/N LOSS = -1.14 dB
0.13 μS
f_d = 0
-1.0 μS
0
1.0 μS
DELAY SENSITIVITY

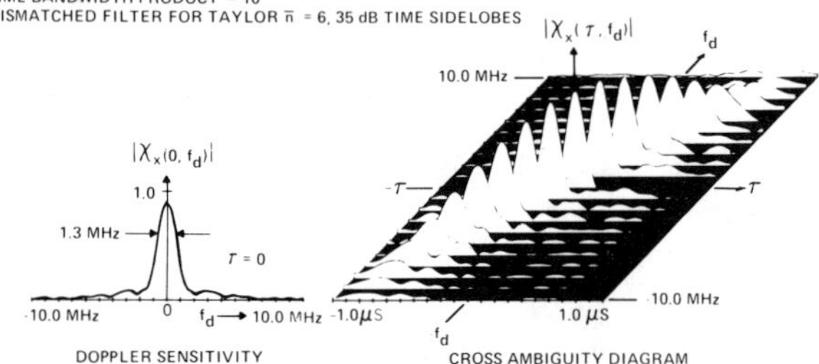

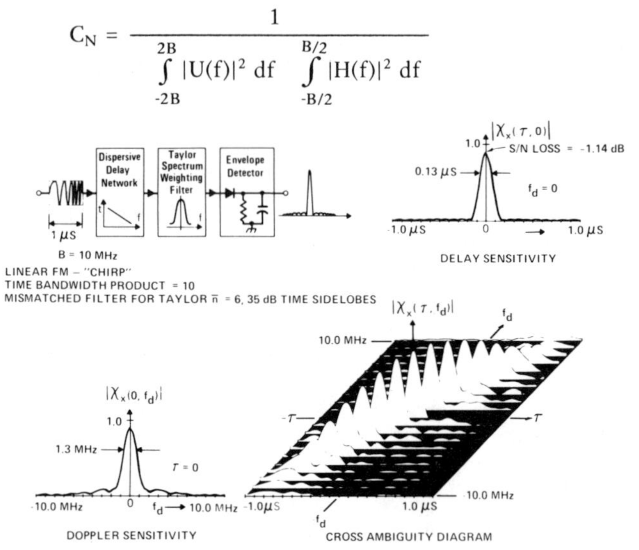

Figure 19 **Cross Ambiguity Diagram for Pulse of Linear FM**

The procedure outlined above requires several comments in order for the reader to understand the rationale for the method used. The filter function, H(f), is such that the small time bandwidth linear FM spectrum is match filtered U*(f), normalized to unity magnitude by dividing by $|U(f)|^2$, and multiplied by the Taylor weights, T(f). This procedure for $f_d = 0$ results in:

$$U(f)\,H(f) = \frac{U(f)\,U^*(f)}{|U(f)|^2}\,T(f) = T(f)$$

which, in turn, results in:

$$|X_x(\tau, 0)|^2 = C_N \left| \int T(f)\, e^{j2\pi f\tau} df \right|^2$$

which is simply the Fourier transform of T(f); by definition, the desired low sidelobe output thereby is produced. This technique is particularly useful for low time bandwidth waveforms since the signal spectrum U(f) has severe Fresnel ripples which, if not normalized before weighting, would produce large range lobes. Implementation of this technique is quite difficult with an analog filter, but it is accomplished readily with a digital processor using the FFT algorithm to do the necessary transforms. This technique is described in more detail in Reference 13.

It should be noted that the cross-ambiguity function was normalized with the factor C_N rather than normalizing U(f) and H(f) prior to the computation. This procedure insures the proper normalization of $X_x(\tau, f_d)$. The value of C_N is unity if the correct normalization of U(f) and H(f) has been done initially.

The limits of integration on $\int |U(f)|^2\, df$ were confined to ±2B since most of the spectral energy is within that region. All of the transforms were done using a high sampling rate discrete Fourier transform since the integrals involved do not have simple closed form solutions.

This technique is applicable to any type of waveform and is implemented readily with the digital signal processor in the frequency domain. The penalty for the low sidelobes is a loss in signal to noise ratio and a broadening of the uncertainty region of the cross-ambiguity function.

Figure 20 compares the square root of the ambiguity function of the match filtered linear FM pulse to the square root of the cross-ambiguity function for the spectrum weighted case. The mismatched signal experiences a 1.14 dB signal-to-noise loss measured directly from the normalized cross-ambiguity function. The near interference level on the τ and f_d axes are reduced below the match filtered levels. The main lobe response width of the mismatched function is increased by about 45% along both the τ and f_d axes.

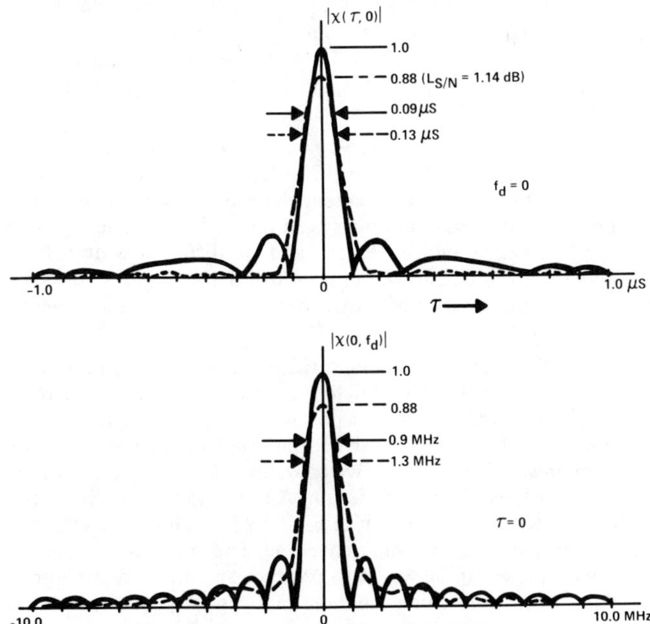

Figure 20 **Frequency Weighting for Sidelobe Reduction**

Barker Phase Code

Figure 21 graphs $X(\tau, f_d)$ for the 13-bit Barker binary phase coded pulse. The ambiguity function was computed from a modified version of Equation (25). The technique for computation is outlined below:

1) $u(x) = \sum_{k=0}^{12} q_k \,\text{rect}\,(x - kT/T)$, $T = 1/13\ \mu s$ (code segment)

   $q_k = 1, 1, 1, 1, 1, -1, -1, 1, 1, -1, 1, -1, 1$ for $k = 0, 1, 2 \ldots 12$

2) $u(x) = \text{rect}\,(t/T) * \sum_{k=0}^{12} q_k\, \delta(x - kT)$

This alternate form of the binary coded waveform suggests a simple method of computing the ambiguity function.

Let $\text{rect}\,(t/T) = u_1(x)$ and

$$\sum_{k=0}^{12} q_k\, \delta(x - kT) = u_2(x)$$

so that

$$u(x) = u_1(x) \cdot u_2(x)$$

Now:

$$X(\tau, f_d) = u(x)\, e^{j2\pi f_d x} * u^*(-x)$$

Substituting for u(x) and rearranging the order of convolution

$$X(\tau, f_d) = \left[u_1(x)\, e^{j2\pi f_d x} * u_1^*(-x)\right] * \left[u_2(x)\, e^{j2\pi f_d x} * u_2^*(-x)\right]$$

The equation indicates that our desired result can be obtained by first convolving our rectangle function, rect (x/T), with a Doppler shifted version of itself, then convolving the result with the set of 25 delta functions produced by convolving the 13 coded delta functions with a Doppler shifted version of themselves.

The resulting equation is:

$$X(\tau, f_d) = \frac{1}{13} \sum_{m=-12}^{12} X_2(\tau - mT)\, X_1(mT)$$

$$X_1(mT) = \sum_{k=0}^{12-m} q_k\, q_{k+m}\, e^{j2\pi f_d kT} \quad 0 \leqslant m \leqslant 12$$

$$X_1(mT) = \sum_{k=-m}^{12} q_k\, q_{k+m}\, e^{j2\pi f_d kT} \quad -12 \leqslant m < 0$$

$$X_2(y) = \frac{\sin \pi f_d(T - |y|)}{\pi f_d(T - |y|)} \left[1 - \frac{|y|}{T}\right],\ |y| \leqslant T$$

$$X_2(y) = 0 \qquad |y| > T$$

Summary of Simple Waveform Results

These five examples should serve to demonstrate the computational techniques used in computing the waveform filter response functions. Once the basic principles are understood, some of these functions can be visualized mentally by simply sliding and multiplying the waveform across the filter impulse response. Some practice permits the radar engineer to roughly assess a waveform and its filter mentally, thereby saving many hours of calculation.

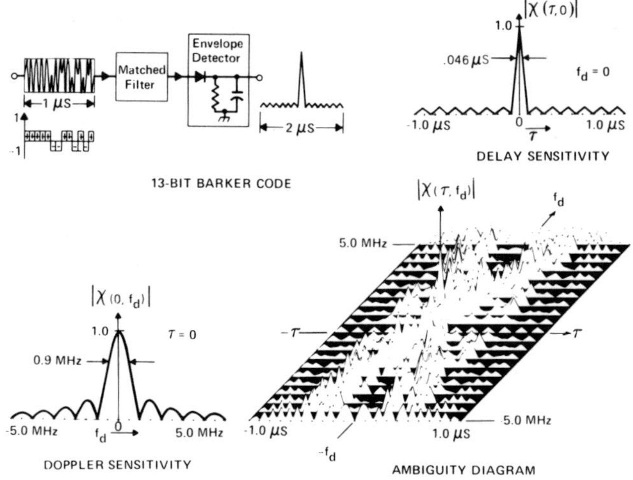

Figure 21 **Ambiguity Diagram for 13-Bit Barker Code**

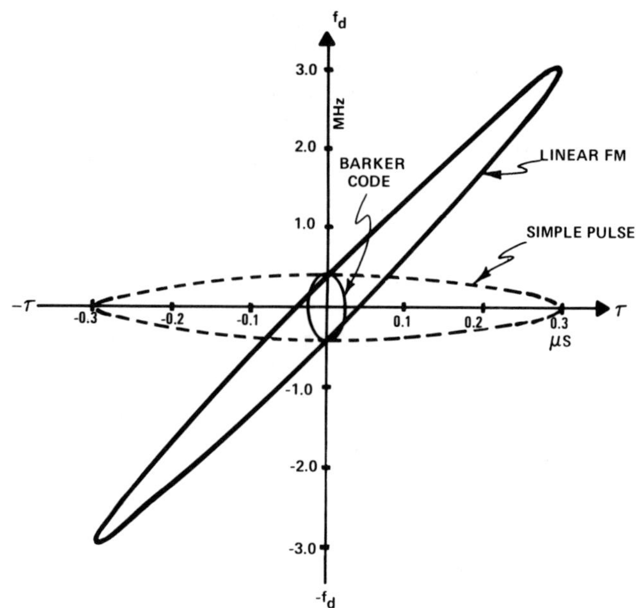

Figure 22 **Contour Comparison**

All three of the match filtered waveforms just analyzed are 1 microsecond long; they can be thought of as having equal energy and, therefore, equal signal-to-noise ratios at the filter output. Figure 22 graphs the 3 dB contour of the ambiguity functions of each waveform. Note that all three waveforms have the same ambiguity width on the Doppler axis, because they all have the same time envelope – a 1 microsecond rectangular pulse – which results in equal t_o values. The 3 dB Doppler axis width is about equal to the reciprocal pulse length. For a 1 μs pulse, the Doppler width is about 1 MHz. On the delay axis, the story is quite different. The Barker code has the narrowest τ dimension because it has the largest bandwidth. Each of its coded segments is only 1/13 μs; so its total bandwidth is on the order of 13 MHz. The linear FM pulse comes next with a 10 MHz bandwidth, followed by the 1 μs simple pulse with an inherent 1 MHz bandwidth. The FM pulse has a strong Doppler-time delay coupling ($\alpha \neq 0$) resulting in the characteristic skewing of the 3 dB contour. The slope of this contour, $df_d/d\tau$, is equal to the ratio B/T, or 10 MHz/1 μs in the example. Thus, it is evident that three waveforms with the same time duration can produce quite different response functions.

The applications of these waveforms can be summarized as follows:

Simple Pulse

1) Widely used in older generation radars for search and track functions (magnetrons are not easily phase modulated).
2) Used where range accuracy and resolution requirements can be met with a pulse wide enough to provide sufficient energy for detection.
3) Used in inexpensive radars where signal generation and processing costs must be minimized.
4) Has the minimum ratio of time sidelobe extent to compressed pulse width.

Linear FM

1) Commonly used to increase range accuracy and resolution when long pulses are required to get reasonable signal-to-noise ratios (+10 to 20 dB).
2) Used to search for targets having unknown velocity since Doppler sensitivity of the waveform is low. Permits the equivalent of a multiple Doppler cell search.
3) Variety of hardware is available to generate and process this waveform type.

Barker Phase Code

1) Used in search radars to increase range accuracy and resolution while maintaining pulse lengths necessary for detection.
2) Used to improve subclutter visibility in typical search radars in conjunction with MTI.
3) The 13-bit code represents the highest perfect code known and results in about -22 dB time sidelobes.
4) The 0-180° phase code is implemented easily and can be processed simply in the time domain with digital techniques, lumped constant delay lines, or the more recent surface wave acoustic device.

Other Waveforms

There are an infinite variety of waveforms from which the radar designer can choose; the ones mentioned above are the most common and are generally adequate for most jobs. It should be noted that most radar waveforms consist of pulses having a rectangular envelope, because most transmitter devices operate most efficiently at one fixed power level. Listed below are some other types of waveforms which can be used for special applications:

1) *Amplitude Modulated Pulse* – For example, a Gaussian shaped pulse reduces the Doppler lobes of the ambiguity diagram.
2) *Non-Linear FM Rectangular Pulse* – Various frequency modulation schemes such as "V" FM can be used to reduce the delay-Doppler coupling. Quadratic and Cubic FM are in this category also [2, 14, 15].
3) *Contiguous Phase Coded Pulses* – The 13-bit Barker code is just the tip of an iceberg in this category; there are a variety of ways to phase code a rectangular pulse. A basic code can be repeated, or random phase can be used. The code segments can be of equal width or they can have different widths. Phase codes do not have to go between 0 and 180 degrees but can use any imaginable phase quantization increment; see Chapter 8 by Cook.
4) *Burst Waveforms* – The burst waveform or pulse train is probably the most flexible vehicle for molding the desired cross-ambiguity function. We purposely say cross-ambiguity here since most bursts make use of some type of time or frequency domain weighting on receive to reduce the sidelobe responses in the cross-ambiguity function. A burst waveform generally can be described as shown in Figure 23. Each pulse in the burst typically has a rectangular envelope. The amplitude of each successive pulse can be varied to produce an amplitude weighted burst, an action which generally reduces the processed peak Doppler response lobes and which is used to sort between targets closely spaced in velocity. Stepping the frequency of each pulse permits high range resolution due to increased bandwidth;

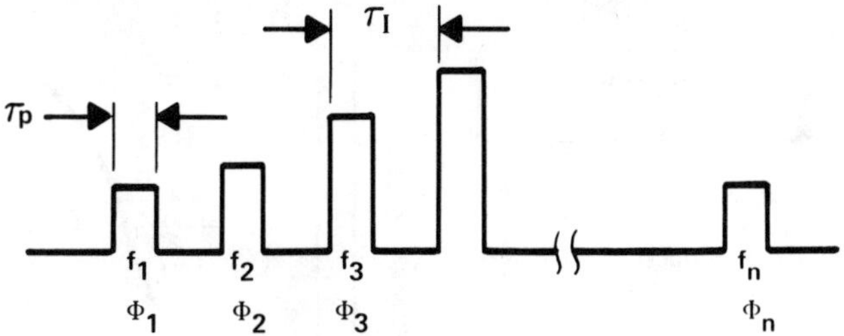

VARIABLE PARAMETERS:

1. INDIVIDUAL PULSE MODULATION
   WIDTH, AMPLITUDE, PHASE/FREQUENCY MODULATION
2. INTERPULSE FREQUENCY STEPPING
3. INTERPULSE PHASE
4. INTERPULSE SPACING

Figure 23 The General Pulse Burst

staggering the interpulse spacing can reduce peak interference levels. Signal generation and processing equipment capable of handling individual pulses in the burst can usually be modified simply to permit burst generation and processing.

Figure 24 shows a typical 3 dB contour representation of a burst of coherent simple pulses of sine wave; Figure 25 shows a similar representation for a stepped frequency burst.

Generation and Processing Techniques

This section should give the reader some feel for the types of generation and processing techniques that are used to implement the more commonly used waveforms. The techniques have been summarized in graphic form in Figures 26 through 38.

Waveform Generation Techniques

Figures 26 and 27 show several means for waveform generation.

1) *Passive Analog Filter* – In theory, a filter can be designed that will produce any desired waveform when impulsed; thus, any $h_r(t)$ can be designed. Practical realization of this theory is limited to certain types of waveforms. The most important application of this technique is in the generation of linear FM by impulsing a dispersive delay line and then bandlimiting and gating the output pulse.
2) *Memory Readout* – Any waveform can be generated digitally by simply computing values of the waveform at equal time intervals and then clocking these samples out of a shift register. The samples are converted to an analog signal which can be up converted and transmitted.
3) *Active Generation* – One method of generating various FM pulsed waveforms is to apply a programmed control voltage to a *Voltage Controlled Oscillator (VCO)*. In this way large bandwidth signals (several hundred megahertz) can be generated.
4) *Active Phase Coder* – This method often is used to produce a simple binary phase code. A code control signal directs a pulse of sine wave between a 0 and 180 degree phase shifter. More complicated phase codes can be implemented in a similar fashion.
5) *Burst Generator* – This scheme permits transmission of an arbitrary burst of pulses. Each pulse can be at a different center frequency and have a different phase applied. The frequencies usually are generated coherently.

   (See Chapter 13 for a novel way of generating linear FM waveforms.)

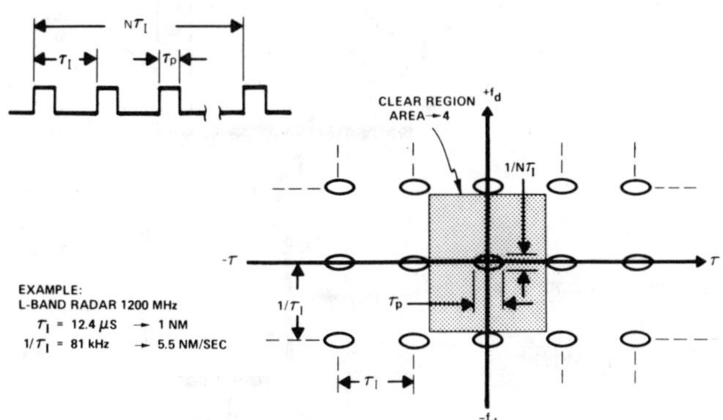

Figure 24 Ambiguity Contour for Coherent Constant Frequency Pulse Burst

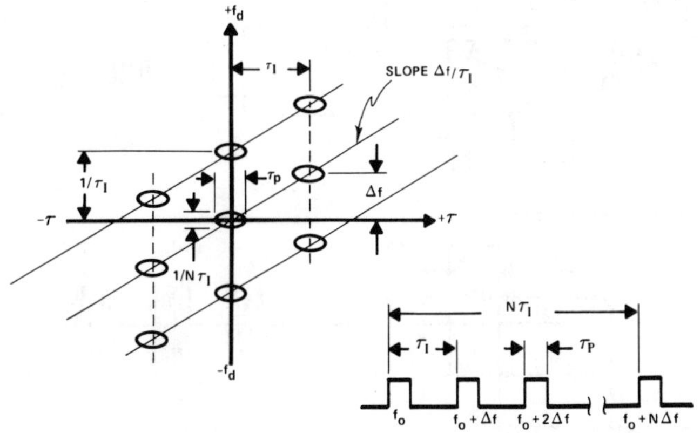

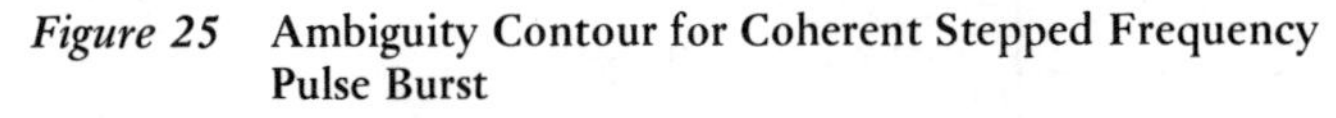

Figure 25 Ambiguity Contour for Coherent Stepped Frequency Pulse Burst

Waveform Processing Techniques

Figures 28 through 38 summarize a number of processing schemes and some special devices used in conjunction with these schemes.

1) *Passive Analog IF Waveform Processing* – Figure 28 shows a typical passive (all range) filter scheme. The analog IF filter can, in theory, have any transfer function, $H_r(f)$; in practice, though, limits do exist. Filters can be of the lumped constant (L, R, C) variety, which are used primarily for bandpass and spectrum weighting applications in conjunction with dispersive acoustic delay lines.

   The dispersive delay line can take several forms, as indicated in Figures 29, 30, 31, and 32. In general these devices convert electrical to sonic waves, perform the necessary filtering in the sonic medium, and then convert back to electrical signals. The

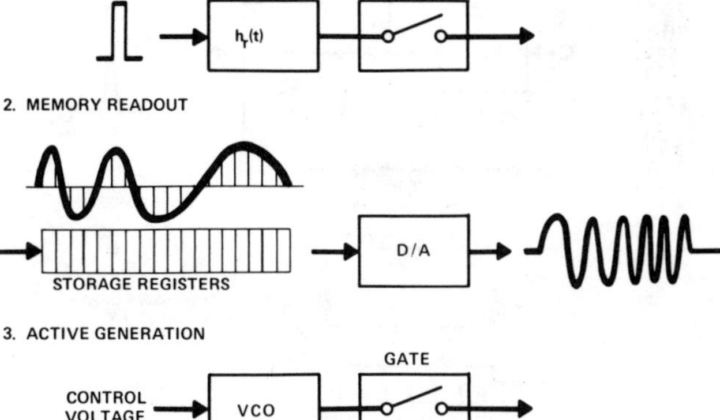

Figure 26 Waveform Generation Techniques

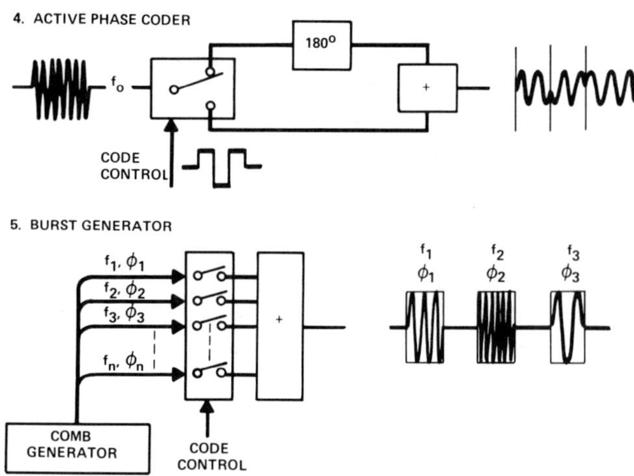

Figure 27 Waveform Generation Techniques

electrical to sound transducers are lossy; therefore about 30 to 60 dB of gain must be provided ahead of these devices to prevent signal-to-noise loss.

a) Metallic Strip Dispersive Delay Line – These units are generally made of steel or aluminum and are thin ribbons of metal, coiled and placed in temperature controlled ovens. They have an input and output transducer resulting in 30 to 60 dB of loss. The transfer function of the metal strip has a nearly linear delay versus frequency characteristic, as shown in Figure 29. It is used with linear FM only. Typical operating constraints are indicated in the figure.

b) Diffraction Grating Delay Line – These units make use of a diffraction grating to achieve the desired transfer function. Figure 30 shows the general shape of the wedge and perpendicular diffraction grating configurations. Electrical IF signals are launched into the non-dispersive quartz medium. The gratings cause different frequencies to experience different delays by directing the sound wave across different paths. The grating spacing is varied across the coupling boundary so that different portions of the grating are resonant to different frequencies.

c) Diffraction Grating on a Strip – This unit is made of a sheet metal delay medium with a diffraction grating etched into its surface. The transfer function characteristics result from the grating shape. Sound waves are reflected from different portions of the grating, depending on frequencies. These units usually are designed for linear FM. Figure 31 outlines the general shape and characteristics of these devices.

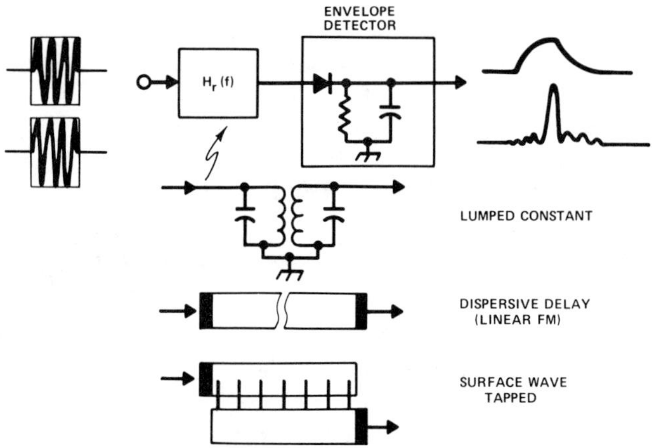

Figure 28 Passive Analog IF Waveform Processing

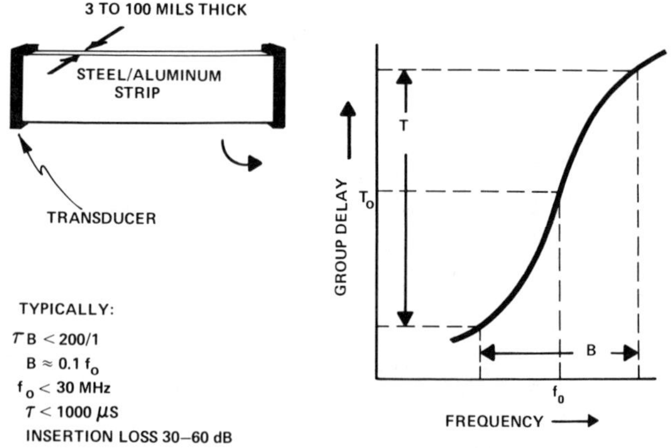

Figure 29 Metallic Strip Dispersive Delay Line

d) Surface Acoustic Wave Delay – These units are currently in the development stage. Some specialized units have been successfully put into production. In particular, Westinghouse has developed a small and relatively inexpensive 13-bit Barker code device. These units make use of sonic surface wave propagation so that taps can be placed wherever desired, thereby making them much more flexible than the bulk wave devices mentioned above. The possibilities for designing different transfer functions into the gratings are unlimited and bandwidths in the hundreds of megahertz are possible. Figure 32 shows a linear FM implementation.

[*Editor's Note* – Figure 3 of Chapter 10 summarizes the state of the art of dispersive delay devices. By cascading lines, even larger time-bandwidth products than indicated in that figure can be obtained. Anderson has built and delivered a 25,000:1 linear FM system using the cascade of twenty 500 μs IMCON lines for a total dispersion of 10 ms. Its bandwidth is 2.5 MHz, its close-in peak sidelobes are 30 dB down, and the far sidelobes fall below that value. Anderson also has produced a 3300:1 linear FM system consisting of four 550 μs IMCON lines for a total of 2.2 ms. Its bandwidth is 1.5 MHz and the peak sidelobe is 40 dB down.]

(See Chapters 10 and 12 and the "Present and Future Trends" section of Part 2 for further details on surface wave acoustic devices.)

2) *Digital Processing* – Within the past several years, digital processing of radar signals has become quite popular. Recent advances in A/D converters, integrated circuits technology, and the emergence of the Fast Fourier Algorithm have made prac-

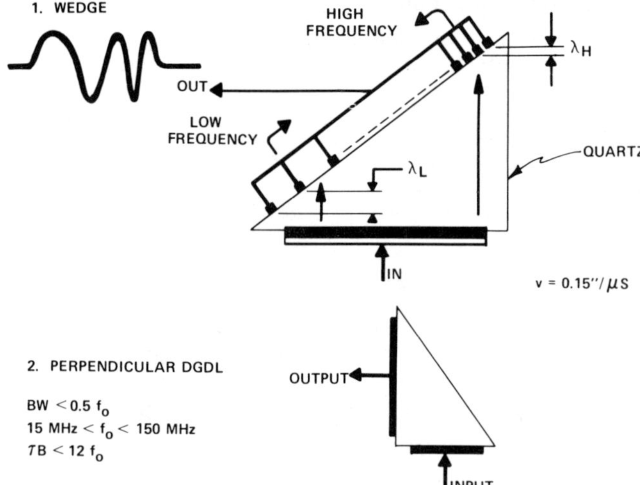

Figure 30 Diffraction Grating Delay Line

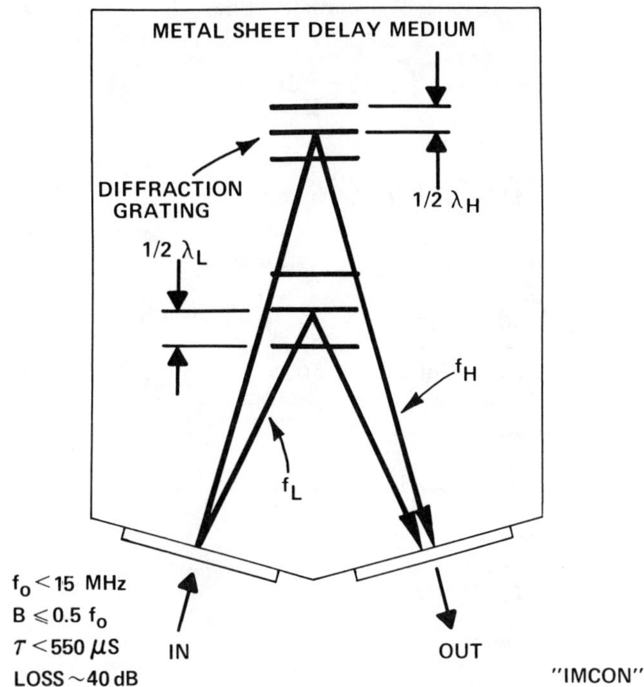

Figure 31 Diffraction Grating on a Strip

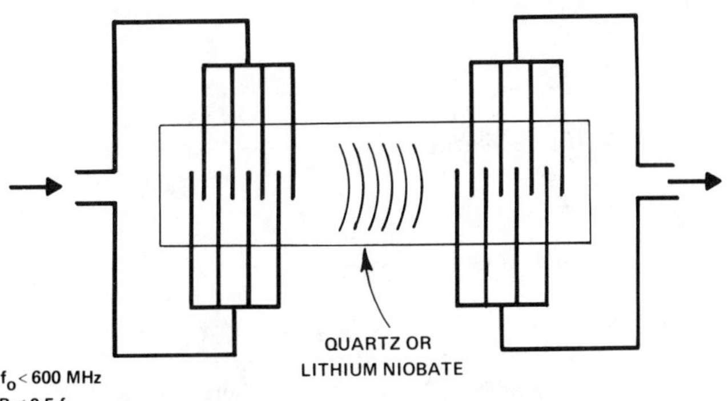

Figure 32 Surface Wave Acoustic Delay

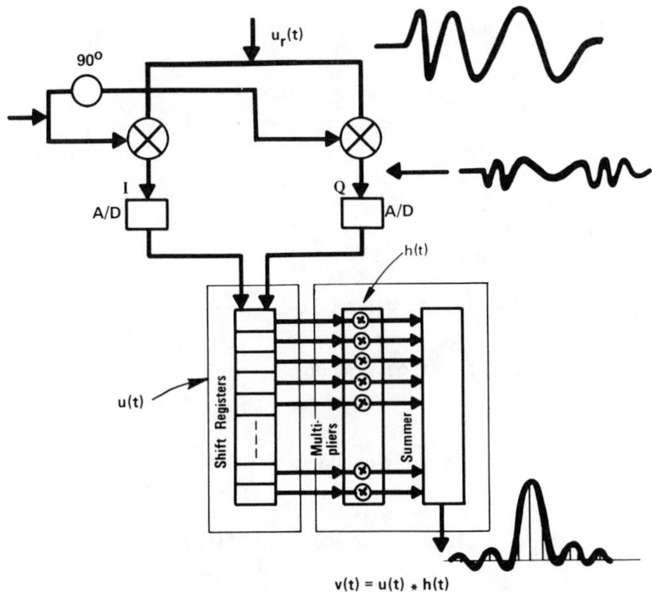

Figure 33 Digital Processing — Time Domain

tical the digital generation and filtering of signals. The advantages are in stability, reproducibility, and flexibility. The types of waveforms and filter functions which can be simulated are, in theory, unlimited. Figure 33 shows a typical I&Q channel time domain processor. The convolution process is performed directly in digital format. Figure 34 shows the frequency domain equivalent. Signal bandwidths up to 10 MHz and unlimited time duration can be handled in either configuration. (See Chapters 9, 10, and 11.)

As an example of the extreme flexibility of digital techniques, Figure 35 depicts an L-band radar configured to transmit and process a 256 MHz bandwidth waveform making use of a burst waveform. Figure 36 describes the waveform processing concept and Figure 37 approximates the cross-ambiguity contours.

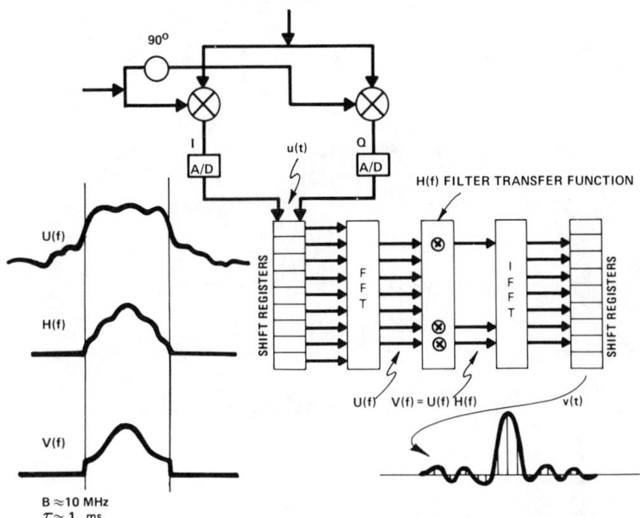

Figure 34 Digital Processor — Frequency Domain

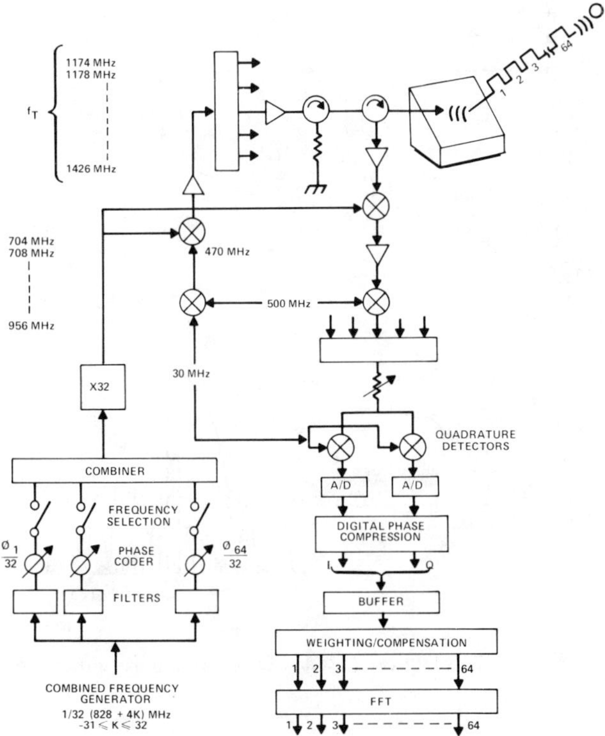

Figure 35 Burst Waveform Implementation

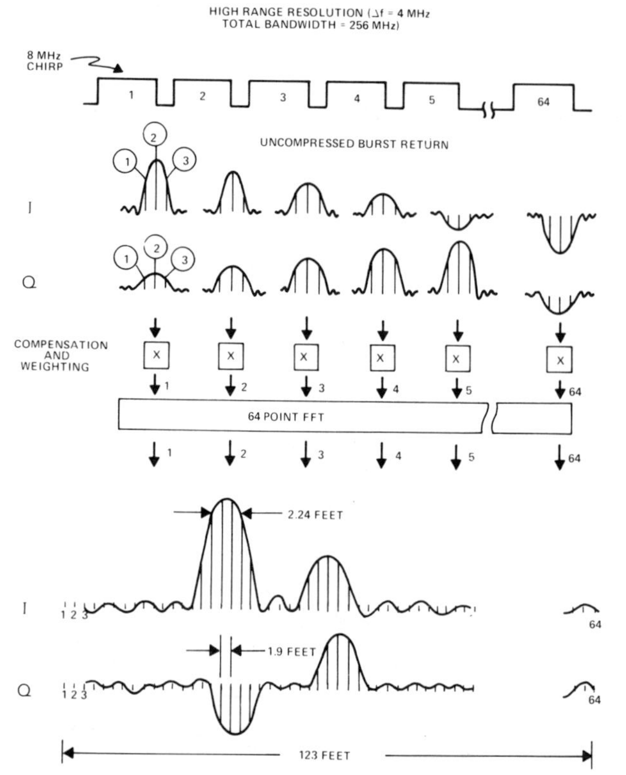

HIGH RANGE RESOLUTION (Δf = 4 MHz
TOTAL BANDWIDTH = 256 MHz)
8 MHz CHIRP
UNCOMPRESSED BURST RETURN
I
Q
COMPENSATION AND WEIGHTING
64 POINT FFT
2.24 FEET
1.9 FEET
123 FEET

Figure 36 **Burst Waveform Processing**

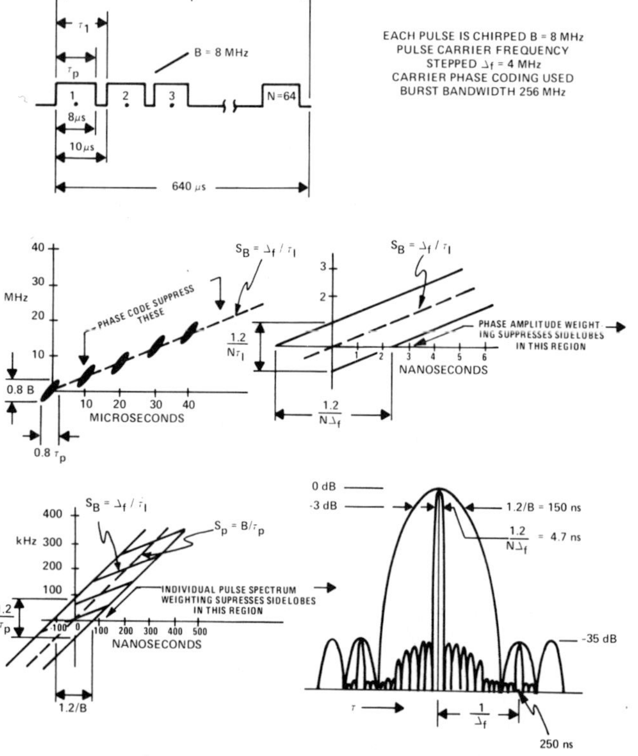

Figure 37 **Typical Detailed Cross Ambiguity Function**

Appendix I Fourier Transform Identities

Definition $U(f) = \int_{-\infty}^{\infty} u(t)\, e^{-j2\pi ft}\, dt$

1) Inverse

$$u(t) = \int_{-\infty}^{\infty} U(f)\, e^{j2\pi ft}\, df$$

2) Delta Function Transform

$$1 = \int_{-\infty}^{\infty} \delta(t)\, e^{-j2\pi ft}\, dt \qquad 1 = \int_{-\infty}^{\infty} \delta(f)\, e^{j2\pi ft}\, df$$

$$\delta(t) = \int_{-\infty}^{\infty} 1\, e^{j2\pi ft}\, df \qquad \delta(f) = \int_{-\infty}^{\infty} 1\, e^{-2\pi ft}\, dt$$

3) Superposition

$$a\,u(t) + b\,v(t) <=> a\,U(f) + b\,V(f)$$

4) Average Value

$$U(o) = \int_{-\infty}^{\infty} u(t)\, dt$$

5) Time Reversal

$$u(-t) <=> U(-f)$$

6) Delay

$$U(t-\tau) <=> e^{-j2\pi f\tau}\, U(f)$$

$$U(f-f_o) <=> e^{j2\pi f_o t}\, u(t)$$

7) Stretch

$$u(kt) <=> |1/k|\, U(f/k)$$

8) Differentiation

$$u^n(t) <=> (2\pi jf)^n\, U(f)$$

$$U^n(f) <=> (-2\pi jt)^n\, u(t)$$

Where $u^n(t)$ and $U^n(f)$ are the nth derivatives of u and U, respectively.

9) Convolution

If $W(f) = U(f)\,V(f)$

then $w(t) = \int_{-\infty}^{\infty} u(\alpha)\, v(t-\alpha)\, d\alpha$

or $w(t) = \int_{-\infty}^{\infty} u(t-\alpha)\, v(\alpha)\, d\alpha$

If $w(t) = u(t)\,v(t)$

then $W(f) = \int_{-\infty}^{\infty} U(\alpha)\, V(f-\alpha)\, d\alpha$

or $W(f) = \int_{-\infty}^{\infty} U(f-\alpha)\, V(\alpha)\, d\alpha$

10) Conjugates

$u^*(t) <=> U^*(-f)$ Corollary: if $u^*(t) = u(t)$, $U(f) = U^*(-f)$

$U^*(f) <=> u^*(-t)$ Corollary: if $U^*(f) = U(f)$, $u(t) = u^*(-t)$

11) Power Theorem

$$\int_{-\infty}^{\infty} u(t)\, v^*(t)\, dt = \int_{-\infty}^{\infty} U(f)\, V^*(f)\, df$$

$$\int_{-\infty}^{\infty} |u(t)|^2\, dt = \int_{-\infty}^{\infty} |U(f)|^2\, df$$

12) Conjugate and Delay

$$u^*(t-\tau) <=> e^{-j2\pi ft}\, U^*(-f)$$

$$U^*(f-f_o) <=> e^{j2\pi f_o t}\, u^*(-t)$$

13) Schwartz's Inequality

$$\int_{-\infty}^{\infty} u(t)\, u^*(t)\, dt \quad \int_{-\infty}^{\infty} v(t)\, v^*(t)\, dt \geqslant \left| \int_{-\infty}^{\infty} u^*(t)\, v(t)\, dt \right|^2$$

Appendix II Complex Form of the Convolution Integral

We want to prove that:

$$v_c(t) \approx \int u_c(x)\, h_c(t-x)\, dx$$

$$v(t) \approx \int u(x)\, h(t-x)\, dx$$

When we are told that

$$v_r(t) = \int u_r(x)\, h_r(t-x)\, dx$$

The proof is outlined below:

$$V_r(f) = U_r(f)\, H_r(f)$$

$$V_c(f) + V_c^*(-f) = [U_c(f) + U_c^*(-f)] \times [H_c(f) + H_c^*(-f)]$$

$$= U_c(f)\, H_c(f) + U_c^*(-f)\, H_c^*(-f)$$

$$+ U_c^*(-f)\, H_c(f) + U_c(f)\, H_c^*(-f)$$

But the last 2 terms on the right side of the last equation are negligible for small time bandwidth since, from Figure 3, it can be seen that $U_c^*(-f) \approx 0$ for $f > 0$. The same is true of $H_c^*(-f)$ when $f > 0$.

$$V_c(f) + V_c^*(-f) \approx U_c(f)\, H_c(f) + U_c^*(-f)\, H_c^*(-f)$$

It follows immediately that

$$V_c(f) \approx U_c(f)\, H_c(f) \quad \text{and} \quad V_c^*(-f) \approx U_c^*(-f)\, H_c^*(-f)$$

Since $V_c(f) \approx 0$ when $f < 0$

and $V_c^*(-f) \approx 0$ when $f > 0$

The corresponding situations apply for $U_c(f)$, $U_c^*(-f)$, $H_c(f)$, and $H_c^*(-f)$.

So $$V_c(f) \approx U_c(f)\, H_c(f) \qquad (1)$$

and $$V_c(f+f_o) \approx U_c(f+f_o)\, H_c(f+f_o)$$

which corresponds to

$$V(f) \approx U(f)\, H(f) \qquad (2)$$

since $V(f) = V_c(f+f_o)$, as evident from Figure 3.

From Equations (1) and (2) above, it follows, from Appendix I, that

$$v_c(t) \approx \int u_c(x)\, h_c(t-x)\, dx$$

and

$$v(t) \approx \int u(x)\, h(t-x)\, dx$$

Appendix III Signal-to-Noise Loss of Mismatched Filter

We want to prove that:

$$|X_x(\tau_o, f_{d_o})|^2 = L_{S/N}$$

where: X_x is in its normalized form and $L_{S/N}$ is the loss in S/N ratio resulting from use of the non-matched filter. Note that τ_o and f_{d_o} are not necessarily at (0, 0).

Proof:

The S/N ratio for the non-matched filter is

$$(S/N)_{NM} = |\int u(x)\, w^*(\tau_o + x)\, e^{j2\pi f_{d_o} x}\, dx|^2 \Big/ N_o \int |W(f)|^2\, df \qquad (1)$$

For the matched filter

$$(S/N)_M = |\int u(x)\, u^*(x)\, dx|^2 \Big/ N_o \int |U(f)|^2\, df$$

$$(S/N)_M = \int |u(x)|^2\, dx \Big/ N_o \qquad (2)$$

Taking the ratio $(S/N)_{NM} \big/ (S/N)_M = L_{S/N}$

from Equations (1) and (2) we get:

$$L_{S/N} = |\int u(x)\, w^*(\tau_o + x)\, e^{j2\pi f_{d_o} x}\, dx|^2 \Big/ \int |W(f)|^2\, df \int |u(x)|^2\, dx$$

In the normalized case, both integrals in the denominator of the right side of the equation above are unity. Thus

$$L_{S/N} = |\int u(x)\, w^*(\tau_o + x)\, e^{j2\pi f_{d_o} x}\, dx|^2 \equiv |X_x(\tau_o, f_{do})|^2$$

References

[1] Woodward, P.M.: "Probability and Information Theory, with Applications to Radar," 1953.

[2] Rihaczek, A.W.: *Principles of High-Resolution Radar,* McGraw-Hill, 1969.

[3] Rihaczek, A.W.: "Radar Signal Design for Target Resolution," *Proceedings of the IEEE, Vol. 53, No. 2,* pp. 116-128, February 1965.

[4] Gabor, D.: "Theory of Communication," *Journal of the IEE, Vol. 93 (Part III),* pp. 429-457, November 1946.

[5] Helstrom, C.W.: *Statistical Theory of Signal Detection,* Pergamon Press, New York, 1960.

[6] Berkowitz, R.S.: *Modern Radar,* John Wiley & Sons, Inc., 1965.

[7] Di Franco, J.V. and Rubin, W.L.: *Radar Detection,* Prentice-Hall, Inc., Englewood Cliffs, New Jersey, 1968.

[8] Blau, W.: "Synthesis of Ambiguity Functions for Prescribed Responses," *IEEE Transactions on Aerospace and Electronic Systems,* July 1967.

[9] Rihaczek, A.W.; and Mitchell, R.L.: "Design of Zigzag FM Signals," *IEEE Transactions on Aerospace and Electronic Systems,* September 1967.

[10] Wolf, J.D.; Lee, G.M.; and Suyo, C.E.: "Radar Waveform Synthesis by Mean-Square Optimization Techniques," *IEEE Transactions on Aerospace and Electronic Systems,* July 1969.

[11] Nathanson, F.E.: *Radar Design Principles,* McGraw-Hill, 1969.

[12] Sinsky, A.I.; and Wang, C.D.: "Standarization of the Definition of the Radar Ambiguity Function," *IEEE Transactions on Aerospace and Electronic Systems,* July 1974.

[13] Powell, T.H.; and Sinsky, A.I.: "A Time Sidelobe Reduction Technique for Small Time-Bandwidth Chirp," *IEEE Transactions on Aerospace and Electronic Systems,* May 1974.

[14] Skolnik, M.: *Radar Handbook* (Chapter 3), McGraw-Hill, 1970.

[15] Cook, C.E.; and Bernfeld, M.: *Radar Signals – An Introduction to Theory and Applications,* Academic Press, Inc., New York, March 1967.

[16] Brookner, E.: "Multidimensional Ambiguity Functions of Linear, Interferometer, Antenna Arrays," *IEEE Transactions on Antennas and Propagation, Vol. AP-12,* pp. 551-561, September 1964.

[17] Brookner, E.: "Optimum Clutter Rejection," *IEEE Transactions on Information Theory, Vol. IT-11,* pp. 597-599, October 1965.

Chapter 8 Cook

Large Time-Bandwidth Radar Signals

Basic Properties of a Matched Filter Radar System

Figure 1 illustrates, in a very general way, the key elements of a matched filter radar system. In this diagram, a driving impulse (upper left) is fed into the signal generation filter, designated as H*(f). The characteristics of this filter are such that its response, or output, to being impulsed is a dispersed signal extending over the time interval T. This signal contains, as a result of the inherent characteristics of H*(f), a phase or frequency modulation referenced to a nominal center frequency, f_o, of the amplitude response of the filter. The bandwidth of H*(f) is W, as indicated by the spectrum response characteristic shown at the bottom of the figure. The time-dispersed output of H*(f) is amplified (and, in most practical implementations, time-truncated to eliminate long leading and trailing edges), fed to the antenna, radiated, reflected from a target, and returned (in greatly attenuated form) to the receiver. In order not to complicate the discussion, the receiver can be considered to have a bandwidth greater than W, the signal bandwidth. After suitable amplification, the output of the receiver is fed into the matched filter, H(f). The filters H*(f) and H(f) form a *Matched Filter Pair,* the combined response of which is given by $|H(f)|^2$. Since the bandwidth of H(f) is also W, this filter establishes the basic bandwidth of the radar receiver. The matched filter (and either H*(f) or H(f) can be considered as fitting this description) is usually a tapped delay line or a dispersive delay line. Various examples of these filter types are described in later portions of this chapter.

Elementary Fourier analysis indicates that the output of the receiver filter H(f) is related only to the squared magnitude response, $|H(f)|^2$, of the matched filter pair, and not to the time-dispersed characteristics of the output of the transmitter filter H*(f). Thus, the pulse at the output of the radar receiver is a bandpass filtered version of the original driving impulse. In principle, one can think of the action of the filter H*(f) as distorting the driving impulse in a prescribed way in order to produce the time-dispersed signal that is transmitted, and one can conceive of the action of H(f) as exactly offsetting the distorting effects of H*(f) in order to recover the original form of the signal. Since we have defined H*(f) and H(f) as having the same amplitude response characteristics, it is clear that the only differences between these two filters lies in their phase response characteristics; indeed, from the definition of conjugate filters, the phase response of one is the negative of that of the other. The output of the receiver matched filter is given by

$$h(t) = \int_{-\infty}^{\infty} |H(f)|^2 \exp\,[j2\pi ft]\; df$$

As shown in Figure 1, this output may contain associated time sidelobe signals. In some cases these signals may be considered objectionable, e.g., small-target masking effects. To remedy this situation, a mismatch filter, W(f), often follows the matched filter. The purpose of this filter is to shape the overall receiver response so as to lower the range sidelobes. This action is analogous to the taper associated with antenna illumination functions for reduction of antenna pattern sidelobes. When a mismatch or sidelobe reduction filter is used, the radar receiver output is described by

$$h_w(t) = \int_{-\infty}^{\infty} |H(f)|^2\, W(f) \exp\,[j2\pi ft]\; df$$

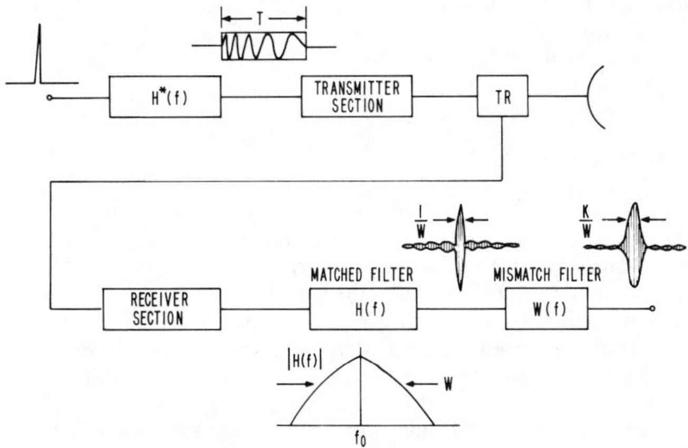

Figure 1 Basic Large Time-Bandwidth Radar System TW ≫ 1

The penalty for employing sidelobe reduction in this way is a loss in signal resolution (by the factor K shown in Figure 1), and a loss in signal-to-noise ratio that may be on the order of 1-3 dB. In all cases, only the matched filter combination provides the best signal-to-noise ratio at the receiver output when the interference is white gaussian noise.

One might think that the synthesis of a matched filter pair is an intricate design problem. There are two special cases of matched filter waveforms in which the filter design problem is relatively straightforward, in both analytic and practical terms. One such situation is the *maximum length sequence,* or *M-sequence*, binary phase code signal for which the matched filter signal generation is achieved with a tapped delay line; the receiver matched filter is realized by an identical filter in which the input and output connections have been reversed. The second signal of interest is the *linear FM,* or *chirp,* waveform. The signal generation filter has a linear delay versus frequency (or parabolic phase) characteristic. The conjugate of this filter can be achieved by using the identical filter preceded by a sideband inversion circuit to reverse the sense of the linear FM slope. Both of these signals have found extensive use in modern radar systems.

Some of the theoretical characteristics and attributes of large time-bandwidth signals are summarized below.

Signal

$$s(t) = a(t) \cos\,[2\pi f_o t + \phi(t)],\ -T/2 < t < T/2$$

Energy

$$E = \int_{-T/2}^{T/2} s^2(t)\,dt = \frac{1}{2} A^2 T \text{ (rectangular pulse)}$$

Matched Filter Output Signal-To-Noise Ratio

$$R_{max} = \frac{2E}{N_o} \text{ (independent of modulation parameters)}$$

Matched Filter Frequency Response

$$H(f) = S^*(f)$$

Matched Filter Impulse Response

$h(t) = ks(-t)$

Effective Signal Duration, α

$$\alpha^2 = \frac{4\pi^2}{2E} \int_{-T/2}^{T/2} t^2 |S(t)|^2 dt$$

$$= \frac{\pi^2 T^2}{3} \text{ (rectangular pulse); or } \alpha = 1.81 \text{ T}$$

Effective Signal Bandwidth, β

$$\beta^2 = \frac{8\pi^2}{E} \int_0^\infty (f-f_o)^2 \; |S(f)|^2 df$$

$$= \infty \text{ (rectangular pulse)}$$

Time-Bandwidth Product

Either TW or $\alpha\beta$

The signal s(t) is the output of the transmitter filter shown in Figure 1, and may have a phase function that is either continuous or discrete in nature; for example, the waveform could contain binary phase modulation. It is worthwhile to emphasize the matched filter output signal-to-noise relationship — it is completely independent of all of the signal parameters, except the energy contained in the waveform. This fact tells us that, with a large time-bandwidth (or matched filter) signal, duration T can be chosen to satisfy the detection requirements of the system in spite of any limitations imposed by transmitter peak power. Then, the signal modulation characteristics derived from the matched filter can be chosen to realize a signal bandwidth consistent with the desired resolution capability of the system. This design option is one that is not available in most simple pulsed systems.

The large time-bandwidth attributes shown above are primarily of interest to those concerned with the statistical analysis of signals. As indicated by the examples chosen, the definitions indicated may or may not be reasonably close to what one would intuitively surmise as practical figures for signal durations and bandwidths. When used with care, these definitions can be helpful in analyzing a variety of problems and in providing some measure of comparison for various types of large time-bandwidth signals.

So far, we have not touched explicitly on why the radar designer would consider using large time-bandwidth, or, as they are more commonly called, pulse compression signals. Some of the major reasons are listed below.

- Simultaneous Range Resolution and Detection Performance
  - Detection, resolution, and measurement accuracy can be independent design parameters
- Power Management
  - More efficient use of average power
  - Avoiding peak power limitations
- Reduction of Vulnerability to Certain Kinds of Interference
  - Clutter
  - Intentional Jamming

Large TW signals can be designed to have responses that are sensitive to the Doppler shift of the received signal. Thus, they can be used to provide combined range and Doppler measurement capability. In a large TW matched filter system, a dispersed pulse is transmitted and a narrow pulse is reconstituted by the action of the receiver matched filter. In the use of a matched filter pair, one can invoke the principle of conservation of energy to establish the fact that there must be an increase in the peak power level of the matched filter output compared to that of its input. In fact, this increase in the effective peak power of the matched filter time-compressed output is by the factor TW. Clearly, the burden on the actual peak power that need be generated at the transmitter is eased, allowing more efficient use of the average power capability of most available high power transmitter tubes. Large TW signals also result in a reduction in the vulnerability to certain kinds of interference or clutter signals. In the latter situation the radar designer is concerned with either the velocity or range resolution characteristics that can be obtained with large TW signals as a means for discriminating against unwanted target reflections. In the case of intentional interference, the use of wide-band modulation can be used to force the intentional jammer to radiate wide-band noise, which then becomes the optimum mode of jamming. The fact that the matched filter signal contains a fairly sophisticated type of modulation tends to inhibit a jamming source from duplicating this signal for the purposes of providing false information to the radar receiver.

Some of the disadvantages associated with the use of pulse compression waveforms are listed below.

- Minimum Range Set by the Transmitted Pulse Width, T
- Waveform Generated Self-Clutter and Range Sidelobes
- Ambiguous Range-Doppler Indications at Matched Filter Output
  - Can be minimized by careful choice of signal parameters
- More Complex Receiver and Transmitter
  - Signal generation and processing
  - Sensitivity to distortion and mismatching factors

The most obvious one is that the minimum range capability of the radar system is set by the transmitted pulse length, T. For example, if the radiated signal has a time-bandwidth (TW) product of 100 and the signal bandwidth is 1 MHz (500 foot resolution), then the minimum range is on the order of 8 miles (100 microseconds). In some applications that penalty may not be acceptable. Thus, the system designer should consider quite carefully the intended application before deciding what kind of a large TW signal, if any, is to be used. The fact that large TW signals often result in the existence of range sidelobe signals (as shown in Figure 1) can be a disadvantage when one is trying to detect small targets located near large targets, or when one is trying to detect signals in clutter. (This issue is discussed further at a later point.) For certain signals, these range and Doppler sidelobes can be nearly as large as the desired matched filter output. This situation can lead to an ambiguity, particularly in the case of noisy reception, when attempting to determine the true range and Doppler shift of the signals being received. Normally, these large values of ambiguous responses are associated with signals that have a high degree of periodicity in either their time or frequency characteristics. The nature of these responses is described by the behavior of the *Radar Ambiguity Function,* which has many interesting mathematical properties that are beyond the scope of this discussion. Chapter 4 of Reference 1 provides some insight into the nature of the ambiguity function and how it is applied to radar waveform design; see also Chapter 7 by Sinsky.

Although we have mentioned that the design of a matched filter system need not be unduly complicated, it does result in a more complex receiver and transmitter arrangement than that of a simple pulsed radar. The design of the matched filter must be approached with some care; some of the important details are illustrated in later sections. Finally, when a large TW signal is used, there is a greater sensitivity to different forms of system distortions, such as in amplitude and phase, and the distortion caused by the action of system nonlinearities (such as hard limiting) on overlapping signals before they are processed by the receiver filter. These distortions normally result in additional ambiguities in the form of range sidelobes that appear at the receiver output. The topic of the effects of distortion on large TW signals is treated in considerable detail in Chapter 11 in Reference 1.

Matched Filter Waveform Examples

With this background in hand, let us proceed to the consideration of some specific examples of pulse compression (large time-bandwidth or matched filter) signals. Although we have carried out this discussion under the assumption that there is a pair of matched filters used by the system – one in the transmitter and one in the receiver – we could easily postulate the use in the receiver of a single filter that is matched to a large TW signal produced by an electronic signal generator in the transmitter.

Linear FM

The linear FM pulse compression signal has proved to be one of the most useful waveforms for radar application. The example illustrated in Figure 2 shows an increasing frequency sweep. The instantaneous frequency is given by $f_o + \mu t/2\pi$, where μ is defined by 2π W/T. The matched filter for this signal is a dispersive delay line that has a linear delay versus frequency characteristic. The matched filter output has the approximate (sin x)/x shape, and the peak amplitude buildup is given by $(TW)^{1/2}$. The spectrum shown is approximately rectangular in shape, and this approximation improves as the value of TW increases. One of the advantages of the linear FM signal is that the output of the matched filter remains fairly well correlated over a range of Doppler shifts that are a large fraction of the signal bandwidth. When the received signal undergoes a Doppler shift, there is a displacement of the matched filter output from the true time location by an amount $\pm(T/W)f_d$, with f_d being the Doppler shift. This occurrence is often cited as a disadvantage of the linear FM signal, since it results in the requirement of having to know the value of Doppler shift before the true time location of the target can be determined. The solution to this problem is touched on at the bottom of the following list of properties of linear FM pulse compression.

- Has Relatively Large In-Close Range Sidelobes, but Overall Extended Sidelobe Interference is Very Low
- Closely Approximates the Optimum Band Limited Signal for Stationary or Slowly Moving Clutter Interference
- Optimum Range Measurement Accuracy Obtained by Using Range Prediction Property

  Matched filter output for positive slope FM shows locations of moving and stationary reflectors Δt seconds in future

$$\Delta t = \frac{T}{W} f_o$$

L-Band Example $T = 100\ \mu s$, $W = 10$ MHz, $\Delta t = 10^{-2}$ s
X-Band Example $T = 10\ \mu s$, $W = 10$ MHz, $\Delta t = 10^{-3}$ s
$T = 10\ \mu s$, $W = 100$ MHz, $\Delta t = 10^{-4}$ s

To insure that Δt is a positive number (i.e., prediction in the future), a positive slope linear FM signal must be transmitted. In most cases Δt is small enough so that it can be used in range tracking equations without having to be concerned over the effects of acceleration during the interval Δt.

As has been mentioned, the linear FM matched filter output has a (sin x)/x shape, thus implying relatively large range sidelobes adjacent to the peak output. These sidelobes can be reduced by the technique of receiver response mismatching illustrated in Figure 1. In actuality, the total integrated energy of all of the range sidelobes of this signal is relatively small; its value is related to the waveform figure of merit for signal detection in stationary range-distributed clutter. The optimum signal for this situation has a rectangular spectrum (exact (sin x)/x waveform). Thus, the unweighted linear FM compressed pulse signal is nearly optimum for use in the case of clutter that is stationary or moving very slowly with respect to the target velocity.

Barker Codes

The Barker Code waveforms are short binary or m-ary phase sequences that have the property of unit sidelobe level at the matched filter output. The peak response is N, the length of the code. Figure 3 presents the signal parameters and major waveform features of the binary Barker codes, where "+" indicates a zero phase shift reference and "–" denotes a 180 degree phase shift reference. The longest binary Barker sequence that is known is of length 13. This limitation is the major reason that this type of signal is not too practical for most large TW signal applications. Some effort has gone into the investigation of polyphase (4 or 6 equally arranged phase shift values possible) sequences with Barker properties (unit sidelobes). The longest lengths that have been uncovered for these polyphase Barker codes is 15 or 16; thus, it does not appear that very long sequences having the Barker sidelobe property will be available in the foreseeable future. The matched filter for a Barker signal can be either a tapped delay line or an appropriate digital shift register.

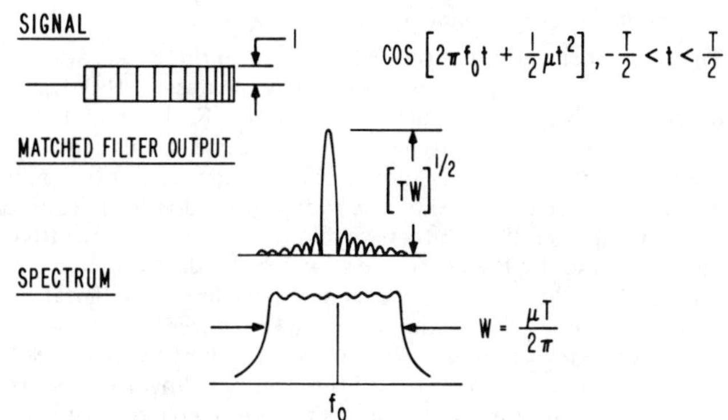

Figure 2 Linear FM Pulse Compression

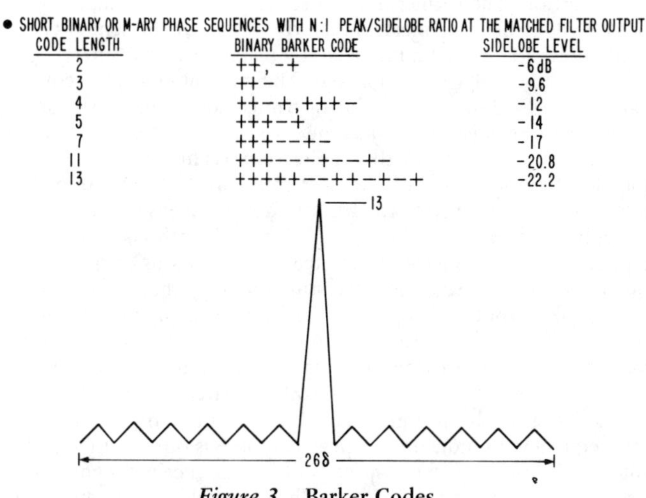

- SHORT BINARY OR M-ARY PHASE SEQUENCES WITH N:1 PEAK/SIDELOBE RATIO AT THE MATCHED FILTER OUTPUT

CODE LENGTH	BINARY BARKER CODE	SIDELOBE LEVEL
2	++, −+	−6dB
3	++−	−9.6
4	++−+, +++−	−12
5	+++−+	−14
7	+++−−+−	−17
11	+++−−−+−−+−	−20.8
13	+++++−−++−+−+	−22.2

Figure 3 Barker Codes

- LONG BINARY PHASE SEQUENCES OF LENGTH $N = 2^n - 1$, n = GENERATING SHIFT REGISTER LENGTH
- BANDWIDTH SET BY CLOCK RATE OF SHIFT REGISTER
- MATCHED FILTER OUTPUTS

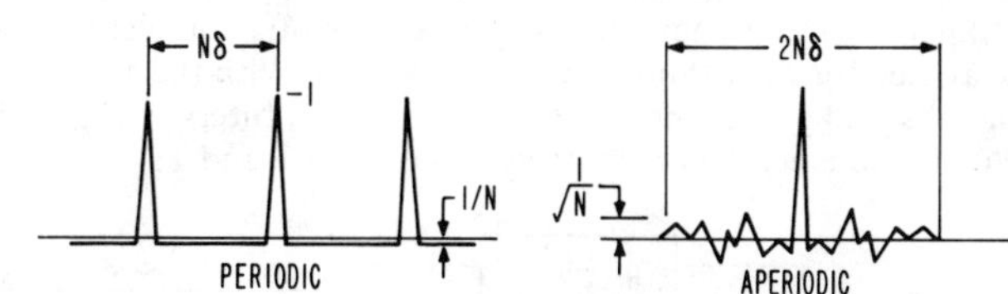

- TOTAL SIDELOBE CLUTTER RELATIVELY LARGE FOR TRUNCATED (APERIODIC) SEQUENCE
- A DOPPLER SHIFT OF 1/Nδ REDUCES THE PEAK RESPONSE TO THE SIDELOBE LEVEL
  - M-SEQUENCE CAN BE USED FOR SIMULTANEOUS ONE-HIT RANGE AND VELOCITY MEASUREMENT (PARAMETER ESTIMATION)

Figure 4 Maximal Length Sequences

Maximum Length Sequences

This class of signals is conceptually related to the Barker waveforms. They are long binary phase shift sequences (0-180 degree) of length N, N being as indicated in Figure 4. Rather than being generated by one of a matched filter pair, these signals can be generated with a relatively short feedback shift register of length n. The sequence produced by the shift register does not begin to repeat until after $2^n - 1$ shifts have taken place. The bandwidth of the signal is set by the clock rate at which the shift register is shifted. The matched filter is a tapped delay line, or a digital version of the same. The maximum length sequences can be used in two ways. One of these is a periodic repetition of the basic sequence; the result is the matched filter output shown at the left of Figure 4. This output has the desirable property of very low sidelobe levels, but has the disadvantages of ambiguous peak outputs and the fact that the system must always be transmitting during reception. The former is of no consequence if the unambiguous range of the system is less than the length of the sequence, while the latter can be minimized if a bistatic transmitter/receiver with very low leakage is employed. The maximum length sequence can also be used in an aperiodic, or truncated, format which involves the transmission of one sequence of length $N = 2^n - 1$. The matched filter output for this application has much higher sidelobe levels, as indicated on the right in Figure 4. The total integrated sidelobe energy of the truncated sequence is relatively large, but can be reduced by straightforward methods of sidelobe reduction. Both the periodic and truncated versions of the M-sequence waveform have the interesting property that the matched filter peak output reduces to the sidelobe level when the Doppler shift is $1/N\delta$, $N\delta$ being the duration of the basic sequence. This fact provides a basis for performing simultaneous range and Doppler measurement by means of a bank of matched filters, as shown in Figure 5a. Each filter in this bank is tuned to a different center frequency, as indicated. A peak response is obtained from that filter which matches the Doppler shift of the received signal; this peak is at the true range location. The other filters, for this same signal input, have outputs similar to the sidelobe formations of the matched output. This situation can lead to mutual interference among targets if several are under surveillance in the same range interval. A Doppler correction matrix, illustrated in Figure 5b, can be used to simulate a bank of matched filters. This arrangement makes use of one tapped delay line plus a Doppler correction matrix that is composed of a combination of resistors – relatively inexpensive items compared to the expense of a complete tapped delay line for each Doppler value of interest. With the advances in digital signal processing, the bank of matched filters of Figure 5a can be implemented easily; see Chapters 9, 10, and 11.

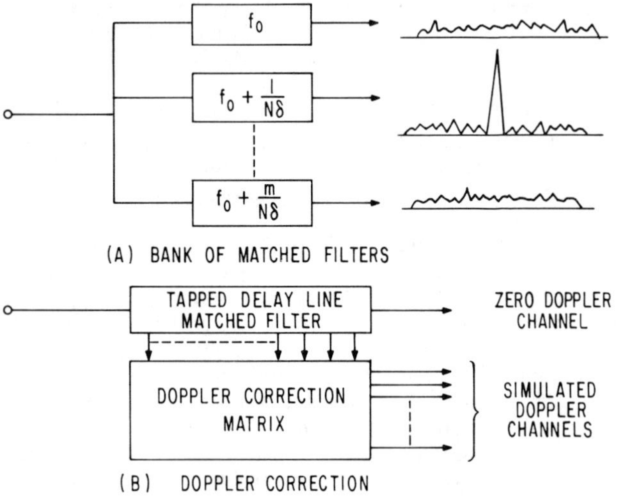

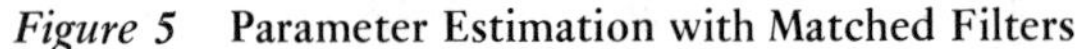

Figure 5 **Parameter Estimation with Matched Filters**

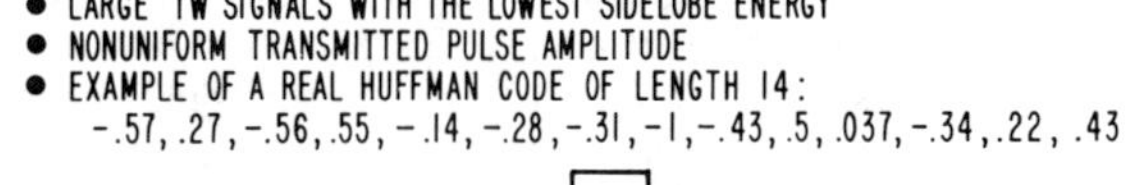

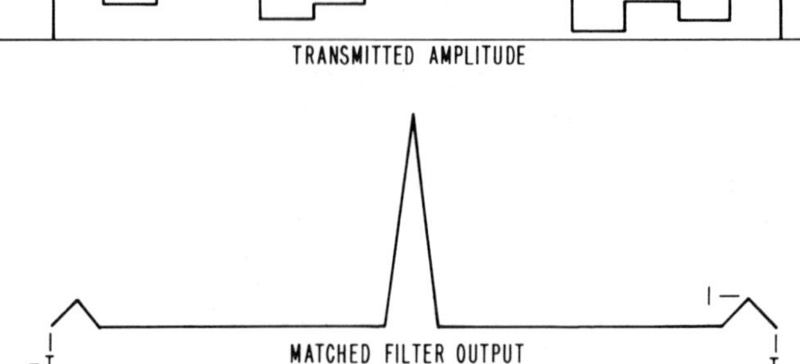

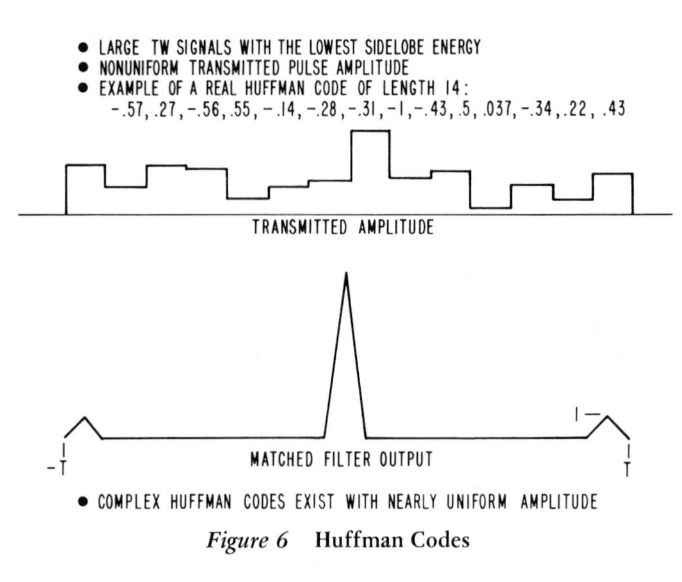

Figure 6 **Huffman Codes**

Huffman Codes

Huffman codes have the properties, shown in Figure 6, of a matched filter output that has zero sidelobe levels everywhere except for unit sidelobes at the beginning and end of the waveform coming out of the matched filter. The obvious disadvantage of this type of signal also is shown in the figure – the large variation in the amplitude of the transmitted pulse envelope (29 dB in the example shown). If the intended application can support the use of a linear amplifier transmitter, then the Huffman codes can be used as a signal that generates very little sidelobe interference. The example shown makes use of binary phase shifts. Other forms of Huffman codes with more uniform amplitude levels require the phase shifts to take on any possible value, which would complicate the design of the matched filter. The Huffman codes do not stay well correlated under Doppler shift conditions; however, since the known Huffman sequences are relatively short, it would take a relatively large Doppler shift to degrade the matched filter output.

Pulse Train Signals

Pulse train signals have a fairly restricted but important application. Because of their periodic nature, they can be used to create signals that exhibit what are known as *Doppler Clear Areas.* That is, in these regions there are no, or very low, range or Doppler sidelobe signals in the near vicinity of the matched filter peak output. This feature can be an important one when it is desired to examine many targets that cover a broad range of sizes, and which are contained in a compact region with potentially interfering targets at different velocities. The simple pulse train shown in Figure 7 has relatively large range and Doppler ambiguities, which can be reduced or moved around by adding phase or frequency coding to the various pulses in the pulse train. Generally, the penalty for this procedure is the reduction of the portion of the Doppler-clear area in which targets may be viewed. This limitation might be acceptable if it were to result in orienting the remaining Doppler clear area to better match the range and Doppler distribution of the targets of interest.

Other Large TW Waveforms

The features of several other large TW signals that have found application at one time or another in restricted applications are summarized below.

- *Coded And/Or Weighted Pulse Trains* – can achieve minimum self-clutter in specified intervals of range and Doppler; are sensitive to signal parameter variations
- *Stepped-FM* – approximates linear FM signal; it has large sidelobe response under Doppler shift conditions

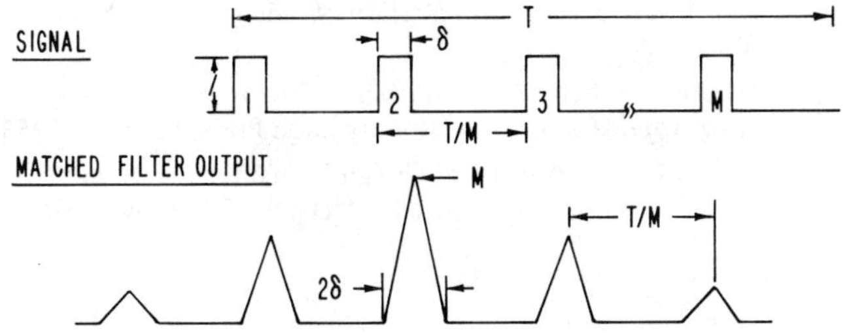

Figure 7 Pulse Train Signals

- *Frank Polyphase Code* – also approximates linear FM, but with much lower range sidelobes (matched filter). Doppler mismatched response has very marked range sidelobe ambiguities; it should not be used except under very low Doppler shift conditions

Waveform Design Criteria

How does one go about choosing a suitable large time-bandwidth signal for implementation in a particular system? The choice depends greatly on the application involved and the desired type of information that the system is expected to supply, in terms of range resolution, range and Doppler measurement capability, numbers of targets, the percentage of time a target might be expected to be masked by clutter, dynamic range of target responses, etc. Sometimes subjective factors can enter, such as biases about certain types of waveform characteristics or about forms that the matched filter implementation might take. Ideally, the system designer would be left free to make a choice on the basis of clearly stated system requirements. That list should include the following points:

- Does the signal chosen make efficient use of bandwidth? Does it achieve resolution and measurement accuracy in the most efficient manner, given the bandwidth constraints of the system? How should the problem of signal sidelobes be treated? Is it really important to reduce them to a gnat's eyebrow? This determination depends on whether the primary application is to isolated target measurement, to groups of isolated targets, or to targets in clutter.
- If sidelobe reduction is called for, the designer should be concerned with the total sidelobe energy contained in the processed output, and not just that in the peak levels.
- What are the expected distributions in range and velocity for targets that must be seen at the same time? Is it better to try to solve the range-velocity measurement problem strictly through matched filter techniques, or is it better to use the matched filter to achieve one measurement (such as range) and rely on other methods for extracting the remaining parameter?

Among the criteria listed above, that of the range sidelobe levels of the matched filter output can be very important. Figure 8 illustrates the sidelobe structures of an unweighted linear FM processed output (approximate (sin x)/x shape) and a Hamming weighted compressed pulse (-43 dB peak sidelobes). Also shown are the average sidelobe levels of aperiodic M-sequence compressed pulses for TW products of 100 and 1000. One can see that the initial sidelobes of the unweighted pulse are comparatively high; however, they fall off relatively rapidly below the sidelobe levels of the other signals illustrated. A fairly simple conclusion can be drawn – if the system is required to detect a small number of signals (each having a large dynamic range) and the possibility exists that any two might be within a few range resolution cells of each other, then the processor that has the (sin x)/x output is not the best one to use. However, if there are a large number of uniformly distributed signals existing over a large range interval (relative to the uncompressed pulse width duration), then the (sin x)/x output would be the most desirable of those shown. In the latter instance, the combined sidelobe energy interference of the other signals illustrated is greater than that of the (sin x)/x signal; thus, it has a greater effect in regard to small signals spread through the range interval of interest. This example is one in which a system specification solely of low range sidelobe levels would be incomplete, thus returning us to the initial criterion of the most effective use of bandwidth. In many cases, achieving low range sidelobe levels is costly in terms of bandwidth, and the system designer must be alert to the fact that, in many applications, range resolution is one of the better means by which to discriminate among signals under certain interference states. A logical choice might be to give the radar operator the option of switching the sidelobe reduction filter of Figure 1 in or out, depending on the specific circumstances.

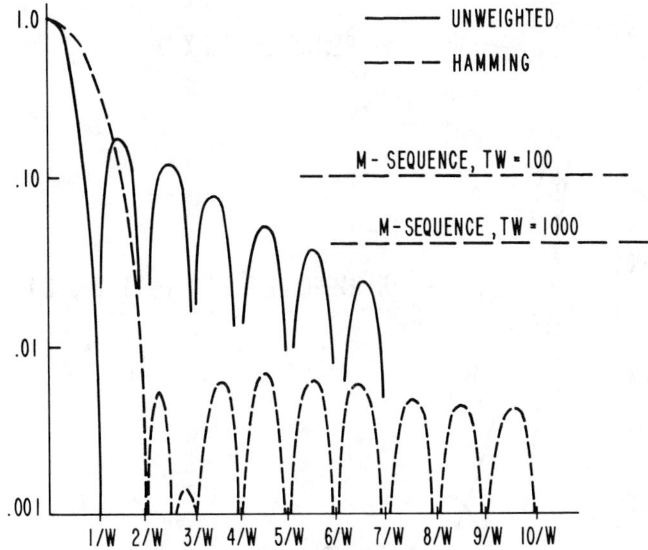

Figure 8 Comparative Sidelobe Levels

The comparison of the potential interference effects of waveform sidelobe structures (or self-clutter) was formalized by P.W. Woodward [2] many years ago using what he called the *Equivalent Rectangle of Resolution.* This concept is illustrated in Figure 9. The dotted line rectangle shown contains the equivalent amount of energy to that contained in the (sin x)/x waveform. Intuitively, one would conclude that a wider rectangle is required to match the total energy in the Hamming weighted pulse. These equivalent rectangles become a measure of the total interference that can be generated by each matched filter signal in a densely distributed target environment. The mathematical definition of this criterion is shown below and some comparative calculated results for signals having various spectrum power density functions are listed in Table 1.

$$T_r = \frac{\int_{-\infty}^{\infty} |S(f)|^4 \, df}{\left[\int_{-\infty}^{\infty} |S(f)|^2 \, df\right]^2}$$

Obviously, this criterion is most useful when the interfering signals have little or no relative velocity compared to the signals of interest. Woodward's *Equivalent Rectangle of Resolution,* or *Time Resolution Constant,* can be extended to the more general case in-

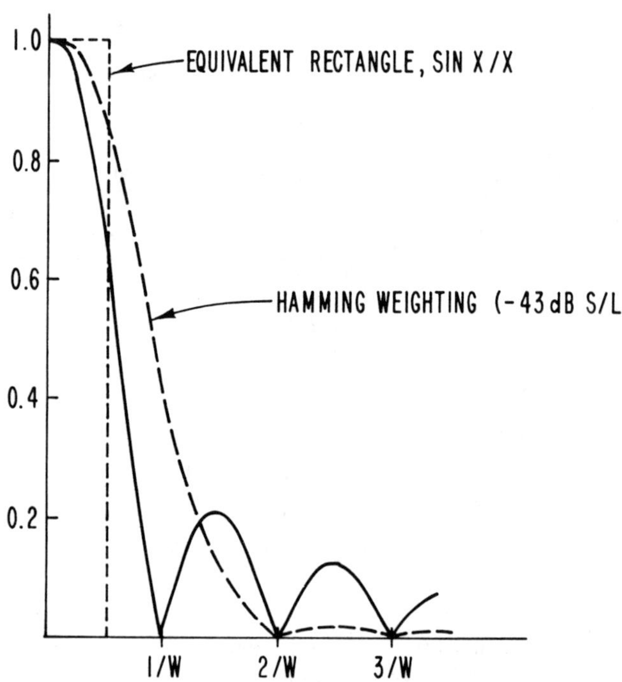

Figure 9 **Waveform Self-Clutter Criterion Equivalent Rectangle of Resolution**

Spectrum Power Density	*Band Limited Signals, W* *Sidelobe Level*	*Normalized T_r*
rect (f/W)	-13 dB	1.0
cos (πf/W)	-23	1.23
$\cos^2$ (πf/W)	-32	1.50
Hamming	-43	1.36

Table 1 **Time Resolution Constants for Various Spectrum Power Densities**

volving relative motion among various target returns of interest. This expansion is discussed in greater detail in Chapter 10 of Reference 1 and in Reference 3. Other criteria for the design of signals for clutter rejection may be found in the literature*. In one way or another, the ambiguity function properties of the matched filter signal are involved, with the objective being to maximize the desired matched filter output in certain velocity bands while minimizing the output from the undesired signals at other velocities. This problem has been, and continues to be, an interesting one for the radar designer.

*See first Editor's Note in Chapter 7 for example.

References

[1] Cook, C.E.; and Bernfeld, M.: *Radar Signals – An Introduction to Theory and Application,* Academic Press, New York, 1967.

[2] Woodward, P.M.: *Probability and Information Theory, with Application to Radar,* Pergamon Press, London, 1953.

[3] Deley, G.W.: "Waveform Design," Chapter 3 in *Radar Handbook* (M.I. Skolnik, ed.), McGraw-Hill, New York, 1970.

Chapter 9

Sheats

Fast Fourier Transform

The Fourier transform is a concept that is quite familiar to most engineers.

$$F(w) = \int_{-\infty}^{\infty} f(t)\, e^{-j\omega t}\, dt$$

This formula would be used, for example, in computing the far field pattern of an antenna from the aperture illumination function as illustrated in Figure 1.

A related concept, the Fourier series,

$$f(t) = \sum_{n=-\infty}^{\infty} C_N e^{jnx},\ C_N = a_N + jb_N$$

is likewise useful in many engineering problems.

The spectrum of a rectangular radar pulse (illustrated for an ideal case in Figure 2) can be found readily using the Fourier series. The series has coefficients given by

$$a_N = \frac{2}{T}\int_0^T f(t) \cos nx,\ b_N = \frac{2}{T}\int_0^T f(t) \cos nx$$

which may be identified with the harmonics of the waveform f(t).

Just how the Fourier integral (or transform) enters the radar problem is not so clear. Consider a typical situation as illustrated in Figure 3.

A radar transmitter transmits a known signal, s(t), which is reflected by a target and received by a receiver. The scintillations of the target and the thermal noise of the receiver both corrupt the signal, often to such an extent that an observer has trouble deciding whether the received signal is just noise or noise plus a signal. North in 1963 at RCA described a method for detection which is now known as the *Matched Filter* receiver (see Figure 4 and Chapters 7 and 8.)

The received signal, r(t), is correlated with the known transmitted signal, s(t). It can be shown readily that this structure is equivalent to the one shown in Figure 5, in which a bandpass filter, rather than a correlator, is used.

The bandpass filter is matched to the known form of the spectrum of the transmitted signal. Radar engineers, recognizing that the correct frequency of the received signal may not be known (due to an unknown Doppler shift), use a bank of filters (centered at different frequencies) hoping that the signal falls in at least one of them, as shown in Figure 6.

The structure begins to look like a spectrum analyzer, for it is in this manner that such a device actually is built. It can be seen that this device outputs a set of voltage levels (or numbers) which indicate the relative strength of discrete frequency components of the unknown signal.

This discussion may seem to be getting away from the subject of the Fourier transform; recall, though, that the Fourier transform can be used to find the spectrum of a signal, and that a spectrum analyzer actually forms a part of the radar in the fashion illustrated above. The problem now is how to process the received signal in such a way that the Fourier transform can be used.

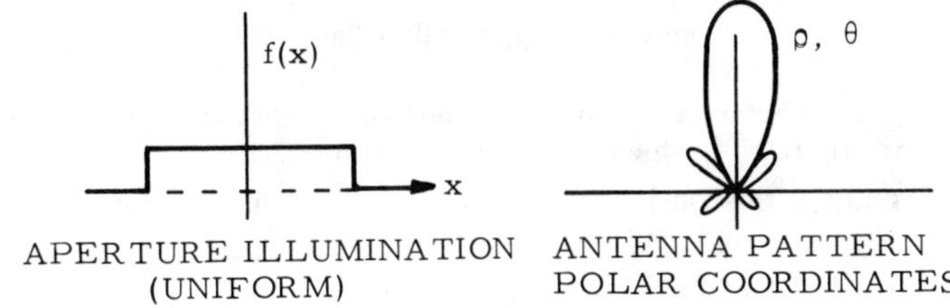

Figure 1 **Illumination Function**

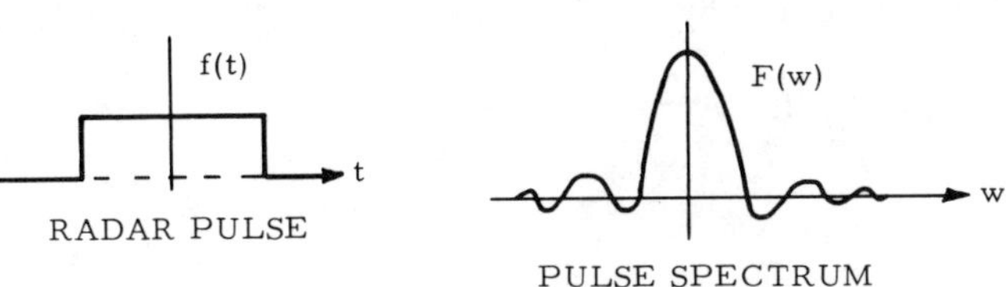

Figure 2 **Radar Pulse Spectrum**

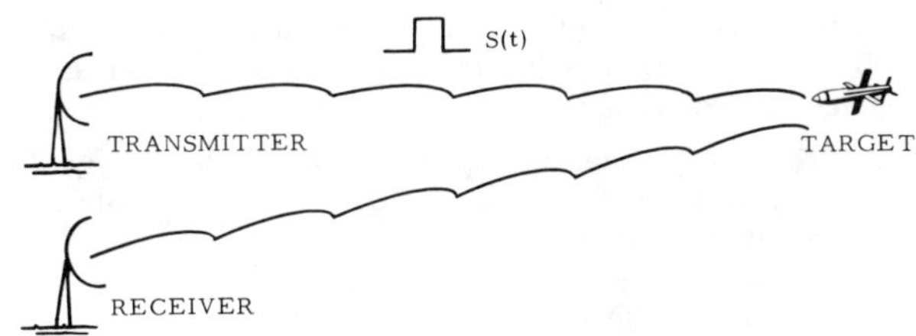

Figure 3 **Simplified Radar**

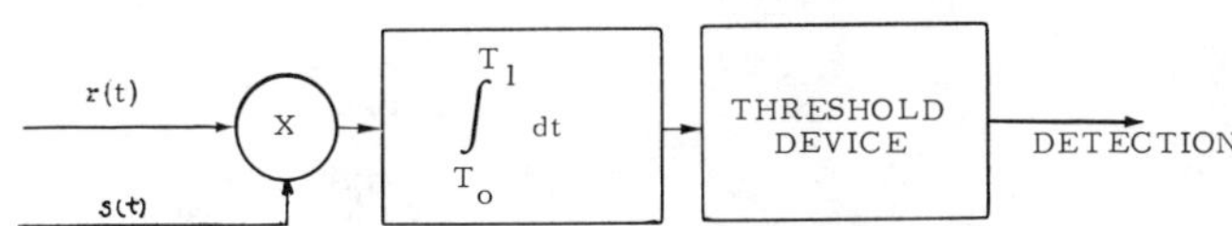

Figure 4 **Correlator Receiver**

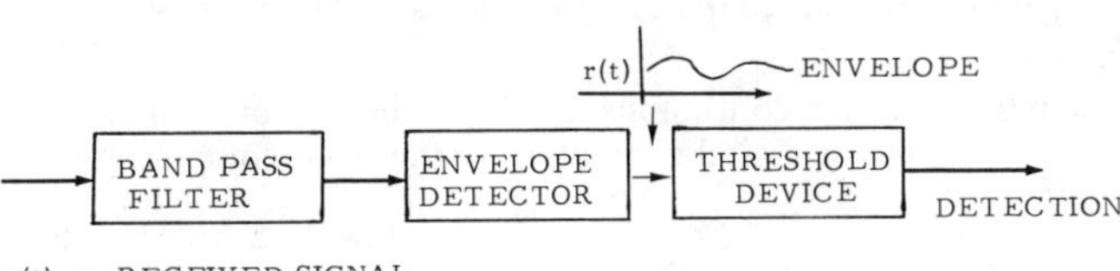

Figure 5 **Matched Filter**

With the advent of the digital computer, engineers have recognized that continuous signals can be dealt with effectively by sampling. Thus, if some signal is given (such as the output of a radar receiver), the signal could be sampled at some suitable rate and, from the set of samples, the signal could be reconstructed at any time.

Furthermore, the sampling theorem guarantees that if the signal is confined to a bandwidth of B Hz, no information is lost by sampling at 2B samples/s. Recognizing this fact, engineers began, in the mid-50's to consider computing spectra from sets of samples

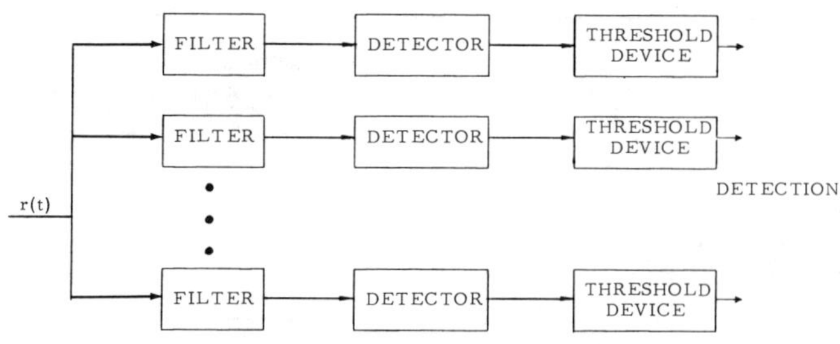

Figure 6 **Doppler Filter Bank**

of signals. This research led to the study of the Discrete Fourier Transform (DFT), which is illustrated in Figure 7.

Thus, just as the continuous signal f(t) $0 \leqslant t \leqslant T$ has the transform

$$F(w) = \int_{-\infty}^{\infty} f(t)e^{-j\omega t}\, dt$$

The discrete signal, f(nt), shown in Figure 7, has the discrete transform, F(kw), given by

$$F(kw) = \sum_{n=0}^{N-1} f(nT)e^{-j\omega T \cdot n \cdot K}$$

where:

$\omega = 2\pi/NT$

N = number of samples

T = spacing of samples

K = index over frequency

This definition is consistent, it is mathematically rigorous, and, like all good scientific formulas, it is believed because it gives results that agree with experimental data.

Note that the integral has been replaced by a summation sign, because discrete, rather than continuous, data is being considered. Changing notation a bit, let

$$F(k) = F(kw),\ f_n = f(nT)$$

and

$$W = e^{-j\,(2\pi)/N}$$

Then

$$F(k) = \sum_{n=0}^{N-1} f_n W^{nk}$$

This expression is the DFT — a formula used for computing spectra.

When going from a continuous signal to a discrete or sampled signal in order to use digital computing methods for computing the

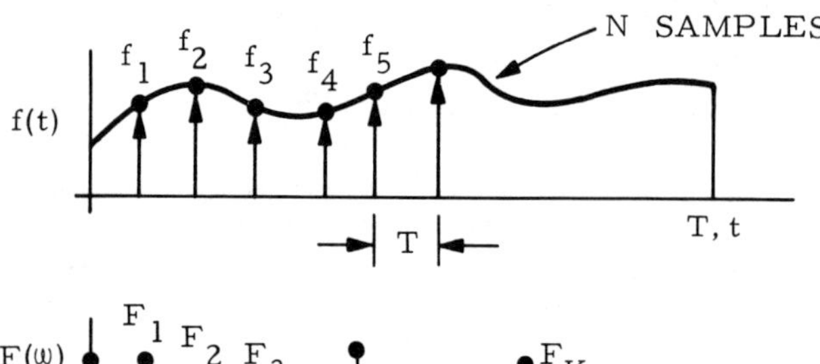

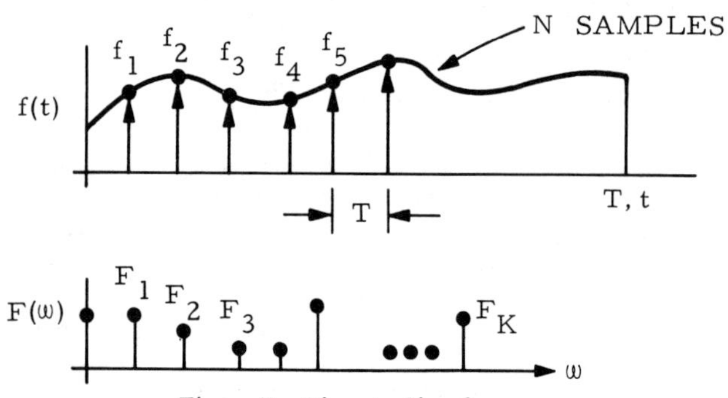

Figure 7 **Discrete Signals**

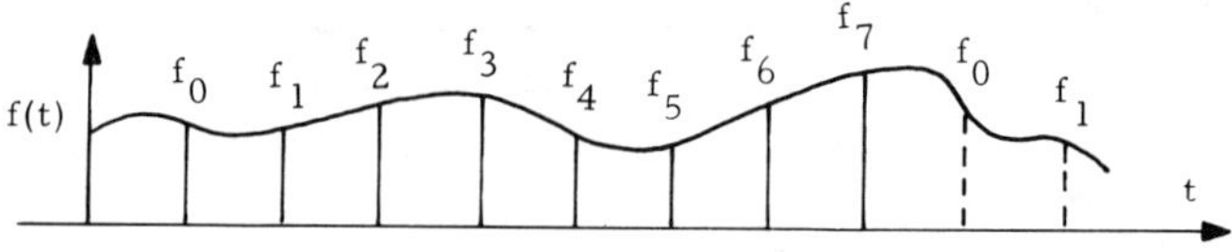

Figure 8 **Discrete Signal Batch Processing**

spectrum by means of the discrete Fourier transform, the fact that a finite memory is available for storing digitized samples forces chopping the signal into sections or batches of samples. These sections are contiguous blocks, as illustrated in Figure 8a. Figure 8b assumes that eight samples are taken in one block, and labeled f_0 through f_7.

To emphasize the extent of calculation involved for this trivial example, all eight equations are written out.

$$F_0 = f_0 + f_1 + f_2 + f_3 + f_4 + f_5 + f_6 + f_7$$
$$F_1 = f_0 + f_1W^1 + f_2W^2 + f_3W^3 + f_4W^4 + f_5W^5 + f_6W^6 + f_7W^7$$
$$F_2 = f_0 + f_1W^2 + f_2W^4 + f_3W^6 + f_4W^0 + f_5W^2 + f_6W^4 + f_7W^6$$
$$F_3 = f_0 + f_1W^3 + f_2W^6 + f_3W^1 + f_4W^4 + f_5W^7 + f_6W^2 + f_7W^5$$
$$F_4 = f_0 + f_1W^4 + f_2W^0 + f_3W^4 + f_4W^0 + f_5W^4 + f_6W^0 + f_7W^4$$
$$F_5 = f_0 + f_1W^5 + f_2W^2 + f_3W^7 + f_4W^4 + f_5W^1 + f_6W^6 + f_7W^3$$
$$F_6 = f_0 + f_1W^6 + f_2W^4 + f_3W^2 + f_4W^0 + f_5W^6 + f_6W^4 + f_7W^2$$
$$F_7 = f_0 + f_1W^7 + f_2W^6 + f_3W^5 + f_4W^4 + f_5W^3 + f_6W^2 + f_7W^1$$

This exercise is carried out to point out that there are N^2 (= 64) operations to be performed.

A note about W-terms — they are complex numbers; they all have magnitude unity; and they can be represented as points in the complex plane on the unit circle. Angular separation is $2\pi/N$ radians (in this case, $\pi/4$ or 45°). This definition is best demonstrated by referring to Figure 9.

Note — all terms reduce modulo 8; i.e., $W^9 = W^1$, $W^{10} = W^2$, etc.

The equations written above can be written more conveniently in matrix form.

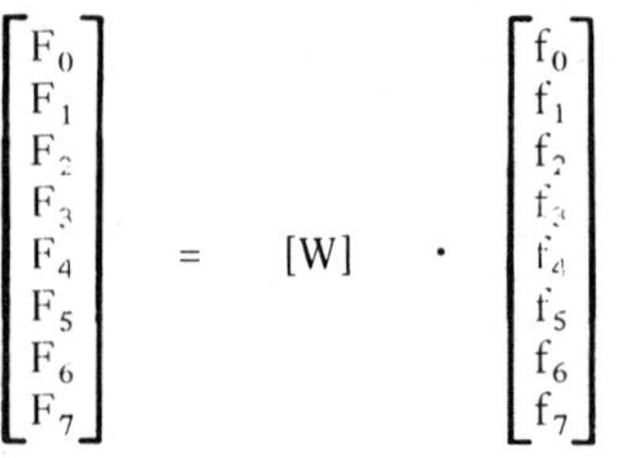

$$\begin{bmatrix} F_0 \\ F_1 \\ F_2 \\ F_3 \\ F_4 \\ F_5 \\ F_6 \\ F_7 \end{bmatrix} = [W] \cdot \begin{bmatrix} f_0 \\ f_1 \\ f_2 \\ f_3 \\ f_4 \\ f_5 \\ f_6 \\ f_7 \end{bmatrix}$$

If the W-matrix is written out it is

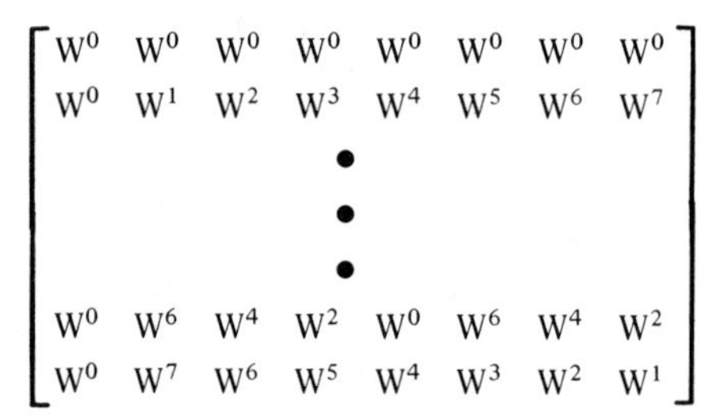

$$\begin{bmatrix} W^0 & W^0 & W^0 & W^0 & W^0 & W^0 & W^0 & W^0 \\ W^0 & W^1 & W^2 & W^3 & W^4 & W^5 & W^6 & W^7 \\ & & & & \bullet & & & \\ & & & & \bullet & & & \\ & & & & \bullet & & & \\ W^0 & W^6 & W^4 & W^2 & W^0 & W^6 & W^4 & W^2 \\ W^0 & W^7 & W^6 & W^5 & W^4 & W^3 & W^2 & W^1 \end{bmatrix}$$

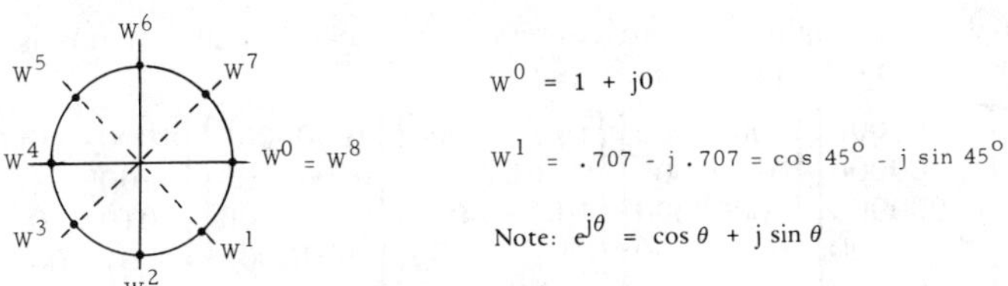

Figure 9 W-Terms

To save some labor in writing out this matrix, just the exponents are included.

$$
W = \begin{bmatrix}
0&0&0&0&0&0&0&0\\
0&1&2&3&4&5&6&7\\
0&2&4&6&0&2&4&6\\
0&3&6&1&4&7&2&5\\
0&4&0&4&0&4&0&4\\
0&5&2&7&4&1&6&3\\
0&6&4&2&0&6&4&2\\
0&7&6&5&4&3&2&1
\end{bmatrix}
\begin{matrix} 000\\001\\010\\011\\100\\101\\110\\111 \end{matrix}
$$

To the right of the matrix are the row numbers, 0 through 7, in binary form. The reason for this row numbering is shown shortly.

As has been noted, considerable computation is required even in this trivial case. From the equations above, there are N^2 multiplications and N^2 adds (each complex; note that a complex multiply is 4 multiplications and 4 additions).

In many problems of interest there are typically 1000 f_n's — the number of (complex) operations required to compute a spectrum is $10^{3 \cdot 2} = 10^6$ — using a reasonably good computer with a 4 μs multiply time, 16 s is required. Add to this period time for the additions and memory access, and the total is a full minute more — hardly a real-time process.

An amazing labor-saving device called the "Cooley-Tukey"* algorithm can be employed. This algorithm has been renamed the FFT — it is not in itself a transform — just an efficient algorithm for computing a transform. The trick involved is to manipulate the W-matrix so as to introduce as many ones or zeros as possible and thereby reduce the number of operations (additions and multiplications). At the same time, the processing is arranged in a sequence of repeated operations.

The first step is to re-arrange the rows of W in "bit-reversed" order; that is, the index number of each row has its digits reversed. The row, index 001, now has index 100 and is interchanged with Row 4.

$$
W = \left[\begin{array}{cccc:cccc}
0&0&0&0&0&0&0&0\\
0&4&0&4&0&4&0&4\\
0&2&4&6&0&2&4&6\\
0&6&4&2&0&6&4&2\\
\hdashline
0&1&2&3&4&5&6&7\\
0&5&2&7&4&1&6&3\\
0&3&6&1&4&7&2&5\\
0&7&6&5&4&3&2&1
\end{array}\right]
\begin{matrix} 000\\100\\010\\110\\ \\001\\101\\011\\111 \end{matrix}
$$

A binary number written to the right of the rows indicates the new order. Note that this procedure does not change the matrix (as an operator). Now the matrix can be factored, after first partitioning. Note that:

1) The top left sub-matrix $\begin{bmatrix}0&0&0&0\\0&4&0&4\\0&2&4&6\\0&6&4&2\end{bmatrix}$ is repeated.

2) The matrix $\begin{bmatrix}4&5&6&7\\4&1&6&3\\4&7&2&5\\4&3&2&1\end{bmatrix}$ is the negative of $\begin{bmatrix}0&1&2&3\\0&5&2&7\\0&3&6&1\\0&7&6&5\end{bmatrix}$

i.e., $W^4 = -W^0$
$W^5 = -W^1$
$W^6 = -W^2$ etc.

3)
$$
\begin{bmatrix}0&1&2&3\\0&5&2&7\\0&3&6&1\\0&7&6&5\end{bmatrix} = \begin{bmatrix}0&0&0&0\\0&4&0&4\\0&2&4&6\\0&6&4&2\end{bmatrix} \cdot \begin{bmatrix}0&0&0&0\\0&1&0&0\\0&0&2&0\\0&0&0&3\end{bmatrix}
$$

Remember, these numbers are exponents — $1 \times 2 = W^1 \times W^2 = 3$.

Thus, the original W-matrix can be written in the form

$$
[W] = \begin{bmatrix} T & T \\ T \cdot K & -T \cdot K \end{bmatrix}
$$

where: $T = \begin{bmatrix}0&0&0&0\\0&4&0&4\\0&2&4&6\\0&6&4&2\end{bmatrix}$ $K = \begin{bmatrix}0&0&0&0\\0&1&0&0\\0&0&2&0\\0&0&0&3\end{bmatrix}$ $I = \begin{bmatrix}1&0&0&0\\0&1&0&0\\0&0&1&0\\0&0&0&1\end{bmatrix}$

Then

$$
[W] = \begin{bmatrix} T & T \\ T \cdot K & -T \cdot K \end{bmatrix} = \begin{bmatrix} T & \phi \\ \phi & T \end{bmatrix} \begin{bmatrix} I & I \\ K & -K \end{bmatrix}
$$

or

$$
[W] = \begin{bmatrix} T & \phi \\ \phi & T \end{bmatrix} \cdot \begin{bmatrix} I & \phi \\ \phi & K \end{bmatrix} \cdot \begin{bmatrix} I & I \\ I & -I \end{bmatrix}
$$

which shows the matrix factoring step by step. Note that the first "iteration" produces three matrices

$$
\begin{bmatrix}
0&0&0&0&0&0&0&0\\
0&4&0&4&0&0&0&0\\
0&2&4&6&0&0&0&0\\
0&6&4&2&0&0&0&0\\
0&0&0&0&0&0&0&0\\
0&0&0&0&0&4&0&4\\
0&0&0&0&0&2&4&6\\
0&0&0&0&0&6&4&2
\end{bmatrix} \cdot
\begin{bmatrix}
1&0&0&0&0&0&0&0\\
0&1&0&0&0&0&0&0\\
0&0&1&0&0&0&0&0\\
0&0&0&1&0&0&0&0\\
0&0&0&0&0&0&0&0\\
0&0&0&0&0&1&0&0\\
0&0&0&0&0&0&2&0\\
0&0&0&0&0&0&0&3
\end{bmatrix} \cdot
\begin{bmatrix}
1&0&0&0&1&0&0&0\\
0&1&0&0&0&1&0&0\\
0&0&1&0&0&0&1&0\\
0&0&0&1&0&0&0&1\\
1&0&0&0&-1&0&0&0\\
0&1&0&0&0&-1&0&0\\
0&0&1&0&0&0&-1&0\\
0&0&0&1&0&0&0&-1
\end{bmatrix}
$$

Care must be taken to note that the I-matrix contains the number "1," not W^1 (which is .707 - j .707) and the null matrix, ϕ, contains the number zero, not W^0 (which is 1 + j0). The T-matrix, however, still has as its elements the exponents of the W-terms.

The first matrix can be partitioned again

$$
\begin{bmatrix}
0&0&0&0&0&0&0&0\\
0&4&0&4&0&0&0&0\\
0&2&4&6&0&0&0&0\\
0&6&4&2&0&0&0&0\\
0&0&0&0&0&0&0&0\\
0&0&0&0&0&4&0&4\\
0&0&0&0&0&2&4&6\\
0&0&0&0&0&6&4&2
\end{bmatrix} \rightarrow
\left[\begin{array}{cccc:cccc}
0&0&0&0&0&0&0&0\\
0&4&0&4&0&0&0&0\\
0&2&4&6&0&0&0&0\\
0&6&4&2&0&0&0&0\\
\hdashline
0&0&0&0&0&0&0&0\\
0&0&0&0&0&4&0&4\\
0&0&0&0&0&2&4&6\\
0&0&0&0&0&6&4&2
\end{array}\right]
$$

*Co-discovered in 1965 by J. Cooley and J. Tukey of BTL.

The matrix $\begin{bmatrix}0&0&0&0\\0&4&0&4\\0&2&4&6\\0&6&4&2\end{bmatrix}$ can then be factored as follows

$$\begin{bmatrix}0&0&0&0\\0&4&0&4\\0&2&4&6\\0&6&4&2\end{bmatrix} = \begin{bmatrix}T & T'\\ T'\cdot K & -T'\cdot K'\end{bmatrix}$$

$$= \begin{bmatrix}T' & \phi\\ \phi & T'\end{bmatrix}\cdot\begin{bmatrix}I' & \phi\\ \phi & K'\end{bmatrix}\cdot\begin{bmatrix}I' & I'\\ I' & -I'\end{bmatrix}$$

where:

$$T' = \begin{bmatrix}0&0\\0&4\end{bmatrix} \quad K' = \begin{bmatrix}0&0\\0&2\end{bmatrix} \quad I' = \begin{bmatrix}1&0\\0&1\end{bmatrix}$$

Note $\begin{bmatrix}0&0\\0&4\end{bmatrix}\cdot\begin{bmatrix}0&0\\0&2\end{bmatrix} = \begin{bmatrix}0&2\\0&6\end{bmatrix}$ *and* $\begin{bmatrix}4&6\\4&2\end{bmatrix} = -\begin{bmatrix}0&2\\0&6\end{bmatrix}$

This expression is an iteration of the procedure already used. Once again, it should be noted that the T-matrix contains exponents while the null and identity matrices contain literal 1 or ϕ.

The sub-matrices T′, K′, and I′ are now 2 x 2, and no further factoring is possible. The final expression for W is

$$[W] = \begin{bmatrix}T'&\phi&\phi&\phi\\ \phi&T'&\phi&\phi\\ \phi&\phi&T'&\phi\\ \phi&\phi&\phi&T'\end{bmatrix}\cdot\begin{bmatrix}I'&\phi&\phi&\phi\\ \phi&K'&\phi&\phi\\ \phi&\phi&I'&\phi\\ \phi&\phi&\phi&K'\end{bmatrix}\cdot\begin{bmatrix}I'&I'&\phi&\phi\\ I'&-I&\phi&\phi\\ \phi&\phi&I'&I'\\ \phi&\phi&I'&-I\end{bmatrix}\cdot\begin{bmatrix}I&\phi\\ \phi&K\end{bmatrix}\cdot\begin{bmatrix}I&I\\ I&-I\end{bmatrix}$$

The development of the FFT given here is known as the *sparse matrix* method. It is presented quite elegantly in an article by Pease [1]. Note that the first "iteration" produces three matrices

$$\begin{bmatrix}0&0&0&0&0&0&0&0\\0&4&0&4&0&0&0&0\\0&2&4&6&0&0&0&0\\0&6&4&2&0&0&0&0\\0&0&0&0&0&0&0&0\\0&0&0&0&0&4&0&4\\0&0&0&0&0&2&4&6\\0&0&0&0&0&6&4&2\end{bmatrix}\cdot\begin{bmatrix}1&0&0&0&0&0&0&0\\0&1&0&0&0&0&0&0\\0&0&1&0&0&0&0&0\\0&0&0&1&0&0&0&0\\0&0&0&0&0&0&0&0\\0&0&0&0&0&1&0&0\\0&0&0&0&0&0&2&0\\0&0&0&0&0&0&0&3\end{bmatrix}\cdot\begin{bmatrix}1&0&0&0&1&0&0&0\\0&1&0&0&0&1&0&0\\0&0&1&0&0&0&1&0\\0&0&0&1&0&0&0&1\\1&0&0&0&-1&0&0&0\\0&1&0&0&0&-1&0&0\\0&0&1&0&0&0&-1&0\\0&0&0&1&0&0&0&-1\end{bmatrix}$$

The second iteration splits the left-hand matrix into 3 matrices by partitioning and factoring the sub-matrix.

$$\begin{bmatrix}0&0&0&0\\0&4&0&4\\0&2&4&6\\0&6&4&2\end{bmatrix}$$

Thus the left matrix

$$\begin{bmatrix}0&0&0&0&0&0&0&0\\0&4&0&4&0&0&0&0\\0&2&4&6&0&0&0&0\\0&6&4&2&0&0&0&0\\0&0&0&0&0&0&0&0\\0&0&0&0&0&4&0&4\\0&0&0&0&0&2&4&6\\0&0&0&0&0&6&4&2\end{bmatrix}$$

splits into the following three terms

$$\begin{bmatrix}0&0&0&0&0&0&0&0\\0&4&0&0&0&0&0&0\\0&0&0&0&0&0&0&0\\0&0&0&4&0&0&0&0\\0&0&0&0&0&0&0&0\\0&0&0&0&0&4&0&0\\0&0&0&0&0&0&0&0\\0&0&0&0&0&0&0&4\end{bmatrix}\cdot\begin{bmatrix}1&0&0&0&0&0&0&0\\0&1&0&0&0&0&0&0\\0&0&0&0&0&0&0&0\\0&0&0&2&0&0&0&0\\0&0&0&0&1&0&0&0\\0&0&0&0&0&1&0&0\\0&0&0&0&0&0&0&0\\0&0&0&0&0&0&0&2\end{bmatrix}\cdot\begin{bmatrix}1&0&1&0&0&0&0&0\\0&1&0&1&0&0&0&0\\1&0&-1&0&0&0&0&0\\0&1&0&-1&0&0&0&0\\0&0&0&0&1&0&1&0\\0&0&0&0&0&1&0&1\\0&0&0&0&1&0&-1&0\\0&0&0&0&0&1&0&-1\end{bmatrix}$$

while the right two matrices remain unchanged. Thus when W is written out in full it appears as follows

$$[W] = \begin{bmatrix}0&0&0&0&0&0&0&0\\0&4&0&0&0&0&0&0\\0&0&0&0&0&0&0&0\\0&0&0&4&0&0&0&0\\0&0&0&0&0&0&0&0\\0&0&0&0&0&4&0&0\\0&0&0&0&0&0&0&0\\0&0&0&0&0&0&0&4\end{bmatrix}\begin{bmatrix}1&0&0&0&0&0&0&0\\0&1&0&0&0&0&0&0\\0&0&0&0&0&0&0&0\\0&0&0&2&0&0&0&0\\0&0&0&0&1&0&0&0\\0&0&0&0&0&1&0&0\\0&0&0&0&0&0&0&0\\0&0&0&0&0&0&0&2\end{bmatrix}\begin{bmatrix}1&0&1&0&0&0&0&0\\0&1&0&1&0&0&0&0\\1&0&-1&0&0&0&0&0\\0&1&0&-1&0&0&0&0\\0&0&0&0&1&0&1&0\\0&0&0&0&0&1&0&1\\0&0&0&0&1&0&-1&0\\0&0&0&0&0&1&0&-1\end{bmatrix}\begin{bmatrix}1&0&0&0&0&0&0&0\\0&1&0&0&0&0&0&0\\0&0&1&0&0&0&0&0\\0&0&0&1&0&0&0&0\\0&0&0&0&0&0&0&0\\0&0&0&0&0&1&0&0\\0&0&0&0&0&0&2&0\\0&0&0&0&0&0&0&3\end{bmatrix}\begin{bmatrix}1&0&0&0&1&0&0&0\\0&1&0&0&0&1&0&0\\0&0&1&0&0&0&1&0\\0&0&0&1&0&0&0&1\\1&0&0&0&-1&0&0&0\\0&1&0&0&0&-1&0&0\\0&0&1&0&0&0&-1&0\\0&0&0&1&0&0&0&-1\end{bmatrix}$$

This procedure may be used to obtain the FFT of any sequence of length 2^N; with some variation, it may be used for mixed Radix FFTs (i.e., combinations of bases other than base 2) as well.

This process may appear at first to increase, rather than reduce, the computations because there are now 5 matrices instead of just one to evaluate. But note that the new matrices are now either diagonal or unity-quasi-diagonal. The diagonal matrices require N/2 (complex) operations; the quasi-diagonal ones require N additions. There are $\log_2(N)$ diagonal matrices and, hence, $N/2 \log_2(N)$ operations in all. For a large value of N (e.g., 10^3), $N/2 \log_2 N = 500 \times 10 = 5000$ operations — a considerable saving in computation compared to the 10^6 operations required if the FFT algorithm is not used.

These operations can be done in sequence, starting from the right, as follows

$$\begin{bmatrix}1000 & 1000\\ 0100 & 0100\\ & \\ \text{etc.} & \\ & \end{bmatrix}\begin{bmatrix}f_0\\ f_1\\ \cdot\\ \cdot\\ \cdot\\ \cdot\\ f_7\end{bmatrix} = \begin{bmatrix}A_0\\ A_1\\ \cdot\\ \cdot\\ \cdot\\ \cdot\\ A_7\end{bmatrix}$$

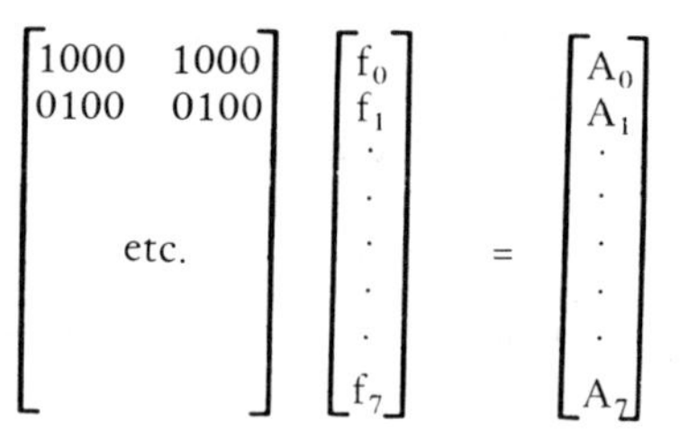

That is, the first quasi-diagonal matrix operating on the f-column vector produces a new column vector, A. That is

$A_0 = f_0 + f_4$, $A_1 = f_1 + f_5$, etc.

$A_4 = f_0 - f_4$

A Mason-type flow chart may be used to depict more graphically the matrix operations. For example, A_7 is computed as shown below

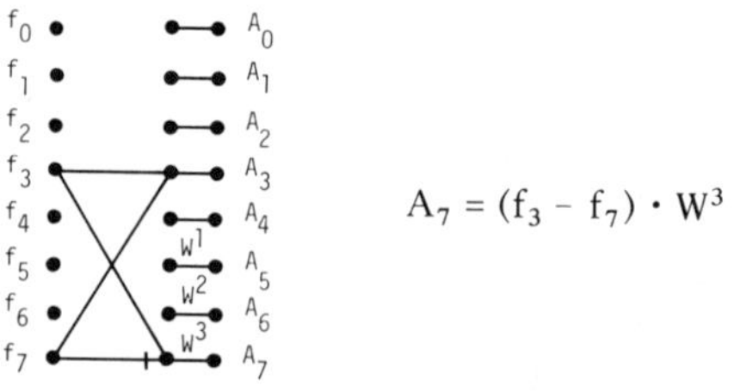

$$A_7 = (f_3 - f_7) \cdot W^3$$

The operation flow-charted by

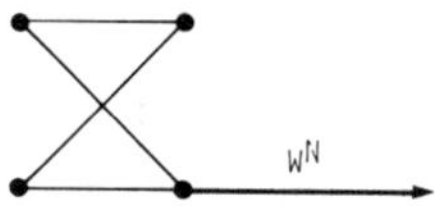

is called a *butterfly* and the W^N factor is called the *twiddle* factor. Many general-purpose computers now have as a hardware option a "butterfly" unit for doing FFTs; it is clear that getting through an FFT really only involves going through a butterfly operation repeatedly with different twiddle factors.

Note that F's come out in "Bit-Reversed" order.

This is Cooleys' first algorithm and is called-

Decimation in Time

NOTE ●—+—● = ●——— -1 ———●

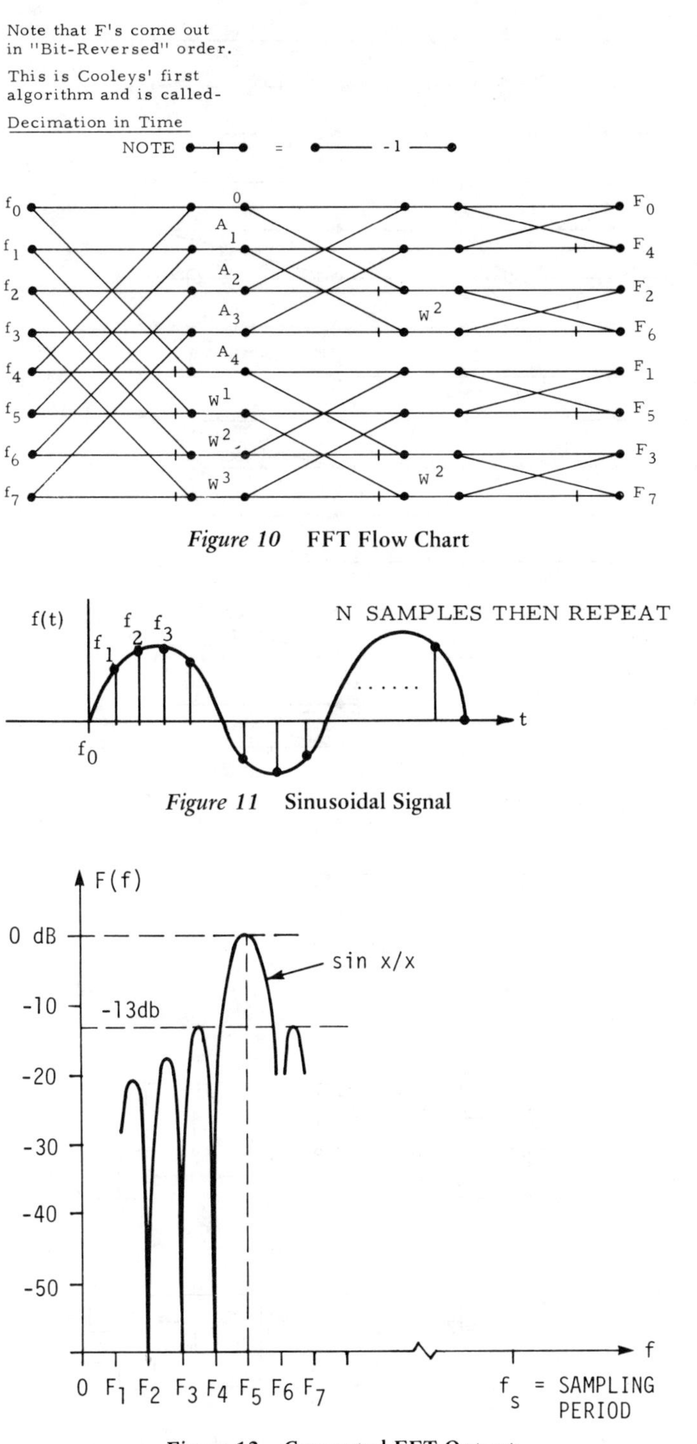

Figure 10 **FFT Flow Chart**

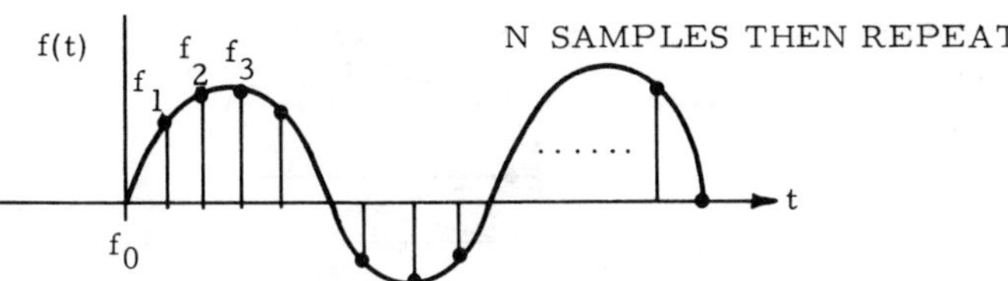

Figure 11 **Sinusoidal Signal**

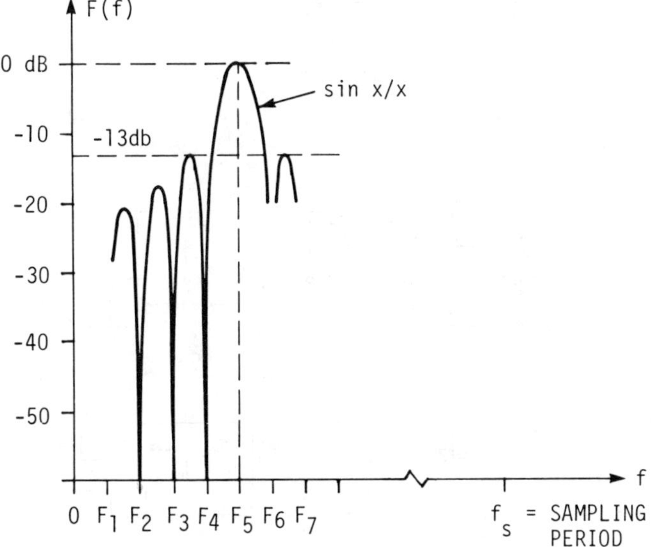

Figure 12 **Computed FFT Output**

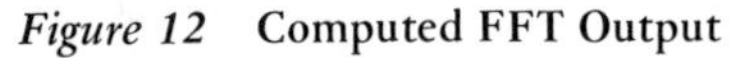

The full flow chart (Mason-type) is written out as shown in Figure 10.

The amazing amount of labor saved by this algorithm enabled Fourier transforms to be computed in real time by digital computers. Special-purpose machines (with complex multiply hardware and special address computing hardware) make the process even faster.

Consider the sinusoid shown in Figure 11; let this graph be sampled at a fixed rate to generate the f's as shown, resulting in a fixed number of samples. Now select a spectral coefficient (say F_5) and compute its magnitude with the FFT; then, change the period (or frequency) of the sinusoid and repeat, plotting magni-

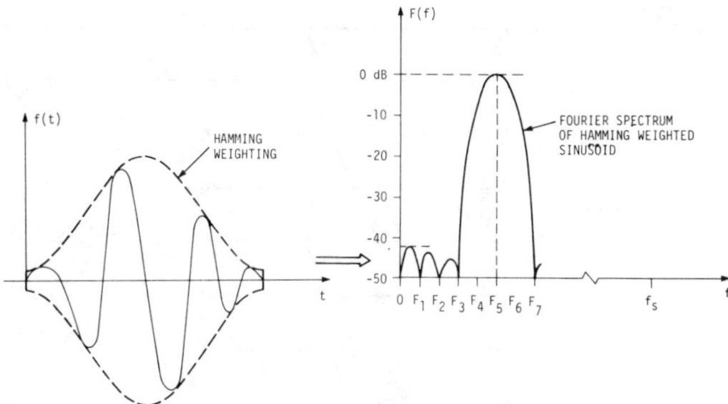

Figure 13 **Weighted Transform**

tude of F_5 (a complex number) against W (the radian frequency of the "input" sinusoid). The result is as shown in Figure 12.

Note that F_5 has a "response" to a changing input frequency which is similar to that of a filter. The stop-band response is poor, but it can be improved by "weighting" the input. Note that what is being computed is a Fourier transform of two functions multiplied together (i.e., sinusoid and rectangular function).

As a result, the two transforms are convolved (i.e., an impulse and a (sin x)/x function). If a tapered function is used to time limit the sequence (which must be finite), the result is the transform of this modulating function (see Figure 13). This action reduces the "sidelobe" response of the filter in exactly the same way that tapering an antenna array reduces far-field pattern sidelobes.

Note that the filters being discussed are realized digitally with an FFT – returning to the radar problem mentioned earlier. One of the important uses of the FFT is to realize digitally a bank of contiguous filters and thus produce a spectrum analyzer or Doppler filter bank.

The usual approach to implementing a FFT is the "double buffer" method. Referring to the flow chart in Figure 10, the buffer is not cleared in the order that data arrive; hence, to do "in place" computation, two buffers are needed – one to accommodate new data coming in and one to hold the block being worked on. The storage required is thus twice the block length.

A remarkable Raytheon invention (by G. Works) gets around this difficulty in an extremely clever way – by using Shift Register storage. The manner in which the FFT is obtained is understood most easily by referring to the flow chart (Figure 14) which shows diagrammatically how a decimation in time Cooley-Tukey process is carried out for a simple case (N=16). The W_0 to W_8 terms correspond to the trig weighting terms, W^N; the f_0 to f_{16}, input data points; the A's and B's, intermediate calculations; and F_0 through F_{15}, the discrete Fourier transform of the input data samples. To implement this process with shift registers, the scheme shown in Figure 15 is used. Note that data are entered into the shift register in normal order. After eight-word counts, the first register (length = 8 data words) is full. The clock advances one more count as the switches activate to the position shown in Figure 15b and the arithmetic unit forms the quantities A_0 and A_8 using the equations given above. These results are stored in the shift register assembly, as shown in Figure 15b. Processing continues in this fashion. Figure 15c shows the state of the unit after five additional clock counts – note that quantities A_0 and A_4 are forming the quantities B_0 and B_4 in the second AU; A_4 was forwarded from the first AU as the result of processing f_4 and f_{12}. One binary counter controls the entire unit. The switches and arithmetic units are activated by the 1's in the clock count word. The same word is bit-reversed and used as the address for the look-up table of sines and cosines required in computing the Fourier coefficients.

This unit is described in References 2, 3, and 4.

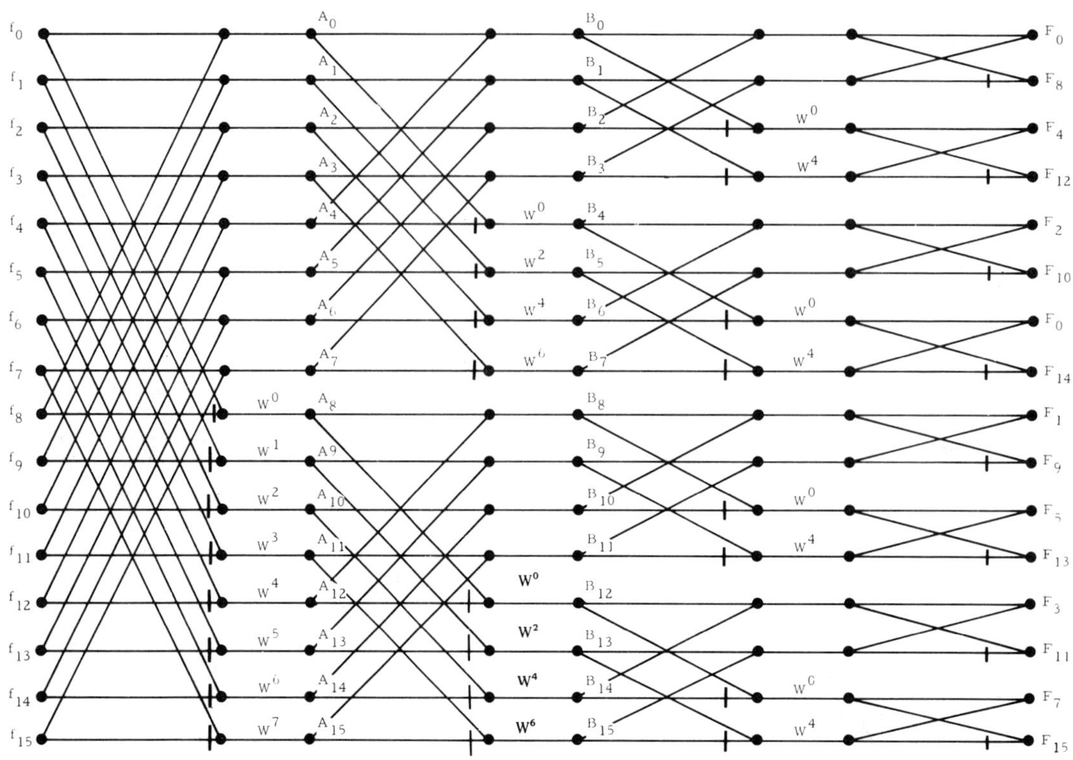

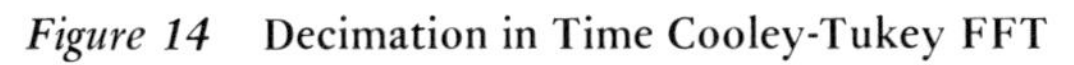

Figure 14 Decimation in Time Cooley-Tukey FFT

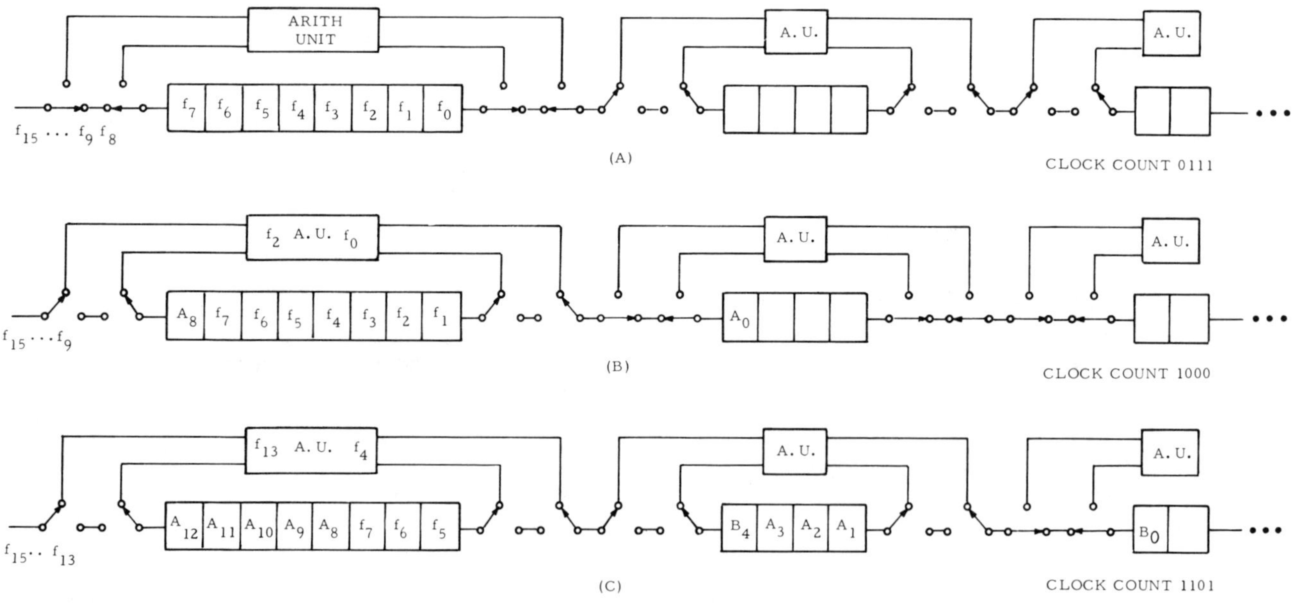

Figure 15 Pipeline FFT

References

[1] Pease, M.C.: "Adaptation of FFT for Parallel Processing," *Journal of the Association of Computing Machinery,* April 1968.

[2] Groginsky, H.; and Works, G.: "A Pipeline Fast Fourier Transformer," *EASCON,* 1969.

[3] Sheats, L.; and Vickers, H.: "Implementation of a Pipeline FFT," *NEREM,* 1970.

[4] Works, G.: "Data Manipulating System," *US Patent No. 3,816,729,* 11 June 1974.

Chapter 10

Purdy

Signal Processing Linear Frequency Modulated Signals

Introduction

Linear Frequency Modulated (LFM) radar signals have been the workhorse of radar signals for about as long as radars have existed. These signals have achieved preeminence for a variety of reasons — they are easy to generate, they provide both good range resolution and high energy, and they are easy to process. Despite this ease of processing — or perhaps because of it — many diverse techniques and devices have been developed to provide the pulse compression processing required by these signals. This chapter is a brief qualitative overview of some of the more well-known techniques. While the focus is on LFM compression, the reader should keep in mind that most devices and techniques discussed here have much wider applicability and can be used to process other modulations; some of the devices and systems can be used with burst waveforms. LFM processing is used only as a vehicle to describe the basic signal processing concepts.

This chapter is divided into three sections. The first discusses existing analog pulse compression devices and indicates their capabilities. The second contains brief summaries of the processing techniques used in the ALCOR* system. (This system is particularly interesting since it combines, in one stage of development or another, three distinct types of LFM pulse compression techniques, each of which is discussed.) The last section presents one general technique for digital pulse compression; known as fast convolution, this procedure allows the matched filtering of arbitrary waveforms and thus is far more useful than just as an LFM matched filter.

LFM signals have the characteristic that the instantaneous frequency increases (or decreases) linearly over the duration of the signal. Thus, the general LFM signal may be written as

$$s(t) = A(t) \cos 2\pi (f_o + \mu t/2)\, t \tag{1}$$

where:

$A(t) = 1$ for $-T/2 < t < T/2$

$= 0$ for $t \geqslant |T/2|$

f_o = carrier frequency

$\mu = W/T$

W = total bandwidth excursion

T = total time duration

Like all radar signals, the given LFM signal of Equation (1) has a unique matched filter which has an impulse response that is the time reversed complex conjugate of the given signal. Thus, the matched filter to s(t) has an impulse response h(t) given by

$$h(t) = A(t) \cos 2\pi(f_o - \mu t/2)t \tag{2}$$

Filters with this characteristic are generally known as dispersive filters since different instantaneous frequencies have different

*ALCOR is a C-band radar on the Kwajalein Atoll in the Pacific.

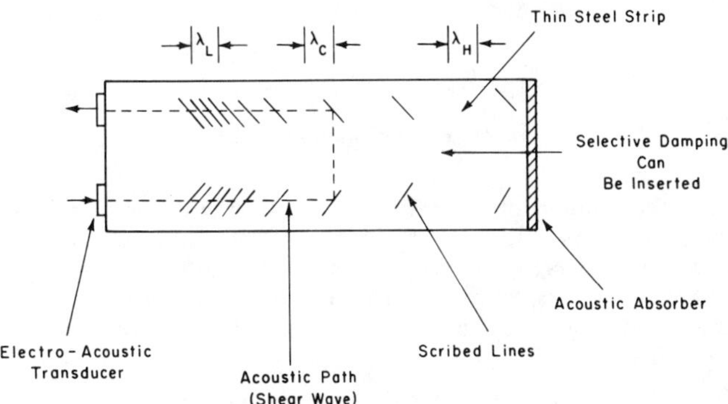

Figure 1 IMCON Line

delays. Thus, when the LFM signal of Equation (1) is processed by the filter above, the lower frequencies, which occur first in time, have long delays while the higher frequency at the end of the signal have shorter delays. The result is that all components of the input arrive at the output at the same time. The result is an integrated and compressed signal. This "pulse compression" is the basic processing required of LFM signals.

Analog Pulse Compression

The first analog device, shown in Figure 1, is an IMCON line [1]. It is an ultrasonic bulk-wave device which operates in the following fashion. The signal enters the device at the input transducer and generates an acoustic shear wave which propagates through the medium (steel). This wave travels through an oblique grating that consists of grooves the spacing of which increases (for an up-chirp) as a function of distance from the input transducer. The wave is reflected strongly at a right angle in a region where the groove spacing in the propagation direction matches the wavelength of the wave. A second reflection in the symmetrically placed mirror-image grating sends the wave to the output transducer. The groove positions are established such that the surface wave travels from input to output along a path the length and delay of which are linearly proportional to wavelength. Thus, for the device shown, short wavelengths are reflected over short total delays and long wavelengths have long delays. The result is a dispersive device.

This class of devices is able to process signals with fairly substantial time durations but there are some limitations in bandwidth, primarily as a result of the thickness of the typical unit. If it is made thin, though the bandwidth capability is increased, it can become mechanically unstable.

Surface wave devices [2] are discussed in Chapter 12 by Worley; thus, only one device is covered here — a Reflective Array Compressor (RAC), shown in Figure 2, which was developed at Lincoln Laboratory. This unit is a surface wave implementation of the IMCON principle discussed above. Etchings on the surface again

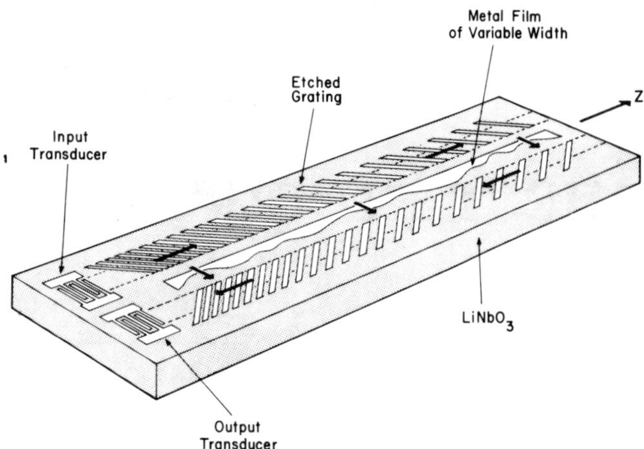

Figure 2 Phase-Compensated RAC

allow the coupling of different wavelengths over different delays and produce the dispersive characteristic. These surface wave RACs can support substantially higher bandwidths than similar bulk wave devices. At the moment, the durations (delays) are smaller, but substantial progress is being made to increase this capability.

Figure 3 summarizes the time durations (dispersion) and bandwidths (swept frequency) that can be achieved with several different devices. In all cases, only a single device is presented. Generally, there is a wide spectrum of devices to select from depending on the duration and bandwidth required. However, IMCON lines and Surface Acoustic Wave (SAW) devices provide the most advanced analog processing.

Wideband Pulse Compression

As mentioned above, all the devices in Figure 3 are single devices. However, many situations exist where a single device does not have sufficient capability to process the required signals. Several techniques exist for overcoming these limits, two of which follow. The idea for the first technique is expressed in Figure 4 — the frequency region is broken into narrow sections and a path delay provides a gross dispersion which is the staircase portion of the response. The resulting incremental differences between the staircase function and the true LFM response are then compensated for by dispersive elements in each section of the delay. This incremental

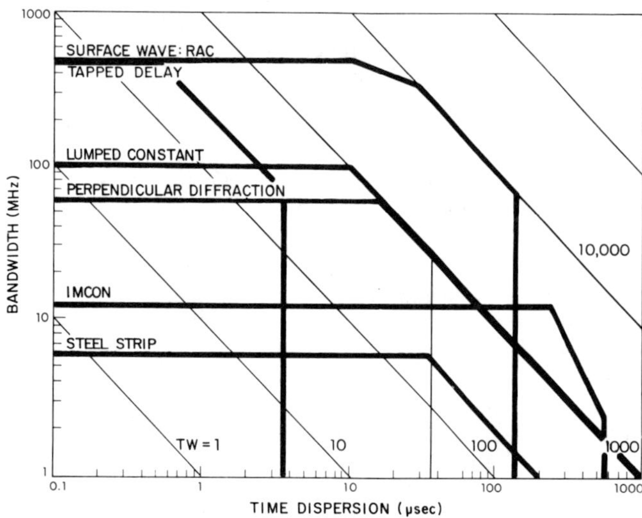

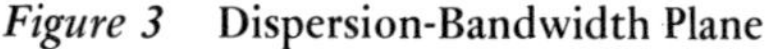

Figure 3 Dispersion-Bandwidth Plane

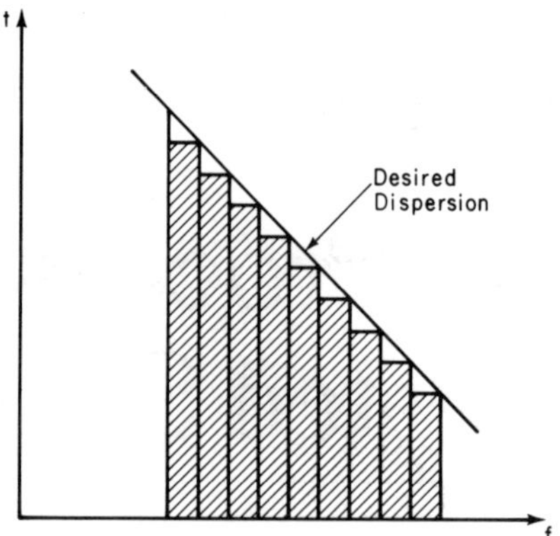

Figure 4 Bandwidth-Partitioned Matched Filter

dispersion is a very small percentage of the total dispersion of the signal and thus may be implemented and replicated easily.*

The second technique [3] is the LFM correlation and spectrum analysis technique, illustrated in Figure 5. Assume an LFM signal is transmitted and suppose that a target is located at one of the three different time positions indicated by A, B, and C. Assume that the received signal is multiplied by an LFM ramp with the same slope as the transmitted signal and generated at a fixed, known time. If the target were at "A," the result after multiplication would be a constant frequency, A. If the target were at "B," the result would be frequency B; the situation is similar for "C." Thus, if the frequency of the signal after multiplication is measured, it indicates the range (i.e., whether the target is at A, B, or C). It also can be shown that the inherent resolution of the transmitted signal is preserved. The advantage of this technique

*See second Editor's Note of Chapter 7 for the results obtained by Anderson Laboratories using an alternate way to increase the time-bandwidth product (TB) by cascading single delay devices. Lines having the same bandwidth are placed in cascade so as to increase only the time dispersion of the pulse compressor (from T to NT for N lines). The technique depicted in Figure 4 increases both the time dispersion and bandwidth by the number of lines used (from T to NT and B to NB).

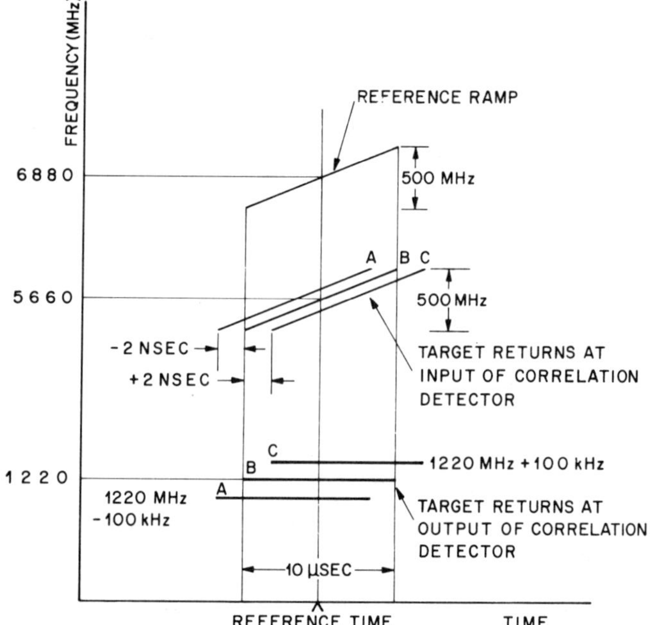

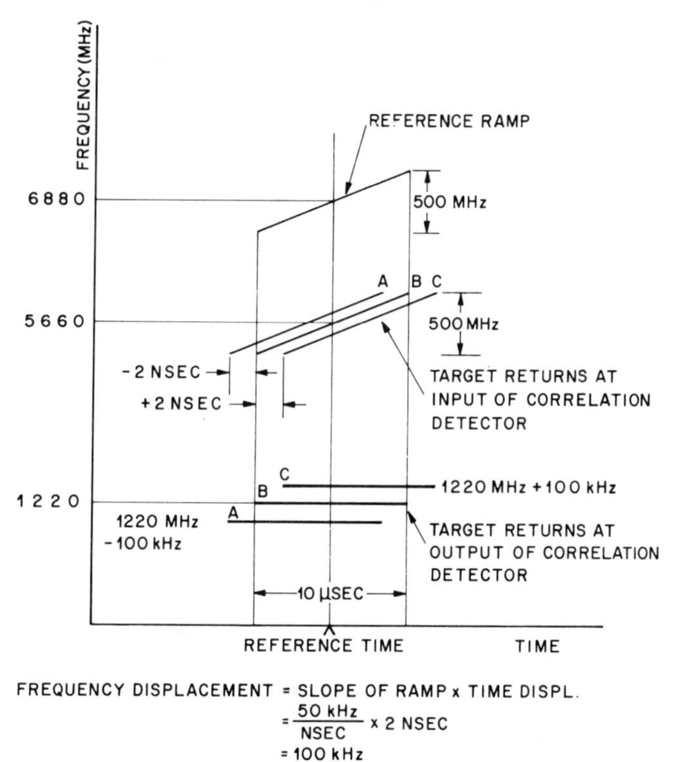

Figure 5 ALCOR Wideband Signal Processing

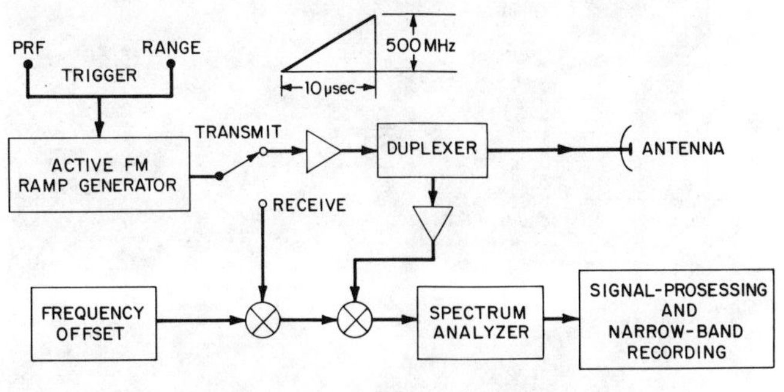

Figure 6 **ALCOR Pulse Compression**

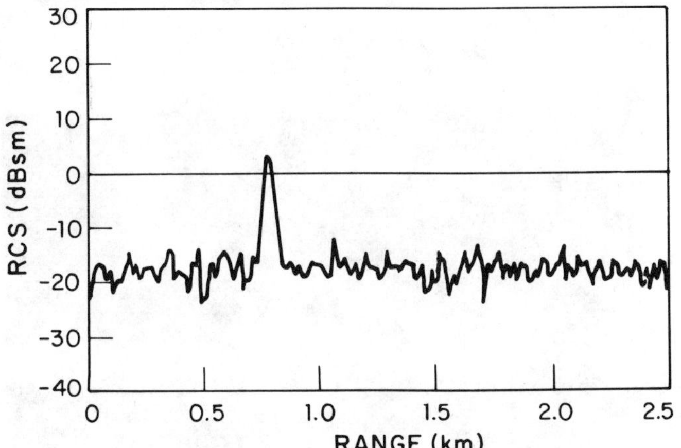

Figure 7 **Narrowband (6 MHz) Response**

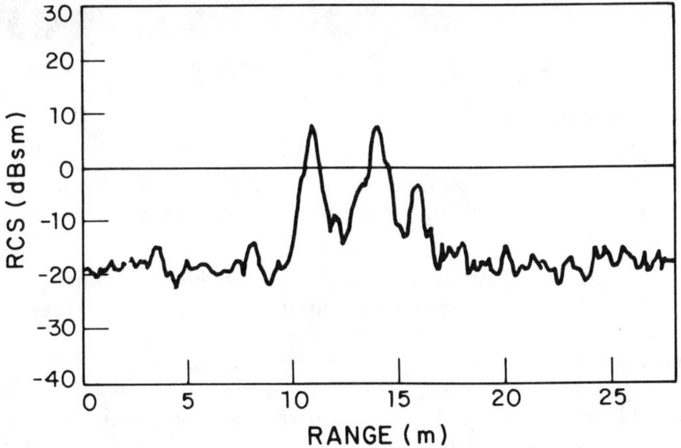

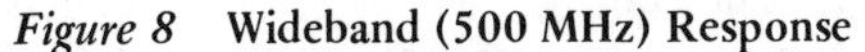

Figure 8 **Wideband (500 MHz) Response**

is that, after the multiplication process, all of the signals are at low bandwidths. Thus, the multiplication process can be performed early in the signal processing chain where large bandwidths are handled easily. This technique greatly simplifies the signal processing; for this reason, it is often used to process very wide band signals when other techniques may be impossible to apply. However, the penalty is a limited range window since the target must be reasonably close to the reference ramp. Otherwise, either some of the return energy is lost or extra noise is fed into the processor. In either case, a degradation of the signal-to-noise ratio and performance occurs. Figure 6 shows its implementation.

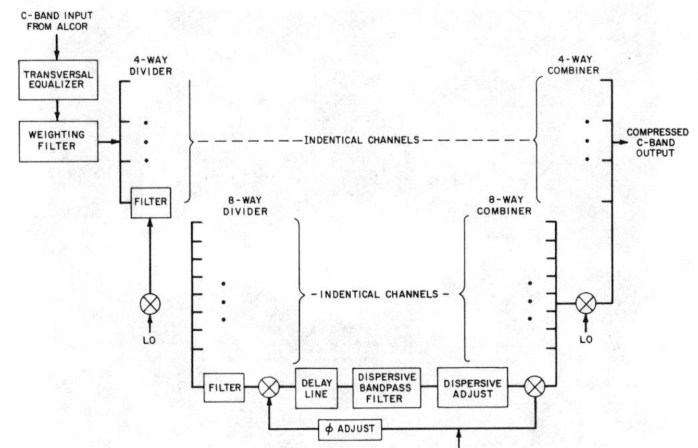

Figure 9 **All Range Processor (ARP)**

Three of the above techniques outlined above have been implemented on the ALCOR radar. This C-band radar has a 500 MHz, 10 μs signal; the 500 MHz provides 0.5m of range resolution (after weighting). The bandwidth is exceptionally large and, as such, strains the signal processing. The first procedure implemented on ALCOR was the LFM correlation and spectrum analysis technique described above (Figure 6). In this case, the range of reference frequencies after multiplication by the extended ramp is ±5 MHz; the bandwidth thereafter is only 10 MHz.

As an example of the quality of data generated by this system, Figures 7 and 8 show the return from the same target as observed (virtually simultaneously) by a 6 MHz ALCOR signal and by the 500 MHz signal as processed above. The target is only a point target as viewed by the narrow-bandwidth signal (Figure 7) but the extended nature of the target becomes clearly visible with 500 MHz (Figure 8). Note the units of the range axis in each case.

The second processor implemented on ALCOR was the All Range Processor. This system is based on the frequency partition technique described earlier. The processor system diagram is shown in Figure 9. The incoming signal is first divided into four distinct frequency channels, then each is subdivided into eight different frequency intervals for a total of 32 different frequency channels. Each then consists of a prefilter, some phase adjustments, and the primary processing components – the delay and the dispersive filter. The delay is different for each channel. All dispersive filters are of the same design and provide for the full signal duration, but each handles only 1/32 of the bandwidth. Eight of the outputs of each of these 32 channels are first combined; the result is four outputs which subsequently are combined to yield the complete output. The All Range Processor consists of seven racks of carefully tuned equipment; despite this fact, it works very well, as indicated by the sample outputs shown in Figure 10.

Figure 10a shows the response when only eight of the channels are combined. When two sets of eight are combined the response is improved (Figure 10b). The full 32 channel output (Figure10c) shows the very narrow pulse characteristic of 500 MHz bandwidth signals. This output signal is the familiar (sin x)/x with 13 dB sidelobes. When the signal is weighted, the result, shown in Figure 10d, is a compressed 500 MHz signal with 32 dB sidelobes.

The third processor used on ALCOR is a Reflective Array Compressor (RAC) device built by Lincoln Laboratory. This device, shown in Figure 11, provides the same matched filtering operation that is provided by the seven racks of equipment in the All Range Processor. This reduction in size, together with the fact that the signal processing is equivalent, indicates the capabilities and potential of surface wave devices.

The signal generator is an important part of the ALCOR radar which is required by all the processing techniques. Because of the

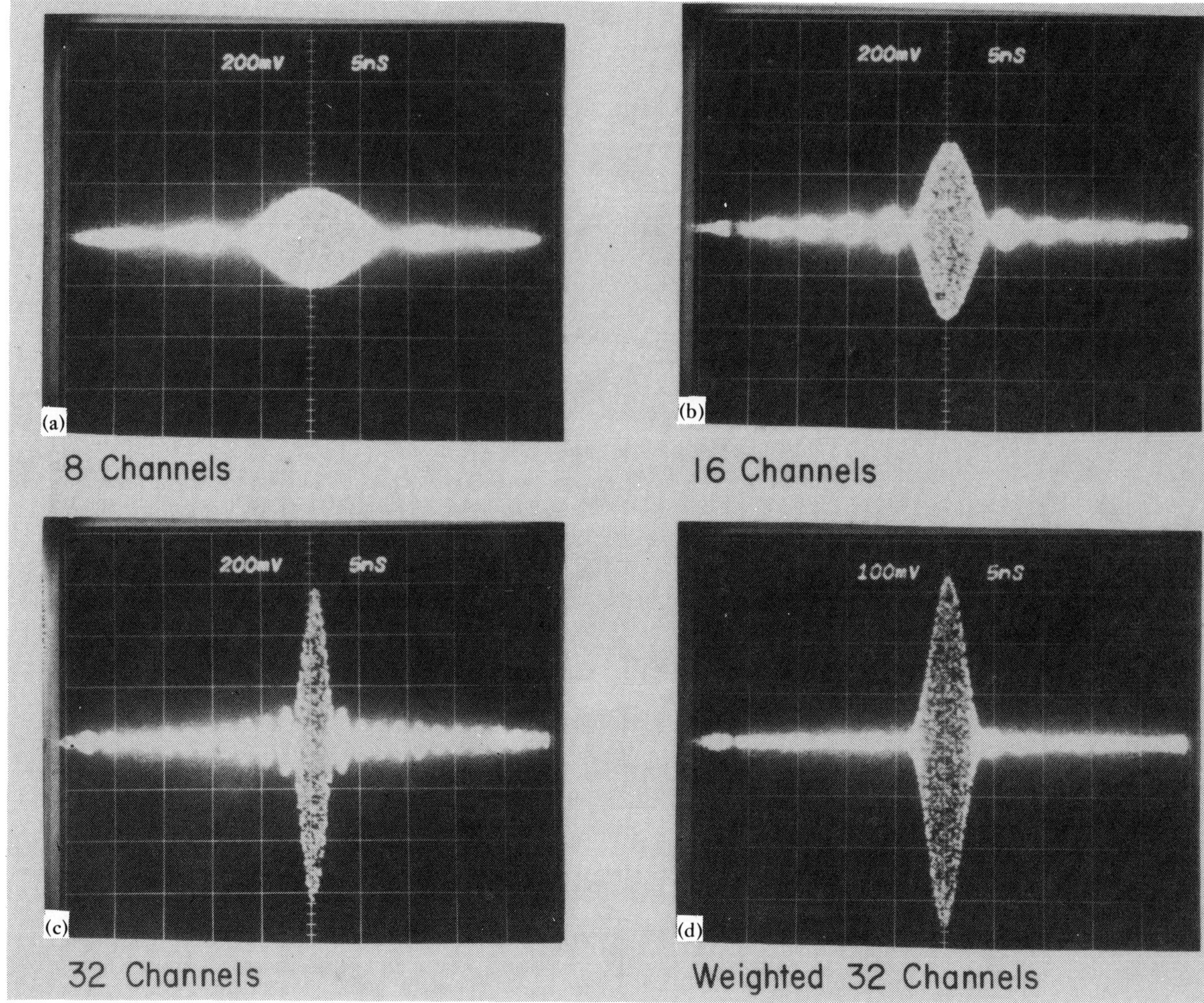

Figure 10 **ARWBC Time Response**

wide bandwidth, it is important to have a very good linearity throughout the signal. This end is achieved by a rather carefully controlled feedback system (see Figure 12). A video sweep generates a ramp that controls the Voltage Controlled Oscillator (VCO) which, in turn, generates the transmitted signal. If the ramp were perfect, an LFM signal with perfect linearity would be generated.

However, errors and drift occur. These effects are corrected by first delaying the signal slightly, then multiplying the signal by its delayed version. If no deviations exist, this action yields a perfectly constant frequency with no phase distortions. However, errors and drift do exist, so phase distortions are present. When the product signal is compared to a very carefully controlled sine wave, this phase distortion is evident at the output of the phase detector. The result then is used to correct the voltage sweep generated by the VCO.

Digital Pulse Compression

Digital systems offer the opportunity to spread the processing load over the available time. Thus, after the analog to digital conversion process, the resulting data can be placed in memory and the processing can proceed at a non-real time rate – either faster or slower. All that is required is that the processing be completed before the new data arrive. Moreover, it is almost always possible to increase the processing rate by increasing the amount of hardware. Thus, by increasing the hardware through parallelism, it is possible to reduce the processing time.

Digital processing also offers the advantage of high flexibility, arbitrary time bandwidths, and arbitrary control of the processing errors (i.e., computation noise). The time bandwidth product can be made arbitrarily large if the available processing time is long enough. The processing errors can be reduced simply by increasing word length. However, increased TW and increased precision imply increased hardware, either in terms of memory, computation units, or word length. Consequently, while large TW products, high precision, and short processing times can be achieved, this high performance requires large digital systems. These parameters – the available time, the TW product, the processing precision, and the physical size of a processor – are nearly always the fundamental ones that must be considered in selecting between an analog and a digital processing system.

Since any digital approach would use one or more integrated circuit technologies, it is worthwhile to review some of them. In all cases, only those circuit families which are widely available (circa early 1976) with reasonably complete product lines are considered. Several additional technologies are emerging (i.e., Integrated Injection Logic (IIL) and large Charge Coupled Device (CCD) memo-

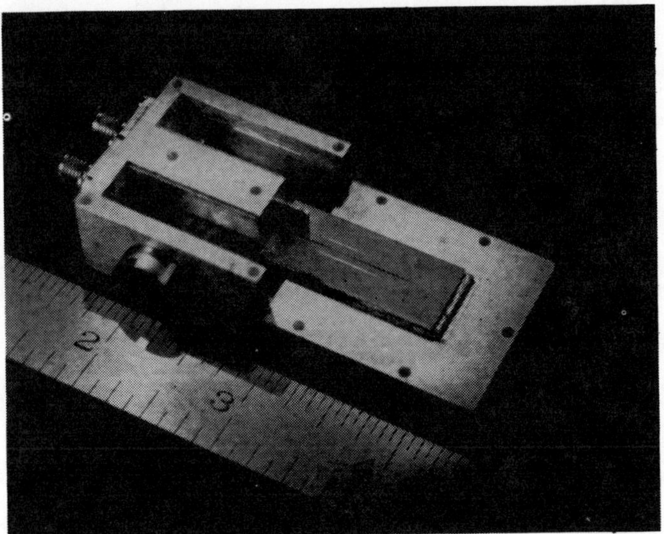

Figure 11 **RAC Pulse Compressor for 500 MHz, 10 μs LFM Signals**

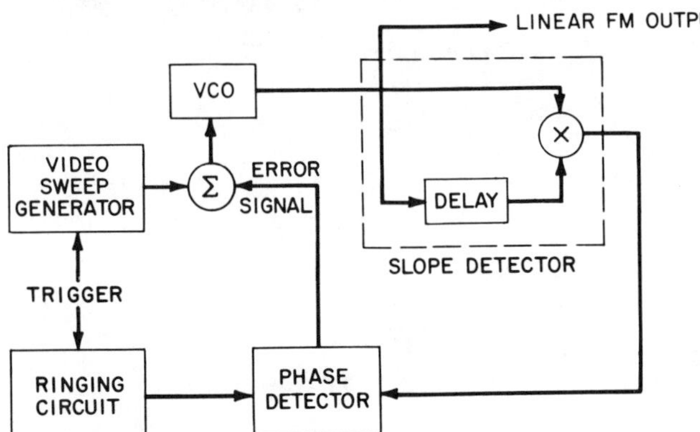

Figure 12 **ALCOR Signal Generator**

ries (65k)) but these are not discussed here since, at present, they either do not have a full product line or they are not available in quantity. Similarly, many high performance specialized digital components have been developed for particular applications, but these also are not discussed, again since they are not generally available.

Digital components fall into two main categories – memory and logic. The characteristics of representative devices from these two groups are shown in Table 1. The memory sizes selected are the largest presently available in each of the technologies. It should be noted that smaller sizes in each technology generally have faster speeds.

The propagation delay for the gates is indicative of the relative speeds. The devices selected (either a NOR or a NAND) are comparable in complexity for all the technologies. All except the ECL 100k have four to a package; the 100k has five. The power in both the memory and gate areas is for a single complete integrated circuit.

The most widely available and used technology is Transistor-Transistor Logic (TTL). This family (in its low power, high power, and Schottky versions) provides good speed with medium power dissipation. The various forms of Metal-on-Silicon (MOS) memories provide good levels of integration since the power levels are low. This technology finds wide applicability in memory circuits; at present, 4k bits per package are available.* These devices can be either dynamic or static – they may or may not have to be refreshed. In general, the static units are preferable since control is greatly simplified; however, the dynamic MOS devices have higher levels of integration. Charge Coupled Devices (CCDs) have recently been applied to digital memory circuits and dynamic 16k bit ICs are presently available. This technology offers the promise of being able to achieve even higher densities (65k ICs are being developed). Lastly, the Emitter Coupled Logic (ECL) family provides high speed and concurrent high power, as well as having the potential advantages of complementary outputs and nearly constant power supply drain. The 10k line is the most complete and is being used in an increasing number of systems. The ECL III and 100k series are new, so the product lines are not extensive.

*16k bit chips should be available soon.

MEMORIES

Technology	Size	Type	Read or Write Cycle Time (ns)	Milliwatts/Package
CCD	16k	SERIAL	4 MHz*	200
CMOS	1k	RAM	450**	15***
MOS (static)	4k	RAM	400	450
MOS (dynamic)	4k	RAM	300	430
TTL	1k	RAM	90	550
ECL 10k	1k	RAM	35	520

GATES

Technology			Prop Delay (ns)	Milliwatts/Package
CMOS	QUAD	NAND	30****	6*****
TTL	QUAD	NAND	10	40
TTL (Low Power)	QUAD	NAND	33	4
TTL (Schottky)	QUAD	NAND	3	76
TTL (Low Power Schottky)	QUAD	NAND	10	8
TTL (High Power)	QUAD	NAND	6	88
ECL 10k	QUAD	NOR	2	100******
ECL III	QUAD	NOR	1	240******
ECL 100k	QUINT	NOR	0.75	250******

*Transfer rate

**With V_{dd} = 5V and C_l = 50 pF

***Quiescent power = 0.5 μW

****With V_{dd} = 10V and C_l = 15 pF

*****At 2 MHz clock

******Exclusive of Terminator

Table 1 **Digital Components (circa early 1976)**

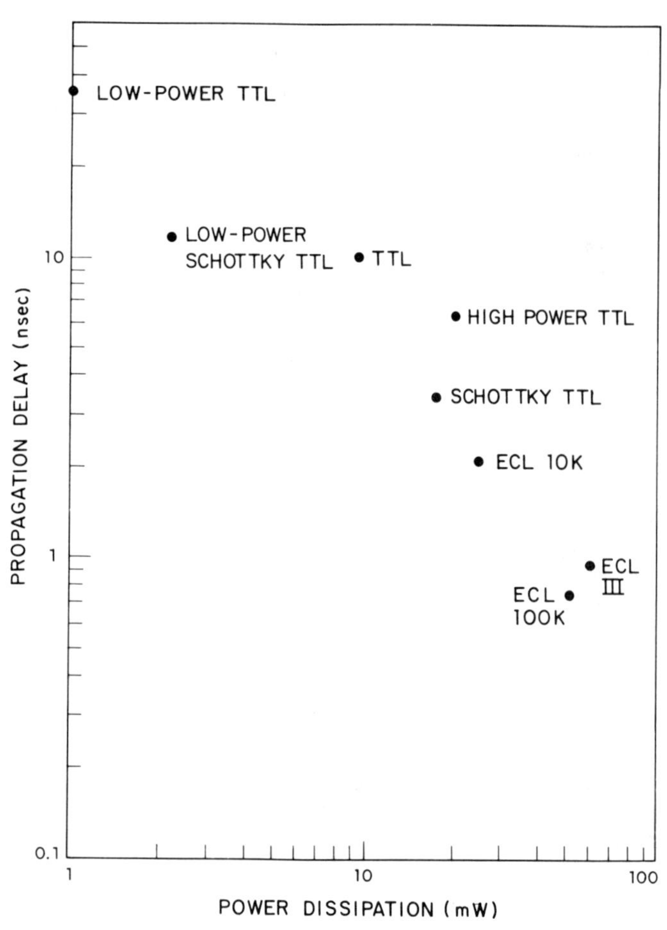

Figure 13 Delay-Dissipation Plane

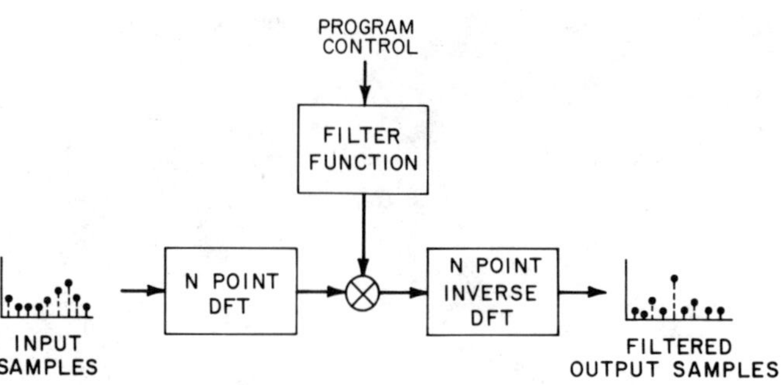

Figure 14 Convolution by Discrete Fourier Transform

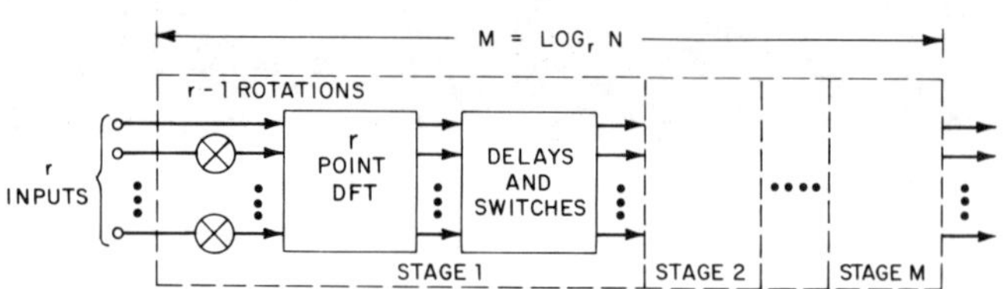

Figure 15 Pipeline FFT

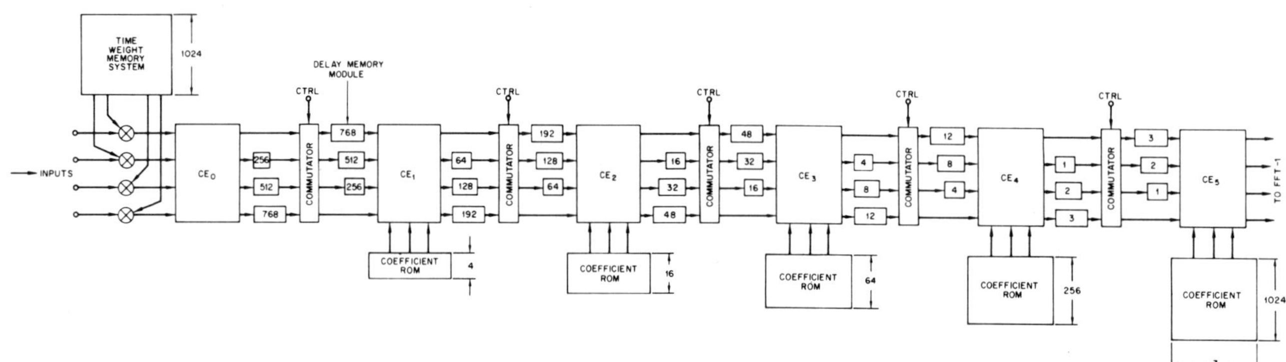

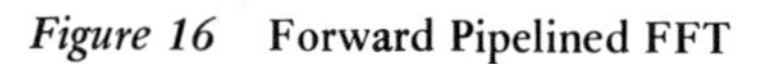

Figure 16 Forward Pipelined FFT

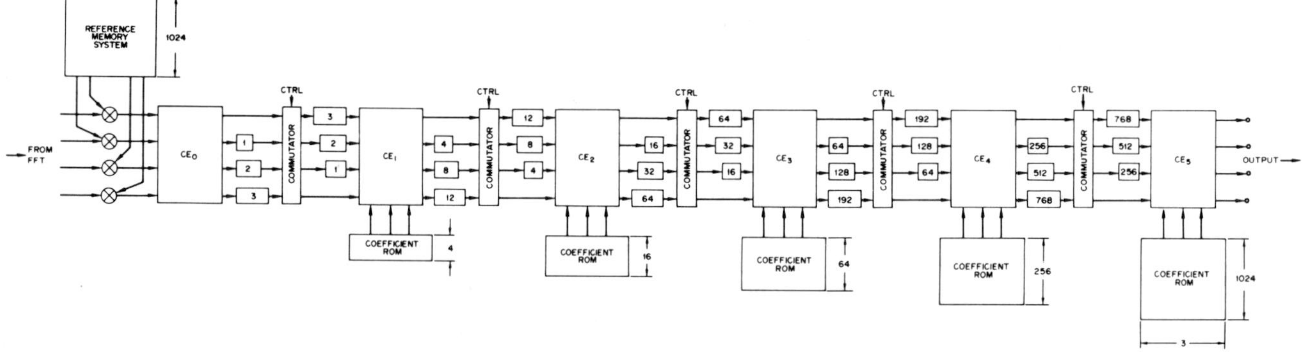

Figure 17 Inverse Pipelined FFT

One traditional way of comparing technologies is by examining the speed power product, as is done for the gates (excluding CMOS) in Figure 13 with a graph of propagation delay time vs. power per gate [4, 5] (as opposed to power per package in Table 1). Power and speed can often be traded directly, and the power levels strongly influence the amount of integration that can be achieved in a given IC. In Figure 13, the points are only representative and the designer should examine individual specification sheets for precise maximum, minimum, and typical performance values.

All of the above components are available for use in digital systems. However, the particular form of the implementation of a matched filter for a LFM signal can take two different forms – either direct convolution in the time domain or fast convolution. The latter is preferable when time bandwidth products are large (greater than approximately 100), which is almost always the case with radar signals. The technique also has vast flexibility and is applicable to the processing of arbitrary signals within certain time duration and bandwidth constraints. The technique, shown in Figure 14, is simple in concept. It consists of taking an input sequence of samples (of the received signals), performing a Fourier transform to yield samples in the frequency domain, multiplying these samples by a stored filter function (the spectrum of the desired filter), and then performing an inverse transform to yield the time domain response. Thus, any filter function can be implemented by storing it in memory. It is also possible to arrange the flow of information in such a fashion as to provide a continuous flow through the filter even though the transform size is finite.

In principle, the above technique is clear. While implementations of the Fourier transform may take several different forms, nearly all are based on the Fast Fourier Transform (FFT) algorithm. The particular implementation discussed here is the pipeline FFT shown for an arbitrary radix [6] in Figure 15.* For an N-point transform, there are $\log_r N$ stages and each stage has r inputs of N/r samples. As r grows longer, the number of stages decreases, but the complexity of each stage increases. Radix 2 is the most familiar form, but the higher radices are more efficient. As an example of a radix 4 convolver system [7], Figures 16 and 17 respectively show a forward and an inverse 4k transform. The six elementary computations (ECs) in each transform are identical but there are slight differences in delay and coefficients. These ECs consist of four inputs, three complex multipliers, and eight complex adds, as shown in Figure 18.

Regardless of the implementation of the Fourier transform, the fast convolution technique implies restrictions only on the time duration and bandwidth of the signals. Figure 19 shows the time-bandwidth plane for fast convolution. The straight lines are lines of constant time-bandwidth product. The curved lines show the limits for each transform size (N). Implicit in these curves is an assumption of a constant range window of 10 km. Thus, for example, a 16k transform allows the processing over a 10 km window of arbitrary signals (including LFM signals) with time duration and bandwidth anywhere under the 16k line.

The other parameter of interest is the amount of time for this processing. A system using a single FFT is presently being fabricated with commercially available high-speed digital circuits (ECL 10k) which should provide complete matched filtering operations within the times shown in Figure 20. The number of Doppler channels refers to successive filtering of the data with different presumed Doppler shifts. For an LFM signal, only one filtering operation is required. The initial delay through the pipeline is included in these curves. The through-put rate can be obtained by examining the slope.

Since all digital processing requires a conversion from analog signals to digital data, it is appropriate to review available analog to digital converters (A/D's). A/D's generally have the form shown in Figure 21. The analog input at each sample time is held for comparison with an encoder section. The encoder compares the analog voltage with a finite set of fixed reference voltages. The one particular reference (of the finite set) which is the closest to the analog input is noted and the corresponding digital word is then read-out. Often, this word is reformatted into alternative digital codes.

As a summary of the presently available A/D's, Figure 22 shows the bit sample rate plane. The line indicates the approximate limit of sample rates and word lengths that are available from industry. Some devices with capabilities outside this line are being developed presently but are not currently available.

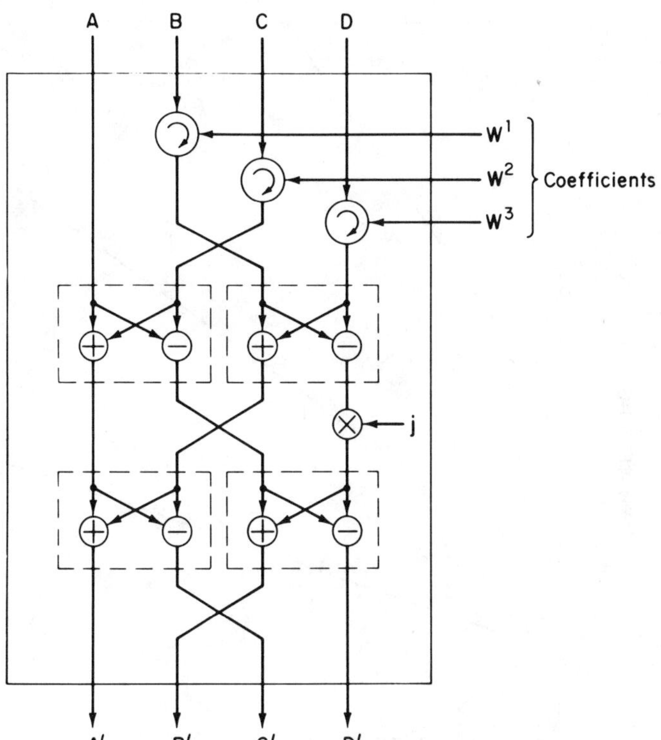

Figure 18 Radix 4 Complex Computation Element

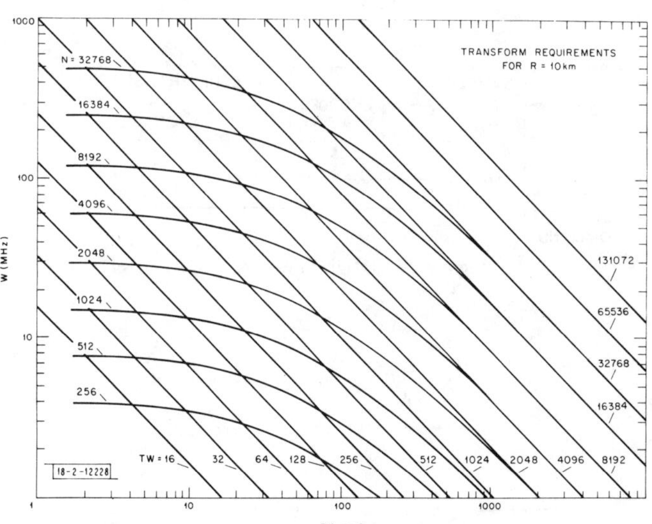

Figure 19 Time-Bandwidth Plane

*See Chapter 9 for a detailed discussion of this device.

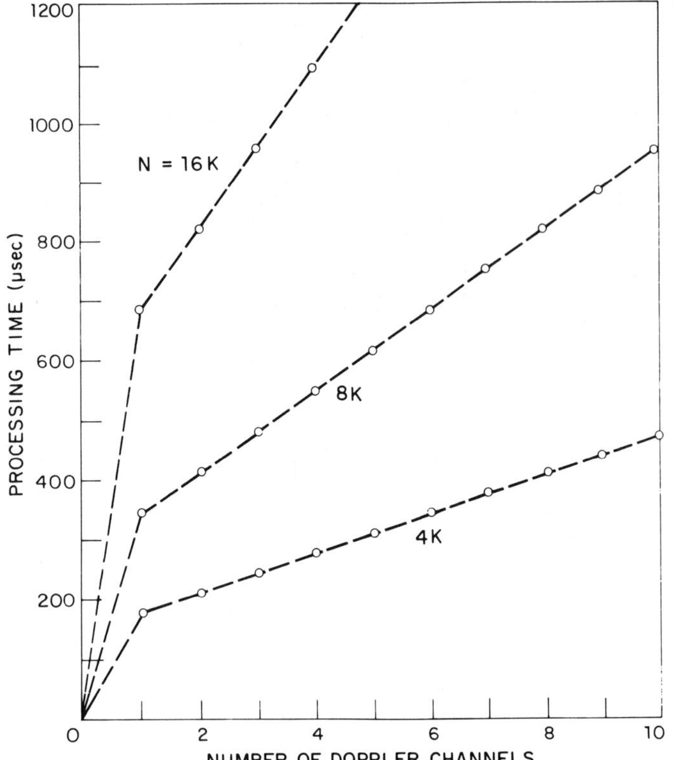

Figure 20 Convolver Processing Time

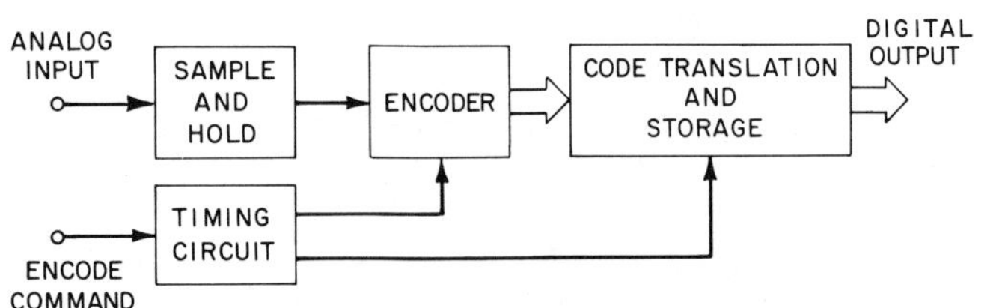

Figure 21 High-Speed A/D Converter

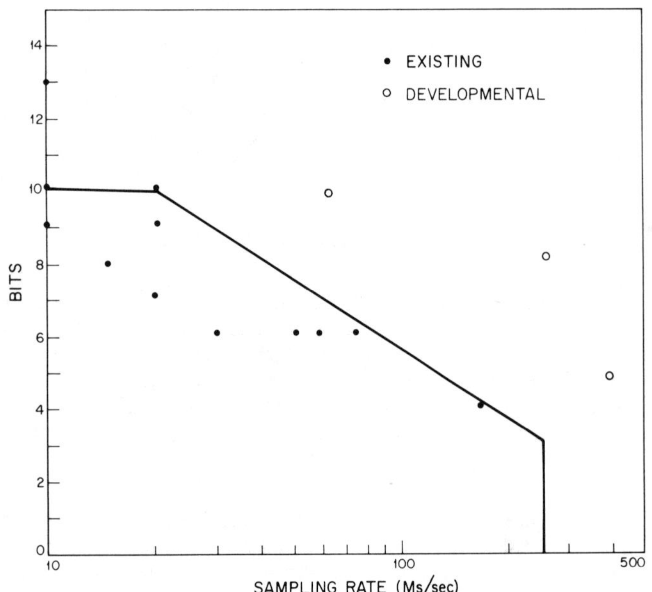

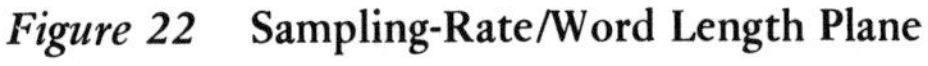

Figure 22 Sampling-Rate/Word Length Plane

Acknowledgements

The material in this chapter is a summary of the work of many people. Those individuals at Lincoln Laboratory who have contributed substantially to the Signal Processor Systems discussed here include W.W. Camp, J.M. Frankovich, A.H. Huntoon, J. Margolin, J.P. Perry, E. Stern, J.O. Taylor, and R.C. Williamson. The efforts described here were supported by the Department of the Army and the Department of the Air Force.

References

[1] Martin, T.A.: "The IMCON Pulse Compression Filter and Its Applications," *IEEE Transactions on Microwave Theory and Technique, Vol. MTT-21, No. 4,* pp. 186-194, April 1973.

[2] *Proceedings of the IEEE, Vol. 64, No. 5,* May 1976.

[3] Camp. W.W.; Axelbank, M.; Lynn, V.L.; and Margolin, J.: "ALCOR – A High Sensitivity Radar with 0.5m Range Resolution," *IEEE 1971 International Convention Digest,* pp. 112-113, March 1971.

[4] Texas Instruments, Inc.: *The TTL Data Book for Design Engineers,* 1973.

[5] Motorola Semiconductor Products Inc.: *Semiconductor Data Library* (Volume 4, MECL Integrated Circuits), 1974.

[6] Gold, B.; and Bially, T.: "Parallelism in Fast Fourier Transform Hardware," *IEEE Transactions on Antennas and Propagation, Vol. AP-21, No. 1,* pp. 5-16, February 1973.

[7] Purdy, R.J.; et al.: "Digital Signal Processor Designs for Radar Applications," *TN 1974-58,* Lincoln Laboratory, 31 December 1974.

Radar Matched Filtering

Introduction

An important application of digital processing has been in the area of radar matched filtering. The classical approach to digital matched filtering has employed a tapped delay with a weighted summation. This method recently has been replaced by the use of a Fast Fourier Transform (FFT) to perform circular convolution. Developments of pipeline FFT techniques and floating point implementations have enabled large dynamic range, high speed matched filter processors to be implemented.

A new method of matched filter processing, called the *step transform,* uses less than half the hardware of a FFT convolution processor. The step transform uses subaperture processing to reduce the size of the FFT. This algorithm also has application to Synthetic Aperture Radar (SAR) in which quadratic phase focusing is required.

Radar Matched Filters Using Fourier Transform Techniques

The Impact of the Fast Fourier Transform

The classical method of designing a digital matched filter for radar consists of a simple tapped delay line with each tap output multiplied by the appropriate weighted replica sample of the transmitted waveform. All products thus formed are then summed as indicated in Figure 1 to provide the desired convolution, Y_n, of the input sequence, S_i, with the reference sequence, R_i.

$$Y_n = \sum_{i=1}^{N} R_i S_{n+i-N}$$

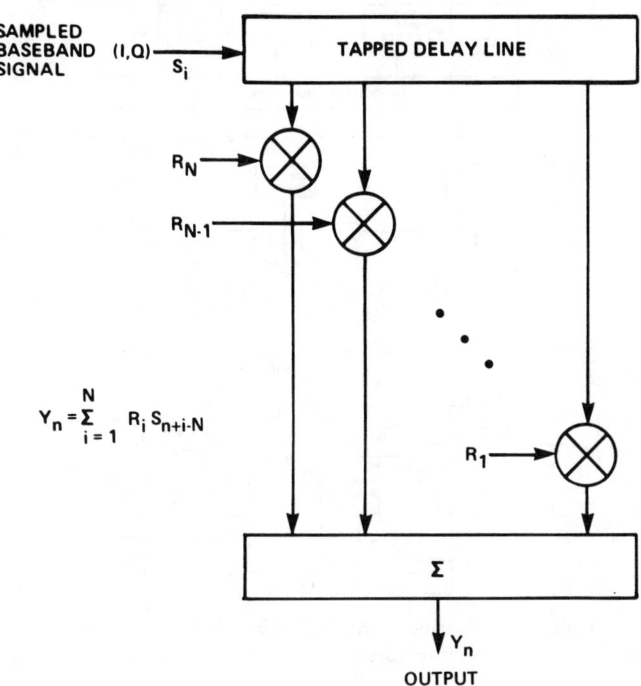

Figure 1 Classical Tapped Delay Line Matched Filter

The computation requirements of convolution implemented in this direct manner are very large from either a computer software or special purpose hardware point of view. The number of complex number multiplications is approximately equal to N^2 where N is the number of sample points in the signal and reference function. Similarly, with a multiplier at each tap, there are N multipliers in a high speed hardware implementation. An implementation of a digital matched filter therefore was given very little serious consideration prior to the development of the Fast Fourier Transform (FFT) algorithm by Cooley and Tukey [1, 2]. The number of multiply-add operations for computation of a discrete Fourier transform was reduced from N^2 to $N \log_2 N$ by the FFT algorithm. With a simplified method available for transforming a signal to the frequency domain, use of the Borel convolution theorem becomes practical. That is, the product of the Fourier transform of two functions is the Fourier transform of the convolution of the two functions. Stated mathematically

$$F^{-1}[S(f)R(f)] \equiv \int_{-\infty}^{\infty} s(\tau)r(t-\tau)d\tau = y(\tau)$$

where $S(f) = F[s(t)]$, and $R(f) = F[r(t)]$.

The operation with sampled data functions and the discrete Fourier transform is illustrated in Figure 2. Sampled input and reference data are periodic at the sequence length. Thus, if a convolution of N sample signal and reference points is desired, the transform aperture must be greater than N as regards the number of convolution points. For N convolution points, transform apertures of length 2N are necessary. The convolution of sampled data in this manner is often referred to as a *circular convolution.*

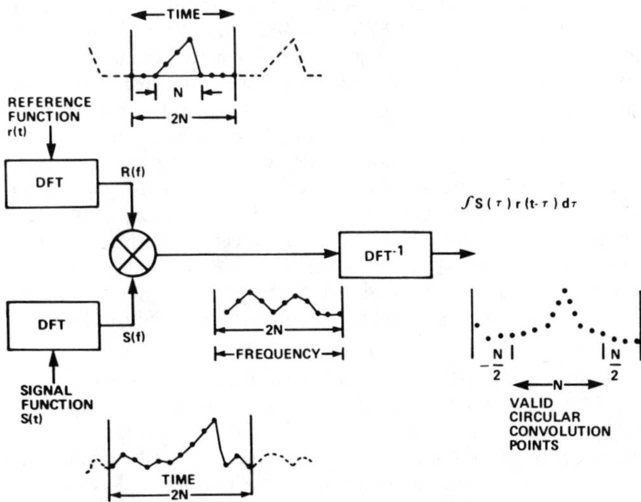

Figure 2 Convolution Via the Discrete Fourier Transform

The FFT algorithm permits computation of N convolution points of two N point sample sequences with $2N(\log_2 2N+1)$ radix-2 arithmetic operations. This reduction is by a factor of $N/(4\log_2 2N+1)$ compared to the classical tapped delay line approach.

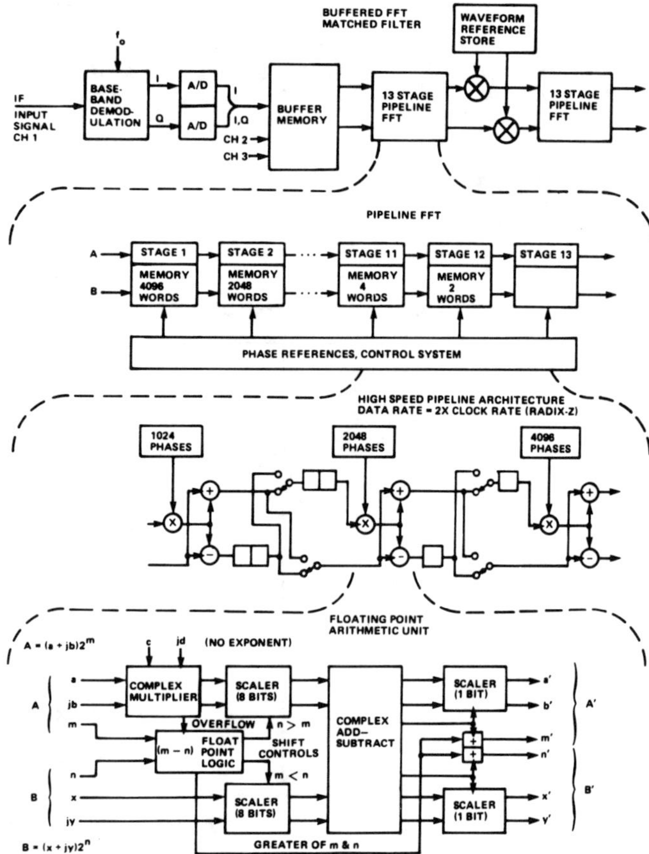

Figure 3 **Floating Point Buffered FFT Matched Filter Architecture**

Multi-Channel-Buffered FFT

A composite diagram of an FFT matched filter architecture which is useful for multichannel-multimode radar applications is shown in Figure 3. The buffered FFT matched filter takes advantage of limited range window coverage required in actual radar operation compared to the total range sweep of a pulse. Storage of the signals over the range window permits the matched filter processor to either operate at a lower speed or process additional channels. Use of this approach in a multimode radar impacts the radar pulse scheduling procedure as an additional constraint with transmitter duty cycle and the target environment. Since memory costs are relatively low compared to a large high speed FFT, the buffered FFT matched filter approach generally is the most cost effective architecture. The speed capability of the FFT can be set to match exactly the maximum signal processing rate required by the radar target environment.

In the example, a 8192 point (13 stage) pipeline FFT is shown. This large size FFT is inefficient for the processing of short track-mode pulses. To accommodate mode variations of this type, the buffer organizations and the length of the FFT can be made adaptable to waveform and range window length, or multiple short pulses can be processed in a single large FFT filter aperture.

High Speed Pipeline FFT

Radar applications typically require matched filters with bandwidths of 10 MHz or more. There may be a system requirement for all range coverage for which the buffered FFT must operate in a manner such that the transform points are outputted at the sampling rate of the signals as represented functionally by the tapped delay line diagram of Figure 1. In computing an N point FFT, data points are processed $\log_2 N$ times. The high speed FFT processing rates can be accommodated efficiently with a pipeline architecture in which an FFT arithmetic processing stage is inserted for each pass through the FFT "butterfly." The general stage-to-stage pipeline form for a radix-2 system is shown in the middle two diagrams of Figure 3. Various pipeline architectures have been designed for achieving real time FFT processing rates [3-5]. A key element is the "shuffle" memory in each FFT stage which permits the signal coefficients to flow efficiently through the pipeline with minimum memory storage and control.* The radix-2 arithmetic unit consists of a complex multiplier, adder, and subtracter. The complex multiply operation is, in reality, a phase shift; other techniques, such as the CORDIC [6] algorithm, can be used for this function.

Quantization Levels and Arithmetic

The performance level of an FFT matched filter depends on the number of quantization bits used in the process, the rounding-truncation rules, and whether fixed or floating point arithmetic is employed. The arithmetic unit shown for the FFT of Figure 3 uses a simplified floating point algorithm [7] which provides high performance levels with a minimum of hardware complexity for signal processing applications.

The features of the floating point approach are that:

1) Floating point numbers are not assigned to the phase reference factors.
2) The exponents are incremented in the positive direction only.
3) No left-justification of a mantissa is performed.
4) A common exponent is shared by the *In-phase* (I) and *Quadrature* (Q) components of a complex sample.

In this approach, the mean square error (mse) level of the range sidelobe structure of a compressed pulse relative to the peak output is (in dB) approximately

$$\text{mse} = 6(B-1) + 10 \log TW$$

where B is the number of mantissa quantization bits (excluding sign) and TW is the time-bandwidth product of the waveform. Other factors affecting performance are the size of waveform relative to the FFT aperture and the number of bits in the A/D converter.

Fixed point arithmetic has the principal disadvantage of causing small signal suppression unless the number of quantization bits is large enough to avoid overflow. This problem occurs because of the need for block scaling (division by two) at selected points in the FFT process. This division has the effect of shifting weak targets and noise out of the quantization range as a large target is integrated.

Step Transform Algorithm for Linear FM Pulse Compression

The step transform algorithm provides an efficient method to perform pulse compression of linear FM type waveforms. This algorithm can reduce the hardware requirements by up to 50% of conventional FFT approaches.

A block diagram of the step transform is shown in Figure 4. A short duration reference is mixed with the incoming signal and the output is fed to a Fast Fourier Transform (FFT) processor the time aperture of which is equal to the duration of the reference waveform but is less than the total waveform length. The output of the FFT is stored in a memory adequate in size to hold the full waveform. The samples are reordered and then passed to a second short aperture FFT the output of which represents the pulse compression of a linear FM signal. High speed FFT processing can be obtained using pipeline processing techniques and high-speed digital logic to provide a continuous pulse compressed output.

*See Chapter 9 by Sheats for a detailed description of the pipeline FFT "shuffle" memory.

Step Transform Technique

The concept of the step transform can be understood by first referencing to Figure 5. This figure shows a time-frequency plot of a linear FM signal received at an arbitrary time, Δ, relative to a fixed reference linear FM. Mixing these two waveforms results in a CW signal the frequency of which is proportional to Δ and which exists over the period of time that the signals overlap. As Δ increases, the amount of overlap decreases, resulting in an output energy loss, a decrease in resolution, and an increase in collapsing loss.

These losses can be reduced by using additional reference signals which are spaced closely together, as in Figure 6. Each reference covers only a small region of incremental range delay and is mixed with the signal to produce output frequencies proportional to the time displacement of the references and the signal. During any increment of time that two references overlap the outputs of the mixers are identical with the exception that they are displaced in frequency.

The multiple references can be replaced by a single sawtooth reference (shown in Figure 7) with no loss in information. When the sawtooth frequency reference is mixed with the linear FM signal, a step frequency function results (Figure 8). A short aperture FFT is used to measure the frequency of each step. This spectrum analysis starts at the beginning of the short reference ramp and stops at its end. The resolution of the spectrum analyzer corresponds to one over the time duration of the short ramp. The corresponding resolution is less than that of one range bin for a mixed linear FM signal. However, full resolution can be achieved by coherently combining successive spectrum analysis outputs over the waveform duration.

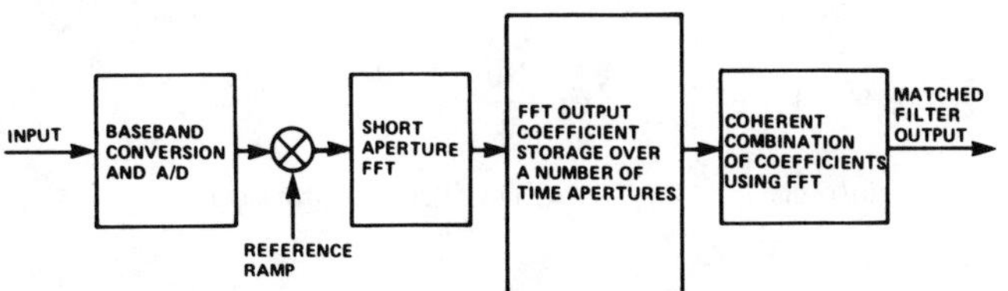

Figure 4 Implementation of Step Transform Matched Filter

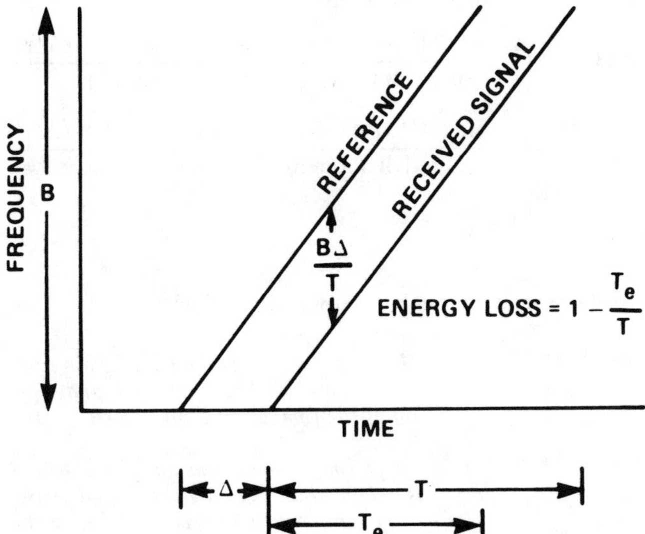

Figure 5 Correlation of Linear FM Signal and Reference

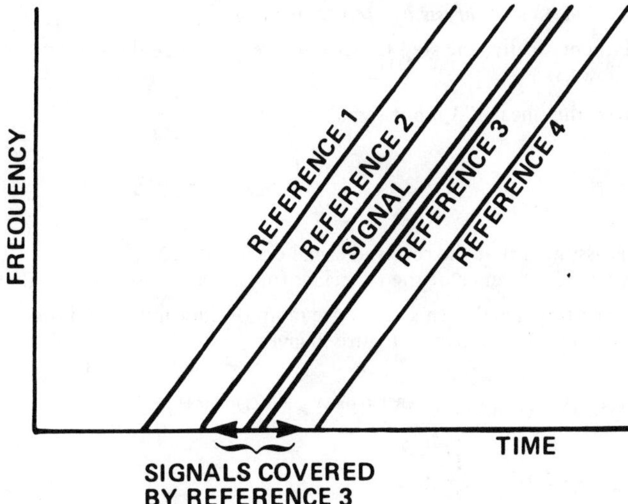

Figure 6 Correlation of Linear FM Signal with Multiple References to Minimize Energy Loss

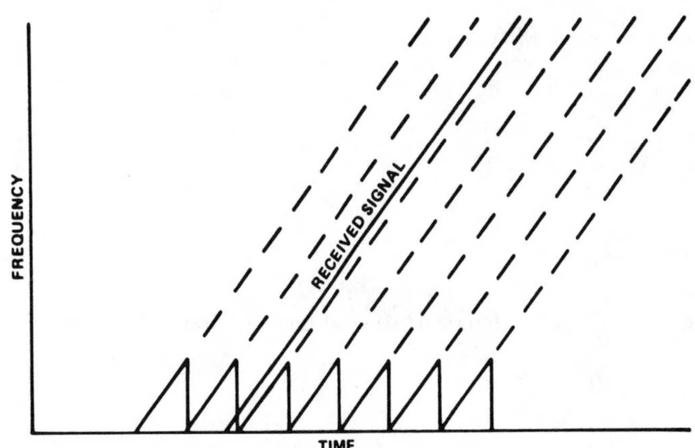

Figure 7 Reduction of Redundant Correlations

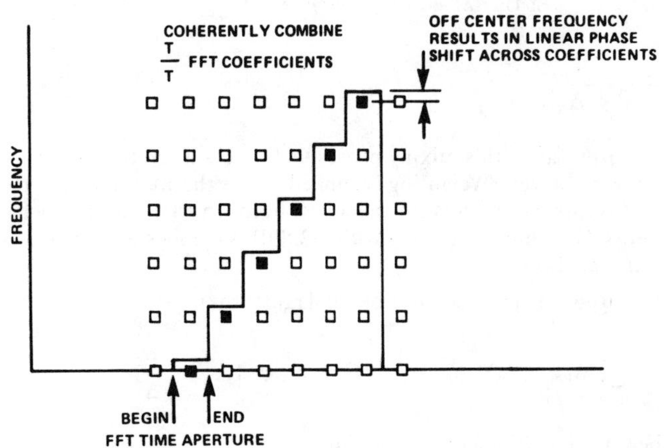

Figure 8 Repeated Spectrum Analysis of Short Time Apertures

Mathematical Analysis of Step Transform

Mathematically, the step transform process can be derived as follows.

Given the linear FM input signal

$$S(t) = e^{j(\pi B/T)t^2} \qquad -\frac{T}{2} \leqslant t \leqslant \frac{T}{2} \tag{1}$$

it is assumed that the received signal occurs at mΔ relative to a zero time reference of the processor (m can be a non-integer value).

Mixing this signal with a reference ramp of duration T′ and displaced by -nΔ (shown in Figure 9) gives

$$S(t+m\Delta)S^*(t-n\Delta) = e^{j(\pi B/T)(t+m\Delta)^2} e^{-j(\pi B/T)(t-n\Delta)^2}$$

$$-\frac{T'}{2} + n\Delta \leqslant t \leqslant \frac{T'}{2} + n\Delta \tag{2}$$

If we sample this output at interval Δ, letting k be the sample number,

$$t = -\frac{T'}{2} + n\Delta + k\Delta$$

where

$$0 \leqslant k \leqslant \frac{T'}{\Delta} - 1.$$

Then

$$S_{k,m}S_{k,n} = e^{j(\pi B/T)[(k+n)\Delta-(T'/2)+m\Delta]^2} e^{-j(\pi B/T)[(k+n)\Delta-(T'/2)-n\Delta]^2},$$

$$0 \leqslant k \leqslant \frac{T'}{\Delta} - 1 \tag{3}$$

$$S_{k,m}S^*_{k,n} = e^{j2(\pi B\Delta^2/T)[m+n]k} e^{j(\pi B/T)[n^2\Delta^2-n\Delta T']} \cdot e^{j(\pi B/T)[m^2\Delta^2-m\Delta T']} e^{j(\pi B/T)[2mn\Delta^2]} \tag{4}$$

$$0 \leqslant k \leqslant \frac{T'}{\Delta} - 1$$

The output after this mixing process is fed to a discrete Fourier Transform device. Weighting is applied across the FFT aperture to reduce cross talk effects (or sidelobes) between FFT output coefficients. Hamming weighting with -42.8 dB sidelobes was chosen for this analysis.

The output of the Discrete Fourier Transform is

$$S_r = \sum_{k=0}^{(T'/\Delta)-1} W_k \; S_{k,m}S^*_{k,n} \; e^{-2\pi j(rk/N)}, \quad r = 0, 1, \ldots, \frac{T'}{\Delta} - 1 \tag{5}$$

where W_k is the weighting function.

Assuming an even number of samples for a symmetrical sampled Hamming weighting function

$$S_r = e^{j\phi} \sum_{k=0}^{(T'/\Delta)-1} 0.54 - 0.46 \cos 2\pi \frac{(\Delta/2) + k\Delta}{T'} \; e^{j(2\pi B/T)(m+n)k\Delta^2} \cdot e^{-2\pi j(rk\Delta/T')}, \quad r = 0, 1, \ldots, \frac{T'}{\Delta} - 1 \tag{6}$$

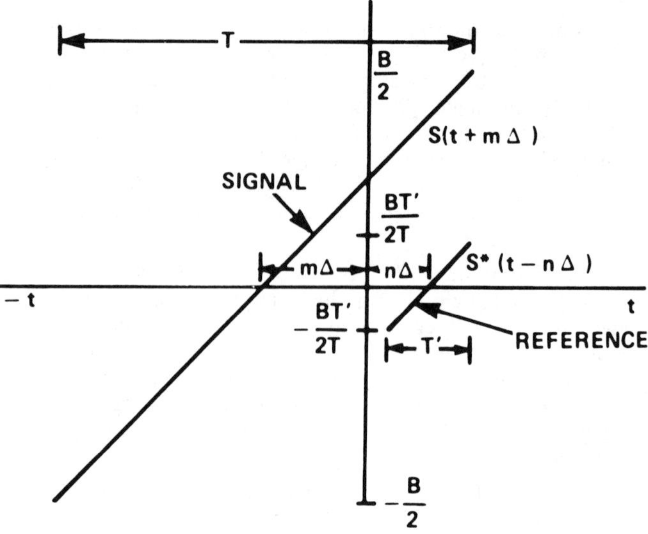

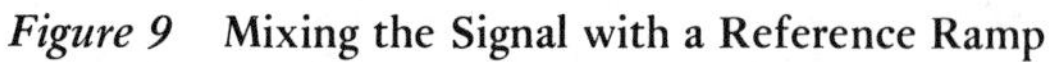

Figure 9 **Mixing the Signal with a Reference Ramp**

$$S_r = e^{j\phi} \sum_{k=0}^{(T'/\Delta)-1} 0.54\, e^{j2\pi k[(B/T)(m+n)\Delta^2-(r\Delta/T')]} - \sum_{k=0}^{(T'/\Delta)-1} 0.23 \cdot e^{j2\pi(k[(B/T)(m+n)\Delta^2-(r\Delta/T')+(\Delta/T')]+(\Delta/2T'))}$$

$$-\sum_{k=0}^{(T'/\Delta)-1} 0.23\, e^{j2\pi(k[(B/T)(m+n)\Delta^2-(r\Delta/T')-(\Delta/T')]-(\Delta/2T'))} \tag{7}$$

Using the series

$$\sum_{k=0}^{N} e^{jk\theta} = 1 + e^{j((N+1)\theta/2} \sin\frac{N\theta}{2} \csc\frac{\theta}{2} \tag{8}$$

$$S_r = e^{j(\pi B)/T(2nm\Delta^2)} e^{j(\pi B)/T(m^2\Delta^2-m\Delta T')} e^{j(\pi B)/T(n^2\Delta^2-n\Delta T')}$$

$$\cdot \Bigg\{ 0.08 + e^{j(T'/\Delta)\pi[(B\Delta^2/T)(m+n)-(r\Delta/T')]}$$

$$\cdot \Bigg[0.54 \frac{\sin((T'/\Delta)-1)\pi[(B\Delta^2/T)(m+n)-(r\Delta/T')]}{\sin\pi[(B\Delta^2/T)(m+n)-(r\Delta/T')]}$$

$$+0.23\, e^{j(\pi\Delta/T')} \frac{\sin((T'/\Delta)-1)\pi[(B/T)(m+n)\Delta^2-(r\Delta/T')-(\Delta/T')]}{\sin\pi[(B/T)(m+n)\Delta^2-(r\Delta/T')+(\Delta/T')]}$$

$$+0.23\, e^{-j(\Delta\pi/T')} \frac{\sin((T'/\Delta)-1)\pi[(B/T)(m+n)\Delta^2-(r\Delta/T')-(\Delta/T')]}{\sin[(B/T)(m+n)\Delta^2(\gamma\Delta/T')-(\Delta/T')]} \Bigg] \Bigg\}$$

$$r = 0, 1, \ldots \frac{T'}{\Delta} \tag{9}$$

The above equation can be interpreted as follows:

$e^{j(\pi B/T)[2nm\Delta^2]}$ — *This term represents the phase change to which the second FFT responds and separates the fine range resolution elements.*

$e^{j(\pi B/T)[m^2\Delta^2-m\Delta T']}$ — *This term is one of constant phase, which is independent of n; it gives a fixed phase out of the second FFT which is lost when the absolute value is taken.*

$e^{j(\pi B/T)[n^2\Delta^2-n\Delta T']}$ — *This term is known phase function dependent on the reference displacement which must be corrected prior to processing in the second FFT.*

The rest of the terms are the filter transfer function. The center of the filter is at $(B\Delta^2/T)(m+n)-(r\Delta/T')$.

In Equation (9), indexing S_r, the Fourier coefficient which corresponds to the stepping increment that a received target would produce, produces a rotary vector the frequency of which corresponds to the difference frequency between the center of the filter (S_r) and the actual frequency. This equivalence occurs, providing the known quadratic phase term which is a function of n is first subtracted. The equation also shows that, when the filter function is stepped by an integral number of Fourier coefficients, there is no change in the filter phase function.

Applying the corrected signal to a second FFT resolves each range resolution cell into T/T′ elements, providing the full waveform resolution. Weighting is also required on the second FFT input to reduce adjacent bin sidelobe levels; it is selected to establish the radar sidelobe levels.

There are some important factors which have to be considered in order to minimize sidelobe level:

1) Input spectral aliasing
2) Sidelobes of the spectral coefficients (short aperture)
3) Time sample aliasing of first FFT coefficients
4) Quadratic phase change from step to step
5) Straddling loss due to time sampling
6) Second FFT sidelobes
7) Amplitude variations due to coefficient amplitude versus frequency characteristic in both the first and second FFTs.

The first item above involves the time sampling of the input baseband signal. Since the sampling is complex, it should be at least equal to the bandwidth of the signal. Since there is no perfect band limiting filter, a guard band is necessary to avoid spectral foldover losses. Therefore, sampling rates on the order of 1.1 to 1.5 times Nyquist provide low sampling losses due to aliasing (depending upon the band limiting filter characteristics). The output of the first spectrum analyzer includes the full coverage of the spectrum up to the sampling frequency. Those coefficients outside the bandwidth of the signal can be dropped before going into the storage memory.

A very important point which must be considered in order to minimize system sidelobe levels is that time weighting must be applied across the aperture of the first FFT. If time weighting were not applied, each Fourier coefficient would have a (sin x)/x sidelobe response which would result in unacceptable sidelobes for most radar signal processing. Weighting increases the main frequency lobe width for each coefficient, and lowers the sidelobes according to the function used.

The second FFT receives sequential samples derived from adjacent apertures of the first; the sampling rate from the first FFT output must be sufficient to prevent aliasing in the second. To increase the sampling rate out of the first FFT, it is necessary to recompute the spectral coefficients by sliding the time aperture in steps, each of which correspond to the desired frequency coefficient sampling time. For adequate performance of the matched filter, this time should be at least equal to twice the period of the aperture. Since the weighting function in the first FFT widens the bandwidth covered by each spectral coefficient, there is a large amount of overlap in frequency. Not all frequencies covered by a coefficient in the first FFT are extracted from that coefficient. Therefore, some time sampling aliasing can be tolerated provided that aliased signals do not fall in the frequency region from which the fine resolution frequencies are to be extracted. The useful region for a low resolution coefficient is equal to the frequency difference between coefficients, or one over the time aperture of the first FFT.

Another factor which must be applied is a quadratic phase correction to the input of the second FFT to remove the effect of simultaneous stepping in time and frequency. The mathematical analysis of the previous section has shown why this factor exists and why the deterministic correction is required.

The output sampling rate of the matched filter should be higher than that of Nyquist sampling to minimize the effects of straddling loss. The sampling rate can be increased without changing the clock rate, since the Fourier coefficients of the first FFT (which represent the "guard band" and have been discarded) can be replaced by "zeros" in the second FFT. This alteration increases the output sampling rate without changing the output waveform shape or bandwidth.

Close-in sidelobes which result from the second FFT operation can be reduced by amplitude weighting on the input to that FFT, thereby reducing the sidelobes according to the weighting function selected.

Since each FFT coefficient has an equivalent bandpass characteristic which is not completely flat, the slope of the linear FM must be such as to have the same off-center frequency for each coefficient in order to prevent a large unwanted amplitude modulation from appearing as an input to the second FFT. There are several ways to assure that this condition is met; these requirements are the same as for linear phase compensation:

1) Select the size of the first and second FFTs, ensuring that step frequency shifts ahead of the first are an integral multiple of the number of coefficients.
2) Use several reference functions on the input so that the step frequency is shifted to always coincide with a coefficient number.
3) Allow small deviations from integer steps but keep the harmonic modulations small.

A very large degree of waveform flexibility can be achieved using these options.

The achievable matched filter time-bandwidth product for a practical implementation of the step transform process can be expressed as:

$$TW = \frac{(\text{number of points in first FFT}) \times (\text{number of points in second FFT})}{\left(\dfrac{\text{time steps of aperture}}{\text{period of aperture}}\right) \times \left(\dfrac{\text{sampling rate}}{\text{Nyquist rate}}\right)^2}$$

Variations in the time-bandwidth product can be obtained by varying the above factors.

A last correction factor is applied to correct for the amplitude of the fine resolution coefficients as they deviate off the center frequency of the coefficients of the first FFT. This change is incorporated as a gain correction on the output of the filter.

Implementation of Processor – Step Transform Example

The primary functional elements of the step transform processor are highlighted in Figure 10, which also indicates the data flow through the processor. The input is in the form of an intermediate frequency (IF) signal of bandwidth W Hz. This signal is demodulated to baseband in-phase (I) and quadrature (Q) components by the baseband demodulator system; these signals each have a bandwidth of W/2 Hz and a minimum sampling rate (Nyquist frequency) of W Hz.

In the example shown, the I and Q signals are sampled at 1.23 times the Nyquist frequency. This factor is affected by the minimum sidelobe level required in the pulse compression system; the ratio of 1.23 is sufficient for -35 dB sidelobe levels.

After sampling, the input data are organized into overlapping signal samples in the buffer and switch unit. A portion of the signal is repeated corresponding to an overlapped aperture. An 18-sample point overlap is used. In the example, with a time-bandwidth product (TW) equal to 592, the input signal is demodulated with a linear FM sawtooth, 32 samples in length. Thus, the FFT must process 32 sample apertures. To avoid aliasing in the frequency domain, the time samples represented by successive ramps must be increased because of the overlap in successive ramps, as shown in

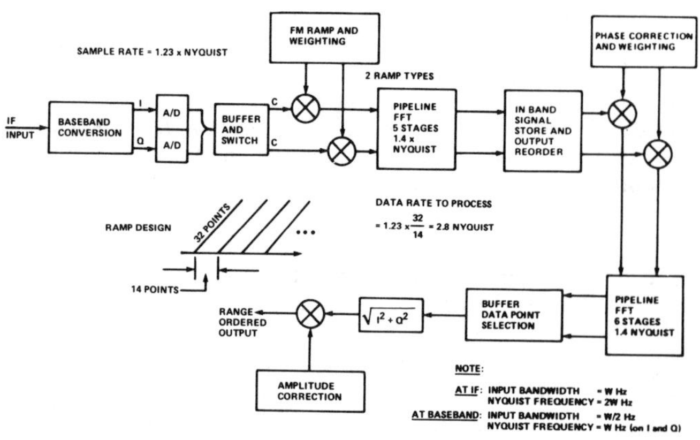

Figure 10 Single Channel Step Transform System (TW = 592)

the diagram. Time weighting is applied across the input sample interval together with the FM demodulation to reduce the sidelobe levels in the frequency domain. Fourteen sample intervals separate the starting points of the demodulating 32-point ramps. The net data rate necessary to process the input signal is then 1.23 x (32/14) = 2.8 times the input bandwidth (Nyquist frequency for baseband signals).

Since the architecture of the pipeline FFT provides processing of two parallel input data streams at the clock rate of the processor, the clock rate required for real-time processing in this case is 1.4 times Nyquist; 16 sample intervals are required for a 32-point FFT. The frequency coefficients from the first pipeline FFT are stored in a memory unit which also reorders the data for fine range analysis in the second FFT. Prior to insertion in the second FFT, a phase correction term is applied to the samples. The desired weighting function for range sidelobe reduction is applied in conjunction with the phase correction function. In the example, the second FFT has 6 stages (64 points); the first FFT has 5 stages (32 points). Each 64-point aperture of the second FFT would ideally have one-to-one correspondence with two sets of 32 coefficients from the first FFT. However, because the sampling rate is greater than Nyquist, all frequencies above the 32/1.23 = 26th point are in the aliasing guard band and therefore invalid. Zeros are inserted for these samples.

The output range ordered data rate can be effectively equal only to the input sampling rate. Thus, the desired terms must be selected from the second FFT, a function of the final buffer.

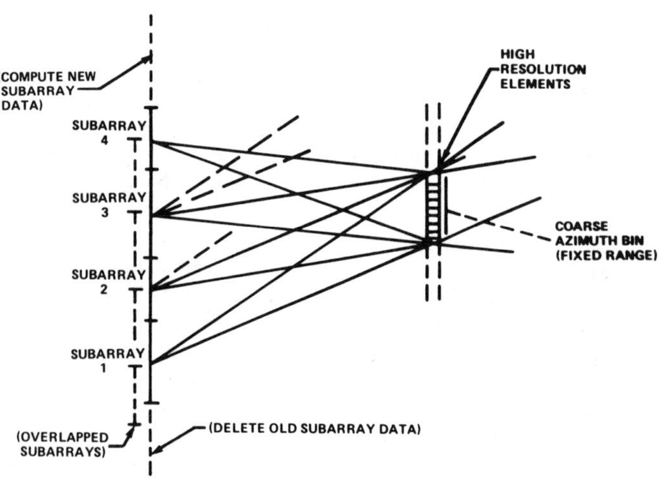

Figure 11 Subarray Synthetic Aperture Processing

The amplitude of the signal is obtained by calculating an approximation to the square root of the sum of the squares of the I and Q output samples. Prior to thresholding and other post-processing functions, a final amplitude adjustment is made to correct for roll-off of the weighting function applied ahead of the first FFT.

Step Transform Algorithm Applied to Synthetic Aperture Radar

The step transform processor is efficient for the strip map mode type synthetic aperture described in detail in Chapters 16, 17, and 18. As discussed in Chapters 16 and 18 continuous linear FM processing is required for focused strip map SAR systems.

In performing the azimuth processing for such a focused synthetic aperture system, the range (or phase) pattern of pulse returns are "matched" to that which an azimuth element would follow due to the motion of the radar vehicle. The more pulses which are combined or the larger the length of the synthetic array, the higher the azimuth resolution.

The step transform process can be considered as a subarray approach to SAR processors, as shown in Figure 11. Each wide beam subarray can be focused toward the same azimuth element; the subarrays then can be combined to obtain the high angular resolution corresponding to that of the full aperture. In performing this operation, it is also necessary to overlap subarrays to avoid the grating lobes that would result if the subarrays were end to end.

As the full aperture is moved in space, each subarray changes its relative position in the aperture and a different beam position is required for combining with the other subarrays.

Since all subarray beam positions are used as it moves through the aperture, these beam positions can be computed as soon as the subarray data have been accumulated.

Figure 11 shows how the subarray beam positions are selected to cover a common azimuth coarse resolution element. Beam positions from subarrays 1, 2, 3, and 4 are shown overlapping at the same coarse azimuth bin. Each subarray requires a different beam position. If the new data are being obtained by subarray #3, its bottom beam position would be used while the other beam positions for subarray #3 are stored until needed. After a beam position has been used for each subarray, that beam position is not used again; therefore, it is not necessary to continue to store that data.

Figure 11 also shows the combining of the four subarrays to resolve the coarse resolution element to a number of high resolution elements. The above process then can be repeated, shifting the subarrays, to obtain the next set of contiguous high resolution elements.

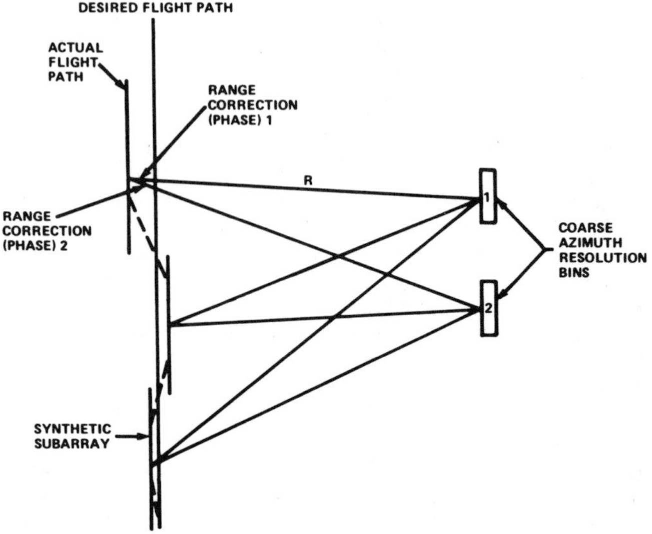

Figure 12 Platform Motion Compensation

Platform motion of the aircraft and range walk can be compensated for using the subarray approach by incorporating a range correction when combining the subarray outputs. This procedure is shown in Figure 12. As new subarrays are added and each subarray moves along the full aperture distance, the correction factor for the subarray is modified to match the actual flight geometry.

A synthetic aperture processing procedure using subarrays can be summarized as follows:

1) Focus data over short subarray time
2) Store subarray data over large array length
3) Combine subarray outputs to form large array
4) Compute new subarray data entering large aperture
5) Drop old subarray data leaving large aperture

The implementation of the process is identical to the range pulse compression except that pulse-to-pulse samples of the same range bin are applied for the azimuth processing.

Comparison of SAR Processors

A comparison of four approaches to SAR processing is shown in Table 1. The common convolution approach performs synthetic aperture processing using discrete time weighting and summation. For a large number of pulses in the array, the number of taps and complex multipliers can become quite large since a complex multiplier is required for every sample in the array.

A second approach to synthetic aperture processing is a FFT batch processing implementation. This method makes use of the fact that convolution can be performed by taking the Fourier transform of the input signal and multiplying that spectrum by a reference spectrum which is the Fourier transform of the convolving function. The inverse Fourier Transform then results in the time convolution at the output. The FFT batch processing is efficient in that fewer complex multipliers are required than for the common convolution approach. Because of the nature of the FFT process, the convolution using a discrete FFT is cyclic or periodic; therefore, some additional processing must be done beyond the length of the convolving function. In most cases an analyzing aperture greater than twice the length of the convolving function (2BT) is used.

A third approach is a deramping process over the full length of the aperture, followed by an FFT. The requirement to overlap the deramping reference by as much as 75% to avoid losses makes the process less efficient.

The subarray processing technique uses much less computational hardware and reference memory, and the minimum data memory size. It also has the advantage of the ability to compensate for undesired base motion.

The data memory corresponds to the minimum required to perform the azimuth focusing. Both the number of complex multipliers and reference size are substantially less than those required for shift register convolution and FFT batch processing.

The classical optical processor for strip mapper SAR systems is described in Chapter 16 by Curtis. An efficient real-time digital processor for spotlight mapper SAR systems is presented in Chapter 18 by Brookner.

References

[1] Cooley, J.W.; and Tukey, J.W.: "An Algorithm for the Machine Calculation of Complex Fourier Series," *Mathematics of Computation, Vol. 19, No. 90,* pp. 297-301, 1965.

[2] Cochran, W.T.; et al.: "What is the Fast Fourier Transform?" *IEEE Transactions on Audio and Electroacoustics, Vol. AU-15, No. 2,* June 1967.

[3] Halpern, H.M.; and Perry, R.P.: "Digital Matched Filters Using Fast Fourier Transforms," *EASCON Proceedings,* pp. 222-230, 1971.

[4] Groginsky, H.A.; and Works, G.A.: "A Pipeline Fast Fourier Transform," *EASCON Proceedings,* pp. 22-29, 1969.

[5] Gold, B.; and Bially, T.: "Parallelism in FFT Hardware," *IEEE Transactions on Audio and Electroacoustics, Vol. AU-21, No. 3,* February 1973.

[6] Volder, J.E.: "The CORDIC Trigonometric Computing Technique," *IRE Transactions on Electronic Computers, Vol. EC-8, No. 3,* pp. 330-334, September 1959.

[7] Martinson, L.W.; and Smith, R.J.: "Digital Matched Filtering with Pipelined Floating Point FFTs," *IEEE Transactions on Acoustics, Speech, and Signal Processing, Vol. ASSP-23, No. 2,* April 1975.

Approach	Data Memory	Complex Multipliers	Reference Memory	Remarks
Shift Register Convolution (Common Approach)	BT	BT	BT	Accommodates aircraft maneuvers, automatic focusing
Batch Processing (Convolution)	4 BT	$2 \log_2 2$ BT +2	2 BT	Efficient fixed parameter processing, cannot respond rapidly to aircraft maneuvers
Long Chirp Deramping (75% overlapped)	4 BT	$2 \log_2$ BT +4	4 BT	Cannot respond to aircraft maneuvers
Subarray Processing	BT	$\log_2$ BT +4	$4\sqrt{BT}$	Provides efficient process along with ability to accommodate aircraft maneuvers and variable focusing

Table 1 **Comparison of Strip Map SAR Processors**

Chapter 12 Worley

Surface Acoustic Wave Devices for Radar Systems

This chapter deals with one current method of generating and processing waveforms used in radar systems – Surface Acoustic Wave (SAW) devices. There has been much activity in this area in the last several years with a great amount of funding from the Department of Defense. The great interest in these devices owes to the fact that they allow the generation and processing of complex waveforms with devices which are orders of magnitude smaller, far more reliable than preceding hardware, easily reproduced, and relatively inexpensive.* This chapter is divided into two parts – the first deals with the physical and electrical characteristics of these devices and a description of what they can do; the second deals with two specific devices.

We are all familiar with one or more types of surface waves. The water wave is a form of surface wave. The motion of a drop of water is the same (elliptical) as the motion of an atom on a solid surface as a surface wave passes under each. Earthquakes are another form of surface wave. In fact, the origin of the theory used today to describe the mechanics of SAW devices comes from physicists in the 1800's attempting to describe the transmission of earthquakes over and through the earth.

There are many types of surface waves, depending upon whether the surface is free or an interface exists between two layers and whether the substrate is piezoelectric, magnetic, or neither. From a device standpoint, Rayleigh waves are the most interesting.

Figure 1 shows a diagram of a Rayleigh wave. This mode is composed of two coupled modes – a compressional component causing particle motion in the direction of propagation, and a shear component causing particle motion perpendicular to the surface. The amplitude of the Rayleigh wave disturbance decays exponentially into the surface and penetrates to a depth of only about two wavelengths. The propagation of the wave is dispersionless, a point which is of utmost importance for making devices with a linear phase response. The velocity of a Rayleigh wave on a typical substrate is 3×10^5 cm/sec, or five orders of magnitude less than the velocity of electro-magnetic waves. Thus, if one were to make EM and SAW travelling wave devices with equivalent properties, the SAW device would be five orders of magnitude smaller. The confinement of the SAW to the surface allows the acoustical signal to be manipulated (tapped, guided, coupled to, etc.) as it travels along. The planar nature of the devices means that they can be fabricated by the standard photo-resist techniques of the semiconductor industry and so can be mass produced easily. The planar nature also necessitates that the device be hermetically sealed, since surface damage would seriously affect its response.

The acoustic signal is generated from an electric signal via the piezoelectric effect with either a piezoelectrically active substrate or overlay. If an electric field is placed across a piezoelectric material, a mechanical strain is induced; this strain will propagate if there is a gradient to the field. By reciprocity, the mechanical disturbance is detected by the same electrode configuration which

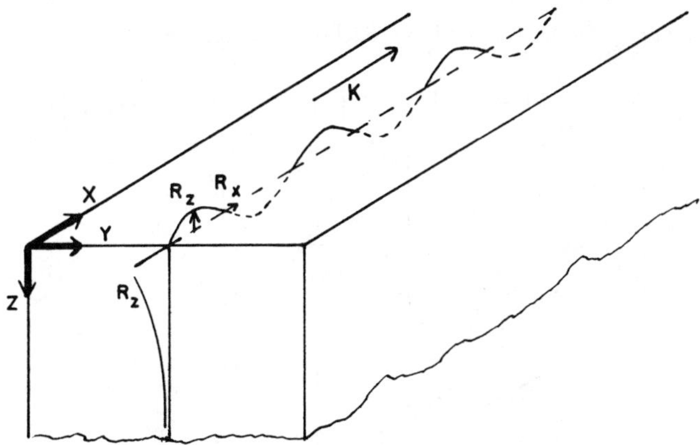

Figure 1 Rayleigh Surface Acoustic Wave

would generate it. To generate a SAW, a pair of interdigital electrodes with a center-to-center spacing of one half wavelength at the desired center frequency of operation is photolithographically etched out of a thin film (≅ 2000 Å) of aluminum or gold which is vacuum deposited on the surface. The two fingers are then connected to opposite electrical poles. This transduction is schematically shown in Figure 2. The SAW generated in this manner travels in the forward and reverse directions along the surface perpendicular to the fingers. Usually some damping material must be placed in the path of the reverse wave to prevent spurious echoes. As is shown later, filter structures are synthesized by placing many finger pairs of various spacings and lengths in acoustical series and electrical parallel on the surface. Since the fingers usually must be one-half wavelength apart, the maximum frequency at which a SAW device can be expected to operate is determined by the fineness with which the fingers can be fabricated. Practically, the maximum frequency is approximately 800 MHz, at which point the finger width is 1 μm. With optical techniques, it is difficult to fabricate fingers of less than this width, though devices have been fabricated up to 3 GHz using electron beam methods.

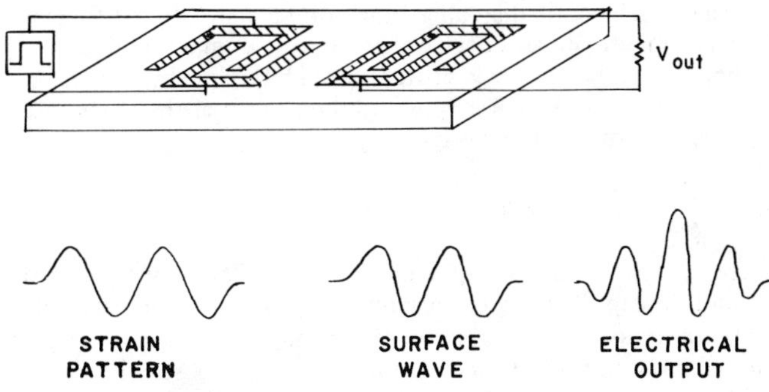

Figure 2 Diagram of the Generation and Detection of a Surface Acoustic Wave

*Chapter 10 by Purdy gives an example involving the replacement of seven racks of equipment with the RAC device of Figure 11 in Chapter 10.

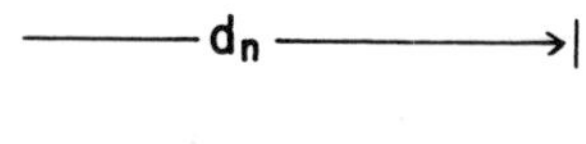

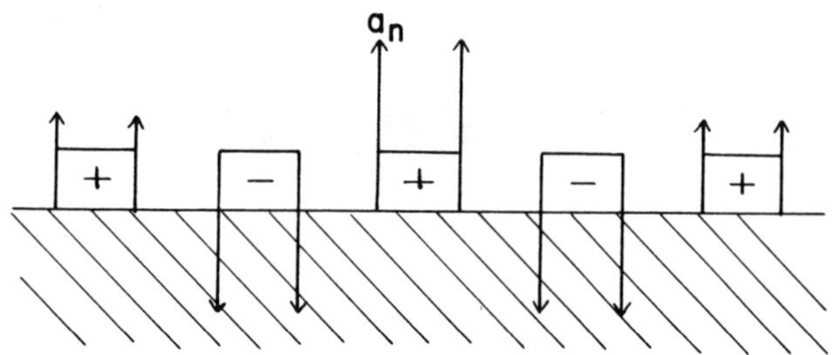

Figure 3a Delta Function Model of a Surface Wave Device

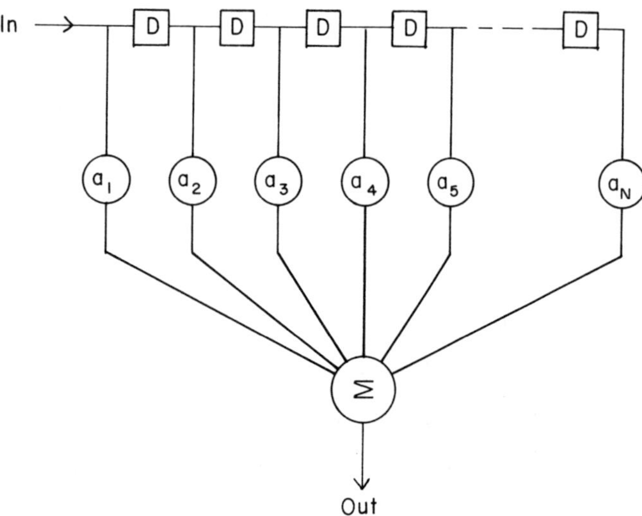

Figure 3b Non-Recursive Digital Filter

A first order model of the SAW device assumes that each edge of each finger is an infinitely narrow source of strain [1], as shown in cross-section in Figure 3a. The amplitude of the strain is given by the relative amount of overlap of each finger in the interdigital array. The response of a device is given by

$$H(f) = \sum_{n=1}^{N} a_n \exp(j2\pi d_n f/v)$$

where a_n is the relative overlap of the appropriate finger pair and d_n is the distance from some reference point to the nth edge.

A SAW device is nothing more than a time domain tapped delay line. Because of this fact, it is analogous to other types of devices, most notably the non-recursive digital filter and the end-fire array antenna. A schematic diagram of the non-recursive digital filter is shown in Figure 3b; this device also is a time domain tapped delay line where the energy is shifted through the line by the clock. D is the unit delay which is typically the inverse of the clock rate. The response of this filter is given by

$$H(f) = \sum_{n=1}^{N} a_n \exp(j2\pi f T_n)$$

where a_n is the weight at tap n and T_n is the delay between the input and the tap. Because the response equations are the same, all the literature on designing non-recursive digital filters may be applied directly to SAW devices. A similar analogy exists with the end-fire array antenna and that knowledge may also be applied directly [2].

A more accurate model than the one given earlier for predicting the electrical properties of the device is shown in the form of an equivalent circuit in Figure 4 [3].

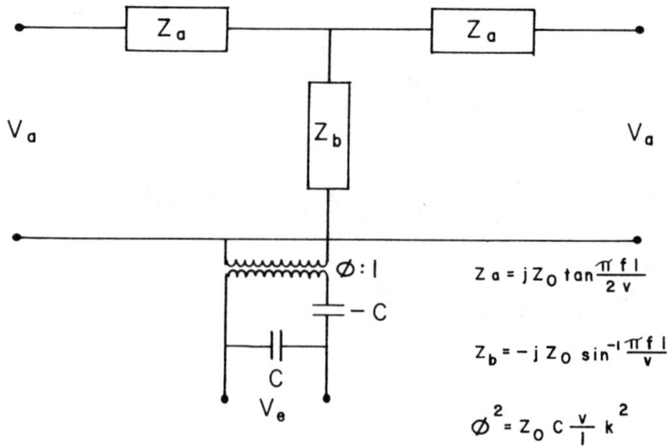

Figure 4 Equivalent Circuit Model of One Finger Pair of a Surface Wave Device

This model is able to include many second order effects, such as electrical and mass loading of the fingers on the SAW and detection and regeneration of the signal as it passes under its own transducer. Z_o is the characteristic impedance of the substrate, V_e is the electrical input or output, and V_a is the acoustic output or input (the mechanical force on the lattice which is the acoustic equivalent of a voltage). As can be seen, the device looks like a coupled transmission line. The transformer represents the electrical to mechanical conversion. Its value is dependent upon the piezoelectric constants of the material; i.e. how much strain is induced for a given voltage. A transducer consisting of a uniform array of interdigital fingers looks like a resistor and capacitor in series at the center frequency of the array. The capacitance is just that of the interdigital fingers; the resistance is given by

$$R = \frac{4}{\pi} k^2 (1/\omega_o C_s)$$

where k is the electromechanical coupling coefficient of the material and C_s is the capacitance of one finger pair. The electrical Q of the input is

$$Q = \pi/(4k^2 N)$$

where N is the number of finger pairs.

Some of the possible devices which can be made from SAW delay lines are Resonators, Simple Delay Lines, Memories, Oscillators, Bandpass Filters, Pulse Compressors, Spread Spectrum Correlators, Transform Signal Processors, Nonlinear Convolvers, Couplers, and Isolators. The remainder of this chapter deals with two of these devices – Pulse Compression Filters and Phase Code Correlators.

Figure 5 shows a drawing of a transducer pair used in an FM pulse compression filter. Notice that the center-to-center spacings of the fingers in the large array vary from one end to the other. Let us assume that each finger pair generates one half cycle of the impulse response of the delay line. Upon impulsing the large array, one would generate a cycle with a short wavelength (high frequency) at the left of the array; at the other end, the cycle would have a long wavelength (low frequency). The high frequency must travel only the separation between the two transducers before it is detected while the low frequency must travel the separation plus the length of the array before its presence is realized. The signal which is output at the second transducer varies with frequency in time since time and distance are related by a constant. Since there exists the freedom to grade the finger spacing in any fashion, it is possible to make a SAW delay line which has almost any FM characteristic. This device can then be used to generate the FM signal for a pulse compression radar. In addition, if one

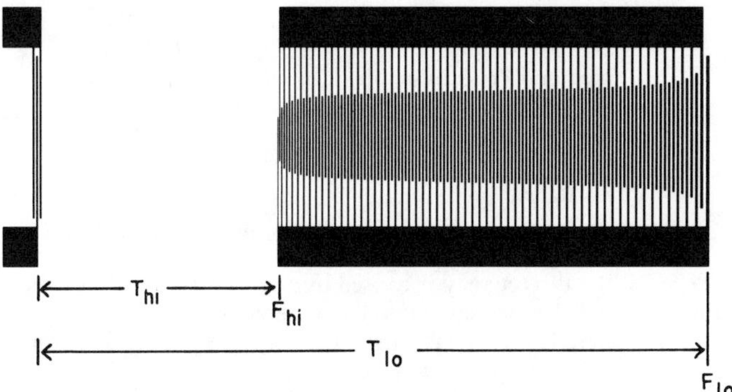

Figure 5 FM Transducer

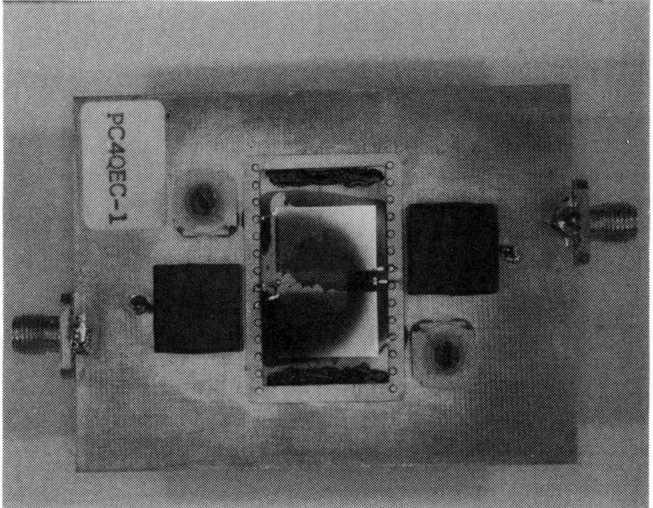

Figure 6 Picture of FM Pulse Compression Filter. Rectangle in the Center of the Surface Wave Device

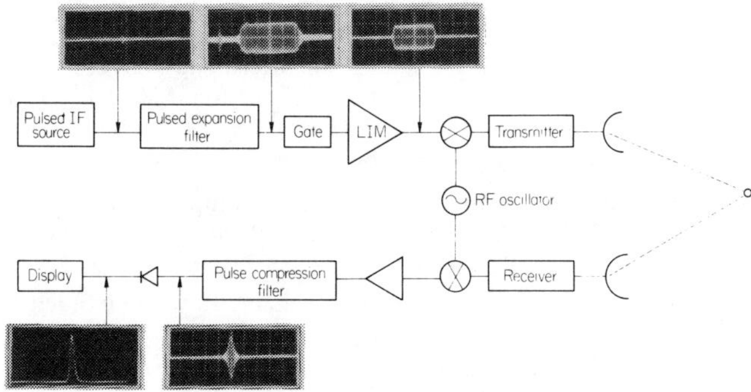

Figure 7 Pulse Compression Signal as it Passes Through System

Center Frequency	5 to 1000 MHz
Bandwidth	to 100%
Delay Time	to 150 μs
Time Bandwidth Product	10 to 10,000
Spectrum Phase Ripple	0.5° rms
Spectrum Amplitude Ripple	0.5 dB rms
Time Signal Amplitude Ripple	0.5 dB rms
Insertion Loss	25 to 50 dB
Sidelobe Level	40 dB

Table 1 Properties of Current SAW FM Pulse Compression Filters

makes a delay line with the opposite slope and puts it in the receiver, one can compress the long original FM signal.* Besides grading the spacings of the fingers, it is also possible to grade the relative overlap of the fingers to obtain a desired amplitude response. A picture of a device is shown in Figure 6, with dimensions 0.75″ x 1.25″ x 0.06″ – representing a volume and weight reduction of 500:1 over the equipment it was designed to replace.

Figure 7 shows a schematic of the signal as it travels through a radar system. The signals shown are from a SAW filter which had a center frequency of 30 MHz, a dispersion slope of 1.25 MHz/μs, and a pulse length of 6 μs. The expansion filter was pulsed with a 0.1 μs burst at 30 MHz. The output of the filter which shows an amplitude ripple of 0.5 dB was then limited and gated to 3 μs with a 0.25 μs rise and fall time. This signal was then fed into the receiver; the output of the compressor was .350 μs wide at the -4 dB points, the sidelobes were 25 dB down, and spurious responses were more than 40 dB down. The CW insertion loss of each filter at the center frequency is 27 dB. In these filters, an FM array for both input and output is used and each array has 140 fingers. Also, weighting is used in the compression filter to decrease the sidelobes from the theoretical -13 dB for autocorrelation to -25 dB.

Table 1 lists the present capabilities of SAW pulse compression filters, focusing on the properties of existing devices rather than on expected results from future devices. All these characteristics are not possible with any one filter. For example, for low time-bandwidth product filters, it is difficult to achieve very low sidelobes, due to the Fresnel ripple in the spectrum.** Narrow bandwidth filters generally have lower insertion losses than do wide bandwidth filters.

A second type of device worth a close look is the phase coded variety. Figure 8a shows a drawing of a delay line the output response of which is that of a seven bit Barker code. The transducer on the left has as many finger pairs as there are half cycles in each bit of the code. The single finger pairs on the right are separated by the width of the input transducer. The phase reversals are obtained by the manner in which each of the fingers of each finger pair is connected to the large bars. For example, it can be seen that the connection of the finger in the second finger pair is the opposite of that of the first finger pair; thus, the second bit is 180° out of phase with the first bit. There is no reason that these taps must be hard wired as they are to the summing bars.

They could be (and some larger codes have been) attached to switches which can connect the finger to either bar to give a switchable code. Figure 8b shows the impulse response of this delay line, clearly showing the phase reversals.

Figure 9a shows a picture of a 127-bit maximal length sequence code synthesizer and correlator made by Hughes [6]. The center frequency is 30 MHz, the bit rate is 5 MHz, and the size of the single crystal quartz substrate is 0.75″ x 5″ x 0.10″. Figure 9b shows the impulse response of the filter; again, one can clearly see the changes in phase. It is so obvious here because there is one cycle too few in the input transducer; thus, the impulse response of the input transducer does not quite fill the width of each bit. The figure also shows how sensitive some SAW devices can be to a broken finger. Unlike the Barker code filter, this device had several fingers at each tap. Some of the bits in the impulse response are considerably smaller in amplitude than the rest, because of just one finger in each tap being disconnected from its summing bar. Figure 9c shows the autocorrelation response of this filter. The width of the correlation peak is exactly what theory predicts. The sidelobe level is -22 dB whereas ideally it should have been

*The same line used to generate the transmitted signal can be used to process the received signal. The frequency spectrum of the transmitted signal is reversed in the receiver, the lowest frequency signal becoming the highest frequency at the SAW IF. The "flipping" is done by selecting the receiver local oscillator (LO) above the echo signal frequency rather than below (or vice versa, depending on how the transmitted signal was generated, the opposite process being executed on receive).

**See References 4 and 5 relative to a technique for removing the sidelobe degradation of low time-bandwidth product filters due to the Fresnel ripple.

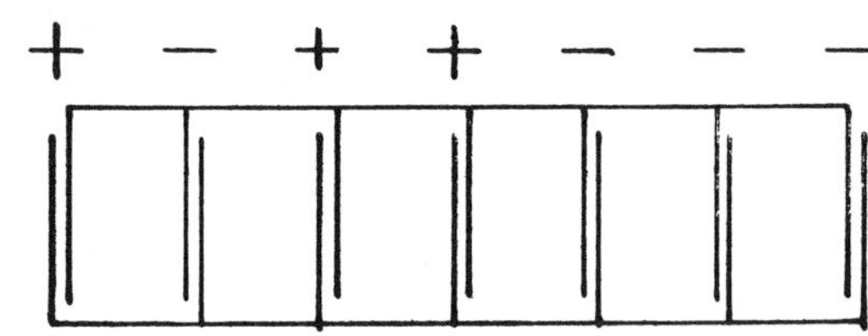

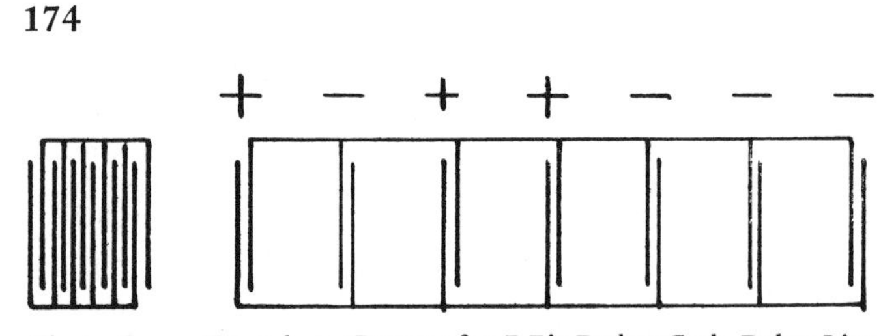

Figure 8a Transducer Pattern for 7-Bit Barker Code Delay Line

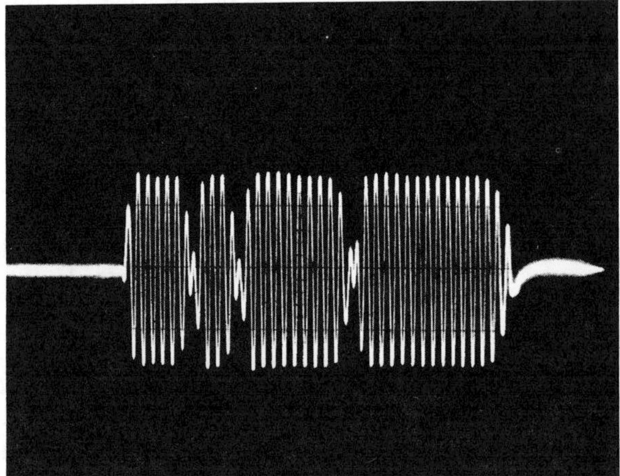

Figure 8b Impulse Response of Device

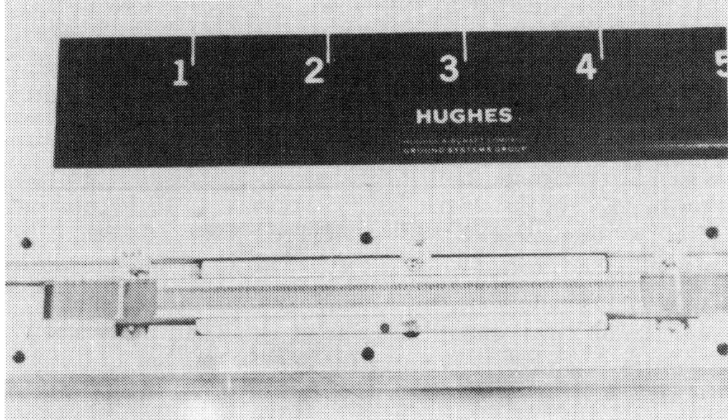

Figure 9a Picture of 127-Bit Synthesizer and Correlator

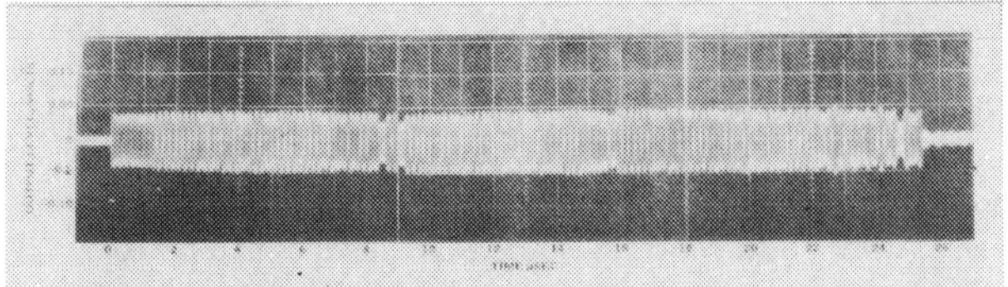

Figure 9b Impulse Response of Device

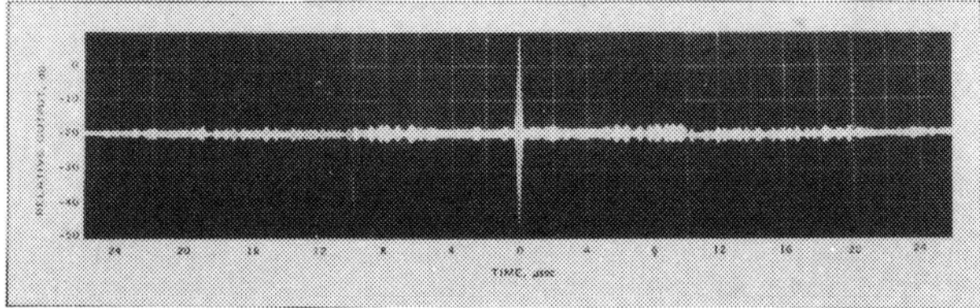

Figure 9c Autocorrelation Response of Device

Center Frequency	5 to 200 MHz
Code Length	7 to 2047
Fixed and Programmable	
Chip Rate	5 to 60 Mbits/s
Insertion Loss	25 to 50 dB
Sidelobe Level	1/N to $1/N^2$
Phase Ripple	0.5° rms

Table 2 Properties of Current SAW Phase Coded Filters

-22.5 dB; the discrepancy is caused by the four bits with broken fingers. The sidelobes show good symmetry, indicative of the fact that there is little phase distortion in the device and, in turn, that all second order effects have been either accounted for or eliminated. The untuned CW insertion loss of this device at its center frequency is 36 dB.

Table 2 lists the current capabilities of SAW phase coded filters. Programmable correlators have been made up to 127 bits by using hybrid techniques of combining a diode array and a SAW tapped delay line. Research is presently underway to produce a monolithic programmable filter on a silicon wafer. A 1016-bit filter has been built using eight 127-bit sequences which could be switched to occur at any of eight time frames within the long code.

In this chapter, an attempt has been made to touch lightly on the basic properties of surface acoustic wave delay lines and examples of the two devices which most directly apply to radar signal processing have been shown. There are other significant devices, such as bandpass filters and oscillators, which can be made from SAW delay lines; there are other areas in which SAW devices are making a significant contribution, such as air traffic control and communications systems. For further details on SAW devices and the recently developed acoustoelectronic devices, see References 7 through 10 and the "Present and Future Trends" section of Part 2.

References

[1] Tancrell, R.H.; and Holland, M.G.: "Acoustic Surface Wave Filters," *Proceedings of the IEEE, Vol. 59, No. 3,* pp. 393-408, March 1971.

[2] Tancrell, R.H.: "Analytic Design of Surface Wave Bandpass Filters," *Proceedings of the 1972 IEEE Ultrasonic Symposium (72-CHO-708-8 SU),* pp. 215-217, October 1972.

[3] Smith, W.R.; et al.: "Analysis of Interdigital Surface Wave Transducers by Use of an Equivalent Circuit Model," *IEEE Transactions on Microwave Theory and Techniques, Vol. MTT-17, No. 11,* pp. 856-864, November 1969.

[4] Bristol, T.W.: "Modern Radar Pulse Compression Techniques," *Record of 1974 NEREM (Radar Systems and Components,* Part 4), pp. 65-72, 28-31 October 1974.

[5] Judd, G.W.: "Techniques for Realizing Low Time Sidelobe Levels in Small Compression Ratio Chirp Waveforms," *1973 Ultrasonics Symposium Proceedings, IEEE Cat. No. 73 CHO 807-8 SU,* pp. 478-481, November 1973.

[6] Judd, G.; et al.: "Acoustic Signal Processing Devices," *First Technical Report Contract DAAB07-72-C-0023,* May 1972.

[7] *Proceedings of 1973 IEEE Ultrasonics Symposium (73-CHO-807-8 SU).*

[8] *IEEE Transactions on Microwave Theory and Techniques, Vol. MTT-21, No. 4,* April 1973.

[9] *Proceedings of the International Specialist Seminar on Component Performance and System Applications of SAW Devices,* Aviemore, Sootlant, 1973 (Published by the IEE).

[10] , *Proceedings of the IEEE,* Special Issue on Surface Acoustic Wave Devices and Applications, May 1976.

Chapter 13 Haggarty/Meehan/O'Leary

A 1 GHz 2,000,000:1 Pulse Compression Radar–Conceptual System Design

Introduction

This chapter describes a technique for generating, receiving, and processing large time-bandwidth product wideband linear FM pulses. A particular radar application concerned with making coherent high resolution measurements on orbiting bodies is detailed. The basic signal generation technique has features that make it attractive for many applications where a high quality, coherent, linear FM signal source is required. The technique is based on coherent frequency synthesis and digital timing; no dispersive filters or voltage controlled oscillators are employed.

A study performed to identify the critical areas for the prototype that has been designed, implemented, and operated is documented here. The description of the signal generation technique is preceded by a discussion of the overall radar system, including the effect of time dilation on large TW product signals and its correction. The chapter concludes with a summary of the highlights of the signal generation technique.

System Description

The advantages of linear FM pulse compression systems in radar are well known. The range resolution that can be obtained is determined by the bandwidth, W; the single pulse Doppler resolution, by the transmitted pulse length, T. Since the system can be made fully coherent, the fine Doppler resolution is determined by the coherent integration time and the PRF is measured as in a conventional coherent pulsed Doppler radar. Most radars operate with a peak power constraint, but it is signal energy (power x pulse length) that determines the signal-to-noise ratio on reception. The use of pulse coding to obtain long pulse duration permits the signal energy to be large within the peak power constraint, without loss of range resolution.

Linear FM coding results in a range-Doppler coupling – a feature often viewed as a disadvantage since it results in ambiguous range and Doppler measurements. However, the linear frequency-time coupling has the advantage that it simplifies the implementation of the signal processor. A single dispersive filter can be employed to cover a wide Doppler frequency band; for tracking applications, a single correlation receiver can be used to cover a large range window (multiple range cells). In fact, for a given range window, the linear FM slope, W/T, can be chosen to control the bandwidth required in the correlation processor.

The particular signal generator and processor described here employ an X-band, 2 ms duration, 1 GHz bandwidth linear FM pulse. The 1 GHz bandwidth was chosen to obtain 6″ range resolution; the 2 ms pulse duration was chosen since it is the maximum usable pulse length for the intended application, thus minimizing the required peak power-aperture product. The transmission frequency was chosen at X-band so that the percentage bandwidth would be compatible with the need for high quality characteristics in the microwave components.

Antennas with diameters as large as 120 ft have been operated successfully at X-band [1]. At 9 GHz, the combination of a 2 ms

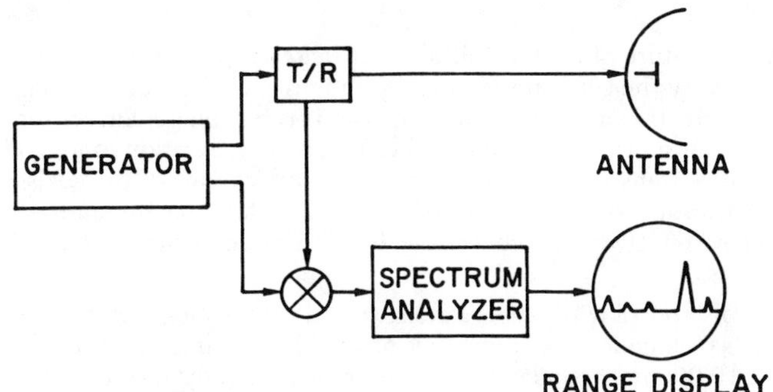

Figure 1 Linear FM Correlation Receiver

pulse, a 5 kW peak power transmitter, and a 120 ft diameter antenna yields a 30 dB signal-to-noise ratio on a 1 m^2 target at 1000 mi range. Transmitter tubes capable of supporting the 1 GHz signal bandwidth with high quality amplitude and phase characteristics are available at these low power levels. In addition, capability achieved with this long pulse technique should act as the catalyst for the development of higher peak power long pulse tubes (which would permit use of smaller apertures).

An additional advantage of the X-band frequency is that cross-range resolution comparable to the 6″ range resolution can be achieved via coherent multiple-pulse Doppler processing for small, slowly rotating (nearly stabilized) objects. Atmospheric absorption and ionospheric dispersion are also acceptable at X-band; see Figure 28 of Chapter 1 and Figure 7 of Chapter 14.

Reception

Because of the linear FM coding, the correlation receiver takes the form illustrated in Figure 1. The received signals are processed by mixing them with a linear FM local oscillator (LO) and spectrum analyzing the resulting output.* The range of the target must be known approximately in order to know when to start the sweeping LO. Targets within a limited range window appear at the output of the mixer in a narrow signal band. If the targets have negligible differential velocity, the spectrum of the sweeping receiver output is the range profile of the target. The scatterers on solid objects of 50 ft length with rotational rates less than 0.1 rev/s have negligible differential velocities in this regard. One mixer and a spectrum analyzer perform the correlation with negligible mismatch for range windows as large as 10% of the transmit pulse duration, T. The output of the spectrum analyzer is the range profile of the target. The band, BW, that must be analyzed for a range window ΔR is equal to $2\Delta R/C \cdot W/T$. The system was designed to measure objects up to 50 ft in length. By employing an auxiliary acquisition and tracking mode, the moderate bandwidth (5 MHz) transmissions of which are interspersed with the high resolution mode, range tracking accuracies of ±50 ft can be ob-

*This method is the same as the LFM correlation and spectrum analysis technique used on the ALCOR; see Figures 5 and 6 of Chapter 10 by Purdy.

tained readily on orbiting bodies. The resulting 150 ft range window combined with 0.5 MHz/μs FM slope yields a 150 kHz bandwidth at the receiver mixer output. This data can be handled in several ways. The narrow bandwidth makes it quite simple to digitize and record the data for later processing on a general purpose digital computer. Such a method is attractive since, not only the required spectrum analysis (or Fourier transformation), but sidelobe weighting (e.g., Taylor weighting) and transversal equalization (to overcome equipment errors) can be carried out simply, exactly, and economically. In fact analog to digital conversion, A/D, and recording rates compatible with 5 MHz bandwidths make it possible to handle range windows as large as 5000 ft in this non-real-time manner. In addition, the rapid advance of integrated circuit technology makes it possible to perform these calculations in real time.

A key feature of digital realization of these operations is the ability to modify the computation rapidly and precisely, thereby making the correction adaptive. In this manner it is feasible to correct for poor transmitter distortion characteristics via calibration loops and adaptive transversal equalization in the receiver. One of the goals of the prototype signal generation experimental program upon which MITRE embarked was the demonstration of these capabilities.

It is also possible to employ available real time analog spectral analysis techniques to obtain range profiles in real time. In particular, the 10,000 time-bandwidth product dispersive filter spectrum analyzer [2] developed at MITRE is capable of handling a 1250 ft range window in combination with the 2 ms, 1 GHz linear FM system.

Time Dilation

The time-bandwidth (TW) product or pulse compression ratio of 2×10^6 requires that the correlation receiver discussed above be flexible (or adaptive) since, even for moderate radial velocities, the Doppler or time dilation effect cannot be modeled adequately as a simple shift of the carrier frequency. As is well known, the true effect of target velocity is a scaling of the time axis of the transmitted signal; the received signal is a time dilated version. If the transmitted signal is f(t), the signal received from a point target, neglecting delay and amplitude scaling, is

$$s(t) = f(at) \tag{1}$$

where the scale factor a is the Doppler scalar

$$a = \frac{c+v}{c-v} \tag{2}$$

and v is the target radial velocity (positive for incoming target, negative for a receding target). The spectrum (Fourier transform) of the received signal, S(f), is related to the spectrum of the transmitted signal, F(f), by the relation

$$s(f) = \frac{1}{a} F(f/a) \tag{3}$$

as can be seen readily by inspection of the Fourier integral. From Equation (3) it is clear that all frequency components of the transmitted signal are Doppler shifted by an amount proportional to their value. Thus, not only is the carrier frequency shifted, but the bandwidth is modified. This change in bandwidth coupled with the change in signal duration (Equation 1) can cause significant distortion for highly coded signals.

There are many ramifications of this phenomenon. In general for large TW products, the receiver – be it a filter or a correlator – must be adaptive. Rihaczek [3] has considered the detrimental effect of this phenomenon for non-tracking linear FM systems; Thor [4] has suggested logarithmic modulation to overcome the problem. Since the application under discussion employs range and velocity tracking, the distortion mechanism is straightforward to analyze and the correction simple to implement. Consider the linear FM transmission

$$f(t) = \text{rect}\left[\frac{t}{T}\right] e^{j2\pi\,[f_o + Wt/2T]\,t} \tag{4}$$

then the received signal is, from Equation (1),

$$S(t) = \text{rect}\left[\frac{at}{T}\right] e^{j2\pi\,[f_o + W/2 \cdot at/T]\,at} \tag{5}$$

which can be rewritten

$$S(t) = \text{rect}\left[\frac{t}{T/a}\right] e^{j2\pi\,[af_o + aW/2 \cdot t/(T/a)]\,t} \tag{6}$$

Comparison of Equations (4) and (6) show that the pulse duration and bandwidth are scaled

$$T \rightarrow T/a \tag{7}$$

and

$$W \rightarrow aW \tag{6}$$

and the linear FM slope becomes

$$\frac{W}{T} \rightarrow a^2 \frac{W}{T} \tag{9}$$

As expected, the carrier frequency is af_o. Significantly, the signal remains a linear FM pulse – it simply has different parameters. Clearly, if the center frequency, af_o, is Doppler tracked in the receiver, the only distortion mechanism arises from the changes indicated in 7 through 9. The signal at the output of the correlation receiver has a slightly different pulse duration due to Equation (7) and a residual linear FM. The frequency modulation out of the difference mixer with center frequency tracking is

$$\text{FM residual} = [1-a^2]\,\frac{W}{T}\,t \tag{10}$$

If a range displacement is considered, there is an additional frequency offset term. However, Equation (10) represents the distortion effect due to correlating the received signal with the transmitted signal rather than with a replica of itself. The magnitude of this residual FM can be seen since, from Equation (2)

$$a = 1 - \frac{2v}{c+v} \tag{11}$$

$$a^2 \approx 1 - \frac{4v}{c} \tag{12}$$

Combining Equations (10) and (12) yields

$$\text{FM residual} = \frac{4v}{c}\,\frac{W}{T}\,t \tag{13}$$

Since the pulse length change shown in (7) is negligible, the resulting time bandwidth product of the residual FM pulse is

$$\text{TW residual} = \frac{4v}{c}\,\text{TW} \tag{14}$$

For typical orbit velocities, $4v/c \approx 10^{-4}$; the maximum distortion effect of time dilation for the 2 x 10^6 TW pulse is

$$\text{TW residual (max)} = 200 \qquad (15)$$

These distortion effects can be corrected exactly by autocorrelating the received signal with a scaled (see Equation (6)) version of the transmitted waveform. The swept frequency LO of Figure 1 simply must be time dilated.

Notice that the accuracy of the correction is not severe – the reduction of a by three orders of magnitude would decrease the residual modulation so that it is well within tolerable limits [5].* This accuracy of estimate can readily be furnished by the interspersed Doppler mode which employs a 2 ms, 5 MHz linear FM pulse. The Doppler shift of the carrier frequency, f_o, is

$$f_D = f_o - af_o$$

$$= (1-a) f_o$$

$$\approx \frac{2v}{c} f_o$$

The maximum of $v/c = 10^{-4}$ yields

$$f_D \text{ (max)} \approx 2 \text{ MHz}$$

Since a 500 Hz single pulse Doppler accuracy can be obtained from the track mode, correction of one part in 4 x 10^3 can easily be achieved. In practice, the high signal-to-noise ratios of the track mode permits estimates well in excess of one part in 10^4.

For a fully coherent system, such as the one under discussion, the correction is simple to implement. Since all signals, both frequency components and timing signals, are generated directly from a master clock, changing the frequency of the master clock produces a scaling in time of the output signal. The scaling compensates exactly for the Doppler effect. The master clock is actually a frequency synthesizer the frequency output of which can be changed in discrete steps over a range of one part in 10^4 – the expected range of v/c encountered for orbiting bodies. For the 10 MHz output of the master clock, this change is of ±1000 Hz in 2^{14} steps. The phase also is controlled digitally at this point to vary the starting time of the received ramp, thereby effecting range tracking. All these changes are controlled automatically by the computer using the data from the interspersed tracking mode.

Actually, two clocks are necessary in a practical system. The first, the fixed clock, runs at the undeviated frequency; it is used to generate the transmit pulse. The second clock, the flexible clock, has the digitally controlled frequency and phase outputs; it is used to generate the receive LO ramp shown in Figure 1. Two separate clocks are used to give the output of the flexible clock time to stabilize after a new command is given. If only one clock were used, it would be necessary for the output of the flexible clock to stabilize in the time between the end of the transmit pulse and the beginning of the return. Providing two clocks allows one to be programmed at the end of a receive interval and have the entire interpulse period to stabilize.

System Configuration

Figure 2 shows a block diagram of the entire system. The frequency standard, chosen for its excellent short term frequency stability, feeds the fixed and flexible clocks. These clocks, in turn, are gated at the appropriate times to a frequency ramp generator (which generates the subpulses) and to a ramp extender (which produces the 2 x 10^6 TW pulse). The pulse is sent to the transmitter in the transmit mode, and to the receive mixer in the receive mode.

*See Figures 8 and 9 of Chapter 14 by Brookner.

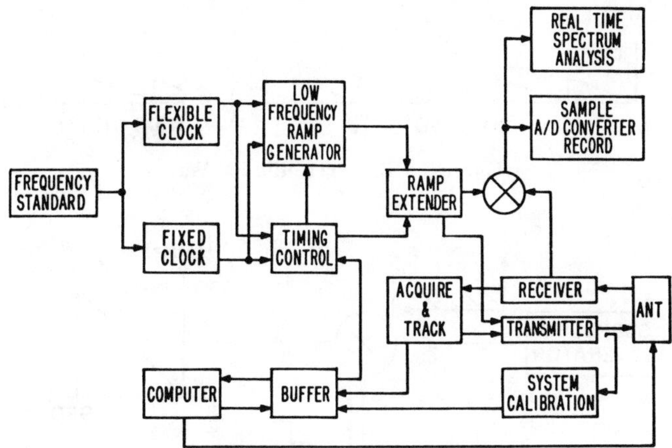

Figure 2 **Block Diagram of 2,000,000:1 Pulse Compression Radar System**

The acquire and track mode operates with a lower bandwidth signal, but it uses the same transmitter. The data from the acquire and track mode is fed to the computer to generate the settings for the receive ramp generation. That mode is interspersed with the wideband mode so that the trajectory of the object can be updated continuously. Notice that, since the received frequency, af_o, of the Doppler shifted pulse is estimated in the track mode and this information used to scale the frequency of the clock, no conversion of velocity frequency relation is required. The frequency estimate is applied directly by the computer to the flexible clock.

Signal Generation

The signal generation described here is optimized to the application. Since the system is designed to make measurements on tracked objects less than 100 range cells (50 ft) in length, the 2 ms pulse is composed of 192 subpulses. Consequently, by construction, range sidelobes due to waveform deficiencies tend not to occur in the first 192 range resolution cells.*

The signal is generated in two steps. A periodic train of linear FM subpulses is synthesized by adding 65 individual coherent Fourier harmonics with proper amplitude and phase. Each resulting subpulse has 5 MHz bandwidth and 10 μs duration (TW = 50). The 2 x 10^6 TW pulse is generated by sequentially mixing 192 subpulses with an appropriate set of 15 local oscillators. Figure 3 illustrates the basic periodic FM wave. Figure 4 is a conceptual block diagram of how such a periodic FM function could be extended in swept bandwidth and period. The required stepped local oscillator is shown in Figure 5. The practical problems associated with this basic method of extending the FM ramp are overwhelming. Not only is the mixer design extremely difficult, but 192 coherent frequencies also must be synthesized, precisely phase and amplitude equalized, and gated into the mixer in proper time sequence through a low loss combining network. An alternate procedure is illustrated in Figure 6.

The periodic train of linear FM subpulses are fed into a mixer the local oscillator of which changes by 5 MHz at the precise time when the subpulse ramp starts. The effect of this first stage is to extend the subpulse FM ramp bandwidth and period from 5 MHz and 10 μs to 10 MHz and 20 μs, respectively. At the next stage, the resulting periodic FM ramp is extended in three steps so it becomes a 30 MHz ramp with 60 μs period. This process continues until the full 960 MHz, 1920 μs signal is achieved.

*The sidelobes referred to here are those caused by signal generator errors and are *not* the system sidelobes which also depend on the exciter, transmitter, plumbing, etc. It is this latter class of sidelobes that the previous section addressed under the topics of digital transversal equalization.

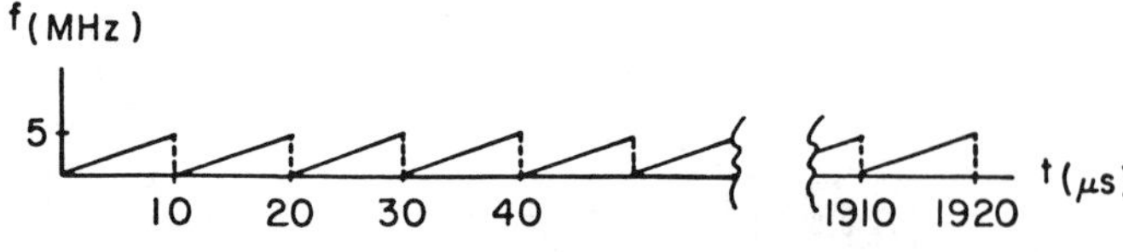

Figure 3 Basic Periodic FM Wave

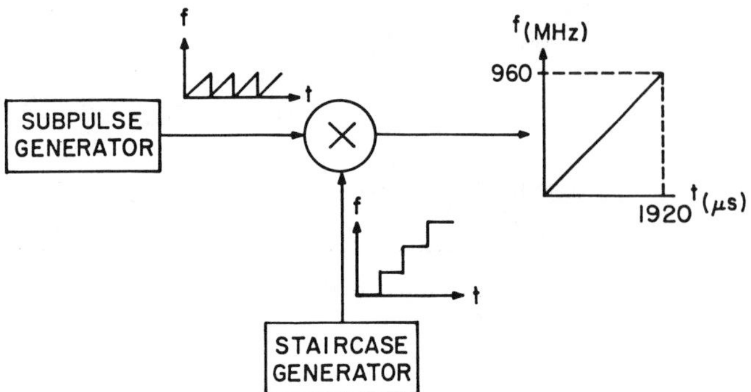

Figure 4 Conceptual Block Diagram of Ramp Extender

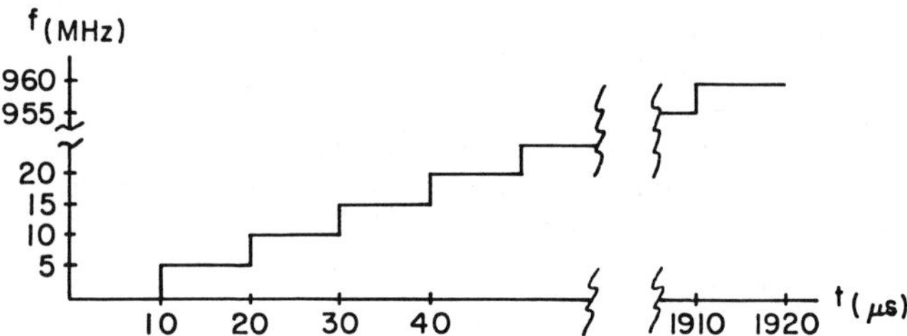

Figure 5 Stepped Local Oscillator Required for Ramp Extender

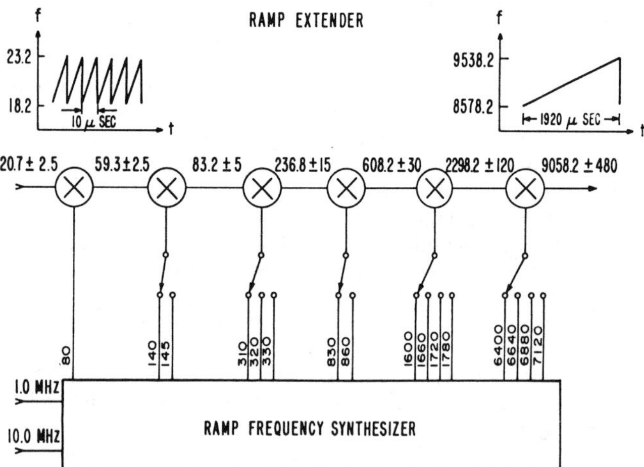

Figure 6 Block Diagram of Ramp Extender

The mixing and switching signals used in the ramp extender also are synthesized directly from the master system clock so that the system maintains complete phase coherence.

Note that by choosing the mixing frequencies properly it is possible to make the signal come out of the final stage at the desired transmit frequency. The conversion frequencies are selected to maintain the percentage bandwidth at approximately a constant value throughout. The basic increment is chosen so that the switching transients are less than one percent of the subpulse period. This choice insures that range sidelobes due to switching are at least 40 dB below the compressed pulse height.

With the exception of the second stage, which employs three local oscillators, the ramp extender is a binary ladder. The advantage of the binary sequency is that frequency doublers can be employed extensively in the synthesis of the microwave local oscillators, resulting in simple, efficient, solid state equipment. Since the last eight oscillators can be obtained with four nearly identical frequency multiplier chains, the delay equalization problem is minimized. Delay equalization is required for the receive mode during which the flexible clock frequency is changed to effect time dilation correction. However, the maximum frequency change of one part in 10^4 makes delay equalization simple to attain.

Since the FM subpulse is translated and not multiplied as it proceeds through the ladder, the phase errors are not magnified. In fact, because a multiplicity of local oscillators are simultaneously changed at each switching interval, the errors of the latter stages tend to be masked. Consequently, the overall errors are not additive but increase only slowly. This masking or shadowing acts to preserve the basic feature of the signal – the lack of sidelobes in the first 192 range cells – which is caused by the periodic nature of the dominant amplitude and phase errors which repeat 192 times within the pulse. The three major causes of such periodic errors are switching transients, variations in the phase of the local oscillators, and errors in the basic periodic saw-tooth FM waveform that is the input of the ramp extender.

Sub-Pulse Generation

Any periodic function can be represented by a Fourier series defined as

$$f_P(t) = \sum_{-\infty}^{\infty} f_N \, e^{j2\pi \cdot nt/T_P} \tag{16}$$

where

T_P = the period of the pulse train

f_N = the nth Fourier coefficient of the expansion

The Fourier coefficients are given by the usual formula

$$f_N = \frac{1}{T_P} \int_{-T_P}^{T_P} f_P(t) \, e^{-j2\pi \cdot nt/T_P} \, dt$$

The periodic pulse train in the system is synthesized directly as the sum of sinusoids with the appropriate amplitude and phase weights. The magnitudes of the Fourier coefficients of the ideal waveform of Figure 3 are shown in Figure 7. Since only a finite number of terms can be used in the synthesis procedure, it is necessary to truncate the series in Equation (16). It can be seen in Figure 7 that most of the energy is contained in the lines between -25 and +25.

In fact, it was shown by a computer simulation that, if the 65 lines between -32 and +32 are included, there is only slight degradation in the system performance. If there were no system errors,

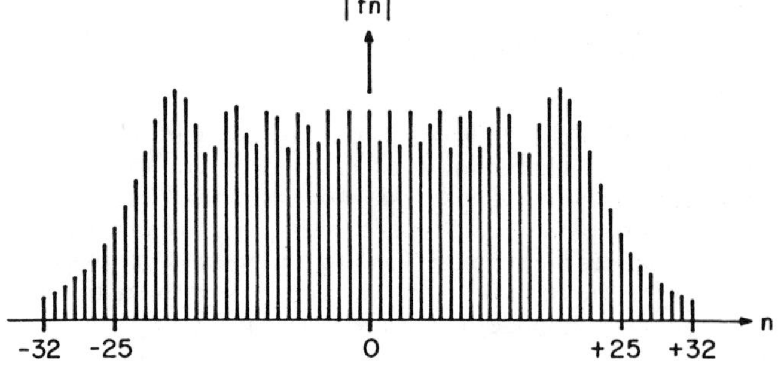

Figure 7 Magnitudes of the Fourier Coefficients of the Waveform of Figure 3

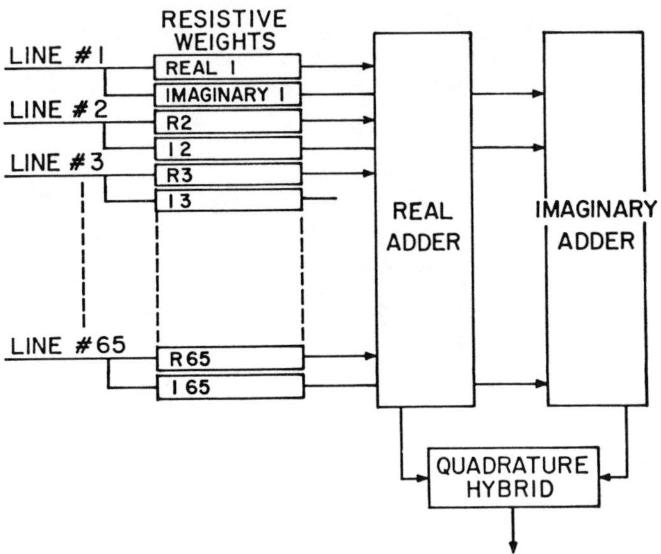

Figure 8 Block Diagram of Weighting Network

the error range sidelobes in the correlation function at the receiver output due to this truncation effect would be 46 dB down from the peak. Even assuming reasonable component tolerances, the worst error sidelobe could be expected to be below -40 dB for 90% of the error distributions.

The 65 lines spaced by 100 kHz and of equal amplitude and phase are synthesized directly from a frequency standard which is the basic system clock. The frequencies are then given the correct real and imaginary (in phase and quadrature) weights by resistive pads, as shown in Figure 8. The real and imaginary components are added separately, then combined in a broadband quadrature hybrid, a device which has a constant 90° phase shift across the band of interest. The output of the hybrid is the desired periodic signal; 192 periods can be gated to obtain the desired subpulse train. The actual frequencies range from 17.5 to 23.9 MHz.

Flexible Clock

The flexible clock may be considered as the frequency standard of the radar since all other signals are derived from it. Its output is $F = \cos 2\pi\ [10^7(1+B)\ t + \phi]$, where B and ϕ are computer controlled. The range of B is 0 to $\pm 10^{-4}$ in 16,384 increments; ϕ is variable in 100 increments from 0 to 1. This function is performed while maintaining substantially the spectral purity of the frequency standard.

Figure 9 is a block diagram of the flexible clock. The number generator in the flexible clock is a frequency synthesizer capable of generating frequencies from 9 to 11 MHz. These frequencies are produced in the following manner. Consider that, at any divider input in the number generator, the difference between the frequency at that point and 10 MHz is divided by some power of two. (This power is equal to the number of remaining divide-by-two counters before the number generator output.) Also consider that a binary signal, 0 or 1 (.5 MHz is equal to binary 1), may be injected at any divider via the mixer preceding that divider. Therefore, 10 MHz plus the weighted sums of the negative powers of 2 multiplied by a scale factor of .5 MHz is realized at the output of the number generator. This output frequency is (in MHz)

$$F_N = 10 + .5N\left[M + \frac{L}{2} + \frac{K}{2^2} + \frac{J}{2^3} + \ldots + \frac{A}{2^{12}}\right]$$

where F_N is the output frequency. Inputs A through M are 0 or 1, and N is ±1 as directed by the computer. This output is fed to the phase compressor.

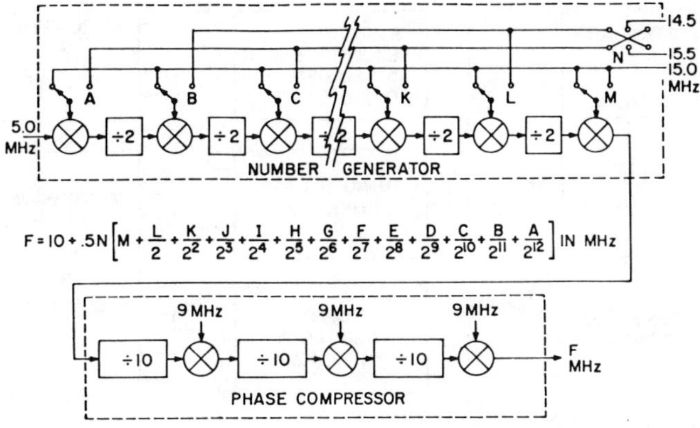

Figure 9 Block Diagram of Flexible Clock

The phase compressor divides the difference between its input frequency and 10 MHz by 10^3. Since the input frequency range is 10 ± 1 MHz, the output range is 10 ± .001 MHz. This compression of the phase also reduces the spurious signals arising from the mixing operations in the number generator by 60 dB. The three decade dividers in the phase compressor allow 1000 possible values of output phase, depending on the states of the decade dividers at t = 0. The initial setting of the second and third dividers are computer controlled; the first divider is always set to the zero state at t = 0. Therefore 100 phase values actually are realized. The output phase is

$$\phi = \left[\frac{\phi_N}{1000} + \frac{N_2}{100} + \frac{N_3}{10}\right] 2\pi$$

where ϕ_N is the random phase of the signal F_N from the number generator, and N_2 and N_3 are the settings of the initial states of the second and third dividers; N_2 and N_3 are decimal integers.

The 100 possible values of ϕ allow the zero crossing of the flexible clock output to be set to any of 100 values of time. This feature is equivalent to 100 values of delay of 1 ns increments, where 100 ns is equal to the 10 MHz period.

The 10 MHz from the flexible clock is frequency divided further by 100 in the low frequency synthesizer to obtain the 100 kHz required by the basic ramp. By computer controlling the initial setting of these additional decade dividers, 10,000 increments of delay (with each increment equal to 1 ns) are possible. Ten thousand nanoseconds is the period of the basic ramp.

Summary

A large time bandwidth linear FM signal generation technique and its application to a particular radar design has been presented. The radar (at X-band) would have two modes of interspersed operation – an acquisition and tracking mode, and a high resolution mode. In the high resolution mode, a 2 ms duration, 1 GHz bandwidth linear FM pulse is employed. A correlation receiver would be employed with all signal processing but the initial mixing operation performed digitally.

The signal generator is configured so that the major sidelobes appear outside of the range window. The basis for the signal generation techniques is the process of frequency synthesis (i.e., the production of a multiplicity of phase coherent sinusoids from a common oscillatory standard). A block diagram of the 2 x 10^6 TW signal generator and associated clock is shown in Figure 10. The system frequency standard is operated upon to produce a coherent frequency with the same high quality short term phase stability but with a variable (digitally commandable) phase and frequency. The remainder of the figure represents the signal generator and up-converter. The low frequency synthesizer produces 65 fre-

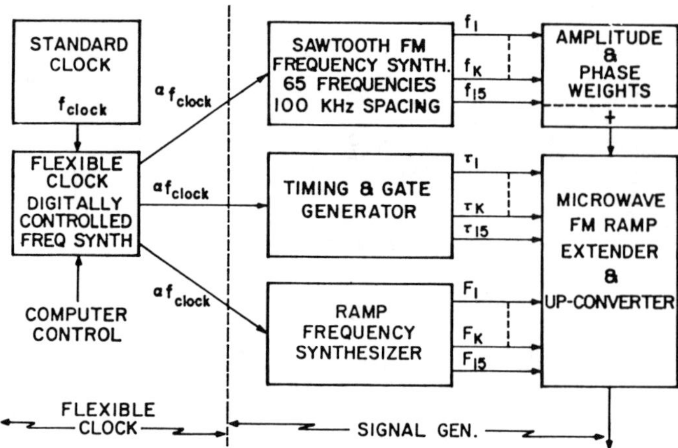

Figure 10 Block Diagram of 2,000,000:1 Signal Generator and Clock

quencies which are then weighted and combined to produce a high quality, periodic, coherent linear frequency modulated signal. This signal is up-converted to X-band and extended into a 2 ms 1 GHz linear FM signal, simultaneously, by a series of 6 mixing operations. The bandwidths and time duration of all signals involved are related directly to the output of the flexible clock. Thus, a change in the clock frequency changes the parameters of the output 2 x 10^6 TW pulse to account rigorously for the effects of target motion. A change in the same clock phase causes the pulse to move in time and is used for the range tracking function.

References

[1] Weiss, H.G.: "A Description of a High-Performance Microwave Experimental Facility," *Radar Techniques for Detection, Tracking, and Navigation* (W.T. Blackband, ed.), Gordon and Breach Science Publisher, New York, 1966.

[2] Haggarty, R.D.; Hart, L.A.; and O'Leary, G.C.: "A 10,000:1 Pulse Compression Filter Using a Tapped Line Filter Synthesis Technique," *1968 EASCON Convention Record,* pp. 306-314.

[3] Rihaczek, A.W.: *Principles of High-Resolution Radar,* McGraw-Hill, New York, 1969, p. 262.

[4] Thor, R.C.: "A Large Time-Bandwidth Product Pulse-Compression Technique," *IRE Transactions on Military Electronics, Vol. MIL-6,* pp. 169-173, April 1962.

[5] Lynch, J.T.: "A Straightforward General Analysis of Signal Distortion with Applications to Wideband Ionospheric Dispersion," ESD-TR-67-623, The MITRE Corporation, March 1968. (This document has been approved for public release and sale; its distribution is unlimited and may be obtained through ESD or DDC; see also Chapter 14.)

Present and Future Trends in Signal Waveform Design and Processing Technology and Techniques

Waveform Design

More complex waveforms are employed by radar systems which must have antijam features. With the development of digital processing, it is possible to generate and process most any type of complex waveform. In the future it is expected that there will be extensive use of coded phase spread spectrum waveforms which are generated and processed digitally. Such waveforms have the disadvantage of being Doppler sensitive. If $T\Delta f_D > 1$, where T equals the waveform duration and Δf_D equals the target echo Doppler shift, then a bank of filters is necessary to process the signal — creating a problem which does not exist for the simple linear FM chirp waveform.

In Chapter 8, Hamming weighting is used for the linear FM chirp waveform in order to achieve a peak sidelobe level of 43 dB. This weighting results in a mismatch loss in the receiver output signal-to-noise ratio of 1.3 dB. The availability of SAW filters and digital circuitry makes it possible to generate easily a nonlinear FM waveform which can be processed by either a matched SAW filter or digital processor in the receiver. The output of this matched filter provides very low sidelobes [1]. In this way, low range sidelobes are achieved without incurring a mismatch loss. For applications where $T\Delta f_D < 1$, this waveform is very attractive since it can be easily generated and processed by SAW lines. It is expected that much wider use will be made of this type of waveform in the future.

Amplitude weighted waveforms, such as the Huffman waveform of Figure 6 in Chapter 8, also may be used in the future if linear power amplifiers become practical.

Winograd Fourier Transform Algorithm (WFTA)

Recently, a new FFT algorithm (the Winograd Fourier Transform Algorithm or WFTA) has been developed which promises a considerable reduction in the number of multiplies needed over what has been achieved already with the standard FFT discussed in Chapter 9 [2, 40]. The WFTA algorithm reduces the number of multiplies by a factor of between 3 and 10 over that required with the standard FFT, while requiring about the same number of adds [3,4]. The algorithm has been programmed on the IBM 360/91 and results in a 40% reduction in the processing time [5]. A prime factor version of the algorithm programmed on an IBM 370/155 in FORTRAN resulted typically in a 50% reduction for transforms consisting of 60 to 2520 points [6]. To fully realize the potential of this algorithm, it is felt that a hardwired, parallel signal processor designed specifically for implementation of the algorithm should be built. Thereby, considerable overhead required for indexing the data in carrying out the WFTA would be eliminated [3, 6].

Digital Components [78]

Each year the number of devices which can be realized on an integrated circuit approximately doubles. Advanced chips are expected to have 10^6 components in 1980 as compared to 1977 levels of 60,000 to 100,000. The logic clock rate also is expected to increase (to 10 GHz in some specialized integrated circuits), although very high speeds and high densities can not be achieved simultaneously. These developments should exert a strong positive influence on radar systems. It may, in fact, become possible to A/D convert the received signal without the need for down conversion.

Memories

Read Only Memory (ROM)

The cycle time for various types of ROMs are as follows:

ROM	35-50 ns cycle time (2048 x 8)
PROM	70 ns (2048 x 8)
EAROM and EEROM	2 μs
EPROM	0.5 μs (up to 16K)

For ROMs, the binary information is stored in the memory during manufacturing. PROMs are programmable ROMs in which the programming can be done by the user by electrically burning out specific diode junctions or polysilicon fuses. EAROM and EEROM are electrically erasable ROMs (a 28 V pulse of 100 ms duration is used for erasure; a pulse at the same voltage for 10-20 ms is used for writing). EPROMs are also electrically programmable ROMs, but are erased by ultraviolet radiation of the chip through a transparent quartz window on the package. They are also erasable by X- and γ-rays, and hence not hardenable. The Intel 2716 2048 x 8 bit* EPROM requires 50 ms to read in an 8-bit word (byte) and 100 seconds for all 16,384 bits. The chip requires only a 5 V supply for normal read only operation and 25 V for programming. Its access time is 0.45 μs, maximum active power dissipation is 525 mW, and standby power is 132 mW.

A 1024 bit high speed (15 ns cycle time) EAROM using amorphous semiconductors should be available soon [42]. These amorphous semiconductor memories have been coined "Ovionic" memories after S.R. Ovshinsky, the unconventional, controversial, and mostly self-taught scientist who did pioneering work with amorphous semiconductors and zealously championed them. Amorphous memories are less volatile and faster than MNOS (metal-nitride-oxide semiconductor) EAROMs [42].

Random Access Memory (RAM)

Texas Instruments (TI) has announced plans to introduce a 64K** RAM by the end of 1977. A large number of vendors are supplying 16K RAM memories (into which, as well as out of, data can be read) having access times between 150 and 400 ns [7, 8]. For example, a 16K NMOS Mostek chip having a 375 ns read or write cycle time is now available which requires 462 mW while active and 20 mW during standby [9]. 4K static*** transistor-transistor

*11 parallel input address bits locate one of 2^{11} = 2048 8-bit words which are read out in parallel.

**Capital K is used to represent 1024 as is the standard in the computer industry; lower case k is used for 1000, per the IEEE standard.

***Static memories hold their setting indefinitely (as long as the power does not fail), flip flop circuits being used. Dynamic memories store the information as charges on capacitive structures; because these charges leak in time, periodic refreshing of the capacitors is needed for dynamic memories.

logic (TTL) and integrated injection logic (I^2L) RAMs are available which have maximum cycle times of less than 80 ns and respective dissipations of 850 mW and 500 mW [41, 75]. Using emitter coupled logic (ECL), 1024 x 1 and 16 x 4 bit RAM units can now provide 15 ns and 6 ns, maximum access times, respectively. Hughes has built several 512 x 1 bit RAM chips with 4 ns access times, but the power dissipation is several watts.

Block Addressed Memories

A 92k bit bubble memory device (25 grams, 1.0 x 1.1 x 0.4 inches) is available commercially from TI with an average time to access the first bit of 4 ms (this time also is called the latency time) and a read-write operation rate of 100 kbit/s. The device sells (in 1977) for $200 (about 0.2 cents/bit) and consumes 0.5 W. Random access to any bit position is not available, unlike with a RAM or ROM.

This 92 kbit memory stores the data on 157 minor loops, each having 641 bit positions. These minor loops feed adjacent bit positions of a major output loop having 640 bit positions. The bits on the major loop can be read out only sequentially (shift register style); the bits on the minor loop similarly are read out sequentially to the major loop. The loops operate at 100 kHz read/write operation, i.e., an average loop moves one bit position in 10 μs. Thus, the maximum time to read out a bit is (639 x 10 μs) + (156 x 10 μs) = 7.95 ms. Hence, the average latency time is 7.95/2 = 3.98 ms. To enhance production, only 144 of the minor loops are guaranteed useful, yielding 144 x 641 = 92,304 bits of memory. TI has announced plans to market a 256 kbit bubble memory by the end of 1977; read/write speeds should increase to 1 MHz.

A 65kb, 5 MHz charge-coupled device (CCD) chip is available from TI. CCDs typically have an order of magnitude faster access time than do magnetic bubble memories. An available 16 Kbit LARAM-organized CCD chip provides a 4 bit parallel output, with an average latency time of 12.8 μs and a 20 Mbit/s data rate [10].

EBAM (or BEAMOS) [10] is an electron-beam-addressed storage system which uses a two-dimensionally deflected electron beam to obtain the binary data stored on a metal-oxide semiconductor (MOS) target having PN junctions in silicon. The latency time is up to 30 μs with a data transfer rate of 10 Mbit/s [11]. The first such systems are expected to have a 10^8 bit storage capacity.

Figure 1 [12] gives the price-performance of existing memories (circa 1976).

Magnetic-bubble and EBAM, both of which have the advantage of being nonvolatile, and CCD memories are suitable for storing the clutter map of a surveillance radar (in place of a disc system). Such a memory is used with the moving target detector (MTD) [103] radar processor described in Chapter 1. CCDs also are attractive for simple tap delay line processing and for pipeline FFT shift register memories.

Josephson Junction Technology

Memories with very fast cycle times may be in the offing through cryogenic (4.2K) Josephson Junction technology. 6 ns has been achieved [43] in the laboratory for a nondestructive memory; studies have been performed for a 2 ns cycle time [10]. A cycle time of about 0.2 ns is expected for a destructive-read memory system [10]. These devices will consume orders of magnitude less power than today's devices in configurations that eventually may contain hundreds of thousands of memory elements on a chip [71]. An energy of 10^{-17}J, five orders of magnitude above kT, has been reported [72].

Logic

8, 12, and 16-bit single chip (130, 150, and 160 ns, respectively) parallel multipliers are available from TRW which require 1.8, 3.5, and 5 watts, respectively. 16 bit 3.8 ns single chip multipliers have been built by Hughes. Using digital CCD (DCCD) technology,

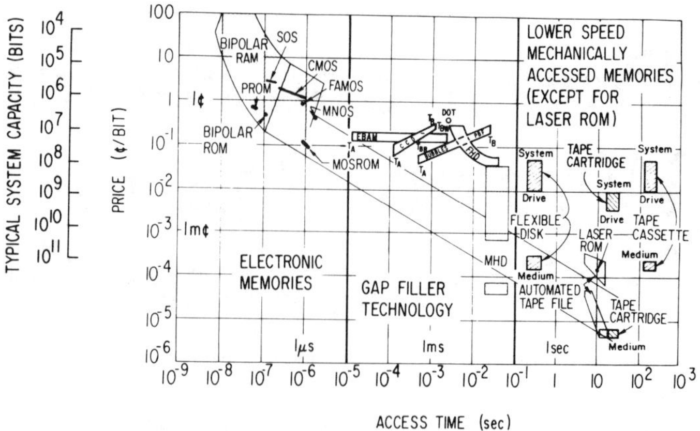

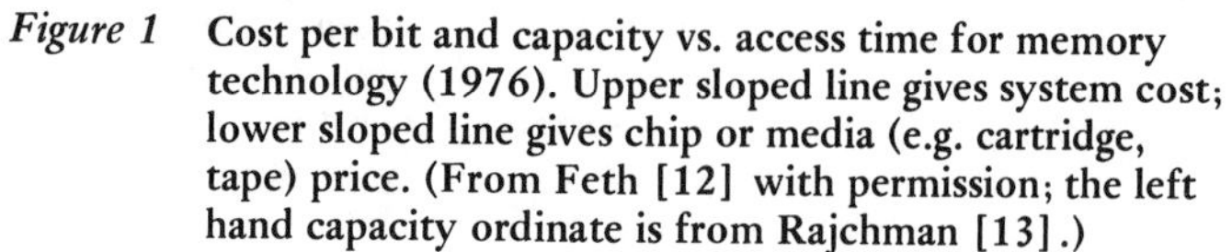

Figure 1 **Cost per bit and capacity vs. access time for memory technology (1976). Upper sloped line gives system cost; lower sloped line gives chip or media (e.g. cartridge, tape) price. (From Feth [12] with permission; the left hand capacity ordinate is from Rajchman [13].)**

it should be possible in a few years to put a butterfly (see Chapter 9) of the FFT algorithm on a single chip, with this chip operating at a 5 MHz rate (butterflies per second) and requiring 1.8 W of power [44-46]. This procedure involves using the CCD to do conventional logical operations such as AND (A•B), OR (A+B), NOR ($\overline{A+B} = \overline{A}\cdot\overline{B}$), exclusive-OR ($A\cdot\overline{B} + \overline{A}\cdot B$), and basic 2-bit half adder and full-adder operations [44, 45]. The basic butterfly chip (consisting of three time-multiplexed 16 bit adder/substractors and four 16 x 16 bit multipliers for which the 16 most significant bits are used) could be made modifiable by user-generated control signals and gates so as to perform other functions (2-pole recursive digital filtering, single-pole digital filtering, and serial correlation [44-46]).

By 1985, it should be possible to build a 2 ns 16 x 16 multiplier with dielectrically-isolated emitter coupled logic (D-ECL) operating with 5 W. Other technologies exist (GaAs metal-semiconductor field effect transistor [MESFET] and transferred-electron device [TED]) which should provide logic circuits having logic gate speeds of up to 7 to 10 GHz in the near future [47, 48]. A 2:1 divider operating from dc to 4 GHz has been built using GaAs MESFETs [14, 85]. The Josephson Junction technology has the promise of both high speed (10 ps delays for gates operating at 4.2K) and low power (approximately one ten thousandth that for I^2L or CMOS*/SOS). The propagation-delay times power-dissipation product for Josephson Junctions is expected to be on the order of 10^{-5}pJ; see Figure 2. This power consumption advantage is lost partly in the power required to provide the cryogenics. Research is proceeding to find materials that provide superconductance at temperatures where closed-cycle refrigeration systems can be used ($\geqslant$ 10K), superconductance at 14.5K being demonstrated [74]. A 4-bit Josephson Junction 27 ns (external test equipment limit; 12 ns is expected without this limit) multiplier has been built which required 1.6 mW [73].

A/D Converters

400 MHz A/D converters are presently operating in the laboratory at TRW, four bits being achieved with four cells. Hughes has built several monolithic 7 bit, 250 MHz A/D converters. Using transferred-electron and field-effect devices (TEDs and FETs) a 5 GHz A/D converter is under development for the Office of Naval Research (ONR) by TRW [47, 48].

*Complementary metal-oxide semiconductor

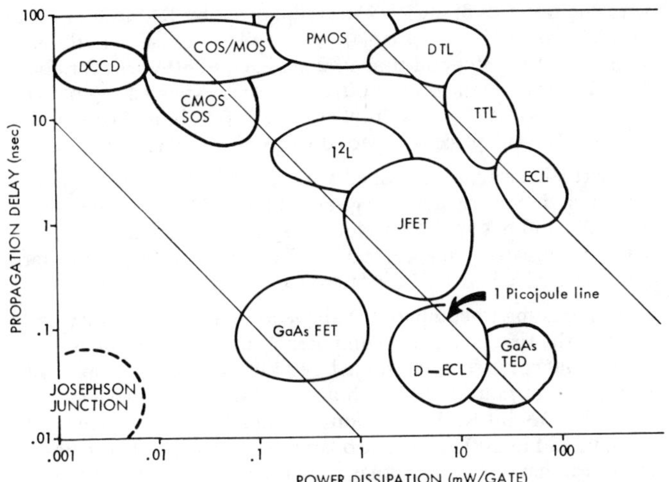

Figure 2 **Propagation delay vs. power dissipated per gate for today's technology. The diagonal lines give the lines of constant energy to switch a single gate.** ***Note*** **– DTL = Diode Transistor Logic, PMOS = P-channel MOS, JFET = Junction FET, D-ECL = Dielectrically-isolated ECL, COS/MOS = Complementary-oxide on silicon/metal-oxide semiconductor; other terms defined in text.**

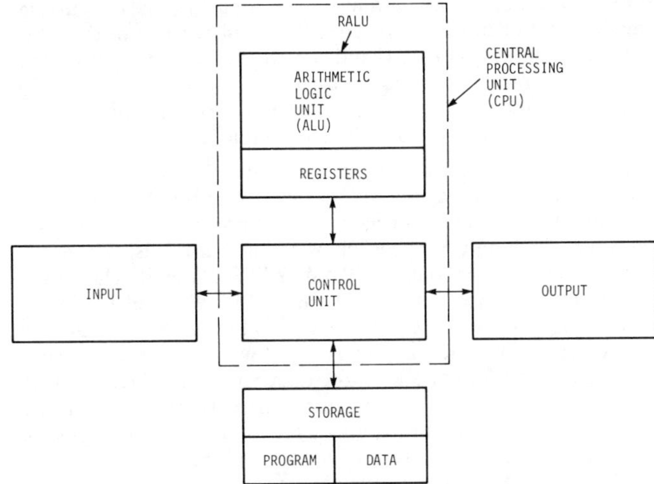

Figure 3 **Five basic components of a computer. A microcomputer (μC) has all five components on a single chip; the microprocessor (μP) usually has only the central processing unit (CPU) portion on a single chip [76].**

Microprocessors (μP) and Microcomputers (μC)

μP and μC can be used for CFAR processing, pulse Doppler processing, FFT processing, target detection logic (such as coincidence detectors), and beam splitting algorithm computations instead of specially wired hardware or a general purpose computer [15].

A μP is part of a whole computer all on a chip while a μC is a whole computer on a chip; see Figure 3.

To better understand what a μP and a μC are, brief explanation of the fundamental computer block diagram of Figure 3 is given. Assume that an FFT algorithm is to be carried out. The user could write his FFT algorithm for the computer in an easy-to-program high-order language such as BASIC. This language then would be converted to a lower-order machine-oriented language called an assembly language. This conversion can be carried out on another computer using a program called the compiler.

Assembly language computer instructions are in the form of a short string of alphabetic and numerical symbols. For example, ADD A,B could mean add contents of registers A and B; MUL A,B could mean multiply contents of A and B. Another type of instruction would be load register A with the contents of memory location 901.

Letters and numbers are used for the assembly languages as a convenience for the programmer, as it is easier to remember strings of alphabetical and numerical numbers than binary numbers. Also, it is easier to correct errors in such instructions.

These assembly language instructions next are converted into instructions that the machine can understand – binary sequences called machine language instructions. This conversion is carried out by a program called the assembler if done on the same computer (if done on another computer, the cross-assembler). For each assembly language instruction there is one machine language instruction, a one-to-one mapping. The mapping from the higher order BASIC language to assembly language is one-to-many. Some compilers convert directly from the higher order language to the binary machine language.

Once the machine language program (also called the macroprogram) has been developed, it is put into the program part of memory of Figure 3 in successive locations in the order to be executed. When the macroprogram is executed in the computer, it is stepped through starting with the first instruction in the first location of the macroprogram memory. A program counter is used to indicate which program step is next. This counter could reside in the register, arithmetic, and logic unit (RALU) or in the control unit; see Figure 3.

The arithmetic and logic unit (ALU) (see Figure 3) cannot carry out a complex macroprogram instruction (called a macroinstruction), such as multiply the contents of registers A and B unless a hardware multiplier is available in the ALU (which is not the case in 1977) for μPs and μCs; it is possible though, to incorporate the TRW multiplier-on-a-chip into the architecture of the signal processor. The ALU can carry out simple instructions, such as add the contents of registers A and B or shift the contents of register A one position to the left (which is equivalent to multiplying by two).

The multiplication of the contents of registers A and B is achieved by successive additions. Moreover, a program exists in the control unit which is used to tell the RALU what sequence of addition steps are to be carried out when the machine language instruction "multiply the contents of A and B" is observed. Each RALU instruction of this program is called a microinstruction; the whole sequence of microinstructions which forms the machine language multiply instruction is called a macroinstruction [77]. Another sequence of microinstructions exists for the machine language divide instruction. Moreover, for each machine language instruction there is a set of microinstructions called a macroinstruction. These sets of microinstructions (called microprograms) reside in the control unit. The control unit receives the next machine language instruction and then executes the set of microinstructions which implements this machine language instruction in the RALU. Thus the control unit interprets the machine language instructions for the RALU, reducing the commands to sequences of simpler instruction that the RALU can understand.

Not all machine language instructions require more than one microinstruction. Requiring one microinstruction is the machine instruction to add the contents of registers A and B or to load register A with the contents of memory location 901. Nonetheless, the control unit is used to translate these machine language commands to signals that cause the RALU to perform the desired operation.

The μC has all the basic parts of the computer shown in Figure 3 – the RALU, the control unit, the machine language memory, possibly a data memory, and means for inputting and outputting the data and/or instructions. The μP has only certain of these elements, often just the RALU and control unit which together are called the central processing unit (CPU) of the computer; see Figure 3. Separate memory and input-output hardware chips must be used in addition to form a complete computer.

If the signal processor and its μC are to be produced in large volume, the macroprogram would be included in the ROM on the μC chip during manufacturing, i.e., mask-programmed. For systems for which only a few μC chips will be built, it is more economical for the user to have a RAM or EPROM on it. The new (1977) Intel 8748 μC has a 1024 x 8 bit EPROM available for macroprogram storage as well as a 64 x 8 bit RAM. It has 96 macroinstructions and can be purchased with a minimum cycle time of either 2.5 μs or 5 μs. More than 50% of the macroinstructions require a single cycle (i.e., microinstruction). This μC is useful for applications requiring only a few units, where changes are expected in the algorithms to be used, or where the algorithms need experimentation and development.

The memory is expandable to 4096 x 8 bits by the use of external memory units (such as the Intel 2716 2048 x 8 EPROM discussed previously). Intel supplies a version of the 8748 μC (the 8048) which uses a 1024 x 8 bit ROM in place of the EPROM. This unit is suitable for volume applications. It can be used after having developed the algorithms with the 8748. No compilers are available at present for these units and, hence, the programs must be written in assembly language. The popular Intel 8080 and 8080A μPs (the 8080A is a refined version of the 8080 μP to be replaced with the faster 8085 in 1977) are supported by BASIC and PL/M (a subset of PL/1) higher language compilers. These two μPs have a register-to-register add time of 2 μs. Each 8080A sells for about $20 (1976) in quantites of 100; the price is expected to go down to $10 in 1977 [79]. The Intel 8048 sells for about $100 in volume (1977); the price is expected to go down to $25 (1978).

Hewlett-Packard has developed a static CMOS/SOS 16-bit parallel μP which does a register-to-register add in 0.875 μs, draws about 350 mW, and has an 8 MHz clock, 34 16 bit instructions, and 10,000 transistors on a single 5.7 x 5.9 mm chip [80]. Appropriately, it is called an MC^2 for Micro-CPU chip; see Figure 3.

It is interesting to compare the above μPs and μCs with the ENIAC (1946), the first electronic computer. It used 18,000 tubes, occupied 3,000 cubic feet, consumed 140 kW, weighed 30 tons, had a mean time to failure of a few hours, a 16K bit ROM (relays and switches), 1K bit RAM (flip-flop accumulators), and performed a 12 digit add in 200 μs [81]. Today (1977), for $100, a system designer can have an ENIAC that is 20 times faster, 10,000 times more reliable, drawing 56,000 times less power, and occupying 1/300,000 the space [81].

For further reading on μPs and μCs, see References 76-79, 82, and 83.

SAW Devices

These devices are seeing much wider use than just for pulse compression. They are being used as bandpass filters [16, 17, 49, 53] (see Tables 1 and 2), to synthesize filter banks, to generate multi-

	Demonstrated	**Production**	**Projected Practical Limits**
Center Frequency	1.0 MHz – 2.75 GHz	10 MHz – 1.5 GHz	10 MHz – 2.0 GHz
Minimum Insertion Loss, IL(3)	0.65 dB	1.0 dB	0.5 dB
Maximum Fractional Bandwidth	100%	50%	100%
Sidelobe Rejection	70 dB	60 dB	90 dB(10)
Minimum Bandwidth, B_3(4)(5)	100 kHz	100 kHz	50 kHz
Minimum Transition Bandwidth(6)	100 kHz	100 kHz	50 kHz
Minimum Shape Factor(7)	1.15	1.2	1.1
Triple Transit Suppression(8)	55 dB	45 dB	60 dB
Amplitude Ripple	± 0.02 dB	± 0.05 dB	± 0.01 dB
Phase Deviation From Linear(9)	± 0.1°	± 2°	± 0.1°

(1) Filters obtained using interdigital transducer simulating tapped delay filter (as in Figures 5 and 8 of Chapter 12 and Figure 6 of this section) as opposed to SAW resonator filters (discussed in text and illustrated in Figure 7 and Table 2).

(2) Capabilities indicated not achieved simultaneously; see footnotes (3), (4), and (7).

(3) Minimum insertion loss (IL) achieved for $B_3 \leq 4.5\%$ for quartz (ST, X) substrate
$B_3 \leq 24\%$ for lithium niobate substrate.

(See footnote (1) of Table 3)

Above minimum achieved if transducer is tuned inductively and impedance-matched, and if new three phase unidirectional transducer [49] is used instead of bidirectional transducer depicted in Figures 2, 5, and 8a of Chapter 12 and Figure 6 of this section.

Bidirectional transducers propagate out both sides of transducer, resulting in a minimum theoretical loss of 3 dB on both transmit and receive. Unidirectional transducers also reject triple-transit signals which are a source of passband amplitude and phase ripple.

For larger B_3 than indicated above for minimum IL (0 dB) the device can be mismatched electrically with the result that IL increases at 12 dB/octave.

(4) Minimum B_3 = 100 kHz achieved only for $f_C \leq 150$ MHz
= 1 MHz for f_C = 500 MHz
= 4 MHz for f_C = 1 GHz

where f_C = filter center frequency [49].

(5) Excludes performance achievable with SAW resonators.

(6) Transition bandwidth equals 3 to 40 dB bandwidth.

(7) Shape factor ≥ 3:1 for B_3 = 100 KHz; low shape factors achieved for large B_3 [49].

(8) Assumes unidirectional transducer; see footnote (3).

(9) Excludes electrical loading effects, which can be compensated by proper design.

(10) 100 dB out of band, 70 dB first sidelobe should be achievable [84].

Table 1 **Interdigital-Transducer(1) Surface Wave Bandpass Filter Capabilities(2) (After Hays and Hartmann [49])**

Type Filter	Center Frequency, f_c (MHz)	Bandwidth B_3 (MHz)	f_c/B_3	Insertion Loss (dB)	Sidelobe Rejection (dB)	Substrate
Simple, two-port, single pole resonator (Figure 7) [53]	60	0.04	1500	6.6	23	Lithium niobate
Two-port, single pole filter using different inductors to vary bandwidth [53]	140.2 140.2 140.2	0.0242 0.0723 0.253	5793 1939 554	10.5 3.2 2.8	>16 >16 —	Quartz
Four-pole Butter-worth [53]	60	0.08	750	15	50	Lithium niobate
Simple two-port, single pole resonator (Figure 7) [27, 59]	34	0.009	3780	7	33	Lithium niobate
Cascade of three of above single pole resonators [27, 59]	34	0.0045	7550	13	80	Lithium niobate
Cascade of two Jaumann circuit bandpass filters [60][(1)]	153	0.043	3558	3	42	ST quartz
Cascade of four band-elimination filters [60][(1)]	153.5		Except for: 1-2 dB, ±5 kHz, Except for ±1.5 kHz	<1 dB, ±150 kHz,	>80 stop band rejection over 60 kHz band	Lithium niobate
Two-port, one-pole (Figure 7) [61]	184	0.0138	13,333	8.6	20 (worst)	Quartz
Two-port, two-pole [61]	184	0.016	11,500	17	48 (worst)	Quartz
Two-port, one-pole (Figure 7) [58][(2)]	157.6	0.0052	30,300	11.6	39	ST quartz
Two-port, one-pole (Figure 7) [58][(3)]	310.4	0.0120	25,900	14.3	29	ST quartz

(1) One SAW resonator per filter (2) $R_1 = 280\Omega$; $R_G = 50\Omega$; $Q_U = 41{,}300$ (3) $R_1 = 416\Omega$; $R_G = 50\Omega$; $Q_U = 32{,}100$.

Table 2 SAW Resonator Filters

ple frequency tones (for frequency-hopped systems), for cross-correlation, as high frequency resonators, as crystal oscillators (10 MHz to 2 GHz), as programmable pseudo-random spread spectrum waveform generators, as moving target indicator circuits (MTI), as burst waveform processors, as Fourier Transform analyzers, and as delay lines; see References 19 and 20.

Delay Lines

A 2.5 GHz SAW delay line having a 210 MHz 3 dB bandwidth, 1.5 μs delay, and 29 dB insertion loss has been built [21]. A low insertion loss (2 dB) lithium niobate delay line centered at 33.8 MHz has been built having a 6.1 MHz bandwidth (18% fractional bandwidth), 0.95 μs delay, and multiple-transit-time spurious responses (for example, triple-transit signals that reflect from the output transducer to the input transducer and then back to the output transducer) over 55 dB down [49]. This low insertion loss delay line uses the new low loss unidirection SAW transducer [49].

A laboratory delay line has been built which has a 0.907 ms delay over a 65 MHz bandwidth with 65 dB insertion loss. It uses a helical delay line (unguided) on a BGO ($Bi_{12}GeO_{20}$) plate [50]. The delay line consists of four lengthwise helical turns on an evacuated 7.25″ long, 1.1″ wide, 0.185″ thick BGO slab; see Figure 4a. BGO is used for long delays because it has the lowest acoustic wave velocity; see Table 3. Quartz is used when excellent temperature stability is required (as for oscillators) and lithium niobate ($LiNbO_3$) is used for wide bandwidth with low insertion loss; see Table 1, footnote 3, and Table 3, footnote 1.

Material and Orientations	v (x 10^5 cm/s)	k^2 (1) (percent)	Delay Time Temperature Coefficient (ppm/°C)
$LiNbO_3$ (YZ)[(2)]	3.48	4.5	91 ± 1
Quartz (YX)	3.17	0.23	-22 ± 1
Quartz (ST, X)	3.15	0.16	0
$LiTaO_3$ (YZ)	3.22	0.74	37 ± 3
$LiTaO_3$ (XZ)	3.22	0.69	36 ± 2
$LiTaO_3$ (ZY)	3.31	0.93	67 ± 2
$Bi_{12}GeO_{20}$ (111) (011)	1.65	1.7	128 ± 2
ZnO (C, any)	2.70	1.0	40 ± 5

(1) k is the electromechanical coupling constant and is a measure of the efficiency of the coupling between electrical and acoustical energy. If the new three-phase unidirectional transducer is used and it is tuned inductively and impedance matched, 100% conversion efficiency can be achieved for fractional bandwidths below $2k/\sqrt{\pi}$ (k given as a fraction) [18]; see footnote (3) of Table 1.

(2) Symbols in parentheses are the orientation of the surface normal and the direction of wave propagation, respectively.

Table 3 Constants of Various Surface Wave Substrates for Differing Orientations (From Holland and Claiborne [16])

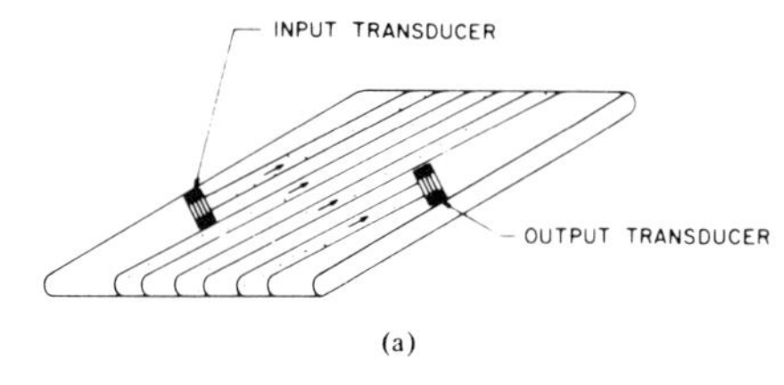

(a)

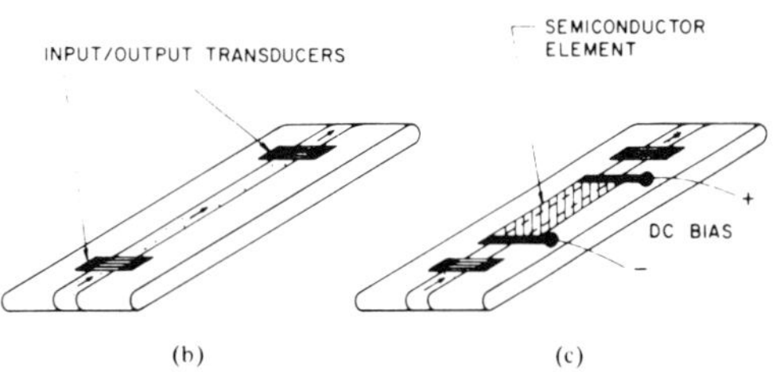

(b) (c)

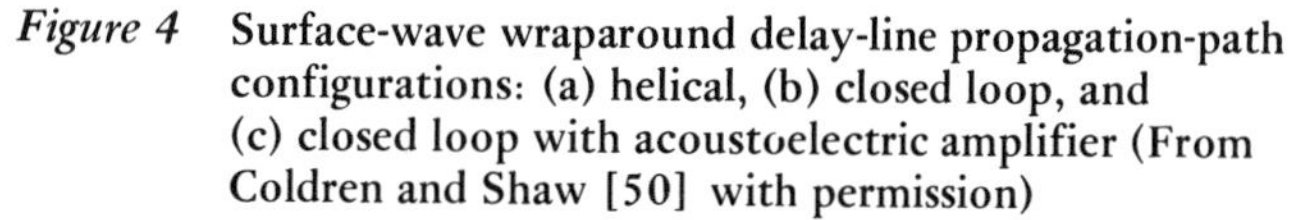

Figure 4 **Surface-wave wraparound delay-line propagation-path configurations: (a) helical, (b) closed loop, and (c) closed loop with acoustoelectric amplifier (From Coldren and Shaw [50] with permission)**

Figures 4b and 4c show closed loop circulating delay lines with and without an internal acoustoelectronic amplifier (to be discussed later) that is used to make up for the insertion loss per loop. A unit using a silicon-on-sapphire (SOS) acoustoelectronic amplifier on a BGO delay line, such as depicted in Figure 4c, has been built which provides unity loop gain (the amplifier making up for the loop losses) [50]. Such circulating memory devices can be used for digital or analog memories with the average latency (access) time equal to half the loop delay. (The CCD and bubble memories described previously also operate using circulating loops, except more than one loop usually is used as for the 92 kbit bubble memory.)

A variable delay line has been implemented using two SAW lines and a voltage controllable oscillator (VCO). Variations in the delay of 30 μs for a 10 MHz bandwidth signal was obtained with a dynamic range of 75 dB [22].

Generation of Chirp Waveforms of Arbitrary Slopes

A technique whereby SAW lines can be used to generate and process linear FM waveforms having different slopes has been devised [23]. This technique consists of the cascade of a SAW line, up-converter mixer, second SAW line, and down-converter mixer, with both mixers fed by the same VCO; see Figure 5. The two SAW lines have a phase which varies as the cube of frequency in addition to nonidentical quadratic phase terms. The cubic phase variation for both SAW lines have the same magnitude but are of opposite sign so as to cancel, leaving a quadratic phase variation the magnitude of which is determined by the VCO frequency.

Fourier Analysis

A chirp SAW line (that is, a SAW line which generates a linear FM signal) can be used to provide the Fourier transform of a signal by first mixing it with a local oscillator having a linear FM slope of magnitude identical to that of the chirp SAW line but of opposite slope. Applying this mixed signal to the chirp SAW line provides a waveform at the output which corresponds to the Fourier transform of the received signal, with time being proportional to frequency. The inverse Fourier transform can be obtained in a similar manner.

The long (0.907 ms) delay line with 65 MHz bandwidth mentioned previously has the potential of generating and providing matched filter processing for large time-bandwidth product (60,000) chirp signals as well as other complex waveforms [50]. A chirp SAW

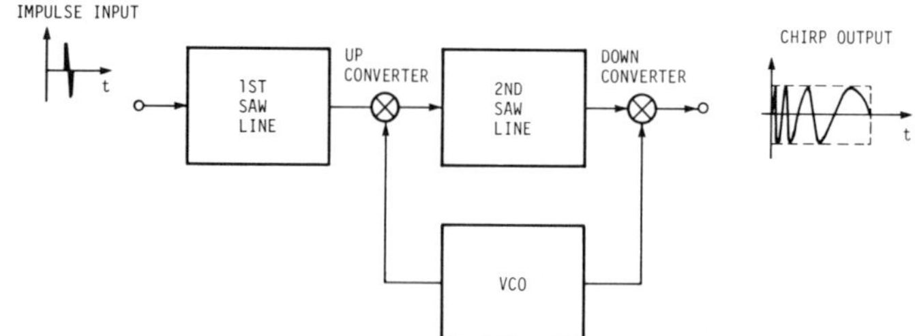

Figure 5 **Technique for obtaining linear FM signal of varying slopes; same circuit can be used for pulse compression (After Newton and Paige [23])**

line having a 1 ms delay, 60 MHz bandwidth, and 60 dB output dynamic range (maximum signal to rms noise level) would be capable of doing a 1.5 x 10^4 point transform in 0.5 ms. It would be equivalent to a digital binary system using 10-bit words and about a 2 ns clock period for multiplies, or about 19 ns per butterfly – a very impressive performance, even when compared to the future DCCD butterfly on a chip which has the potential of doing a butterfly in 100 ns [50, 54].

Implementation of the circular memory filter (CMF) technique of synthesizing a bank of Doppler filters, for both a single channel [24] and a time multiplex version of the CMF invented by Brookner (which is capable of handling multiple channels [25]), becomes easier with SAW lines; see Reference 26.

Pulse Compressor for Arbitrary Waveforms

A technique has been developed whereby a chirp SAW line can be used to pulse compress any arbitrary waveform. One of the chirp SAW devices provides the Fourier transform of the incoming signal; this resulting Fourier transform then is multiplied by a stored replica of the complex conjugate of the Fourier transform of the signal being matched. Next, a second chirp SAW line is used to provide the inverse transform of the resulting product – the matched filter output for the incoming signal.

SAW Resonator Filters

Up to about 1974, the primary method available for implementing SAW bandpass filters was through the use of interdigital transducers to simulate the tapped delay line equivalent to the bandpass filter. This approach is described in Chapter 12. For example, by having the overlap between successive pairs of interdigital electrodes follow a (truncated) (sin x)/x function, a good approximation to rectangular bandpass filter can be synthesized; see Figure 6. This varying overlap is called apodization [16]. The filters of Table 1 are synthesized using this time domain synthesis approach. In 1974 [51] (see also [52]) a new SAW device was developed which permits the synthesis of filters in the frequency domain – the SAW resonator. Its equivalent circuit is similar to that of the bulk crystal resonators used in the past to generate stable oscillators and crystal filters (via poles and zeroes synthesis) [53, 55-57]. The advantage of filters designed using SAW resonators is that much narrower bandwidths (about an order of magnitude narrower) can be achieved. Thus, SAW resonator technology complements the apodized interdigital transducer technology. Table 2 lists some SAW resonator filters that have been built.

The fractional bandwidth (which equals one over the loaded Q) achievable with an apodized interdigital transducer narrowband SAW filter is approximately one over the number of interdigital electrodes. Obtaining fractional bandwidths of 10^{-4} to 10^{-5} (Q = 10^4 to 10^5) with such a filter is not feasible; SAW resonators are needed. SAW resonators provide a simple means for obtaining the large delay paths needed for such high-Q filter construction.

Figure 7 shows the basic two-port SAW resonator and its equivalent circuit. The two surface-wave gratings act as mirrors having

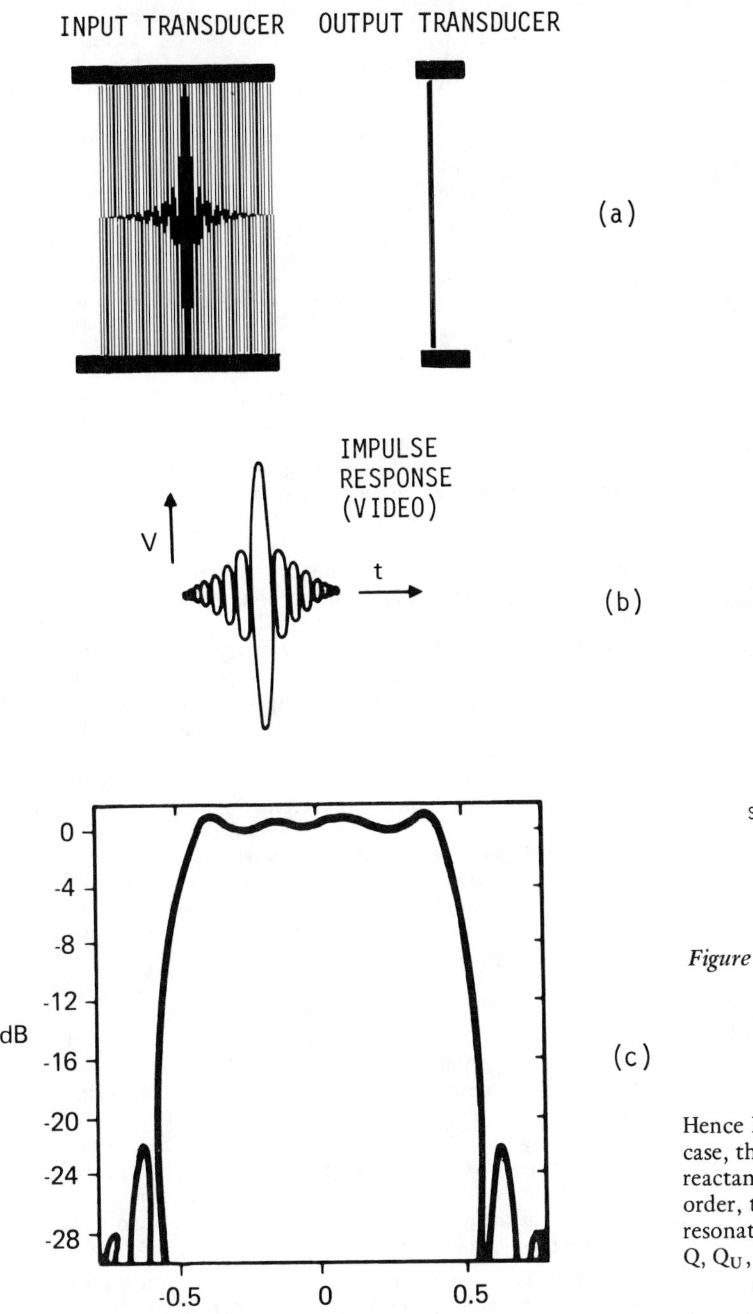

Figure 6 As discussed in Chapter 12, the impulse response of a SAW device using interdigital transducers is defined by the overlap (apodization) between successive pairs of transducer electrodes; (a) truncated (sin x)/x overlap, (b) truncated (sin t)/t impulse response, and (c) resulting nearly rectangular bandpass filter frequency response (From Raytheon Company with permission)

nearly 100% (90 to 99% [27]) reflection. (The dashed lines indicate the positions of the equivalent mirrors.) These two mirrors form the cavity resonator, the signal bouncing between the equivalent mirrors. For the SAW resonator of Figure 7, an unloaded Q nearly equal to that of the substrate material limit can be achieved [58]; see Figure 8. The unloaded Q is obtained when zero resistive loads appear on the input or output terminals of Figure 7b; i.e., the load port is shorted and the source has zero impedance.

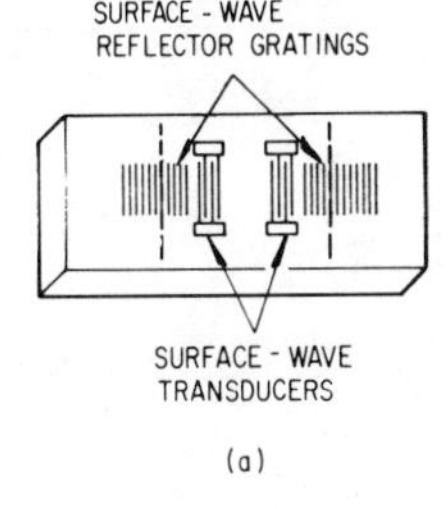

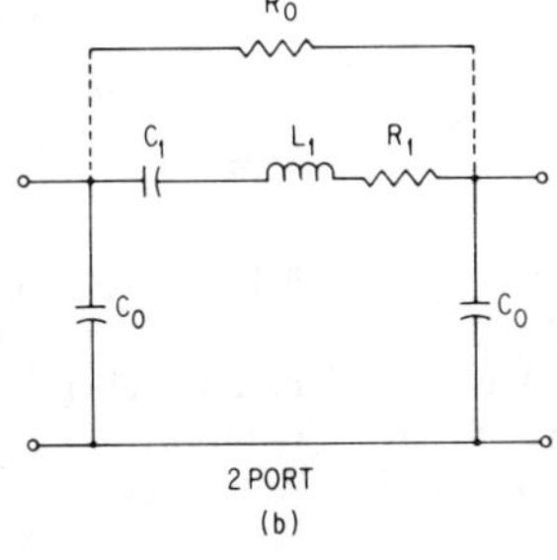

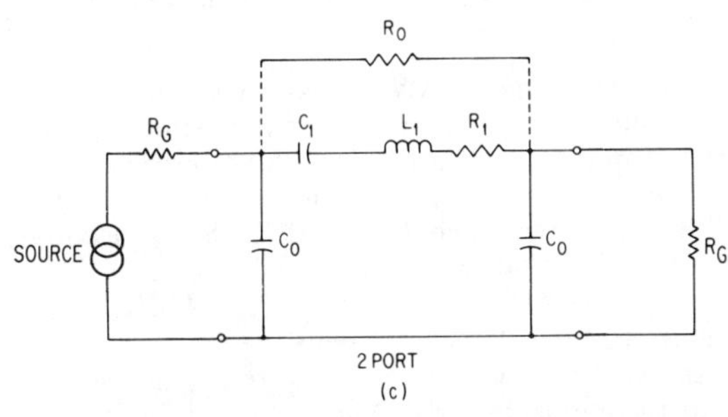

Figure 7 (a) Simple two-port SAW resonator; the dashed lines indicate the positions of two mirrors to which the two gratings are equivalent. (b) equivalent circuit. (c) Complete two-port SAW resonator with source and load R_G. (After Bell and Li [53] with permission)

Hence $R_G = 0$ in Figure 7c for unloaded conditions. (For this case, the circuit has infinite insertion loss.) At resonance, the reactances of C_1 and L_1 of Figure 7b buck each other; to a first order, the circuit reduces to the resistor R_1. When $R_G \neq 0$, the resonator is loaded. The loaded Q, Q_L, is less than the unloaded Q, Q_U, and their ratio is

$$Q_N = Q_U/Q_L = Q_U \frac{B_3}{f_c} \tag{1}$$

where

B_3 = 3 dB bandwidth of loaded SAW resonator

f_c = center frequency of resonator

Q_N is related to the resonator insertion loss IL by [53]

$$IL = -20 \log_{10} \frac{Q_N}{Q_N - 1} = -20 \log_{10} \left[\frac{R_1 + 2R_G}{2R_G}\right] \tag{2}$$

Physically, IL equals the ratio of the total power supplied to the load R_G by the source to the maximum power available from the source of impedance R_G. A plot of Equation (2) is given by the n = 1 curve in Figure 9. Figure 9 also relates IL to Q_N for Butterworth filters formed with n resonators.

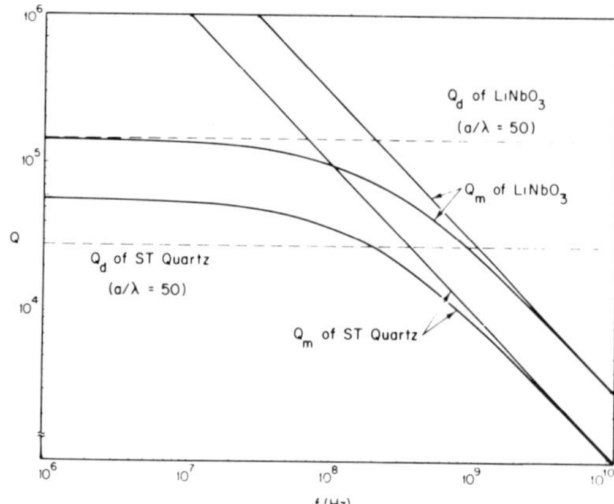

Figure 8 **Q, limited by material (viscous damping and air loading) (Q_m), and limited by beam diffraction (Q_d), for the SAW beamwidth, a, equal to 50λ where λ is the SAW wavelength. (Typically a = 100λ; Q_d is proportional to a^2 [58].) Results given for Y-Z lithium niobate and ST quartz. Curved and straight lines for Q_m apply respectively to air-loaded and evacuated conditions. (From Bell and Li [53] with permission)**

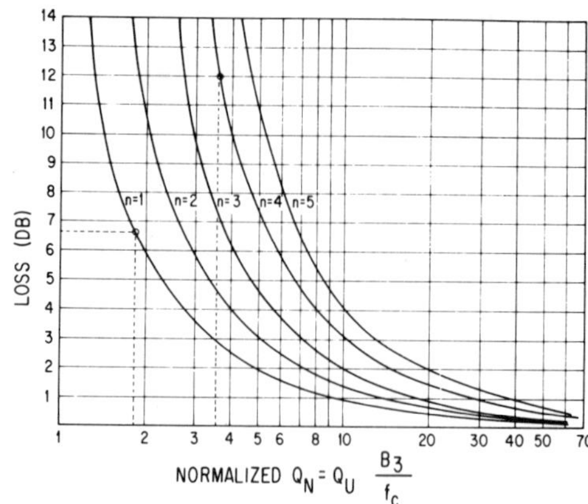

Figure 9 **Insertion loss vs. normalized Q (Q_N) for multipole Butterworth filters using n SAW resonators (From Bell and Li [53] with permission)**

Using the simple two-port SAW resonator configuration of Figure 7, Li of Lincoln Laboratory has achieved an unloaded Q of 32,100 at a center frequency of 310.4 MHz with ST quartz under vacuum [58]. This value is very close to the material Q of 33,000; see Figure 8. In a 50 ohm system ($R_G = 50\Omega$), this value corresponds to a loaded Q of 25,900 and an insertion loss of 14.3 dB, as predicted by Equation (2) and Figure 9 for $R_1 = 416\Omega$. This result is the last entry in Table 2 and represents the best result obtained to date for the SAW resonator filter of Figure 7 with respect to closeness to the theoretical material Q limit, and it is indicative of the type of result to be expected in the future. Recently developed design and fabrication techniques have made this result possible [58]. The two-port resonator for the next to the last entry of Table 2 has a higher unloaded Q (Q_U = 41,300), but it is not as close to the theoretical limit (Q_m = 63,000); it was produced before the latest techniques were developed. It now should be possible to achieve a Q_U of 60,000 at 157 MHz [58].

The SAW resonator filter of Figure 7 has the advantage of small size relative to comparable filters constructed using interdigital transducers. At the frequency of 300 MHz, the SAW resonator of Figure 7 having a Q of 2,000 would be 1 x 2 mm^2 in size; an equivalent interdigital transducer filter having the same Q would be about 3 x 70 mm^2, over two orders of magnitude larger [27, 53]. These resonator filters can be built between 10 MHz and 2 GHz.

A SAW resonator has been developed which provides continuous electronic frequency tuning (via a varactor diode) of up to 0.18 percent of the center frequency [27, 59].

Oscillators

Just as there are two types of SAW filters, there are two types of SAW oscillators – those that use SAW delay lines, and those that use SAW resonators. SAW oscillators which use a quartz delay line in a feedback circuit have the advantage over bulk quartz crystal oscillators of operating to low GHz frequencies without multipliers while providing good short-term stability and ease of modulation. Bulk quartz crystal oscillators can not be built above 50 MHz without either using harmonics (which permits operation to about 250 MHz) or a multiplier chain [27]. The long term stability of SAW quartz delay line oscillators is about $10/10^6$ for the first six months and $2\text{-}3/10^6$ per year after that when mounted in a flat pack [102]. Using a bulk crystal resonator package for the SAW line yields about $3/10^6$ for the first six months and better aging than $2\text{-}3/10^6$ per year after that [102]. In comparison, moderately-good and good bulk quartz crystal oscillators achieve $1/10^6$ per month and $1/10^6$ per year, respectively [18]. In the VHF range, short-term stability comparable to that obtained with a bulk-wave quartz oscillator using a multiplier chain may be obtained in the near future by the use of a SAW resonator [18]. However, in a rugged environment, the delay line SAW oscillator at present has better short-term stability than does the quartz bulk oscillator. When a Y cut quartz SAW delay line oscillator and a well-developed voltage-controlled crystal oscillator (VCXO), both operating at 9.3 GHz, were subjected to 8G rms random vibration, the noise spectrum of the former was essentially unaffected (except below 3 kHz) while that for the latter rose 30 to 40 dB with some resonant spikes 20 dB higher for frequencies below 20 kHz. As a result, the noise spectrum for the VCXO was 10 dB to 50 dB higher than that for the SAW delay line oscillator [62]. The bulk crystal is thin and can be mounted only at a few points for proper operation; the whole SAW delay line bottom surface can be mounted rigidly, the reason for its low sensitivity to vribrations. Such rigid mounting, though, have the disadvantage of rendering the SAW device vulnerable to shifts in frequency due to mechanical stress (through physical distortion during mounting) and temperature stress (due to differential temperature expansions for the mounted SAW device). A poor bond may also degrade the aging characteristics.

Oscillators built using SAW resonators can have short-term stabilities that are over an order of magnitude better than those achieved with SAW delay lines. They can also be built with smaller SAW devices, be less expensive, and have generally lower loss – thus reducing amplifier gain requirements and producing higher stability and greater immunity to external vibrations component [49]. These oscillators can be tuned electronically (0.18% [27, 59]) and can be built to be insensitive to external circuit parameters (e.g., power supply ripple is $<$ 1 ppm/V [59]).

One SAW resonator oscillator built by Lincoln Laboratory to operate at 157 MHz had the same short term stability as a space-qualified temperature-compensated crystal oscillator (TCXO) using a 10 MHz bulk crystal oscillator multiplied up to 150 MHz [58, 63]. Another unit built by Ragan (Texas Instruments [64]) operating at 194 MHz had a short term stability of 1.5×10^{-10} for a 10 ms averaging time. It was also voltage tunable over a 30 kHz range. Based on elevated temperature measurements on a SAW resonator for accelerated life, it was estimated that its long-term stability would be $1/10^6$ per month [65, 66]. It is believed that the long-term stability might be actually an order of magnitude better, $1/10^6$ per year [65, 67].

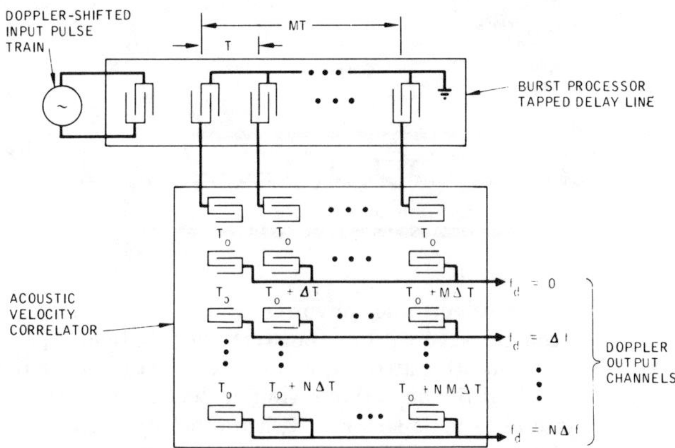

Figure 10 **SAW burst waveform processor called Burst Processor Tapped Delay Line (BPTDL) (After Bristol [28] with permission)**

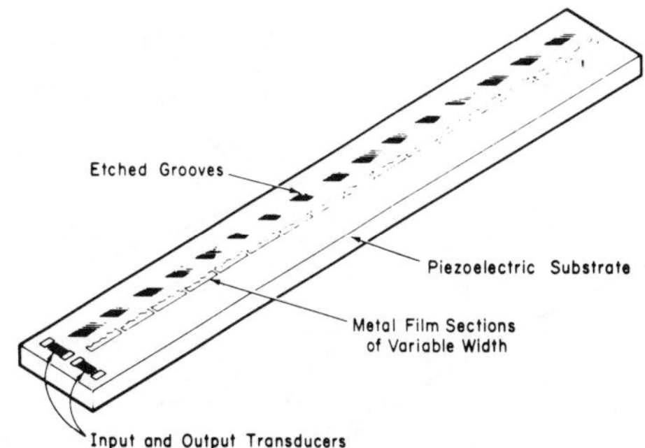

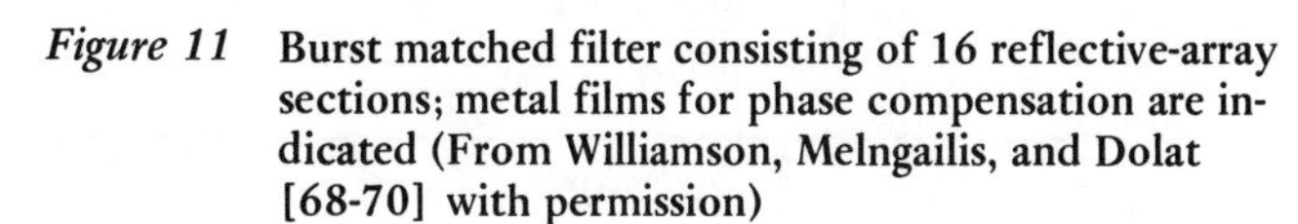

Figure 11 **Burst matched filter consisting of 16 reflective-array sections; metal films for phase compensation are indicated (From Williamson, Melngailis, and Dolat [68-70] with permission)**

Moving Target Indicator (MTI)

SAW devices are being considered for noncoherent MTI radars; the SAW line provides the reference clutter signal in the region where the clutter is not present [21]. In this way, the COHO and stalo are not needed, resulting in a potentially inexpensive radar.

SAW lines also are used for coherent MTI [28]. In a two pulse canceller implemented using SAW lines, the normally undelayed pulse is delayed by 2 μs while the normally delayed pulse is delayed by 2 μs + T_p (where T_p is the pulse-to-pulse period). By delaying the normally undelayed pulse, the same amplitude and phase errors are introduced to both pulses, with the result that broadband cancellation of up to 35 dB can be obtained. Field tests for a multiple PRF system yielded clutter cancellation of 30-35 dB.

Burst Waveform Processors

Figure 10 shows a burst waveform processor called the Burst Processor Tapped Delay Line (BPTDL) [28]. Each short delay, ΔT, is equal to a fraction of the input signal IF frequency period. Thus, different phase shifts are provided to the successive taps T apart on the BPTDL shown at the top of Figure 10. As a result, tap delay lines are synthesized for different Doppler shifts of the received burst waveforms. A 16-tap prototype BPTDL having 50 MHz bandwidth and 80 μs maximum delay has been built for the US Army [28].

Figure 11 shows another type of burst waveform processor, the burst match filter [68-70]. This processor uses the reflective-array-compressor (RAC) configuration with the etched herringbone grooves arranged in 16 groups; see Figure 11. Each group of grooves is used to provide the pulse compression for a single chirped pulse of the burst waveform which consists of 16 3 μs, 60 MHz pulses having 2 μs pulse-to-pulse spacings. The groups of grooves are 2 μs apart. The depth of the grooves is set to provide Hamming weighting within a pulse and from pulse-to-pulse. There is little coupling between the surface wave and the grooves. Consequently, the impulse response of the unit is a burst of 16 pulses having Hamming weighting both on the individual pulses and from pulse-to-pulse, with each pulse chirped 60 MHz, 3 μs wide, and removed by 2 μs from the preceding and following pulses. The metal film sections between pairs of herringbone grooves provide compensation for the phase errors (as was done in Figure 2 of Chapter 10) between and within pulses to within ±20° of ideal. A single SAW device handles one Doppler resolution cell. A bank of M identical devices is used to handle M Doppler cells; all input frequencies are shifted to zero Doppler via appropriate local oscillators signals. With this device, range sidelobes 40 dB down (except for near-in lobes 33 dB down) and Doppler sidelobes 32 dB down were achieved [69].

The burst match filter of Figure 11 has a number of significant advantages over that of Figure 10. First, it has a much greater dynamic range (more than 90 dB maximum peak output signal to output rms noise level) because all the processing is done on one SAW device, i.e., pulse compression of each burst chirped pulse and coherent integration of burst waveforms. The circuit of Figure 10 uses three separate SAW lines, including their added transducer losses. Not shown in Figure 10 is the SAW line used to pulse compress each burst pulse prior to its entering the burst processor. Second, the burst processor of Figure 10 does Doppler matching using true time delay at IF rather than at RF. For large time bandwidth product waveforms, an appreciable mismatch can result.

Monopulse Tracker

Normally, a monopulse tracker requires three channels for processing the received signal – sum, delta-azimuth, and delta-elevation channels. A tracker has been implemented using SAW lines to delay the delta-azimuth and delta-elevation channels by T and 2T, respectively, relative to the sum channel signal where $T > T_u$ (the uncompressed received signal pulsewidth). As a result, only one A/D and digital signal processor is needed for the three received signals [28]. SAW delay lines have the advantage for this application of low distortion and accurate delays (within ±5 ns).

Low Sidelobes for Low Time-Bandwidth Product Chirp Waveform

As indicated in Chapter 12, the versatility of SAW pulse compression lines makes it possible to compensate for the Fresnel ripple sidelobe level limitations obtained with conventional pulse compression lines having low time-bandwidth products [29]. Figure 12 shows the results that can be obtained when this compensation is used compared to the results obtained with conventional lines.

Acoustoelectronic Devices

Convolver

Acoustoelectronic processors use the nonlinear interaction between surface acoustic waves and a semiconductor. One form of acoustoelectronic processor that is under development, the integrating convolver, has the potential to provide match filtering operation for signals 100 ms long with bandwidths of 100 MHz. An existing acoustoelectronic device is the convolver. A version built by Lincoln Laboratory is capable of convolving two 20 μs 100 MHz bandwidth signals – the received signal and a reference signal [30] (see also References 21 and 31 through 34); its dynamic range is about 40 dB. A dynamic range of 50 dB (1 dB compression point to the noise level) was achieved by Lincoln Laboratory with a 100 MHz 10 μs convolver [32]. This convolver consists of a piezoelectric delay line of lithium niobate ($LiNbO_3$), a few 1000 Angstroms above which is placed a silicon (Si) semiconductor; see Figure 13. A metal plate (called the electrode) is placed on top of the silicon;

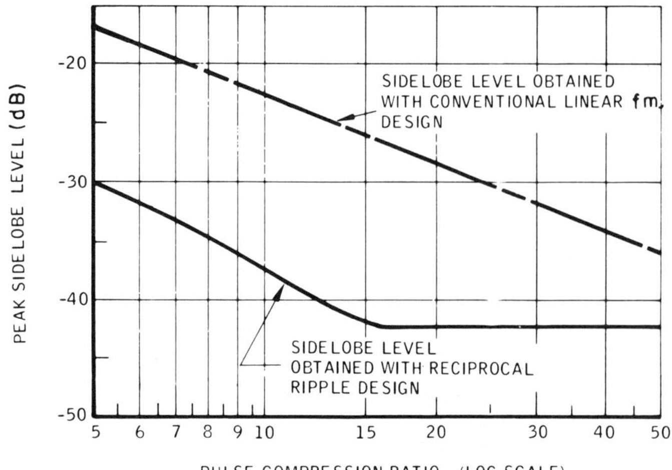

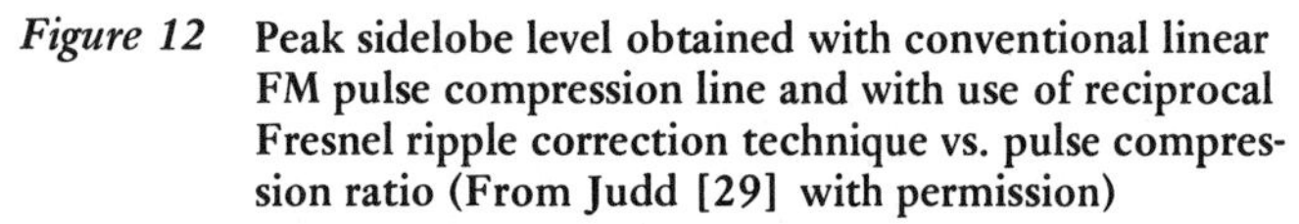

Figure 12 **Peak sidelobe level obtained with conventional linear FM pulse compression line and with use of reciprocal Fresnel ripple correction technique vs. pulse compression ratio (From Judd [29] with permission)**

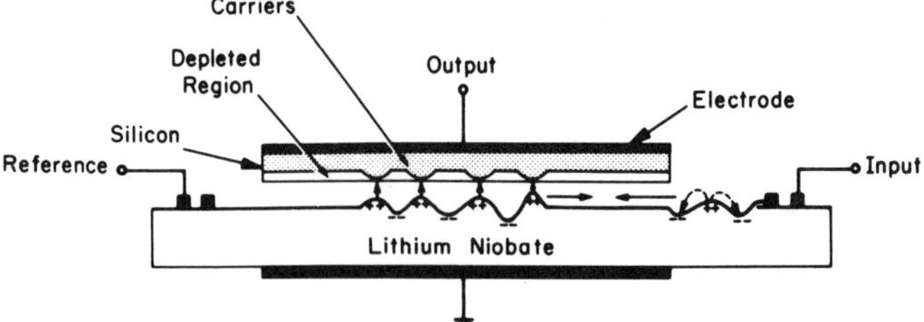

Figure 13 **Acoustoelectronic convolver. Shown is the surface acoustic wave with its transverse electric field (upward pointing arrows), E, and the effect it has on the width of the depletion region of the silicon semiconductor. The departure, $\Delta \ell$, from the equilibrium depletion layer thickness, ℓ_o, is proportional to -E (that is, the thickness decreases for positive E as shown in the figure). The potential across the depletion layer is proportional to the depletion layer thickness squared (that is, $(\ell_o + \Delta\ell)^2$) and hence has a component proportional to E^2. The potential between the electrode and ground is proportional to the integral of the depletion layer potential along the length of the semiconductor, enabling convolution. (From Stern [34] with permission)**

the output terminal is attached to this metal plate. The received signal is fed into one end of the delay line (the input end) while the reference signal is entered simultaneously into the other end. Thus, two surfaces acoustic waves traveling in opposite directions are generated. A nonlinear interaction takes place (the semiconductor responds nonlinearly to the RF electric field of the acoustic waves propagating along the piezoelectric line) which results in the convolution of the two signals appearing at the output terminal. The convolved output signal has a frequency that is equal to the sum of the input received and reference signals, it is time compressed by a factor of two, and it has two times the bandwidth of a conventional convolver output. (The time compression results because the surface waves are traveling in opposite directions [33]). However, the convolved output contains the same information in every respect [21, 33, 34].

The convolver can be used for matched filter radar processing if the matched filter impulsive response is used for the reference waveform. The matched filter signal can be stored digitally in a ROM, from which it is read out at a proper speed and D/A'd. Alternately, the transmitted signal can be stored in a SAW delay line (for the delay time of the line) and then inverted in time to produce the matched filter reference waveform. Time inversion of a transmitted signal can be effected by using the circuit of Figure 14. The linearly varying first local oscillator introduces a frequency variation in time across the pulse. This variation is such that, when it is passed through the first chirp SAW line, the leading edge of the signal becomes delayed by more than is the trailing edge of the signal, to the extent that the leading edge comes out of the dispersive line after the trailing edge and the signal thus is inverted in time. The second linearly varying local oscillator removes the FM modulation introduced by the first local oscillator; the second chirp SAW line removes the time dispersion distortion introduced by the first SAW line. Time reversal can be achieved with a convolver [33] by feeding the signal of frequency f into the input terminal and a narrow pulse of frequency 2f into the electrode. The time-reversed signal then appears at the input terminal [33].

The shape of the matched filter which can be implemented is completely arbitrary except for the bandwidth (100 MHz) and duration (20 μs) constraints. However, the above matched filter implementation is limited to applications in which the time of arrival of the received signal is known *a priori* to a small range window, $\pm(20\ \mu s - T_d)$ for a T_d duration signal. In this respect, the convolver is similar to the stretch processor. A bank of convolvers could be used to handle a wider range window. An alternative is to use the memory correlator described in the next subsection. It is a sliding window correlator, as is the conventional network or SAW-line radar matched filter processor.

Memory Correlator

In the memory correlator, the reference signal is stored spatially along the silicon semiconductor for correlation with the received signal surface acoustic wave. This storage is accomplished through the use of an array of Schottky or PN diodes placed on the under surface of the silicon semiconductor of Figure 13 [33, 35]. The reference signal is stored by first inputting it into the SAW line with the diodes at zero bias. When the reference signal is fully under the diode array embedded in the silicon, a short bias pulse is applied to the electrode to bias the diodes in the forward direction. This action results in each diode contact being charged with a voltage proportional to the amplitude of the reference signal surface acoustic wave below it. The charge is retained after the bias pulse is removed. As a result, the reference signal is stored spatially along the diode array. When the received signal is inputted to the same SAW input terminal, the correlation of the received signal and the reference signal is obtained at the electrode output (Figure 13). For this processor there is no time compression of the correlator output signal. Also, to generate a matched filter receiver reference signal, time inversion of the transmitted signal is not necessary. The reference signal can be stored for 10 to 100 ms if Schottky diodes are used, and up to about 10 s if PN diodes are used. Schottky diodes have the advantage of providing a wider bandwidth correlator, having achieved a 30 MHz bandwidth and projected to provide a 100 MHz bandwidth; the forward direction charge time for Schottky diodes is between 0.1 and 1 ns [33]. The forward charge time for PN diodes is 0.1 μs.

Lincoln Laboratory has built a memory correlator with a bandwidth of 25 MHz and integration time of 10 μs [30]; see also Reference 35. The reference signal can be stored for up to 30 ms without degradation in dynamic range (which is about 40 dB). For a 70 ms storage time, the reference signal amplitude degrades by 3 dB, resulting in a 3 dB dynamic range degradation; for a 100 ms integration time, the degradation is 10 dB. Degradation with increasing delay is in the right direction for a radar system, i.e., the far-out signals are weak and thereby require the lower dynamic range. The storage time can be increased by a factor of two by reducing the operation temperature of the Schottky diodes from 20°C to 10°C.

The above memory correlator also can be used as a storage device for analog signals, having a bandwidth of 25 MHz, duration of 10 μs, and a storage time of 10-100 ms. To read out the stored signal, a short pulse at twice the signal frequency is applied to a terminal at the end opposite to that used for storing the signal. The output is observed at the electrode terminal. The readout is nondestructive; a signal has been scanned experimentally 100,000 times without observing deterioration [35]. Time reversal of the stored signal is achieved by applying the short pulse to a terminal next to the one used for reading in the stored signal (a different terminal is needed because the pulse frequency is twice the signal frequency). The memory correlator can be used as a convolver if the input signal is applied to the terminal opposite to that into which the stored reference signal is read.

The memory correlation has the advantage of being completely adaptive. The reference waveform can be completely arbitrary; it is changed by reading in a new reference waveform and applying a bias pulse, with the old reference waveform being erased in the process. Use of the transmitted waveform as the reference signal eliminates the need for precise control of the transmitter amplitude and phase errors. It is expected that it will be possible to build a memory correlator having a maximum bandwidth of 100 MHz and maximum integration time of 20 μs.

Integrating Convolver (Correlator)

Lincoln Laboratory is developing an acoustoelectronic processor called the integrating convolver which is expected to provide cross correlation of 20 MHz (in the near term, 100 MHz for the far term), 10-100 ms duration signals [30]. This device has essentially the same construction as the memory correlator; its operation, though, is different. No longer is the reference waveform stored in the diodes. The received signal and reference signal are simultaneously inputted at opposite ends of the piezoelectric delay line. The reference and signal surface acoustic waves arrive below a diode in the silicon semiconductor with a specific delay relative to each other, this delay depending on the distance of the diode to the two ends of the piezoelectric delay line. A current term proportional to the product of the amplitudes of the two surface acoustic waves immediately below the diode is generated in the diode. The diode integrates this current for a period of up to 100 ms. Different diodes provide the cross correlation of the two signals for different relative delays. Thus the cross correlation of the two signals is stored spatially in the diode array. The resulting cross correlation signal is read out from the integrating correlator by applying an impulse to the signal input terminal and observing the electrode output. With a 100 MHz bandwidth and 100 ms integration time, a signal processing gain of 70 dB would be achieved.

Like the memory correlator, the integrating correlator (convolver) can be used for matched filter processing if the reference waveform is a delayed replica of the transmitted signal. There is no restriction on the reference signal shape. The device has the limitation that the time of arrival of the received signal is restricted to a small time window, ±20 μs for a 20 μs line. This limitation is similar to that for the convolver except that the time window is independent of the signal duration.

Burst Waveform Processor

A fourth acoustoelectronic device being developed by Lincoln Laboratory is the burst waveform processor [30]. This device is also similar to the memory correlator, except that a polycrystalline Si overlay is used on the diodes. This overlay allows previously stored charges to be retained despite subsequent bias pulsing to enter new charges. As a result, the device can be used to integrate coherently a sequence of identical waveforms which repeat in time, as does a burst waveform.

The operation of the burst processor is as follows. Assume a burst waveform signal consisting of 200 10 μs chirp pulses, having a 50 MHz bandwidth, and having a 50 μs pulse-to-pulse period. Assume a 20 μs piezoelectric line. Initially the diodes have zero charge. The received signal is applied to the input terminal. It is assumed that the time of arrival of the burst waveform is known to within ±5μs. When the first received pulse is wholly within the line, a bias pulse is applied to store the signal spatially in the diode array. Exactly 50 μs later when the next pulse arrives a second bias pulse is applied, with the result that the charge from this second pulse is added to that from the first. This process is repeated until the charges from all 200 pulses have been added coherently in the diodes. At this point, the reference waveform (the transmitted waveform for a single pulse) is fed to the signal terminal of the piezoelectric line, and the matched filter output is read at the electrode terminal. For this system, the output signal is not time compressed (as is the case for the convolver), but rather is in real time (as it is for the memory correlator and a conventional matched filter processor). The above operation assumes a zero Doppler return signal. To handle different Doppler velocities, a bank of burst waveform processors would be used with a separate local oscillator for each to reduce the signal to zero Doppler.

CCD/Memory-Correlator Analog Buffer

Lincoln Laboratory is incorporating CCD technology into their memory correlator device in order to realize a fast-input/slow-output analog buffer [30]. A CCD is connected to each Schottky diode by a FET. The received signal is fed into the input terminal of the piezoelectric line to generate a surface acoustic wave. When the signal is completely under the diode array, a bias pulse is applied to store the signal. The signal charge on each diode is transmitted to its respective CCD via its FET. The CCDs are connected as a shift register; the stored charges can be read out sequentially at a slow rate. For example, a 35 MHz input signal would be read out at a 100 kHz rate.

SAW Amplifier

The above gap-coupled type of silicon/piezoelectric SAW technology used for the memory correlator has been applied to the development of a SAW amplifier [30, 33, 50]; see Figure 4c. Such an amplifier can be used to improve the dynamic range of a long SAW line or as a part of an acoustoelectronic circuit (contained SAW couplers, magic Ts, transformers, dispersive lines, precision attenuators, and oscillators) [16, 21]. Acoustoelectronic amplifiers have been built having gains of 60 dB at frequencies up to 2 GHz with noise figures of 5-10 dB [21]. Other technologies besides the gap technology, such as monolithic structures and stripline systems, are also available for signal processors; see References 19 through 21.

The acoustoelectronic devices described in this section have not advanced to the state of development of the SAW devices described earlier, but they hold a great deal of promise. The most developed of the acoustoelectronic devices are the convolver, memory correlator, and SAW amplifier. The others discussed are still in experimental stages.

Analog CTD (CCD and BBD Technologies)

Charge transfer devices (CTDs) involve the transfer of charge from one capacitor to another. There are two types of CTDs – CCDs and BBDs (bucket brigade devices). BBDs basically consist of a chain of capacitors interconnected with switches and buffer amplifiers; see Figure 15a. By properly cycling the switches open and shut, the charge of one capacitor can be shifted to the next one down the line. Every other capacitor can be used to store charge, with the in-between capacitors used to permit transferring the charges one position down the chain – as with a fireman's "bucket brigade" [86]; see Figure 15b. In 1967, bipolar transistors were used; in 1970, fully integrated metal-oxide-semiconductor field-effect transistors (MOSFETs) were employed [86]; see Figures 15c and 15d.

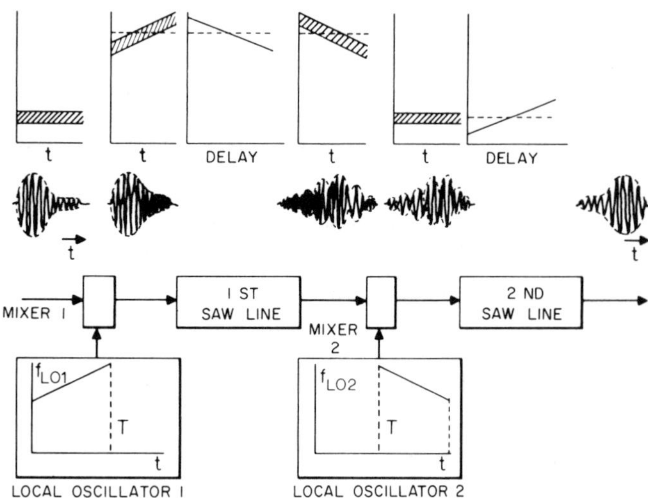

Figure 14 **Circuit for obtaining signal time reversal (After Maines and Paige [18])**

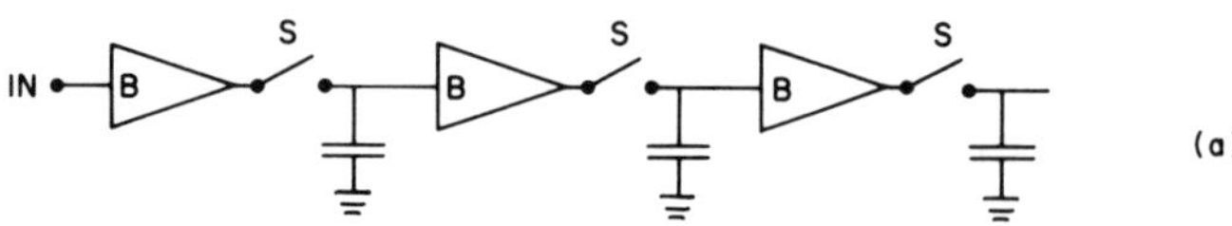

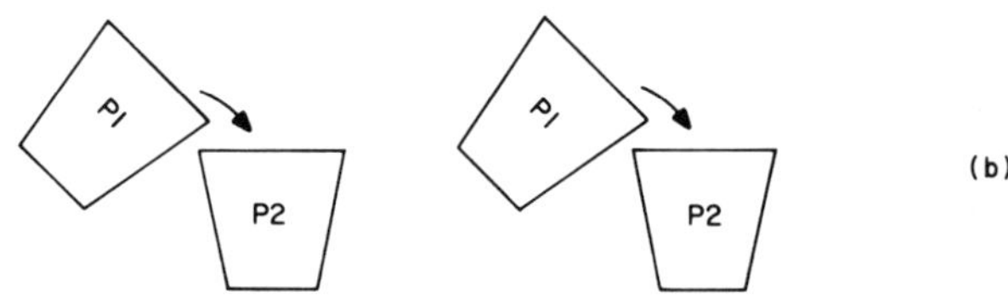

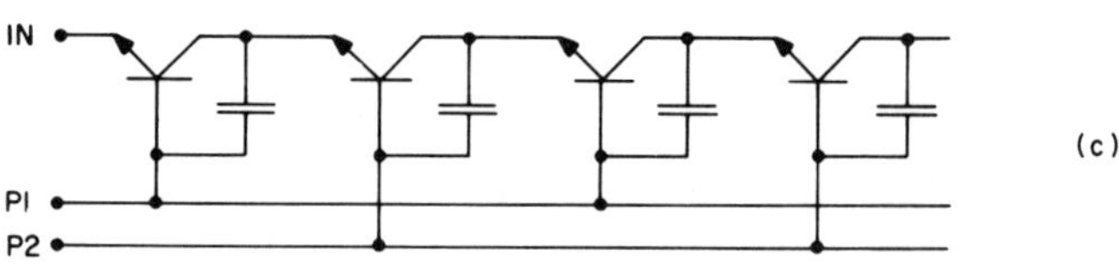

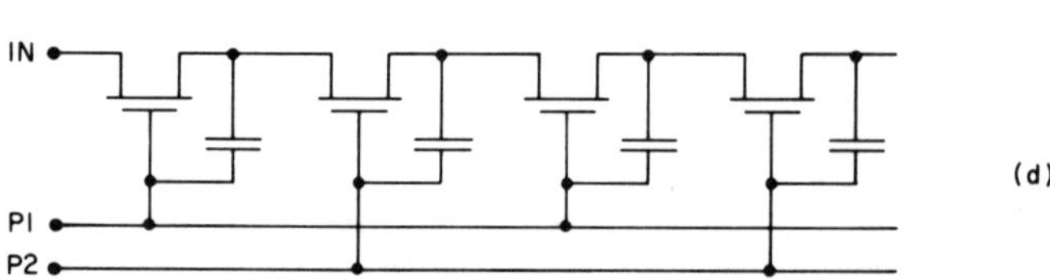

Figure 15 **Bucket brigade circuits: (a) chain of buffer amplifiers, switches, and capacitors, (b) water bucket brigade, two-phase, (c) with bipolar transistor buffer amplifiers, two-phase, and (d) with metal-oxide-semiconductor field-effect transistor (MOSFET) buffer amplifiers (From Séquin and Tompsett, Academic Press [86] with permission)**

CCDs differ from BBDs in that they do not use buffer amplifiers. The charges are stored on a chain of integrated MOS capacitors and moved from one capacitor to the next by proper cycling of the polarity of each capacitor through the use of three transfer electrodes; see Figures 16 and 17. To understand the mechanism of the charge transfer, it is convenient to consider each capacitor (a capacitor is formed below each electrode terminal, P_1, P_2, and P_3) as having a potential well in which charges can be stored (just as a well stores water). The greater the positive charge on the electrode terminal (P_1, P_2, or P_3) the deeper the well (the dashed curves of Figures 16 and 17) and, hence, the greater the charge it can store.

Assume a charge in the well under electrode P_1. P_1 is at a high potential with P_2 and P_3 at low potentials; see Figure 16 a. Now raise the potential of well P_2. The charge then is shared by the wells under P_1 and P_2. Next, lowering the potential of P_1 causes all of the charge to flow to the well of P_2, negative charges going to points of highest potential. Every third electrode is tied together as shown in Figure 17. By varying the potentials on all the P_1, P_2, and P_3 terminals in the manner described above and illustrated in Figures 16 and 17, the charges move to the right from one well to the next. The CCD structure illustrated requires a three-phase control system for charge transfer. The three-phases are needed to isolate charges from adjacent CCD cells, one terminal of a group of three terminals always being at a low potential. A group of three electrodes (P_1, P_2, and P_3) form a single CCD storage cell. Two-phase systems are also possible [86].

The transfer of the charge from the capacitor under P_1 to the capacitor under P_1 in the next cell over does not take place with 100% efficiency. The efficiency with which it actually takes place is called the fractional charge transfer efficiency (CTE). It can be as high as 0.9999 for the type CCD structure shown in Figures 16 and 17, a surface channel CCD (SCCD) [86].

Another type of CCD commonly used is the buried channel CCD (also called bulk channel CCD) (BCCD) [86]. For this device, an n-Si layer is incorporated (via epitaxial growth or ion-implantation) into the surface of the p-Si [86]; see Figure 18. With this device, the electrons are stored in the bulk of the silicon (in the transfer channel of Figure 18). BCCDs have the advantage over SCCDs of higher clock speeds and fractional charge transfer efficiencies; see Table 4 for a comparison of SCCDs and BCCDs.

CCDs can store binary data, in which case they become the digital CCD memory and logic units described previously or the analog signals described in this section. The analog signal processors have the advantage of not requiring A/D converters. CCDs can carry out many of the same functions as can SAW devices. Basically, they can be characterized as analog sampled-data delay lines. They can be used as tapped delay lines, transversal filters, delay lines, pulse compression networks, Fourier transformers, cross correlators, bandpass filters, and analog signal storage devices [36, 86].

Fixed tapped weightings can be built into CCD devices by using a split electrode P_2; see Figure 19 [89]. The corresponding halves of all P_2 split electrodes are connected and applied to a difference amplifier. The difference between the lengths of the two halves determines the weighting for a particular tap; equal halves give zero weighting, and zero length for one half gives ±1 depending on which part has the zero length (see Figure 19). SCCDs have the advantage over BCCDs of better linearity when used as transversal filters with split electrode weighting [90].

Using split electrode weighting, a 500-stage SCCD transversal narrow band filter having Hamming weighting was built which provided a fractional 3 dB bandwidth of 0.0108 (versus 0.0104 predicted), peak sidelobes 38 dB down (versus 42.8 dB predicted), 45 dB total harmonic distortion (for 5V peak-to-peak input signal), and 75 dB dynamic range (5V peak-to-peak input signal to equivalent input rms noise level when output noise is referred to input) [90]. Above results were obtained with a 200 kHz clock rate and

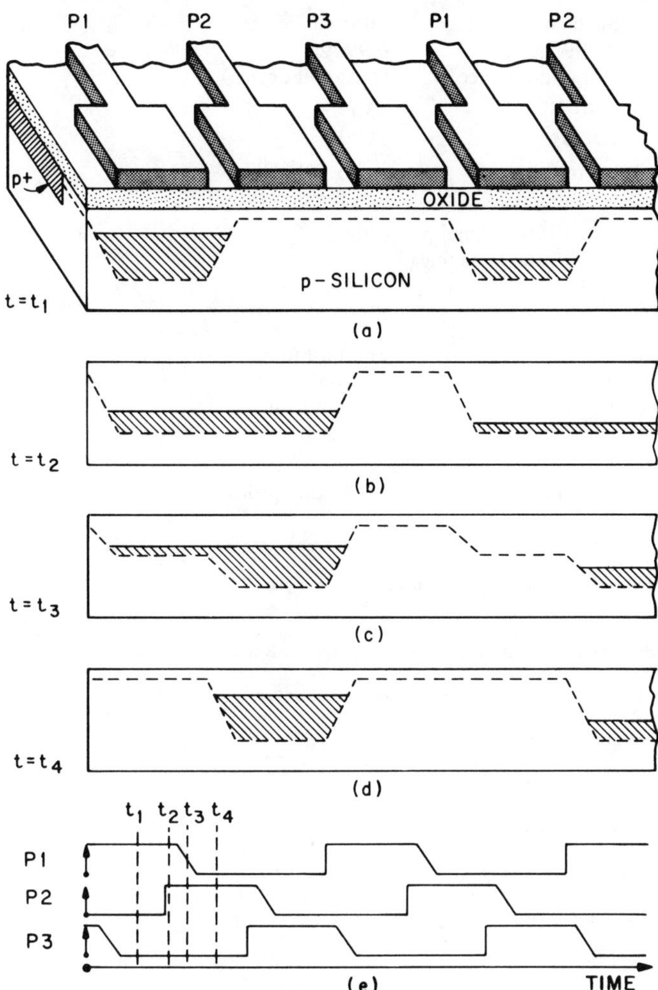

Figure 16 **Illustration of charge transfer in three-phase surface channel CCD system: (a) CCD consisting of chain of metal-oxide-semiconductor capacitors formed by discrete metal contacts (P_1, P_2, P_3) on top of continuous layer of oxide (SiO_2) and p-silicon (p-Si). Dashed lines show potential wells, shaded areas are the amounts of charge stored, and $t = t_1$. (b), (c), and (d) The potential wells and charge distribution at $t = t_2$, t_3, and t_4, respectively. (e) Voltages across P_1, P_2, and P_3 vs. time. (From Séquin and Tompsett, Academic Press [86] with permission)**

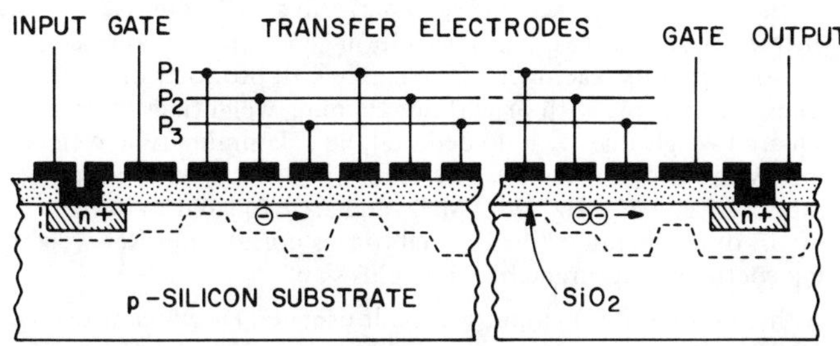

Figure 17 **Three-phase surface channel CCD (SCCD); potential well shown for time $t = t_3$ of Figure 16 (From Tompsett et al., Academic Press [86, 87] with permission)**

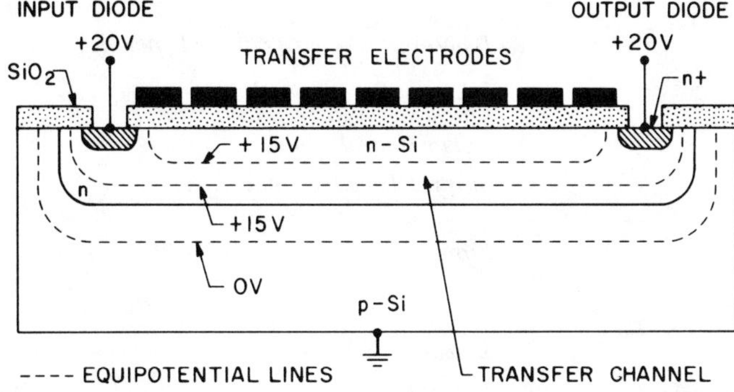

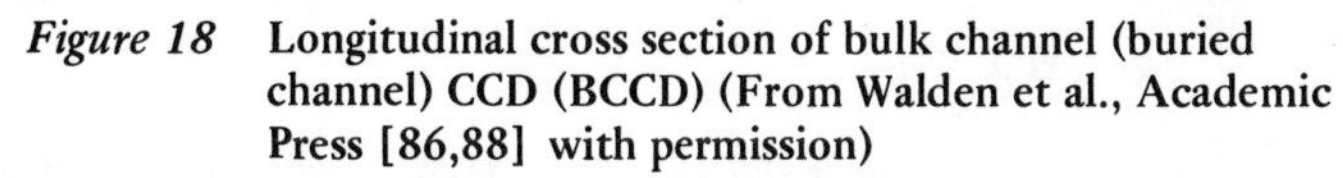

Figure 18 **Longitudinal cross section of bulk channel (buried channel) CCD (BCCD) (From Walden et al., Academic Press [86,88] with permission)**

fractional charge transfer efficiency of 0.9997. The imperfect CTE results in the deviation of resolution from the ideal, and limits the useful transform size for this device to about 1000 points [94]. The measured dynamic range of 75 dB is 25 dB worse than the inherent limit of the device, that limit being set by the sample and hold circuits and differential amplifier in the output circuits. Analysis indicates that this device is capable of a dynamic range of 100 dB while retaining high linearity and having sampling rates from 25 Hz to 10 MHz [90]. The lower limit on the sampling rate (25 Hz) is set by the loss of dynamic range (6 dB) due to dark current (2.0 nA/cm^2) of the device used; the upper limit (10 MHz) is due to the difficulty of providing the higher frequency clock drivers and faster analog circuitry associated with the filter output, as well as the rapid rise in CTE with clock speeds above 10 MHz for the SCCD used [90].

The filter weighting coefficient errors for the above device correspond to 7 bit accuracy for an equivalent digital filter. Weighting coefficient errors limit the Doppler sidelobe rejection achievable (just as antenna weighting errors affect the antenna sidelobe levels). The filter rms sidelobe level is given by [90]

$$\mathrm{SLL} \doteq \frac{\delta/W}{\sqrt{N}} \sqrt{\frac{8}{3}} \tag{3}$$

for the Hamming weighted filter and

$$\mathrm{SLL} \doteq \frac{\delta/W}{\sqrt{N}} \sqrt{\frac{2}{3}} \tag{4}$$

for an unweighted chirp filter implementation where

N = number of taps

δ = quantization error in location of gaps for the split electrodes

W = width of electrodes

For the above filter, N = 500, δ = 0.05 mil and W = 5 mil; thus, SLL = -63 dB. It is possible to make δ = 0.01 mil and W = 30 mil for an equivalent 11½ bit accuracy [90].

Using the same CCD but with a photomask providing a different weighting, a sliding window Fourier transform analyzer (employing the chirp Z transform (CZT) algorithm) was implemented which gave 1 to 200 Hz resolution and 80 dB dynamic range [90]. CZT Fourier transform analysis is achieved with CCDs in a manner analogous to that described previously to obtain Fourier transforms with chirp SAW devices, except that a CCD transversal filter implementation of the chirp SAW device is employed. Specifically, the signal to be Fourier transformed first is multiplied by a linear FM waveform and then passed through a CCD transversal filter

	Surface	Buried
Fractional Charge Transfer Efficiency	0.9999 (for good device)	0.99999 (for good device)
Clock Speed (MHz), Current:	5-50(2)	130(3)
Theoretical Limit:	10-50+	1000
Storage Time (seconds)(4)	up to 50(5) (room temperature)	up to 50(5)(6) (room temperature)
Storage (Electrons/well) [100]	10^7-10^8 (signal processing) 10^5-10^6 (imaging) 10^5 (memory)(7)	
Linearity (%)	0.6	0.6(8)
Harmonics	2nd harmonic 45 dB down for maximum signal(9)	2nd harmonic 45 dB down for maximum signal(10)
Dynamic Range (maximum input rms signal to equivalent input rms noise level when output noise referred to input) (dB)	40-70(11) (~ 90 theoretical) [90]	40-70(12) (~ 90 theoretical)

(1) Note all parameters achieved simultaneously.

(2) High speeds obtained for narrow electrodes (4-5 μm). Speed proportional to one over the electrode width squared. 10 MHz achieved with 10 μm [99, 100].

(3) Philips, Netherlands [91]. Higher values also reported.

(4) Time for well to fill due to dark current. Generally, wells 20% full for normal operation. The well charge rate is constant for constant temperature; it approximately doubles for every 10°C increase.

(5) Laboratory best result. 20 s routinely obtained for these laboratory devices [92]. Similar results obtained at Bell Laboratories [101]. Commercially available units have tens of milliseconds storage time.

(6) More typically buried devices have worse storage time by factors of two to ten.

(7) Projected to go down to 10^3 in future [100].

(8) Compensated internally to the chip. More difficult to achieve 0.6% for buried channel than for surface channel devices.

(9) 60 dB achieved by Bell Laboratories for 0.7V rms, 1 kHz input signal using 20 kHz sampling frequency and 100 μm wide electrode [98, 99].

(10) More difficult to achieve 45 dB down second harmonic for buried channel than for surface channel; at low frequencies, can achieve 60 dB. Good output circuit is needed.

(11) 70 dB achieved for device of Footnote (9) for same conditions. Dynamic range proportional to electrode width [98, 99].

(12) 55 dB available with Fairchild commercial unit.

Table 4 **Parameters for Surface and Buried CCDs(1)**

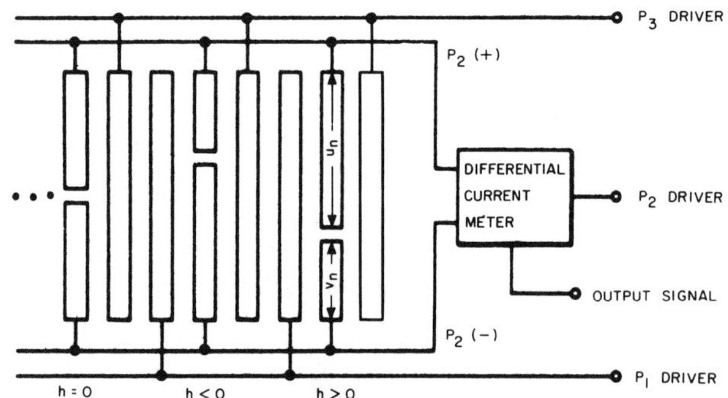

Figure 19 **Top view of three-phase CCD using split electrodes to implement transversal filter weightings. Weighting for n^{th} tap, h_n, is $(u_n-v_n)/(u_n+v_n)$. Signal outputs observed when charges under P_2 electrodes; $t = t_4$. (From Buss et al. [89] with permission)**

which is matched to the modulating linear FM signal. If the Fourier transform phase is desired (in addition to the amplitude), the output has to be multiplied by a linear FM waveform having opposite slope to that used at the input [93].

Buss et al. [93] have made a comparison of performing Fourier transforms digitally (using the FFT algorithm and state-of-the-art I^2L logic (1975) having a clock speed of 2 MHz) and using a CCD CZT. Table 5 shows the specifications to which the general purpose digital FFT was configured. Although this digital FFT is more flexible and provides greater accuracy (16 bits versus 8 bits plus sign, effectively, for the CCD), the CCD CZT requires fewer chips (3 versus 9) and operates at a faster rate (one butterfly per 20 μs versus one per 200 ns). The future DCCDs described in an earlier section also have the potential of one butterfly per 100 ns.

At present, analog CCDs processors are more suitable than digital processors for those applications requiring moderate performance (dynamic range), low cost, high speed, and low power consumption.

Reticon shortly (1977) will have available a commercial CCD chirp which can be used either to perform a 512 point CZT Fourier transform or as two chirp-waveform generators or pulse compression circuits (each with time bandwidth products of 512). These units come with or without Hanning weightings (cosine squared weightings with no pedestal, i.e., Hamming type weightings without the pedestal). The units will operate with sampling rates between 500 Hz and 2 MHz, and are expected to have a 60 dB dynamic range (peak signal to rms noise) and filter weighting coefficient accuracy of 8 bits plus sign.

A three pulse (MTI) circuit was built using CTDs which provided 50 dB of clutter suppression in field testing [37]. A production version of such an MTI circuit could be packaged on a single 5″ x 9″ board.

By marrying CTD and SAW technology, a system has been developed which permits one SAW line to synthesize a Doppler filter bank for many channels or, equivalently, range bins [38, 39]. The technique involves storing the successive pulse returns from a particular range gate in successive locations of one row of a CTD ana-

Algorithm	In place
Cycle Clock	2 MHz
Butterfly Time	4 μs
*Transform Speed**	[(n/2) log (n+1)] x [Butterfly Time]
Transform Lengths	n = 1, 2, 3, . . . , 512 complex points
Arithmetic	16 bit block floating point with rounding
Coefficient Word Length	8 bits
Data Word Length	16 bits

*Includes unscrambling.

Table 5 **Specifications for I²L Digital, General Purpose FFT Processor (From Buss et al. [93])**

log storage memory. The next range gate return is stored in the next row of the CCD memory, and so on. The data is read in from 8 range gates during one pulse interval along a column of the CCD memory and read out along rows of the CTD memory. This arrangement is what is called a corner turning memory. The data from a particular row (consisting of 64 pulse returns from one range gate) is read out in 25.6 μs and processed with a fixed window CTD CZT. The Doppler spectrum for the 8 range gates are read out in an interval of about 204.8 μs via this method. The system can be made to read out in one-fourth this time.

From the above, it should be apparent that CCDs can be used for time compression. By reading in the signal into the CCD memory at one rate and reading it out at another rate, time compression (or time expansion) can be achieved. As a result, it should be possible to use one chirp SAW line to pulse compress chirp waveforms having different time bandwidth products; the chirp waveform is stored in the CCD and read out at a rate which allows its slope to be matched to that of the chirp SAW line used for pulse compression.

Fairchild provides a BCCD delay line consisting of two 455 elements. Its bandwidth is 4 to 5 MHz, dynamic range is about 55 dB, nonlinearity is 3% or less, and sampling rate is 14.3 MHz. Its single unit price is about $95 (1977) and is expected to drop to less than $5 in quantity lots [95].

At present (1977), CCD technology is mainly complementary with that of SAW technology. CCD is capable of providing much larger delays, on the order of seconds, and its bandwidths are limited to 10 to 50 MHz due to limits imposed by the circuitry for getting information and clocking signals onto and off the chip. However, for BCCD, 130 MHz [91] has been achieved and 1 GHz is the theoretical limit; see Table 4.

For further reading, see References 86, 96, and 97.

References

[1] Newton, C.O.: "Signal Processing Aspects of Nonlinear Chirp Radar Signal Waveforms for Surface Wave Pulse Compression Filters," *Proceedings of Signal Processing Conference*, Lausanne, Switzerland, 1975.

[2] Winograd, S.: "On Computing the Discrete Fourier Transform," *Proceedings of the National Academy Science, Vol. 73, No. 4,* pp. 1005-1006, April 1976.

[3] Rader, C.: *Private Communication,* Lincoln Laboratory.

[4] McClellan, J.: *Private Communication,* MIT.

[5] Dixon, N.R.; and Silverman, H.F.: "The 1976 Modular Acoustic Processor (MAP): Signal Analysis and Phonemic Segmentation," *Conference Record of 1976 IEEE International Conference on Acoustics, Speech and Signal Processing,* Philadelphia, Pennsylvania, 12-14 April 1976, pp. 9-14.

[6] Kolber, D.P.; and Parks, T.W.: "A Prime Factor FFT Algorithm Using High Speed Convolutions," to appear in *IEEE Transactions on Acoustics, Speech and Signal Processing,* 1977.

[7] *Electronic Products Magazine, Vol. 19,* pp. 72-81, November 1976.

[8] Torrero, E.A.: "Solid State Devices," *IEEE Spectrum,* pp. 48-54, January 1977.

[9] *Electronic Products Magazine, Vol. 19,* p. 51, December 1976.

[10] Feth, G.C.: "Memories: Smaller, Faster, and Cheaper," *IEEE Spectrum, Vol. 13,* pp. 36-43, June 1976.

[11] Myers, W.: "Key Developments in Computer Technology: A Survey," *IEEE Computer, Vol. 9,* pp. 48-77, November 1976.

[12] Feth, G.C.: "Progress in Memory and Storage Technology," *Paper No. 33-4,* IEEE Electro '76, Boston, Massachusetts, 11-14 May 1976.

[13] Rajchman, J.A.: "New Memory Technologies," *Science, Vol. 195,* pp. 1223-1229, 18 March 1977.

[14] Van Tuyl, R.; Lichti, C.; Lee, R.; and Gowen, E.: "4-GHz Frequency Division with GaAs MESFET ICs," 1977 International Solid State Circuits Conference (ISSCC), pp. 198-199, February 1977.

[15] Daly, R.H.; and Glass, J.M.: "Digital Signal Processing for Radar," *IEEE EASCON '75 Record,* Washington, DC, pp. 215-A to 215-G, 29 September-1 October 1975.

[16] Holland, M.G.; and Claiborne, L.T.: "Practical Surface Acoustic Wave Devices," *Proceedings of the IEEE, Vol. 62,* pp. 582-611, May 1974.

[17] Squire, W.D.; Whitehouse, H.J.; and Alsup, J.M.: "Linear Signal Processing and Ultrasonic Transversal Filters," *IEEE Transactions on Microwave Theory and Techniques, Vol. MTT-17,* pp. 1020-1040, November 1969.

[18] Maines, J.D.; and Paige, E.G.S.: "Surface-Acoustic-Wave Devices for Signal Processing Applications," *Proceedings of the IEEE, Vol. 64,* pp. 639-652, May 1976.

[19] "Special Issue on Surface Acoustic Wave Devices and Applications," *Proceedings of the IEEE, Vol. 64,* May 1976.

[20] *IEEE 1976 Ultrasonics Symposium Proceedings,* 29 September-1 October 1976.

[21] Maines, J.D.; and Paige, E.G.S.: "Surface-Acoustic-Wave Components, Devices and Applications," *Proceedings of the IEE, Vol. 120,* pp. 1078-1110, October 1973.

[22] Dolat, V.S.; and Williamson, R.C.: "A Continuously Variable Delay-Line System," *IEEE 1976 Ultrasonics Symposium Proceedings,* pp. 419-423, 29 September-1 October 1976.

[23] Newton, C.O.; and Paige, E.G.S.: "Surface Acoustic Wave Dispersive Filter with Variable, Linear, Frequency-Time Slope," *IEEE 1976 Ultrasonics Symposium Proceedings,* pp. 424-428, 29 September-1 October 1976.

[24] Brookner, E.: "Synthesis of an Arbitrary Bank of Filters by Means of a Time-Variable Network," *IRE Conventions Record, Part 4,* pp. 221-235, March 1961.

[25] Bickel, H.; and Brookner, E.: "A Delay-Line Synthesized Filter Bank with Electronically Adjustable Impulse Responses," Fourth National Convention on Military Electronics, pp. 489-496, June 1960.

[26] Collins, J.: "SAW Devices Emerging into Commercial and Military Hardware," *MSN,* pp. 15-21, January 1977.

[27] Haydl, W.H.: "Surface Acoustic Wave Resonators," *Microwave Journal, Vol. 19,* pp. 43-46, September 1976.

[28] Bristol, T.W.: "Applications of Surface Acoustic Wave Devices to Radar Signal Processing," *Proceedings of Symposium on Optical and Acoustical Micro-Electronics,* New York, 16-18 April 1974, Volume 23 of Microwave Research Institute Symposia Series, pp. 57-65, J. Fox (ed.), Polytechnic Press, Polytechnic Institute of New York, Brooklyn, New York.

[29] Judd, G.W.: "Technique for Realizing Low Time Sidelobes in Small Compression Ratio Chirp Waveforms," *IEEE Ultrasonics Symposium Proceedings,* pp. 478-481, 1973.

[30] Ralston, R.W.: "Spanning the Gap Between Microelectronic and Surface Wave Signal Processors with Acoustoelectronics," Boston IEEE Sonics and Ultrasonics Chapter Meeting, 13 April 1977.

[31] Reible, S.A.; Cafarella, J.H.; Ralston, R.W.; and Stern, E.: "Convolvers for DPSK Demodulation of Spread Spectrum Signals," *IEEE 1976 Ultrasonic Symposium Proceedings,* pp. 451-455, 29 September-1October 1976.

[32] Cafarella, J.H.; Brown, Jr., W.M.; Stern, E.; and Aluslow, J.A.: "Acoustoelectric Convolvers for Programmable Matched Filtering in Spread-Spectrum Systems," *Proceedings of the IEEE, Vol. 64,* pp. 756-759, May 1976.

[33] Kino, G.S.: "Acoustoelectric Interactions in Acousto-Surface-Wave Devices," *Proceedings of the IEEE, Vol. 64,* pp. 724-748, May 1976.

[34] Stern, E.: "Surface-Wave Devices and Their Applications to Signal Processing," *Proceedings of Symposium on Optical and Acoustical Micro-Electronics,* New York, 16-18 April 1974, Volume 23 of Microwave Research Institute Symposia Series, pp. 19-31, J. Fox (ed.), Polytechnic Press, Polytechnic Institute of New York, Brooklyn, New York.

[35] Ingebrigtsen, K.A.: "The Schottky Diode Acoustoelectric Memory and Correlator – A Novel Programmable Signal Processor," *Proceedings of the IEEE, Vol. 64,* pp. 764-769, May 1976.

[36] Buss, B.D.; Brodersen, R.W.; and Hewes, C.R.: "Charge-Coupled Devices for Analog Signal Processing," *Proceedings of the IEEE, Vol. 64,* pp. 801-804, May 1976.

[37] Lobenstein, H.; and Ludington, D.N.: "A Charged Transfer Device MTI Implementation," *IEEE 1975 International Radar Conference Proceedings,* pp. 107-110, Arlington, Virginia, 21-23 April 1975.

[38] Collins, J.D.: "Frequency Spectrum Analyzer," *US Patent No. 4,005,417,* 25 January 1977.

[39] Collins, J.D.; Scearretta, W.A.; MacFall, D.S.; and Schulz, M.B.: "Signal Processing with CCD and SAW Technologies," *IEEE 1976 Ultrasonic Symposium Proceedings,* pp. 441-450, 29 September-1 October 1976.

[40] Silverman, H.F.: "An Introduction to Programming the Winograd Fourier Transform Algorithm (WFTA)," *IEEE Transactions on Acoustics, Speech and Signal Processing, Vol. ASSP-25, No. 2,* pp. 152-165, April 1977.

[41] "TTL Isoplanar Memory 93470 and 93471 4096 x 1 Bit Fully Decoded Random Access Memory," *Fairchild Camera and Instrument Corporation Data Sheet,* September 1976.

[42] Allan, R.: "Amorphous Semiconductors Revisited," *IEEE Spectrum, Vol. 14,* pp. 41-45, May 1977.

[43] Zappe, Hans H.: *Private Communication,* IBM, Thomas J. Watson Research Center.

[44] Allen, R.A. (ed.); Handy, R.J.; and Sandor, J.E.: "Charge Coupled Devices in Digital LSI," *Technical Digest,* 1976 International Electron Devices Meeting, Washington, DC, pp. 21-26, 6-8 December 1976.

[45] Zimmerman, T.A.; and Barbe, D.F.: "New Role for Charge-Coupled Devices: Digital Signal Processing," *Electronics,* pp. 97-103, 31 March 1977.

[46] Allen, R.A.: *Private Communication,* TRW.

[47] Davis, G.R.: "Gigabit Logic: Real-Time Response for Tomorrow's Threats," *Microwaves, Vol. 15,* pp. 9-12, September 1976.

[48] Claxton, D.: *Private Communication,* TRW.

[49] Hays, R.M.; and Hartmann, C.S.: "Surface-Acoustic-Wave Devices for Communications," *Proceedings of the IEEE, Vol. 64,* pp. 652-671, May 1976.

[50] Coldren, L.A.; and Shaw, H.J.: "Surface-Wave Long Delay Lines," *Proceedings of the IEEE, Vol. 64,* pp. 598-609, May 1976.

[51] Staples, E.J.: "UHF Surface Acoustic Wave Resonators," *Proceedings of the 28th Annual Frequency Control Symposium,* US Army Electronics Command, Fort Monmouth, New Jersey, pp. 280-285, May 1974.

[52] Hartmann, C.S.; and Rosenfeld, R.C.: "Acoustic Surface Wave Resonator Devices," *US Patent No. 3,886,504,* 27 May 1975.

[53] Bell, D.T. Jr.; and Li, R.C.M.: "Surface-Acoustic-Wave Resonators," *Proceedings of the IEEE, Vol. 64,* pp. 711-721, May 1976.

[54] Swartzlander, E.: *Private Communication,* TRW.

[55] Mason, W.P.: *Electromechanical Transducers and Wave Filters (second edition),* Van Nostrand, Princeton, New Jersey, 1948.

[56] Temes, G.C.; and Mitra, S.K. (eds.): *Modern Filter Theory and Design,* Wiley, New York, 1973.

[57] *Reference Data for Radio Engineers (fifth edition),* H.W. Sams and Co., Inc., Indianapolis, Indiana, 1968.

[58] Li, Robert C.M.: *Private Communication,* MIT Lincoln Laboratory.

[59] Cross, P.S.; Haydl, W.H.; and Smith, R.S.: "Design and Applications of Two-Port SAW Resonators on YZ-Lithium Niobate, *Proceedings of the IEEE, Vol. 64,* pp. 682-685, May 1976.

[60] Koyamada, Y.; Ishihara, F.; and Yoshikawa, S.: "Narrow-Band Filters Employing Surface-Acoustic-Wave Resonators," *Proceedings of the IEEE, Vol. 64,* pp. 685-687, May 1976.

[61] Shreve, W.R.: "Surface Wave Resonators and Their Use in Narrow-Band Filters," *IEEE 1976 Ultrasonics Symposium Proceedings* (IEEE Catalog No. 76 CH1120-5SU), pp. 706-713, 29 September-1 October 1976.

[62] Weglein, R.D.; and Otto, O.T.: "Effect of Vibration on SAW-Oscillator Noise Spectra," *Electronics Letters, Vol. 13,* pp. 103-104, 17 February 1977.

[63] Williamson, Richard C.: *Private Communication,* MIT Lincoln Laboratory.

[64] Ragan, L.: "Voltage Controlled Surface Wave Resonator Oscillators," *IEEE 1976 Ultrasonics Symposium Proceedings* (IEEE Catalog No. 76 CH1120-5SU), pp. 252-255, 29 September-1 October 1976.

[65] Shreve, R.W.: *Private Communication,* Texas Instruments.

[66] Bell, D.T.; and Miller, F.P.: "Aging Effects in Plasma Etched SAW Resonators," *Proceedings of the 30th Annual Symposium on Frequency Control,* US Army Electronics Command, Fort Monmouth, New Jersey, pp. 358-362, June 1976.

[67] Bell, D. Jr.: *Private Communication,* Texas Instruments.

[68] Williamson, R.C.; Melngailis, J.; and Dolat, V.S.: "Reflective-Array Matched Filter for a 16-pulse Radar Burst," *IEEE 1975 Ultrasonics Symposium Proceedings* (IEEE Catalog No. 75 CHO 994-4SU), New York, pp. 400-404, 1975.

[69] Melngailis, J.; and Williamson, R.C.: "Surface-Acoustic-Wave Device For Doppler Filtering of Radar Burst Waveforms," 1976 MTT Symposium, Cherry Hill, New Jersey, 14-16 June 1976 (IEEE Catalog No. 76 CH1D87-6MTT).

[70] Williamson, R.C.: "Properties and Applications of Reflective-Array Devices," *Proceedings of the IEEE, Vol. 64,* pp. 702-710, May 1976.

[71] Altman, L.: "Josephson Tunneling Shows Promise," *Electronics,* pp. 55-56, 26 December 1974.

[72] Guéret, P.; Mohr, T.O.; and Wolf, P.: "Single Flux-Quantum Memory Cells," *IEEE Transactions on Magnetics, Vol. MAG-13,* pp. 52-55, January 1977.

[73] Herrell, D.J.: "An Experimental Multiplier Circuit Based On Superconducting Josephson Devices," *IEEE Journal of Solid-State Circuits, Vol. SC-10,* pp. 360-368, October 1975.

[74] Wu, C.T.; and Falco, C.M.: "High-Temperature Nb_3Sn Thin-Film SQUIDs," *Applied Physics Letters, Vol. 30,* pp. 609-611, June 1977.

[75] *Bipolar Microcomputer Components Data Book for Design Engineers (first edition),* Texas Instruments, January 1977.

[76] Ogdin, C.A.: "μC Design Course," *EDN, Vol. 21,* pp. 127-316, 20 November 1976.

[77] Osborne, A.: "Basic Concepts," Volume 1 of *An Introduction to Microcomputers,* Adam Osborne and Associates, Inc., PO Box 2036, Berkeley, California, 1976.

[78] *Electronic Products Magazine, Vol. 19,* p. 17, November 1976.

[79] Cushman, R.H.: "EDN's Third Annual Microprocessor Directory," *EDN, Vol. 21,* pp. 44-89, 20 November 1976.

[80] Forbes, B.E.: "Silicon-on-Sapphire Technology Produces High-Speed Single-Chip Processor," *Hewlett-Packard Journal, Vol. 28,* pp. 2-8, April 1977.

[81] Linvill, J.G.; and Hogan, C.L.: "Intellectual and Economic Fuel for the Electronics Revolution," *Science, Vol. 195,* pp. 1107-1113, 18 March 1977.

[82] Osborne, A.: "Some Real Products," Volume 2 of *An Introduction to Microcomputers,* Adam Osborne and Associates, Inc., PO Box 2036, Berkeley, California, 1976.

[83] "Special Issue on Microprocessor Technology and Applications," *Proceedings of the IEEE, Vol. 64,* June 1976.

[84] Claiborne, L.T.: *Private Communication,* Texas Instruments.

[85] Fawcette, J.: "MSI Logic Reaches 4.5 GHz – Can Application be Far Off?," *MSN, Vol. 7,* pp. 13, 14, 16, and 18, February 1977.

[86] Séquin, C.H.; and Tompsett, M.F.: *Charge Transfer Devices,* Academic Press, Inc., 1975.

[87] Tompsett, M.F.; Amelio, G.F.; and Smith, G.E.: "Charge Coupled 8-Bit Shift Register," *Applied Physics Letters, Vol. 17,* pp. 111-115, 1970.

[88] Walden, R.H.; Krambeck, R.H.; Strain, R.J.; McKenna, J.; Schryer, N.L.; and Smith, G.E.: "The Buried Channel Charge Coupled Device," *Bell System Technical Journal, Vol. 51,* pp. 1635-1640, 1972.

[89] Buss, D.D.; Collins, D.R.; Baily, W.H.; and Reeves, C.R.: "Transversal Filtering Using Charge Transfer Devices," *IEEE Journal of Solid-State Circuits, Vol. SC-8,* pp. 138-146, April 1973.

[90] Brodersen, R.W.; Hewes, C.R.; and Buss, D.D.: "A 500-Stage CCD Transversal Filter for Spectral Analysis," *IEEE Transactions on Electron Devices, Vol. ED-23,* pp. 143-152, February 1976.

[91] Esser, L.J.M.: *IEEE ISSSC Digital Technology Paper, Vol. 18,* p. 28, February 1974.

[92] Claeys, C.L.; Laes, E.E.; Declerck, G.J.; and Van Overstraeten, R.J.: "Elimination of Stacking Faults for Charge-Coupled Device Processing," *Semiconductor Silicon 1977 Proceedings,* sponsored by Electro-Chemical Society, pp. 773-784, May 1977.

[93] Buss, D.D.; Veenkant, R.L.; Brodersen, R.W.; and Hewes, C.R.: "Comparison Between the CCD CZT and the Digital FFT," *Proceedings of 1975 Naval Electronics Laboratory Center International Conference on the Application of Charge-Coupled Devices,* pp. 267-281, October 1975.

[94] Buss, D.D.; Brodersen, R.W.; Hewes, C.R.; and de Wit, M.: "Spectral Analysis Using CCDs," *CCD '76,* Washington, DC.

[95] Derman, S.: "CCDs Reach Practical Stage for Analog Signal Processing," *Electronic Design, Vol. 24,* pp. 34 and 36, 29 March 1976.

[96] Melen, R.; and Buss, D.D. (ed.): *Charge-Coupled Devices: Technology and Applications,* IEEE Press, 1977.

[97] "Special Issue on Charge-Transfer Devices," *IEEE Transactions on Electronic Devices, Vol. ED-23,* February 1976.

[98] Sealer, D.A.; Séquin, C.H.; Ryan, P.M.; Suciu, P.I.; Statile, J.L.; Fuls, E.N.; and Tompsett, M.F.: "A Dual-Differential Analog Shift Register with a Charge-Splitting Input and On-Chip Peripheral Circuits," 1977 IEEE ISSCC, pp. 148, 149, and 248, February 1977.

[99] Sealer, D.A.: *Private Communication,* Bell Laboratories.

[100] Séquin, C.H.: *Private Communication,* Bell Laboratories.

[101] Tompsett, M.F.: *Private Communication,* Bell Laboratories.

[102] Parker, T.E.: "Current Developments in SAW Oscillator Stability," *Proceedings of 31st Annual Symposium on Frequency Control,* US Army Electronics Command, Fort Monmouth, New Jersey, 1-3 June 1977.

[103] Cartledge, L.; and O'Donnell, R.M.: "Description and Performance Evaluation of the Moving Target Detector," *Project Report ATC-69,* Lincoln Laboratory, 8 March 1977; also O'Donnell, R.M.; and Cartledge, L.: "Comparison of the Performance of the Moving Target Detector and the Radar Video Digitizer," *Project Report ATC-70,* Lincoln Laboratory, 26 April 1977.

Part 3
Propagation Effects

Chapter 14 gives a detailed discussion of the subject of ionospheric pulse distortion in simple terms, including the effects of Earth's magnetic field. Figure 7 in that chapter provides an easy-to-use set of curves and nomograph-like axes for determining the distortion due to the ionosphere for various ionospheric conditions, carrier frequencies, and transmitted pulsewidths. A simple equation is given for determining the Faraday rotation as a result of propagation through the ionosphere.

Chapter 15 gives a simple exposition of the effect of atmospheric turbulence on the propagation of optical radar signals. Many of the results given also apply to microwave signal propagation. Also presented are fog and rain attenuation at optical wavelengths, with a comparison to microwave attenuation. The subject of pulse distortion in both clear and inclement weather (including propagation through clouds) is reviewed.

Chapter 14 Brookner

Pulse–Distortion and Faraday–Rotation Ionospheric Limitations

In this chapter, first consideration is given to the pulse shape distortion which results from the ionosphere when Earth's magnetic field can be neglected. Next, pulse shape distortion resulting from the presence of Earth's magnetic field is discussed. Finally, Faraday rotation is treated and related to the time delayed distortion resulting from Earth's magnetic field.

Pulse Distortion Resulting When Earth's Magnetic Field Can Be Neglected

Ionospheric pulse distortion resulting from the presence of Earth's magnetic field can be neglected for carrier frequencies above 100 MHz if a circularly polarized signal is transmitted. However, for a linearly polarized signal, there exist conditions for which Earth's magnetic field must be accounted. These conditions are given in a later section of this chapter. In this section and the next, it is assumed that Earth's magnetic field can be neglected.

Due to the presence of the ionosphere, the increase in group velocity of a radar pulse transmitted through it is proportional to the pulse's RF carrier frequency squared; that is

$$v_G \propto f^2 \tag{1a}$$

Thus

$$\tau_I \propto \frac{1}{f^2} \tag{1b}$$

where

τ_I is the increase in the path delay due to the presence of the ionosphere.

As a result, a narrow pulse having a very wide bandwidth has its lower frequency components arrive later than do its higher frequency components; that is, there is a pulse stretching or, equivalently, a time dispersion of the pulse. This pulse time dispersion is depicted in Figure 1 for transmitted pulses having 3 dB widths, τ_1, varying from 5 to 60 ns at a carrier frequency of 1 GHz. The results are given for the case in which the envelope of the transmitted pulse has a Gaussian shape. As indicated in Figure 1, the pulse received after propagation through the ionosphere also has a Gaussian shape; however, its 3 dB pulse width, τ_2, is wider and its amplitude is lower (by a_P dB). [1, 2] (The parameter N_{TV} is defined in the glossary at the end of this chapter and is discussed later.)

For Figure 1, the assumption is made that the radar is on the ground while the target is above the ionosphere. Thus, the pulse travels twice through the ionosphere. The elevation angle of the satellite as viewed from the radar is assumed to be 10°. The results given in the figure are for a rather severe ionosphere.

The ionosphere behaves like the linear dispersive chirp delay lines discussed in Chapters 7, 8, 10, and 12. The ionospheric dispersion is, however, not linear with frequency. For most radar applications under consideration, though, the ionosphere time dispersion about the radar carrier can be considered to be linear.

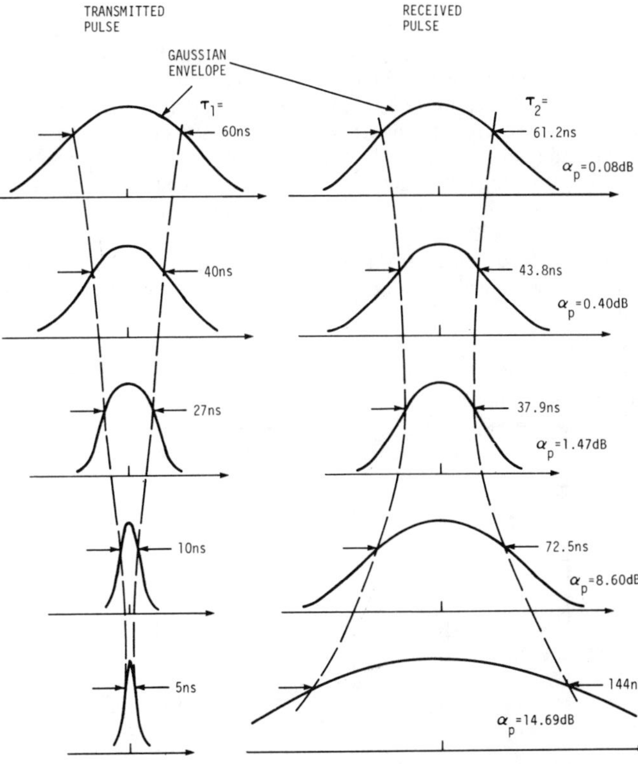

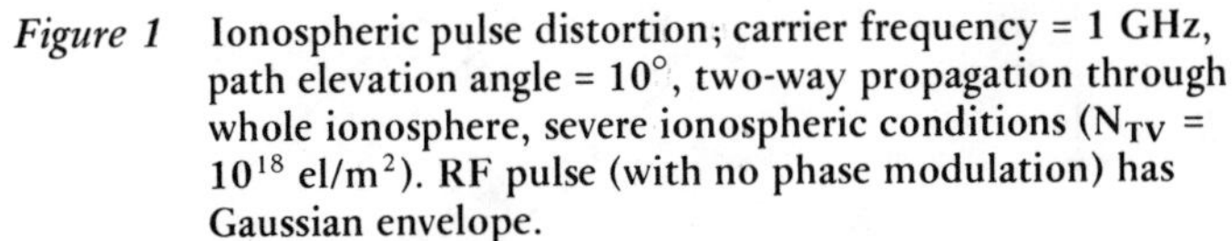

Figure 1 **Ionospheric pulse distortion; carrier frequency = 1 GHz, path elevation angle = 10°, two-way propagation through whole ionosphere, severe ionospheric conditions (N_{TV} = 10^{18} el/m^2). RF pulse (with no phase modulation) has Gaussian envelope.**

The dispersion due to the ionosphere can be compensated for by using a dispersive network in the receiver, the time dispersion of which is the complement of that of the ionosphere. The sum of the ionospheric time dispersion and receiver network time dispersion then provides a constant delay with frequency so that no distortion results; see Figure 2. It is when compensating for ionospheric dispersion for wide bandwidth waveforms that the nonlinearity of the ionospheric dispersion must be taken into account; such as when compensating for the distortion resulting for the 5 ns of Figure 1.

The ionospheric time dispersion results from the electrons present in the ionosphere. The larger the number of electrons present, the greater this time dispersion. Also, the lower the elevation angle of the satellite target, the greater the resulting time dispersion [1, 2].

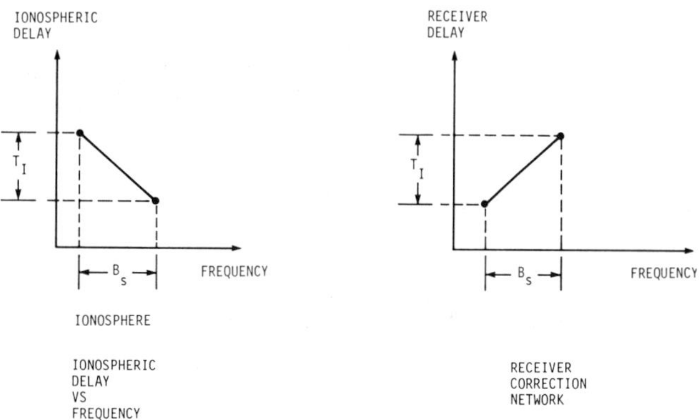

Figure 2 Compensation of Ionospheric Time Dispersion

Figure 3 plots the ionosphere group delay versus carrier frequency for two elevation angles, $\alpha = 10^\circ$ and $= 90^\circ$, and for two ionospheric conditions, severe and medium. This figure indicates that the group delay increases by a factor of 10 when going from the medium to the severe ionospheric conditions assumed. Also, the group time delay increases by a factor of 3 when the target satellite elevation angle changes from 90° to 10°. (The variation with elevation angle does not change as $1/\sin \alpha$ because of Earth's curvature and the elevated position of the ionosphere, its peak being at an altitude of about 300 km. As a result, a ray leaving the earth with a 0° grazing angle penetrates the peak of the ionosphere at a grazing of about 17°.)

Examination of Figure 3 shows that for worst case ionospheric conditions and a 10° elevation angle, the ionospheric group delay is 1000 ns for a carrier frequency of f = 0.9 GHz, and 670 ns for

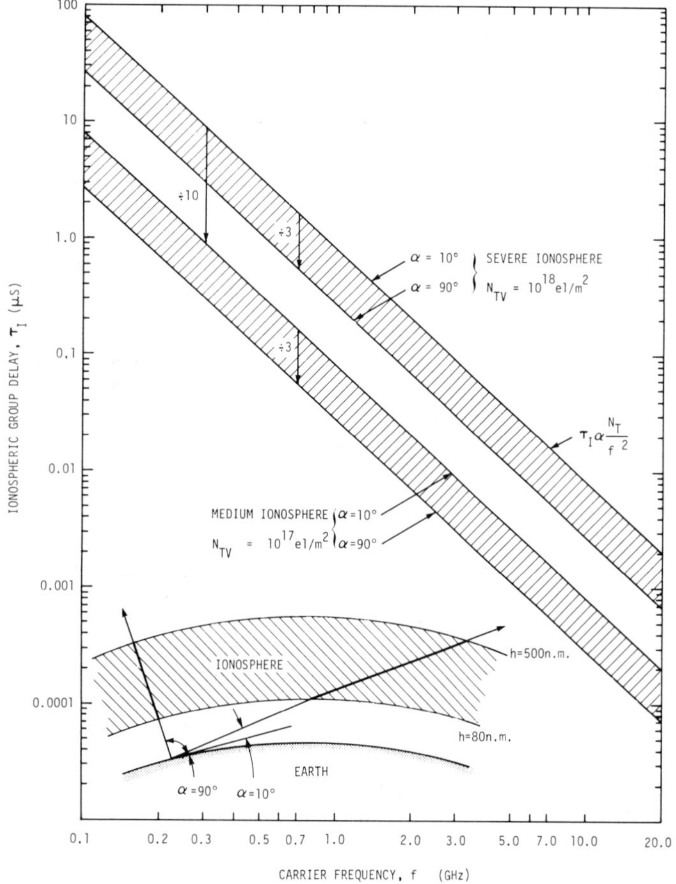

Figure 3 Ionospheric Group Delay vs. Frequency

f = 1.1 GHz – a difference of 330 ns for this 200 MHz bandwidth centered at a carrier frequency of 1 GHz. (Figure 1 shows a dispersion of 144 ns for this bandwidth of 200 MHz at a carrier frequency of 1 GHz. This discrepancy arises from the fact that the 330 ns represents the dispersion between the 16 dB points of the pulse, while the 144 ns dispersion of Figure 1 represents the dispersion between 3 dB points of the pulse.)

The electron content of the ionosphere is given in terms of the number of electrons in a vertical column through the ionosphere, N_{TV} (el/m^2). N_{TV} depends on the solar activity, the time of day, and geographic location [15, 16]. Figure 4 shows the diurnal variation of N_{TV} observed at Stanford, California and Boulder, Colorado (both at about 40°N latitude) in parts of the years 1958 and 1959 near the 1958 sun spot maximum. The electron density is seen to be 5 to 10 times greater at local noon than at local midnight.

The maxima N_{TV} of Figure 4 are characteristic of high sun spot activities. The sun spot maxima (which occur in 11 year cycles) for the years 1947 and 1957 were high, 1957 having the greatest recorded solar activity. For the year 1968, the daily N_{TV} maximum at Hamilton, Massachusetts (also at about 40°N latitude) ranged from about 17 x 10^{16} to 75 x 10^{16} el/m^2 instead of to 170 x 10^{16} el/m^2 as observed in Figure 4 [17]. At Hamilton, the monthly 24 hour daily mean for the worst months of 1968 (January, February, October, and November) was 40 x 10^{16} el/m^2 with the one-sigma standard deviation about that mean being 25%. During strong magnetic storms the maximum can be four-sigma above the mean; i.e., 80 x 10^{16} el/m^2, which is close to the largest value of 87 x 10^{16} el/m^2 observed since 1967 (on March 1970) at Hamilton. For the year 1972, the daily N_{TV} maximum ranged from about 12 x 10^{16} to 50 x 10^{16} el/m^2 [17]. The solar minimum appears to have been reached in 1976 (instead of 1974) with the lowest value observed at Hamilton for N_{TV} since 1968 being 0.3 x 10^{16} el/m^2 (observed at night). N_{TV} peaks just north and south of the geomagnetic equator and reaches its minimum near the geomagnetic poles; the maximum is about two times the value at Hamilton [12, 16].

The geographic variation of N_{TV} can be obtained from ionospheric models (spatial and temporal) which use as an input the world plots of the critical frequency of the F_2 region of the ionosphere, designated as f_0F_2. Ionospheric models have been developed by the Air Force Geophysics Laboratory (AFGL) at Hanscom Field in Bed-

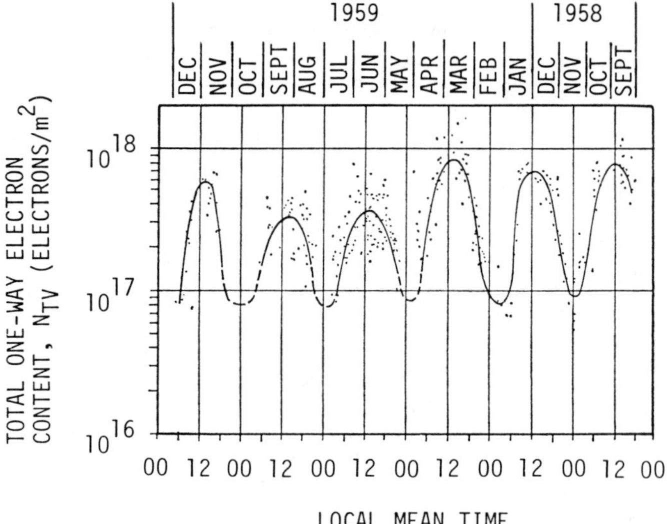

Figure 4 One-way integrated electron density along vertical path through ionosphere. After Lawrence and Posakony [3] (see also Reference 2).

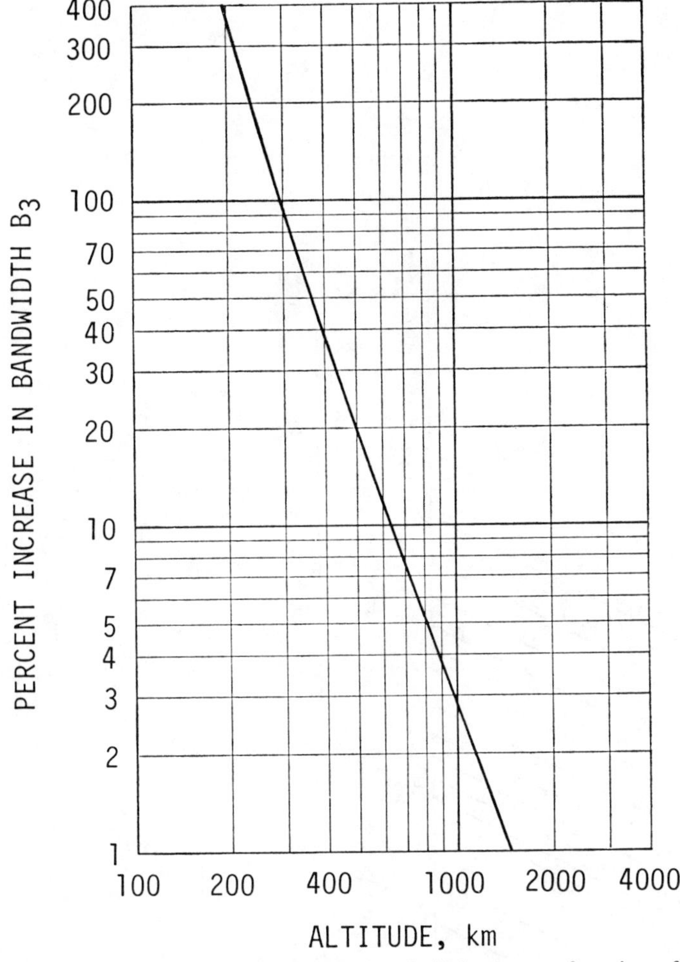

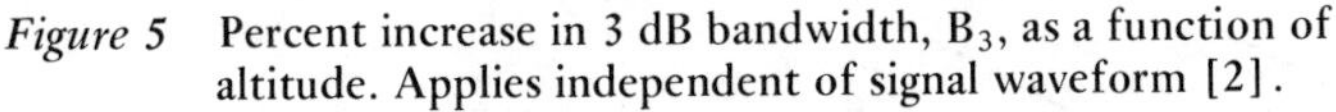

Figure 5 **Percent increase in 3 dB bandwidth, B_3, as a function of altitude. Applies independent of signal waveform [2].**

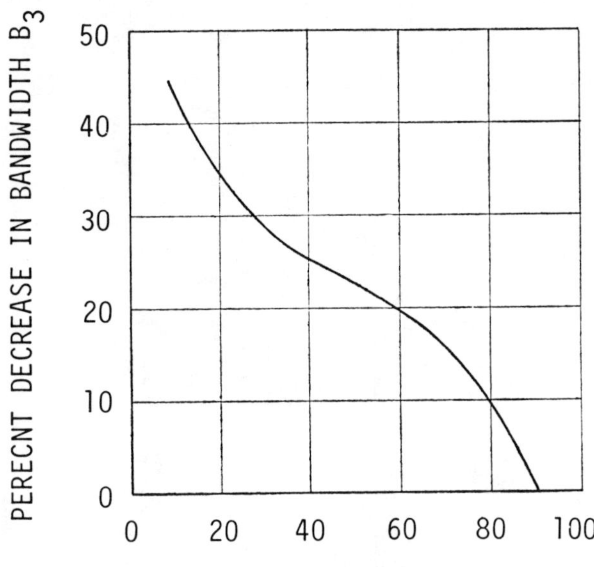

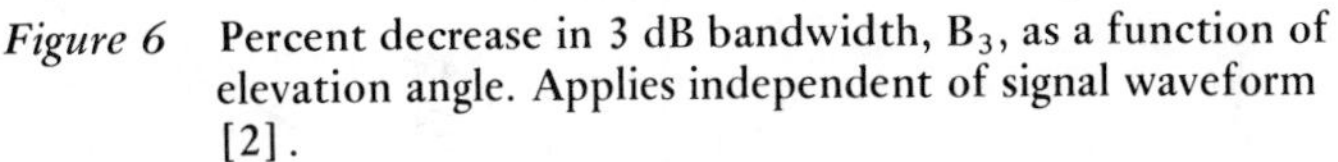

Figure 6 **Percent decrease in 3 dB bandwidth, B_3, as a function of elevation angle. Applies independent of signal waveform [2].**

ford, Massachusetts, the Global Weather Central (GWC) at the Strategic Air Command (SAC) Base in Omaha, Nebraska [4], and at the Raytheon Company in Sudbury, Massachusetts [5]. Global plots of f_0F_2 are provided monthly by the Environmental Science Services Administration (ESSA) in Boulder, Colorado. The electron density, N, (el/m^3) at the peak of the ionosphere's F_2 region is related to the critical frequency, $f_0F_2 = f_P$ (Hz), by [1]

$$f_P^2 = 81N \tag{2}$$

N_{TV} is proportional to $f_P{}^2$ (for a given ionospheric altitude-profile model). For the ionospheric Chapman altitude-profile model of Reference 2, $f_P = 19.59$ MHz when $N_{TV} = 10^{18}$. Accurate predictions of N_{TV} for any geographic location can be obtained directly from GWC, who use worldwide ionospheric sounding data.

The group delay along a path having an elevation angle, α, is proportional to the number of electrons, N_T, in one square meter column along the two-way path of the signal. Specifically

$$\tau_I = \frac{81N_T}{2cf^2} \tag{3}$$

where c is the velocity of light, 3×10^8 m/s. For N_T in electrons/square meters and f in hertz, τ_I is in seconds.

A unit of measure sometimes used for N_T is the TEC (total electron content). One TEC unit equals 10^{16} el/m^2. A convenient relationship to remember between the delay, τ_I, and N_T is that, at a carrier frequency of 1 GHz, when N_{TV} equals one TEC unit, the two-way τ_I equals 2.7 ns (1.328 ft) for a vertical path and 8.1 ns (3.98 ft $\doteq$ 4 ft) for $\alpha = 10°$.

Equation (3) results from the fact that the group velocity is given by (see appendix)

$$v_G = \frac{2cf^2}{f_P{}^2} = \frac{2cf^2}{81N} \tag{4}$$

while

$$\tau_I = \int \frac{1}{v_G} ds \tag{5}$$

Substituting Equation (4) into Equation (5) and making use of

$$N_T = \int N ds \tag{6}$$

yields Equation (3). As indicated, for a path through the whole ionosphere leaving the ground with an elevation angle of 10°, N_T is three times greater than it is for a vertical path through the whole ionosphere.

The signal bandwidth, B_3, which results in a specified allowable pulse distortion, is inversely proportional to (a first order) $\sqrt{N_T}$. Figures 5 and 6 show how B_3 varies with altitude and elevation angle [2].

Plots of the transmitted 3 dB pulse widths, τ_1, which result in 10%, 30%, and 100% 3 dB pulse width broadenings, are given in Figure 7 as a function of N_T for carrier frequencies ranging from 0.1 to 10 GHz. The results are given for the case of an RF pulse having a gaussian shaped envelope. The extension of the results of this figure to other shape pulses is given in the next section. The ordinate scales also give the 3 dB bandwidths, B_3, which are given in terms of τ_1 by [1]

$$B_3 = \frac{0.441}{\tau_1} \tag{7}$$

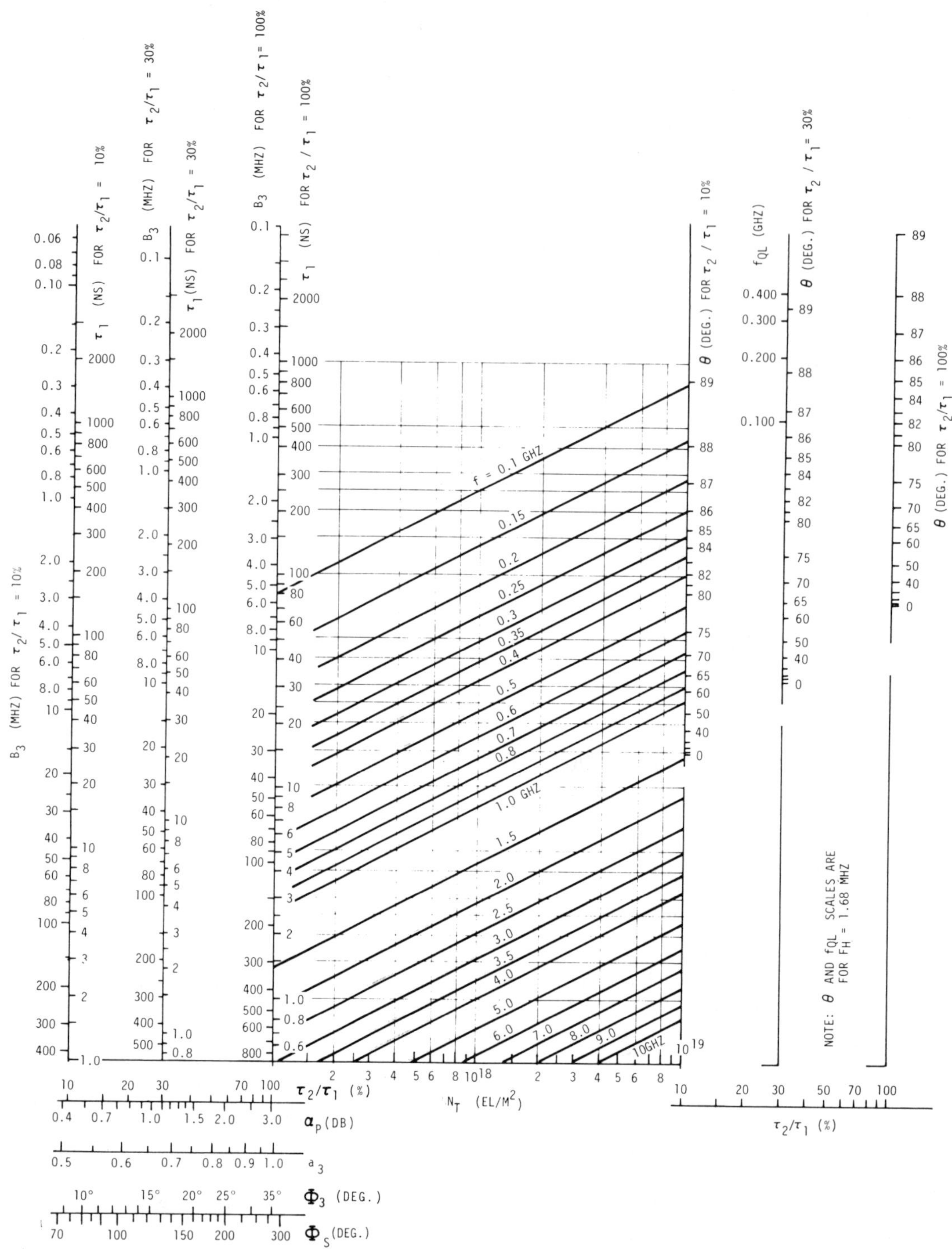

Figure 7 Transmitted 3 dB pulse width, τ_1, which results in 10%, 30%, 100% broadenings of received 3 dB pulse width, τ_2, versus ionospheric path total electron content, N_T (two-way for radar system, one-way for communication system), for carrier frequencies 0.1 to 10 GHz. Also, angle, θ, for which $\Delta\tau_I = \Delta\tau_M$ and minimum frequency, f_{QL}, for which quasi-longitudinal propagation applies for angle θ. Results apply for RF pulses (with no phase modulation) having Gaussian shape. Also apply for chirp pulses having 40 dB Tapered-Taylor ($\bar{n} = 6$) weighting or Hamming weighting, with τ_1 and τ_2 being transmit and received compressed 3 dB pulse widths respectively [6].

τ_2/τ_1*	α_P	a_3	Φ_3	Φ_S
10%	0.414	0.5075	9.10°	73.1°
30%	1.139	0.6833	16.50°	132.6°
100%	3.01	0.9867	34.4°	276.4°

Table 1 **Values for α_P, a_3, Φ_3, and Φ_S when τ_2/τ_1 = 10%, 30%, and 100%**

The ordinates on the right-hand side and the horizontal scales in the lower left- and right-hand corners are discussed later. The pulse amplitude attenuations that result for these distortions are given in Table 1.

The pulse width distortion results given in Figure 7 were derived for the approximation that the ionosphere delay varies linearly with frequency near the RF pulse carrier frequency. This assumption is equivalent to one that the ionosphere is a network having a quadratic phase error with frequency, the higher order (cubic, quadratic, etc.) terms being neglected. The higher order phase errors should have negligible effect on the pulse width dispersion and amplitude degradation results presented in Figure 7 and Table 1.

In Figure 7 and Table 1, Φ_3 represents the quadratic phase error introduced by the ionosphere at the 3 dB bandwidth points of the signal. The parameters Φ_S and a_3 are discussed shortly.

Example 1 – Assume a radar system with two-way propagation such that $N_T = 6 \times 10^{18}$ el/m^2 and f = 1 GHz; from Figure 7, the minimum pulse width that can be transmitted if a 10% pulse width broadening is allowed is τ_1 = 40 ns and B_3 = 11.1 MHz.

Example 2 – For a communications application with one-way propagation, N_T becomes 3×10^{18} el/m^2 in the above example and τ_1 becomes 28 ns (that is $40/\sqrt{2}$ ns); B_3 = 15.8 MHz.

Example 3 – Assume that a 20% pulse width distortion is allowed in Example 2. For distortion other than 10%, 30%, and 100%, the horizontal scales on the lower left-hand side of Figure 7 are used to allow interpolation between the 10%, 30%, and 100% ordinate scales. Interpolating between the 10% and 30% pulse width broadening ordinate scales yields τ_1 = 22.6 ns.

The horizontal scales at the left-hand side also allow the determination of α_P and Φ_3 for values of distortion other than the 10%, 30%, and 100% values given in Table 1. Thus, for Example 3, α_P = 0.79 dB and Φ_3 = 13.2°. These scales also can be used to obtain the parameters a_3 and Φ_S (to be defined later) for arbitrary distortions between 10% and 100%.

The Φ_3 and α_P scales can be used to determine the pulse width broadening and peak amplitude degradation resulting for any network characterized by a quadratic phase shift and constant amplitude across its passband. Assume such a network and assume an IF pulse having a Gaussian shaped envelope. At the 3 dB bandwidth points the network phase shift, Φ_3, is assumed to be 9.10°. Then, from Figure 7, it follows that at the output of the network, this signal has its 3 dB compressed pulse width broadened by 10% and its peak amplitude degraded by 0.414 dB (Example 1 inverted).

The equation used to plot the curves of Figure 7 is [1]

$$\tau_1 = \frac{8.20 \times 10^{-9}}{a_3} \frac{\sqrt{N_T}}{f^{3/2}} \tag{8}$$

where the distortion parameter, a_3, is given in terms of the pulse width spreading factor τ_2/τ_1 by [1]

$$a_3 = \left[\frac{(\tau_2/\tau_1)^2 - 1}{3.168}\right]^{1/4} \tag{9a}$$

which in terms of the signal bandwidth is given by [1, 2]

$$a_3 = 1.86 \times 10^{-17} B_3 \frac{(N_T)^{1/2}}{f^{3/2}} \tag{9b}$$

where, in Equations (8) and (9b), f is in GHz, N_T is in el/m^2, B_3 is in Hz, and τ_1 is in ns.

In terms of Φ_3 (in radians) [1]

$$a_3 = \frac{4}{\pi} \sqrt{\Phi_3} \tag{9c}$$

The equation for the peak amplitude attenuation, α_P, is given by [1]

$$\alpha_P = 5 \log_{10} [1 + 3.168 a_3^4] \tag{10}$$

A convenient equation for the pulse width dispersion, that is, for

$$\Delta\tau_I = \tau_2 - \tau_1 \tag{11}$$

is

$$\Delta\tau_I = \frac{8.20 \times 10^{-9}}{A} \frac{\sqrt{N_T}}{f^{3/2}} \tag{12}$$

where

$$A = \frac{a_3}{\sqrt{1 + 3.168 a_3^4} - 1} \tag{13}$$

and where f is in GHz, N_T is in el/m^2, and $\Delta\tau_I$ is in ns.

Results for Other Shaped Pulses

Figure 7, in addition to applying for an RF pulse having a Gaussian envelope, is relevant for chirped pulses which use a Gaussian type of weighting, such as a 40 dB Taylor weighting or Hamming weighting [1]. Now, τ_1 and τ_2 represent 3 dB compressed pulse widths of the transmitted and received signals, respectively. In terms of B_S, the chirp signal bandwidth, τ_1 is given by [6, 7]

$$\tau_1 = \frac{1.25}{B_S} \tag{14}$$

The parameter Φ_S in Figure 7 and Table 1 represents the quadratic phase error due to the ionosphere at the signal band edges $\pm B_S/2$ away from the carrier frequency.

Figure 8 shows the actual pulse distortion resulting for different quadratic phase errors Φ_S for the case of a 40 dB Tapered-Taylor chirp weighting (having $\bar{n}$ = 6) [2, 7, 8]. This weighting is approximately a Hamming weighting. Specified also in Figure 8 is the parameter a_S to be defined shortly. Note that there is negligible degradation of the sidelobe levels of the weighted, compressed chirp pulse even though there can be significant amplitude degradation and pulse width spreading due to the ionosphere.

Given in Figure 8 is the ionospheric time-bandwidth product, $T_I B_S$, when viewed as a dispersive network in which T_I represents the delayed dispersion of the ionosphere over the bandwidth B_S.

One cannot define a_3 in terms of B_3 using Equation (9b) because B_3 has no meaning for chirped waveforms. Instead, parameter a_S of Figure 8 is used in place of a_3 in Equation (9b), with B_3 replaced by B_S. The distortion parameter, a_3, can be specified for the Tapered-Taylor weighted waveform, but it must be defined by Equation (9a) in terms of τ_2/τ_1 [1]. The distortion parameter a_S is expressed in terms of Φ_S by [1]

$$a_S = \frac{4}{\pi} \sqrt{\Phi_S} \tag{15}$$

* Percent increase in τ_2/τ_1.

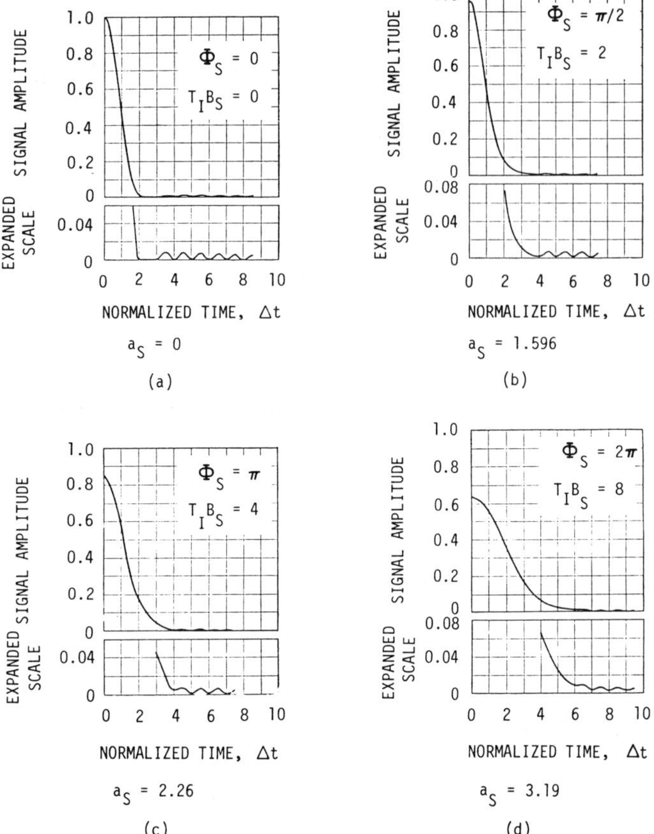

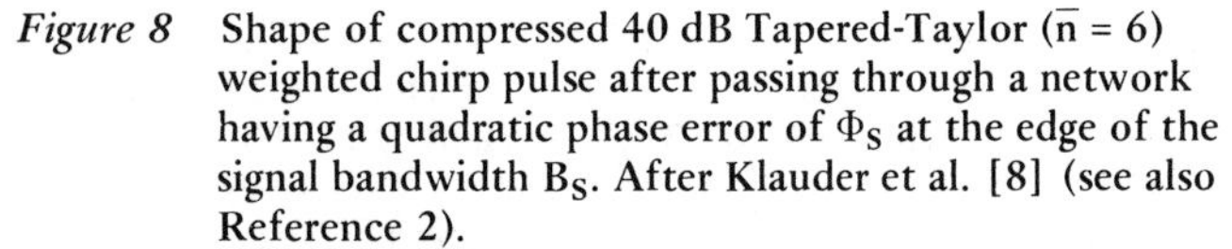

Figure 8 **Shape of compressed 40 dB Tapered-Taylor ($\bar{n}$ = 6) weighted chirp pulse after passing through a network having a quadratic phase error of Φ_S at the edge of the signal bandwidth B_S. After Klauder et al. [8] (see also Reference 2).**

For the unweighted waveform, the distortion of the compressed pulse shape is given in Figure 9 as a function of Φ_S and a_S. Also given in this figure are the ionosphere time-bandwidth products, $T_I B_S$. Note that the pulse width distortion for the unweighted, compressed, chirp waveform is more severe for the same signal bandwidth and ionospheric conditions than it is for the weighted, compressed, chirp waveform; compare Figures 8 and 9.

The results of Figures 7 and 8 also apply to the stepped-frequency waveform approximation of the chirp waveform when a 40 dB Tapered-Taylor or Hamming weighting is used. Correspondingly, the results of Figure 9 apply to the unweighted stepped-frequency waveform approximation of the unweighted chirp waveform. Again, τ_1 and τ_2 represent the compressed pulse 3 dB widths for these cases.

For an RF pulse having a rectangular envelope of duration T and no phase modulation, the pulse distortion due to the ionosphere is as given by Figure 10 [2, 9]. The parameter a_4 in Figure 10 is equal to a_3 of Equation (9b) with B_3 replaced by $B_4 = 1/T$, B_4 being approximately the 4 dB width of the signal spectrum. Also given in Figure 10 is the ionosphere time-bandwidth product in terms of $T_I B_4$. Figure 10 indicates that the sidelobes for a rectangular pulse are degraded more severely due to ionosphere dispersion than they are for the Gaussian pulse or weighted chirp pulses.

Ionospheric Pulse Distortion due to the Presence of Earth's Magnetic Field [6, 11]

Due to the presence of Earth's magnetic field, there exists a difference in the velocities of propagation for right and left circularly

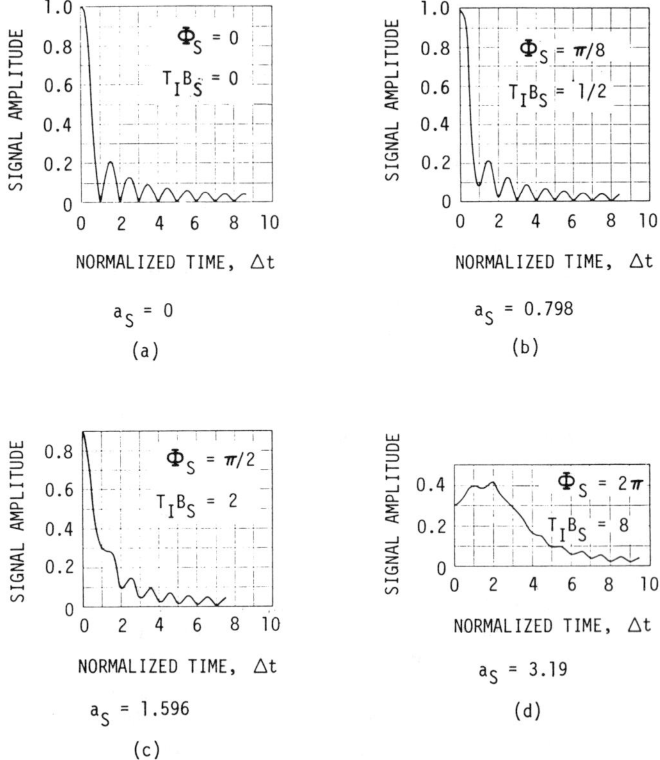

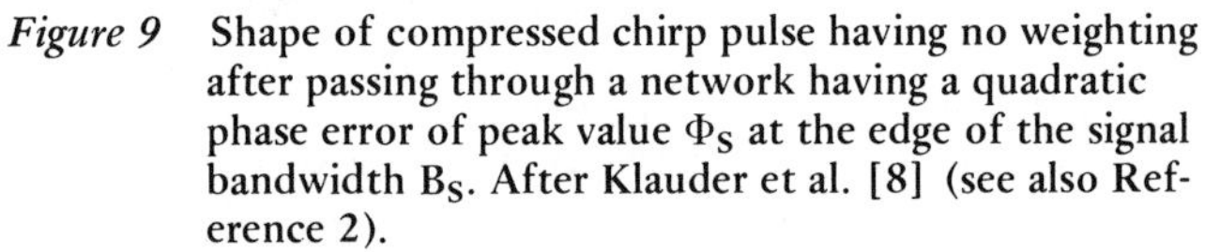

Figure 9 **Shape of compressed chirp pulse having no weighting after passing through a network having a quadratic phase error of peak value Φ_S at the edge of the signal bandwidth B_S. After Klauder et al. [8] (see also Reference 2).**

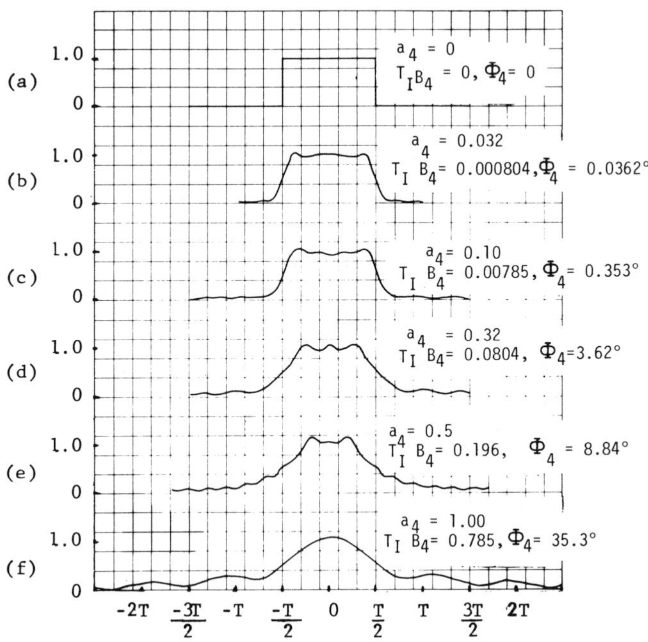

Figure 10 **Shape of rectangular pulse after passing through a network having a quadratic phase error of Φ_4 at the 4 dB bandwidth points which are $\pm 1/2T = \pm B_4/2$ away from the signal carrier frequency. After Knop and Cohn [9] who give corrected versions of Elliott's original curves [10].**

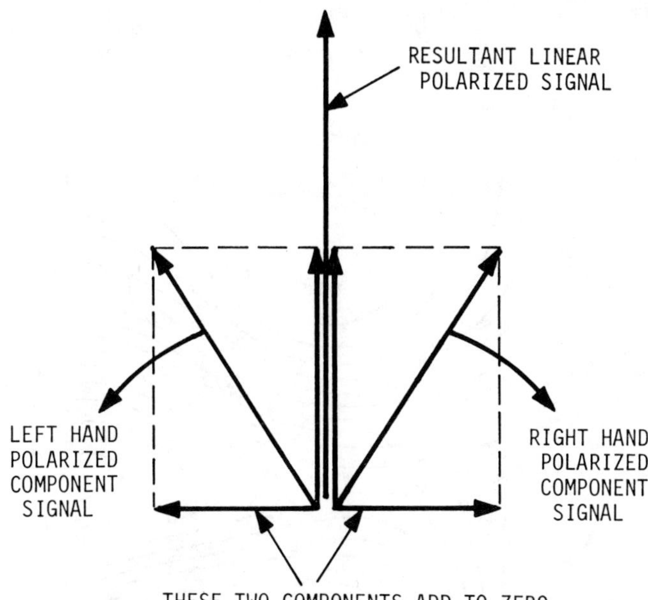

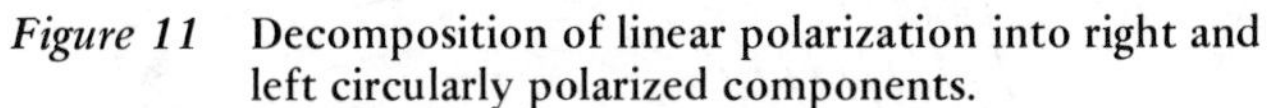

Figure 11 **Decomposition of linear polarization into right and left circularly polarized components.**

polarized signals. A linearly polarized signal is composed of the superposition of a right and left circularly polarized signal (see Figure 11). If a linearly polarized signal is transmitted, its right and left circularly polarized components arrive separated in time by an amount

$$\Delta\tau_M = \frac{162 \times 10^{-18} N_T f_H \cos\theta}{cf^3} \tag{16}$$

where:

θ = is the angle between the direction of Earth's magnetic field and the direction of propagation.

f_H = Earth's gyromagnetic frequency

= 1.68×10^6 Hz at Earth's geomagnetic pole [12-14]

If c, the propagation of the velocity of light in a vacuum, is given in m/s, f_H in Hz, f in GHz, and N_T in el/m^2, $\Delta\tau_M$ is in ns. (An outline of the derivation of Equation (16) is given in the appendix of this chapter.) The right-hand ordinates of Figure 7 give the values of θ for which $\Delta\tau_M = \Delta\tau_I$ for $f_H = 1.68$ MHz. For Example 1, $\Delta\tau_M = \Delta\tau_I$ for $\theta = 43.3°$. For Example 2, $\Delta\tau_M < \Delta\tau_I$ even for $\theta = 0°$.

The condition for which Earth's magnetic field can be neglected is given by

$$\Delta\tau_M \ll \Delta\tau_I \tag{17}$$

It is instructive to see how the θ ordinates were derived for Figure 7. Using Equations (12) and (16), $\Delta\tau_M = \Delta\tau_I$ when

$$\frac{1}{\cos\theta} = 6.58 \times 10^{-17} f_H A \frac{\sqrt{N_T}}{f^{3/2}} \tag{18}$$

where f_H is in Hz, f in GHz, and N_T in el/m^2.

Note that the dependence of $1/\cos\theta$ on N_T and f is the same as that for τ_1 given in Equation (8). Hence, the curves plotted for Equation (8) in Figure 7 can be used to represent $1/\cos\theta$ versus N_T by appropriately labeling the ordinates. Such ordinates are given on the right side of Figure 7 for pulse broadening cases of 10%, 30%, and 100%.

The horizontal scale in the lower right corner of Figure 7 is used to obtain (by interpolation) values of θ for τ_2/τ_1 other than 10%, 30% and 100%, as done earlier in Example 3 for τ_1.

Equation (16) applies when the propagation of the electromagnetic field is not nearly perpendicular to Earth's magnetic field for $f \gg f_H$. This condition is known as quasi-longitudinal propagation. It occurs when

$$\frac{4f^2}{f_H^{\,2}} \gg \sin^2\theta \tan^2\theta \tag{19}$$

The left side of Equation (19) is 50 times or more greater than the right side for f = 100 MHz and $f_H = 1.68$ MHz when $0 \leqslant \theta \leqslant 86.6°$.

Given along the $\tau_2/\tau_1 = 30\%$ θ ordinate is a scale indicating the minimum carrier frequency, f_{QL}, at which the quasi-longitudinal approximation applies for the values of θ on this same ordinate. The frequency, f_{QL}, is obtained from Equation (19) for the restriction that the left side of Equation (19) be 50 times greater than the right side.

Earth's gyromagnetic frequency is proportional to Earth's magnetic field. For the gyromagnetic frequency of $f_H = 1.68$ MHz observed at Earth's magnetic pole, Earth's magnetic field intensity is 0.62 G. Figure 12 gives a plot of Earth's magnetic field intensity as a function of latitude and longitude. A corresponding plot of the magnetic dip angle versus latitude and longitude is given in Figure 13.

For quasi-transverse propagation, the inequality of Equation (19) is reversed and the propagation of the electromagnetic field is nearly perpendicular to Earth's magnetic field. For quasi-transverse propagation, $\Delta\tau_M$ is much smaller. Specifically, using the same units, Equation (16) becomes [6]

$$\Delta\tau_M = \frac{243 \times 10^{-27} N_T f_H^{\,2}}{2\,c\,f^4} \sin^2\theta \tag{16a}$$

Thus, for f = 100 MHz and 1 GHz, $\Delta\tau_M$ is respectively 79.4 and 794 times larger for longitudinal propagation (i.e., $\theta = 0°$) than for transverse propagation (i.e., $\theta = 90°$). For quasi-transverse propagation, the ordinary and extraordinary waves are polarized linearly with the former parallel and the latter perpendicular to the earth's magnetic field [13].

Faraday Rotation

When a linearly polarized signal is transmitted through the ionosphere in the presence of Earth's magnetic field, its direction of polarization rotates. A horizontally polarized signal may become vertically polarized, vice versa, or it may take on some arbitrary linear polarization. The amount of rotation of the linearly polarized signal, θ_R, is related directly (to a first order) to the ionospheric dispersion due to Earth's magnetic field; that is

$$\theta_R = \frac{\pi \Delta\tau_M f}{2} \tag{20}$$

Let $\Delta\tau_M$ equal one over the carrier frequency (for example, let $\Delta\tau_M = 1$ ns for f = 1 GHz); then, $\theta_R = 2\pi/4$ rad or 90° (that is, one quarter of a carrier period). The phase shift of the right polarized signal relative to the left polarized signal is two times θ_R, or 180° for this example.

Using the scales on the right of Figure 7 (or using Equation (16)) together with Equation (20), one can easily obtain the Faraday rotation which results when a linearly polarized signal is propagated through the ionosphere.

Example 4 – Assume a radar located in Florida off the Gulf of Mexico (near the AN/FPS-85) at a latitude of 30°N and

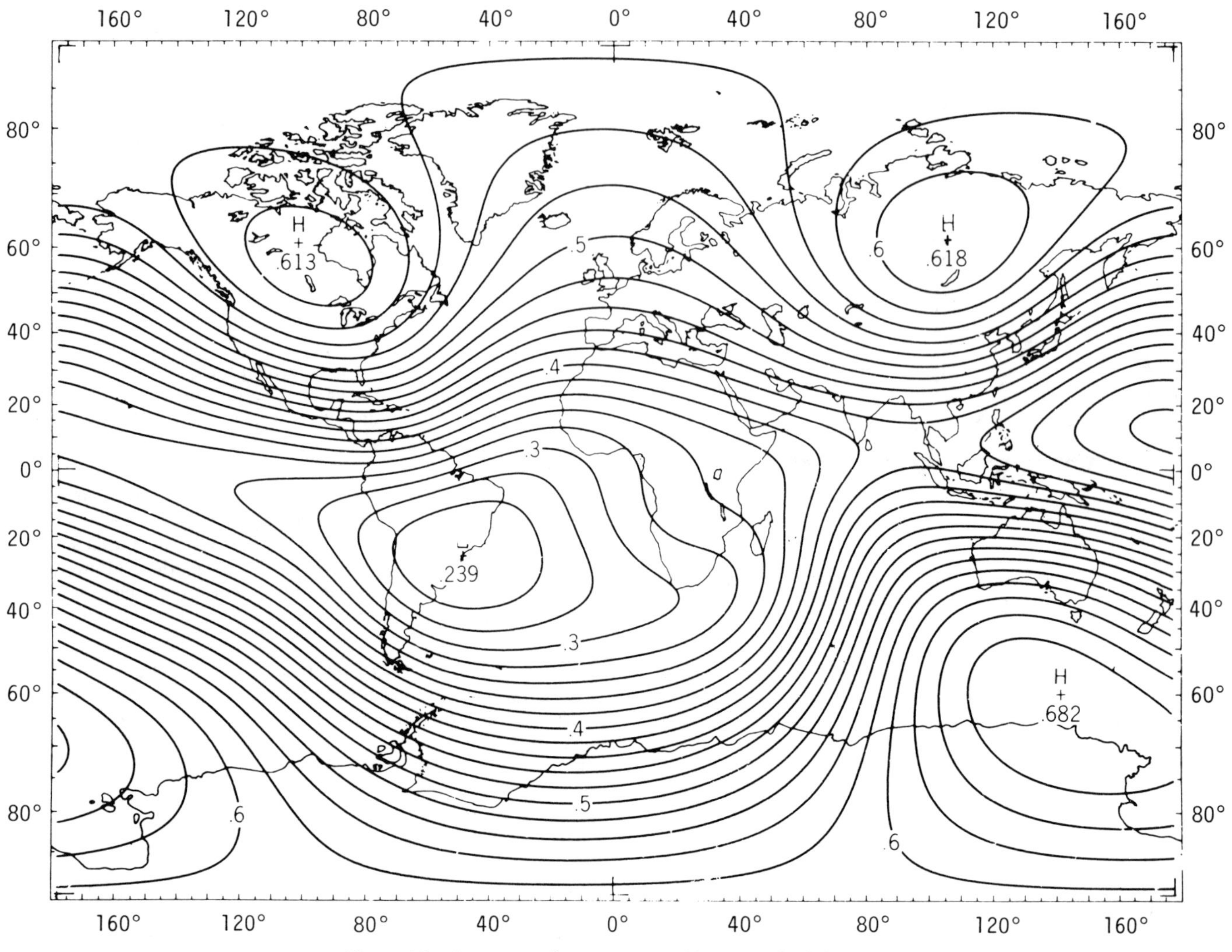

Figure 12 **Contours of constant Earth's magnetic field intensity, F (in Gauss) for epoch 1960. After Cain and Cain [14].**

longitude of 85°W. Assume the radar is looking due south at a target above the ionosphere. Let the radar frequency be 1 GHz (L-band) and assume a linear polarization. Let $N_T = 3 \times 10^{18}$ el/m^2. The radar line of sight pierces the ionosphere peak at 11°N and 85°W, at which point the magnetic field intensity is about 0.4 G and the dip angle is about 40°N (see Figures 12 and 13, respectively). The radar line of sight pierces the peak of the ionosphere with a grazing angle of about 17°. Thus $\theta \doteq 40° - 17° = 23°$ and $f_H = (.4/.62) \times 1.68$ MHz = 1.084 MHz. From Equations (16) and (20), it follows that $\Delta\tau_M = 1.62$ ns and $\theta_R = 145°$.

This example is a rather extreme one. However, it was picked expressly for that reason — it illustrates two things. First, it shows that Faraday rotation can be a problem at L-band for severe ionospheric conditions; second, the dispersion due to Earth's magnetic field must be accounted for if the compensation depicted in Figure 2 is used for a 250 MHz bandwidth L-band system.

Example 5 — The specifications are the same as those above for 100 MHz carrier frequency. Now, $\Delta\tau_M = 1.62$ μs and $\theta = 254$ rad for about 40 rotations. If a circularly polarized array is used, the noncircularity of the radiated field for off-boresite scan angles gives rise to a second pulse 1.62 μs from the first.

Appendix Derivation of Equations Giving Dispersion in the Presidence of Earth's Magnetic Field

The phase velocity, v_P, of an electromagnetic field in a medium having an index of refraction of n is given by [2, 12]

$$v_P = \frac{c}{n} \tag{A1}$$

where, as before, c is the velocity of propagation of light in a vacuum. Due to the presence of Earth's magnetic field, there are two indices of refraction for the ionosphere [12]

$$n_O = 1 - \frac{f_P^2}{2f^2}\left[1 - \frac{f_H}{f}\cos\theta\right] \tag{A2a}$$

$$n_E = 1 - \frac{f_P^2}{2f^2}\left[1 + \frac{f_H}{f}\cos\theta\right] \tag{A2b}$$

for, respectively, left and right circularly polarized signals. The subscripts O and E are used because these two modes of propagation are often referred to as the *ordinary* and *extraordinary* modes of propagation. Equations (A2a) and (A2b) apply when the con-

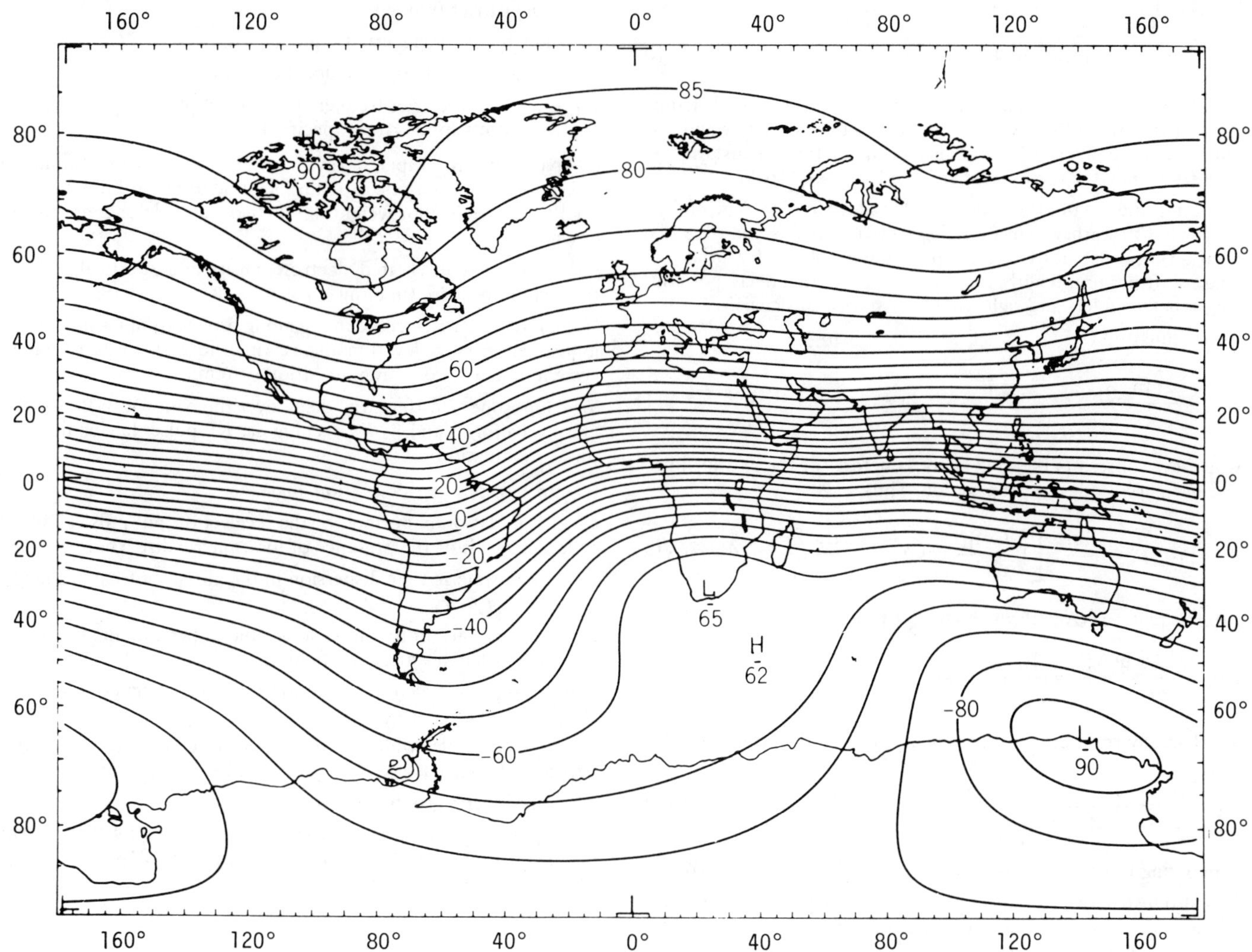

Figure 13 **Contours of constant inclination (dip) of Earth's magnetic field intensity for epoch 1960. After Cain and Cain [14].**

ditions for quasi-longitudinal propagation hold, as given by the inequality of Equation (19).

The group velocity of a dispersive medium is given by [12]

$$v_G = \frac{1}{\frac{1}{2\pi}\frac{dk}{df}} \tag{A3}$$

$$\frac{1}{v_G} = \frac{1}{2\pi}\frac{dk}{df} \tag{A4}$$

where k is the wave number in the medium; that is

$$k = \frac{2\pi}{\lambda} = \frac{2\pi fn}{c} = \text{phase shift per unit length} = \beta \tag{A5}$$

Substituting Equation (A5) in Equation (A4) yields

$$\frac{1}{v_G} = \frac{1}{c}\frac{d(fn)}{df} = \frac{1}{c}\left(n + f\frac{dn}{df}\right) \tag{A6}$$

The group delay of the medium τ_G, is given by

$$\tau_G = \int \frac{1}{v_G}\, ds \tag{A7}$$

The dispersion of the ionosphere due to Earth's magnetic field is given by

$$\Delta\tau_M = \tau_{G,E} - \tau_{G,O} \tag{A8}$$

where $\tau_{G,E}$ and $\tau_{G,O}$ are τ_G for the extraordinary and ordinary modes. Combining Equations (A2a), (A2b), and (A6) through (A8) yields Equation (16) in the text. (The path through the ionosphere over which Equation (A7) is integrated is different for the extraordinary and ordinary modes. However, these two paths can be treated as being the same when $f \gg f_P$, even when very low elevation angles are used.)

The result given by Equation (20) is contrary to what would be expected if the ionosphere differential delay, $\Delta\tau_M$, were due to a delay line having a fixed delay, T_D, independent of frequency. For this latter case, the differential phase shift between the left and right circularly polarized signals is

$$\phi = 2\pi f T_D \tag{A9}$$

In turn, the polarization rotation would be given by

$$\text{Polarization Rotation} = \pi f T_D \quad \text{(A10)}$$

The discrepancies between Equations (20) and (A10) result from the fact that the differential delay, $\Delta\tau_M$, between the left and right circularly polarized components is not actually constant with frequency for the ionosphere (see Equation (16)).

To better understand the reason the Faraday rotation is given by Equation (20) rather than Equation (A10) (which is not a Faraday rotation), it is useful to derive expressions for the differential phase shift per unit length, $\Delta\beta$, and the differential delay per unit length between left and right circularly polarized signals.

From Equation (A5)

$$\beta_{E,O} = \frac{2\pi f n_{E,O}}{c} \quad \text{(A11)}$$

where the subscripts E and O apply as before for the extraordinary and ordinary modes of propagation. Thus

$$\Delta\beta = \beta_O - \beta_E = \frac{2\pi f}{c}\left[2\,\frac{f_P{}^2 f_H}{2f^3}\cos\theta\right] = \frac{2\pi f_P{}^2 f_H}{cf^2}\cos\theta \quad \text{(A12)}$$

From Equations (A4) through (A8), it follows that

$$\Delta\tau_M = \frac{1}{2\pi}\,\frac{d(\Delta\beta)}{df} = \frac{2f_P{}^2 f_H}{cf^3}\cos\theta \quad \text{(A13)}$$

The Faraday rotation per unit propagation length is thus given by

$$\theta_R = \frac{\Delta\beta}{2} = \frac{\pi f_P{}^2 f_H}{cf^2}\cos\theta \quad \text{(A14)}$$

Comparing Equations (A14) and (A13) yields Equation (20).

It is also instructive to indicate how the Faraday rotation for the whole path given by Equation (20) can be derived directly from Equation (A11). From Equation (A11), it follows that

$$\phi_{E,O} = \int \beta_{E,O}\,ds \quad \text{(A15)}$$

The Faraday rotation of a linearly polarized signal over this path is in turn given by

$$\theta_R = \frac{\phi_E - \phi_O}{2} \quad \text{(A16)}$$

Combining Equations (A2), (A11), and (A15), and using Equations (2), (6), and (16), yields Equation (20).

Glossary

A — distortion parameter used in Equation (12) and defined in Equation (13).

a_3, a_4, a_S — distortion parameter for B = B_3, B_4, and B_S in Equation (9b).

B — signal bandwidth

B_3 — 3 dB signal voltage spectrum bandwidth.

B_4 — ~4 dB signal voltage spectrum bandwidth for rectangular RF pulse having no phase modulation

— 1/T

B_S — swept bandwidth for chirp signal

— $1.25/\tau_1$, for 40 dB Tapered-Taylor weighted chirp pulse (equals the 24 dB voltage spectrum bandwidth for unchirped and chirped Gaussian pulses)

f — signal carrier frequency

f_H — Earth's gyromagnetic frequency

— 1.68 MHz at Earth's geomagnetic pole

f_{QL} — minimum carrier frequency at which quasi-longitudinal approximation applies for a specified θ

$\bar{n}$ — parameter for Tapered-Taylor weighting. $\bar{n} - 1$ equals number of sidelobes on each side of main lobe that are 40 dB down for 40 dB Tapered-Taylor weighting; further out, sidelobes are further down. [7]

N_{TV} — one-way integrated electron density along vertical path through ionosphere (in el/m^2).

N_T — integrated electron density along propagation path through ionosphere having elevation angle α (in el/m^2); two-way for radar system, one-way for communication system.

T_I — ionospheric differential delay over the bandwidth B_S for chirp waveforms and B_4 for rectangular RF pulses having no phase modulation.

α — propagation path elevation angle.

α_P — degradation in decibels of the receiver output pulse amplitude as a result of the presence of ionospheric dispersion

$\Delta\tau_I$ — ionospheric time dispersion in the absence of Earth's magnetic field. $\tau_2 - \tau_1$

$\Delta\tau_M$ — contribution to ionospheric time dispersion due to the presence of Earth's magnetic field.

θ — the angle between the direction of Earth's magnetic field and the direction of propagation.

τ_1, τ_2 — 3 dB pulse widths (compressed for chirped waveforms) before and after passage through ionosphere, respectively.

τ_I — ionospheric group delay

Φ_3, Φ_4, Φ_S — phase shift due to ionosphere (or equivalent network having quadratic phase errors) at frequencies $\pm B_3/2$, $\pm B_4/2$, and $\pm B_S/2$ away from the carrier.

References

[1] Brookner, E.: "Ionospheric Dispersion of Electromagnetic Pulses," *IEEE Transactions on Antennas and Propagation, Vol. AP-21, No. 3,* pp. 402-405, May 1973.

[2] Brookner, E.: "Effect of Ionosphere on Radar Waveforms," *Journal of the Franklin Institute, Vol. 280,* pp. 1-22, July 1965.

[3] Lawrence, R.S.; and Posakony, D.J.: "The Total Electron Content of the Ionosphere at Middle Latitudes Near the Peak of the Solar Cycle," *Journal of Geophysical Research, Vol. 68, No. 7,* pp. 1889-1898, April 1963.

[4] Flattery, W.; Allane, A.C.; and Ramsay, C.: "Derivation of Total Electron Content for Real Time Global Applications," *Air Force Global Weather Central, Naval Research Laboratory Symposium, "Effect of Ionosphere on Space Systems and Communications,"* January 1975.

[5] Odom, D.B.; Jones, S.; Hollway, L.; Sforza, P.; Gotto, M.; Tsonis, C.; Press, S.; McDermott, J.; Boak, T.; and Habert, R.: "Effects Simulation Program," *Final Report, Phase II* (AF/ARPA Sponsorship), Raytheon Company, Sudbury, Massachusetts, 01776, June 1970.

[6] Brookner, E.: "Ionospheric Pulse Time Dispersion Including Effects of Earth's Magnetic Field," *IEEE Transactions on Antennas and Propagation, Vol. AP-26,* pp. 307-311, March, 1978.

[7] Brookner, E.: "Antenna Array Fundamentals, Part 1," Lecture No. 2 of *Microwave Journal* Intensive Course Notes, Practical Phased-Array Systems, (E. Brookner, ed.), 1975.

[8] Klauder, J.R.; Price, A.C.; Darlington, S.; and Albersheim, W.J.: "The Theory and Design of Chirp Radars," *Bell System Technical Journal, Vol. 39,* pp. 745-808, July 1960.

[9] Knop, C.M.; and Cohn, G.I.: Comments on "Pulse Waveform Degradation Due to Dispersion in Waveguide," *IEEE Transactions on Microwave Theory and Techniques, Vol. MTT-11,* pp. 445-447, September 1963.

[10] Elliott, R.S.: "Pulse Waveform Degradation Due to Dispersion in Waveguide," *IRE Transactions on Microwave Theory and Techniques, Vol. MTT-5,* pp. 254-257, October 1957.

[11] Brookner, E.: "Ionospheric Time Dispersion Due to Earth's Magnetic Field," *Joint USNC/URSI and PGAP Meeting,* Georgia Institute of Technology, Atlanta, Georgia, June 1974.

[12] Davies, K.: "Ionospheric Radio Propagation," *National Bureau of Standards Monograph 80,* 1965.

[13] Millman, G.H.: "Atmospheric Effect on Radio Wave Propagation," *Modern Radar Analysis, Evaluation, and System Design* (R.S. Berkowitz, ed.), New York, John Wiley & Sons, p. 357, 1965.

[14] Cain, J.C.; and Cain, S.J.: "Derivation of the International Geomagnetic Reference Field [IGRF (10/68)]," *NASA Technical Note NASA TN D-6237,* August 1971.

[15] Klobuchar, et. al.: "Total Electron Content Studies of the Ionosphere," *AFCRL-TR-73-0098,* Hanscom AFB, Massachusetts, 1 February 1973.

[16] Klobuchar, J.A.: "A First-Order, Worldwide, Ionospheric, Time-Delay Algorithm," *AFCRL-TR-75-0502,* Hanscom AFB, Massachusetts, 25 September 1975.

[17] Klobuchar, J.A.; and Allen, R.S.: "Maximum Ionospheric Range Errors for Air Defense Command Radars," *Air Force Geophysics Laboratory Report AFGL-TR-76-0042,* Hanscom AFB, Massachusetts, 2 March 1976.

Chapter 15 Brookner

Effects of the Atmosphere on Laser Radars

Introduction

The degrading effects of the atmosphere are much more pronounced at laser wavelengths than at microwave wavelengths. Many of these effects are familiar to us because our eyes are sensitive to laser wavelengths. When you look at a star, you see it twinkle; a laser signal twinkles similarly. The severe attenuation of optical wavelengths by fog is all too well known. The twinkle phenomenon is a clear weather effect while fog attenuation is an inclement weather effect. In the following discussion clear weather effects are followed by those in inclement weather.

Clear Weather Propagation

Twinkle Phenomenon

Star twinkle or, equivalently, star signal amplitude fluctuations, result from the atmosphere not being uniform, that is, from the atmosphere being inhomogeneous [1]. The atmosphere can be thought of as being composed of many blobs (see Figure 1). Each of these blobs has a different index of refraction. Because of refraction and diffraction by these blobs, the rays of light follow different paths, as indicated in the figure. Upon reaching the receiver, some of these rays add constructively while others add destructively, resulting in the type of blotchy interference pattern shown in the insert of Figure 1 for the one-way path depicted. If the atmosphere were homogeneous, the signal would have had a uniform intensity over the receiver aperture (except for the usual antenna beam pattern variation). Figure 2 shows a blow-up of the interference pattern observed at the receiver aperture. The intensity across the aperture diameter is shown in Figure 3.

Let us consider the signal amplitude seen with a small receiver aperture. The spatial correlation distance for the amplitude fluctuations over the receiver aperture of Figure 2 is about 1 cm, a value in close agreement with the value of 0.7 cm obtained from theory [3, 4] if medium turbulence conditions are assumed (the actual

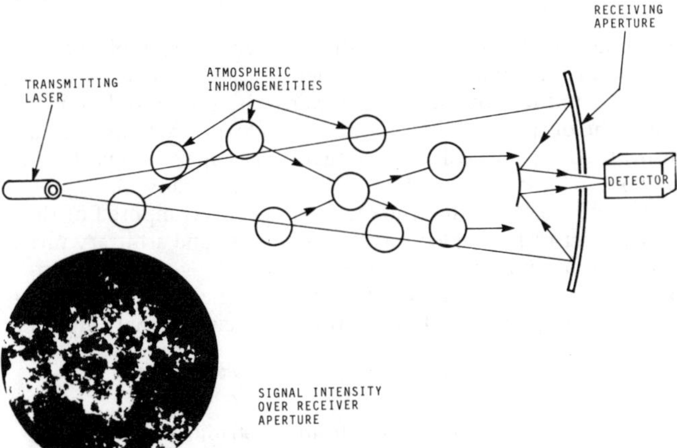

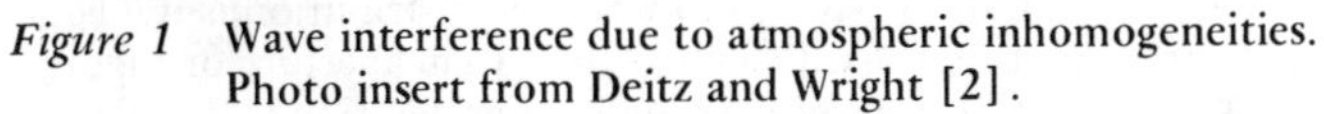

Figure 1 **Wave interference due to atmospheric inhomogeneities. Photo insert from Deitz and Wright [2].**

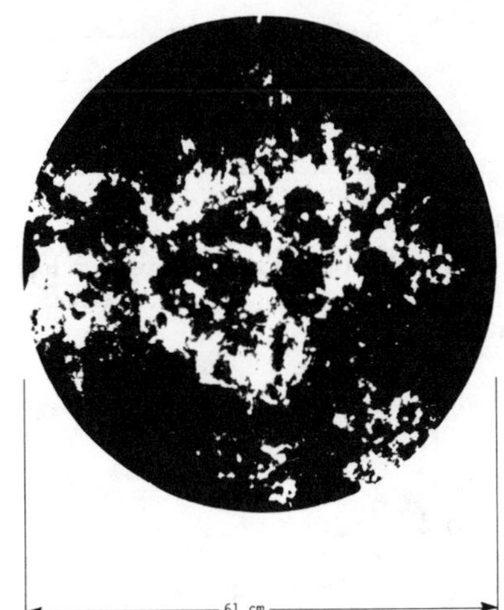

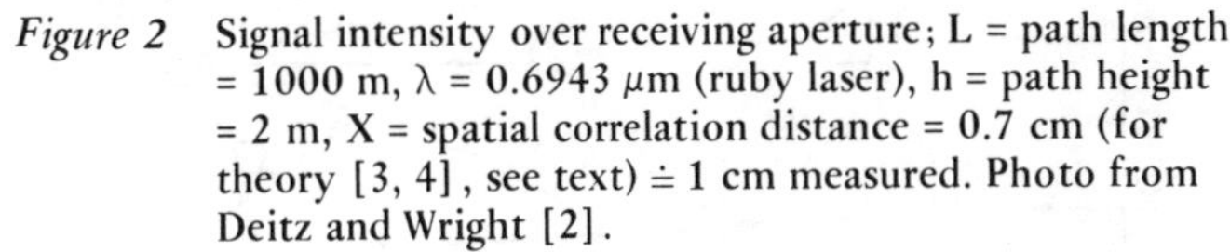

Figure 2 **Signal intensity over receiving aperture; L = path length = 1000 m, λ = 0.6943 μm (ruby laser), h = path height = 2 m, X = spatial correlation distance = 0.7 cm (for theory [3, 4], see text) ≐ 1 cm measured. Photo from Deitz and Wright [2].**

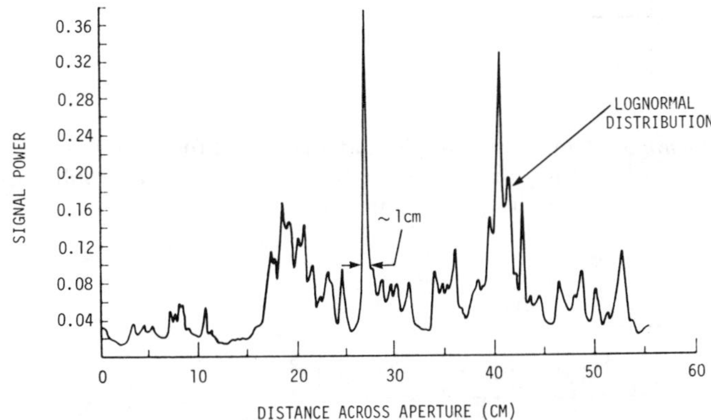

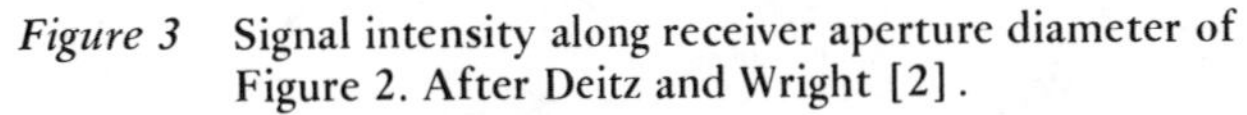

Figure 3 **Signal intensity along receiver aperture diameter of Figure 2. After Deitz and Wright [2].**

turbulence conditions are not given in Reference 2). Consider a receiver aperture which is small compared to the spatial correlation distance of 1 cm, an aperture the size of the eye pupil. Also assume that the atmospheric inhomogeneities are "frozen," that is, that the inhomogeneities do not vary with time. However, assume that there is a wind moving the whole frozen atmospheric inhomogeneities in a direction perpendicular to the line-of-sight between the laser transmitter and the receiver. Assume that the wind is going from right to left when looking at the pattern of Figure 2. As a result, the interference pattern of the figure moves

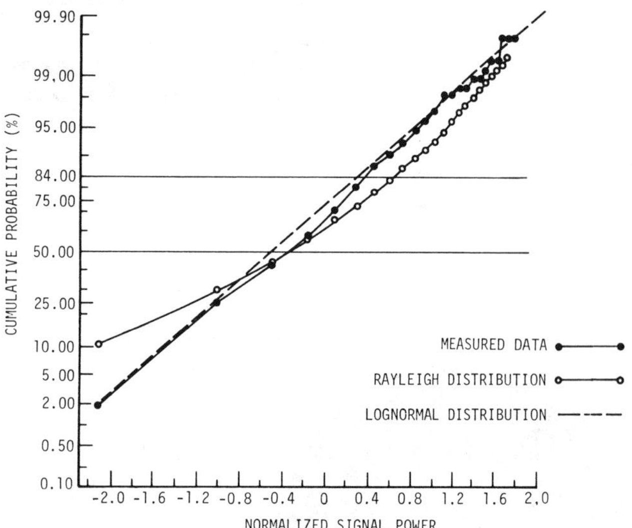

Figure 4 Cumulative probability of received signal power on log-normal paper. Obtained for signal of Figure 3. After Deitz and Wright [2].

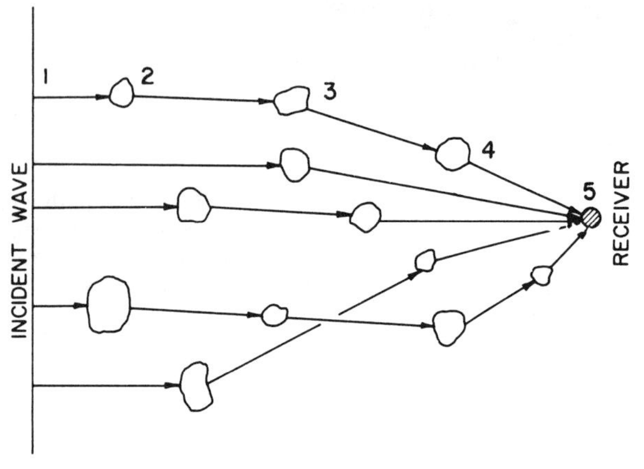

Figure 5 Multipath multiple-scattering model for propagation through inhomogeneous atmosphere. After Strohbehn, Wang, and Speck [6].

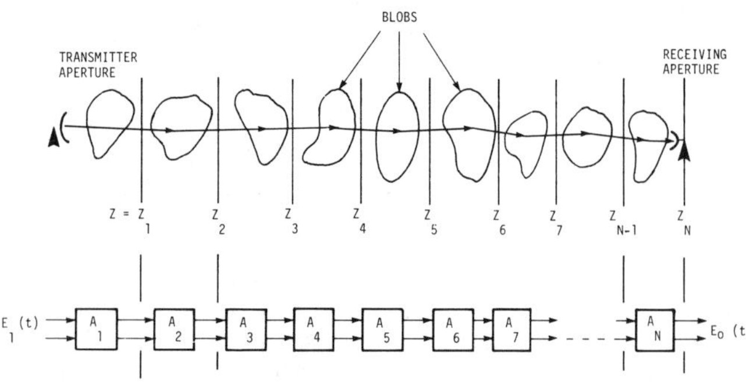

Figure 6 Single path of Figure 5 as cascade of networks having random gain.

from the right to left at the same speed as the wind velocity. If the small receiver aperture were placed at the center of the large aperture shown in Figure 2, the signal seen at this aperture is that shown in Figure 3 with the distance scale replaced by a time scale determined by the wind velocity. Assume a cross wind velocity of 1 nmi/hr (= 0.51 m/s). Then 10 cm becomes about 0.2 s, etc.. The time for independence then is equal to 1 cm/(0.51 m/s) = 19.6 ms with a resultant maximum rate of fluctuation of the signal amplitude of about 50 Hz. The eye would see the lower frequency components of the signal fluctuations. This amplitude scintillation is the same as that of star twinkle. (Star twinkle arises predominantly from an atmospheric inhomogeneous layer at an altitude of about 10 km.)

Amplitude Distribution of Twinkle

Much debate and controversy exists as to the probability density function for the twinkle power fluctuations. Some theoretical and experimental investigators claim a log-normal distribution, others a Rayleigh distribution, still others a Rician distribution, and one a Gaussian distribution [1, 4, 5]. Measurements tend to indicate a log-normal type of distribution or at least one very close to log-normal (see Figure 4). Recently it has been proved by Strohbehn and Wang [6] (see also Reference 5) that it is impossible for the amplitude distribution to be exactly log-normal.

That the distribution is not log-normal can be argued from physical factors. The atmosphere consists of a parallel combination of paths such as those of Figure 5 with each such path being perturbed by its own set of atmospheric inhomogeneities. For the moment, let us consider the amplitude distribution of the signal received after having gone along a path such as that of 1-2-3-4-5, as depicted in Figure 6.

For that path the signal passes successively through many independent atmospheric blobs. The situation is analogous to having the signal pass through a cascade of N networks, as illustrated in the bottom half of Figure 6. Each network has a gain which is random and independent of the gains of the other networks, these gains depending on the index of refraction and size of the atmospheric blob with which they are associated. Consequently, the signal at the output of this cascade of networks is the product of the random gains, A_1, i = 1, 2, . . . , N, of the cascade of networks

$$E_0(t) = A_1 A_2 A_3 \ldots A_N E_1(t) = \prod_{i=1}^{N} A_i E_1(t) \qquad (1)$$

Taking the log of both sides of this equation gives

$$\ln E_0(t) = \sum_{i=1}^{N} \ln A_i + \ln E_1(t)$$

which indicates that the log of the received signal amplitude is the sum of a large number of independent variates (which are the logs of the amplitudes of the network gains). By the central limit theorem, the output amplitude of the cascade of networks should be distributed log-normally. The actual laser propagation path consists of a parallel combination of paths such as those of Figure 6 (see Figure 5). Thus the true received signal is composed of the sum of vectors having log-normal amplitudes and arbitrary phases. Such a sum is not expected to be distributed log-normally. Strohbehn, Wang, and Speck [6] have suggested a combination of log-normal and Rician distributions for the received signal amplitude.

Distribution of Amplitude Fluctuations for Large Receiver Aperture

In the previous section, the amplitude probability density function for the signal fluctuations observed by a small aperture is discussed. The question arises as to what the distribution is if a large collecting aperture is used (such as the 61 cm aperture of Figure 2). The type of optical receiver under consideration is that depicted in Figure 1. The output of this receiver is only proportional to the total energy incident over the collecting aperture. The receiver output energy is not dependent on the phases of the signals impinging on the aperture. This type of receiver is similar to the microwave frequency receiver referred to as the video radiometer receiver. The total received signal power is the sum of the signal energies received over the different parts of the aperture of Figures 1 and 2. From the central limit theorem, one would expect that the signal power fluctuations would approach a Gaussian distribution; however, the signal power fluctuations obtained using large receiving apertures is observed to be closer to log-normal than Gaussian distribution. At first this finding was surprising. However, Mitchell [7] showed that the sum of many independent log-normal variates is closer to a log-normal variant than it is to a Gaussian variate for the conditions of laser propagation.

We now show that because the signal intensity over a small aperture is nearly log-normally distributed, it is reasonable to expect the sum of such signals to be (or at least appear to be) log-normally distributed. Recall that

$$\ln(1 + x) = x \text{ for } x \ll 1 \tag{3}$$

where ln x is the natural log of x. Consequently the log-normal distribution is not much different from the normal distribution when $\sigma/\mu \ll 1$, where σ is the rms of the random variable x and μ is its mean. When a large number of identically distributed independent variates are summed to form a random variable x, $\sigma/\mu \ll 1$. From the central limit theorem, x then can be expected to be approximately Gaussianly distributed and, in turn, approximately log-normally distributed from Equation (3); however, it is closer to a log-normal distribution, as Mitchell's work and the measurements indicate.

Amplitude Distribution of Radar Echo

So far, the discussion has been for the one-way propagation that arises for a communication system. The reason for this limitation is that practically all of the laser signal propagation work in the literature is for this case. For the radar application, the one-way propagation results have to be extrapolated to the two-way radar situation.

If the target is very small in size (smaller than the transverse spatial correlation distance for the amplitude fluctuations), the signal at the target just after reflection has the combination log-normal-Rician distribution mentioned earlier. (This distribution is close to that of a log-normal distribution for the laser propagation situations of interest.) This reflected signal goes through the same atmosphere again and hence is attenuated the same amount as on the outgoing path (reciprocity theorem [55]). Consequently, the radar echo for a small target has the log-normal-Rician distribution. If the target dimensions are much larger than the spatial decorrelation distance, the situation becomes considerably more complicated. The received signal now is composed of the sum of vectors having log-normal-Rician distributed magnitudes and arbitrary relative phases. The distribution for such a signal is not known. However, when the target and receiver are large so that a great number of such vectors are summed, from the central limit theorem the echo signal at the receiver is approximately Gaussianly distributed.

Saturation Phenomena

The interference pattern given in Figures 1 and 2 was obtained using a fairly long propagation path, a path of 1 km. If the path were considerably shorter, the magnitudes of the fluctuations would not be as pronounced. Figure 7 shows the variance of the log of the received signal power fluctuations versus the path length. This figure indicates that the variance of the log of the received signal power fluctuations initially increases linearly with increasing path length as expected. However, at some path length it

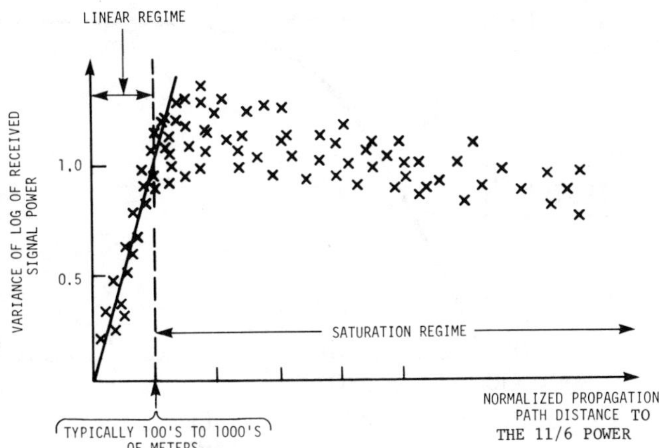

Figure 7 Saturation phenomenon. After deWolf [8].

reaches a peak value after which it decreases a small amount before it eventually levels out to a constant plateau. The region where the variance of the log of the receiver signal fluctuation increases linearly with distance is called the *linear regime*, whereas the region where the variance peaks and flattens out is called the *saturation regime.* This saturation effect first had been noticed experimentally, then explained theoretically although it is as yet not completely understood [1, 4].

Coherent Receiver

For microwave radars the signals received over the various parts of the receiving aperture are added in phase to form a coherent receiving system. The same can be done at laser frequencies. Such a receiver system is illustrated in Figure 8 and is called a *coherent heterodyne receiver* (see Chapter 24 by Jelalian). For this system the signals received over the aperture are added vectorily with their phases being taken into account. If the signals illuminating the different parts of the aperture were in phase, coherent addition would be realized if the aperture were diffraction limited. However, due to atmospheric inhomogeneities the signals over different parts of the aperture are not in phase when a large receiver aperture, such as that of Figure 2, is used, even if the aperture is diffraction limited. For the case of Figure 2, the signal at two points on the aperture separated by 0.7 cm are on the average 1 rad (57.3°) out of phase [1].

The receiver output signal-to-noise ratio versus receiver aperture size is depicted in Figure 9. The receiver signal-to-noise ratio initially increases linearly and then levels off to a constant value even though the aperture is increasing in size and is diffraction limited. In Figure 9, r_0 is the aperture diameter for which the signal-to-noise ratio is 3 dB below that obtained if no atmospheric degradation is present [1, 4, 9], it has a value of about 1.6 cm for the ruby laser propagation path of Figure 2 with medium turbulence conditions.

It is this lack of spatial coherence over the receiver aperture that limits the resolution capability of a star telescope. The parameter r_0 represents the maximum size telescope one can have and still

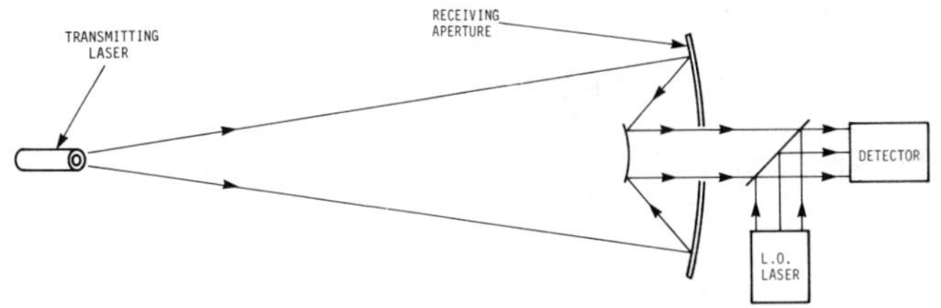

Figure 8 Coherent heterodyne receiver system.

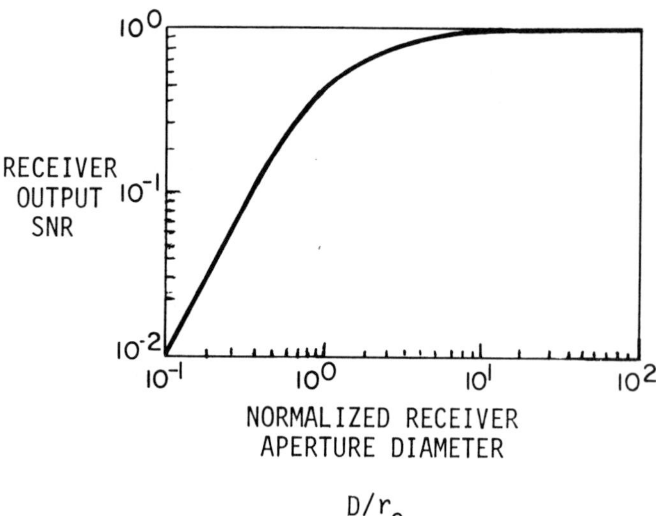

Figure 9 Normalized receiver output power signal-to-noise ratio (SNR) versus normalized heterodyne receiver aperture diameter, D/r_0. After Fried [9]. Example – 1) For L = 1 km and λ = 10.6 μm, r_0 = 0.3 m; 2) For L = 100 km and λ = 0.03 m (10 GHz), r_0 = 300 m [1].

realize its diffraction limited capabilities. The parameter r_0 has a value on the order of 4.5 in at the Ctio Tololo Inter American Observatory in Chile at an altitude of 7,000 ft during good seeing conditions. In the Boston region, it is about 2 in during the same conditions. Hence, if you buy a 6 in diffraction limited Criterion telescope, it is better than the diffraction limited conditions available during good seeing in the Boston region. The larger collecting aperture area does, however, have the advantage of providing a greater sensitivity so that smaller stars can be seen.

Just as it is possible to correct for the incoherence over the receiver aperture observed at the microwave frequencies, it is possible to do the same at laser frequencies. One such possible way was suggested by Brookner [10]. It involves the transmission of the LO signal along with the transmitted signal. As a result, the LO has the same phase distortion across its wavefront as the signal. Amplifying this LO wavefront, narrow-band filtering it, and mixing it together with the amplified signal wavefront results in mixed signals over the mixer receiver aperture that are in-phase and, as a result, add up in-phase. A block diagram for this type of system is depicted in Figure 10. (An alternate approach was suggested by Kompfner [11].

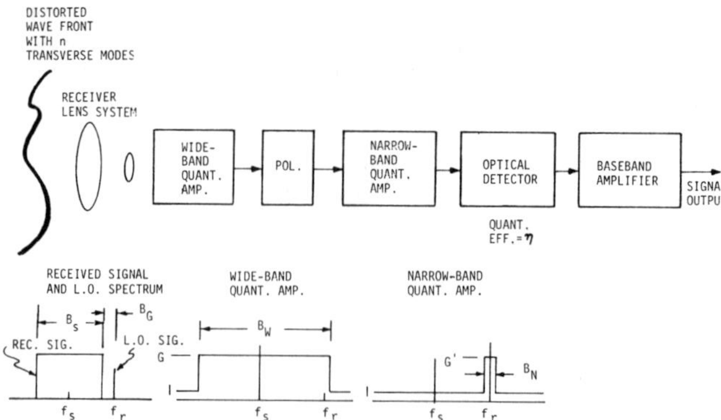

Figure 10 Technique for eliminating receiver output SNR degradation due to atmospheric-inhomogeneity induced spatial incoherence. B_W = bandwidth of wideband quantum amplifier $> B_S + B_G$ where B_G is separation between information signal band and LO signal transmitted along path and B_S is information signal bandwidth. G and G′ are gains of amplifiers. From Brookner [10].

Model for Atmospheric Turbulence

Previously we described the atmosphere as being composed of blobs having different indices of refraction (see Figure 1). Although the blobs shown in Figure 1 are depicted as having the same size, they actually have different sizes. This spherical blob model is just a model of convenience. A more precise model is now developed.

The atmosphere has an index of refraction which continually varies in a random manner in all three directions. Let the index of refraction at an instant of time be given as a function of the position coordinates x, y, and z by

$$n(x, y, z) = n_0(x, y, z) + \Delta n(x, y, z) \tag{4}$$

For simplicity, drop the position variables to obtain

$$n = n_0 + \Delta n \tag{5}$$

where:

n_0 = average atmospheric index of refraction

Δn = deviation of index of refraction n from average value n_0 at location x, y, z.

The severity of the atmospheric inhomogeneities is described by the mean square value of Δn given by [12, 13]

$$< \Delta n^2 > = \frac{1}{2} C_N^2 L_0^{2/3} \tag{6}$$

where:

C_N = index-of-refraction structure constant. (It is this quantity that is tabulated in the literature to give an indication of the strength of the atmospheric irregularities or, equivalently, of the variance of the index of refraction.)

L_0 = the size of the largest inhomogeneity for which the spatial blob size model to be described shortly holds; also called *outer scale of turbulence.*

The parameter C_N^2 and, in turn, $< \Delta n^2 >$ vary as a function of time of day. A typical diurnal 13 hour variation for C_N is given in Figure 11 [1, 14]. The variation over a 24 hour period can be obtained by assuming mirror symmetry of C_N with time about the midday hour.

The outer scale of turbulence, L_0, varies with altitude, h, according to [1]

$$L_0 = \sqrt{4h} \tag{7}$$

where h and L_0 are in meters. For h = 2 m and $C_N^2 = 3 \times 10^{-14} m^{-2/3}$ (medium turbulence, the condition assumed for the experimental results of Figure 2), L_0 = 2.8 m from Equation (7) and $< \Delta n^2 > = 3 \times 10^{-14}$ from Equation (6). The diurnal variation for C_N^2 given in Figure 11 applies for h ≐ 30 m. The index-of-refraction structure constant also varies with altitude (see References 1 and 4, and Figure 27 in Chapter 24).

As indicated, at any instant, the atmospheric turbulence represents a three-dimensional spatially-varying random process. For atmospheric turbulence, as done previously, the time variations often simply are assumed to result from a translational motion of the frozen three-dimensional atmospheric homogeneities. This

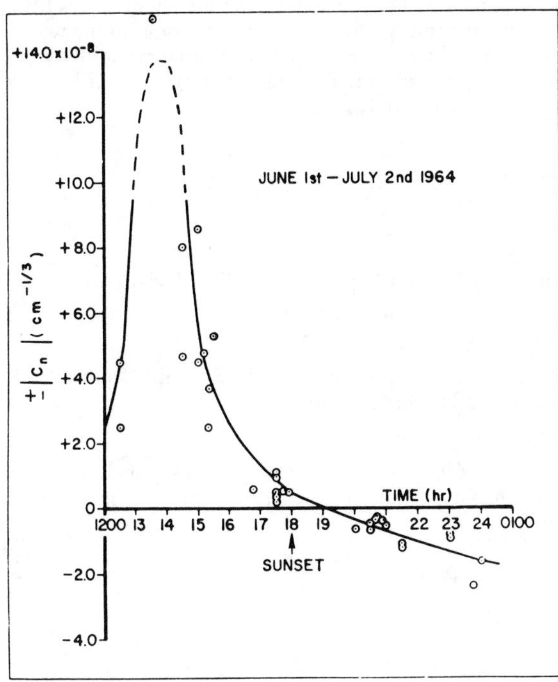

Figure 11 **Diurnal variation of C_N. From Goldstein, Miles, and Chabot [14].**

three-dimensional spatially-varying process is in contrast to the more common one-dimensional random noise process which varies only with time. Usually for a one-dimensional noise process a Fourier transform of the noise voltage is obtained in order to derive the noise power-spectral density of the process. A similar procedure is followed for the three-dimensionally spatially-varying atmospheric turbulence. Now, however, a three-dimensional Fourier transform is obtained. Also, instead of radiating frequency ω, the three-dimensional Fourier transform yields the wave-number K for the three-dimensional power-spectral-density independent variable. Whereas $\omega = 2\pi/\tau$ where τ is the period of the sinusoidal components that compose the noise, $K = 2\pi/\ell$ where ℓ is the spatial period of the sinusoidal components that compose the atmospheric turbulence. The three-dimensional power spectral density of the atmospheric spatial index-of-refraction variations gives the energy in the Fourier component of the atmospheric turbulence having wave number $K = 2\pi/\ell$ or, equivalently, of the turbulence having a sinusoidal period of ℓ. The atmospheric blobs that were spoken of in Figure 1 (where all the blobs were assumed to have the same size ℓ) actually can be interpreted as representing the atmospheric inhomogeneities having a period ℓ and wave number $K = 2\pi/\ell$.

Assume that the character of the turbulence variations is the same in all directions (that is, the turbulence is isotropic [1]); the power spectrum of the atmospheric turbulence in all directions then is given by [4, 13]

$$\Phi(K) \doteq 0.033\, C_N^2 K^{-11/3} \text{ for } \frac{2\pi}{L_0} < K < \frac{2\pi}{\ell_0} \tag{8}$$

where ℓ_0 is the inner scale of the turbulence or, equivalently, the size of the smallest atmospheric inhomogeneities, i.e., "blobs." Empirically [1]

$$\ell_0 = (10^{-9}h)^{1/3} \text{ for } \ell_0 \gtrsim 2\text{mm} \tag{8a}$$

For h = 30m, $\ell_0 \doteq 3$mm.

The above three-dimensional spectrum represents the widely accepted Kolmogorov-Obukhov [1, 4] model for atmospheric turbulence. Designate $\langle \Delta n^2(\ell) \rangle$ as the amount of energy in the inhomogeneities of size ℓ having spectral wave number K. Then, from Equation (8) [15]

$$\langle \Delta n^2(\ell) \rangle = 4\pi K^2 \Phi(K) dK = 4\pi(0.033) C_N^2 K^{-5/3}\, dK$$

for

$$\frac{2\pi}{L_0} < K < \frac{2\pi}{\ell_0} \tag{9}$$

Expressing $\langle \Delta n^2(\ell) \rangle$ in terms of the "blobs" of size ℓ yields

$$\langle \Delta n^2(\ell) \rangle = 4\pi(0.033) C_N^2 \left[\frac{\ell}{2\pi}\right]^{5/3} \frac{2\pi}{\ell^2}\, d\ell$$

for

$$\frac{2\pi}{L_0} < K < \frac{2\pi}{\ell_0} \tag{10}$$

or

$$\langle \Delta n^2(\ell) \rangle = (0.066) \left[\frac{2\pi}{\ell}\right]^{1/3} C_N^2\, d\ell$$

for

$$\frac{2\pi}{L_0} < K < \frac{2\pi}{\ell_0} \tag{11}$$

Using Equation (11) we can derive the received signal carrier phase fluctuations resulting from atmospheric inhomogeneities of size ℓ moving transverse to the line-of-sight. The phase shift of a received signal after it has been propagated a distance L through a medium having constant index of refraction, n_0, is given by

$$\Phi_L = \frac{2\pi}{\lambda} n_0 L \tag{12}$$

where λ is the laser radar wavelength in a vacuum (i.e., for n = 1). The perturbation in the phase shift, ϕ_L, due to an atmospheric "blob" of size ℓ having an index of refraction $n = n_0 + \Delta n(\ell)$ instead of n_0 is, from Equation (12)

$$\Delta\phi_\ell(\ell) = \frac{2\pi}{\lambda} \ell \Delta n(\ell) \tag{13}$$

Obtaining the mean square value of both sides of Equation (13) and using Equation (11) yields

$$\langle \Delta\phi_\ell^2(\ell) \rangle = 0.066\, (2\pi)^{1/3} C_N^2 k^2 \ell^{5/3} d\ell$$

for

$$\frac{2\pi}{L_0} < K < \frac{2\pi}{\ell_0} \tag{14}$$

where

$k = 2\pi/\lambda$

Over a path of length L, the number of inhomogeneities of size ℓ is L/ℓ. Hence, from Equation (14) it follows that the mean square

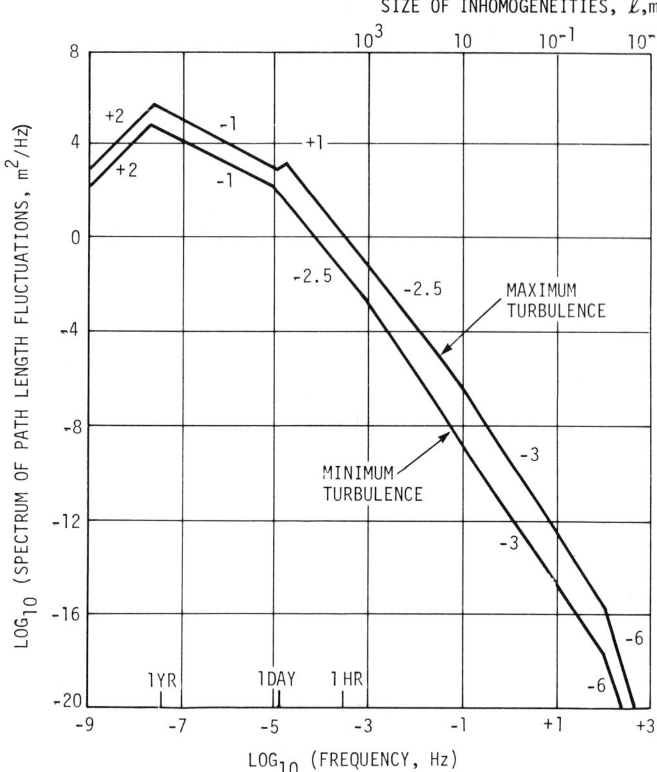

Figure 12 **Spectrum of path-length fluctuations; path length = 15 km, cross-wind speed = 1 m/s and surface refractivity = 313. After Unger [16].**

phase fluctuation over a path of length L due to blobs of size ℓ is given by

$$< \Delta\phi_L^2(\ell) > = 0.066\,(2\pi)^{1/3} C_N^2 k^2 L \ell^{2/3} d\ell$$

for

$$\frac{2\pi}{L_0} < K < \frac{2\pi}{\ell_0} \qquad (15)$$

Rewriting Equation (15) in terms of $K = 2\pi/\ell$ yields

$$< \Delta\phi_L^2(\ell) > = (0.066)\,(2\pi)^2 C_N^2 k^2 L K^{-8/3} dK$$

for

$$\frac{2\pi}{L_0} < K < \frac{2\pi}{\ell_0} \qquad (16)$$

Figure 12 gives a plot of $< \Delta\phi_L^2(\ell) >$ obtained experimentally at microwave frequencies. The top abscissa is in terms of the "blob" size, ℓ. The lower abscissa axis is the logarithm of the frequency of the phase fluctuation resulting from a blob of size ℓ moving transversely to the propagation path with a velocity of 1 m/s. Thus ℓ = 1 m results in a phase component having a frequency fluctuation in the receiver of (1 m/s)/ℓ = 1 Hz; ℓ = 0.1 m, 10 Hz; etc.. The plot of Figure 12 shows that the spectrum density of the phase fluctuations resulting from "blobs" of size ℓ follows approximately the -8/3 = 2.67 power law of Equation (16), the power law ranging from -2.5 to -3.0 in the region where the Kolmogorov-Obukhov model is expected to apply.

As indicated, the data in Figure 12 were obtained at microwave frequencies. The magnitude of the fluctuations is actually lower for laser frequencies, because at laser frequencies there is little dependence of the atmospheric index of refraction on the water vapor content and consequently on the spatial random variation of the water vapor content. The index of refraction at laser frequencies is (approximately) given by

$$N = \frac{77.6}{T}\, p \qquad (17)$$

where

N = index of refraction in parts per million deviation from unity — *refractivity*
= $(n-1) \times 10^6$

p = total atmospheric pressure in millibars (mb); 1 mb = 0.750 millimeters of mercury (mm of Hg).

T = temperature in Kelvin

Typically N ≐ 260. At microwave frequencies N includes a term depending on the partial pressure of the atmospheric water vapor component

$$N = \frac{77.6}{T}\left(p + \frac{4810e}{T}\right) \qquad (18)$$

where

e = partial pressure of water vapor content in millibars.

N ranges from 240 to 400 at microwave frequencies depending on geographic location and weather conditions. Note that it is the spatial variation of the three parameters (p, T, and e) that result in the spatial variations in n.

For microwave frequencies, a more convenient and more accurate expression for N in terms of the relative humidity, H, given as a decimal fraction, is [45]

$$N = 77.6\,\frac{p}{T} + 5.4 \times 10^5\,\frac{H}{T^2}\,\exp\left[\frac{T-273}{3.5} + 2.3\right]^{1/2} \qquad (18a)$$

where p and T are in the same units as in Equations (17) and (18). Equations (18a) gives N to ±1 (i.e., ±1 N unit) for temperatures below 120°F (48.9°C = 322K). The humidity H (as a decimal fraction) is given in terms of the dew point temperature, T_D, (in K) by [46]

$$H = \left(\frac{T}{T_D}\right)^2 \exp\left\{\left[\frac{T_D-273}{3.5} + 2.3\right]^{1/2} - \left[\frac{T-273}{3.5} + 2.3\right]^{1/2}\right\} \qquad (18b)$$

The mechanism for ray interference at the receiver due to diffraction by the atmospheric "blobs" is depicted in Figure 13. The angle over which the rays are diffracted by a "blob" of size ℓ is given in radians by

$$\theta_D = \frac{\lambda}{\ell} \qquad (19)$$

Destructive interference can occur as a result of the addition of diffracted and nondiffracted rays at the receiver. The angle that a ray is diffracted by a blob of size ℓ so as to add 180° out of phase with a nondiffracted ray is designated as θ_F and is given by

$$\theta_F = \sqrt{\frac{\lambda}{\ell}} \qquad (20)$$

(see Figure 14)

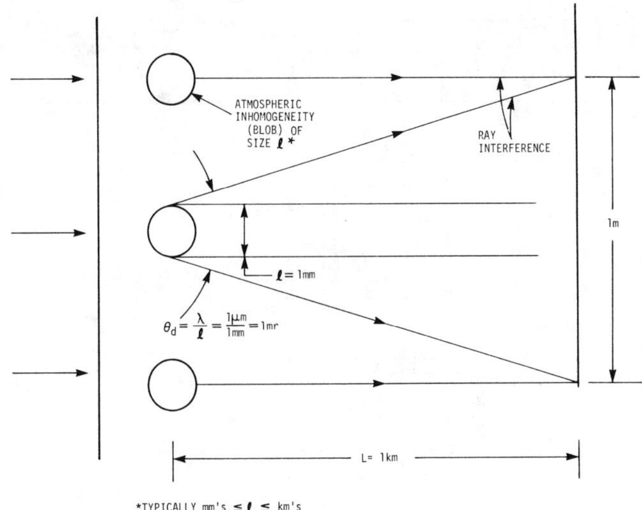

Figure 13 **Ray interference due to diffraction.**

Refraction (bending due to Snell's Law) as well as diffraction can result in ray interference. When the path length is short enough so that one is in the linear region of Figure 7, "blob" refraction is much smaller than θ_F [15]. In the saturation region (see Figure 7) the reverse is true. The path length for the experiment of Figure 1 corresponds to being at the transition between the linear and saturation regions.

Clear Weather Atmospheric Attenuation

Clear weather attenuation results from absorption by the atmospheric constituent gases, aerosol scattering and absorption, and molecular scattering. Figure 15 shows the absorption observed in the ruby laser wavelength region for 17.4° to 19.3° elevation angle paths through the whole atmosphere [17, 18]. The atmosphere constituents forming the absorption lines are indicated. The data were obtained from solar spectrum measurements. The absorption lines labeled as "solar" are due to sun gases as opposed to earth gases.

Figure 16 shows a corresponding plot near the CO_2 P(20) laser line [19, 20]. (The vertical lines represent laser wavelength lines for the gases indicated.) Figure 17 shows a theoretical calculation of atmospheric transmittance [21] (i.e., fraction of power transmitted) for the 10 μm region. Theoretical estimates for transmittance tend to underestimate the absorption [21] (up to as much as a factor of two); hence, care must be taken when using such results. Measurements, however, also have their uncertainties as is indicated shortly. Further work is necessary to allow the accurate prediction of atmospheric absorptions. Extensive compilations of laser atmospheric absorption data are given in References 19, 20, and 22 through 30. The variation of the attenuation coefficient as a function of altitude for CO_2 wavelengths is given in Reference 21 for five atmospheric models.

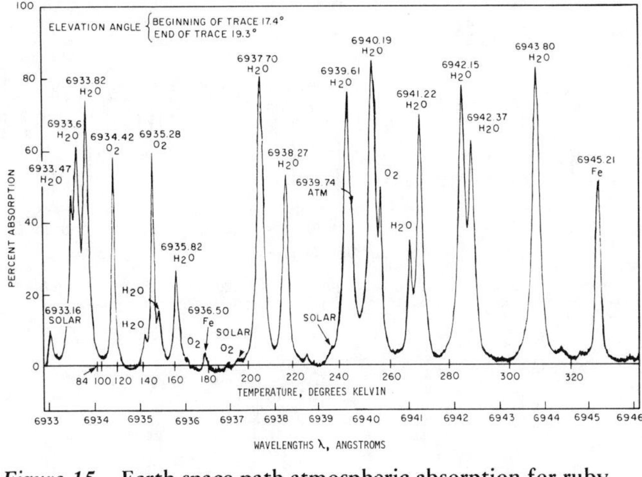

Figure 15 **Earth-space path atmospheric absorption for ruby laser wavelength region. (The absorption lines labeled as "solar" are due to sun gases as opposed to earth gases.) From Long [17, 18].**

A crude estimate of the atmospheric transmittance between 0.5 and 25 μm is obtained from the coarse resolution measurements of Yates and Taylor [31] given in Figures 18 through 21. These data give the atmospheric transmittance for horizontal paths of different lengths having different atmospheric conditions for two altitudes. Note the absence of the fine absorption line structure given in Figures 15 and 17 for the ruby and CO_2 laser regions because of the coarse resolution used.

Also shown in Figure 18 is a theoretically calculated atmospheric transmittance for a coarse resolution based on the LOWTRAN 2 computer program developed at Air Force Cambridge Research Laboratory (AFCRL) [21]. The deviation of the measured results from the theoretical calculations in this figure in the region of 10 μm is believed to be due to the fact that Yates and Taylor artificially set the transmittance to 100% in this window region since they were unable to estimate the water vapor continuum contribution.

Corresponding atmospheric transmittance curves for propagation through the whole atmosphere at various elevation angles are given in Figures 22 through 24 [32].

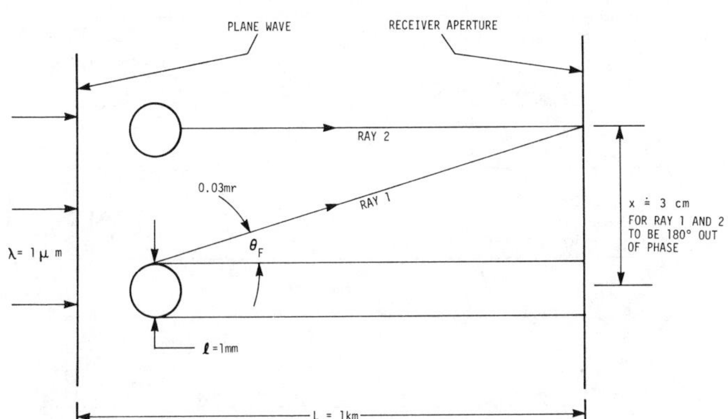

Figure 14 **Blob separation for 180° interference between diffracted and nondiffracted ray.**

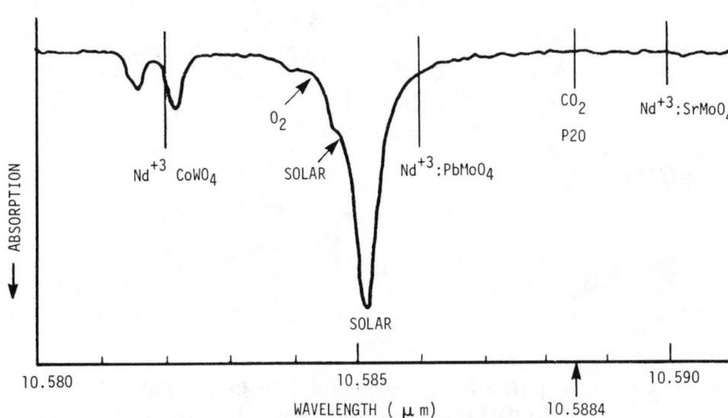

Figure 16 **Earth-space path atmospheric absorption near CO_2 10.5884 μm P(20) line. (The vertical lines represent laser wavelength lines for gases indicated.) From Migeotte [19]; also see Long [20].**

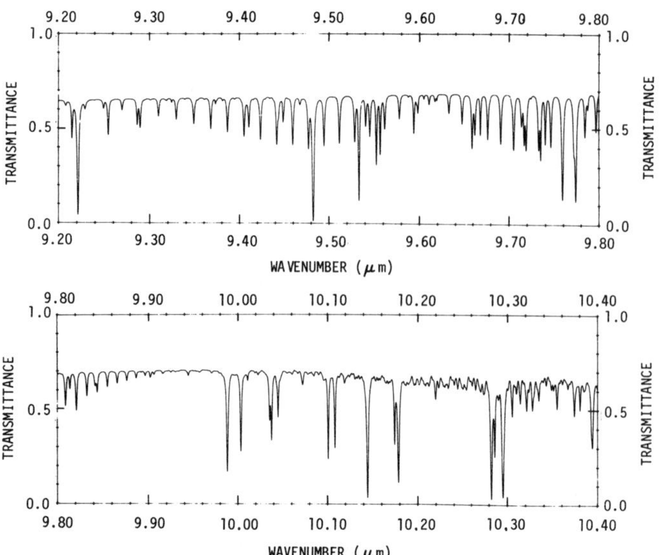

Figure 17 Theoretically computed atmospheric transmittance for 10 μm region; 10 km horizontal path at sea level, mid-latitude winter model for atmosphere. From McClatchey, Selby, and Garing [21].

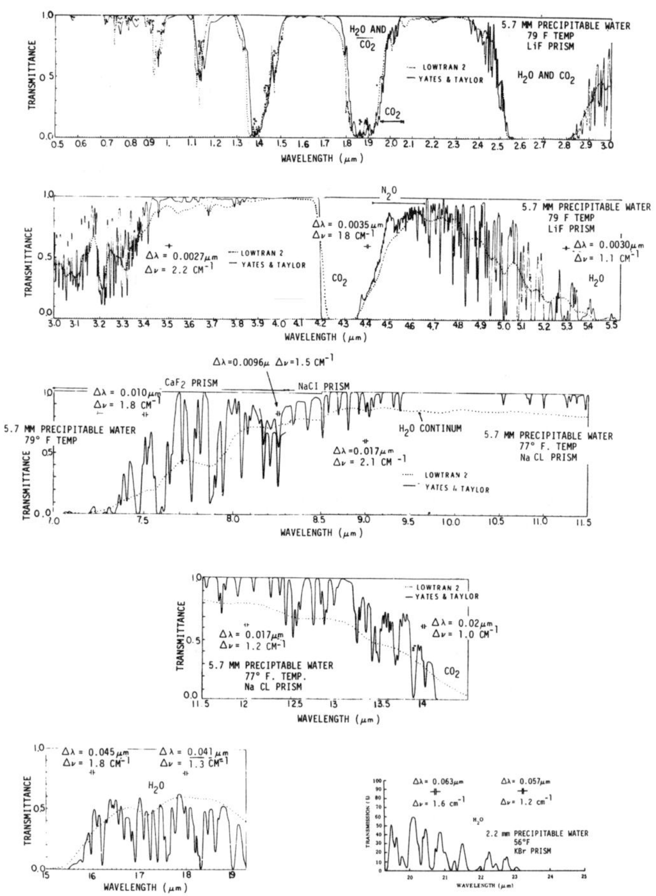

Figure 18 Comparison of Yates and Taylor [31] measurements and LOWTRAN computer predictions [21] of the atmospheric transmittance for a 0.3 km path at sea level for 0.5 μm to 25 μm; coarse resolutions (indicated in figure) used for measurements and theoretical calculations. From McClatchey, Selby and Garing [21].

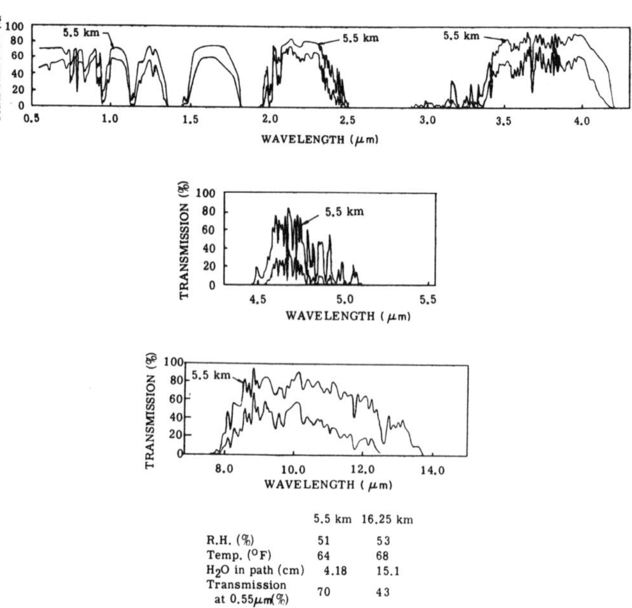

Figure 19 Atmospheric transmittance at sea level for 5.5 and 16.25 km paths when 0.55 μm transmittance 70% and 43% respectively; 0.5 to 14 μm. From Yates and Taylor [31].

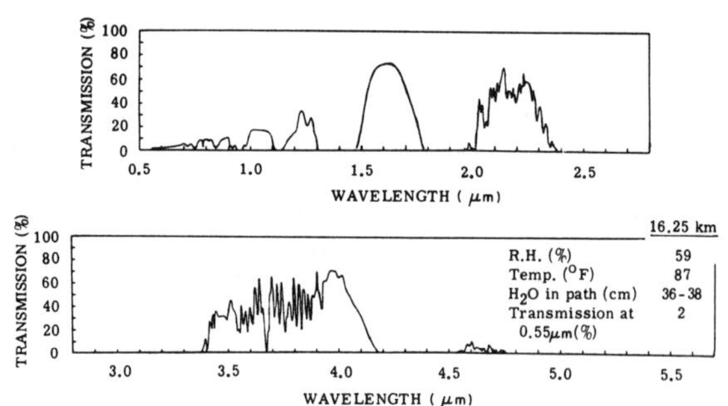

Figure 20 Atmospheric transmission at sea level for a 16.25 km path when 0.55 μm transmission is 2%; 0.5 to 5.7 μm. From Yates and Taylor [31].

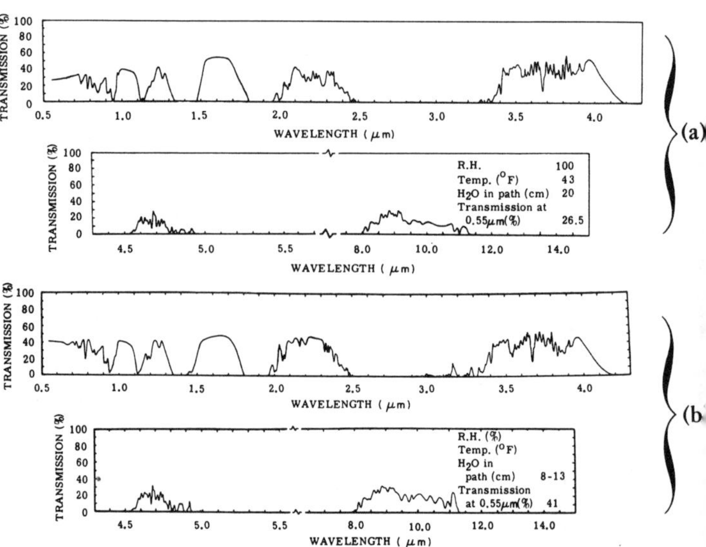

Figure 21 Atmospheric transmission at 10,000 ft altitude for a 27.7 km path – (a) when 0.55 μm transmission is 26.5%; (b) when 0.55 μm transmission is 41%. From Yates and Taylor [31].

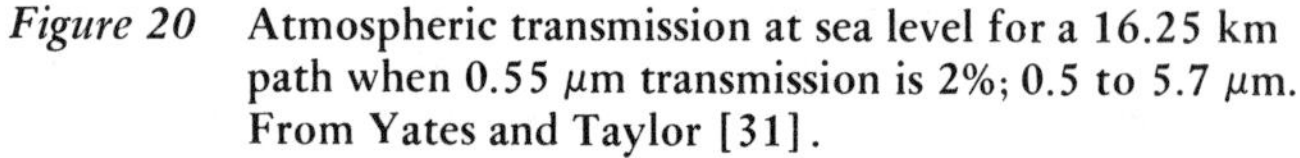

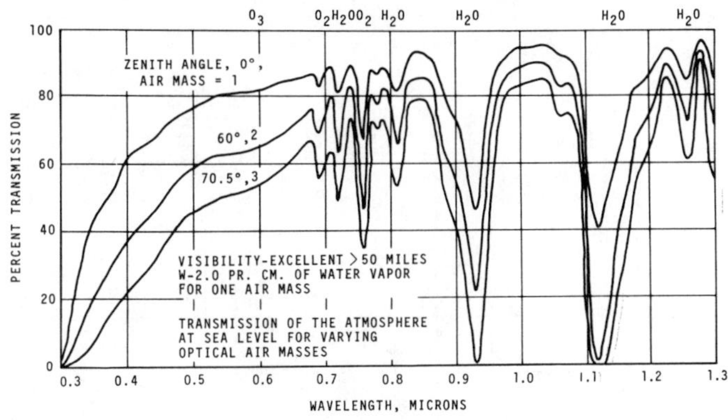

Figure 22 Laser atmospheric transmission through whole atmosphere; zenith angles 0°-70.5°; 0.3-1.3 μm. After Chapman and Carpenter [32].

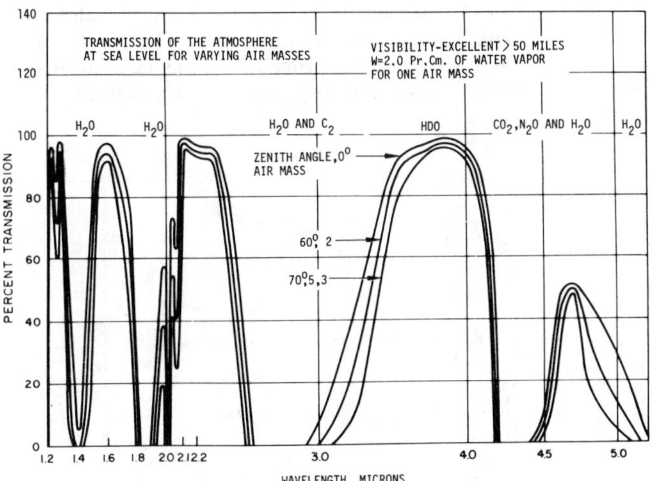

Figure 23 Laser atmospheric transmission through whole atmosphere; zenith angles 0°-70.5°; 1.2-5.0 μm. After Chapman and Carpenter [32].

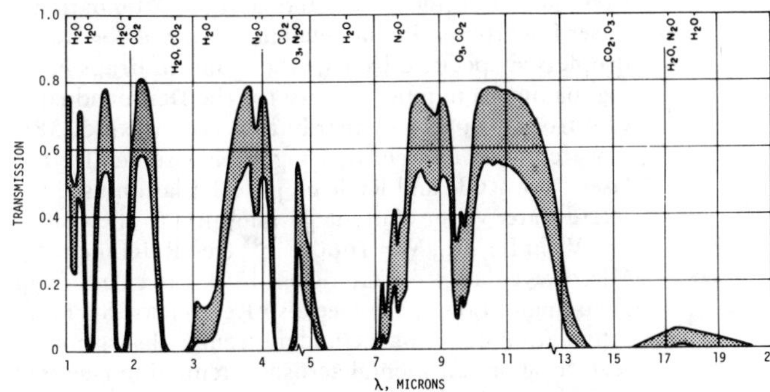

Figure 24 Laser atmospheric transmission through whole atmosphere; zenith angles from 20°-70°; 1-21 μm. Absorbing factors — CO_2, N_2O, CH_4, CO, O_3, and haze (visibility = 32 km). After Chapman and Carpenter [32].

Atmospheric transmittance curves such as those of Figure 18 and 24 depend on the atmospheric conditions. For example at the CO_2 wavelengths the determining factors are the water vapor and CO_2 content. At sea level the H_2O one-way attenuation in dB/km for 10 μm is given by [33]

$$\alpha = 1.43 \times 10^{-2}e + 3.62 \times 10^{-3}e^2 \text{ dB/km} \tag{21}$$

where

e = the partial pressure of water vapor content in Torr or, equivalently, millimeters of mercury (mm of Hg).

The partial pressure of water vapor can be calculated from a knowledge of the atmospheric ambient temperature and dew point temperature or humidity using standard meteorological techniques. A good approximation of e in terms of T and H is obtained by equating the second terms on the right hand sides of Equations (18) and (18a) which yields

$$e = 1.447 \text{ H} \exp\left[\frac{T-273}{3.5} + 2.3\right]^{1/2} \tag{21a}$$

where e is in mb when T is in Kelvin and H is a decimal fraction.

For a relative humidity (RH) of 80% and ambient temperature of 25°C, e = 19 Torr and α = 1.6 dB/km, as compared to 0.37 dB/km for the conditions of Figure 19 for the 5.5 km path (17.8°C, RF = 51%).

For a vertical path through the whole atmosphere at 30°N latitude, the H_2O absorption is 0.59 and 2.4 dB for average January and July humidity conditions, respectively [33]. For other than a vertical path, the attenuation is proportional to the cosecant of the elevation angle. (Equations for the attenuation to arbitrary altitudes are given in Reference 33.)

The sea level absorption at 25°C due to CO_2 for a standard 330 ppm is 0.35 dB/km at the 10.588 μm P (20) line. However, nocturnal CO_2 concentrations near vegetation of up to 900 ppm have been observed [33, 34].

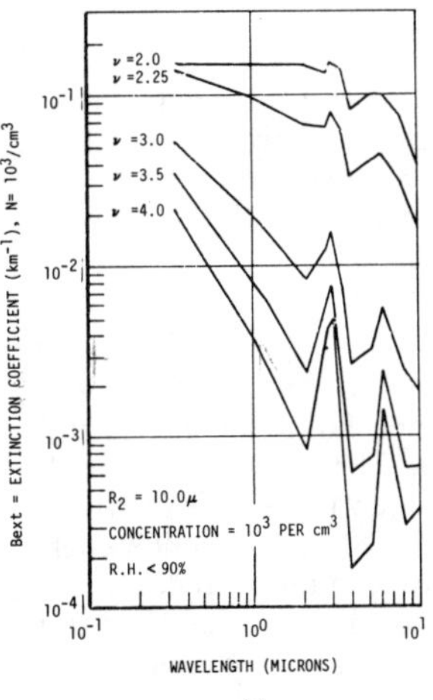

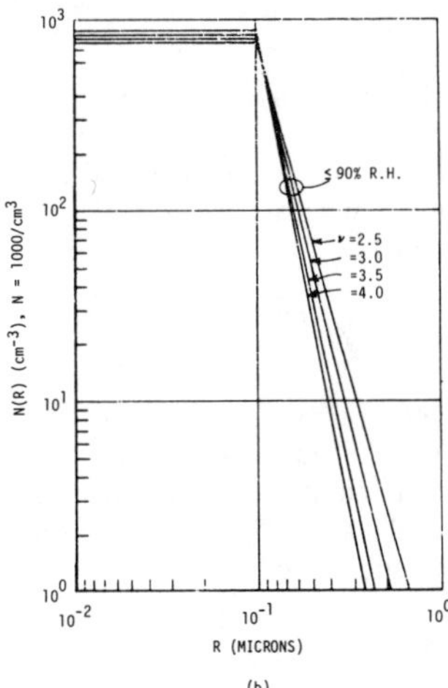

Figure 25 Continental aerosol model and attenuation. (a) Extinction coefficient vs. wavelength for continental aerosol model with relative humidity (RH) $\leqslant$ 90%; concentration of N = 1000/cm^3 is assumed. (b) Normalized Junge's continental aerosol size distribution model for RH $\leqslant$ 90%; concentration of N = 1000/cm^3 is assumed. (Junge's distribution actually has total particle concentration of N = 5 x 10^3/cm^3 instead of 10^3/cm^3 as shown in figures. Hence, extinction coefficient, B_{EXT}, from Figure 25(a) has to be multiplied by 5 for the Junge aerosol model. Example — For λ = 0.63 μm and ν = 3, B_{EXT} = 5.0 x 3.0 x 10^{-2} km^{-1} = 0.15 km^{-1}. Note — Visibility = V $\doteq$ 3.93/B_{EXT} (0.63 μm) = 26 km, for the above example.) From Rensch and Long [35].

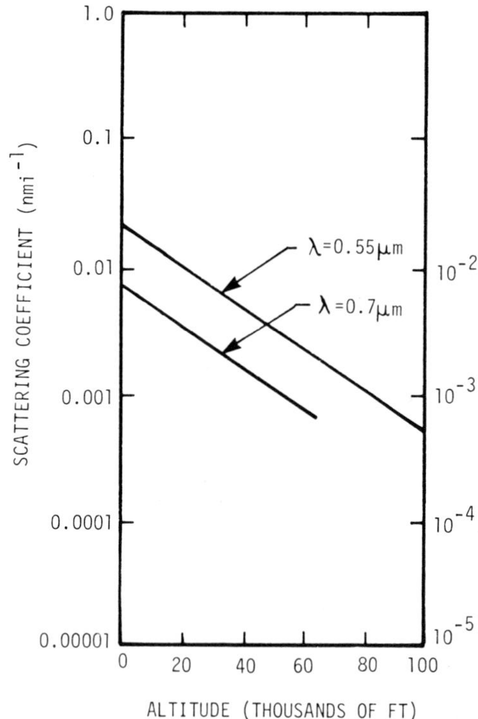

Figure 26 **Atmospheric molecular scattering coefficient, α_S, vs. altitude. (One-way attenuation = 4.34 α_S dB/nmi. One-way power transmittance = $A_S = e^{-\alpha_S L}$ where L is one-way path length in nmi for α_S in 1/nmi.) After Ross [36].**

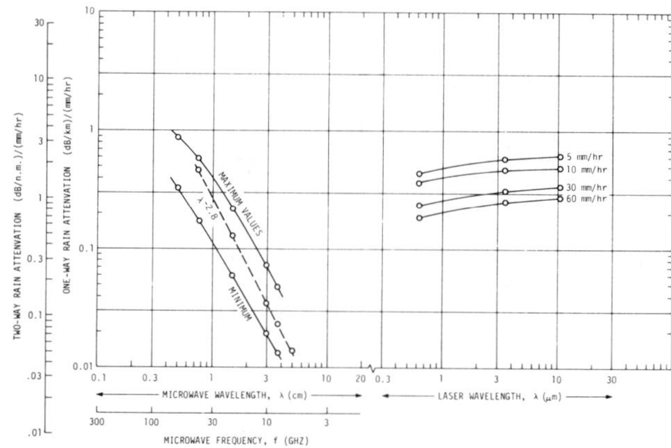

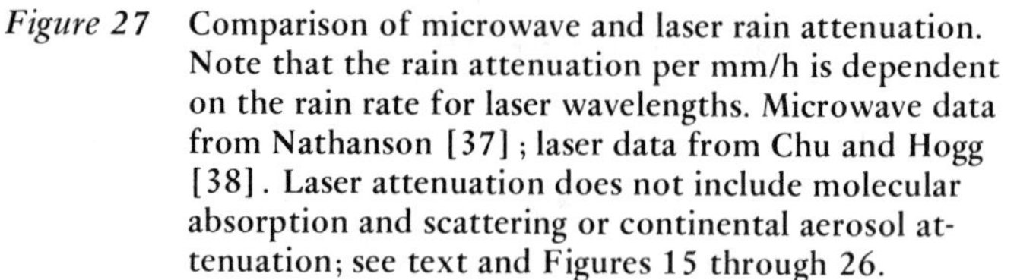

Figure 27 **Comparison of microwave and laser rain attenuation. Note that the rain attenuation per mm/h is dependent on the rain rate for laser wavelengths. Microwave data from Nathanson [37]; laser data from Chu and Hogg [38]. Laser attenuation does not include molecular absorption and scattering or continental aerosol attenuation; see text and Figures 15 through 26.**

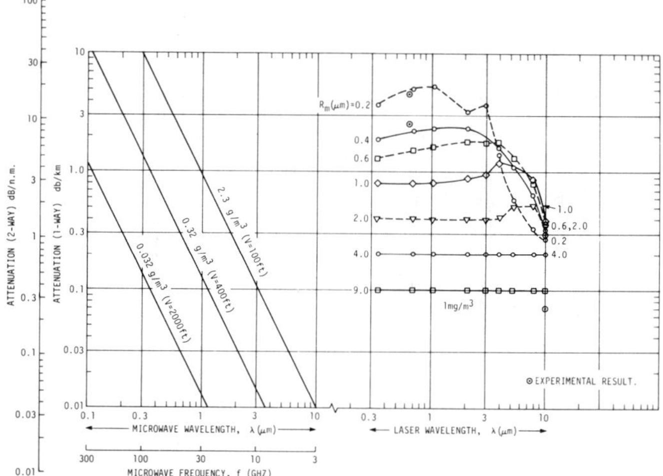

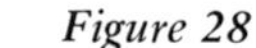

Figure 28 **Comparison of microwave and laser fog attenuation. Laser fog attenuation dependent on size of fog water droplets as specified by R_M, the radius of drops having maximum number density for the Deirmendjian-Chu-Hogg droplet size distribution assumed [35, 38]. A water vapor content of 1 mg/m^3 is assumed for the laser data. Ryde and Ryde empirical relation used to relate water vapor content, M (in g/m^3) to the visibility, V (in ft), i.e., $M = 1660\ V^{-1.43}$ (see Reference 39). Microwave data from Goldstein [39]; laser data from Rensch and Long [35] (see also Reference 38). Laser attenuation does not include molecular absorption and scattering or continental aerosol attenuation (see text and Figures 15 through 26). Note – 1.4 g/m^3 and 4 g/m^3 equivalent to heavy cumulus clouds and very heavy nimbostratus clouds, respectively [37].**

The attenuation due to aerosol absorption and scattering is given in Figure 25(a) for the Junge aerosol model of Figure 25(b) [35]. The signal attenuation, A, for propagation over a distance L is given in terms of the extinction coefficient, B_{EXT}, by the expression

$$A = e^{-B_{ext}L} \tag{22}$$

where, if B_{EXT} is given per unit kilometer, L is in kilometers and A is the fractional power transmittance. The aerosol extinction coefficient varies with altitude; Reference 21 gives this variation for 10.6 μm. In the visible region, the log of the extinction coefficient decreases approximately linearly with the altitude by a factor of two* in going from sea level to an altitude of 20,000 ft [36]; the same fall-off occurs at 10.6 μm for hazy conditions [21].

The atmospheric attenuation due to molecular scattering is given in Figure 26 in terms of the scattering coefficient. At sea level, molecular scattering is generally much less than aerosol scattering [36].

Inclement Weather Propagation

Rain and Fog Attenuation

Figure 27 compares the attenuation due to rain at laser and microwave wavelengths [37, 38]. Note that the laser attenuation is dependent on the rain rate; hence, curves for different rain rates are specified. At 10.6 μm, for a 60 mm/h rain rate the attenuation is 1.0 (dB/nmi)/(mm/h) two-way yielding a total attenuation of 60 dB/nmi. The figure indicates that for rain rates between 5 and 10 mm/h, the attenuation at laser wavelengths between 0.6 and 10 μm is an order of magnitude larger than at the microwave frequency of 10 GHz. One has to go to a microwave frequency of about 60 GHz to obtain equivalent attenuation at laser wavelengths as that obtained at microwave wavelengths.

Figure 28 compares laser and microwave frequency fog attenuation. The fog results were obtained using the Deirmendjian-Chu-Hogg [35, 38] drop size distribution model for different R_m's,

*corresponding to a factor of 100 change for the extinction coefficient.

Location – Crawford Hill, New Jersey
Period – October 1968 to July 1969

Attenuation (dB)	% of Time Exceeded	Description of Conditions
10	48	Thin cloud cover; Diffuse ground shadows.
20	43	Sun disk just disappears.
30	34	–

Table 1 8 to 14 μm Earth-Space Measured Attenuation [40].

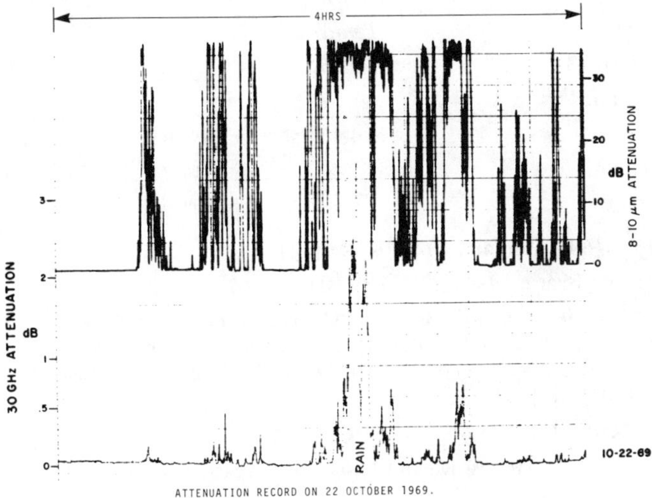

Figure 29 Comparison of 8-12 μm and 30 GHz attenuation for space-to-ground path. Clear sky initially at far left, followed by cumulus clouds and rain shower during middle of track. After Wilson [40].

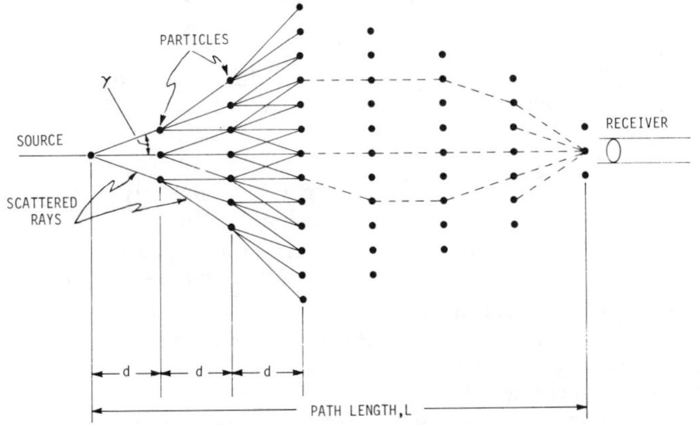

Figure 30 Multiple scattering dispersion from fog, rain, or snow particles. After Chatterton [41].

the radii of the drops having the maximum number density. (The other model parameters used in Figure 28 are $\alpha = 2.0$ and $\gamma = 0.5$ with these parameters as defined in References 35 and 38.) The laser wavelength attenuation results are given for a water droplet content of 1 mg/m^3. The important conclusion from this figure is that the fog attenuation at laser wavelengths of 10 μm is over 1000 times greater than at microwave X-band frequencies. The situation is one or two orders of magnitude worse (depending on the droplet size) at 1 μm. Fog attenuation at the laser wavelength of 10 μm is 30 times or more worse than it is at the microwave frequency of 100 GHz. When molecular absorption is included, the attenuation at 10 μm is ~100 and ~4 times worse than at X-band and 100 GHz, respectively; see Editor's Note on page 340.

Assumptions – 1) 1 km horizontal path 30 m off the ground or path through whole atmosphere with zenith angle 80°
2) Medium Turbulence ($C_N{}^2 = 5 \times 10^{-14}$ m$^{-2/3}$ at altitude of 30 m)
3) Wind Speed = 1 m/s

Maximum Effective Receiver Aperture, r_0	0.32 m
Angular Accuracy for 0.32 m Receiver	22 μr
Coherence Time for Point Receiver	0.094 s
Frequency Spread	
Large Aperture (6 m)	0.15 Hz
Small Aperture (~ 3 cm)	2 Hz
Maximum Available Atmospheric Coherence Bandwidth	~ 1000 GHz#
rms of Received Signal Amplitude*	0.19
Mean of Received Signal Amplitude*	0.98
rms of Natural Logarithm of Signal Amplitude*	0.19

#The absorption line free bandwidth is smaller, probably 250 GHz (see Figure 17).

*Receiver aperture diameter assumed much less than 0.32 m, the maximum effective diameter. Normalized to mean received signal amplitude in vacuum.

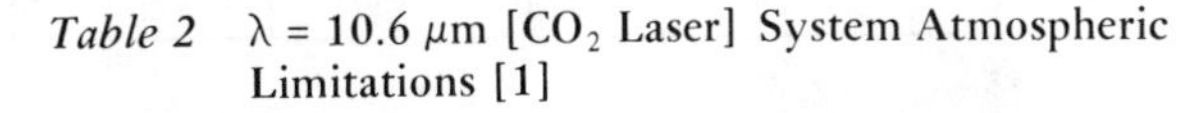

Table 2 λ = 10.6 μm [CO_2 Laser] System Atmospheric Limitations [1]

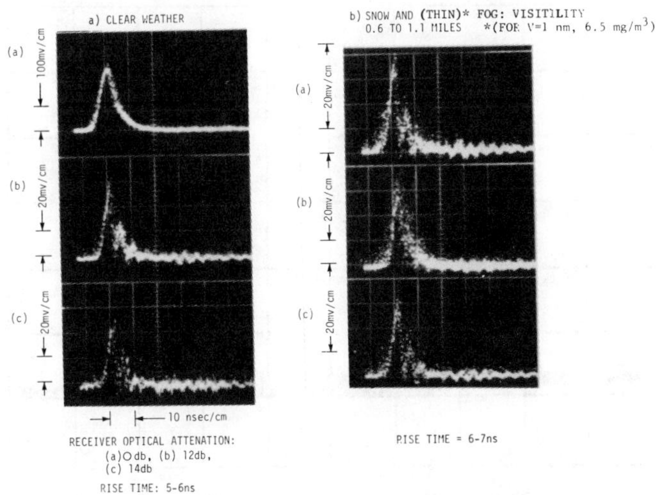

Figure 31 Dispersion through fog and snow; λ = 0.8550 μm (GaAs laser), Path length = 1.8 mi, transmitter beam width = 1°, receiver field of view = 1/3°. After Chatterton [41].

Figure 29 gives a comparison of the attenuation obtained at the optical wavelengths between 8 μm and 14 μm with that obtained at the microwave frequency of 30 GHz for one-way propagation from space to ground through clouds [40]. Table 1 summarizes the frequency of occurrence of various attenuation levels for this 8 to 14 μm band [40].

Pulse Width Dispersion

Of importance in radar (and communications systems as well) is pulse dispersion resulting from fog, rain, and snow. Such dispersion limits the resolution of radars (and the data rates of communications systems). Figure 30 shows how such dispersion comes about as the result of multiple scattering from rain drops, snow particles, or fog particles.

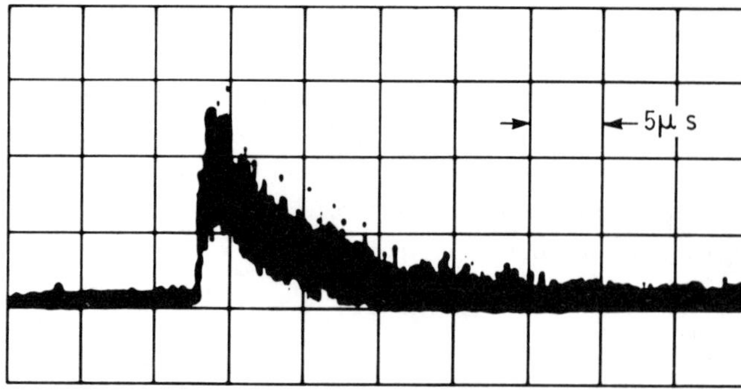

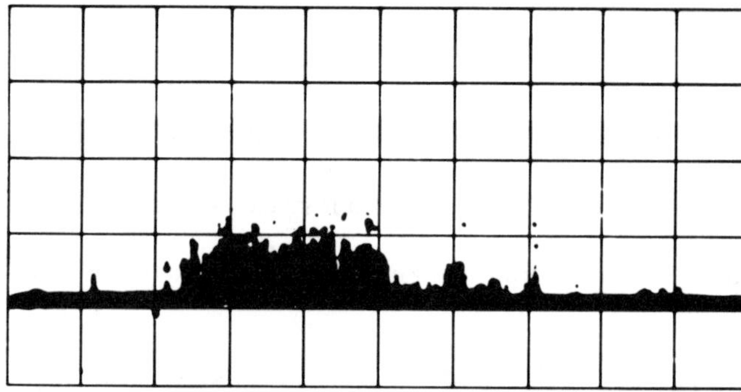

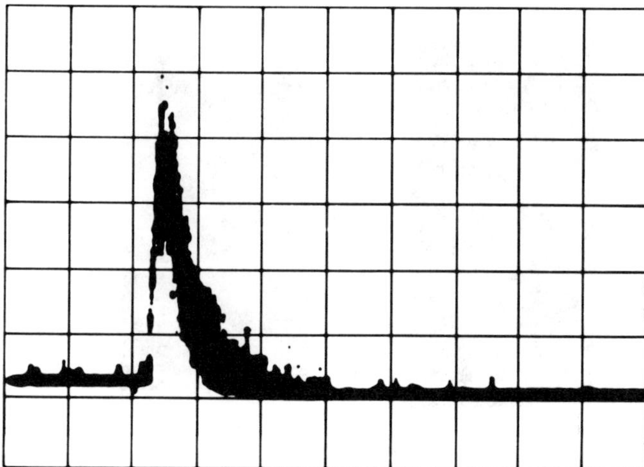

Figure 32 **Transmission through 500 to 2000 meters of cloud; ruby laser λ = 0.6940 μm, transmitted pulse width = 30 ns, fields of view 1° (transmitter) and 1°-4° (receiver). After Bucher [42].**

Results obtained by Chatterton [41] indicate a dispersion of between 4.5 and 5.5 ns due to the combined presence of snow and thin fog (see Figure 31). Thus, if a 1 ns pulse were transmitted through the atmosphere for the conditions of Chatterton's experiment, the received pulse would spread out to a width of about 5 ns.

The dispersion resulting from the propagation of a ruby laser signal through 500 to 2000 m of clouds is indicated in Figure 32 to be between 5 and 15 μs [42].

There is pulse width spreading of laser pulses during clear weather conditions, but such distortion is negligible for most applications of interest. It ranges from picoseconds to tens of picoseconds, depending on the situation [1, 43].

Summary

A summary of the clear weather effects on a CO_2 laser signal is given in Table 2. Table 3 compares the clear weather attenuation at several laser wavelengths with that obtained at microwave frequencies. Finally, Table 4 summarizes some of the main points made in this chapter.

For further reading, the reader is referred to References 1, 4, 12, 13, and 47 to 56.

Ground Paths

Wavelength or Frequency	*dB/nmi (Two-Way)*	
10 GHz (3 cm)	0.05*	
0.63 μm	1.00**	
3.5 μm	4.47****	for 81% RH, 79°F (26.1°C)
10.6 μm	1.49	for 10% RH, 77°F (25°C)***
	7.22	for 80% RH, 77°F (25°C)***

*15°C, 60% humidity; from Blake [44]

**From Figure 19; 17.8°C, 51% RH.

***Assumes standard 330 ppm CO_2. CO_2 concentrations near vegetation of up to 900 ppm have been reported. For such concentration, the attenuations of 1.49 and 7.22 dB/nmi increase by 2.24 dB/nmi [33].

****From LOWTRAN 2 data of Figure 18.

Table 3 **Microwave and Laser Clear Weather Attenuation**

- Received signal amplitude distribution: approximately log-normal for point target independent of receiver aperture size for incoherent system.
- rms of signal amplitude saturates for long ranges.
- There exists maximum effective aperture for coherent heterodyne system.
- Signal Resolution

Clear Weather (path through atmosphere with 80° zenith angle)	~ 1 ps for CO_2 (10.6 μm) ~ 3 ps for GaAs (0.8446 μm)
Snow and Fog	~ 5 ns for GaAs (0.8446 μm) for 1.8 nmi one-way path having V ≐ 1 nmi
Cloud	5-15 μs for ruby laser (0.6940 μm) for 500 to 2000 m thick cloud.

- Rain Attenuation (dB/nmi)
  About same at laser wavelengths as for high frequencies (~ 60 GHz)
- Fog Attenuation (dB/nmi)
  About 100 times worse at 10 μm laser wavelength than at 10 GHz, ~4 times worse at 10 μm than at 100 GHz

Table 4 **Summary of Effects of the Atmosphere on Laser Radars**

References

[1] Brookner, E.: "Atmospheric Propagation and Communication Channel Model for Laser Wavelengths," *IEEE Transactions on Communications Technology, Vol. COM-18,* pp. 396-416, August 1970; also reprinted in *Communications Channels: Characterization and Behavior,* (B. Goldberg, ed.), IEEE Press, 1976.

[2] Deitz, P.H.; and Wright, N.J.: "Saturation of Scintillation Magnitude and Near-Earth Optical Propagation," *Journal of the Optical Society of America, Vol. 59,* pp. 527-535, May 1969.

[3] Brown, W.P.: "Fourth Moment of a Wave Propagation in a Random Medium," *Journal of the Optical Society of America, Vol. 62,* pp. 966-971, August 1972.

[4] Brookner, E.: "Atmospheric Propagation and Communication Channel Model for Laser Wavelengths: An Update," *IEEE Transactions on Communications Technology, Vol. COM-22,* pp. 265-270, February 1974; also reprinted in *Communications Channels: Characterization and Behavior,* (B. Goldberg, ed.), IEEE Press, 1976.

[5] Bissonnette, L.R.: "Log-Normal Probability Distribution of Strong Irradiance Fluctuations: An Asymptotic Analysis," *AGARD Conference Proceedings No. 183 on Optical Propagation in the Atmosphere,* Paper No. 19, Technical Editions and Reproductions, Ltd., Harford House, 7-9 Charlotte St., London, W1P 1HD, May 1976.

[6] Strohbehn, J.W.; Wang, T.; and Speck, J.P.: "On the Probability Distribution of Line-of-Sight Fluctuations of Optical Signals," *Radio Science, Vol. 10,* pp. 59-70, January 1975.

[7] Mitchell, R.L.: "Permanence of the Log-Normal Distribution," *Journal of the Optical Society of America, Vol. 58,* pp. 1267-1272, September 1968.

[8] deWolf, D.A.: "Propagation Through Turbulent Air: New Results," *1973 URSI (International Union of Radio Science) Meeting,* August 21-24, 1973, Boulder, Colorado.

[9] Fried, D.L.: "Optical Heterodyne Detection of an Atmospheric Distorted Wavefront," *Proceedings of the IEEE, Vol. 55,* pp. 57-66, January 1967.

[10] Brookner, E.: "System Combining the Best Features of Heterodyne and Direct Detection Receivers," *Applied Optics, Vol. 10,* pp. 1009-1011, May 1971.

[11] Kompfner, R.: "Light Communication System with Improved Signal-to-Noise Ratio," *US Patent No. 3,532,889,* 6 October 1970.

[12] Hodara, H.: "Effects of a Turbulent Atmosphere on the Phase and Frequency of Optical Waves," *Proceedings of the IEEE, Vol. 56,* pp. 2130-2136, December 1968.

[13] Strohbehn, J.W.: "Line-of-Sight Wave Propagation through the Turbulent Atmosphere," *Proceedings of the IEEE, Vol. 56,* pp. 1301-1318, August 1968.

[14] Goldstein, I.; Miles, P.A.; and Chabot, A.: "Heterodyne Measurements of Light Propagation through Atmospheric Turbulence," *Proceedings of the IEEE, Vol. 53,* pp. 1172-1180, September 1965.

[15] deWolf, D.A.: "Waves in Turbulent Air: Phenomenological Model," *Proceedings of the IEEE, Vol. 62,* pp. 1523-1529, November 1974.

[16] Unger, J.H.W.: "Random Tropospheric Angle Errors in Microwave Observations of the Early Bird Satellite," *Bell System Technical Journal, Vol. 45,* pp. 1439-1474, November 1966.

[17] Long, R.K.: "Atmospheric Attenuation of Ruby Lasers," (Letters), *Proceedings of the IEEE, Vol. 51,* pp. 859-860, May 1963.

[18] Long, R.K.: "Absorption at Ruby Laser Wavelengths for Low Angle Total Atmospheric Paths," *Technical Report 2156-2,* Antenna Laboratory Department of Electrical Engineers, Ohio State University Research Foundation, Columbus, 31 December 1966.

[19] Migeotte, M.: "Annex to Technical Status Report No. 18," *Contract AF61(514)-962,* January-March 1961.

[20] Long, R.K.: "Absorption of Laser Radiation in the Atmosphere," *Technical Report 1579-3,* Antenna Laboratory, Department of Electrical Engineering, Ohio State University Research Foundation, Columbus, May 1963.

[21] McClatchey, R.A.; Selby, J.E.A.; and Garing, J.S.: "Optical Modeling of the Atmosphere," *AGARD Conference Proceedings No. 183 on Optical Propagation in the Atmosphere,* Paper No. 1, Technical Editions and Reproductions Ltd., Harford House, 7-9 Charlotte St., London, W1P 1HD, May 1976.

[22] Migeotte, M.; Neven, L.; and Swenson, J.: "The Solar Spectrum from 2.8 to 23.7 Microns, Part I, Photometric Atlas," *Technical Final Report* (Phase A, Part I), *Contract AF61(514)-432,* 1957.

[23] Migeotte, M.; Neven, L.; and Swenson, J.: "The Solar Spectrum from 2.8 to 23.7 Microns, Part II, Measures and Identifications," *Technical Final Report* (Phase A, Part II), *Contract AF61(514)-432,* 1957.

[24] Minnaert, M.; Mulders, G.F.W.; and Houtgast, J.: *Photometric Atlas of the Solar Spectrum from λ3612 to λ8771,* Schnable, Kampert and Helm, Amsterdam, The Netherlands, 1940.

[25] McClatchey, R.A.; Fenn, R.W.; Selby, J.E.A.; Volz, F.E.; and Garing, J.S.: "Optical Properties of the Atmosphere (Third Edition)," *AFCRL-73-0497,* August 1972.

[26] McClatchey, R.A.: "Atmospheric Attenuation of CO Laser Radiation," *AFCRL-71-0370 (ERP 359),* 1971.

[27] McClatchey, R.A.; and Selby, J.E.A.: "Atmospheric Attenuation of HF and DF Laser Radiation," *AFCRL-72-0312 (ERP 400),* 1972.

[28] McClatchey, R.A.; and Selby, J.E.A.: "Atmospheric Transmittance, 7-30 μm: Attenuation of CO_2 Laser Radiation," *AFCRL-72-0611 (ERP 419),* 1972.

[29] McClatchey, R.A.; Bennedict, W.S.; Clough, S.A.; Burch, D.E.; Colfee, R.F.; Fox, F.; Rothman, L.S.; and Garing, J.S.: "AFCRL Atmospheric Absorption Line Parameters Compilation," *AFCRL-TR-73-0096 (ERP 434),* 1973.

[30] McClatchey, R.A.; and Selby, J.E.A.: "Atmospheric Attenuation of Laser Radiation from 0.76 to 31.25 Micrometers," *AFCRL-TR-74-0003 (ERP 460),* 1974.

[31] Yates, H.W.; and Taylor, J.H.: "Atmospheric Transmission in the Infra-Red," *Journal of the Optical Society of America, Vol. 47,* pp. 223-226, March 1957; also "Infrared Transmission of the Atmosphere," *NRL Report 5453 (AD 240-88),* U.S. Naval Research Lab, Washington, DC, 1960.

[32] Chapman, R.M.; and Carpenter, R.: "Effect of Night Sky Backgrounds on Optical Measurements," *Technical Report 61-23-A,* Geophysics Corporation of America, May 1961.

[33] McCoy, J.H.; Rensch, D.B.; and Long, R.K.: "Water Vapor Continuum Absorption of Carbon Dioxide Laser Radiation Near 10 μ," *Applied Optics, Vol. 8,* pp. 1471-1478, July 1969.

[34] Miller, J.F.; and Coutant, R.W.: "Variations in Atmospheric Carbon Dioxide Concentrations and Their Measurement; Phase I: Status Study," *AD 461 770,* 1965.

[35] Rensch, D.B.; and Long, R.K.: "Comparative Studies of Extinction and Backscattering by Aerosols, Fog, and Rain at 10.6 μ and 0.63 μ," *Applied Optics, Vol. 9,* pp. 1563-1573, July 1970.

[36] Ross, M.: *Laser Receivers,* John Wiley and Sons, Inc., New York, 1966.

[37] Nathanson, F.E.: *Radar Design Principles: Signal Processing and the Environment,* McGraw-Hill Book Co., 1969.

[38] Chu, T.S.; and Hogg, D.C.: "Effects of Precipitation on Propagation at 0.63, 3.5, and 10.5 Microns," *Bell Systems Technical Journal, Vol. 47,* pp. 723-759, May-June 1968.

[39] Goldstein, H.: "Attenuation by Condensed Water" in *Propagation of Short Radio Waves,* D.E. Kerr (ed.), (Vol. 13 in "Radiation Laboratory Series"), pp. 671-692, McGraw-Hill Book Co., 1951.

[40] Wilson, R.W.: "Attenuation on an Earth-Space Path Measured in the Wavelength Range of 8 to 14 Micrometers," *Science, Vol. 168,* pp. 1456-1457, June 19, 1970.

[41] Chatterton, E.J.: "Optical Communications Employing Semiconductor Lasers," *Technical Report 392,* MIT Lincoln Laboratories, 9 June 1965.

[42] Bucher, E.A.: "Light Pulse Propagation Through Clouds – Models and Experiments," *IEEE International Conference on Communications,* Philadelphia, 1972.

[43] Brookner, E.: "Limit Imposed by Atmospheric Dispersion on the Minimum Laser Pulse Width that can be Transmitted Undistorted" (Letters), *Proceedings of the IEEE, Vol. 57,* pp. 1234-1235, June 1969.

[44] Blake, L.V.: "Prediction of Radar Range" (Chapter 2 of *Radar Handbook),* M.I. Skolnik (ed.), McGraw-Hill Book Co., 1970.

[45] Odom, D.B.: *Internal Raytheon Report*

[46] Odom, D.B.: *private communication.*

[47] Lawrence, R.S.; and Strohbehn, J.W.: "A Survey of Clear-Air Propagation Effects Relevant to Optical Communications," *Proceedings of the IEEE, Vol. 58,* October 1970.

[48] Kerr, J.R.; Titterton, P.J.; Kraemer, A.R.; and Cooke, C.R.: "Atmospheric Optical Communications Systems," *Proceedings of the IEEE, Vol. 58,* October 1970.

[49] Barabanenkov, Y.N.; Krartsor, Y.A.; Rytov, S.M.; and Tatarski, V.I.: *Usp. Fiz. Navk, Vol. 102,* 1970; (also *Sov. Phys.-Usp., Vol. 13,* p. 551, 1971).

[50] Ishimaru, A.: "Temporal Frequency Spectra of Multifrequency Waves in Turbulent Atmosphere," *IEEE Transactions on Antennas and Propagation, Vol. AP-20,* pp. 10-19, January 1972.

[51] Brookner, E.: "A Note on the Deterioration of the Coherence Properties of a Laser Beam by Molecular Scatterings," *Radio Science, Vol. 6,* pp. 605-609, June 1971.

[52] Brookner, E.: "Log-Amplitude Fluctuations of a Laser Beam" (Letters), *Journal of Optical Society of America, Vol. 61,* p. 641, May 1971.

[53] Brookner, E.: "Improved Model for the Structure Constant Variations with Altitude" (Letters), *Applied Optics, Vol. 10,* pp. 1960-1962, August 1971.

[54] Brookner, E.; Kolker, M.; and Wilmotte, R.: "Deep Space Optical Communications," *IEEE Spectrum, Vol. 4,* pp. 75-82, January 1967.

[55] Strohbehn, J.W. (ed.): *Laser Beam Propagation through the Atmosphere,* Springer-Verlag, 1977.

[56] "*Optical Communications NSF Grantee-User Meeting*" *Proceedings,* 10-11 November 1976, University of Washington, Seattle, Washington.

Present and Future Trends Relative to Propagation Effects

One of the most exciting areas relative to propagation of a signal through the turbulent atmosphere is adaptive optics. Using adaptive optics, the degradation due to the turbulence of the atmosphere can be compensated for to the extent that diffraction-limited imaging with apertures of 10m or more will be achievable for zenith angles less than 60° [1, 2]. One promising technique for achieving high-quality imagery of satellites is through the use of servo-controlled deformable optics that detect and automatically correct for atmospheric distortion [3].

References

[1] Brookner, E.: "Atmospheric Optics," *Optical Communications National Science Foundation Grantee-User Meeting Proceedings,* University of Washington, Seattle, Washington, pp. 76-77, 10-11 November 1976.

[2] Fried, D.: "Imaging Through Turbulence," *Optical Communications National Science Foundation Grantee-User Meeting Proceedings,* University of Washington, Seattle, Washington, pp. 48-53, 10-11 November 1976.

[3] "Rome Develops New Relations with ESD," *Aviation Week and Space Technology, Vol. 105,* pp. 203-213, 19 July 1976.

Part 4
Synthetic Aperture Radar Techniques

To better understand a subject, one should view it from different aspects. Chapters 16, 17, and 18 do just that for the technique of synthetic aperture processing. In Chapter 16 Curtis introduces strip synthetic aperture mapping from the viewpoints of Doppler processing and also as a long synthetic aperture. He describes in simple terms the classical optical correlator used for processing synthetic aperture maps.

Kovaly in Chapter 17 relates the phenomenon of synthetic aperture processing for improved resolution in the azimuth direction to linear FM waveform pulse compression for improved resolution in the range dimension. He introduces the many different types of synthetic aperture mappings including synthetic aperture mapping from target motion rather than antenna motion. Many examples of actual synthetic aperture maps, including the first ever made, are given. Finally, he gives a very complete and up to date bibliography for those interested in pursuing the subject further. After reading Chapters 16, 17, and 18 and the "Present and Future Trends" section of Part 4, the reader might best start with Kovaly's own book [1].

Chapter 18 introduces synthetic aperture spotlight mapping from basics, that is, without relying on a knowledge of the strip synthetic aperture mapping discussions given in Chapters 16 and 17. It is shown how improved azimuth resolution for spotlight mapping is obtained simply from Doppler resolution. For greater insight, the results obtained for spotlight mapping are related to those obtained in Chapters 16 and 17 for focused and unfocused strip mappers and to standard antenna principles. Doppler-beam-sharpening and partially-focused processing are also discussed. The effects of some system errors on synthetic aperture processing are dealt with and the limit on the size of a map that can be easily processed is given. A simple real-time digital processor is described and, finally, the concept of T-space is introduced.

Reference

[1] Kovaly, J.J. (ed.): *Synthetic Aperture Radar,* Artech House, Inc., Dedham, Massachusetts, 1976.

Part 4
Synthetic Aperture Radar Techniques

Chapter 16 Curtis

Synthetic Aperture Fundamentals

In an effort to keep this discussion of the fundamentals of synthetic aperture radars as simple as possible, though the general cases of Doppler processing are presented, the equations are simplified to the broadside case. That is, the data collection is perpendicular to the flight line of the aircraft so that the equations are simple; thus, it should be easier to grasp the principles of operation. In no sense is this approach meant to indicate the only way to produce synthetic aperture processing. (Brookner discusses nonbroadside synthetic aperture processing in Chapter 18.)

Fundamentals

Consider an aircraft flying along the Z axis with the velocity v and with an omnidirectional beam; see Figure 1. A cone axially centered along the flight path (Z axis) describes a constant squint angle, θ_D. Wherever that cone intersects the earth is a hyperbola (assuming a plane earth); along the hyperbola, the Doppler return has a constant frequency. From the equation for f_D, it is clear that the Doppler frequency is proportional to the velocity multiplied by $\cos\theta_D$ and inversely proportional to the wavelength of the transmitted signal. If f_D is differentiated with respect to θ, the rate of change of Doppler frequency is proportional to the sine of the angle; that is, the Doppler is maximum along the velocity vector and zero at broadside, whereas the rate of change of the Doppler is zero along the velocity vector and maximum at broadside.

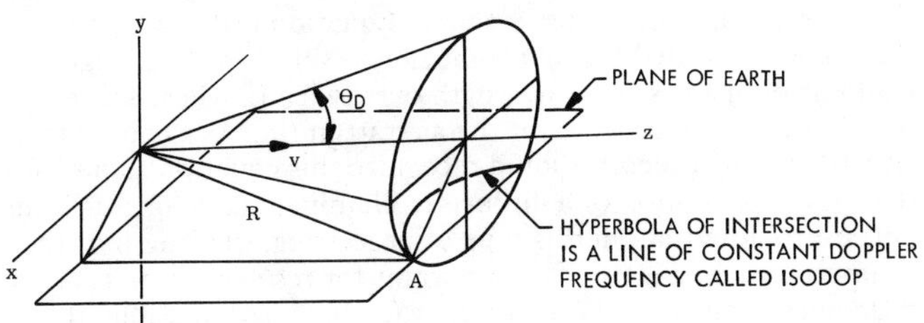

$$f_D = \frac{2v}{\lambda}\cos\theta_D \tag{1}$$

$$\frac{df_D}{d\theta} = -\frac{2v}{\lambda}\sin\theta_D \tag{2}$$

Figure 1 **Doppler Geometry**

Figure 2 shows one quadrant of the lines of intersections of various cone angles of constant Doppler frequency; the cross-hatching represents a radar antenna beam intersecting the ground. With a conventional radar, a target at point A is resolvable only to the beamwidth of the radar antenna. However, since each one of the hyperbolic lines represents a different Doppler frequency, it is possible to resolve targets within the beam with filtering. Considerable resolution improvement can be achieved by this technique; a radar that provides Doppler resolution is called a *coherent* radar.

There are a number of properties of a coherent radar which are deserving of note. One is that its signal is the result of a CW oscillator driving a gated microwave amplifier which is pulsed by a

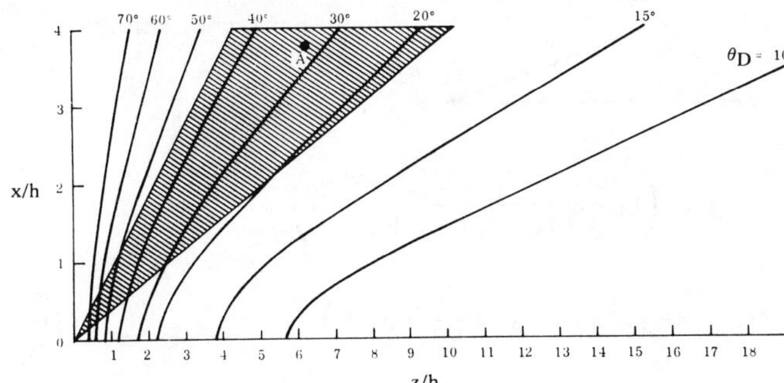

$$z^2\frac{\tan^2\theta_D}{h^2} - \frac{x^2}{h^2} = 1 \tag{3}$$

h = altitude

$$f_D = \frac{2v}{\lambda}\cos\theta_D \tag{1}$$

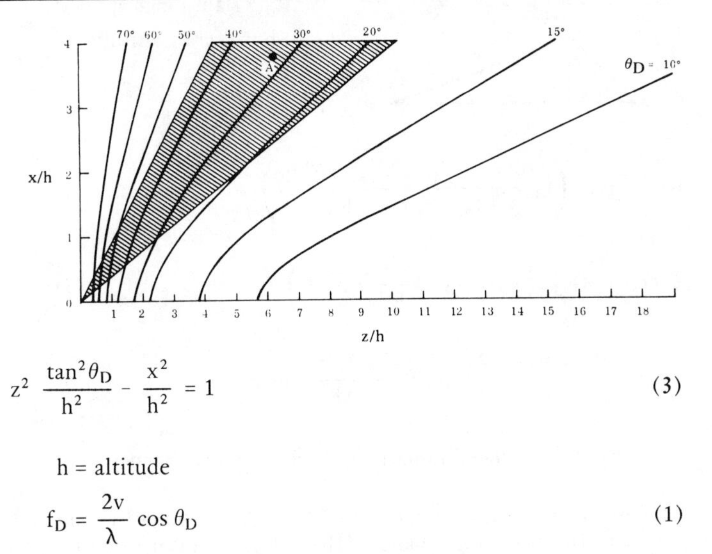

Figure 2 **Isodops**

modulator (see Figure 3). That same signal from the CW oscillator is used for the local oscillator of the mixer in the receiver. In such a radar, there is a phase detector supplied by the oscillator and by the received signal. The output from such a device is proportional to the cosine of the phase of the received signal with respect to the local oscillator. In other words, the oscillator is used as a timing device and measures the time out to the target and back in terms of phase. This measurement is highly ambiguous, but nevertheless, it is a very accurate measure of location of the target.

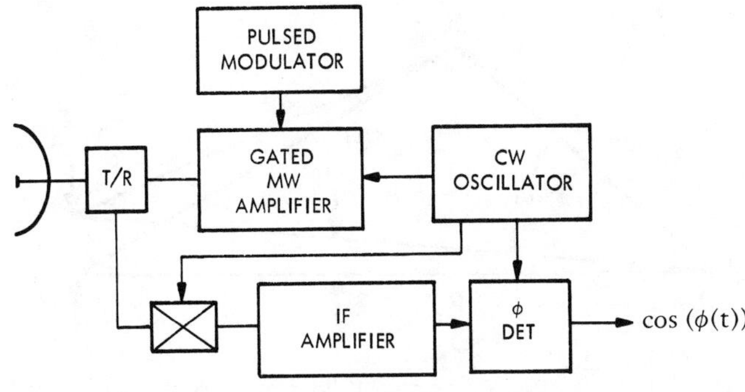

Figure 3 **Coherent Radar**

In Figure 4 is a set of general equations. The aircraft is flying along the line AB with velocity v; vt is the distance traveled during data collection. There is an initial line-of-sight from the aircraft to target C, AC; a later position, B, is a function of vt. In Equation (4), θ_D is the Doppler angle, R(t) is the slant range of the target as a function of time, and h is the altitude of the aircraft above the ground. The plane triangle ABC is useful for finding R(t). It is im-

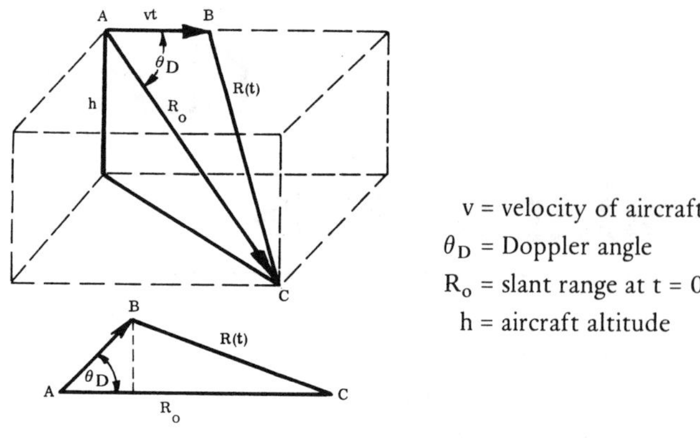

v = velocity of aircraft
θ_D = Doppler angle
R_o = slant range at t = 0
h = aircraft altitude

$$R(t) = R_o \left(1 - \frac{vt\cos\theta_D}{R_o} + \frac{v^2t^2\sin^2\theta_D}{2R_o^{\,2}}\right) \quad (4)$$

$$\phi(t) = \frac{2\pi}{\lambda}\left(2\,vt\cos\theta_D - \frac{v^2t^2}{R_o}\sin^2\theta_D\right) \quad (5)$$

$$f_D = \frac{1}{2\pi}\frac{d\phi(t)}{dt} = \frac{2v}{\lambda}\cos\theta_D - \frac{2v^2t}{\lambda R_o}\sin^2\theta_D \quad (6)$$

Figure 4 **General Equations of Phase and Frequency**

portant to note in Equation (4) that R(t) is a function of vt and v^2t^2. The two-way range change, $2[R(t)-R_o]$, as a function of time times $2\pi/\lambda$ (Equation (5)) indicates the phase, $\phi(t)$, of the signal coming out of the phase detector in the receiver shown in the block diagram of Figure 3. This phase is also a function of vt and v^2t^2. The derivative of the phase times $1/2\pi$ is the frequency, f_D, of the received signal (Equation (6)).

Equation (6) is used to examine the sort of frequency resolution needed to resolve two targets — one at the point C and one at a point some distance, S_a, away (see Figure 5). For this resolution, the difference in frequency of these two targets is designated by f_Δ in Equation (7). By expanding the arguments of the sine and cosine and assuming that the angle, Δ, which separates the two targets (as noted in Figure 5) is very small, one can simplify this equation from the general case by dropping terms of Δ^2 and higher, giving Equation (8). One can further simplify by taking a case where θ_D is 90°, making the target broadside to the flight path of the aircraft, in Equation (9) and Figure 6. Equation (9) is a very simple equation — f_Δ is twice the velocity over wavelength multiplied by the small angle, Δ, which separates the two targets.

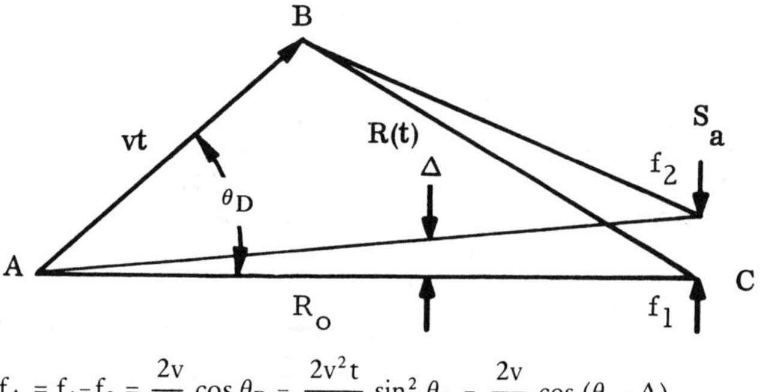

$$f_\Delta = f_1 - f_2 = \frac{2v}{\lambda}\cos\theta_D - \frac{2v^2t}{\lambda R_o}\sin^2\theta_D - \frac{2v}{\lambda}\cos(\theta_D - \Delta) + \frac{2v^2t}{\lambda R_o}\sin^2(\theta_D - \Delta) \quad (7)$$

$$f_\Delta = \frac{-2v}{\lambda}\Delta\sin\theta_D - \frac{2v^2t}{\lambda R_o}\Delta\sin 2\theta_D \quad (8)$$

$$f_\Delta = -\frac{2v}{\lambda}\Delta \text{ For } \theta_D = \frac{\pi}{2} \quad (9)$$

Figure 5 **Resolving Power of Filtering**

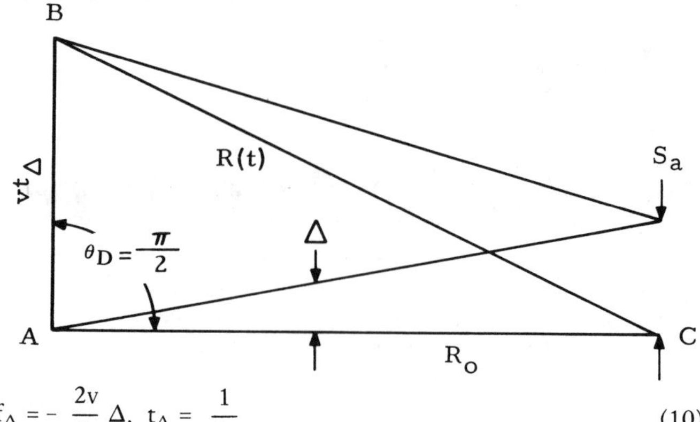

$$f_\Delta = -\frac{2v}{\lambda}\Delta,\ t_\Delta = \frac{1}{f_\Delta} \quad (10)$$

$$\Delta = \frac{\lambda}{2vt_\Delta} = \frac{\lambda}{2L} = \frac{S_a}{R_o} \quad (11)$$

$$S_a = \frac{\lambda R_o}{2vt_\Delta} \quad (12)$$

$$\Delta f_D = -\frac{2v^2t_\Delta}{\lambda R_o} \quad (13)$$

Figure 6 **Doppler Resolution $\theta_D = \pi/2$**

Any filter needs a time of 1/f (where f is the filter bandwidth) for the filter to build up. Substituting in Equation (10) for f_Δ gives the angle, Δ, needed for that frequency resolution. That angle could be defined as the wavelength over twice L, where L is equal to vt_Δ, the length of travel of the aircraft in time t_Δ. Those familiar with antenna theory should recognize this equation as similar to that of the resolution of a uniformly illuminated antenna. It is important to note that, in this type of processing, the length of the synthetic antenna is twice as important for resolving targets. This difference is due to the use of two-way phase in developing the synthetic antenna; with an ordinary antenna, a signal is transmitted to the target, and only the one-way difference in phase helps to resolve targets. The angle Δ can be related to the distance S_a as in Equation (11); solving for S_a yields Equation (12). The change in Doppler frequency over the processing period is given by Equation (13) which follows from Equation (6) for $\theta_D = 90°$.

For a simple example of the necessary width and process of resolution of a filter, Equations (10), (12), and (13) are used. Assume a range of 20 nmi, a resolution of 12 ft (in other words, it is desired to resolve two targets 12 ft apart at 20 nmi), an aircraft velocity of 500 ft/s, and an X-band radar wavelength of 1/10 ft. In this case, since such a tiny angle Δ is created, the necessary frequency resolution, from Equation (10), is one hertz. Also from that equation, $t_\Delta = 1$ and the aircraft would travel 500 feet. During the aircraft travel of 500 feet, the angle to the target changes and the center frequency of the filter would change by 41.7 Hz, according to Equation (13). Circuit designers would realize that such a device would be a rather tough analog filter to build. It has been done, and it can be done, but when you do it for a large

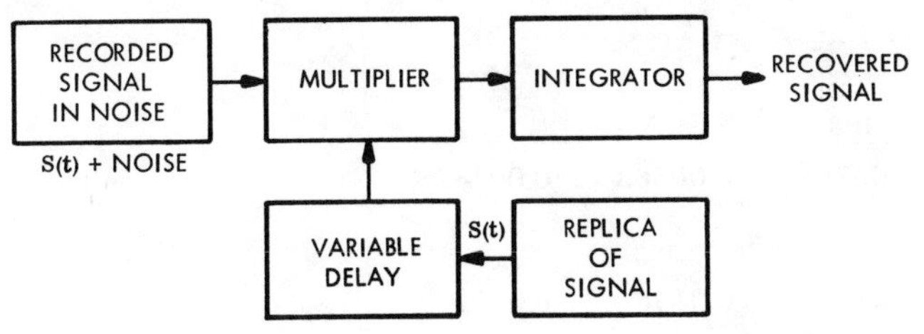

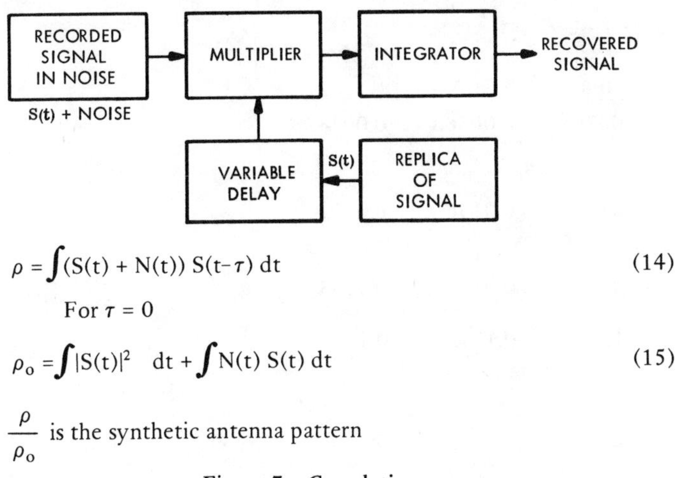

$$\rho = \int (S(t) + N(t))\, S(t-\tau)\, dt \qquad (14)$$

For $\tau = 0$

$$\rho_o = \int |S(t)|^2 \, dt + \int N(t)\, S(t)\, dt \qquad (15)$$

$\frac{\rho}{\rho_o}$ is the synthetic antenna pattern

Figure 7 Correlation

$$\cos\left(\frac{2\pi v^2 t^2}{\lambda R_o}\right) = \text{reference function for sidelooking case} \qquad (16)$$

$$\rho = \int_{-T}^{T} \cos\left(\frac{2\pi v^2 t^2}{\lambda R_o}\right) \cos\left(\frac{2\pi v^2 (t-\tau)^2}{\lambda R_o}\right) dt \qquad (17)$$

$$\rho_o = \int_{-T}^{T} \cos^2\left(\frac{2\pi v^2 t^2}{\lambda R_o}\right) dt \qquad (18)$$

$\frac{\rho}{\rho_o}$ = synthetic antenna pattern

Figure 8 Focused Antenna

number of range elements, the process gets difficult. The detailed calculations and results for this example are listed below.

R_o = 20 nmi (range to target)

S_A = 12 ft (separation of two resolvable targets)

v = 500 ft/s (aircraft speed)

λ = 0.1 ft (X-band)

$$f_\Delta = \frac{2v\Delta}{\lambda} = \frac{2 \times 500 \times 12}{.1 \times 1.2 \times 10^5} = 1.0 \text{ Hz}$$

$$vt_\Delta = \frac{\lambda R_o}{2S_A} = \frac{.1 \times 1.2 \times 10^5}{2 \times 12} = 500 \text{ ft}$$

$$\Delta f_D = -\frac{2 \times 2.5 \times 10^5 \times 1}{.1 \times 1.2 \times 10^5} = 41.7 \text{ Hz}$$

To get a device which does essentially the sort of thing discussed above, one builds the correlator shown in Figure 7. It produces the same filter that moves in center frequency with the signal. Basically, the correlator consists of a multiplier through which the recorded signal is multiplied against a delayed replica of the transmitted signal, then integrated (Equation (14)). But, where does one get the replica of the signal for use in the correlator? The received signal is known when one knows the aircraft velocity and the target location; the phase and frequency history can be computed as functions of time from Equations (5) and (6). For every target on the ground, the expected return can be computed – the results are what is meant by a replica of the signal.

In such a system, true signal goes from the receiver phase detector to a recorder. The problem that arises, though, is that it is not known when the signal will be received. Therefore, it is necessary to utilize a variable delay to control the time at which the replica enters the multiplier. This process is repeated a number of times until the signal is found. When there is a single target, say the planet Venus, the process can be programmed on a computer.

For the delay of the replica to be equal to the radar signal delay due to range (in other words, when the signal and the replica are in the same time phase, or τ in Equation (14) is equal to zero), there is a signal squared term related to the power of the signal (Equation (15)). Since N(t) is noise, it is not a replica of S(t); the integral of the product of the two disjoint signals is very close to zero in the second term of Equation (15). If one takes the ratio of the correlator output for τ varying from -T to +T, to that for τ equal to 0, the result is the synthetic antenna pattern.

At θ_D equal to 90°, the second term in Equation (5) predominates, and the output of the phase detector of the radar receiver is $\cos \phi(t)$. In the side-looking case, the reference function replica of the signal is the cosine function shown in Equation (16) of Figure 8. The correlation function, ρ, would be written as shown in Equation (17), where the first term of the integrand is the reference function, representing the computed replica of a target return at 90° to the aircraft flight path and at a range of R_o. The second term of the product represents the return from a target which begins forward of broadside to the aircraft and ends aft. The correlation function is maximum when the target return is properly aligned in time with the signal replica (Equation (18)). This happens when the alignment delay τ is zero. The ratio of ρ to ρ_o varies with τ and represents the synthetic antenna pattern. These integrals cannot be solved in closed form due to the cosine squared term.

Figure 9 shows the computer runs for the ratio of ρ to ρ_o, the antenna pattern. Note that the amplitude response of any curve is a function of the delay, τ. The narrowness of the response is a function of the limits of integration, T. The larger T, the more radar target data is collected and the longer the synthetic aperture. In this particular case, the return from the target is assumed to be uniform; as with a real antenna with uniform illumination, the sidelobes are between -13 and -14 dB.

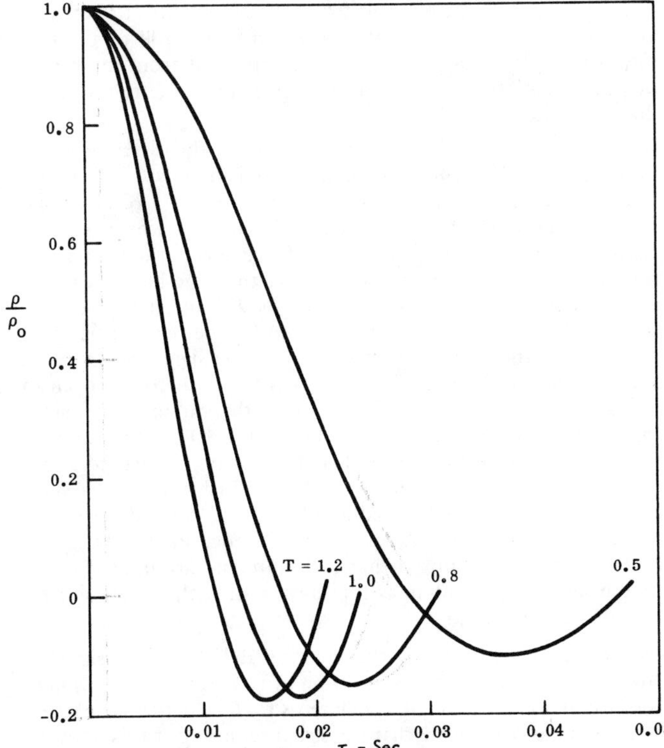

Figure 9 Synthetic Antenna Pattern

	T (seconds)			
	0.5	0.8	1.0	1.2
$\frac{\rho}{\rho_o} = 0.707$	$\tau = 0.01225$	0.00745	0.00588	0.004875
$\frac{\rho}{\rho_o} = 0$	$\tau = 0.0283$	0.0171	0.0136	0.0112
1st Sidelobe	$\tau = 0.0366$	0.0231	0.0186	0.0155
	Amplitude = -0.1075	-0.1535	-0.171	-0.1769

Example:

R_o = 20 nmi (range to target)
v = 500 ft/s (Aircraft Speed)
λ = 0.1 ft (X-band)

For

T = 0.5 and τ = 0.01225
2vT = 2 x 500 x 0.5 = 500 ft (synthetic antenna length)
$2v\tau$ = 2 x 500 x 0.01225 = 12.25 ft resolution

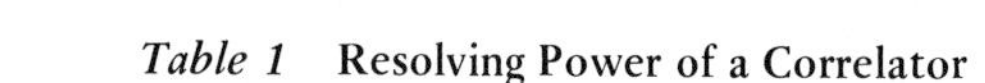

Table 1 **Resolving Power of a Correlator**

Table 1 uses the same example for the correlator that was used previously for the filter; the various figures are called off from the curves in Figure 9 for different limits of integration or, equivalently, time of travel of the aircraft. The first line is the half-power point; the slippage — the time delay between the reference function and the true signal — is 12 ¼ milliseconds. The first zero is at 28 milliseconds; the first sidelobe peak is at τ equal to 37 milliseconds with an amplitude of -0.1075. The example specifications are as noted earlier — R_o is 20 nmi, the velocity of the aircraft is 500 ft/s, and the wavelength is 0.1 ft at X-band. The synthetic antenna length is still 500 ft, the resolution is 12.25 ft, a quarter of a foot different from the previous case. Essentially, this correlator gives the same results as does the filter.

Before proceeding, it is necessary to highlight the differences between a focused antenna and an unfocused antenna (see Figure 10). Observe in the upper half of Figure 10 that a received signal from the target has a path delay at the edge which is an eighth of a wavelength greater than the delay at the center. If the antenna is any longer, the signal at the edge has a phase, with respect to the center, that exceeds 90° and it begins to subtract from the summation collected by the balance of the antenna. As the antenna is made longer, the signal becomes smaller instead of larger. This length is the limit for a linear antenna, but it can be avoided by curving the antenna and focusing it, as shown in the lower half of Figure 10. The problem with this solution, though, is that the antenna is focused at only one range. But a processor can be built which produces a synthetic antenna that is focused at all ranges. Most radar applications deal with a linear antenna which is focused at infinity; often, it is not recognized that such antennas are limited in length because the antennas typically dealt with are far short of this maximum length.

At the top of Figure 11 are the equation and plot of the signal from a broadside target as a function of time. The time at which the aircraft is broadside of the target is t = 0. The fortuitous assumption has been made that the signal returning at this point is of such a phase that the detector response is maximum, but that

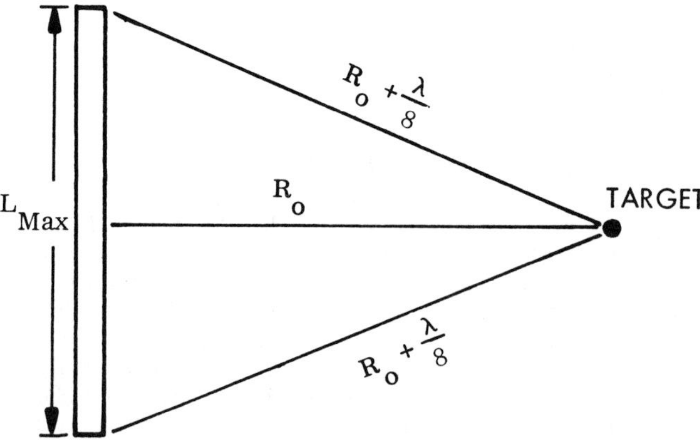

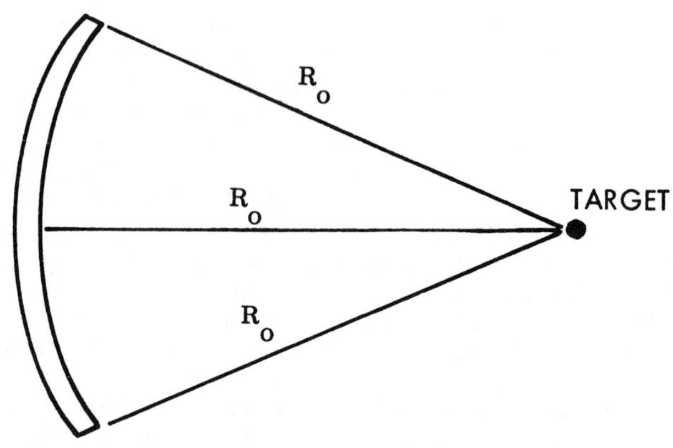

Figure 10 Focused and Unfocused Antennas

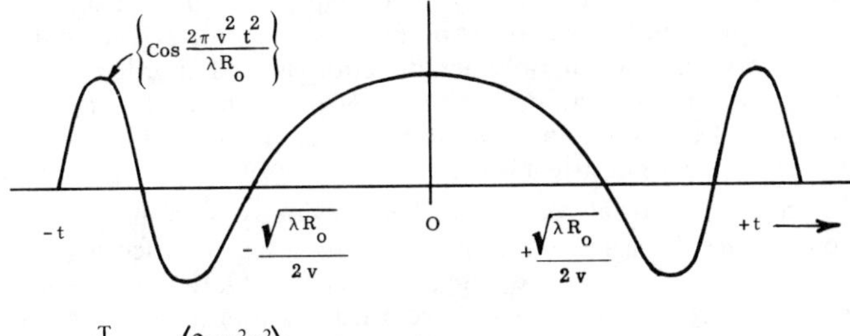

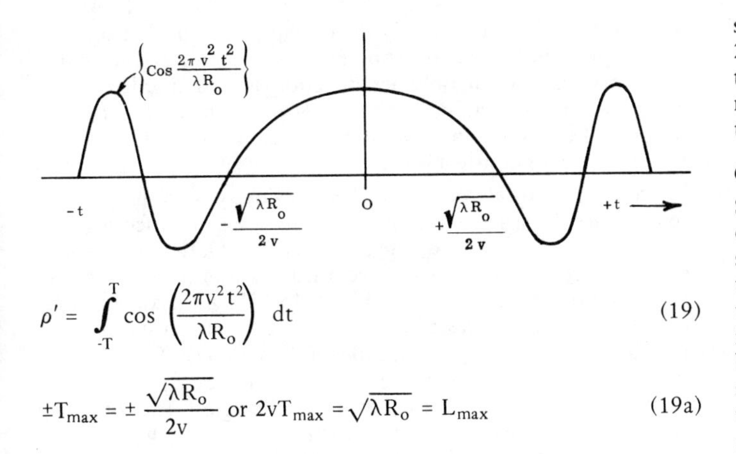

$$\rho' = \int_{-T}^{T} \cos\left(\frac{2\pi v^2 t^2}{\lambda R_o}\right) dt \qquad (19)$$

$$\pm T_{max} = \pm \frac{\sqrt{\lambda R_o}}{2v} \text{ or } 2vT_{max} = \sqrt{\lambda R_o} = L_{max} \qquad (19a)$$

Example:

R_o = 20 nmi (Range to Target)

λ = 0.1 ft (X-band)

S_{max} = 55 ft

L_{max} = 110 ft

Figure 11 Unfocused Antenna

assumption is not necessarily true. However, there are ways of handling this problem, though there is no need to complicate the situation with those issues at this point. Again, the two first zeroes of the curve represent the 90° points – the usable limit of integration for an unfocused linear synthetic antenna.

One way to implement an unfocused antenna is to use a correlator with a constant reference function. For instance, if the reference function is the constant one (which is omitted), the correlation integral is that shown in Equation (19). The limits, $\pm T_{max}$ of Equation (19a), are the values of t for which the argument of the cosine in Equation (19) equals $\pm\pi/2$. As in the examples for R_o = 20 nmi and λ = 0.1 ft, the maximum resolution that can be obtained is 55 ft because the length of the antenna in the unfocused mode is limited, in this case to 110 ft. In the other example, though, L was 500 ft – an arbitrary value, and not a limiting one.

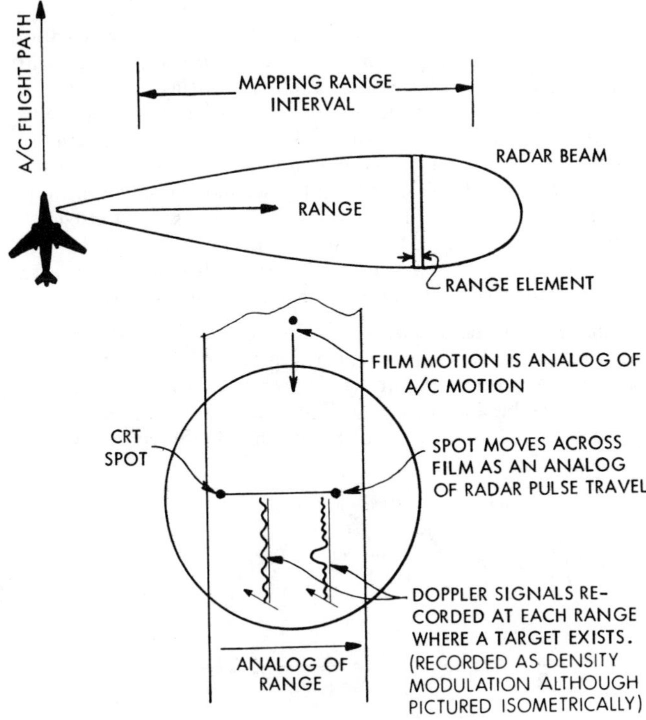

Figure 12 Signal Recording

Optical Correlation

Suppose one wanted to record Doppler radar signals on film – consider the true situation. An aircraft flies along a flight path as shown in Figure 12, with a broadside radar beam illuminating the terrain. The beam is wide due to a small real antenna. A pulse is transmitted which slides out along the radar beam. The mapping interval is as indicated. An analog recording of the echo signal can be made on film. Suppose that there is a cathode ray tube with a spot which moves across the tube face in the time during which the radar pulse is required to cover the range interval. The film moves in an analog fashion in relation to the aircraft velocity. If at every analog range on the film where a target exists, the radar signal amplitude modulates the CRT, there is a density function on the film that represents the signal (see Figure 12). All the signals can be recorded in range provided that there is enough resolution in the CRT and enough light to modulate the film. A record of all the target Doppler data from each range interval thus is produced on the film. A correlator can be built using this data.

Before discussing optical correlators, it is necessary to explore optical operators which involve some very simple ideas. Suppose a plane wave of light falls on a grating (at the left in Figure 13) which alternately passes and stops the light. In the vertical direction, one could conceptualize this situation as a space distribution of a square wave of light. If this distribution is at the front focal plane of a lens, then there is a diffraction grating at the rear focal plane. In electrical engineering terminology, the light at the rear focal plane is more readily thought of as the Fourier transform of the grating; in other words, the grating is the typical time domain and the diffraction pattern is the frequency domain. If the intensity of light is examined at the rear focal plane, the vertical linear distribution of the light is that which one would find in an electrical case with the proper scaling. Thus, the rear focal plane is the frequency domain and the front focal plane is the time domain – one signal is the Fourier transform of the other, as represented in Equation (20).

It is obvious that if filtering certain frequency lines at the rear focal plane were desired, slits could be utilized to pass desired signals and to block out all others; thus, a frequency filter would be produced in this optical system. Also, since the lens is circular, only a cut through its center is shown; the same kind of operation can be executed in two dimensions, and some other topics to be discussed involve operations in two dimensions. However, the latter two concerns are independent. Another point is that this lens system can be used as an integrator. In the Fourier transform in Equation (21), f(x) is multiplied by exponential $-j\omega x$. Suppose

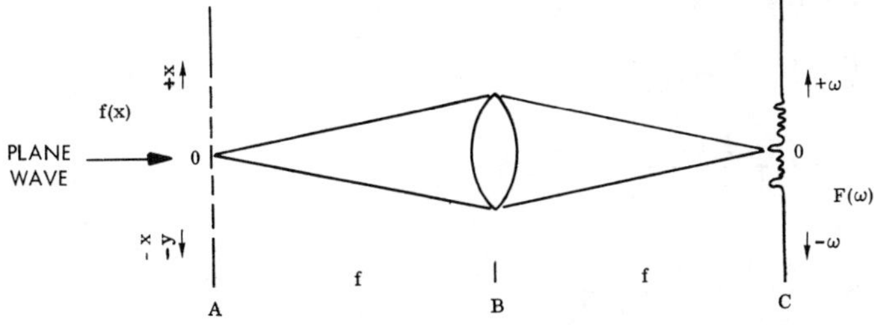

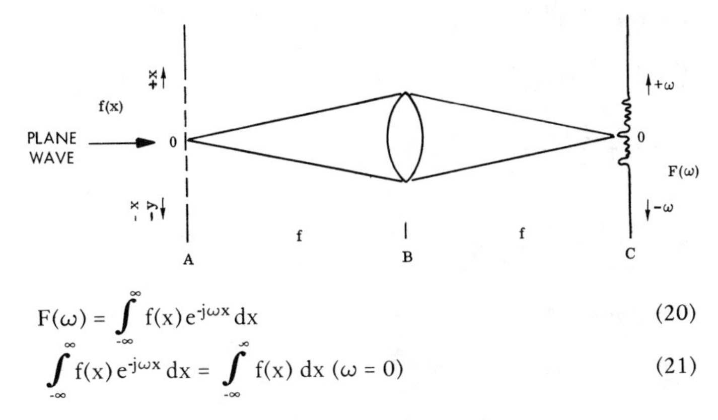

$$F(\omega) = \int_{-\infty}^{\infty} f(x) e^{-j\omega x} dx \qquad (20)$$

$$\int_{-\infty}^{\infty} f(x) e^{-j\omega x} dx = \int_{-\infty}^{\infty} f(x)\, dx \quad (\omega = 0) \qquad (21)$$

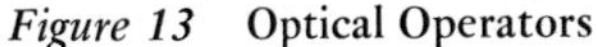

Figure 13 Optical Operators

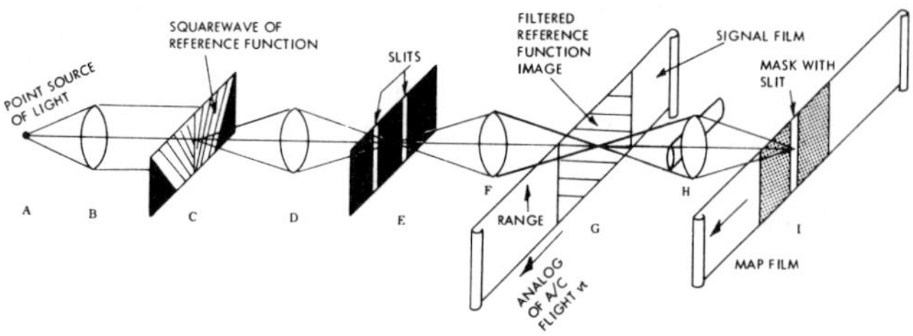

Figure 14 **Optical Correlator**

that ω equals zero. Then e is raised to the power zero and is equal to one. The result is a simple integrator. If one examines the signal light on the axis through a very narrow slit at the rear focus, the result is the integral of the signal, f(x).

A complete optical correlator is shown in Figure 14. If there is a point source of light at A (the front focus of a lens at B), parallel light leaves the lens, producing a plane wave of light. On the transparency at C is scribed the reference function at every analog range. From Equation (16) it follows that the reference function is an inverse function of range; thus, the frequency is higher the closer the target. It is very difficult to produce a sine wave function on a transparency, but it is very easy to make a square wave function. What is done is to make a square wave of the computed Doppler signal that should be received from the target at every range. Then the Fourier transform of the square wave is taken with the lens at D and passing only the fundamental component of the signal spectrum at E; the filtered spectrum is then inversely transformed with the lens at F. At G is an image of the true reference function; in other words, the filter at E has eliminated all the square wave components except the fundamental. If that reference function shines on a piece of film which carries the actual received signals recorded as shown in Figure 14, the light passing through the reference function on film is the multiplication of the reference function with the signal that is received. As the film moves along, the variable delay of the correlation process is created; thus, the signal is multiplied by the reference, and they slide past each other at point G. The result is an analog representation of the integrand of Equation (17); an integrator must follow. The lens, H,

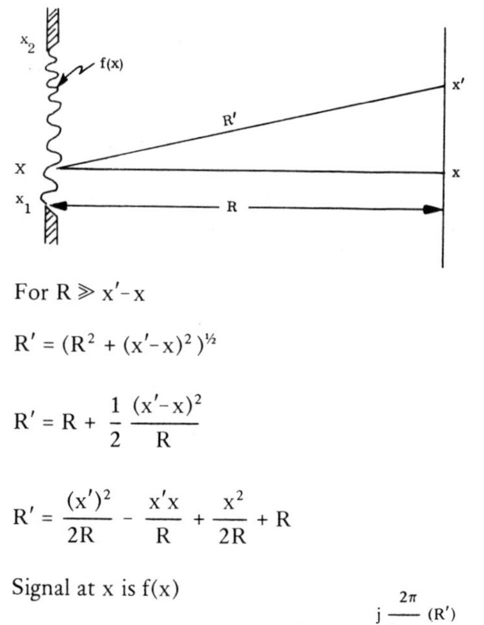

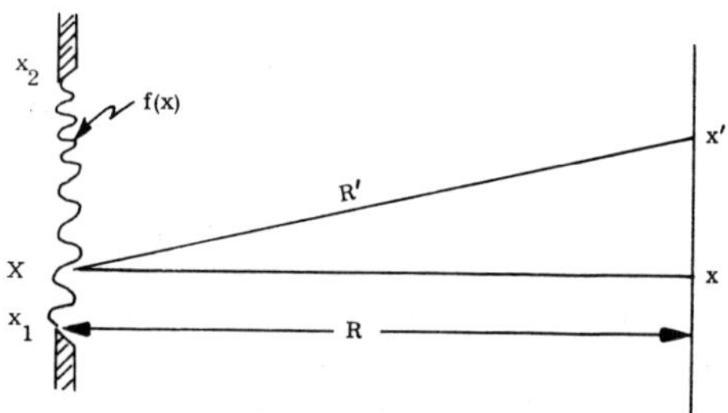

For $R \gg x'-x$

$$R' = (R^2 + (x'-x)^2)^{1/2}$$

$$R' = R + \frac{1}{2}\frac{(x'-x)^2}{R}$$

$$R' = \frac{(x')^2}{2R} - \frac{x'x}{R} + \frac{x^2}{2R} + R$$

Signal at x is f(x)

Signal at x′ due to signal at x is $f(x)\, e^{j\frac{2\pi}{\lambda_L}(R')}$ (22)

Figure 15 **Zone Plate**

serves as this integrator; if only the output is examined at the on-axis slit at point I, the correlation process is complete. The cylinder lens, with power in only one direction, is used in order to image the analog range at G to the range on the map film at I, one for one. The lens has no power in the horizontal direction and, therefore, does not affect our spectrum analysis.

In actuality, a correlator as described previously is not necessary to correlate the signal – the signal film itself has self-focusing properties. A signal f(x) (of the form cosine ax^2), the type of signal received from a target, is recorded as a transparency on film for the correlator (at the left in Figure 15). If light passing through the film at point x is described as f(x), and travels a distance R′ to x′ at the right, it can be described in Equation (22) as $f(x)e^{j2\pi R'/\lambda_L}$, where λ_L is the light wavelength. The algebraic calculation of R′ is shown at the bottom of the figure. The total light at x′ on the right has contributions from the full length of the transparency and must be summed vectorially; this process is discussed below.

Zone Plate Equations:

$$F(x') = \int_{x_1}^{x_2} f(x)\, e^{j\frac{2\pi}{\lambda_L}(R')}\, dx \qquad (23)$$

$$R' = \frac{(x')^2}{2R} - \frac{x'x}{R} + \frac{x^2}{2R} + R$$

$$\text{Let } f(x) = \cos ax^2 = \frac{e^{j\,ax^2}}{2} + \frac{e^{-j\,ax^2}}{2}$$

$$F(x') = e^{j\frac{2\pi}{\lambda_L}R}\, e^{j\frac{2\pi}{2}\frac{(x')^2}{2R}}$$

$$\cdot\left\{\frac{1}{2}\int_{x_1}^{x_2} e^{j\,ax^2}\, e^{j\frac{2\pi}{\lambda_L R}\frac{x^2}{2}}\, e^{-j\frac{2\pi}{\lambda_L R}x'x}\, dx + \frac{1}{2}\int_{x_1}^{x_2} e^{-j\,ax^2}\, e^{j\frac{2\pi}{\lambda_L R}\frac{x^2}{2}}\, e^{-j\frac{2\pi}{\lambda_L R}x'x}\, dx\right\} \qquad (24)$$

This group of expressions is formidable; however, the implications must be understood. Equation (23) is the vector sum of the light passing through the transparency and contributing to the light at point x′. When the substitutions shown for R′ and f(x) are made, the rather unpalatable equation for F(x′) results. Consider the first integral – if a in the first exponential is made equal to and the negative of the coefficient of x^2 in the second exponential, the terms in x^2 drop out. In a similar manner in the second integral the term in x^2 can be made to cancel, when a is equal to the negative of its value for the first integral. The result is an easy integration in exponential x, the essence of the equations above.

The above simplification of the first integral of Equation (24) occurs when either R is negative or positive for corresponding a's that are of opposite sign but equal magnitude. Thus, there must be foci – one behind the signal film at -R; one in front at +R. Consider the second integral of Equation (24) shown below as Equation (25).

Zone Plate Focus Equations:

$$I_2 = \frac{1}{2}\int_{x_1}^{x_2} e^{-j\,ax^2}\, e^{j\frac{2\pi}{\lambda_L R}\frac{x^2}{2}}\, e^{-j\frac{2\pi}{\lambda_L R}x'x}\, dx \qquad (25)$$

$$\text{Let } a = \frac{\pi}{\lambda_L R},\quad R = \frac{\pi}{a\lambda_L};\quad x_\mu = \frac{x_1 + x_2}{2},\quad \Delta x = x_2 - x_1 \qquad (25a)$$

$$I_2 = \frac{1}{2} \int_{x_\mu - \frac{\Delta x}{2}}^{x_\mu + \frac{\Delta x}{2}} e^{-j2ax'x} dx = e^{-2\,ax'x_\mu} \cdot \frac{1}{2} \Delta x \frac{\sin ax'\Delta x}{ax'\Delta x} \quad (26)$$

I_2 is max at $x' = 0$

$$I_2 \text{ at 3 dB points, } x' = \pm\frac{0.445\,\pi}{a\Delta x} = \pm .445 \frac{\lambda_L R}{\Delta x} \quad (26a)$$

To simplify the integral, a is made equal to the values in Equation (25a); R is positive. Carrying out the integration, note that the light focuses at the right at a distance R with the amplitude shown in Equation (26). The amplitude is a sin $\Delta x/\Delta x$ distribution which is maximum at x' equal to zero, which is on the center line of the distance from x_1 to x_2. The focus is a very small spot of light of dimension given by Equation (26a).

The results of the equations above are illustrated in Figure 16. There is a virtual focus on the left and a real focus on the right. The virtual focus gives a highly defocused spot of light (spread function) at the right focal plane; it interferes somewhat with the real focus and fogs the film. If there were a slit at the right focal point when the signal is of the form cos (ax^2), light would pass through the slit. Some of the light would be the defocused spread function of the virtual focus.

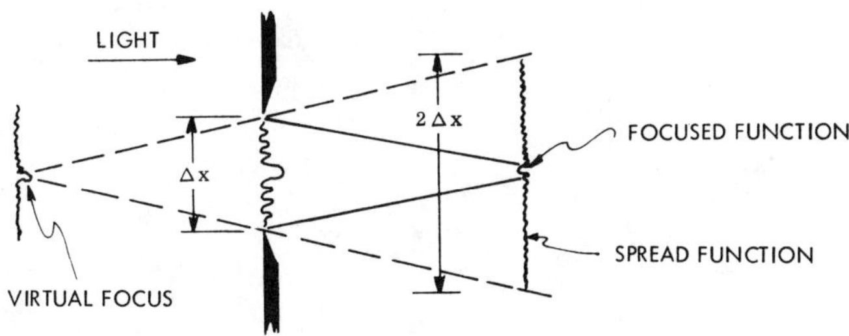

Figure 16 On Axis Focus

If the signal is on a carrier, the focus is centered about the carrier frequency and both foci are shifted so that the spread function is out of the slit, as shown in Figure 17. In this manner, the spread function is prevented from interfering with the signal focus. There is one difficulty with this approach – the focal lengths are on the order of hundreds of feet, so a lens must be inserted in the light path to shorten the distance. In principle, only one lens is needed to accomplish this reduction. This correlation is simpler and better than the type described previously. Needless to say, a computer could be used instead of an optical correlator.

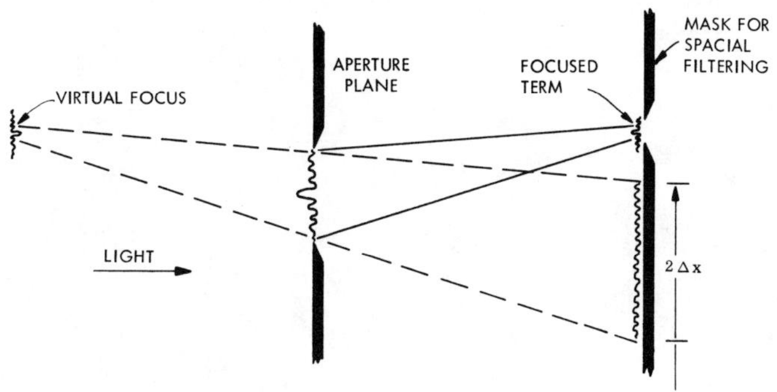

Figure 17 Off Axis Focus

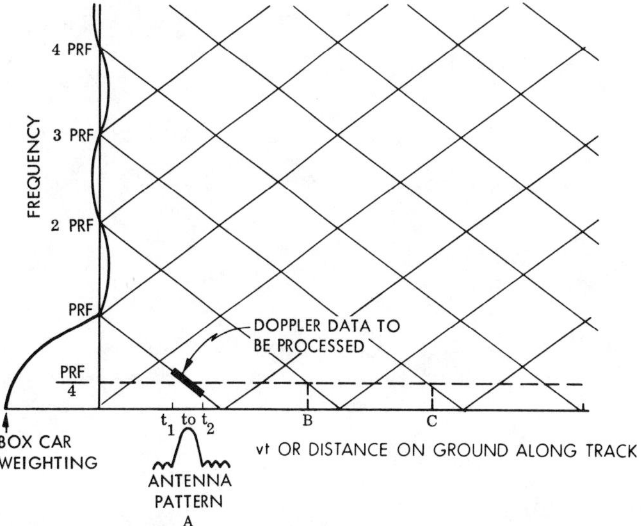

Figure 18 Ambiguities

The synthetic aperture radar is highly ambiguous if great care is not employed in the design. With a pulsed radar, there are spectral lines at the pulse repetition frequency about the carrier. Suppose that the pulse repetition frequency is 4,000 Hz – in the spectrum, the carrier is surrounded on both sides by many RF lines spaced 4.000 hertz apart. Figure 18 shows these repeated spectral lines at base band; there is a dc term and a series of PRF lines, shown respectively as the first, second, etc., PRF frequencies. The desired radar data modulates each PRF line; that is, there is an identical signal spectrum about each PRF line. The diagonal lines drawn from each PRF show the signal spectrum associated with each line for a single target fore or aft of broadside. Note that the IF spectrum is folded at video. It is usual practice to box car the video signals so that the signal energy around each PRF line is folded about the carrier. The box car weighting shown on the left is the filter characteristic. If targets at point A are being processed, the real antenna pattern should illuminate only this area. If the antenna should illuminate the ground at point B, this data would be processed by the correlator as readily as the data at point A. To prevent the processing of ambiguous data, care must be taken not to use too wide a real beam and to direct the beam to the area to be processed.

Examples of high resolution maps resulting from synthetic aperture processing are shown in Chapter 17.

Chapter 17 Kovaly

High Resolution Radar Fundamentals (Synthetic Aperture & Pulse Compression)

Radar had its biggest impact during the second World War. Since that time, the major innovations in the field of radar have been limited – the synthetic aperture, pulse compression, and digital signal processing. This chapter is on Synthetic Aperture Radar (SAR), a class of high resolution radar which obtains fine angular resolution by coherent processing of backscattered Doppler histories. This chapter serves as a synopsis of the fundamentals which are covered in more detail by the references.

Figure 1 defines the geometry of a side-looking radar system; the assumption is that an airborne vehicle carries on-board a coherent radar. The physical aperture of the radar generates an RF beam whose beamwidth is β. It illuminates the shaded patch on the ground with RF energy as the aircraft moves in the direction of the flight path. A point, P, on the ground, is shown just entering the beam; its range at that time is R. As the aircraft moves, the point P traverses the RF beam. It arrives at the point of closest approach when it is broadside at a range R_o. It continues on through the beam and has its original range, R, as it exits from the beam.

A block diagram of the coherent radar in the aircraft is shown in Figure 2. The coherent radar remembers the phase difference from transmission of a pulse to reception of a pulse; its stable local oscillator generates the carrier frequency, f_C. The transmitted pulse has a frequency modulation about the RF carrier which varies linearly from some high frequency (at the beginning of the pulse) to some lower frequency (at the end). The physical aperture radiates this energy, in the form of a beam, and receives the backscattered signal. The backscattered signal contains the carrier frequency, with a Doppler shift in frequency which is, in reality, its phase as a function of time. The receiver section conditions the signal. In the coherent detector the Doppler received signal is compared with the sum of a replica of the carrier frequency and a recording offset frequency, f_o. The recording frequency, f_o, is used to shift the Doppler spectrum away from zero frequency. The output of the coherent detector is the Doppler spectrum centered about f_o. After coherent detection the information enters a correlator and finally is applied to a display for visual observation. The top view of Figure 3 shows, from the Pythagorean theorem, that the slant range to point P is time varying. The Doppler shift in frequency is equal to the time varying range multiplied by the ratio of $2/\lambda$. After going into the detector and making the proper substitutions, the detected frequency is the offset frequency plus a time varying frequency. The coefficient of the time variable is to be recognized from the physics of circular motion as the centripetal acceleration of the point along the radar beam.

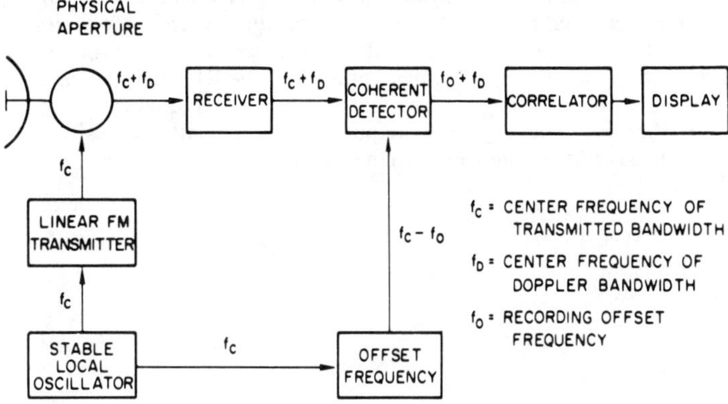

Figure 2 Coherent Radar System

The output of the coherent detector is called the Doppler frequency history; its characteristics are described in Figure 4. This frequency history for each scatterer is generated from the time when

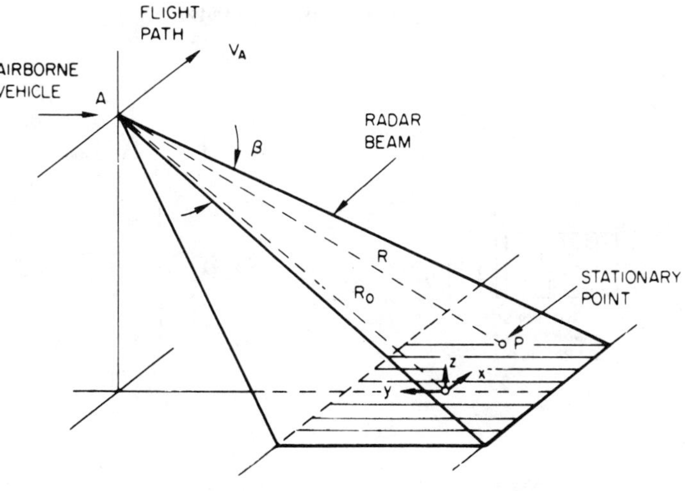

Figure 1 Side-Looking Radar System Geometry

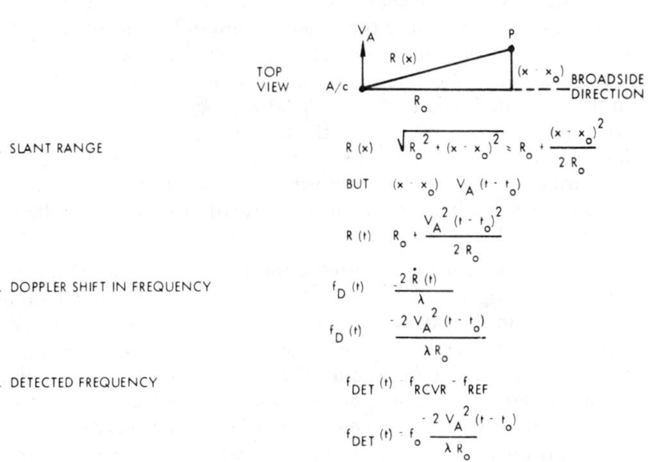

TOP VIEW
V_A
P
R (x)
A/c
R_o
$(x - x_o)$
BROADSIDE DIRECTION
1. SLANT RANGE
$R(x) = \sqrt{R_o^2 + (x - x_o)^2} \approx R_o + \frac{(x - x_o)^2}{2 R_o}$
BUT $(x - x_o) = V_A (t - t_o)$
$R(t) = R_o + \frac{V_A^2 (t - t_o)^2}{2 R_o}$
2. DOPPLER SHIFT IN FREQUENCY
$f_D(t) = \frac{-2 \dot{R}(t)}{\lambda}$
$f_D(t) = \frac{-2 V_A^2 (t - t_o)}{\lambda R_o}$
3. DETECTED FREQUENCY
$f_{DET}(t) = f_{RCVR} - f_{REF}$
$f_{DET}(t) = f_o + \frac{-2 V_A^2 (t - t_o)}{\lambda R_o}$

Figure 3 Doppler-Frequency of Stationary Ground Object (Broadside)

the backscatterer enters the beam to the time when it leaves the physical aperture beam — an interval of time, T. Each scatterer has the same Doppler history which goes from some high frequency to some low frequency. The total bandwidth which is generated over that time is the centripetal acceleration of the ground point multiplied by the ratio of T/λ.

Thus far, the discussion has described the received radar signal characteristics when the radar's beam is at a right angle to the aircraft's velocity vector. Similar signal characteristics are obtained when the radar beam is directed at a squint angle, θ_S, relative to the aircraft's velocity vector. Figure 5 describes the Doppler frequency history of a stationary ground object located at a squint angle.

Similar concepts are used to describe the signal characteristics of the transmitted pulse; they are summarized in Figure 6. Conventional pulse radars generate a pulse τ microseconds long with amplitude A. To obtain fine range resolution, the RF carrier incorporates a linear change in frequency from some frequency, f_1, to another frequency, f_2, over the pulse length, τ. Mathematically, the linear FM transmitter pulse has the following waveform:

$$f(t) = \cos\left(2\pi f_c t - \frac{2\pi\alpha t^2}{2}\right)$$

$$-\frac{\tau}{2} \leqslant t \leqslant +\frac{\tau}{2}$$

Elsewhere, $f(t) = 0$

Where:

f_C = Carrier Frequency

α = Rate of Frequency Sweep $(\Delta f)_R/\tau$

$(\Delta f)_R$ = Swept Spectrum Bandwidth

τ = Uncompressed Pulsewidth

The Detected Frequency is:

$$f_{DET}(t) = f_o - \alpha t$$

The detected frequency on a pulse-to-pulse basis, neglecting Doppler shifts for the time being, has the same form as the previously described Doppler frequency history. When this signal is passed through a correlator or a pulse compression device (see Figure 7) the output of the one dimensional correlator is the envelope function shown (with its characteristic sin x/x function). Because of the frequency coding of the pulse and the proper phase matching in the correlator, an accentuated response is formed where the height of the mainlobe is the original amplitude multiplied by the square root of the time bandwidth product. The width of this main pulse at its 4 dB point and the distance between the first two nulls are related to the time bandwidth product. For uniform weighting of the amplitude of the transmitted pulse, the first sidelobe of sin x/x function is 13.3 dB down. Sophisticated amplitude weighting techniques in the transmitter or receiver are used to reduce these sidelobe levels at the expense of broadening of the mainlobe width.

High resolution through synthetic aperture generation (Doppler frequency histories) and pulse compression (FM transmitted pulse) are linearly independent operations which can be combined, as in Figure 8. The three orthogonal coordinates are range (pulse), azimuth (Doppler) and amplitude. The figure shows that the output of the coherent detector appears as a series of pulses; each pulse has a duration which may be on the order of microseconds and a bandwidth which is on the order of megahertz. These pulses are the backscattered signals received by the radar and are stored in the correlator over a period of time, T, which is on the order of

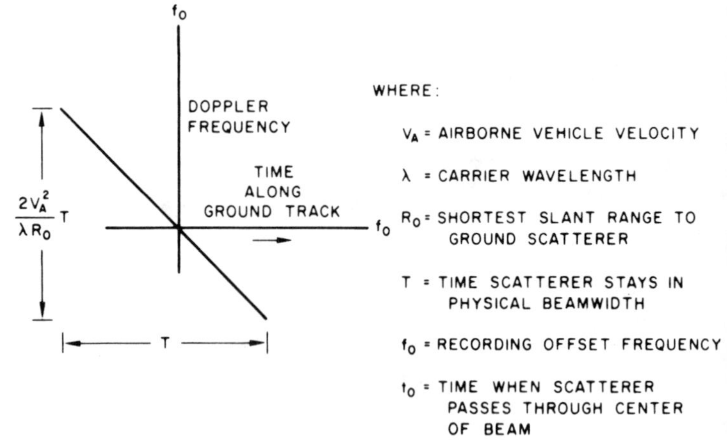

Figure 4 Doppler Frequency History

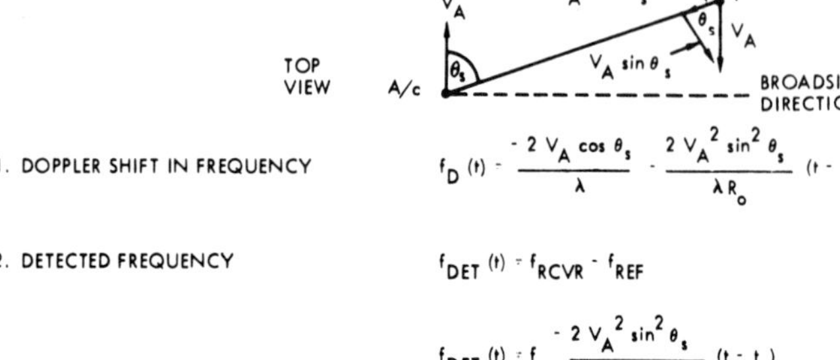

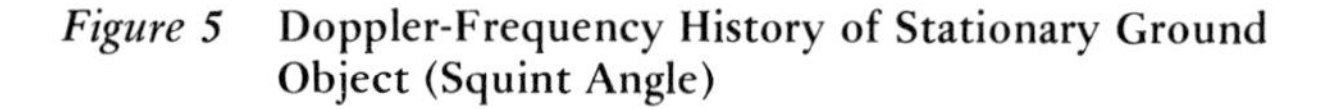

Figure 5 **Doppler-Frequency History of Stationary Ground Object (Squint Angle)**

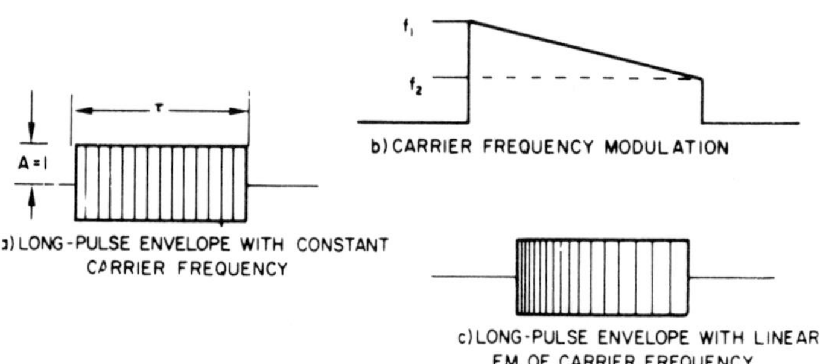

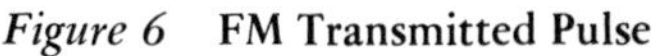

Figure 6 **FM Transmitted Pulse**

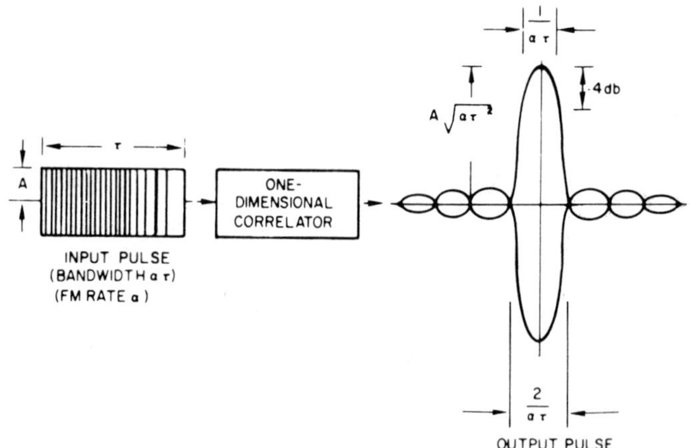

Figure 7 **One-Dimensional Correlator**

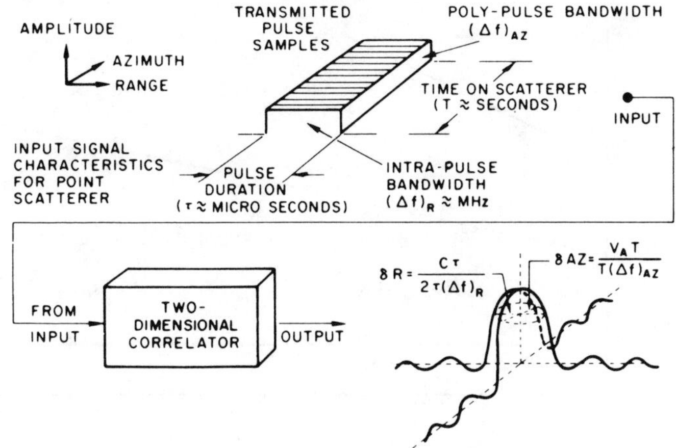

Figure 8 Two-Dimensional Correlator

seconds. Over T a Doppler bandwidth is generated. With this bulk of information as the input to a two-dimensional correlator, the output is a bi-variant sin x/x distribution. In the range dimension, the resolution is δR; in the orthogonal direction, the azimuth resolution is δA_z. Of interest are the similar forms of resolution in the two directions. Each has a numerator which is related to a velocity – in range, the velocity of light; in azimuth, the velocity of the aircraft. Each resolution function is inversely proportional to a bandwidth – in the range dimension, the RF bandwidth; in the azimuth dimension, the Doppler bandwidth. Fundamentally, then, resolution is obtained in both dimensions by generating a bandwidth. At this point in the system – when this resolution function has been manifested – a synthetic aperture has been generated. Comparisons of the two degrees of resolution are shown in Tables 1 and 2. Workers in the field have used the expression "Beam-Sharpening" for the increased resolution provided by the synthetic aperture.

	BEAM-SHARPENING	PULSE COMPRESSION
CODING	PULSE-TO-PULSE	WITHIN A PULSE
DIMENSION	AZIMUTH	RANGE
TIME SCALE	SEVERAL SECONDS	MICROSECONDS
BANDWIDTH	FEW HUNDRED HZ/S (DOPPLER BANDWIDTH)	MHZ/S
CODE GENERATION	LINEAR FREQUENCY MODULATION (AIRCRAFT FLIES STRAIGHT PATH)	LINEAR FREQUENCY MODULATION (ELECTRONICALLY)

Table 1 Comparison of Beam-Sharpening with Pulse Compression

RANGE DIRECTION	AZIMUTH DIRECTION
1. TIME WIDTH τ_R OF IMAGE AT -4 dB LEVEL $\tau_R = 1/(\Delta f)_R$ WHERE $(\Delta f)_R$ IS TRANSMITTING FM BANDWIDTH	1. TIME WIDTH τ_{Az} OF IMAGE AT -4 dB LEVEL $\tau_{Az} = 1/(\Delta f)_{Az}$ WHERE $(\Delta f)_{Az}$ IS DOPPLER BANDWIDTH
2. MULTIPLY BOTH SIDES OF ABOVE EQUATION BY C/2 WHERE C IS VELOCITY OF LIGHT (FACTOR OF 2 REQUIRED FOR TWO-WAY RANGE): $\frac{C\tau_R}{2} = C/2\,(\Delta f)_R$	2. MUTLIPLY BOTH SIDES OF ABOVE EQUATION BY AIRBORNE VEHICLE VELOCITY V_A: $V_A\ \tau_{Az} = V_A/(\Delta f)_{Az}$
3. HERE $\frac{C\tau_R}{2}$ IS DEFINED AS THE RANGE RESOLUTION δR: $\delta R = C/2\,(\Delta f)_R$	3. HERE $V_A\ \tau_{Az}$ IS DEFINED AS THE AZIMUTH RESOLUTION δAz: $\delta Az = V_A/(\Delta f)_{Az}$

Table 2 Comparison of Derivations for Range Resolution (δR) and Azimuth Resolution (δAz)

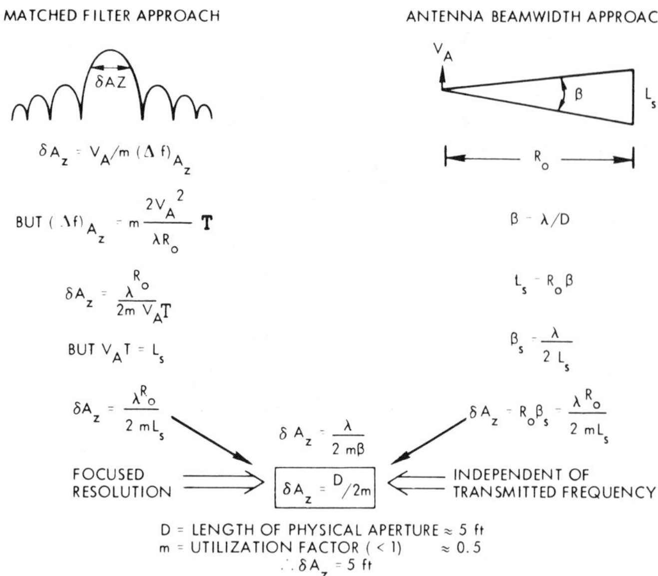

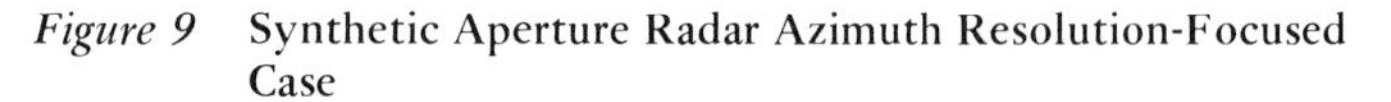

Figure 9 **Synthetic Aperture Radar Azimuth Resolution-Focused Case**

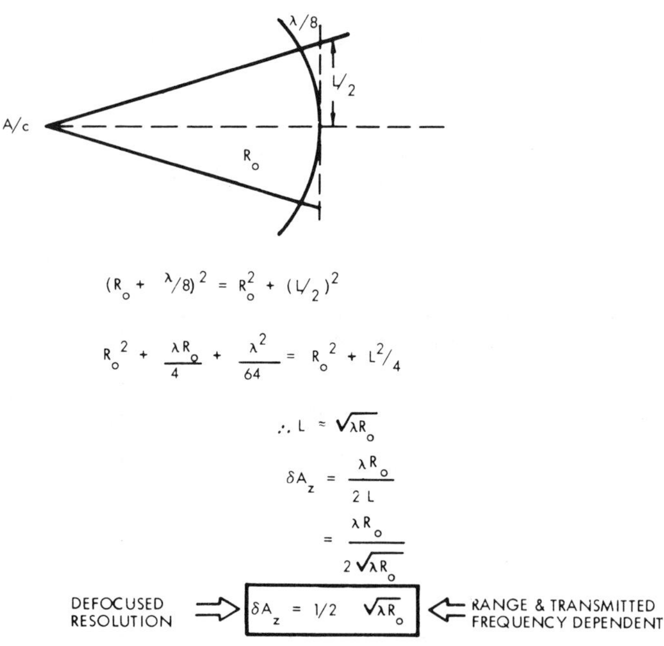

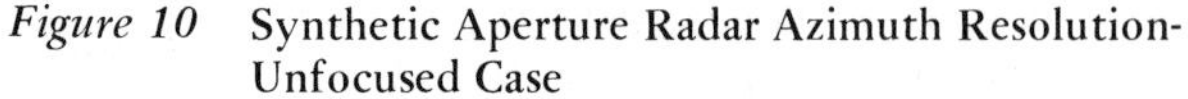

Figure 10 **Synthetic Aperture Radar Azimuth Resolution-Unfocused Case**

Depending upon the system requirements, the full degree of resolution can be obtained by processing all of the Doppler bandwidth and by compensating for the phase difference on a pulse-to-pulse basis. If the phase variations are not compensated for, an unfocused or lower level of resolution is obtained. Two approaches for describing the focused SAR azimuth resolution are shown in Figure 9. The more erudite matched filter approach starts with the relationships of vehicle velocity and Doppler bandwidth. An alternate approach begins with the resolution of the physical antenna. A constant term, m, (called the utilization factor) is introduced in both of these approaches. It can have a value between 0 and 1, depending upon the amount of Doppler bandwidth to be processed. The focused azimuth resolution is independent of transmitted frequency and is dependent only on the size, D, of the physical

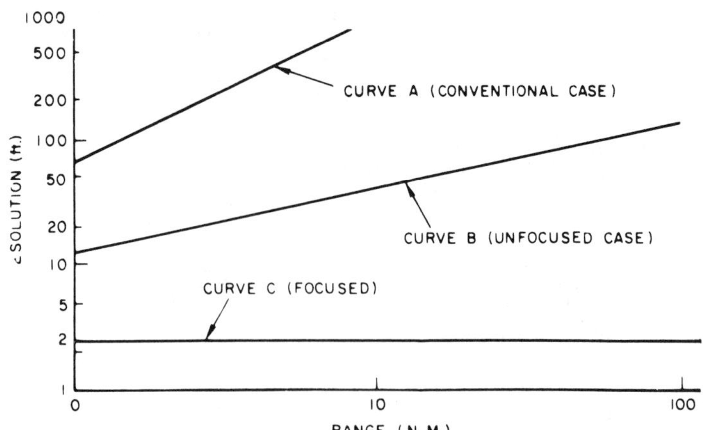

Figure 11 **Azimuth Resolution for Three Cases (A) Conventional, (B) Unfocused, and (C) Focused**

aperture and the utilization factor, m. If D were 5 feet and m were 0.5 (or 50% utilization), then the azimuth resolution would be 5 feet.

On the other hand, Figure 10 derives the SAR azimuth resolution for the unfocused case. Here the data are stored without compensating for phase differences. The phase is permitted to run out until the phase difference at the edge of the integration time (relative to the middle of that time) is 90°. Simple geometry shows that, in this case, the azimuth resolution is one half of the square root of the product of λ and R_o.

A mathematical comparison of the three degrees of fine azimuth resolution is as follows:

Conventional Case

$$\Delta X_{CONV} = \frac{\lambda R}{D}$$

Unfocused Synthetic Antenna

$$\Delta X_{UNF} = \frac{1}{2}\sqrt{\lambda R}$$

Focused Synthetic Antenna

$$\Delta X_{FOC} = \frac{D}{2m}$$

A numerical comparison for a 5 foot aperture is plotted in Figure 11. Figure 12 shows the resolution functions for both the focused and unfocused case. In the focused case, the sin x/x variation has nulls and sidelobes which are very well defined; in the unfocused case, the nulls are filled in, and the first sidelobe level is about 9 dB down from the main lobe (compared to 13.3 dB down in the focused case).

Various forms of high resolution maps can be configured; they are depicted in Figures 13 and 14. In the early days of synthetic aperture mapping, the investigators pointed their physical aperture at an angle 90° from the velocity vector where the largest Doppler bandwidth is obtained; this method is the most convenient for generating surveillance maps.

Over the years, in line with the need for obtaining improved resolutions in other applications, different forms of mapping have been devised. One relatively old type, the strip map in Figure 13, operates in a squint mode which is characterized by the squint angle (the angle between the velocity vector and the direction in which

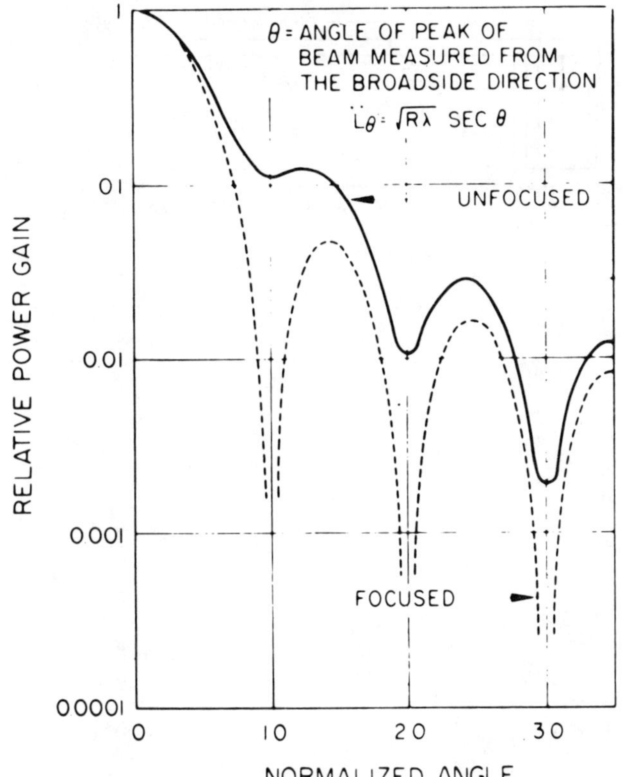

Figure 12 Comparative Power Gain Patterns for Focused and Unfocused Synthetic Arrays

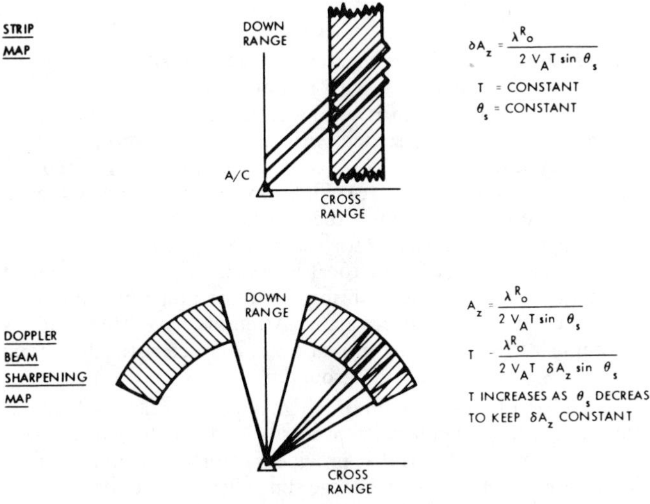

Figure 13 Mapping Modes

the physical beam is pointing). The advent of phased arrays and digital processing and a better understanding of what composes a synthetic aperture have enabled mappers to use the squint mode of operation. To the right of the mapping type in the figure is the fundamental azimuth resolution equation with the constant parameters identified. Pulse compression in the range direction is assumed in all mapping modes. The second type of mapping mode, characterized as Doppler beam sharpening, is akin to the Planned Position Indicator (PPI) type of mapping. The Doppler beam sharpening map can be made on either side of the velocity vector but not directly over it. In this case, the squint angle is varied;

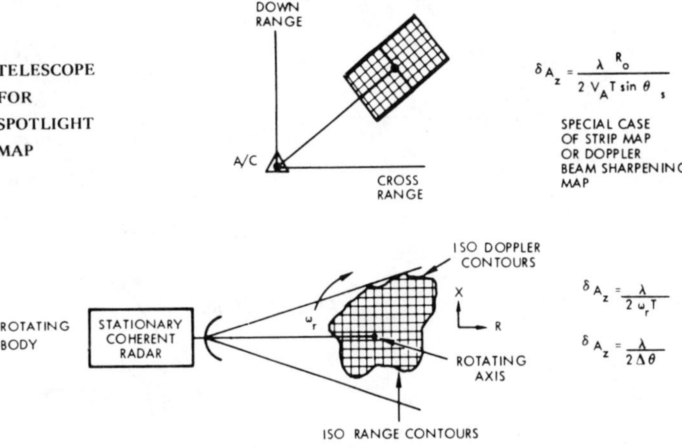

Figure 14 Mapping Modes

however, as the beam gets closer to the velocity vector and the squint angle decreases, the integration time increases. Some systems use two physical apertures, one on each side of the vehicle. If a phased array is used and employs its high beam switching rate capability, strip maps and Doppler beam sharpening maps can be made simultaneously. Strip mapping is discussed in detail in Chapter 16.

Figure 14 shows another type of mapping mode, called the telescope or spotlight map. In this case, a very high resolution map of a particular patch on the ground over an extended period of time is produced. It is similar to the strip map; the physical aperture diameter, time, etc., are adjusted to give the right size map. Spotlight mapping is discussed in detail by Brookner in Chapter 18.

Early investigators knew that the aircraft's motion in sweeping the physical beam across the ground was paramount to generating a synthetic aperture. However, there is a more fundamental point to be made, as shown in the lower part of Figure 14 – the coherent radar remains stationary while the object to be mapped rotates within the physical beam. As the object rotates, it changes its aspect angle relative to the radar. This type of mapping has been used for mapping Venus and the Moon; it is particularly useful when dealing with a Venusian planet, where the surface is hidden by an atmosphere too difficult to penetrate optically. In creating a radar map in such a case, the range coordinate is orthogonal to the beam and the isodops are shown parallel to the axis of rotation. Azimuth resolution is inversely proportional to the angular velocity of the rotating object; conversion of the angular velocity provides a more fundamental form of the azimuth resolution which is dependent upon the change in aspect angle of the object with respect to the radar. Thus, it is not necessary to have the aircraft move in order to generate a Doppler bandwidth which contains the information for improved azimuth resolution.

Signal processors are the "heart" of mapping systems which capitalize on the use of the synthetic aperture. The more common processor types are listed below:

- Range-Gated Filter Bank
- Recirculating Delay Line
- Integration on Photographic Film
- Electronic Storage Tube Integrator

} *Unfocused*

- Optical System*
- Digital Signal Processing**

} *Focused*

In early developmental efforts, great emphasis was placed on trying to determine the best manner in which to store the returned data required to process the Doppler bandwidth. The first four

*see Chapter 16
**see Chapters 11 and 18

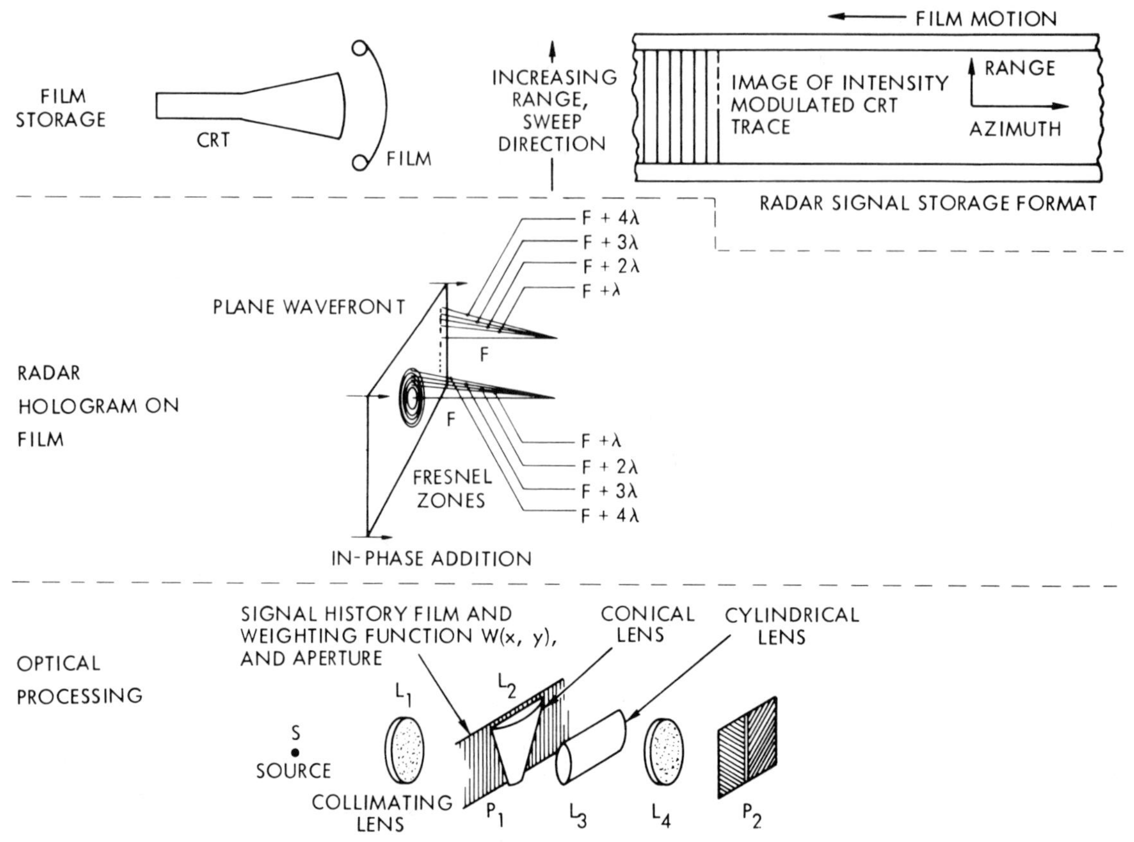

Figure 15 **Simultaneous Pulse Compression and Beam Sharpening (Optical Processing)**

types of processors can be focused but, historically, range-gated Doppler filter banks, circulating delay lines, integration on film, and electronic storage tubes were commonly unfocused systems. Since 1960, primarily focused systems of the optical and digital signal processing types have been developed.

Figure 15 both summarizes the essentials of optical type processing and introduces new terminology. A single sweep is generated across the cathode ray tube; the output of the radar's coherent detector is used to modulate the intensity of this sweep. The film moves across the tube; the speed of the film is directly proportional to the aircraft's velocity. The lines are packed on the film so that there is some overlap. The phase history of the backscattered energy is stored on this film; the phase history within the pulse is stored in the range dimension. When stored on the film, these phase histories have been characteristically called radar holograms, owing to their similarity to optical holograms. A signal film from the image film is produced in the optical processor, which typically is constructed with the elements shown in the lower diagram of Figure 15. A coherent source of light (collimated to form a plane wave) shines on the signal film which contains the radar hologram; the conical lens has a varying focal length to compensate for the phase variations in range. Because there is a different focal length for each range element, a cylindrical lens is used to bring all of the range information into focus in one particular plane. Attempts have been made to accomplish the processing in real time; characteristically, though, the film is processed on the ground.

Figure 16 is a block diagram of a high resolution radar which employs a real time digital signal processor and which incorporates two compensation loops that provide motion compensation for corrections for azimuth and range slip. These corrections are required in ultra high resolution radars in order to keep the individual points resolvable over long integration times and at squint angles other than 90°. Figure 17 is an expanded block diagram of the real time digital signal processor. In-phase and quadrature

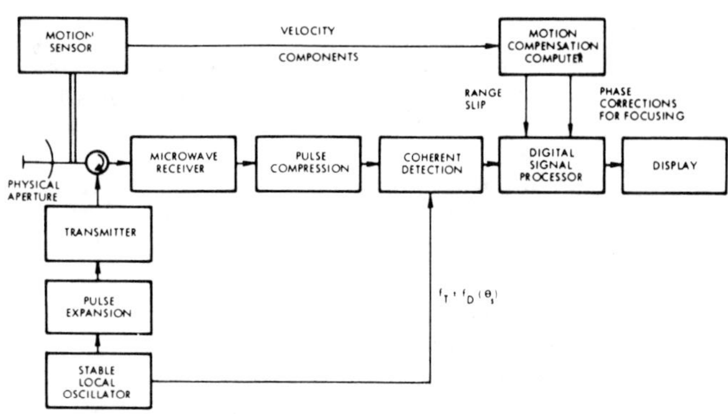

Figure 16 **High Resolution Radar**

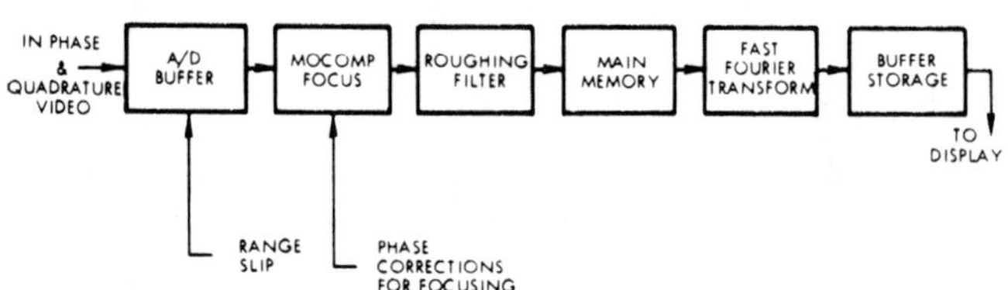

Figure 17 **Digital Signal Processor (Real Time)**

ERROR SOURCE	PERFORMANCE	LIMITATIONS
TROPOSHERIC FLUCTUATIONS	LENGTH OF SYNTHETIC APERTURE	< 1000 ft
MOTION COMPENSATION	FOCUSING	> FEW MILLI g's
OSCILLATOR INSTABILITY	RANGE	> 50 nmi

Table 3 **Sources of Phase Errors**

DISTORTIONS	CAUSES	MOTION LIMITS	DISTORTION SPECIFICATIONS
NON-UNIFORM BRIGHTNESS ACROSS MAP	OFFSET BETWEEN MAP CENTER AND REAL BEAM CENTER	LOW-FREQUENCY VELOCITY ERROR	BRIGHTNESS VARIATIONS ≤ 6 dB
DEFOCUS OF MAP CENTER	"PAIRED ECHOES" NEAR FULL NULL OF AZIMUTH RESOLUTION CELL	MID-FREQUENCY VELOCITY ERROR	AMPLITUDE OF SPURIOUS SIGNAL ≥ 26dB BELOW PEAK RESPONSE
INCREASE IN PEAK AND RANDOM FAR-OUT SIDELOBE LEVELS	"PAIRED ECHOES" BEYOND NULL OF AZIMUTH RESOLUTION CELL	HIGH-FREQUENCY VELOCITY ERROR	DISCRETE RESONANCES 26 dB BELOW RANDOM ERRORS 30dB BELOW

Table 4 **Motion Compensation**

video is converted into digital format with A/D converters; correct strobing of the A/D converters compensates for range slip and phase corrections, computed by the motion compensation computer, focus the azimuth phase histories. A roughing or pre-summing filter is used to reduce the amount of data which enters the main memory. The Doppler bandwidth may be larger than what is needed; to save storage in the main memory, the roughing filter passes only the Doppler band of interest. After storage, a Fast Fourier transform is used to accomplish the frequency analysis. Real time processing is available today. (See Chapters 11 and 18 for further details on digital processing.)

The main sources of error which limit the performance of high resolution radars are shown in Table 3. An expanded description of the errors (and their causes) which limit azimuth resolution, motion limits, and distortion specifications are listed in Table 4.

The more common uses of synthetic aperture radars have been combat surveillance, precision navigation, aerospace surveillance, and missile guidance. The advantages of SAR are capability to be used at either day or night; atmospheric or cloud penetration; surface penetration; photographic quality; large area coverage; physical aperture and spacecraft (size) compatibility; and its usefulness for collecting geological, geophysical, and geomorphological information.

Some examples of the type of imagery obtained by a Side-Looking Airborne Radar (SLAR) are shown in Figure 18. This airborne system has two physical apertures which map to both sides of the aircraft. The dark horizontal lines are unmapped areas directly beneath the aircraft. Clearly visible are the third-order basin and range provinces; fourth-order individual mountains and valleys are discernible. The fifth- and sixth-order micro-reliefs and soil types are also indicated, as is vegetation.

The first experimental demonstration of the synthetic aperture radar concept was performed by a group at the University of Illinois. In a report dated 8 July 1953, Kovaly and co-workers produced a strip map (see Figure 19) of a section of Key West, Florida. With this early system, a 4.13-degree physical aperture beamwidth was used to make an effective synthetic beamwidth of 0.4 degree. Even in this first synthetic aperture radar imagery, the land-water boundaries, with correct geometrical forms, were clearly visible, and individual buildings were resolved.

The radar system that produced this early imagery did not employ pulse compression for improved range resolution. Furthermore; the azimuth processing was unfocused; that is, there was no phase correction to compensate for the changing Doppler frequency over the integration time.

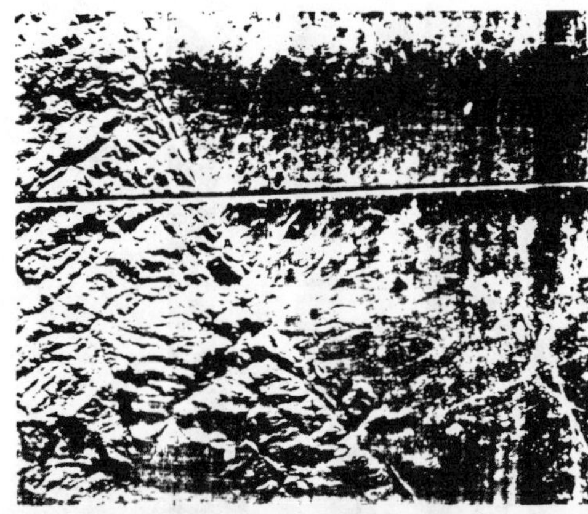

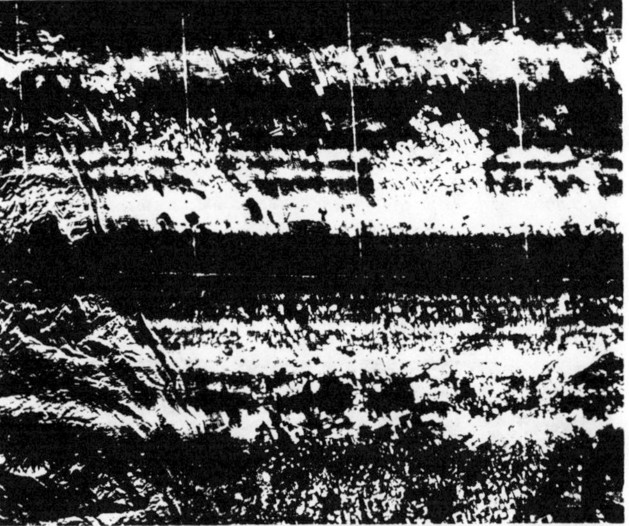

By permission of Autometric Operation, Equipment Division, Raytheon Company

Figure 18 **Radar imagery made by a Side-Looking Airborne Radar (SLAR) – North Plateau, Wyoming (top); Denver, Colorado (bottom).**

Since the time when the first SAR map was created in 1953, many significant advances have been made in the design of SAR systems. The group at the Radar and Optics Laboratory at the University of Michigan in Ann Arbor, Michigan, have been the forerunners in advancing the state-of-the-art in the design of this elegant electronic system. An example of the high-resolution SAR imagery produced by this group is shown in Figure 20 – a map of the Detroit, Michigan area. The water areas of Lake St. Clair and the Detroit River, with the Ambassador and MacArthur bridges, are identified easily. Man-made structures, like buildings and roads, as well as natural terrain, are recognized easily.

The Goodyear Aerospace Corporation has contributed significantly to the production of high-resolution imagery. Some examples of their radar maps are shown in Figure 21. Compare this SAR image of southwestern Arizona with a similar geological map made with SLAR in Figure 18. The well-defined shadowed areas of the mountains show the highly eroded nature of this ridge. Some other geologically interesting areas are the parallel drainage region and the sand dunes. The cultured patterns of barren and irrigated fields are vivid. Figure 21 (bottom) shows much geological structure in the Meteor Crater in Winslow, Arizona. The numerous lineaments were not included

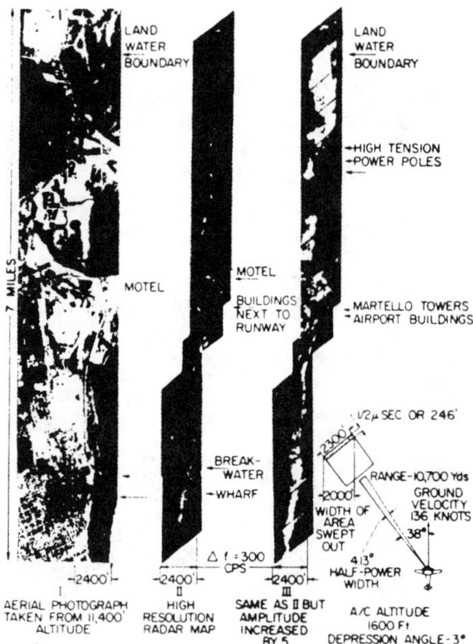

By permission of the Institute of Electrical and Electronics Engineers, Inc.

Figure 19 **First SAR map, an example of early synthetic aperture radar imagery.**

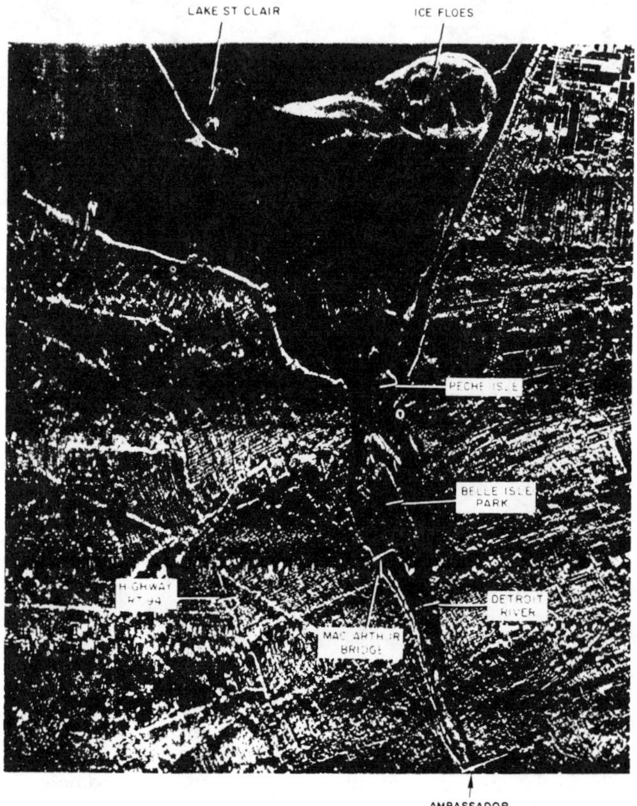

By permission of Radar and Optics Laboratory, University of Michigan, Ann Arbor, Michigan

Figure 20 **Synthetic aperture radar imagery of Detroit area.**

initially on the generalized Arizona geologic map. The structured signatures from this radar image were analyzed in comparison with the known structures, and a revised map, containing more detail than before, was prepared.

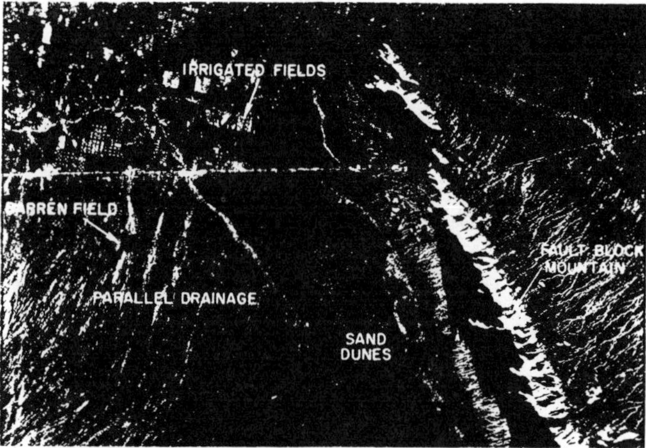

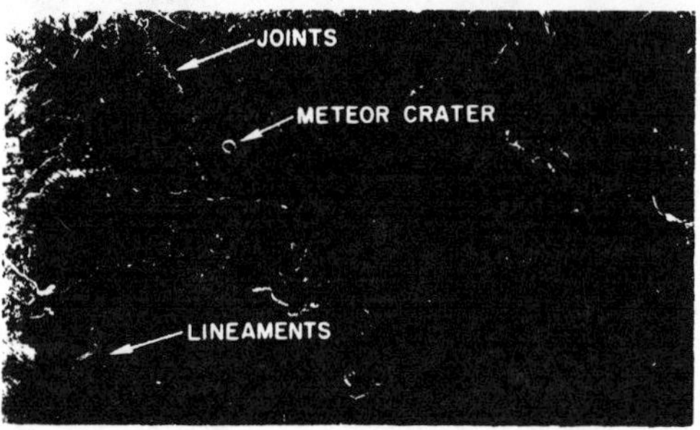

By permission of Goodyear Aerospace Corp., Arizona Division

Figure 21 **Recent synthetic aperture radar imagery of rural areas – southwestern Arizona (top) and Meteor Crater area, Arizona (bottom).**

Figure 22 is a comparison of a photographic image (left) of the moon's surface taken by the NASA Lunar Orbiter I spacecraft and the radar image (right) of the Meteor Crater Area, Arizona. This comparison emphasizes the photo-like quality of the radar image that could be obtained by a future vehicle orbiting one of the unmapped planets that cannot be seen by an optical imaging system.

Radar imagery of Venus has been produced with Earth-based radars by employing the delay-Doppler (synthetic aperture) imaging of a rigid rotating body. Figure 23 is a Venus radar map prepared by the Cornell University astronomers from data obtained with the radio telescope located at Arecibo, Puerto Rico. Locations of rough and smooth surfaces can be pinpointed. First-order bulge, second-order platform, and third-order ranges with fourth-order mountain masses are present. Here is a clear indication of how a radar system can penetrate through the thick cloud cover around Venus, a task than an Earth optical telscope cannot do. Similar radar maps of Venus have been imaged by the group at MIT's Lincoln Laboratory. From these radar maps, radar astronomers can postulate that the surface of Venus is twice as rough as the average surface of the moon, and that the average reflectivity of Venus is higher than that of the moon. In all probability, then, Venus's surface is composed of a denser material than is that of the moon.

Included with this chapter is a bibliography of papers and books that have been written on the subject of synthetic aperture radars, as well as those related to the subject of high resolution imagery. This bibliography of SAR literature can be used by those individuals who want to examine in more detail the topics that have been discussed within this chapter.

MOON CRATER

METEOR CRATER

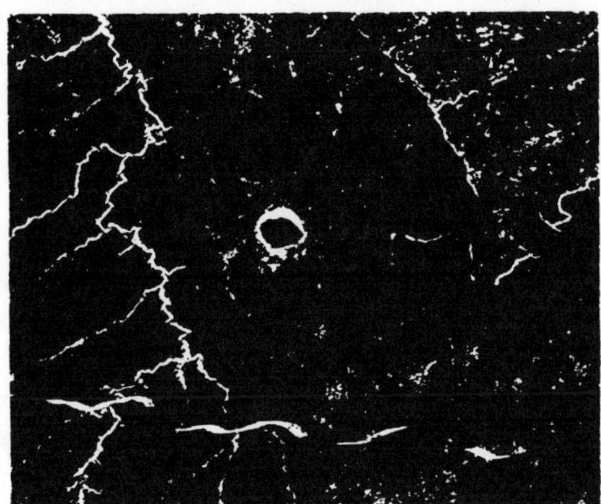

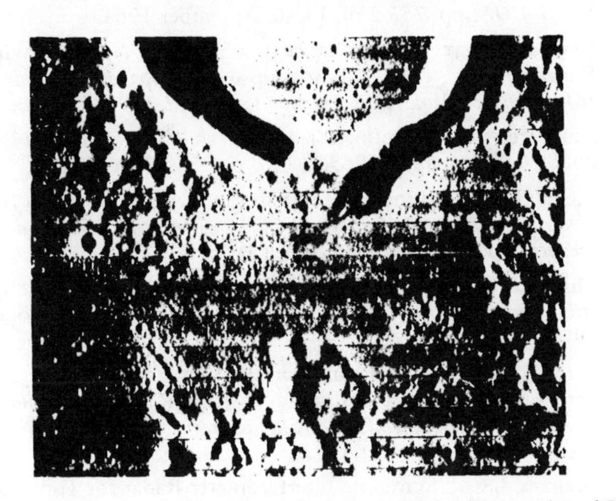

Left, courtesy of National Aeronautics and Space Administration; right, by permission of Goodyear Aerospace Corp., Arizona Division.

Figure 22 Comparison of photographic (left) and radar imagery (right) of similar types.

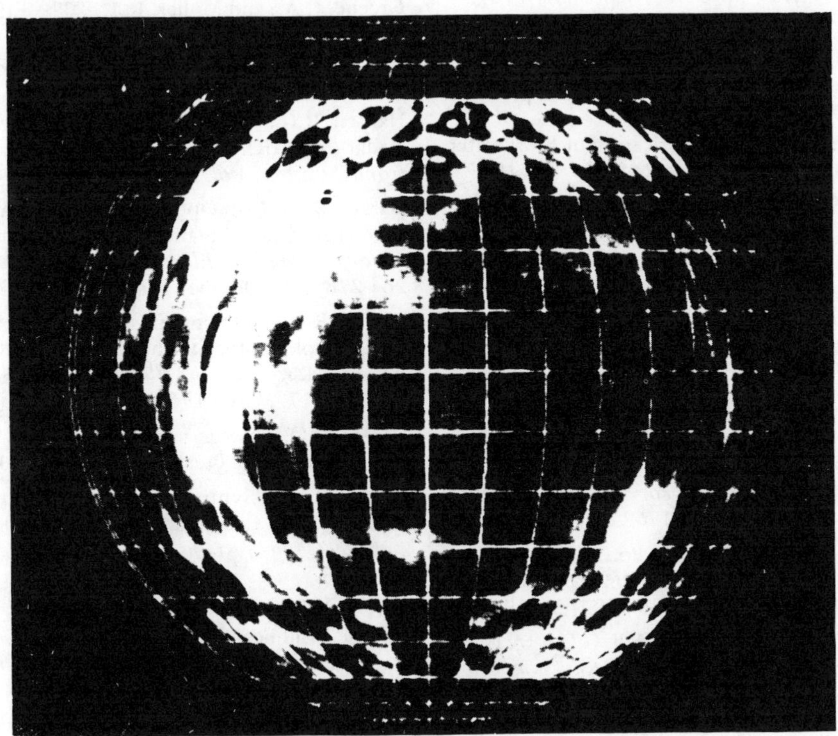

Courtesy of Cornell University Center for Radiophysics and Space Research, Ithaca, New York

Figure 23 Delay-Doppler radar map of Venus

Bibliography

Ausherman, D.A.; et al.: "Radar Data Processing and Exploitation Facility," *The Record of the IEEE 1975 International Radar Conference,* pp. 493-498, 21-23 April 1975.

Bayma, R.W.; and McInnes, P.A.: "Aperture Size and Ambiguity Constraints for a Synthetic Aperture Radar," *The Record of the IEEE 1975 International Radar Conference,* pp. 499-504, 21-23 April 1975.

Bickmore, R.W.: "Time Versus Space in Antenna Theory," Chapter 4 in Vol. III of *Microwave Scanning Antennas* (R.C. Hansen, ed.), Academic Press, New York and London, 1966.

Brown, W.M.; and Palermo, C.J.: "Theory of Coherent Systems," *IRE Transactions on Military Electronics, Vol. MIL-6, No. 2,* pp. 187-196, April 1962.

Brown, W.M.; and Palermo, C.J.: "Effects of Phase Errors on Resolution," *IEEE Transactions on Military Electronics, Vol. MIL-9, pp. 4-9,* January 1965.

Brown, W.M.: "Synthetic Aperture Radar," *IEEE Transactions on Aerospace and Electronics Systems, Vol. AES-3, No. 2,* pp. 217-229, March 1967.

Brown, W.M.; and Fredricks, R.J.: "Range-Doppler Imaging with Motion through Resolution Cells," *IEEE Transactions on Aerospace and Electronic Systems, Vol. AES-5, No. 1,* pp. 98-102, January 1969.

Brown, W.M.; and Porcello, L.J.: "An Introduction to Synthetic Aperture Radar," *IEEE Spectrum,* pp. 52-62, September 1969.

Brown, W.M.; Houser, G.G.; and Jenkins, R.E.: "Synthetic Aperture Processing with Limited Storage and Presumming," *IEEE Transactions on Aerospace and Electronic Systems, Vol. AES-9, No. 2,* pp. 166-176, March 1973.

Bryan, M. Leonard; and Larson, R.: "Classification of Freshwater Ice Using Multispectral Radar Images," *The Record of the IEEE 1975 International Radar Conference,* pp. 511-515, April 1975.

Cary, R.H.J.; and Thraves, J.: "Antennas for Sideways Looking Airborne Radar," *IEEE Conference Publication 77,* pp. 37-42, 1971.

Cheng, D.K.: "Degradation Effects of Arbitrary Phase Errors on High-Resolution Radar Performance," *Proceedings of the IEE, Vol. III, No. 8,* pp. 1375-1378, August 1964.

Clarke, J.: "Reduction of Storage Requirement in Synthetic Aperture Radar," *IEEE Transactions on Aerospace and Electronic Systems, Vol. AES-7, No. 6,* pp. 1213-1215, November 1971.

Collin, R.E.; and Zucker, F.J.: *Antenna Theory, Part II,* Chapter 27, pp. 626-643, McGraw-Hill, New York, 1969.

Cutrona, L.J.; Leith, E.N.; Palermo, C.J.; and Porcello, L.J.: "Optical Data Processing and Filtering Systems," *IRE Transactions on Information Theory, Vol. IT-6, No. 3,* pp. 386-400, June 1960.

Cutrona, L.J.; Vivian, W.E.; Leith, E.N.; and Hall, G.O.: "A High Resolution Radar Combat-Surveillance System," *IRE Transactions on Military Electronics, Vol. MIL-5, No. 2,* pp. 127-131, April 1961.

Cutrona, L.J.; and Hall, G.O.: "A Comparison of Techniques for Achieving Fine Azimuth Resolution," *IRE Transactions on Military Electronics, Vol. MIL-6, No. 2,* pp. 119-121, April 1962.

Cutrona, L.J.; Leith, E.N.; Porcello, L.J.; and Vivian, W.E.: "On the Application of Coherent Optical Processing Techniques to Synthetic Aperture Radar," *Proceedings of the IEEE, Vol. 54, No. 8,* pp. 1026-1032, August 1966.

Cutrona, L.J.: "Synthetic Aperture Radar," Chapter 23 in *Radar Handbook* (M.I. Skolnik, ed.), McGraw-Hill, New York, 1970.

Develet, J.A. Jr.: "The Influence of Random Phase Errors on the Angular Resolution of Synthetic Aperture Radar Systems," *IEEE Transactions on Aerospace and Navigational Electronics, Vol. ANE-11, No. 1,* pp. 58-65, March 1964.

Develet, J.A.: "The Influence of Random Phase Errors on the Edge Response of Synthetic Aperture Mapping Radar Systems," *MIL-E-CON,* pp. 255-258, 14-16 September 1964.

Develet, J.A. Jr.: "Performance of a Synthetic Aperture Mapping Radar System," *IEEE Transactions on Aerospace and Navigational Electronics, Vol. ANE-11, No. 3,* pp. 173-179, September 1964.

Dulberger, L.: "Targeting for Air Attack," *Space/Aeronautics,* pp. 84-95, November 1965.

Eichel, L.A.: "An Inexpensive Side-Looking Radar with a Novel Display," *The Record of the IEEE 1975 International Radar Conference,* pp. 522-526, 21-23 April 1975

Ellis, A.B.E.: "Parallel Processing for a Synthetic Aperture Radar," *IEE Conference Publication 105,* pp. 311-317, 23-25 October 1973.

Farrell, J.L.; Mims, J.H.; and Sorrell, A.: "Effects of Navigation Errors in Maneuvering Synthetic Aperture Radar," *IEEE Transactions on Aerospace and Electronic Systems, Vol. AES-9, No. 5,* pp. 758-776, September 1973.

Graham, L.C.: "Synthetic Interferometer Radar for Topographic Mapping," *Proceedings of the IEEE, Vol. 62, No. 6,* pp. 763-768, June 1974.

Graham, L.C.: "Stereoscopic Synthetic Array Application in Earth Resource Monitoring," *Proceedings of the IEEE 1975 NAECON,* pp. 125-132.

Greene, C.A.; and Moller, R.T.: "The Effect of Normally Distributed Random Phase Errors on Synthetic Array Gain Patterns," *IRE Transactions on Military Electronics, Vol. MIL-6, No. 2,* pp. 130-139, April 1962.

Hansen, R.C.: "The Segmented Aperture Synthetic Aperture Radar (SASAR)," *IEEE Transactions on Aerospace and Electronic Systems, Vol. AES-10, No. 6,* pp. 800-804, November 1974.

Harger, R.O.: "An Optimum Design of Ambiguity Function, Antenna Pattern, and Signal for Side-Looking Radars," *IEEE Transactions on Military Electronics, Vol. MIL-9, Nos. 3 and 4,* pp. 264-278, July/October 1965.

Harger, R.O.; and Crimmins, T.R.: "The Solution of a Nonlinear Design Problem for Synthetic-Aperture Radars," *IEEE Transactions on Aerospace and Electronic Systems, Vol. AES-4, No. 2,* pp. 316-317, March 1968.

Harger, R.O.: *Synthetic Aperture Radar Systems: Theory and Design,* Academic Press, New York, 1970.

Harger, R.O.: "Synthetic Aperture Radar System Design for Random Field Classification," *IEEE Transactions on Aerospace and Electronic Systems, Vol. AES-9, No. 5,* pp. 732-740, September 1973.

Harger, R.O.: "Harmonic Radar Systems for Near-Ground In-Foliage Nonlinear Scatterers," *IEEE Transactions in Aerospace and Electronic Systems, Vol. AES-12, No. 2,* pp. 230-245, March 1976.

Heimiller, R.C.: "Theory and Evaluation of Gain Patterns of Synthetic Arrays," *IRE Transactions on Military Electronics, Vol. MIL-6, No. 2,* pp. 122-129, April 1962.

Hermann, G.; Kerr, D.; and Tammy, M.: "Interactive SAR Interpretation Facility," *The Record of the IEEE 1975 International Radar Conference,* pp. 488-492, 21-23 April 1975.

Hildebrand, B.P.: "Statistics of Focused and Defocused Radar Maps," *IEEE Transactions on Aerospace and Electronic Systems, Vol. AES-10, No. 5,* pp. 615-621, September 1974.

Holahan, J.: "Synthetic Aperture Radar," *Space/Aeronautics,* pp. 88-93, November 1963.

Johansen, E.L.: "The Synthetic-Array Radar Image of a Flat Plate," *IEEE Transactions on Aerospace and Electronic Systems, Vol. AES-6, No. 3,* pp. 395-398, May 1970.

Kelly, P.M.; et al.: "Data Processing for Synthetic Aperture Radar," *Proceedings of the IEEE 1964 NAECON,* pp. 194-200.

Kirk, John C. Jr.: "Digital Synthetic Aperture Radar Technology," *The Record of the IEEE 1975 International Radar Conference,* pp. 482-487, 21-23 April 1975.

Kirk, John C. Jr.: "A Discussion of Digital Processing in Synthetic Aperture Radar," *IEEE Transactions on Aerospace and Electronic Systems, Vol. AES-11, No. 3,* pp. 326-337, May 1975.

Kirk, John C. Jr.: "Motion Compensation for Synthetic Aperture Radar," *IEEE Transactions on Aerospace and Electronic Systems, Vol. AES-11, No. 3,* pp. 338-348, May 1975.

Kock, W.E.: "Holography Can Help Radar Find New Performance Horizons," *Electronics,* pp. 80-88, 12 October 1970.

Kock, W.E.: "A Holographic (Synthetic Aperture) Method for Increasing the Gain of Ground-to-Air Radars," *Proceedings of the IEEE, Vol. 59, No. 3,* pp. 426-427, March 1971.

Kovaly, J.J.: "Satellite High Resolution Radar Mapping Techniques," *Annals of the New York Academy of Sciences, Vol. 163,* pp. 154-170, 4 September 1969.

Kovaly, J.J.: "Radar Techniques for Planetary Mapping with Orbiting Vehicle," *Annals of the New York Academy of Sciences, Vol. 187,* pp. 154-176, 25 January 1972.

Kovaly, J.J.: *Synthetic Aperture Radar,* Artech House, Dedham, Massachusetts, 1976.

Larson, R.W.; Zelenka, J.S.; and Johansen, E.L.: "A Microwave Hologram Radar System," *IEEE Transactions on Aerospace and Electronic Systems, Vol. AES-8, No. 2,* pp. 208-217, March 1972.

Leith, E.N.: "Optical Processing Techniques for Simultaneous Pulse Compression and Beam Sharpening," *IEEE Transactions on Aerospace and Electronic Systems, Vol. AES-4, No. 6,* pp. 879-885, November 1968.

Leith, E.N.; Friesem, A.A.; and Funkhouser, A.T.: "Optical Simulation of Radar Ambiguities," *IEEE Transactions on Aerospace and Electronic Systems, Vol. AES-6, No. 6,* pp. 832-840, November 1970.

Levine, D.: *Radargrammetry,* McGraw-Hill, Inc., New York, 1960.

MacPhie, R.H.: "The Circular Synthetic Radar," *IEEE Transactions on Aerospace and Electronic Systems, Vol. AES-9, No. 4,* pp. 608-611, July 1973.

McCord, H.L.: "The Equivalence Among Three Approaches to Deriving Synthetic Array Patterns and Analyzing Processing Techniques," *IRE Transactions on Military Electronics, Vol. MIL-6, No. 1,* pp. 116-119, January 1962.

Mims, J.H.; and Farrell, J.L.: "Synthetic Aperture Imaging with Maneuvers," *IEEE Transactions on Aerospace and Electronic Systems, Vol. AES-8, No. 4,* pp. 410-418, July 1972.

Mitchell, R.L.: "Models of Extended Targets and their Coherent Radar Images," *Proceedings of the IEEE, Vol. 62, No. 6,* pp. 754-758, June 1974.

Mitchell, R.L.: *Radar Signal Simulation,* Artech House, Dedham, Massachusetts, 1976.

Moore, R.K.; and Rouse, J.W. Jr.: "Electronic Processing for Synthetic Aperture Array Radar," *Proceedings of the IEEE (Letters), Vol. 55, No. 2,* pp. 233-234, February 1967.

Nuttall, J.A.: "Dynamic Range Study of Correlators used in Conjunction with Side-Looking Synthetic Radar Systems," *Supplement to IEEE Transactions on Aerospace and Electronic Systems, Vol. AES-2, No. 6,* pp. 385-394, November 1966.

Pilon, R.O.; and Purves, C.G.: "Radar Imagery of Oil Slicks," *IEEE Transactions on Aerospace and Electronic Systems, Vol. AES-9, No. 5,* pp. 630-636, September 1973.

Porcello, L.J.: "Turbulence-Induced Phase Errors in Synthetic Aperture Radars," *IEEE Transactions on Aerospace and Electronic Systems, Vol. AES-6, No. 5,* pp. 636-644, September 1970.

Porcello, L.J.; et al.: "The Apollo Lunar Sounder Radar System," *Proceedings of the IEEE, Vol. 62, No. 6,* pp. 769-783, June 1974.

Raney, R.K.: "Synthetic Aperture Imaging Radar and Moving Targets," *IEEE Transactions on Aerospace and Electronic Systems, Vol. AES-7, No. 3,* pp. 499-505, May 1971.

Rawson, R.; Smith, F.; and Larson, R.: "The ERIM Simultaneous X- and L-Band Dual Polarization Radar," *The Record of the IEEE 1975 International Radar Conference,* pp. 505-510, 21-23 April 1975.

Reutov, A.P.; and Mikhaylov, B.A.: *Sidelooking Radar,* Radiolokatsionnyye Stantsii Bokovogo Obzora, Moscow, Soviet Radio, 1970. (Available in translation AD787070 from US Department of Commerce Clearing-house for Federal Scientific and Technical Information).

Revillon, G.: "Synthetic Aperture Antennas and their Applications to Side-Looking Radars," *L'Onde Electrique 45, No. 458,* pp. 561-567, May 1965. (In French)

Rihaczek, A.W.: "An Example: Synthetic Aperture Radar," *Chapter 13 in Principles of High-Resolution Radar,* McGraw-Hill, New York, pp. 441-483, 1969.

Rogers, E.E.; and Ingalls, R.P.: "Radar Mapping of Venus with Interferometric Resolution of the Range-Doppler Ambiguity, *Radio Science, Vol. 5, No. 2,* pp. 425-433, February 1970.

Rondinelli, L.A.; and Zeoli, G.W.: "Evaluation of the Effect of Random Tropospheric Propagation Phase Errors on Synthetic Array Performance," *Eighth Annual Radar Symposium Record,* University of Michigan, pp. 235-256, June 1962.

Ruina, J.P.; and Angulo, C.M.: "Antenna Resolution as Limited by Atmospheric Turbulence," *IEEE Transactions on Antennas and Propagation, Vol. AP-11, No. 2,* pp. 153-161, March 1963.

Shapiro, I.I.: "Planetary Radar Astronomy," *IEEE Spectrum,* pp. 70-79, March 1968.

Sherwin, C.W.; Ruina, J.P.; and Rawcliffe, D.: "Some Early Developments in Synthetic Aperture Radar Systems," *IRE Transactions on Military Electronics, Vol. MIL-6, No. 2,* pp. 111-115, April 1962.

Shuchman, R.A.; Davis, C.F.; and Jackson, P.L.: "Contour Strip Mine Detection and Identification with Imaging Radar," *The Record of the IEEE 1975 International Radar Conference,* pp. 516-521, 21-23 April 1975.

Shuchman, R.A.; Rawson, R.F.; and Drake, B.: "A Dual Frequency and Dual Polarization Synthetic Aperture Radar System and Experiments in Agriculture Assessment," *Proceedings of the IEEE 1975 NAECON,* pp. 133-140.

Steinberg, B.D.: "The Effects of Relative Source Strength and Signal-to-Noise Ratio on Angular Resolution of Antennas," *Proceedings of the IEEE, Vol. 62, No. 6,* pp. 758-762, June 1974.

Steinberg, B.D.: "Hard Limiting in Synthetic Aperture Signal Processing," *IEEE Transactions on Aerospace and Electronic Systems, Vol. AES-11, No. 4,* pp. 556-561, July 1975.

Tyler, G.L.: "The Bistatic, Continuous-Wave Radar Method for the Study of Planet Surfaces," *Journal of Geophysical Research, Vol. 71, No. 6,* pp. 1559-1567, 15 March 1966.

Chapter 18

Brookner

Synthetic Aperture Radar Spotlight Mapper

Introduction

This chapter assumes no prior knowledge of synthetic spotlight mapping; no knowledge is assumed of even ordinary strip-mapping synthetic apertures of the type discussed by Curtis and Kovaly in the preceding two chapters. Here, synthetic aperture mapping is explained from a different point of view; at the same time, though, the results are related to those obtained by Curtis and Kovaly. A physical feel is given for how and why cross-range resolution is produced with the spotlight mapper.

Why Spotlight Mappings?

For some applications, only a map of a small region is needed. The requirement is not to make a map over a large region continuously, such as the type discussed by Curtis and Kovaly. For the application under consideration here, only a short time is needed to make a map, which may be a synthetic aperture map if the resolution so demands. A small synthetic aperture map of this sort is called a spotlight map; this mapping mode would be used in a multi-function array radar in which other functions must be satisfied such as air-to-air search and track modes, and terrain following and avoidance modes. These other modes are interwoven with the spotlight mapping mode.

In multifunction array radars, the spotlight mapper can be used for two purposes. One use is the updating of the navigation system, in particular, a map would be made of a certain region where a particular landmark is known to be. If the navigation system indicates that the landmark is located other than where it should be, the system is off and can be corrected for the discrepancy. The other application is weapon delivery. A map is made of the target area so that the weapon can be directed. Spotlight mapping is not for reconnaissance; that job is for strip mapping. Figure 1 depicts a multifunctional array radar having the functions of spotlight mapping, air-to-air search and track, and terrain avoidance and following.

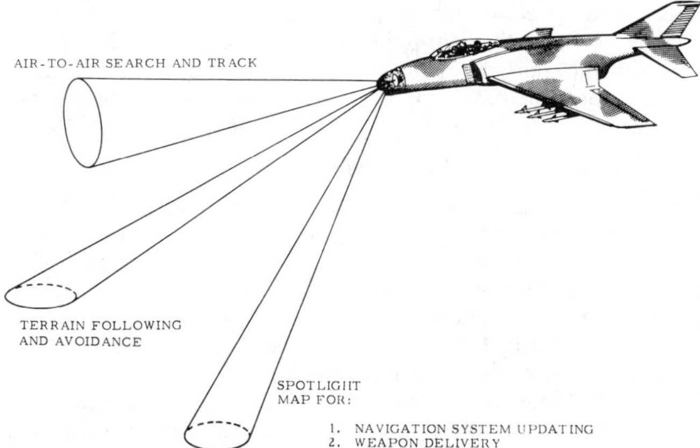

Figure 1 Multifunction Array Radar

Fundamental Difference Between Spotlight and Strip Mappers

For the purpose of comparison, a spotlight mapper and a strip mapper are depicted in Figure 2. To simplify the discussion of how they differ with regard to signal processing, initial attention is given only to how resolution is obtained in the cross-range direction (that is, in a direction perpendicular to the radar line-of-sight or, equivalently, perpendicular to the radar slant range). The question of how resolution in the slant range direction is obtained is addressed later. The process of obtaining resolution in the cross-range direction is unique to the synthetic aperture map; hence, it is appropriate to concentrate on cross-range resolution first. Therefore, the assumption is made that a CW signal is transmitted even though, when using such a signal, one does not obtain resolution in the slant range direction. To get around this problem, for the time being, it is assumed that there are scatterers only at one slant range (equivalently, only in one range interval or gate). The resulting problem is one of obtaining cross-range resolution for these scatterers. The scatterers assumed are labeled $X_1, X_2, \ldots X_N$ for the strip mapper and spotlight mapper in Figure 2. For the strip mapper, these scatterers are at a fixed offset distance, R_S, from the aircraft flight path. For the assumption of a narrow real antenna beam, the slant ranges to the scatterers are about equal to the offset distance, R_S. Hence, the strip mapper scatterers can be considered to be at a fixed slant range, R_S. (See Figure 2)

The strip mapper antenna beam is depicted as pointing straight out to the side perpendicular to the aircraft. For this type of system, signal processing begins when the signal enters the beam, perhaps defined by its one-way 3 dB points. All points along any line parallel to the flight path enter the beam at different times. As a result, the processing and integration start at different times for the scatterers.

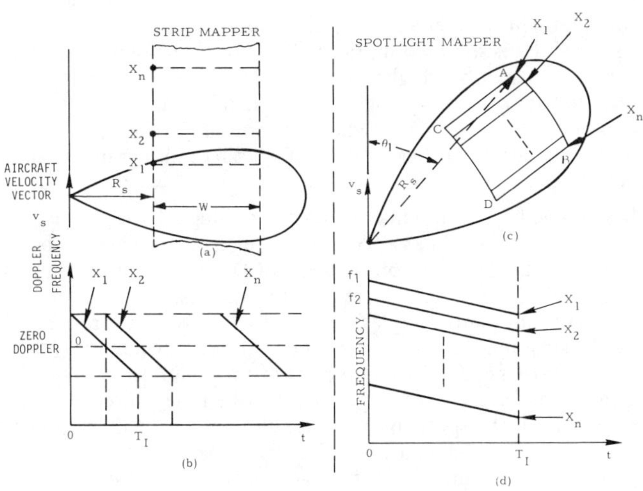

Figure 2 Spotlight Versus Strip Mapper

Let us follow the Doppler history for one of these points. The time at which point X_1 enters the strip mapper real beam is illustrated in Figure 2(a). At this time, the return from point X_1 has a positive Doppler. The point X_1 progresses through the beam; eventually it is broadside to the aircraft and, equivalently, the point X_1 is at the peak of the beam. At this instant, the Doppler velocity is zero. After passing the beam midpoint, the Doppler velocity of the scatterers goes negative. The point exits the beam at the one-way 3 dB point on the other side of the antenna pattern, and the processing and integration terminate at this time. Figure 2(b) shows the Doppler history for the point X_1.

A strip mapper for which the synthetic aperture processing of the scatterers starts and ends as the scatterers enter and leave the 3 dB one-way real antenna beam pattern points is referred to as a focused synthetic aperture mapper (m = 1 in Figure 9 in Chapter 17). Such a synthetic aperture mapper has the best possible processing time or, equivalently, synthetic aperture length. The cross-range resolution for the focused synthetic aperture strip mapper is about equal to half the dimension of the real antenna aperture along the aircraft's axis. As Curtis and Kovaly indicate, it is possible to integrate for a much shorter time to achieve an unfocused (or defocused) synthetic aperture antenna. Specifically, the start and end of the integration time are those instants for which the distance to the scatterer is $\lambda/8$ greater than when the scatterer is abeam (at the real beam peak) where λ is the radar wavelength; see Figure 10 in Chapter 16 and Figure 10 in Chapter 17. With this assumption, the phase variation of the received signal is quadratic with values of 90° at the beginning and end of the integration time and a value of 0° at the middle. For the unfocused case, the integration generally starts and ends much after the scatterer enters the 3 dB one-way real antenna beam points; the cross-range resolution is about $(1/2)\sqrt{\lambda R_S}$. It is possible, of course, to have integration times which fall between the two extremes and to realize a mid-value cross-range resolution. The tendency still is to call this in-between type of synthetic aperture processing focused processing; the distinction made here is to call it partially focused, to differentiate it from fully focused processing which employs the maximum possible integration time. Shortly, we relate focused and unfocused synthetic aperture processings to the spotlight mapper processing.

Consider a point of the strip mapper that is further down range — point X_2. It has exactly the same Doppler history as X_1, except that it enters the beam a little later; this Doppler history is depicted in Figure 2(b). The important thing to be emphasized is that the frequency-time histories of the successive scatterers, $X_1, X_2, \ldots, X_N$, are all exactly the same, but they start at different times (because they arrive at different times).

In a spotlight mapper, the antenna beam may or may not be squinted. (It is squinted in Figure 2(c)). It can be, in fact, actually perpendicular to the flight path; more generally, though, it is squinted at an arbitrary angle. The angle of squint depends on a number of factors, such as when the spotlight mode can be multiplexed in with the other modes and whether more than one spotlight map is to be made of a region.

During the processing time required for a single spotlight map, the antenna beam is held at a fixed squint angle, θ. The area to be mapped is the small region labeled ABCD. This size is in contrast to the very long (much longer than the antenna beam width, possibly 100's of nautical miles) region of width W that is processed by the strip mapper (see Figure 2(a)). Another difference is that, for the strip mapper, different parts of the region to be mapped arrive at different times, depending on when they enter the antenna beam; for the spotlight mapper, all the points of the region are illuminated simultaneously (that is, the integration time for all the points in the region ABCD starts and ends at the same time). The whole region, ABCD, to be processed remains within the antenna 3 dB one-way width during the whole processing time. The processing time is short enough that the aircraft does not move sufficiently for the region ABCD to leave the beam.

At the start of the integration time, t = 0, X_1 has a positive Doppler frequency. This Doppler frequency is designated as f_1 in Figure 2(d). As time goes on, the angle between the aircraft and point X_1 increases. As a result, the Doppler velocity for point X_1 decreases. For illustrative purpose in the figure, the variation of the Doppler of X_1 with time is assumed (as it was by Curtis) linear. For short times, such as those used for synthetic aperture spotlight mapping, this assumption is a good one. The conditions for which this assumption is true are given later.

Point X_2 of the spotlight map has a slightly larger squint angle than does X_1. Therefore, at the start of the processing time, (t = 0), this point has a lower Doppler velocity (f_2 in the figure) than does X_1. For the case of the short integration time, the Doppler history variation with time for point X_2 is essentially the same as for X_1, except that it is displaced downward by the difference, $\Delta f = f_1 - f_2$, that existed at t = 0. Thus, the difference in Doppler frequency between the points X_1 and X_2 remains fixed; also, the Doppler histories for X_1 and X_2 have the same slope. The only difference between their frequency time histories is that they start at different initial Doppler frequencies. The same situation holds for the other scattering points $X_3, \ldots, X_N$ on the map. These points have successively larger squint angles. Therefore, they have successively lower Doppler velocities.

For a spotlight mapper, in order to perform resolution in the cross-range direction (that is, in order to resolve X_1 and X_2), it is necessary only to have two Doppler filters which separate the Doppler-separated echo signals. What is needed is one filter that tracks the Doppler frequency of point X_1 and another one that tracks that of point X_2. The outputs of these filters then can be read to resolve these signals from each other and, equivalently, to resolve X_1 from X_2. The minimum Doppler-frequency separation, Δf_D, between two echo signals for them to be resolved is equal to about one over the processing time or integration time, T_I. If T_I equals a half second, Δf_D must be 2 Hz for X_1 and X_2 to be resolved.

Determination of the cross-range resolution is clear-cut. For example, when $T_I = 0.5$s, the resolution equals the difference in cross-range direction that must exist between X_1 and X_2 in order for the difference in their Doppler-velocity frequencies to equal 2 Hz. This statement essentially sums up synthetic spotlight mapping.

Next is a comparison of the method for obtaining cross-range resolution with the strip mapper, as described by Curtis, with the resolution method described here for the spotlight mapper. Curtis indicates that, to separate returns of the strip mapper, one cross correlates them with an optical reference signal (see Figure 14 in Chapter 16). Because the returns arrive at different times, the outputs of the correlator peak at different, resolvable times for the returns from the different points such as X_1 and X_2 of Figure 2(a). For the spotlight mapper, on the other hand, all the returns arrive at the same time. Doppler filtering must be used to resolve the scatterers such as X_1 and X_2 in Figure 2(d).

Processing of Spotlight Mapper Signal

How is the spotlight mode signal processed? Shown in Figure 2(d) are the frequency histories of the signals coming back from the points $X_1, X_2, \ldots, X_N$, all located at the same slant range from the radar. The processor for these return signals is shown in Figure 3. The RF return, made up of the superposition of the echos from scatterers X_1 through X_N, is heterodyned by a local oscillator having a linearly varying frequency. The slope of the LO is made equal to the slope of the Doppler echos from the scatterers X_1 through X_N. As a result, the echo signals at the output of the mixer all have constant frequency histories, as shown in the figure. The separation in frequency between the echos from points X_1 and X_2 is still the same, Δf. Likewise, the echos from the other scatterers

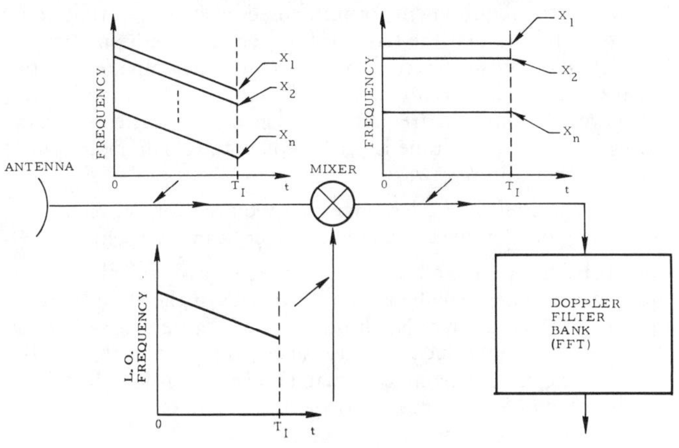

Figure 3 **Receiver Processor**

maintain the same difference in frequency that they had prior to the mixer, only now the echos are made up of CW tones of duration T_I instead of linearly varying frequencies of the same duration.

To process CW IF echo signals, they are put through a bank of Doppler filters. For the assumption $T_I = 0.5$ s, the filters have a 0.5 s integration time or, equivalently, a frequency bandwidth of 2 Hz. Figure 3 shows this filter bank as a digital Fast Fourier Transformer (FFT) which does the processing in real time. This digital FFT filter bank implementation represents a simple way of doing the processing and getting the desired resolution.

Cross-Range Resolution — The Doppler Resolution Viewpoint

To equate the 2 Hz Doppler resolution to cross-range distance resolution, a determination is made of the cross-range separation, r_θ, between X_1 and X_2 which yields a 2 Hz Doppler frequency differential, Δf (see Figure 4).

Let v_S be the aircraft velocity. The Doppler velocity for scatterer X_1 at squint angle θ is then

$$v_D = v_S \cos\theta \tag{1}$$

Hence, the Doppler frequency shift is

$$f_D = \frac{2v_D}{\lambda} = \frac{2v_S \cos\theta}{\lambda} \tag{2}$$

Differentiating Equation (2) yields

$$df_D = \frac{-2v_S \sin\theta \; d\theta}{\lambda} \tag{3}$$

where df_D is the differential change in f_D for a differential change, $d\theta$, in the scatterer squint angle, θ. Replacing $d\theta$ by $\Delta\theta$ (the difference in squint angles of scatterers X_1 and X_2) yields the difference in Doppler velocities Δf_D of the scatterers X_1 and X_2; that is

$$\Delta f_D = \frac{-2v_S \sin\theta \; \Delta\theta}{\lambda} \tag{4}$$

For an integration time of T_I, the Doppler filters actually have a resolution, Δf_R, equal to

$$\Delta f_R = \frac{1.35}{T_I} \tag{5}$$

instead of $1/T_I$. The factor 1.35 allows a loss of resolution due to a time-dependent amplitude weighting placed on the return signals in order to achieve low Doppler sidelobes, and due to system phase errors. (A Tapered-Taylor weighting [1] for 40 dB Doppler sidelobes, two-way quadratic phase error of 90° peak, and two-way cubic phase error of 180°/8 = 22.5° peak are assumed.)

Setting Equation (4) equal to Equation (5) and solving for $\Delta\theta$ yields

$$\Delta\theta = \frac{1.35\,\lambda}{2T_I v_S \sin\theta} \tag{6}$$

In Equation (6), $\Delta\theta$ represents the minimum squint angle separation that X_1 and X_2 can have and still be resolved when the integration time is T_I.

Multiplying $\Delta\theta$ by R_S, the slant range to X_1 and X_2, yields r_θ, the minimum cross-range separation that X_1 and X_2 can have and still be resolvable; that is,

$$r_\theta = R_S \cdot \Delta\theta = \frac{1.35\,\lambda R_S}{2T_I v_S \sin\theta}\;; \tag{7}$$

see Figure 4.

Carrying out a sample calculation using the same assumptions Curtis used (i.e., $R_S = 20$ nmi, $v_S = 500$ ft/s, $\theta = 90°$, $T_I = 1.0$ s, and $\lambda = 0.1$ ft; see Table 1 in Chapter 16), yields $r_\theta = 16.4$ ft. Curtis obtained a result of 12.25 ft. The difference is due to the weighting factor 1.35 in Equation (5). In this chapter, a $T_I = 0.5$ s example is used, with $R_S = 53$ nmi, $v_S = 2000$ ft/s, $\theta = 60°$, and $\lambda = 0.1$ ft. For these assumptions, Equation (7) gives $r_\theta = 25$ ft.

(The above equations were derived for the assumption of zero depression angle ϕ from the aircraft to the region being mapped. In a later section of this chapter, the extension to the case of nonzero depression angle is given.)

Partially-Focused, Unfocused, and Doppler-Beam-Sharpening Processing

Kovaly in Chapter 17 discussed both focused and unfocused synthetic aperture strip mapper processing, and gives expressions for the cross-range resolutions for each in Figure 9 and 10 in Chapter 17. Where does the spotlight mapper synthetic aperture fit? Is it a focused or unfocused synthetic aperture mapper, or does it fall in between and form the partially focused mapper mentioned earlier? The answer generally is in between the focused and unfocused synthetic apertures — it is usually a partially focused synthetic aperture mapper. There are instances, however, where the processing can be such that it could be classified as an unfocused synthetic aperture mapper, such as when the application permits the use of the coarse resolution achieved with an unfocused synthetic aperture mapper.

Kovaly's focused synthetic aperture processing involves integrating for a time equal to the time it takes for the signal to enter and leave the real antenna 3 dB points in the beam pattern. As pointed out for the spotlight mapper, the integration time is much smaller

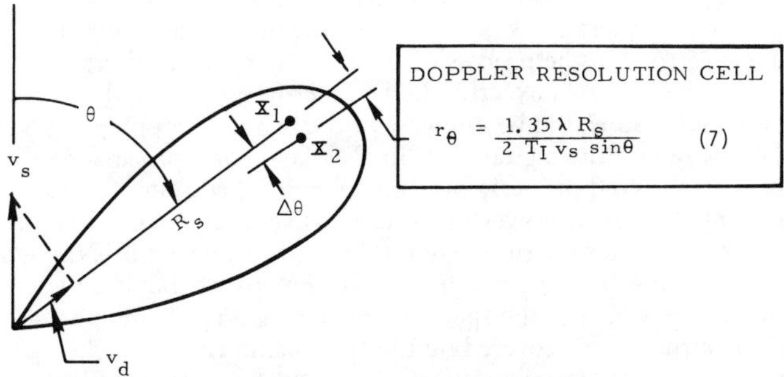

Figure 4 **Cross Range Resolution**

than this period; in fact, it is so small that all the scatterers of the region being mapped do not leave the antenna pattern one-way 3 dB points.

The integration time for the spotlight mapper though, is generally larger than for unfocused processing. Before proceeding, though, it is desirable to elaborate on the conditions for unfocused processing and its physical significance. Consider first the case where $\theta = 90°$ – the broadside case. For this situation, as indicated previously, unfocused processing occurs when the radial distance to the target at the beginning and end of the integration period is $\lambda/8$ greater than at the middle (at which time the target is abeam). The phase of the return at the beginning and end of the integration time is then 90° out of phase with the return received at the middle of the integration. This condition is the same as that for being in the near-to-far-field transition of an antenna. In terms used for the near-to-far-field transition of a synthetic aperture array of length L is the equivalent statement – unfocused conditions are achieved when the distance from the synthetic aperture to the scatterer is L^2/λ. Usually for a real antenna the near-to-far-field transition is defined as occurring when $R_S = 2L^2/\lambda$. This difference results from two factors. First, for a real antenna a 22.5° phase difference (equivalent to a $\lambda/16$ path length difference) between the signals from the ends of the antenna relative to the center of the antenna is used (instead of 90°). Second, whereas for a real antenna a $\lambda/16$ path difference between the ends of the aperture and the center produces a 360°/16 = 22.5° phase difference, for a synthetic aperture twice that phase difference, 45°, is produced. As mentioned by Curtis in Chapter 16, the signal propagates two ways – up and back for each synthetic aperture array element. The unfocused synthetic aperture has a length of $L = \sqrt{\lambda R_S}$ (see Figure 11 in Chapter 16); hence, $L^2/\lambda = \lambda R_S/\lambda = R_S$, as indicated. The unfocused aperture resolution is

$$r_\theta = \frac{1.35\sqrt{\lambda R_S}}{2} \tag{8}$$

For nonbroadside conditions ($\theta \neq 90°$), the conditions for unfocused processing are identical to those above if applied to the equivalent synthetic aperture obtained by projecting the synthetic aperture onto the perpendicular to the radar line of sight. Thus the unfocused synthetic aperture length for arbitrary θ becomes

$$L = \frac{\sqrt{\lambda R_S}}{\sin\theta} \tag{9}$$

For the nonbroadside synthetic aperture, unfocused conditions occur when the difference in the path lengths from the target to the center of the array and either end is $\lambda/8$ greater than obtained if the target were at infinity.) Substituting L as obtained from Equation (9) into Equation (7) for $T_I v_S$ yields Equation (8). Thus, Equation (8) applies independent of θ.

Applying the values for the example case to Equation (9) (i.e., R_S = 53 nmi, λ = 0.1 ft, θ = 60°) yields L = 207 ft so that, for v_S = 2000 ft/s, T_I = 0.104 s for unfocused processing. Thus, the integration time of 0.5 s used for the previous example spotlight mapper is longer than that for unfocused conditions. Using Equation (7) or (8) yields r_θ = 121 ft for unfocused conditions, versus the 25 ft obtained for the partially focused case.

A convenient equivalent condition for unfocused processing can be given in terms of the change in Doppler velocity of the target during the integration time. For unfocused processing, the target's Doppler velocity changes by exactly $2/T_I$ during the integration time, T_I. (This conclusion follows directly from the quadratic variation of phase with time which has peak values of 90° at the ends of the integration time and zero at the center.) This figure is approximately equal to the width of one receiver filter, its 3 dB width being $1.35/T_I$. It is exactly equal to the 6.6 dB width.

Physically, the requirement for unfocused processing is thus equivalent to requiring that the target return remain chiefly in one Doppler-velocity filter during the whole integration period. With this restriction, there is no need for the ramped LO in the receiver of Figure 3 because the frequency change of any scatterer's echo during the integration time is small enough to remain mainly in one individual filter.

If the integration time is less than that required for unfocused processing, one has what is called Doppler beam sharpening.

For the partially focused spotlight mode processing, if the size of the spotlight map is small enough, the slopes of the Doppler returns from the scatterers X_1 through X_N of Figure 2 are all about the same; thus, only one LO ramp is needed to obtain the CW IF signals of Figure 3. The map is small enough if it is less than $S_{\theta,D}$ in azimuth and $S_{R,D}$ in ground range R_G, where

$$S_{\theta,D} = \frac{2K\, r_\theta^2 \tan\theta}{f_w^2 \lambda} \tag{10}$$

and

$$S_{R,D} = \frac{4K r_\theta^2}{f_w^2 \lambda} \tag{11}$$

Equation (10) gives the value of $S_{\theta,D}$ for which the peak two-way quadratic phase error is less than or equal to K·45° when a single LO oscillator is used, the slope of which is matched to the slope from the scatterer at the center of the spotlight map. The parameters f_W represents a weighting factor. For 40 dB Tapered-Taylor weighting, two-way quadratic phase error of 90°, and two-way cubic phase error of 22.5°, f_W = 1.35, the value used in Equation (5). Equation (11) applies for the same assumptions as does Equation (10) – a two-way quadratic phase error less than or equal to K·45° when a single LO oscillator is used. If the spotlight map is $S_{\theta,D}$ by $S_{R,D}$ in size with K = 1, the maximum two-way quadratic phase error is 90° at two opposite corners of the map.

The largest integration time one can have and still consider the Doppler variation to be linear with time, as assumed for Figures 2 and 3, is

$$T_I = \sqrt[3]{\frac{3Q\lambda R_S^2}{2v_S^3 \sin^2\theta \cos\theta}} \tag{12}$$

When Equation (12) applies the deviation from a linear Doppler history is a peak two-way cubic phase error of Q·45°. For Q = 1/2, the peak cubic phase error is 22.5°. For the example, R_S = 53 nmi, λ = 0.1 ft, v_S = 2000 ft/s, and θ = 60°, Equation (12) yields T_I = 1.374 s for Q = 1/2 which is greater than the 0.5 s assumed in the example.

Required Aircraft Velocity Accuracy

So far, the discussion has focused on how the resolution can be obtained in the cross-range dimension by simply conceptualizing it as Doppler resolution in the cross-range dimension across the real beam. A question that arises concerns the effect of aircraft velocity errors on the processing of the system. If the aircraft velocity were known exactly, the Doppler frequency of the center of the map also would be known and the bank of Doppler filters (synthesized by the digital Fast Fourier Transform processor) could be centered properly over the Doppler spectrum of the region to be mapped. However, if the exact velocity of the aircraft is unknown, this spectrum may not be centered where it is thought to be. As a result, if the width of the Doppler filter bank is just equal to the width of the spectrum of the region to be mapped, this spectrum is not covered by the filter bank. Hence, in such a situation, essentially no velocity errors of the aircraft could be tolerated. The solution is to have a bank of filters wider than the region that is to be processed.

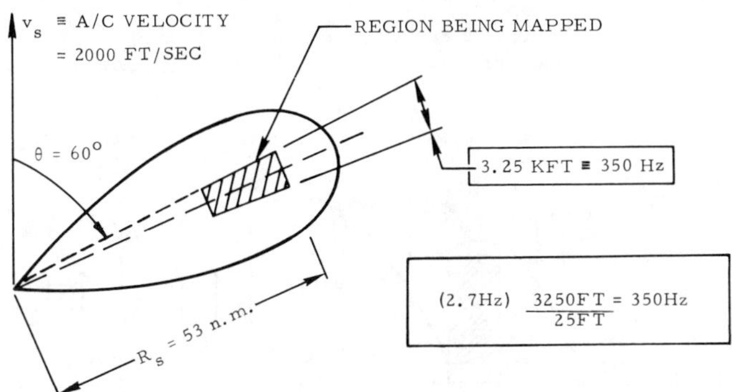

Figure 5 Region to be Mapped

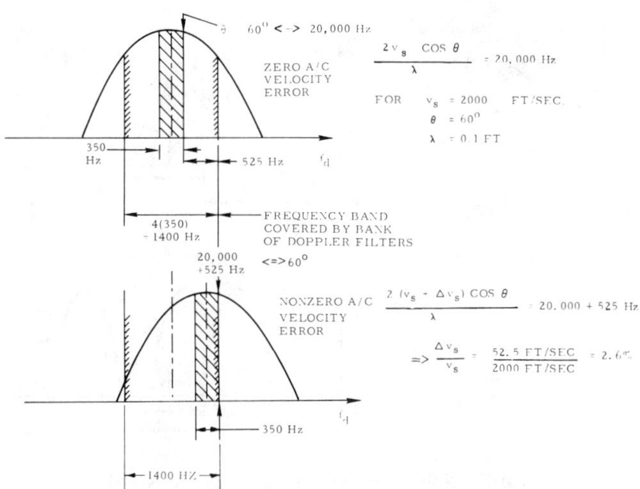

Figure 6 Velocity Accuracy Needed for Mapping

Assume that the area it is desired to map is a 3250 ft wide region, as shown in Figure 5. From Equation (4) it follows that the width of the Doppler spectrum of the region to be mapped is 350 Hz. Assume a Doppler filter bank four times as wide, or 4(350) = 1400 Hz. Figure 6 shows the position of the signal spectrum relative to the Doppler bank of filters used to process the received signal for both a zero aircraft velocity error and an aircraft velocity error which places the signal spectrum just at the edge of the filter bank. (Note that the azimuth width of the map selected for Figure 5 is well within the limits specified by Equation (10), with $S_{\theta,D}$ equaling 11900 ft for K = 1. The maximum ground range extent, $S_{R,D}$, permitted by Equation (11) for K = 1 is 13700 ft.)

For the latter case one has the maximum allowable aircraft velocity error because any larger error results in the signal spectrum falling outside the filter bank coverage. For the example assumptions made the maximum aircraft velocity error allowable is 52.5 ft/s, or equivalently, a 2.6% aircraft velocity error. This accuracy is easy to achieve. You could halve the width of the bank of filters and still have no problem. The required accuracy becomes 1.3% or 25 ft/s (about 15 nmi/h) which is still easy to achieve.

For the type system for which a spotlight mapper is intended, the aircraft velocity is expected to be known to an accuracy of 1 nmi/h so that the filter bank coverage only need be slightly (about 10%) larger than the map Doppler spectrum width.

The application for which the velocity of the aircraft must be known to high accuracy is where precise target ground location is needed, say a weapon delivery system. Table 1 gives the requirements for this case.

For the example used in this chapter (and assumed for Table 1), the range resolution is 25 ft, which is equivalent to a 2.7 Hz resolution in Doppler. From Equation (2), this value is in turn equivalent to an aircraft velocity change of 0.27 ft/s. Hence, to determine the position of the target to within 25 ft in cross-range the aircraft velocity must be known to 0.27 ft/s or roughly 0.01%. This requirement is a very difficult one. If these numbers are multiplied by a factor of six for an accuracy of 150 ft for the location of the target, the aircraft velocity must be known to 1.62 ft/s or about 1 nmi/h — an achievable figure. Thus the position of a target on the ground can be located to within roughly 150 ft with a reasonable Doppler navigation system.

Range Resolution

Up to this point, no consideration has been given to how range resolution is obtained. A CW transmitted signal has been assumed, and the problem of range resolution has been circumvented by assuming scatterers at only one slant range. For the assumption of a CW system and the region to be mapped as shown in Figure 7(a), the receive signal spectrum is as shown in Figure 7(b). Assume that the spectrum is zero outside the bandwidth, Δf_d. Then from the Nyquist sampling theorem, the signal spectrum indicated in Figure 7(b) can be sampled at a rate equal to the signal bandwidth, Δf_d, without loss of information (complex samples of the signal amplitude and phase being assumed). Sampling the received signal is equivalent to sampling the transmitted signal; it can be accomplished by gating on the CW transmitter periodically for a short interval of time τ. As a result, a coherent train of pulses (of width

$\theta = 60°$, $\lambda = 0.1$ ft, $R_S = 53$ nmi, $v_S = 2000$ ft/s, $T_I = 0.5$ s

Position Accuracy Required (ft)	Equivalent A/C Doppler Velocity Accuracy Required (Hz)	Equivalent A/C Velocity Accuracy Required (%)	Aircraft Velocity Error Allowable (ft/s)	(nmi/hr)
25	2.7	0.0135	0.27	0.16
100	4(2.7)	0.054	1.08	0.64
150	6(2.7)	0.081	1.62	0.96
250	10(2.7)	0.135	2.7	1.6

Table 1 Velocity Accuracy Needed for Accurate Target Location

Note: The above computed allowable aircraft velocity errors assume that all of the position error is due to the aircraft velocity error. Actually, there are other sources of errors; an appropriate error budget must be allotted to each contribution to the overall system position error desired.

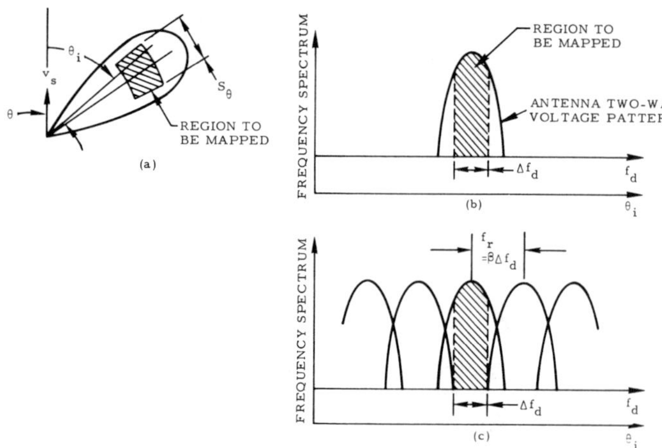

Figure 7 **Spectrum Aliasing and Range Ambiguity**

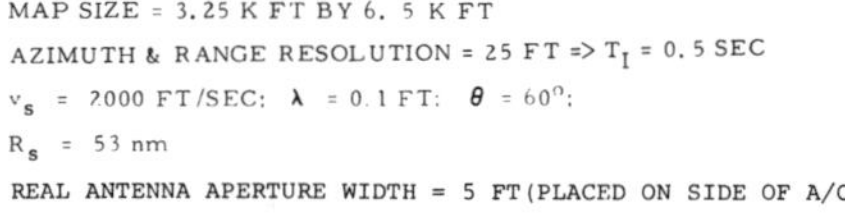

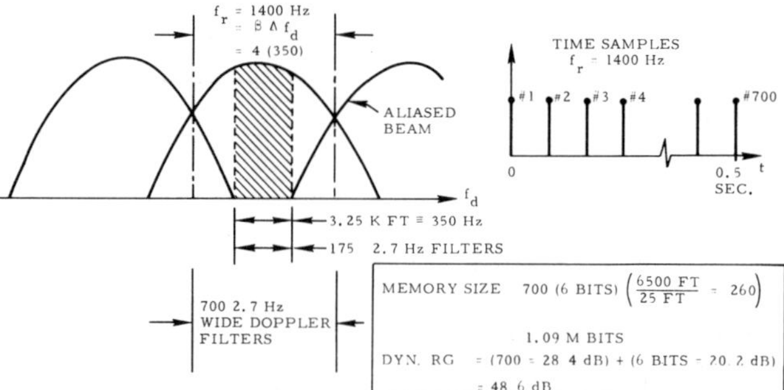

Figure 8 **Real Time Digital Processor Size Requirements**

τ) can be transmitted instead of a CW waveform without any loss of the cross-range resolution capabilities discussed previously. It is necessary only that the pulse repetition rate equal the required Nyquist sampling rate.

The transmitted signal pulses are reflected from the scatterers at range R_S and are received $2R_S/c$ s after transmission, where c represents the velocity of light. These received pulses from the scatterers at the range R_S can be processed in essentially the same manner as is the received signal for the case where a CW signal has been transmitted. The pulses received $2R_S/c$ s after transmission are heterodyned with the same LO used for the CW received signal shown in Figure 3. Because the received signal now is sampled, the LO also can be sampled. Thus, a constant frequency LO, the phase of which is shifted at successive pulse-to-pulse intervals, can be used. The phase of this LO is shifted from pulse-to-pulse in a quadratic manner corresponding to the frequency variation of the CW LO. Successive pulses at the output of the mixer are given the weightings necessary to provide a 40 dB tapered-Taylor weighted sidelobe level at the output of the digital Fast Fourier Transform processor which synthesizes the receiver filter bank.

If there were scatterers at a range $R_S + \tau$ as well as at R_S, their echos would be received a pulse width, τ, apart. Pulses received τ apart can be resolved. Pulses received from slant range $R_S + \tau$ can be processed independently and in the same manner as those from range R_S, thus providing a cross-range resolution at both ranges. Similarly, the returns from any other slant range can be resolved and processed independently to provide cross-range resolution for that slant range. Fourier transforming the returns from successive range cells τ apart provides cross-range resolution for successive range resolution cells; as a result, a two-dimensional spotlight map of the region is created.

A sampled CW waveform or, equivalently, a pulsed CW waveform such as that used above is referred to as a pulse Doppler waveform. By using a pulse Doppler waveform instead of a CW signal, slant range resolution as well as cross-range resolution is achieved. Furthermore, instead of a simple train of narrow unmodulated RF pulses, a train of phase-coded pulses can be transmitted (specifically, chirped pulses) as done in Chapter 17, pulse compression being then used to achieve fine range resolution.

Choice of Pulse Repetition Rate

The question of how rapidly the signal must be sampled or, equivalently, how high a PRF must be used for the pulse Doppler signal must be reexamined. Above it was assumed that the signal would be sampled at a rate equal to the received signal bandwidth. Actually the received signal has a very large bandwidth, since the antenna sidelobes are not zero. However, as long as the sidelobe level is low enough, the sidelobe signal portion of the spectrum can be neglected – for low enough sidelobes, the aliasing of the sidelobe signal frequency components onto the signal spectrum of the region being mapped produces negligible degrading effects. For this situation, a sampling which results in the situation depicted in Figure 7(c) can be used. The sampling here is such that the nulls of the main antenna lobe are aliased to the edge of the region to be mapped. This type of sampling is depicted in Figure 8 for the example of Figures 5 and 6. The sampling rate is chosen to be 4 times the width of the spectrum of the region to be mapped (that is, $\beta = 4$ for Figures 7 and 8). The width of the spectrum of the region to be mapped for the example under consideration is 350 Hz (that is, $\Delta f_d = 350$ Hz for Figure 7) so that the sampling rates (or equivalently PRF) is 1400 pps (that is, $f_R = 1400$ pps for Figures 7 and 8).

To this point, the discussion has dealt with the requirements for the PRF which prevents degradation due to the aliasing. The system PRF must also be chosen to avoid range ambiguities – returns from second time around ranges falling on top of the returns from the range being mapped. For the example given above the PRF was chosen to be 1400 pps, corresponding to an unambiguous range interval $R_A = 57.8$ nmi. The spotlight map is being made at a range of about 53 nmi. Scatterers at the ambiguous range of $53 + 57.8 \doteq 111$ nmi must be either beyond the horizon or not illuminated by the antenna main lobe.

If for the above example, scatterers at ranges further than 111 nmi limited the PRF to lower than 1400 pps, it follows from Figure 8 that the map is in turn limited to a width less than 3250 ft for $\beta = 4$. Thus the range ambiguities set a limit to the maximum map width. The maximum spotlight map width, as limited by range ambiguities at the ambiguous range $R_A + R_S = AR_S + R_S = (A + 1)R_S$, where $A = R_A/R_S$, is given by

$$S_{\theta,A} = \frac{\lambda c}{4Av_S\beta \sin\theta} \tag{13}$$

where c is the velocity of light. The radar PRF, f_R, is assumed to be

$$f_R = \frac{c}{2R_A} = \frac{c}{2AR_S} \tag{14}$$

for Equation (13). If $R_A = 53$ nmi (that is, $A = 1$), for the parameters of Figure 8 it is necessary that $f_R \leqslant 1528$ Hz and $S_{\theta,A} = 3550$ ft so that the PRF of $f_R = 1400$ Hz and map width of 3250 ft are satisfactory; these values even provide some margin.

Processor Memory Requirements

It has been shown that a pulse Doppler signal is transmitted to achieve resolution in the slant-range dimension. The returns from successive range cells are processed sequentially, using the Fast Fourier Transform, to provide cross-range resolution. For each range resolution cell, T_I seconds of pulses are integrated, which for the example of Figure 8 is 700 pulses. If the Fast Fourier Transform processing is done by a general purpose computer, all of the 700 samples must be received before the Fourier Transform processing can be initiated. Hence, the amplitudes and phases of the received pulses must be stored in a memory before the Fourier Transform process is initiated; this processing is done immediately after all the data is received so that essentially real time processing operation is achieved. For the example under consideration in Figures 5 and 8, assume that the region to be mapped is 6500 ft long in the slant range direction, with the slant range resolution equalling the cross-range resolution of 25 ft. (This range length is smaller than the maximum of 13700 ft allowed by Equation (11) for K = 1.) Thus, there are 260 range resolution cells for which 700 pulse amplitude and phase samples must be stored in memory. Assume 6-bit quantization per sample — 1 bit for sign for each quadrature of the received pulses and 2 bits for amplitude. Thus, the total is

$$(\text{700 pulses per range cell})\ (\text{6 bits})\left(\frac{6500\ \text{ft}}{25\ \text{ft}} = \text{260 range cells}\right)$$

$$= 1.09\ \text{M bits}$$

which must be stored in the memory for the processing of a 3250 ft by 6500 ft map having a 25 ft resolution. The size of such a memory is about 1/2 ft^3 (1976). Much smaller sizes will be required in the near future — it is predicted that a single charge coupled device (CCD) chip capable of storing 1 to 10 million bits should be available in a few years [2].

Less memory would have been required if the real antenna pattern had an ideal shape, specifically, if it were perfectly rectangular and had a width equal to the region to be mapped. Under such circumstances, the receive signal spectrum would be equal to the spectrum of the region to be mapped, that is, it would be 350 Hz wide so that a Nyquist sampling rate of 350 pulses per second could be used. This sampling rate results in a reduction of the number of complex samples to be stored per range resolution cell by a factor of four, reducing 700 samples to 700/4 = 175. Hence, the memory size could be reduced by a factor of four to about 0.25 Mbits.

MAP SIZE = 3.25 K FT BY 6.5 K FT

AZIMUTH & RANGE RESOLUTION = 25 FT ⇒ T_I = 0.5 SEC.

v_s = 2000 FT/SEC; λ = 0.1 FT; θ = 60°;

R_s = 53 nm

REAL ANTENNA APERTURE WIDTH = 5 FT (PLACED ON SIDE OF A/C)

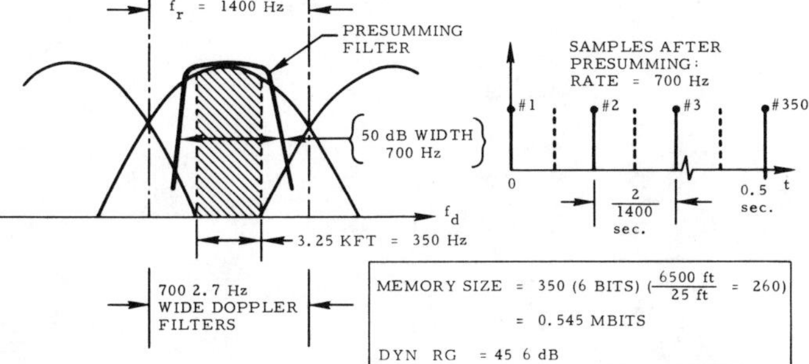

Figure 9 **Processor Size Requirements with Presumming**

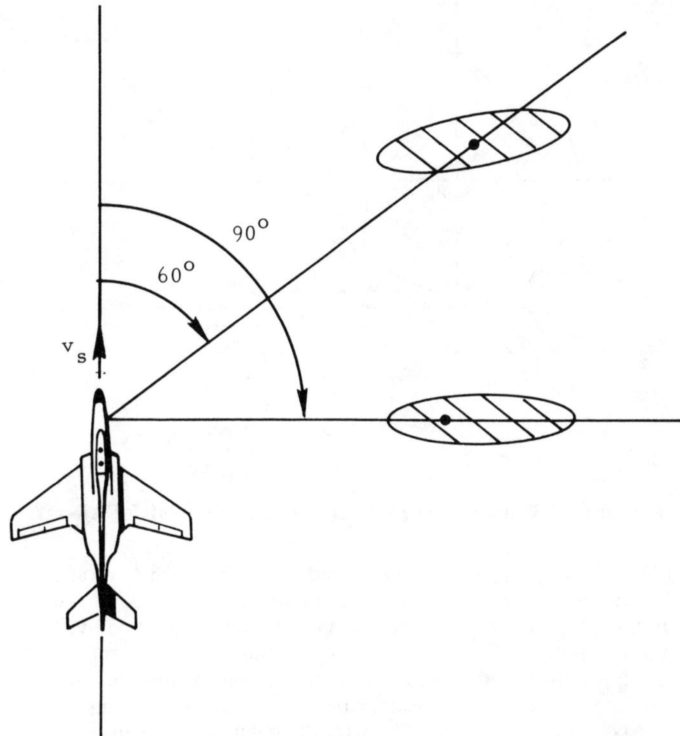

Figure 10 **Real Beam (Nonsynthetic) Ground Contour for Phased Array Antenna**

An ideal rectangular antenna pattern for the real antenna beam is an impossibility, but something close to it can be achieved by using receiver filtering or, equivalently, smoothing just prior to the storage of the data in the memory. This filtering removes the part of the received signal spectrum of the main beam that is not part of the region being mapped. The filtering is done digitally. Figure 9 shows the signal spectrum at the output of such a filter, which is called a presumming filter. The spectrum of the signal after this filter is narrow enough to allow a sampling rate of 700 pulses per second to be used without running into degradation from aliasing. Thus, only half the samples at the output of the presumming filters need be stored, resulting in a reduction by a factor of 2 in the main memory storage; the storage now requires only about 0.5 Mbits and its volume is roughly 1/4 ft^3 (1976). This savings is achieved at the expense of the requirement of a presumming filter, which, in turn, necessitates a memory unit. However, this filter's memory generally is much smaller, on the order of 50 kbits for the above example.

The processing performed by the presumming filter can be looked at from another viewpoint; in particular, it can be thought of as synthetic aperture mapping over a relatively short interval, generally on the order of 20 pulses. The antenna pattern for this short synthetic aperture is given by the frequency characteristics of the presumming filter indicated in Figure 9.

The step transform processor applied in Chapter 11 to the strip mapper requires more hardware than the straightforward FFT processor described in this chapter, if used for the spotlight mapper task. The step transform processor is the best solution when used to do sliding window processing as is required for the strip mapper application.

Real Beam Ground Contour

Next for consideration is the shape of the region on the ground illuminated by the real antenna beam. As indicated earlier, the spotlight mapping mode can be used in a multifunction array

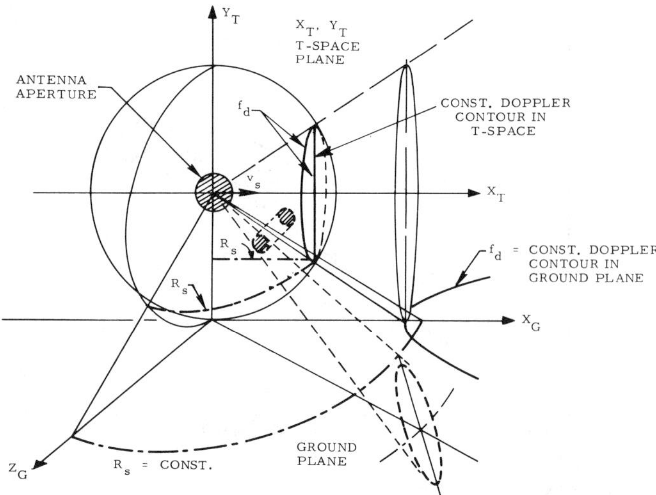

Figure 11 **T-Space – Phased Array Located on Side of Aircraft**

radar, with the array antenna located any place on the aircraft. For this example, assume that the antenna is on the side of the aircraft. Figure 10 shows the patch of ground illuminated by the antenna for two different squint angles, the shaded area represents the approximately elliptical region illuminated within the 3 dB two-way contour of the beam. Normally, it might be thought that the axis of this ellipse lines up with the radar line-of-sight, but such is not the case for squint angles other than 90°.

T-Space

T-space is used to determine the shape, size, and orientation of the patch illuminated on the ground by the real phased-array antenna beam. T-space originally was developed by Van Aulock [3] for phased array analysis; it also is referred to as sin α-sin β space, direction-cosine space, and u-v space [1]. In addition to allowing the determination of the region illuminated on the ground by the real antenna beam, T-space has the additional advantage that contours of constant Doppler in this space become straight lines when the antenna array is mounted on the side of the aircraft and become circles when the array is mounted on the aircraft nose. The following properties are independent of whether the array is mounted on the side or nose of the aircraft. Contours of constant slant range become straight lines in T-space. The size and shape of the pattern illuminated on the ground are independent of the squint angle when plotted in T-space; the shape is given by the far field antenna pattern [3, 4]. Figure 11 illustrates how the T-space projection is obtained.

Nonzero Depression Angles

The equations derived in this chapter are for the assumption of zero depression angle, ϕ, to the target. Some of the equations given apply independently of the depression angle, but some have to be modified for nonzero depression angles. Equation (1) through (4) should be multiplied by cos ϕ*; Equation (6), divided by cos ϕ; Equation (8), multiplied by $\sin \theta_D / \sin \theta$; Equations (9) and (12), θ replaced by θ_D; Equation (10) divided by cos ϕ; Equation (11), multiplied by $\cos \phi \sin^2\theta / \sin^2\theta_D$; and Equations (5), (7), (13), and (14) remain unchanged. The angle θ_D represents the angle between the radar line of sight and the aircraft velocity vector (see Figure 4 in Chapter 16); the angle θ for nonzero ϕ represents the angle between the aircraft velocity vector and the projection of the radar line of sight onto the horizontal plane containing the aircraft velocity vector.

For nonzero depression angle the slant range resolution of the radar r_S is no longer equal to the ground resolution; rather, the ground range resolution becomes $r_G = r_S / \cos \phi$.

For further discussions on sidelookers, see References 5, 6, and 7 and the "Present and Future Trends" section of Part 4.

*Alternatively, for Equations (1) through (4) replace θ by θ_D.

References

[1] Brookner, E.: "Antenna Array Fundamentals, Part 1," in *Practical Phased-Array Systems, Microwave Journal* Intensive Course (E. Brookner, ed.), Dedham, Massachusetts, 1975.

[2] *Computer Design,* pp. 42-43, July 1976.

[3] Van Aulock, W.H.: "Properties of Phased Arrays," *Proceedings of the IRE, Vol. 48,* pp. 1715-1727, October, 1960; also in Hansen, R.C. (ed.): *Significant Phased Array Papers,* Artech House, Inc., Dedham, Mass., 1973.

[4] Lewis, L.R.: "Phased Array Elements, Part 1," in *Practical Phased-Array Systems (Microwave Journal* Intensive Course Notes – E. Brookner, ed.), Dedham, Massachusetts, 1975.

[5] Cutrona, L.J.: "Synthetic Aperture Radar," in *Radar Handbook* (M.I. Skolnik, ed.), McGraw-Hill, New York, 1970.

[6] Kovaly, J.J.: Bibliography at end of Chapter 17.

[7] Kovaly, J.J. (ed.): *Synthetic Aperture Radar,* Artech House, Dedham, Massachusetts, 1976.

Present and Future Trends in Synthetic Aperture Radar Systems and Techniques

The development of Synthetic Aperture Radar (SAR) is summarized in Table 1. The following subsections briefly describe some of the most recent developments in SAR systems and techniques.

Synthetic Interferometer Radar for Topographic Mapping [2]

A synthetic aperture mapping system capable of making topographic maps (maps providing 3-dimensional information, that is, terrain altitude as well as location) has been built. The principle of the system is illustrated in Figure 1. Two side looking antennas are put one above the other in the aircraft. The lower antenna produces a standard SAR strip map. The upper antenna also provides a strip map; however, it has the signal from the lower antenna added to it at RF so that interferometric action takes place. When the phase of the return from the lower antenna bucks that of the upper antenna, the signal disappears for the upper receiver system (that is, the scatterer is in the null of the interferometer formed by the two antennas). For such scatterers a dark strip is formed on the SAR map that is produced by the interferometric receiver system. These strips on the interferometer map permit first the determination of the elevation angles of the scatterers and, in turn, the altitude of the scatterers through the following equations:

$$y = y \quad (1)$$

$$x = R\cos\theta \quad (2)$$

$$z = R\sin\theta \quad (3)$$

where: R = the slant range to the scatterer

θ = elevation angle

A system was built under joint Army-Air Force sponsorship and operated by the Air Force Charting and Geodetic Squadron. The radar operated at a wavelength of 3 cm (7.5 GHz) and used two 4 x 6 ft antennas separated vertically by 4 ft. It was capable of a 1-σ accuracy in elevation of 0.2 mr over a combination of rough (slopes exceed 10%) and smooth (slopes less than 10%) terrains. This accuracy is more than adequate to permit compilation of 1:250,000 scale class B maps. This type system has the advantage of being able to operate in adverse weather and at night. Figure 2 shows a sample strip map of Arizona south of Phoenix. The images extend about 2.5 nmi in slant range.

Date	Development
1951	Carl Wiley of Goodyear postulates Doppler beam-sharpening concept.
1952	University of Illinois demonstrates beam-sharpening concept.
1953	Project Wolverine formulates SAR radar development program.
1957	Project Michigan produces first SAR imagery using optical correlator.
Mid 1960s	Analog electronic SAR correlation demonstrated in non-real time.
Late 1960s	Digital electronic SAR correlation demonstrated in non-real time.
Early 1970s	Real-time digital SAR demonstrated with motion compensation.

Table 1 **Summary of SAR Developments (From Kirk [1])**

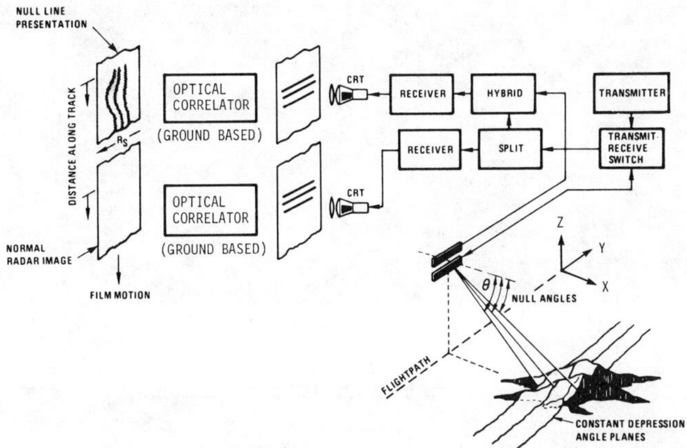

Figure 1 **Synthetic Interferometer Radar for Topographic Mapping (After Graham [2])**

Harmonic Synthetic Aperture Radar for Detection of Metallic Objects in Foliage (METRRA) [3]

A synthetic radar based on the nonlinear scattering action of metallic man-made objects has been built. For such scatterers the third harmonic of the incident signal (as well as the fundamental) are returned to the radar receiver. The natural clutter in which the object is embedded returns only the fundamental of the incident signal. Consequently, metallic objects can be detected in the presence of severe clutter. Furthermore, by radiating with a low frequency that could penetrate the foliage, one could detect man-made targets with a high angular and range resolution by observing the third harmonic of the reflected signal — something not achieved if the fundamental frequency is observed.

For this third harmonic synthetic aperture metal detection radar (designated as *ME*tal *T*arget *ReRA*diation (METRRA)) the cross range resolution r_θ is given approximately by:

$$r_\theta = \sqrt{\frac{\lambda R_S}{3}} \quad (4)$$

where: λ = wavelength of the transmitted signal

R_S = slant range to the target

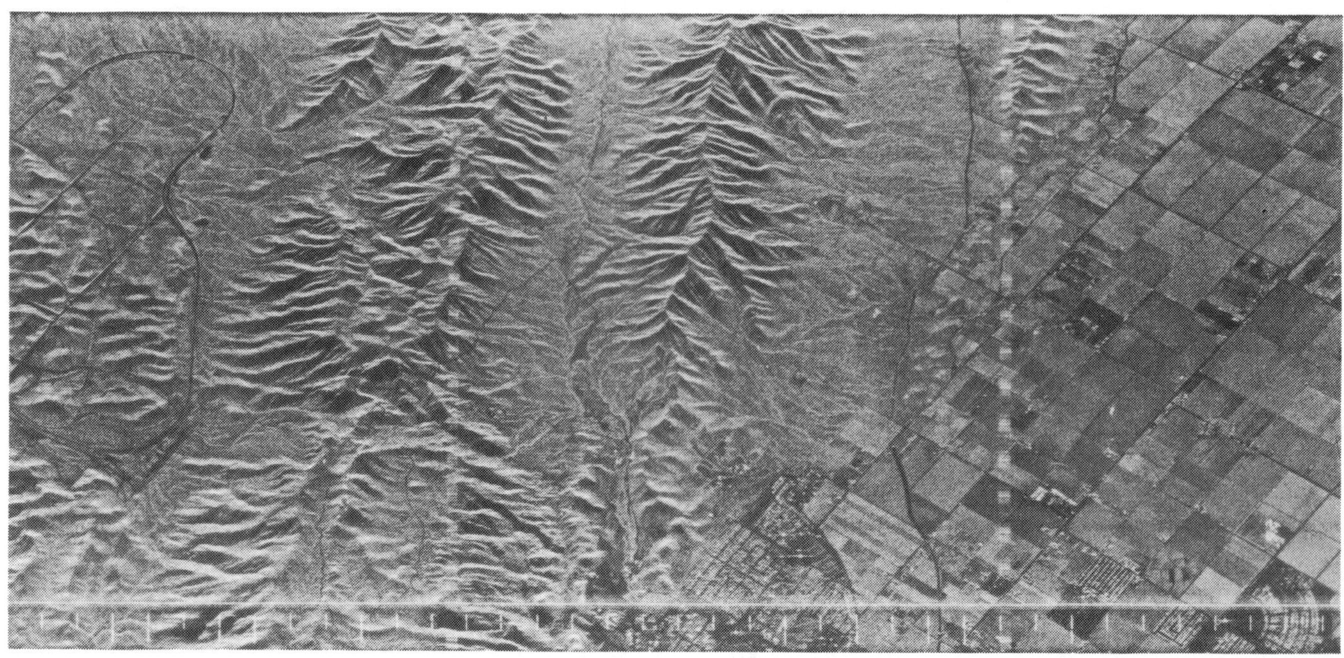

a) Normal High Resolution Image

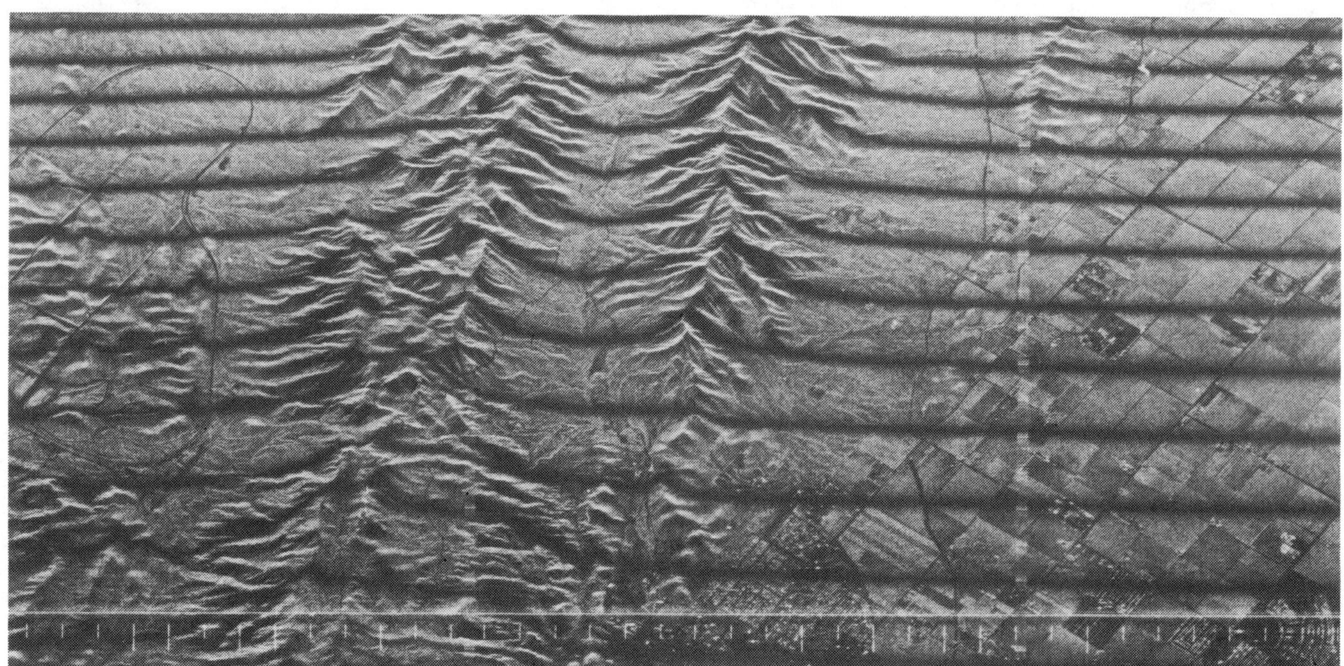

b) Interferometer Image

Figure 2 Radar Imagery of South Mountain Area, Phoenix, Arizona. Strip is approximately 2.5 nmi slant range; near range edge depression is approximately 50°; and flight altitude is 30,000 ft Mean Sea Level (MSL). (Compliments of Goodyear)

The slant range resolution of the system is dependent on the waveform transmitted. For a rectangular RF pulse having no phase modulation, the resolution is the same as would be obtained if the fundamental were received. For an RF pulse having a Gaussian envelope with no phase modulation, the resolution is $\sqrt{\nu}$ better (for $\nu \geqslant 3$), where ν is the nonlinear power law of the target (i.e., the nonlinearity affected component of the received signal is proportional to $E_i^{\nu}(t)$ if the incident wave equals $E_i(t)$); and one-third better for a chirp waveform having a time bandwidth product much greater than one (the third harmonic being proportional to $A^{\nu}(t)\cos[3\omega_o(t) + 3\Psi(t)]$ for a νth law target with $\nu \geqslant 3$ if the transmitted signal is $A(t)\cos[\omega_o t + \Psi(t)]$). (Typically $\nu = 3$ for small $E_i(t)$ and $\nu = 1$ for large $E_i(t)$.)

Apollo Lunar Sounder Radar System [4]

For the Apollo 17 Lunar Sounder Experiment (ALSE), three frequencies were used – 5, 15, and 150 MHz (60, 20, and 2 m), designated as HF1, HF2, and VHF, respectively. The VHF radiation provided standard synthetic aperture imaging of the lunar surface. The HF frequencies provided profile data of the moon's surface and subsurface lunar sounding profile information. The HF antenna was pointed straight down below the Apollo Command Module while the VHF antenna beam was pointed out to the right to avoid left-right image return ambiguities. The VHF radar also provided lunear profile and some sounding data; however, for sounding the HF signals had the advantage of a greater penetration (see Table 2). An interesting feature of the system is that synthetic aperture processing was used for the HF signals to obtain high resolution along track for the sounding and profile data (see Table 2, which also shows the major parameters for the ALSE). Because of the very large range and angle compression ratios, the dynamic range of the SAR mapper processed data can be 50-60 dB*; however, the film dynamic range is limited to 20-25 dB. To circumvent this problem a holographic imaging film is produced using the same optical correlator used to generate the conventional SAR images of the lunear surface. Using this film, the SAR image is viewed in a holographic viewer to nearly its full dynamic range (40-50 dB).

Table 3 shows the parameters of the film recorder used aboard the Command Module. Detailed results on the analysis of the lunear sounding data are to be published in the *Journal of Geophysical Research* (Redbook) [6].

Hologram Matrix Radar [7]

The hologram matrix radar is a novel synthetic aperture type radar which permits depth sounding of ice (to 0.5 to 5 m thickness) even though it uses a CW transmitter. It makes use of the spatial distribution of the scattered waves rather than time delay (as is done with conventional systems, such as the Apollo Lunar Sounder) to determine slant range distance. This radar, called the Holographic Ice Surveying System (HISS), is expected to achieve superior performance at short ranges compared to that of a conventional sounding radar using a moderate bandwidth waveform. The system is designed to use a moderately low carrier frequency (3 GHz used for helicopter system) which has the advantage of penetrating media with high loss, such as sea ice. The system uses focusing to different depths so as to eliminate the usual transmitter beam $1/R^2$ divergence loss.

ERIM Simultaneous X- and L-Band Dual Polarized SAR [8-10]

The Environmental Research Institute of Michigan (ERIM) has built a two-frequency SAR (see Table 4). This system is capable of producing four synthetic maps simultaneously – two at L-band (one having HH, the other having HV polarization; that is, horizontal polarization is transmitted and horizontal and vertical polarizations are received simultaneously) and two at X-band (HH and HV polarizations). This SAR can be used for monitoring strip mine activity and large area enforcement. Of 63 mine areas mapped in Kentucky, 60 were correctly identified as active, reclaimed, or unreclaimed (orphan) strip mines [10]. The system also has the potential of correctly classifying ice and, in some cases, measuring its thickness (to 1 m) [9].

SEASAT

NASA has under construction an earth resources SAR for the SEASAT satellite scheduled for launch late 1977. The parameters are given in Table 5. The spare SEASAT SAR is expected to be flown on the shuttle in about 1979 [6].

*For some systems a dynamic range of up to 90 dB is achievable [5].

	HF Mode		VHF Mode
	HF1	*HF2*	
Frequency (MHz)	5	15	150
Wavelength, λ, (m)	60	20	2
Estimated depth of penetration (m)*	1300	800	160
Center frequency (MHz)	5.266	15.8	158
RF bandwidth (MHz)	0.5333	1.6	16.0
Pulsewidth (μs)	240	80	8.0
Time-bandwidth product	128	128	128
Range resolution, free space (m)**	300	100	10
Transmitter peak power (W)	130	118	95
Transmitter average power (W)	12.4	3.7	1.5
Effective antenna gain (dB one-way, including efficiency)	−0.8	−0.7	+7.3
Noise figure (dB)	11.4	11.4	10.0
Pulse repetition rate (s^{-1})	397***	397***	1984
Recording duration (μs)	600	600	70
AGC gain range (dB)	12.1	12.1	13.9
Echo tracker	no	no	yes

*Based on model analysis.
**Best along track resolution equals 5λ, which is same as above slant range resolution.
***Interlaced on HF1 and HF2.

Table 2 Lunar Sounder System Characteristics (From Porcello, et al. [4])

Parameter (Units)	Channel HF1	HF2	VHF
Film speed (mm/s)	5	5	5
CRT sweep speed (mm/s)	4.03×10^4	4.03×10^4	7.14×10^5
Sweep duration (μs)	620	620	70
Calibration frequency (cycles/mm)	3.75	3.75	3.6
Dynamic range (dB)	30	25	20
Temporal range bandwidth (MHz)	0.533	1.6	16
Range spatial frequencies (cycles/mm)	5.62 to 18.55	21.8 to 60.6	4.32 to 27.4
Number of 5λ range resolution cells	331	992	1120
Temporal along-track (Doppler) bandwidth (Hz)	±43	±100	±100
Along-track spatial frequencies recorded (cycles/mm)	±8.6	±20	±20

Table 3 ALSE Recorder-Signal-Storage Parameters (From Porcello, et al. [4])

Parameter	X-Band	L-Band
Center Frequency	9.450 GHz	1.315 GHz
Resolution	Variable from 5 ft x 7 ft	to classified
Transmitter (Peak) @ 2% duty cycle	1.2 kW	6 kW
Antenna Gain	28 dB	16.5 dB
Ant. Beamwidth	1.1°	7°
Polarization Isolation	23 dB	19 dB
Polarization	HH, HV	HH, HV
Depression Angle	0° to 90°	0° to 90°
Maximum Range	13 nmi	13 nmi
CRT	WX30267P11	WX30267P11
No. of Spots/Scan	8,000	4,000
Film Capacity	100 ft	100 ft
Film (Kodak)	Microfile 5460	Microfile 5460

Table 4 Parameters of ERIM X- and L-Band Dual Polarized SAR (From Rawson, Smith, and Larson [8])

Frequency	1.275 GHz ± 9 MHz
Cross track resolution	25 m
Along track resolution	6.25 m (minimum)*
Uncompressed pulse width	32 μs
Signal bandwidth	18 MHz
Antenna gain	33 dB
Peak power	800 W**
Average power	64 W
Pulse repetition frequency	1200-2000 Hz
Swath width	100 km
Angle from vertical of beam axis	20°
Satellite altitude	800 km
No. of pulses in flight	about 8

*In order to remove speculars from image it normally is planned to average four adjacent range cells, yielding a resolution of 25 m.
**Can be increased to 1100 W when spacecraft has power available.

Table 5 **SEASAT SAR Parameters [6, 13]**

Frequency	1.757 GHz
Cross track resolution	15 km
Along track resolution	about 15 km
Uncompressed pulse width	128 μs
Compressed pulse width	1 μs
Along track Doppler resolution	8 kHz
Aperture diameter	15 in (0.38 m)
Peak power	20 W
Noise figure	6.5 dB

Table 6 **Venus Probe SAR [6, 11]**

SAR for Mapping Venus Surface [11]

NASA has under construction a SAR system for a Venus probe scheduled for launch in 1978 with mapping in 1979 [6]. The system is expected to have a ground range resolution perpendicular to the ground track of about 15 km [6]. The along track resolution is expected to be about the same [11]. The radar parameters are given in Table 6. Plans are also underway for a 2-frequency, 2-polarization (VV and HH) synthetic aperture mapper for Venus [6]. One of the frequencies would be at X-band while the other remains to be determined [6]. This system is expected to fly in about six years.

Land Surveying [12]

Tests are underway to demonstrate the feasibility of using a SAR to locate ground corner reflectors to an accuracy of 3-4 m with promise of 1-2 m. These tests are being planned by the Bureau of Land Management in Denver.

Airborne Electronically Agile Radar (EAR) [14-17]

Westinghouse is building an X-band multifunction phased-array radar for possible use aboard USAF/Rockwell B-1, General Dynamics FB-111, or the B-52. Its modes consist of spotlight synthetic aperture mapping, terrain following and avoidance, ground speed measurement to improve accuracy of on-board inertial-navigation system, and beacon locating for ground support operations. It uses an 86 cm antenna diameter (placed in the nose of the aircraft) having 1818 radiating elements and ferrite phase shifters (the latter accounting for 10% of the total radar cost) and ±60° cone angle coverage. Dual TWT and solid-state driver transmitters are used for redundancy. An interesting study finding was that, contrary to what might be expected, a reliability of 130-150 hr Mean-Time-Between-Failures (MTBF) is more cost-effective than the reliability level of 325 hr originally specified. This result is obtained when cost of ownership over the life of the radar is considered. Currently, maintenance costs over the life of a radar about equal acquisition cost. Flight tests are planned aboard a B-52 in late 1978.

References

[1] Kirk, Jr., J.C.: "Digital Synthetic Aperture Radar Technology," *Record of the IEEE 1975 International Radar Conference,* Arlington, Virginia, pp. 482-487, 21-23 April 1975.

[2] Graham, L.C.: "Synthetic Interferometer Radar for Topographic Mapping," *Proceedings of the IEEE, Vol. 62, No. 6,* pp. 763-768, June 1974.

[3] Harger, R.O.: "Harmonic Radar Systems for Near-Ground In-Foliage Nonlinear Scatterers," *IEEE Transactions on Aerospace and Electronic Systems, Vol. AES-12, No. 2,* pp. 230-245, March 1976.

[4] Porcello, L.J.; Jordan, R.L.; Zelenka, J.S.; Adams, G.F.; Phillips, R.J.; Brown, Jr., W.E.; Ward, S.H.; and Jackson, P.L.: "The Apollo Lunar Sounder Radar System," *Proceedings of the IEEE, Vol. 62, No. 6,* pp. 769-783, June 1974.

[5] Heimiller, R.C.: *Private Communication,* Environmental Research Institute of Michigan.

[6] Brown, Jr., W.E.: *Private Communication,* Jet Propulsion Laboratory.

[7] Iizuka, K.; Ogura, H.; Yen, J.L.; Nguyen, V.; and Weedmark, J.R.: "A Hologram Matrix Radar," *Proceedings of the IEEE, Vol. 64, No. 10,* pp. 1493-1504, October 1976.

[8] Rawson, R.F.; Smith, F.; and Larson, R.: "The ERIM Simultaneous X- and L-Band Dual Polarization Radar," *Record of the IEEE 1975 International Radar Conference,* Arlington, Virginia, pp. 505-510, 21-23 April 1975.

[9] Bryan, M.L.; and Larson, R.: "Classification of Fresh Water Ice Using Multispectral Radar Images," *Record of the IEEE 1975 International Radar Conference,* Arlington, Virginia, pp. 511-515, 21-23 April 1975.

[10] Shuchman, R.A.; Davis, C.F.; and Jackson, P.L.: "Contour Strip Mine Detection and Identification with Imaging Radar," *Record of the IEEE 1975 International Radar Conference,* Arlington, Virginia, pp. 516-521, 21-23 April 1975.

[11] deLeon, J.C.: "Synthetic Array Radar to Map the Surface of Venus," *Microwaves, Vol. 16,* pp. 12, 14, January 1977.

[12] *Aviation Week and Space Technology, Vol. 104,* p. 53, 12 July 1976.

[13] Moore, Richard K.: *Private Communication,* University of Kansas.

[14] *Aviation Week and Space Technology, Vol. 102,* p. 59, 27 May 1974.

[15] Jones, W.S.: *Private Communication,* Westinghouse Electric Company.

[16] Klass, P.J.: "Airborne Role Planned for Agile Radar," *Aviation Week and Space Technology, Vol. 104,* pp. 45, 47, 21 June 1976.

[17] Pettecs-Snider, E.: "EAR Spells Versatility," *Aerospace International,* pp. 43-46, May/June 1977.

Part 5
Radar Systems and Components

The early junction transistor (invented in 1948 by John Bardeen, Walter Brattain, and William Shockley of Bell Telephone Laboratories) could operate only up to a few megahertz. Modern diffused bipolar transistors and GaAs FETs can operate up to tens of gigahertz.

The magnetron remained a laboratory device until the invention of the resonant cavity in 1939; see Table 1. Now, megawatts of peak power can be produced with an efficiency of over 60% by using magnetrons.

The magnetron is basically a dc pulsed oscillator which is not coherent from pulse to pulse. As a result, this device has been applied to systems which require no pulse to pulse coherence or to MTI systems which require coherence between two or among three pulses (which is achieved by retaining the phase of the transmitted pulse, through the use of what is called a COHO oscillator, and measuring the phase of the received pulse relative to that reference phase). There is now, however, also the potential of achieving coherent operation of magnetrons through the technique of locking coaxial magnetrons; see Chapter 23 by Smith.

The reentrant beam, continuous cathode, crossed-field amplifier (frequently called the Amplitron and invented by W.C. Brown in 1953) physically looks like a magnetron oscillator but is actually an RF amplifier. It is characterized by having low gain (6-15 dB), very high efficiency (40-80%), moderate bandwidth (8-15%), and reasonably low noise characteristics (although not as low as those obtainable with the klystron and TWT). The tube operates in a saturated mode, hence amplitude modulation is not possible. However, pulse code modulation of the type described in Chapters 7 and 8 can be used.

The tube can be used for coherent pulse Doppler systems. It often is employed as a power booster output tube following a TWT or klystron. When used in this way, low- and high-power modes of operation for the radar are possible because, when the anode-to-cathode voltage of the CFA is removed, the signal passes through as if the tube were not present (except for an insertion loss ranging between 0.2 and 2 dB). The AN/TPN-25 (the PAR of the AN/TPN-19; see Table 1f of Chapter 1) operates in this manner. The high power mode is used only for penetration through heavy rain. A cascade of CFAs can be used for the transmitter power amplifier, allowing different power levels to be used for different target ranges by selective turning on of one or more of the CFAs.

The CFA discussed above operates as a backward wave amplifier; that is, the group velocity of the RF electromagnetic field in the tube is opposite in direction to the space-charge in the tube. A CFA which operates with a forward wave (that is, where the group velocity of the RF electromagnetic field travels in the direction of the space-charge) has also been developed in France in 1950. This tube, however, has not been used as widely as the backward-wave CFAs (Amplitrons). It has a lower power efficiency generally (30-60%) and a slightly greater insertion loss when the tube is turned off (1.5-3 dB). Its gain is about the same (10-15 dB), it has the potential advantage of wider bandwidth, and its ability to operate across a dc power supply (constant voltage) is utilized in a few systems.

The klystron and TWT are linear beam power amplifiers and can be used in coherent systems such as pulse Doppler systems. They also permit the use of phase or frequency coding of the transmitted signals, such as that described in Chapters 7 and 8. One of the major differences between the klystron and TWT is bandwidth. Klystrons generally have bandwidths of a few percent of the car-

Device	*Date*	*Comment*
Magnetron	1921	A.W. Hull invents magnetron for use as diode switch. He also observes oscillations at 30 kHz with a power output of 8 kW and 69% dc to RF conversion efficiency.
CW Split Plate Magnetron	1924	H. Yagi of Japan (inventor of the Yagi antenna) invents device which produces a few watts of CW microwave power. A. Zacek of Prague also obtains CW microwave power by cyclotron oscillations at about the same time.
Resonant Cavity Magnetron	1939	J.T. Randal and H.A.H. Boot of England invent the resonant cavity traveling wave type magnetron with internal resonators. They obtain 400 W of CW and 10 kW of pulsed power at 9.8 cm, order of magnitude better than what had been achieved before.
Klystron	1939	W.W. Hansen, R.H. Varian, and S.F. Varian develop the klystron, which becomes the first operational high-frequency tube. Initially, though, it is relegated to role as local oscillator in superheterodyne radar receivers.
Traveling Wave Tube (Helix Type)	1940	Invented by R. Knompfer. The first tube is built in 1943 for operation at 9.1 cm with gain of 6 dB and a noise figure of 13 dB.
Transistor	1948	Invented by John Bardeen, Walter Brattain, and William Shockley.
Forward Wave Crossed-Field Amplifier	1950	Invented by R.R. Warnecke, W. Kleen, A. Lerbs, O. Döhler, and H. Hubers in France.
Amplitron (Reentrant beam, continuous cathode, crossed-field amplifier)	1953	Invented by W.C. Brown.

Table 1 **Some Significant Early Device Developments**

rier frequency although bandwidths of up to 10% have been achieved for high power klystrons. TWTs (helix type) have octave bandwidths. Both have high gains, typically 30 to 60 dB, and both can be used in "unsaturated operation" to amplify amplitude-modulated signals (though with poor power efficiency).

The klystron (invented in 1939; see Table 1) uses a slow wave structure to achieve microwave power amplification. The circuit wave velocity is made approximately equal to the velocity of the electrons by proper design of the tube structure. In this way, the electric field of the wave travels with the electrons and accelerates them. These tubes use a cascade of resonant cavities; it is the narrow bandwidth of each cavity that results in the device's narrow bandwith.

The TWT (invented in 1940; see Table 1) amplifies RF power by continuous interaction of the electron beam with an interaction structure. One type of interaction structure is the helix. Because the helix is not strongly resonant at any one frequency (unlike the cavities), helix TWTs can have octave bandwidths. Another type of interaction structure is the coupled-cavity which, because of its narrow bandwidth, typically limits the TWT bandwidth to the order of 10% (though 30% is possible).

Because of their low noise characteristics, TWTs are also used as low noise receivers; noise figures of 4.5 dB have been achieved at S-band, for example. For applications in which the TWT's very wide bandwidth is not required, FET amplifiers are being used as replacements for low noise receiver TWT amplifiers.

Described in Part 5 are the latest developments in solid state transmitter technology, antenna arrays, high power linear-beam tubes, and crossed-field tubes. In the "Present and Future Trends" section at the end of this part, new solid state transmitters (e.g., that of the Transportable Surveillance Radar), new antenna systems (e.g., the dome antenna), new tube techniques (e.g., the thin-film field emission cathode), and new tubes (e.g., the extended interaction oscillator, millimeter wave gyrotron, and relativistic-electron beam tube) are described.

Chapter 19 Dodson

L Band Solid State Array Radar Overview*

If one were to tabulate all of the details which might be covered in a discussion on solid state radar technology, a volume similar to Skolnik's *Radar Handbook* would be required [1]. Therefore, this discussion is limited to several general areas – definitions, advantages, system constraints, possible uses, examples of typical active element modules with their specifications, methods of power generation, low-noise considerations, and device reliability.

A solid state radar is a system composed of solid state modules, functioning collectively to satisfy some particular system requirement or threat. A module is defined as a miniaturized subsystem or package that contains two or more semiconductor circuits (either microwave or digital) that jointly function to support an overall system requirement.

Why use solid state? Proponents respond with reasons based on size, density, weight, ease of reproducibility, low "projected" volume production cost, and high "potential" reliability. Note that projected and potential are in quotation marks – herein lie some of the major problems.

What are some of the inherent system constraints and considerations? First, the radar designer is faced with the relatively low output power capability of transmitters which use semiconductor devices. In many instances, system performance requirements preclude a suitable array operation for solid state. In addition to high elemental power needs, specific requirements for antenna gain, beam width, numbers of elements, etc. may pose serious problems for solid state. Therefore, in order to employ solid state technology, certain compromises may be necessary.

Module reliability is another consideration. Measured data are somewhat minimal. *Module* production experience is very limited, resulting in questionable confidence in the learning curve theory supporting the low projected volume cost.

Possible module applications obviously include electronically steerable arrays. Such arrays could generate single or multiple beams and be functionally adaptable. Electronic steering may also be a potential approach in antenna beam stabilization for ships, airplanes, or space vehicles. Similar modules are being used in manpack and handheld systems. Finally, there are some possible civilian applications for microwave integrated circuit module technology. Auto collision avoidance, commercial air traffic control, and beacon tracking for emergency vehicles (fire, police, etc.) are but a few examples.

There are basically two types of steerable array radars. The active element type is one employing an active transmitter and/or receiver module per radiating antenna element. The passive element variety makes use of a beam steering lens, and requires some kind of high power phase shifter per antenna element.

One example of an active element type of phased array is illustrated in Figure 1. Shown is an artist's concept of a long range ground based radar system with one face partially cut away. Each face might comprise tens of thousands of solid state transceiver modules. A cart for maintenance personnel is employed to replace defective modules. If the modules are as reliable as solid state proponents claim, the cart may not be necessary.

Figure 1 Artist's Conception of a Large Ground Based System

Figure 2 shows a tactical, field-transportable radar employing solid state modules. This sketch is an artist's concept of the TPS-59 Marine Corps system developed by the General Electric Company. The TPS-59 system makes use of a mechanically rotated antenna for azimuth scanning, plus electronic scanning in elevation. This system is currently undergoing qualification and field testing.

Figure 3 is a block diagram of a typical active element module. As noted earlier, array radars do not necessarily require a transmitter and/or receiver on each radiating element. Nevertheless, this particular transceiver model is helpful for understanding module operation. A single RF input is fed through a digitally controlled

Figure 2 Mobile System, The Marine Corps TPS-59

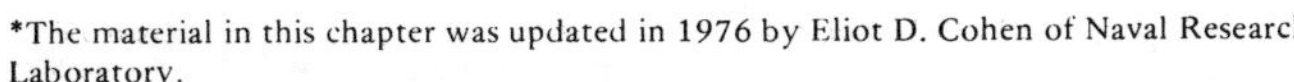

*The material in this chapter was updated in 1976 by Eliot D. Cohen of Naval Research Laboratory.

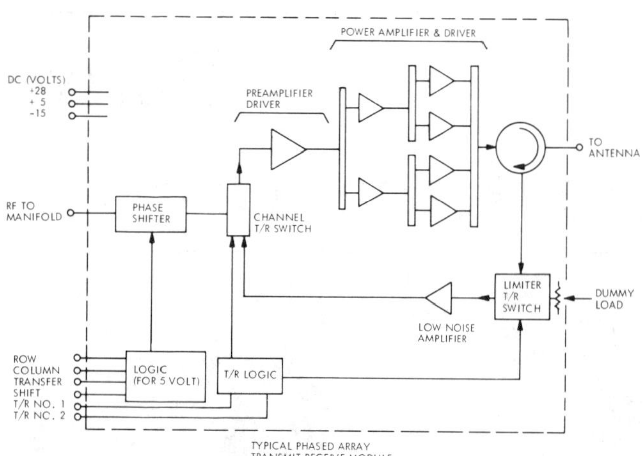

Figure 3 Typical Phased Array Transmit Receive Module

phase shifter (for beam steering) to a channel selector switch. This switch can be positioned in either the transmit or receive mode. In the transmit mode, a multi-stage bi-polar transistor power amplifier drives the elemental antenna through a circulator-duplexer. In the receive mode, the antenna feeds a low noise amplifier through a duplexer and T/R switch-limiter. The receiver output is phase shifted in a reciprocal manner and fed back into the system RF manifold network.

Some typical active element module requirements are given below:

Bandwidth
10 to 20%

Peak Power Output
25 to 1000 watts

Duty Factor
1% to approximately 40%

Pulse Width
Tens of microseconds to several milliseconds

Noise Figure
< 4 dB overall

Phasor Bits
4:22-1/2, 45, 90, 180°

Tracking (Unit-To-Unit)
Phase – between 10 and 15° rms
Amplitude – < 1 dB

Size
< λ/2 in any dimension

Weight
0.5 to 3 lbs

Efficiency
25 to 45% depending on Duty Factor

Size is optional but usually must be less than 1/2 wavelength in the plane of the antenna in order to meet the usual system grating lobe requirements. Weight specifications vary depending on the application and heat sink requirements. Overall module efficiency currently achievable is on the order of 25-45% and is related directly to duty factor. In low duty factor applications, low module efficiency results because proportionately greater time is spent in the receive mode. As duty factor increases, module efficiency tends to approach that of the transmitter.

Figure 4 shows an L-band module developed at RCA in Moorestown, New Jersey. The block diagram includes the T/R switch, phase shifter, low level class A transmitter drivers, and the six-transistor hybrid-combined power amplifier. Also shown is the three-transistor low-noise receiver circuit, with T/R switch and load. In the transmit mode, this load is switched to the receive arm of the circulator duplexer to terminate any reflected RF power resulting from antenna mismatch during beam steering.

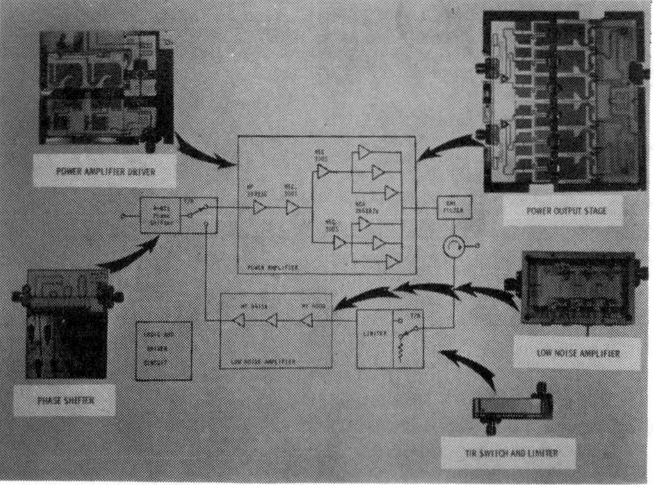

Figure 4 RCA Solid State T/R Module Subassemblies Block Diagram

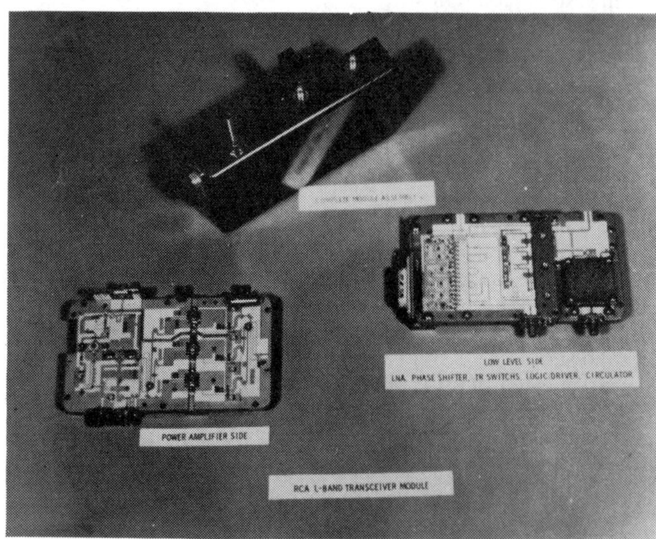

Figure 5 RCA L-Band Transceiver Module

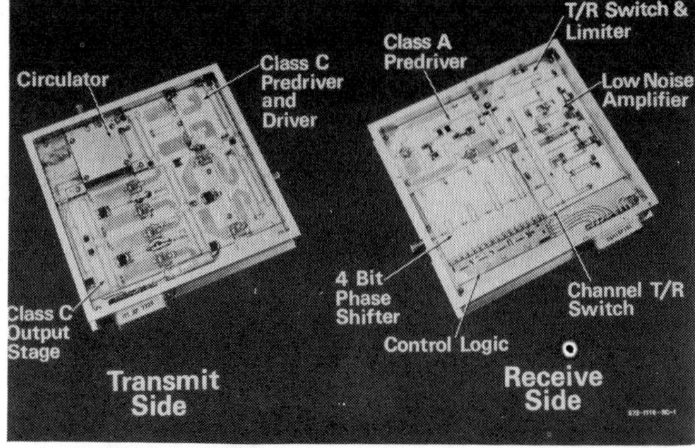

Figure 6 Westinghouse Completely Integrated Modules

Another module is pictured in Figure 5 – an L-band transceiver suitable for long-range ground-based search radars. This particular module contains three cascaded power transistor stages. Also shown are the circulator, phase shifter, and logic circuitry (which are discussed later). This module is designed for wide pulse width, high duty factor operation. Thus, a good thermal interface must be provided with the system. Another version of an L-band T/R module is shown in Figure 6. This design makes use of chip carrier type circuits for power transistors.

What are the key problem areas encountered in the design and use of these modules? First, a high RF power generation capability is required for some radars. Next, meeting spectral purity requirements may involve a trade-off between device noise performance and power generation capability. In addition, module reproducibility and optimum packaging techniques must be considered carefully.

The limiting factor in the generation of solid state RF power is obviously the capability of the semiconductor devices themselves. Key parameters of concern include thermal time constant, thermal impedance, and efficiency. It has been shown that device failure is a function of peak junction temperature; therefore, this parameter is of greatest concern. Availability of the devices is also important, to preclude module designs employing single source transistors or diodes.

Both TRAPATT diodes and silicon bipolar transistors are candidates for generating L-band power. TRAPATTs are capable of generating significantly higher peak output powers than are transistors. For example, a peak power of 1200 W with 24% efficiency has been achieved with five series-connected TRAPATTs at 1.9 GHz [2]. However, at L-band, transistors have demonstrated much higher average power outputs. In addition, peak power outputs of up to 300 W have been achieved recently with transistors operating with 10 μs pulse widths at 1% duty factors. This performance, coupled with the relative ease of circuit design, the ready availability of devices, and the ability of transistors to easily operate in the Class C mode, precludes the use of TRAPATT devices below 2 GHz except for applications requiring very high peak power at low duty factors. Above 2 GHz, TRAPATTs are an attractive candidate for phased array applications, particularly since both average and peak powers available from transistors decrease dramatically at S-band [3].

Table 1 lists the characteristics of those transistors which are suitable for use in the final stage of transmitter amplifiers. Some devies with these characteristics are available in packaged chip carrier configurations which provide a certain amount of input and output matching. In addition to the highest power devices there are, of course, a large number of lower power devices for use in driver stages. The four leading American manufacturers of microwave frequency power transistors are Microwave Semiconductor Corporation; Power Hybrids, Inc.; TRW, Inc.; and Communications Transistor Corporation.

Figure 7 is included to illustrate chip carrier technology. Depicted is an MSC 2100, rated at ≈ 100 W when operating at 900 MHz. The device-circuit size advantage is obvious. A 2100 chip is alloyed down and matching networks are placed on dielectric substrates

Frequency Range (MHz)	Duty Cycle (%)	Pulse Width (μs)	Power Output (W)	Gain (dB)	Collector Efficiency (%)	V_{CC} (V)
1090	1	10	300	6.3-6.6	45	48-50
960-1215	1	10	225	-----	----	45
1025-1150	1	10	250	5.5	----	50
1200-1350	----	1000	60-65	7-7.5	45	35
1000-1400	100	-----	35	7.0	50	28

Table 1 Characteristics of the Highest Power L-Band Transistors Currently Available

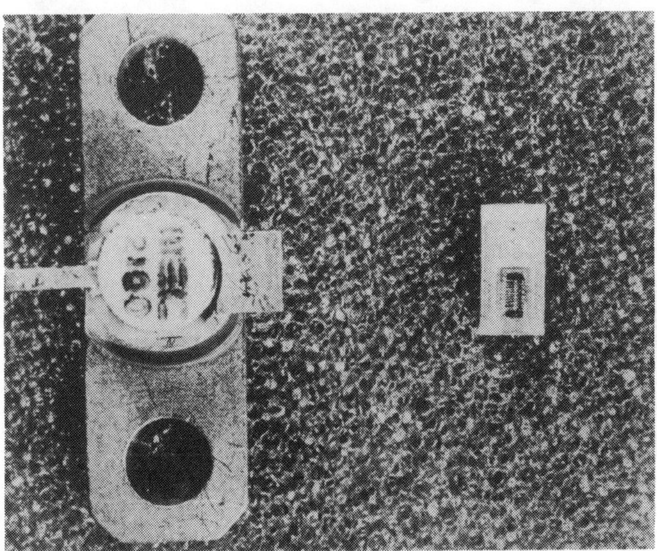

Figure 7 MSC-2100 Package and Chip Carrier Designs

Typical 40 Watt Peak Performance using MSC 80135 Chip

	Package	*Chip Carrier*
Frequency	L-Band	L-Band
Input Impedance	0.7 + j4.0	8.0 + j5.0
Gain	8 dB	8 dB
Bandwidth	100 MHz (3 dB)	100 MHz (1 dB) 250 MHz (3 dB)
Collector Efficiency	55%	55%

Table 2 Advantages of Chip Carrier Design

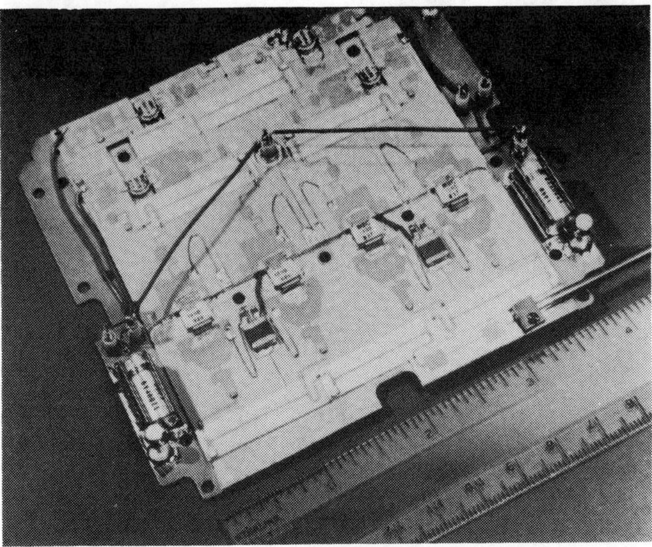

Figure 8 MA/MSC Power Amplifier Subassembly

to form the chip carrier circuit. In this case matching networks for the transistor input and output are included in the package. The entire unit can be hermetically sealed. By doing so, the external matching circuitry is simplified and device bandwidth is extended. The advantages of a typical chip carrier design are summarized in Table 2. For the packaged MSC 80135 without internal matching, the input impedance is predominantly inductive in L-band. Hence, it is very difficult to obtain broadband operation; the chip carrier approach is preferable. O'Reilly, *et al.* [4] at the Naval Research Laboratory and Pitzalis, *et al.* [5] at the US Army Electronics Command have demonstrated very wideband performance using

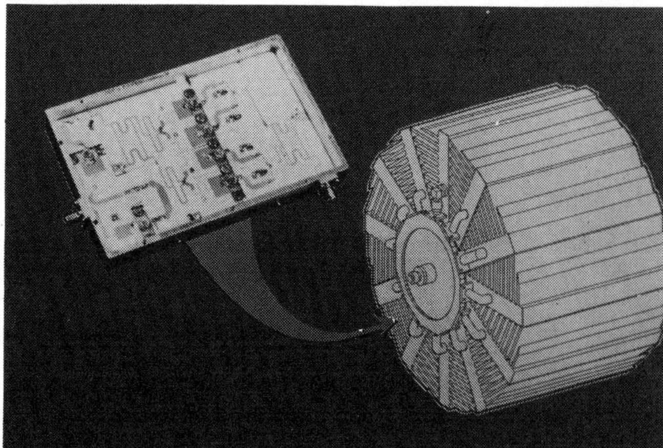

Figure 9 **Westinghouse 1 kW L-Band Amplifier**

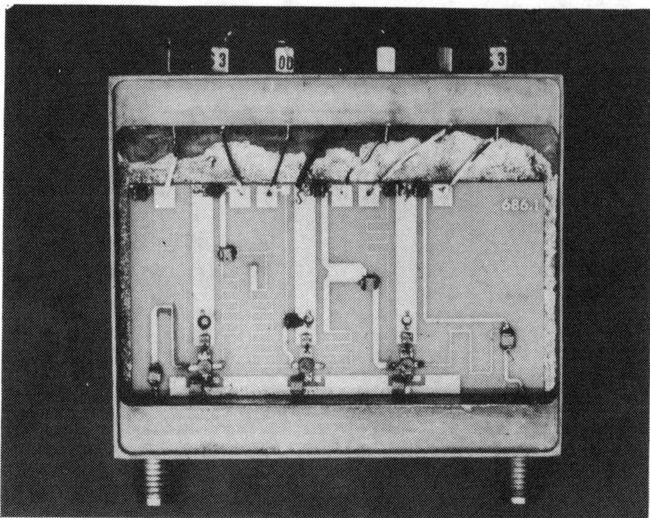

Figure 10 **Microwave Associates Low Noise Amplifier**

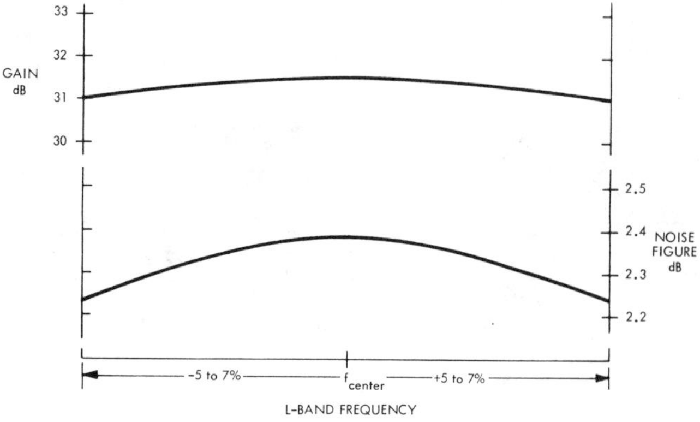

Figure 11 **Typical 3-Transistor Cascaded Low Noise Amplifier Performance**

chip carrier techniques. A recent article by Belohoubek, *et al.* [6] describes a new BeO carrier with very low parasitics. Use of this carrier provides better power sharing between transistor cells and better high frequency performance.

A carrier plate is shown in Figure 8; it contains four MSC 80135 transistors and is centered at L-band. The capacitors provide low voltage, high current energy storage during the pulse to minimize collector voltage droop. Chip carrier devices have been employed to simplify external circuitry. Note the lack of external components in the matching networks.

Figure 9 is a 1 kW L-band amplifier developed by Westinghouse in 1972 for the Rome Air Development Center. It contains twelve 100 W amplifiers. These amplifiers are placed around the hub of a radial line combiner, and the entire package is air cooled. The complexity of an amplifier capable of generating 1 kW at L-band is, of course, largely dependent on the duty factor and pulse width requirements. The Westinghouse amplifier pictured in Figure 9 is capable of operating with pulse widths up to 1 ms and at a duty of 1%. If pulse widths of only 10 μs are acceptable, a 1 kW output stage can be developed using only four 250 W transistors of the type tabulated in Table 1.

The state-of-the-art of low noise receiver devices is the next consideration. At L-band, silicon bipolar transistors still are used in most receivers although at higher frequencies GaAs FETs are attractive. At present, silicon bipolars are available with noise figures ranging from 1.1 dB at 1 GHz to 1.8 dB at 2 GHz; associated gains are 15 dB and 10.5 dB, respectively. These figures are for chip noise, not amplifier noise. Two examples of transistors which operate at approximately this performance level are the Hewlett-Packard HXTR-6101 and the Avantek AT-4611.

Figure 10 shows a typical low noise receiver which makes use of three Avantek transistors. The dc bias lines are to the side of circuit, thereby, reducing the reactive effects of the distribution line. Nearly all receiver designers today make use of computer-aided design techniques such as the DEMON program developed by Gelnovatch and Chase [7]. This receiver was designed using such techniques. Figure 11 illustrates the performance expected from an amplifier of this type. Achievement of gain of 40 dB and very wide bandwidth is possible in a well designed receiver; thus good phase and amplitude tracking are assured from unit to unit. The noise figure shown in Figure 11 is for the receiver amplifier part of the module only, not for the overall module. Approximately 1 dB must be added to the receiver noise figure to take into account the insertion loss effects of the circulator, limiter, and T/R switch.

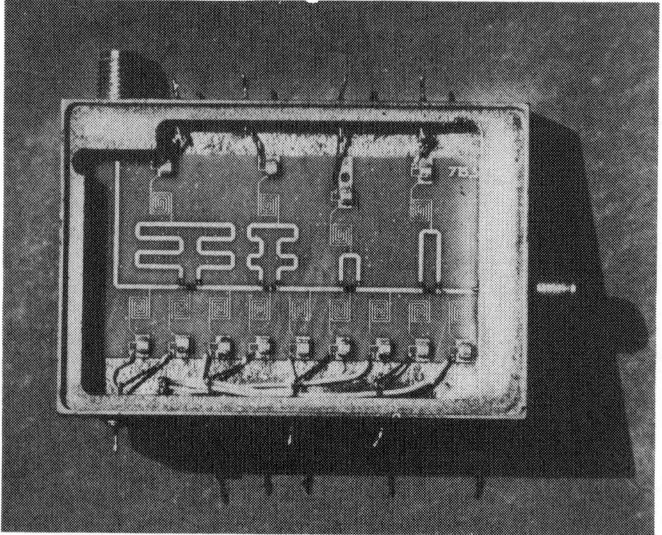

Figure 12 **Microwave Associates 4-Bit Phase Shifter**

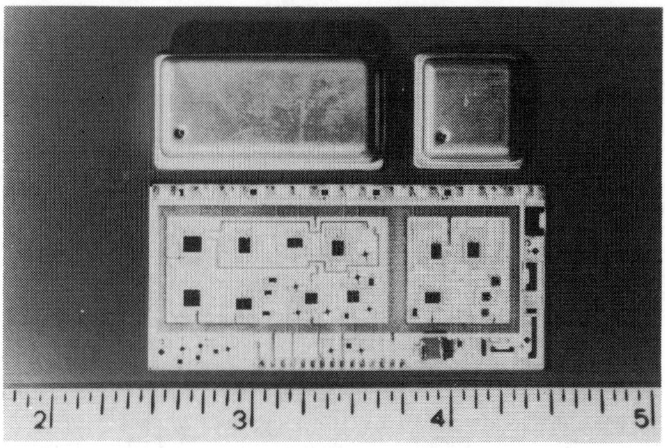

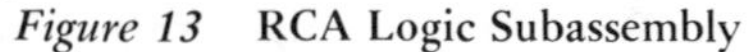

Figure 13 RCA Logic Subassembly

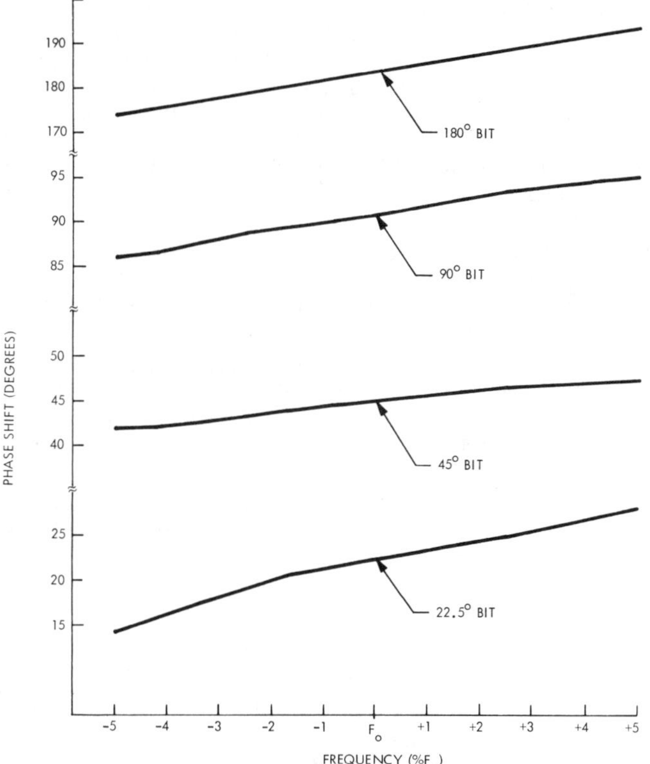

Figure 14 L-Band Phase Shifter Phase Response

A line length phase shifter (that might be used in modules) is shown in Figure 12. For most applications, designers use switched line circuits for phase shifters which require 4 diodes per bit. One advantage of the line-length switch, aside from its linear phase vs. frequency relationship, is that all of the diodes are alike. The designer does not have to consider device matching circuitry unique to a particular bit position, which normally would be required with a loaded line type of phase shifter. Operation of the loaded line phase shifter is based on changing velocity of propagation down the structure. The switching times for the diode phase shifters can be as small as 100 to 200 ns, if desired.

An RCA version of the phase shifter logic circuit is shown in Figure 13. This particular logic is complementary MOS. The key advantage of CMOS is low power dissipation – an order of magnitude less power is required for CMOS logic than for standard TTL (see Table 1 of Chapter 10). The row and column bit rate to the modules is typically limited to 500 kHz for a large array if a serial data flow is used. For a small array, rates two to five times faster can be achieved.

Figure 14 indicates the typical phase vs. frequency relationship of a 4-bit digitally controlled phase shifter. There should be a nearly linear relationship between phase and frequency. However, the system computer can compensate for certain discrepancies provided that all modules are alike. Figure 15 gives the loss of a phase shifter with the characteristics shown in Figure 14. A typical loss is 3 dB, depending on diode biasing current. Also shown in Figure 15 is the VSWR at the input and output of the phase shifter as a function of bit size and frequency.

No overview of phased array radar systems would be complete without a discussion of reliability. To begin, consider still another L-band module with power transistor stages, balanced couplers output, and drivers, which is shown in Figure 16. A parts count of this and the opposite side circuitry shown in Figure 17 (which consists of the circulator, phase shifter, and logic) would result in the listing shown in Table 3. Use of these particular components is not necessarily recommended; nevertheless, they do serve as a reliability model, making use of the MTBF predictions of the *RADC Reliability Handbook.* Using these particular devices and the estimated device λ's, a 40,000 hour MTBF is forecast for the module. This prediction is based on *random* failures expected.

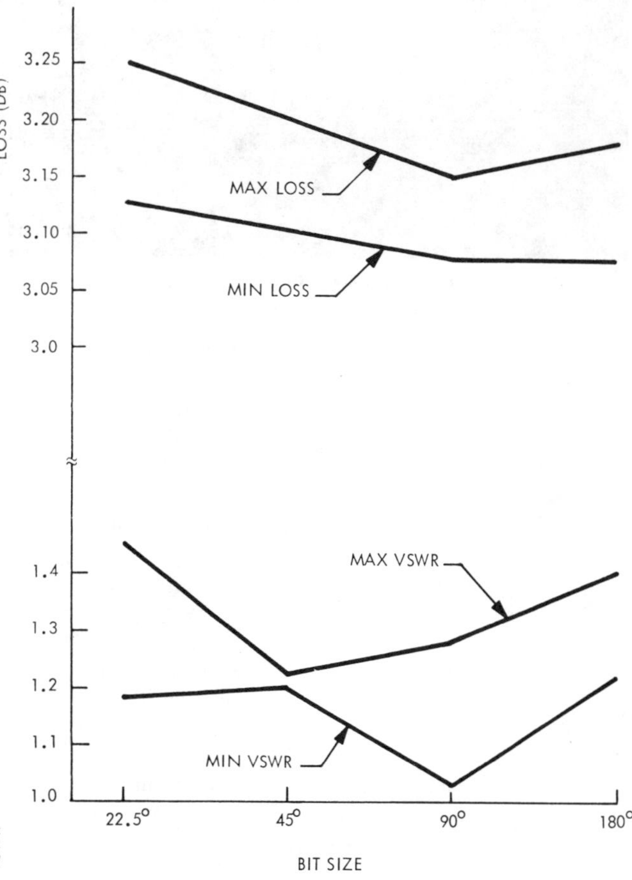

Figure 15 L-Band Phase Shifter – Range of Loss and VSWR over the Frequency Band vs. Bit Size

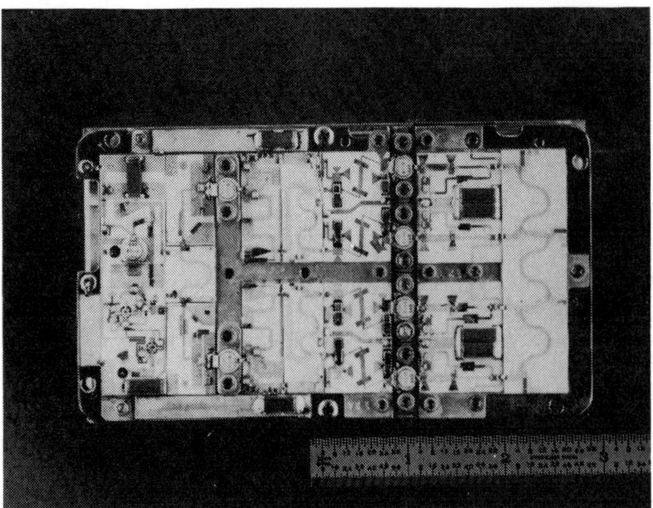

Figure 16 Camel Module, High Power Side

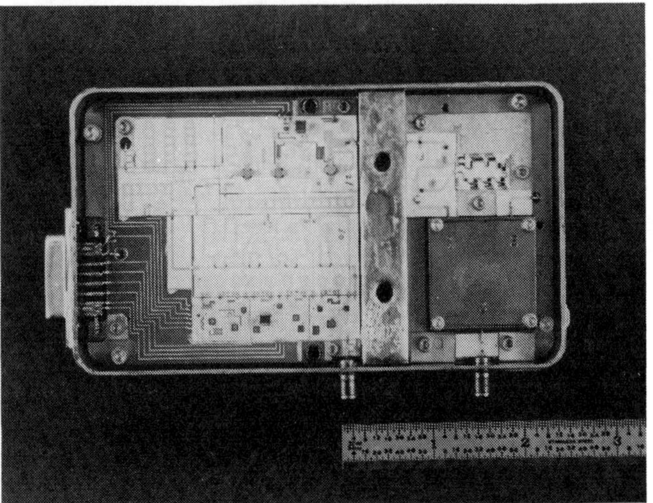

Figure 17 Camel Module, Low Power Side

Recently, several studies of power transistor device reliability were undertaken by the Naval Research Laboratory. One contract program with Microwave Semiconductor Corporation is of particular interest because it resulted in the generation of a substantial amount of data under pulsed RF step-stress conditions. Pulsed RF step-stress testing requires very complex testing facilities compared to oven testing and/or dc fixed or pulsed testing. However, pulsed RF step-stress is greatly preferred over all other types of reliability testing because it most nearly simulates system operating conditions. Data generated using the other types of testing usually show very poor correlation with RF test data.

In the early stages of the MSC reliability program, the devices tested were emitter-ballasted Microwave Semiconductor Corporation MSC-1330A and MSC-1330B devices. These devices are aluminum metallized devices capable of generating 30 W CW with 8.5 dB gain and 50% collector efficiency, or 70 W peak with 10 μs pulse width, 10% duty cycle operation at 1.3 GHz. Devices were tested with RF pulsewidths of 1.5 ms and 120 μs at 15 and 30% duty factors, respectively. Sequential evaluations were made at three peak junction temperatures – 340°C, 280°C, and 250°C. Usually a group of eight devices was tested at each temperature. The results of the tests are described in detail by Dodson and Weisenberger [8] and by Poole and Walshak [9]. Later, the aluminum metallized transistors were tested at temperatures as low as 190°C. In addition, tests were made on special MSC-1330/A transistors with gold-refractory metallizations, as well as two other devices – the AMPAC 1214-30 and the MSC 82010 (a 10 W CW 2 GHz unit). In summary, the following conclusions resulted from this program [10]:

1) MTF values ranging from 10^6 to 10^7 hours (hundreds of years) were predicted for aluminum metallized MSC-1330 transistors operating at normal junction temperatures (75-100°C) under 120 μs, 30% duty pulsed RF conditions.
2) The projected MTF for the gold-refractory metallized MSC 1330/A compared to the same device with aluminum metallization is at least 21 times greater at a junction temperature of 280°C and at least 6 times greater at a junction temperature of 230°C.
3) Projected MTF for the MSC 82010 with the refractory metallization at a junction temperature of 280°C is at least 32 times greater than for the MSC-1330/A or MSC-1330/B with aluminum metallization (at the same temperature).

Part Type	Description	λ (Failures per 10^6 Hours)	N (Number)	N λ	ΣNλ (Total for Part Type)
Transistor	RF Power	1.0	12	12.0	
	RF Small Signal	0.1	3	0.3	
	Switching	0.1	4	0.4	
					12.7
Integrated Circuit	Silicon Monolithic	0.1	5	0.5	
					0.5
Diode	Graded Junction Varactor	0.1	2	0.2	
	PIN Switching	0.1	21	2.1	
					2.3
Capacitor	Solid Tantalum Electrolytic	0.1	6	0.6	
	TA_2O_2 Thin Film	0.1	62	6.2	
					6.8
Resistor	Ta Thin Film	0.01	22	0.22	
					0.22
Interconnections	Thermal-Compression and Weld	0.01	108	1.08	
					1.08
				Total	23.6 Failures per 10^6 Hours

Table 3 L-Band Typical Module Failure Rate Breakdown by Part Type

Date of Failure	SNR	Type Failure	Total Number Failures	Total Number Modules	Total Module-Hours	MTBF (Hours)* Transmitter	Receiver	Logic	Other	Complete Module
6/2/72	3	Transmitter	1	8	41	14	25	25	25	14
8/23/72	37	Receiver	2	50	27,583	9,213	9,213	17,074	17,074	6,454
9/20/72	4	Other	3	78	51,001	17,034	17,034	31,570	–	9,231
9/25/72	57	Transmitter	4	78	55,936	13,089	18,683	34,624	18,683	8,334
5/10/73	45	Transmitter	5	98	183,881	33,282	61,416	113,822	61,416	23,169
5/10/73	79	Receiver	6	98	–	–	43,028	–	–	20,043
5/10/73	84	Other	7	98	–	–	–	–	43,028	17,947
5/10/73	101	Receiver	8	98	–	–	33,282	–	–	16,145
5/14/73	12	Transmitter	9	98	184,805	27,536	33,450	114,394	43,244	14,766
5/14/73	41	Other	10	98	–	–	–	–	33,450	13,528
5/14/73	95	Logic	11	98	–	–	–	61,725	–	12,493
5/14/73	8	Logic	12	98	–	–	–	43,244	–	11,624
5/15/73	96	Receiver	13	98	–	–	27,536	–	–	10,867
6/8/73	33	Receiver	14	98	215,000	32,035	27,090	50,310	38,915	11,847
6/20/73	14	Receiver	15	98	227,000	33,823	24,743	53,118	41,087	11,781
7/25/73	43	Other	16	98	242,036	36,063	26,382	56,636	36,063	11,884
8/1/73	48	Receiver	17	98	270,000	40,230	26,352	63,180	40,230	12,582
8/1/73	85	Receiver	18	98	–	–	23,706	–	–	11,961
11/25/73	*Final Total*		18	98	374,580**	55,814	32,888	87,652	55,812	16,594

*Interim MTBF computed on basis of termination of test at that point.

** Last 104,580 hours were failure-free

Table 4 **Fractional Array – Module Failure Analysis Summary MTBF – 80% Confidence**

Figure 18 is a L-band test bed for one of the modules shown previously. This test bed (located at RCA in Moorestown) contains 72 elements. Recall that a 40,000 hour MTBF was projected for modules in this array; in fact, how well did they perform? The module life test summary is shown in Table 4. When testing ended on 25 November 1973, the total operating time was 374,580 hours, of which the last 104,580 hours were failure-free. Eighteen modules failed. There were four failures in the transmitter circuitry, and fourteen failures in the small signal circuitry. Receiver failures were not attributed to circuit design problems. Seven early failures were related directly to workmanship and lack of assembly process control. It is not surprising that so many quality control failures occurred early in production; such results are typical of the learning curve process experienced in the manufacture of any product. Improvements in quality control, processing techniques, and screening methods certainly reduce this effect.

The test indicates that the MTBF of the transmitter circuit is 55,814 hours (with an 80% confidence level). If all of the failures are taken into consideration, the composite module MTBF is 16,594 hours with an 80% confidence level. This figure is much less than the forecast 40,000 hours. However, excessive failures in the receiver circuits and those attributed to quality control were not originally anticipated; thus, extreme care must be exercised in making use of reliability handbook data. The values of device failures per million hours, used for reliability predictions, must be anticipated and learning processes recognized early in any program. Confidence levels can only be established with quantity and long term measurements.

It is important to note that the failure rate of the transmitters was significantly lower than predicted. This improvement has been attributed to the 500 hour burn-in performed on the transmitter before it was placed in the module. No burn-in was performed on either the logic or receiver circuits. Thus, the infant mortalities for these circuits occurred during array system operation.

Figure 18 **RCA Test Bed Array**

Figure 19 is the Westinghouse low noise receiver; it is used in an L-band system. The significant performance parameters are shown in the figure. Improved reliability is exemplified by the combination receiver/limiter, shown in Figure 20 with the unit it replaces. About sixty of these units were built and placed in the field. Over 500,000 operating hours have been logged on the receiver and switch units without encountering failures. This type of data provides the basis for subsequent reliability predictions. The final circuit example is the AN/FPS-27 S-band preamp and limiter module shown in Figure 21. Westinghouse reports over 60,000 hours of successful operation with this circuit while experiencing only one diode failure.

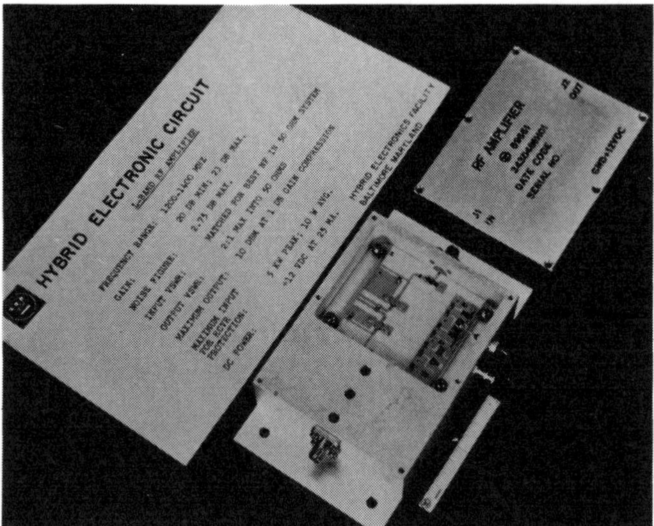

Figure 19 Westinghouse Hybrid L-Band RF Amplifier

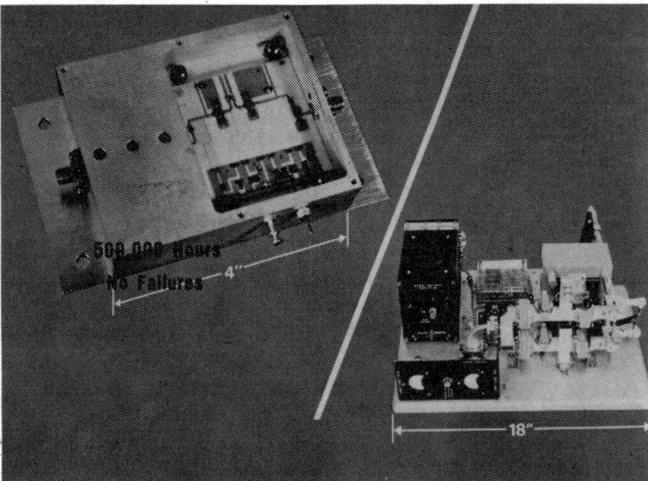

Figure 20 Westinghouse Receiver AN/FPS-107 (500,000 operating hours logged without failure)

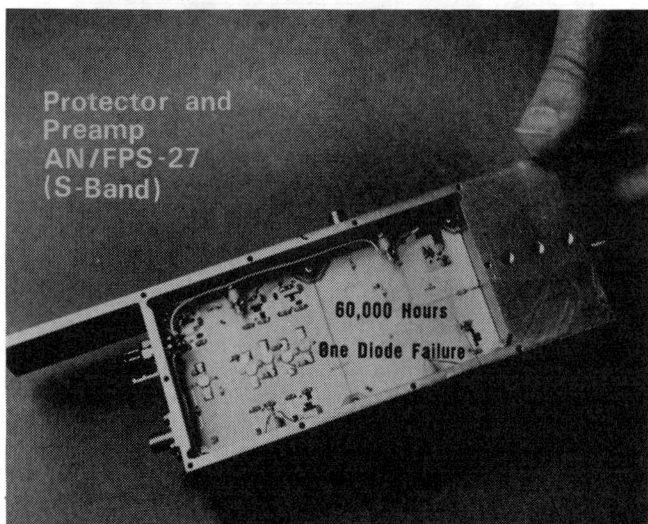

Figure 21 Westinghouse Protector and Preamp for AN/FPS-27 (S-Band) (60,000 operating hours logged with one diode failure)

Module reproducibility and quality control is another area of concern. The first major factor to be considered is the availability of semiconductor devices. One problem is related to the requirement that all devices should be alike. Alike to what extent? There have been insufficient volume requirements to justify capital investments in production and test equipment. This factor is a key one. Furthermore, it should also be noted that no established military specification exists (to date) which specifically covers Microwave Integrated Circuits (MICs). MIL-STD-883 is often referred to, but it is primarily a specification governing digital logic circuits. It was never intended to apply to MICs although it is often used as a basis for their qualification testing and screening.

Still another factor affecting manufacturing cost is module packaging. This could remain a major cost item after "bottoming" out

Figure 22 Raytheon (Camel) Module Package

Figure 23 General Electric Power Amplifier Package for the AN/TPS-59 (see Figure 2 of this Chapter and Figure 90 of Chapter 1)

on a production learning curve. Module packages, similar to the ones previously shown, presently cost about $50. This cost is related directly to the hermeticity requirement which arises when unpassivated semiconductor devices are used. The optimum package size and number of circuits per enclosure remains to be established.

Module packaging is illustrated in Figure 22 which shows a Raytheon assembly. This type of module would be suitable for use in the ground based array shown earlier. The flange provides a heat sink, the module is clamped to an external cold plate, and the lid is welded to the box. Figure 23 is an example of a rather unique packaging concept. This General Electric design contains a two-stage L-band amplifier delivering 50 W peak output power at approximately 20% duty factor. It has microstrip launched input and output terminals; thus, connectors are eliminated. (Hermetic connections are costly and always represent a potential source of failure.) Collector voltage is applied to the circuit lid by means of a contact connection. These units can be stacked like dominoes in a manner similar to the digital computer "mother-board" concept. Outputs are combined to provide an RF power level of several kilowatts.

In conclusion, substantial progress has been made during the last few years toward achieving a practical all solid-state phased array radar system. In fact, the state-of-the-art and economics are such that at UHF a solid state phased array is less expensive than a tube system – the reason that the PAVE PAWS radar is going solid state (see Figure 18 in Chapter 1 by Brookner).

Acknowledgment

The author wishes to acknowledge the many contributors to the material presented in this chapter. Included are the Air Force, Army Electronics Command, Marine Corps Electronics Command, The Naval Research Laboratory Electronics Technology Division, and the industrial firms noted.

References

[1] Skolnik, M.I. (ed.): *Radar Handbook,* McGraw-Hill, New York, 1970

[2] Liu, S.G.: "2000 W GHz Complementary TRAPATT Diodes," *1973 International Solid-State Circuits Conference Digest of Technical Papers,* pp. 124-125

[3] Cohen, E.D.: "TRAPATTs and IMPATTs: Current Status and Future Impact on Military Systems," *1975 Eascon Record,* Paper 130

[4] O'Reilly, G.T.; Neidert, R.E.; and Wilson, L.K.; "A Computer Aided Design of L-Band Transistor Power Amplifiers," *1974 IEEE-MTT Symposium Digest of Technical Papers,* pp. 135-137

[5] Pitzalis, O. Jr.; and Gilson, R.A.: "Broad-Band Microwave Class-C Transistor Amplifiers," *IEEE Transactions on Microwave Theory and Techniques, Vol. MTT-21, No. 11,* pp. 660-668, November 1973.

[6] Belohoubek, E.F.; Presser, A.; and Veloric, H.S.; "Improved Circuit-Device Interface for Microwave Bipolar Power Transistors," *IEEE Journal of Solid-State Circuits, Vol. SC-11, No. 2,* pp. 256-263, April 1976.

[7] Gelnovatch, V.G.; and Chase, I.L.: "DEMON – An Optimal Seeking Computer Program for the Design of Microwave Circuits," *IEEE Journal of Solid-State Circuits, Vol. SC-5, No. 6,* pp. 303-309, December 1970.

[8] Dodson, B.C. Jr.; and Weisenberger, W.H.: "Reliability Testing of Microwave Transistors for Array-Radar Applications," *IEEE Transactions on Microwave Theory and Techniques, Vol. MTT-22, No. 12,* pp. 1239-1246, December 1974

[9] Poole, W.E.; and Walshak, L.G.: "Median-Time-to-Failure (MTF) of an L-Band Power Transistor Under RF Conditions," *Twelfth Annual Proceedings on Reliability Physics 1974,* pp. 109-115, April 1974

[10] Walshak, L.G.: "Long Term Reliability Investigations of the MSC-1330 Microwave Power Transistor and the AMPAC 1214-30 Internally Matched Device," *Final Report on Contract N00014-74-C-0362,* October 1975

Chapter 20 — Harwell

Airborne Solid State Radar Technology

Introduction

The architecture of an active element airborne phased array radar differs very little from that for ground based systems (see Figure 1). Basic elements are the array, a receiver/processor, a signal generator to supply excitation signals for the RF modules, and a computer. Prime differences in the design of an airborne system are the higher operating frequencies and the emphasis that must be placed on reducing weight and power to minimize the impact on the airframe.

The vast majority of modules for airborne arrays have been designed to operate in the 8 to 10 gigahertz region. This focus leads to the major distinction between ground based (low frequency) modules and the X-band modules — the requirement for frequency multiplication in transmit mode and down conversion in receive mode. The block diagram, Figure 2, is typical of an X-band module. A single phasor is used to properly phase S-band transmitter and LO excitation signals. An S-band power amplifier is followed by a frequency quadrupler in the transmit channel. A small-signal quadrupler is used in the LO line to derive the X-band drive for a balanced mixer. Down conversion of the radar returns to an intermediate frequency in the 500 to 1500 megahertz range is the usual case.

The motivation for airborne solid state radars is probably best examined by reviewing the mission profile for an advanced military aircraft. If a mission is profiled (Figure 3), it usually begins with a navigation and weather avoidance requirement. To minimize the probability of detection, a terrain following/terrain avoidance operation is required for a low altitude penetration mode. When hostile territory is entered, a self defense mode may be of prime importance. In the target area a precision map, most likely synthetic aperture mapping, and air-to-ground ranging are required for weapon delivery. As the aircraft exits the target area contour flight modes again are used and frequently may terminate with a rendezvous requirement calling for beacon compatibility.

In summary, the entire spectrum of functions that can be performed by a forward-looking radar includes special purpose equipment for terrain following/terrain avoidance and air-to-ground ranging functions; navigational and precision map equipment; and the very high average power fire control systems for search, track, and missile illumination functions.

Figure 4 shows the array of equipments required. On examining this array of equipments, several problems are immediately evident. First, there exists a basic question as to whether or not this amount of equipment can fit into the forward compartment of a practical aircraft. The likely answer is *no!* Second, system complexity has been increased enormously with the end result of decreased reliability and increased maintenance and procurement costs.

These difficulties lead to the phased array. As Figure 5 indicates, the phased array radar provides the opportunity to implement, with a single piece of equipment, all the forward-looking radar functions required to fulfill the missions of Figure 3. Instantaneous beam shape and pointing angle controls permit this equipment to provide all the radar modes, either individually or simultaneously, required to satisfy the projected mission profile.

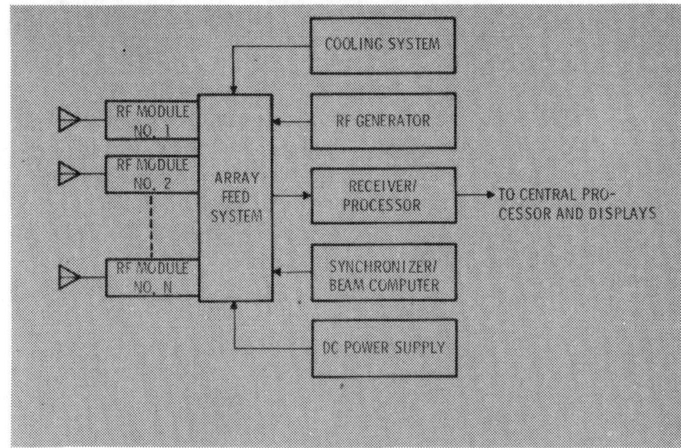

Figure 1 Active Element Phased-Array Radar System

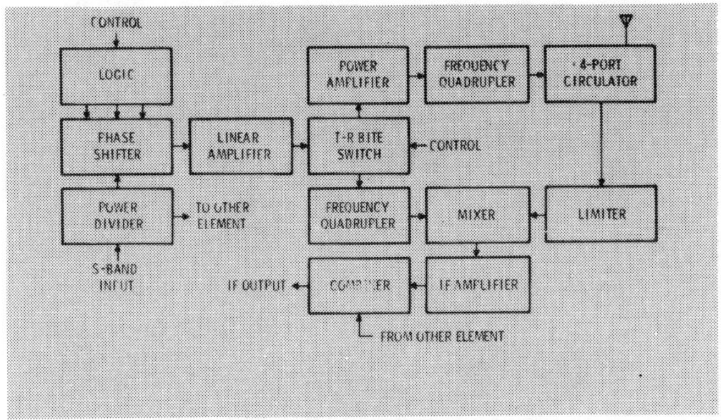

Figure 2 X-Band Phased-Array Module

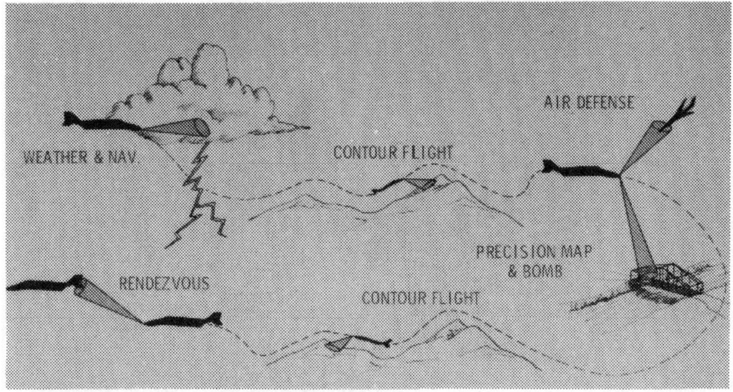

Figure 3 Possible Mission Profile for Solid State Phased Array Radar System

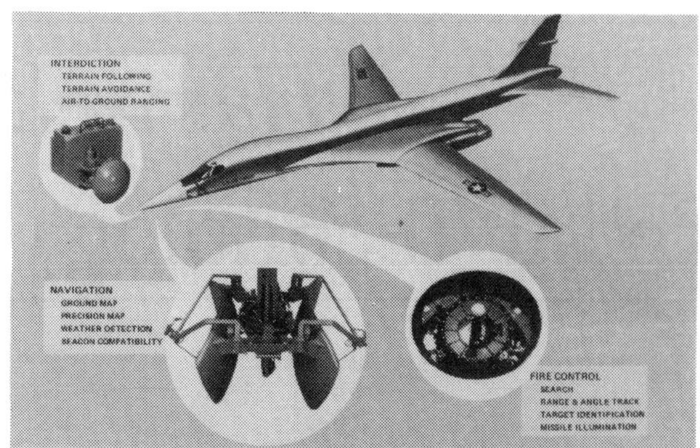

Figure 4 **The Expanded Airborne Radar Equipment. Advanced system mission analyses forecast a much expanded requirement for their radar sensors. System definitions are no longer segregated into air-to-ground and air-to-air functions as each system now requires the capabilities of both to meet mission demands.**

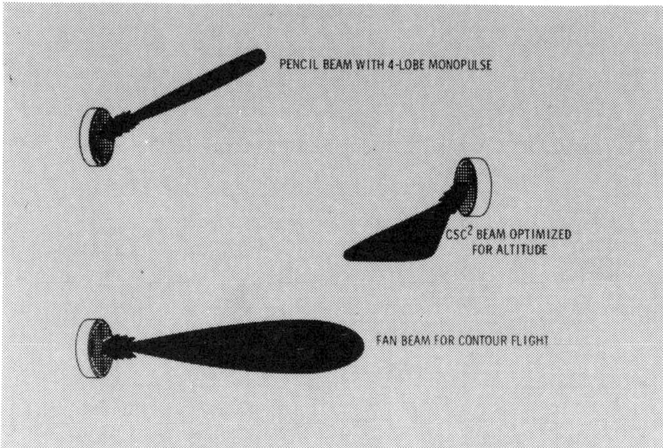

Figure 5 Airborne Phased Array Radar Beamshapes

However, solid state is certainly not the only way to implement a phased array radar. The Department of Defense has responded to the alternatives by sponsoring several major programs; probably the two most noted are the RARF programs. Figure 6 is a photograph of the Emerson array, which was in development for several years and has been integrated successfully with Hughes Aircraft's FLAMR equipment. This arrangement has produced remarkable SAR imagery during flight evaluations. (Figure 19 of Chapter 1 shows the Raytheon RARF reflectarray antenna which was fed by a TWT transmitter.)

The transmission or reflective arrays can employ either diode or ferrite phasors. The system in Figure 7 utilizes ferrites. There is an antenna on each port of the phasor (as it is a transmissive system) and all elements are fed from a single transmitter. The transmitter is a coherent, high average power source. It is this particular point – the liability of a single high power transmitter – that the solid state radar addresses.

At Texas Instruments, 90 percent of all radar system failures are traced to about 10 percent of the components. The 10 percent are those components that deal with microwave power generation or distribution and antenna scanning mechanisms (see Figure 8). The resolution of this problem is through the redundancy inherent in an active element array. With redundancy, element functions can be lost without losing system function.

Incidentally, the most sensitive array parameter is usually sidelobe structure. The first modes of operation to feel the impact of a failure thinned array is either terrain following or the very low rms sidelobe requirements of the air-to-air modes.

The failure limit of Figure 9 is based on the terrain following sidelobe requirement of the RASSR array. In RASSR there are 824 dual modules or 1648 independent transmit/receive functions. Analysis predicts that a malfunction in approximately 107 of the modules can be tolerated before the performance becomes substandard.

Using this information and the predicted module MTBF of about 30,000 hours, the useful life of the array can be computed. The chart in Figure 9 indicates that approximately 4,000 hours of maintenance free operation is expected.

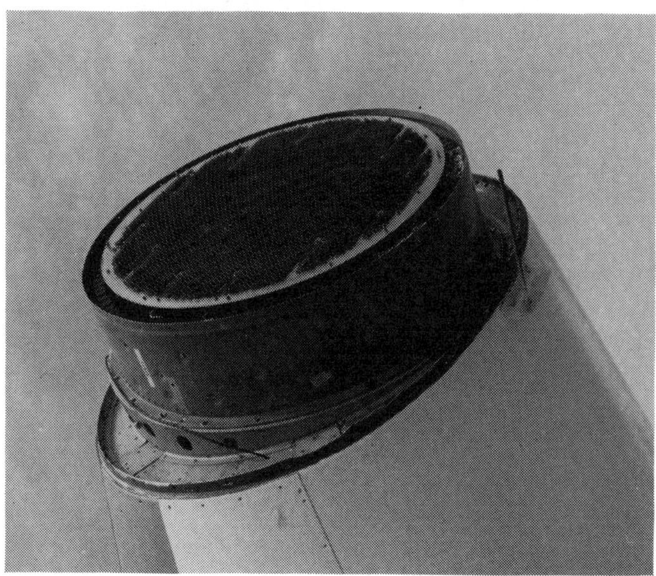

Figure 6 Radome Antenna RF (RARF) Circuitry; TWT Transmitter Used

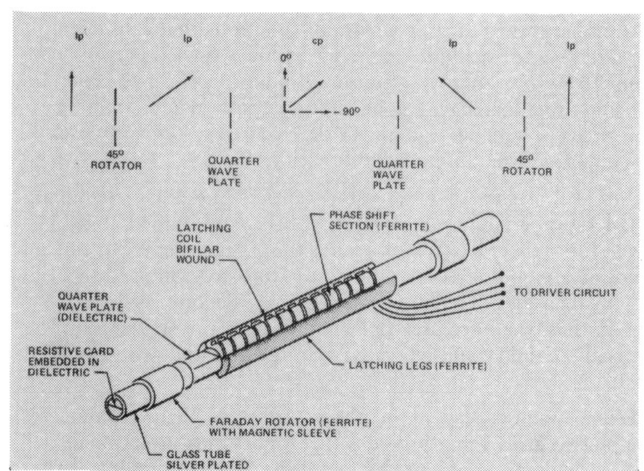

Figure 7 Phase Shifter Concept

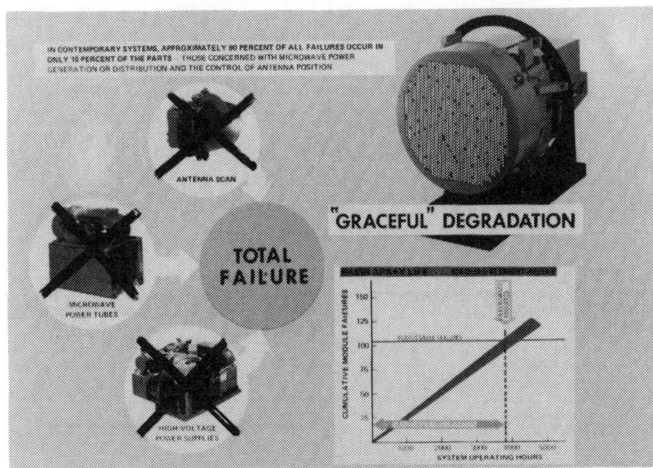

Figure 8 Solid-State Phased-Array Radar Reliability

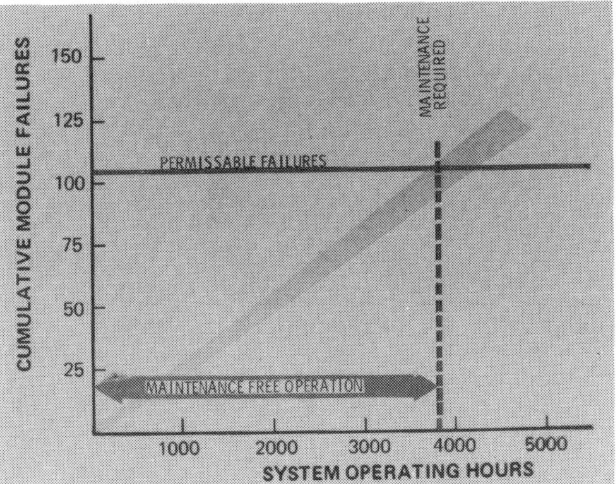

Figure 9 Array Operating Life

Figure 10 The MERA Array

The MERA Radar

The MERA (Molecular Electronics for Radar Applications) system shown in Figure 10, was the first of the solid state phased array radars. It was conceived in 1964 with the initial objective of simply advancing the state of the art in microwave integrated circuits. After the program was underway, the emphasis was broadened to include the demonstration of this new technology. The outcome was the development and successful testing of a laboratory model of an all solid state phased array radar.

This demonstration radar took on the modes listed in Table 1 – ground mapping, terrain following, terrain avoidance, and air-to-ground ranging. The system was a fixed frequency radar containing 604 transmit and receive elements, each of which had an output of approximately 1/2 watt.

The design of the MERA module is shown in Figure 11. This module set the configuration for all future transmit/receive modules operating at X-band. The module elements are preamplifier and phase-shift network, followed in the transmit mode by a pulsed high power amplifier, a frequency multiplier, and a TR mechanism. On receive, the LO excitation signal is used to phase the received signal. The mixer output is through a 500 megahertz IF amplifier.

The MERA system is shown in Figure 12. Elements are the array structure, the signal generator (which provides the transmit and the LO excitation to the modules), synchronizer/computer, receiver/processor, and the TF and ranging computers. The power supply equipment and the display equipment are solely for evaluation purposes.

Among the most significant MERA contributions was the successful demonstration of the feasibility of a solid state radar. MERA also developed the basic concepts that are used on today's solid state radar designs; it demonstrated the feasibility of microwave integrated circuits (MIC); and most importantly, it identified those areas in devices, circuits, and systems where additional development was needed to build useful systems.

The RASSR Radar

The RASSR (Reliable Advanced Solid-State Radar) program (Figure 13) followed the MERA; its purpose was really twofold – to demonstrate that a practical system, one which solves some operational requirements in an operational environment, can be built; and to show that the reliability improvement promised by the technology is achievable. RASSR is an air-to-ground radar. The modes of operation were modeled after the AN/APQ-122 system – the Air Force wanted to have a conventional technology radar system for a one-on-one comparison with the performance achieved by the RASSR. The RASSR modes are ground mapping, terrain avoidance/terrain following, weather avoidance, beacon, and a very elementary station keeping mode. It is also a coherent system with this capability to be evaluated in air-to-air and precision map modes.

The reliability forecast for the total RASSR system is about 650 hours MTBF. An advanced thermal design is employed, a very important part of a solid state radar design which normally does not receive much attention. The RASSR system is capable of simultaneous multimode operation and is tunable over a band from 9.2 to 9.5 gigahertz. There are four coherent frequency channels that the pilot can select via push button controls. Frequency agility has been implemented using 31 basic channels with a pseudo-random pulse-to-pulse selection. Polarization is horizontal – indicative of the system after which it's modeled, the APQ-122. The array contains 1648 elements putting out one watt each. Utilizing a 5% duty cycle, the system provides about 80 watts of average power. The system noise figure is 10 dB.

Modes	
Ground Mapping	Terrain Avoidance
Terrain Following	Air-to-Ground Ranging

Specifications	
Frequency	9.0 GHz
Antenna Gain	
Transmit	32 dB
Receive	30 dB
Antenna Polarization	Vertical
Number of RF Elements	604
Transmitter Power (Peak)	352 W
Pulse Compression Ratio	113:1
System Noise Figure	12.5 dB
IF Frequency	500 MHz

Table 1 **MERA System Parameters**

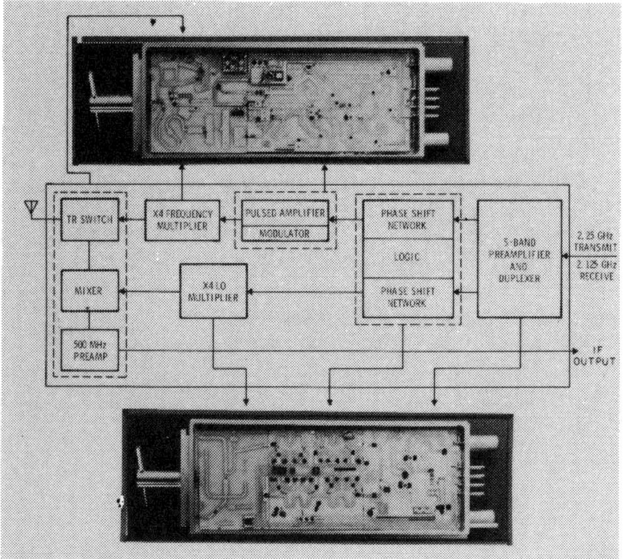

Figure 11 **MERA Module Block Diagram**

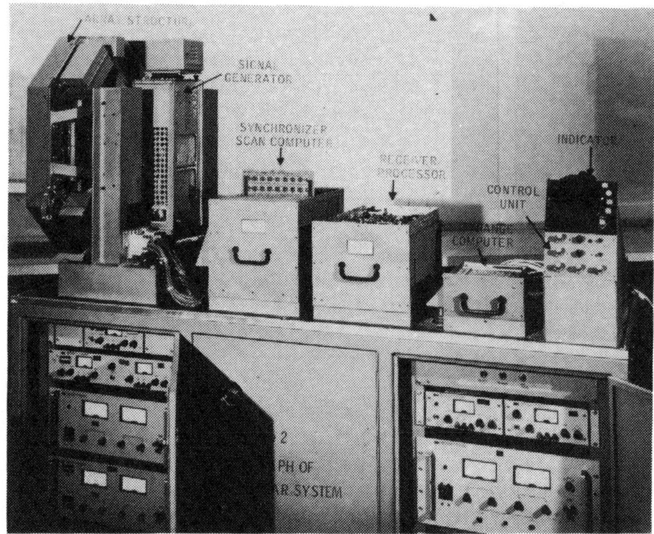

Figure 12 **The MERA System**

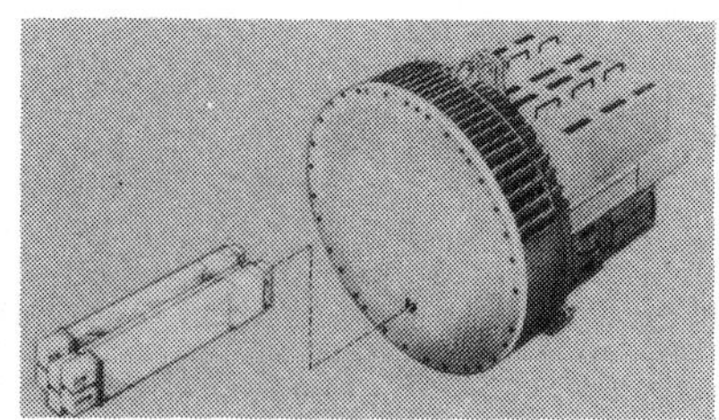

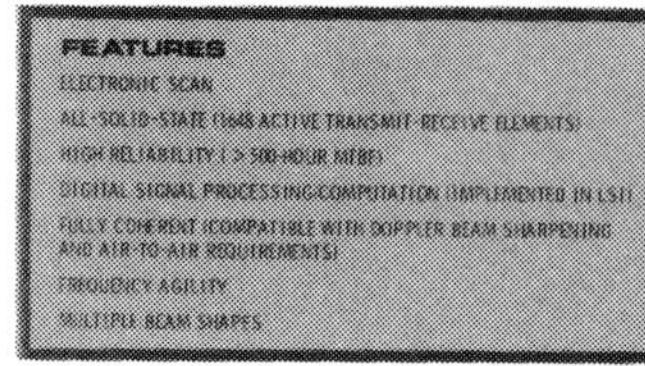

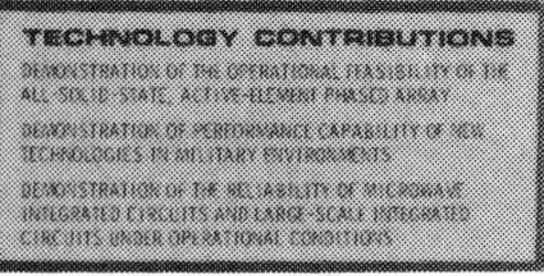

Figure 13 **RASSR (Reliable Advanced Solid-State Radar) An all solid-state, electronically-scanned, phased-array utilizing microwave integrated circuits and Large-Scale Integration (LSI).**

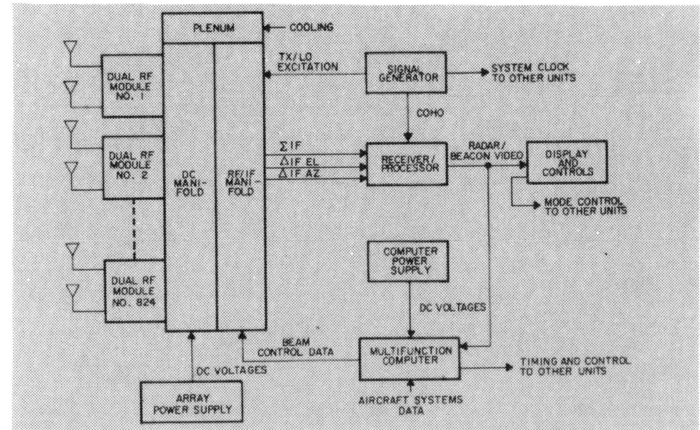

Figure 14 **RASSR System Functional Block Diagram**

Figure 15 **RASSR Forward Assembly**

The block diagram (Figure 14) follows the classic example set by MERA. Figure 15 is a photograph of the RASSR forward assembly. Unfortunately, the photo was taken with the signal generator removed; normally it is located in the area under the forward equipment group mount. Other units are the Receiver/Processor and a Regulator Line Replacement Unit (LRU) to provide the final power conditioning for the array. The array is 36 inches in diameter with about 32 inches of active aperture.

The RASSR module (Figure 16) again looks very similar to the MERA block diagram. Two transceivers were packaged per module, and loaded waveguide elements were used for antennas. A single phase shifter was employed to do the phase shifting of both transmit and receive signals. The IF output frequency is 480 MHz.

Figure 17 is the first prototype module. Two series of prototype configurations preceded a run of production units. 1650 production modules were completed during the first half of 1974. In this photograph the functions just oulined in the block diagram can be seen clearly. The phasor and the associated logic circuitry, the preamplifier, the three stage power amplifier, transmit multiplier, a 4-port circulator, and the EMI filter can be identified. Proceeding down the receive channel, there is an LO multiplier, a limiter-protected mixer, and an IF amplifier. Near the connector are circuits for power combining and dividing – division of the excitation signals and combining of the IF output from each of the receivers. Also included is a thermostat for protection in the event of a thermal control system malfunction.

The test results listed in Table 2 are indicative of the level of performance achieved from a RASSR module. The specifications stated are those developed when system errors were budgeted. This level of performance became the program's design goals. Peak power is typically 1.4 watts at mid-band and 1.3 watts at band-edge. The module has a useful bandwidth of 500 MHz. The transmitter efficiency runs about 8 to 9 percent. Noise figure results were lower than expected. Control of phase errors is really the major challenge when implementing a phased array radar in which there must be precision sidelobe control.

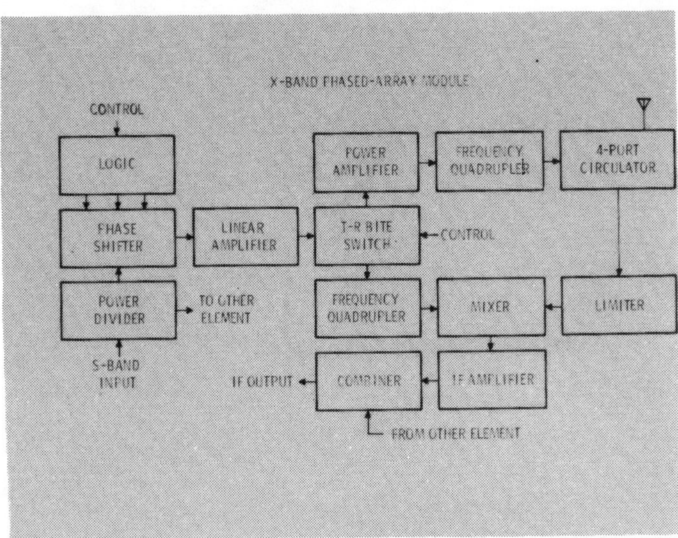

Figure 16 **RASSR Integrated Microwave Module**

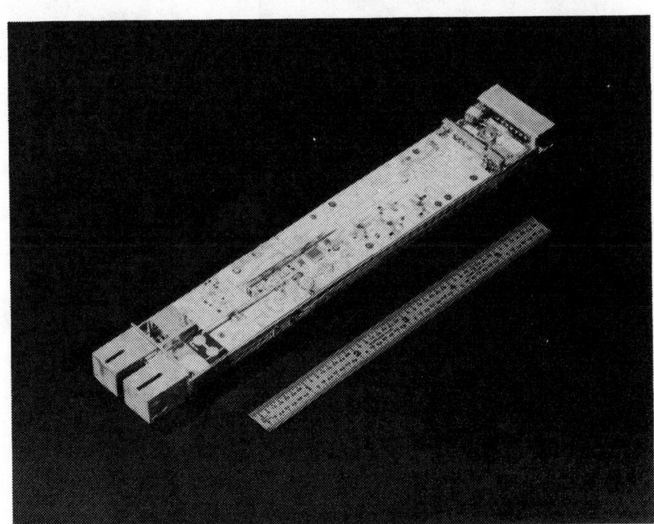

Figure 17 **RASSR Dual Module**

Parameter		Performance* Measured	Specified
Peak Power (watts)	F_o	1.4	1.25 min
	F_1	1.3	1.0 min
Transmit Bandwidth (megahertz)		500	300 min
Transmitter Efficiency (percent)		9	8 expected
Noise Figure (decibels)	F_o	9.5	8.5 max
	F_1	9.7	9.0 max
Phase Error (degrees)			
Transmit	F_o	±12*	±13* max
	F_1	±17*	±14* max
Receive	F_o	±10*	±13* max

*Referenced to antenna terminals, at +25°C

Table 2 **RASSR Module Test Results**

Module Design Considerations

Some common design constraints applicable to all circuits are minimum size and weight, low power dissipation, high MTBF, and compatibility with high volume production.

The phasor (Figure 18) is a four-bit loaded line phase-shifter. A loaded line configuration was selected because it could provide adequate bandwidth and is a less complex design. The major design consideration is phase precision, which involves two factors – the characteristics of the diodes and the precision with which they can be repeated; and the quality of the match in and out of the phasor. A Medium Scale Integration (MSI) chip was used for the control logic. There has been a lot of discussion about logic failures. RASSR's solution to the problem is to replace the six IC chips and seven discrete transistors, which would be ball bonded in place, with a single MSI chip (a fully passivated beam lead device) that can be mounted automatically in a manner that is fully protected from subsequent damage to the active areas.

A low loss, low bias beam lead diode has been developed to minimize dissipation in the modules and to simplify assembly of the phasor circuit. In all circuits that are operating CW, every element must be reviewed in terms of dissipation.

The preamplifier (Figure 19) is probably the most difficult design problem in the module because it carries the burden of compensating for most of the problems encountered in both the transmit and receive chains. This particular design originally was a four-stage preamplifier; it was later converted to three stages, but the design considerations are basically the same. Gain compression is sought to minimize the variations that the LO multiplier and the power amplifier see as a result of varying RF drive levels, either induced

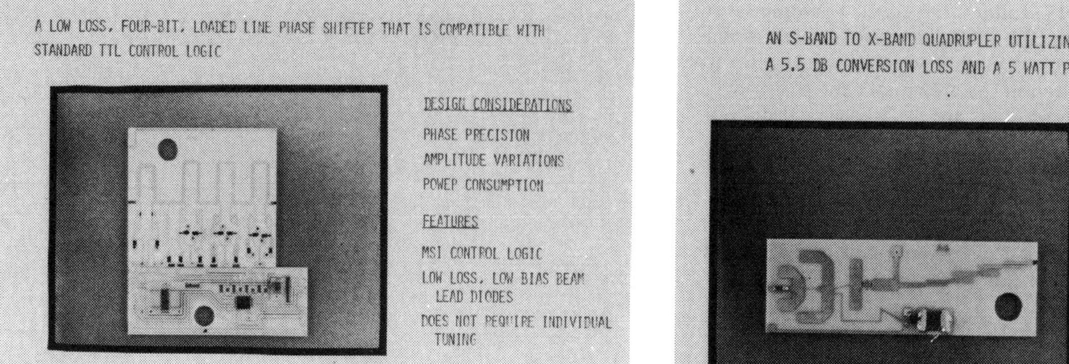

Figure 18 Phasor

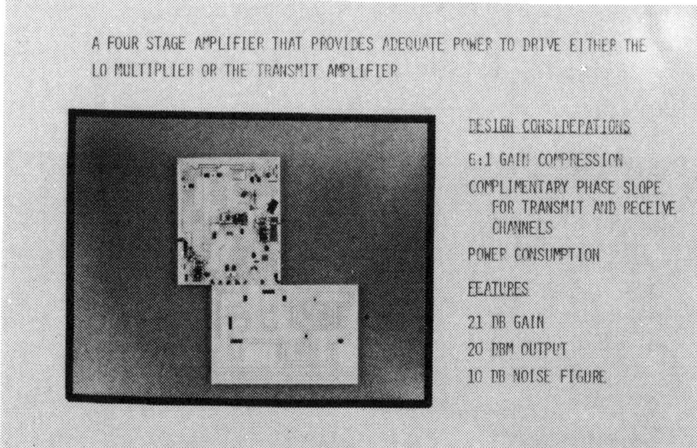

Figure 19 Preamplifier

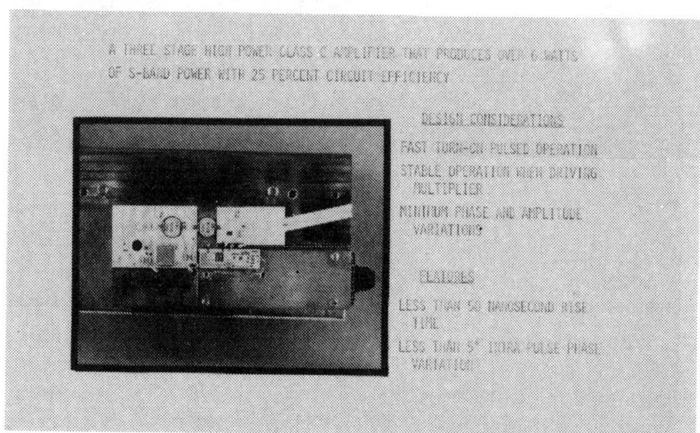

Figure 20 Power Amplifier

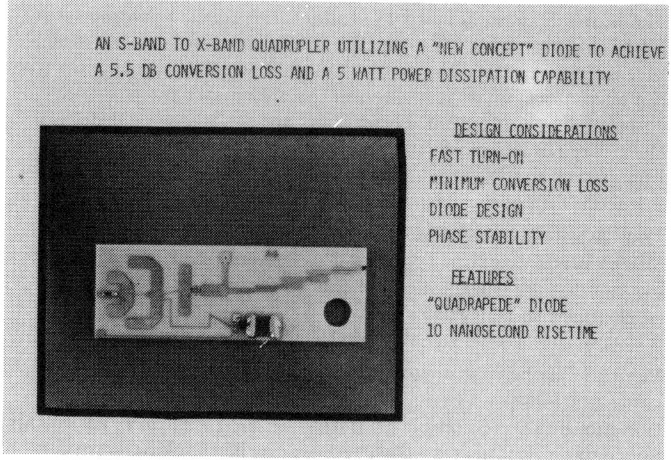

Figure 21 Transmit Multiplier

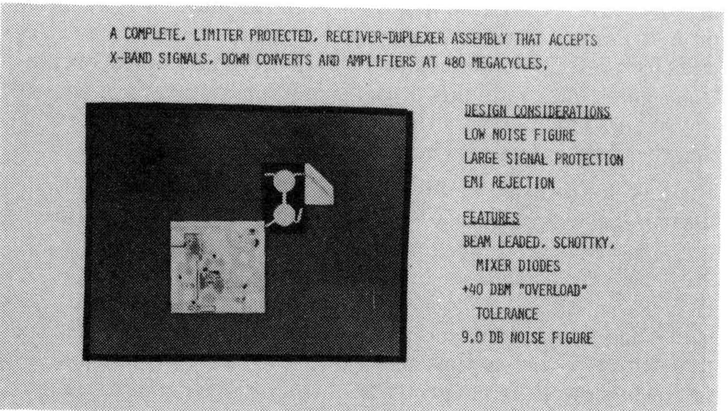

Figure 22 Receiver

by the phasor or coming in from the manifold feed system. It is desired to complement the phase characteristics of both the transmit and receive channels; in other words, to complement the phase sensitivity to RF drive by providing the conjugate of that sensitivity with the preamplifier. Of course, power consumption again must be minimized, as the circuit is operating CW.

The power amplifier (Figure 20) is a three-stage class C amplifier that produces a little over 6 watts of peak power with about 25% circuit efficiency. A primary design consideration is fast turn-on. The unit was specified to be on and stable in 50 ns; it achieved 20 ns. The large variations in input impedance exhibited by a multiplier circuit during turn-on are common knowledge to many. This amplifier must be compatible with the multiplier input impedance. The intrapulse phase variations begin with "turn-on" and run through "turn-off" of the RF pulse. These variations are attributed primarily to the matching circuitry and the thermal time constant of the transistors (the latter probably being the largest contributor). A 5-degree limit (S-band) is placed on intrapulse phase. Circuit performance is comfortably within that boundary.

The transmit multiplier (Figure 21) exhibits a 5.5 dB maximum conversion loss and must handle about 6 ½ watts of incident power. Fast turn-on again is a major consideration; 10 nanoseconds was established as the maximum rise time of this unit. Considerable time was expended developing proper bias for the diode;

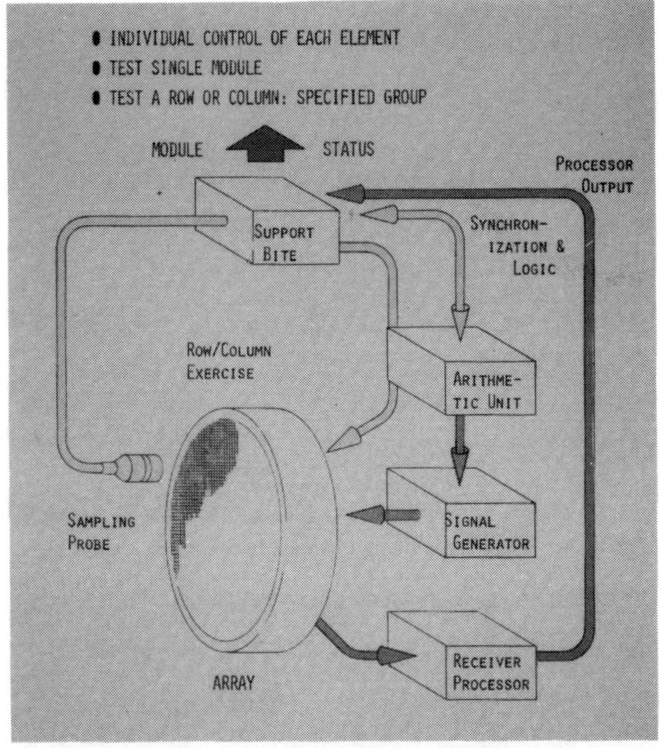

Figure 23 Array Built-In-Test

the bias is a compromise between turn-on characteristics and thermal sensitivity. The diode is a monolithic three-junction structure that was designed specifically for this application.

The receiver (Figure 22) is a complete limiter protected unit; it accepts X-band signals, down-converts, and amplifies at 480 MHz. Of course, low noise figure is a fundamental requirement for any receiver design, but large signal protection is desired, as well. Between the circulator and mixer is a limiter capable of handling +40 dBm overloads for duty cycles up to 0.001. The purpose is to prevent an accidental burn-out of a large number of modules in the event that the finished system is operated in the vicinity of another high power transmitter. The unit provides a 9 dB noise figure at the input to the EMI filter. The EMI filter is to reduce receiver sensitivity to out-of-band signals. Special attention to unwanted receiver responses is demanded when a multiplier is used in the LO chain.

System Design Considerations

An area that warrants some discussion is Built-In Test Equipment (BITE). The BITE operation is implemented as part of the TR switch. A third position of the TR switch provides a termination for the preamplifier and thus excites neither the transmitter nor receiver circuits. Consequently, the module can be switched into a totally inactive state. Now, with a piece of special support equipment (see Figure 23) or with sampling probes around the periphery of the array, it is possible to exercise and evaluate modules in both transmit and receive modes while they remain in the array environment. This process can be accomplished within a few seconds.

Spectral purity is a major consideration for coherent operation. In the photograph of the transmitter there is a dc regulator coupled to the final stage of the power amplifier. The need for the regulator was established by the spectral purity requirements for the system. The four principal contributors to spectral degradation (which results in spurious sidelobes) in solid state radars are the COHO, VCO, frequency synthesizer, and the RF module. These impurities arise, respectively, from vibration, noise, spurs, and power supply ripple. Only the module power supply ripple is really unique. Table 3 summarizes why.

Basically, phase sensitivity is the primary concern. However, note must be taken of gain sensitivities (although it is not the usual driving function) because an amplitude modulation can be converted into phase modulation in subsequent stages. As with any coherent system each component is budgeted with regard to its contribution to spurs or degradation of spectral purity. As an example of the type of requirements that are imposed, the RASSR system power amplifier insertion phase varies at the rate of 12 degrees per volt of B+ change. To satisfy system spectral purity requirements, a total ripple budget for the supply voltage of only 0.04 millivolt rms is allocated. The next most critical circuits are the preamplifier and the LO multiplier. Ripple budgets of 0.08 millivolt rms are needed. A way to relate to the difficulty of obtaining these numbers is by thinking of the basic system power supply converting 400 cycle power to dc power. When the transmitters are pulsed, the power supply has to provide, within a few nanoseconds, more than a thousand amperes of current while meeting the regulation budget. That just cannot happen. The solution is to put localized energy storage in the module, then follow it with a high gain regulator. Each module has two such regulators preceded by some very high quality system supplies.

Thermal control is another system consideration. Thermal phase sensitivity of the module is typically 1° to 2° of net insertion phase change per °C change in the case temperature. For purposes of reliability, junction temperatures should be kept low – less than 125°C is a normal design criterion. For low duty cycle systems, this level presents no problem. The total dissipation of a module is going to range between 6 and 25 watts. Six watts is typical of

Circuit	Supply Voltage	Phase	Amplitude	Ripple/Noise Budget
Power Amplifier	24 V	12°/V	.2 dB/V	.04 mV rms
Transmit Multiplier	60 V	6°/V	.05 dB/V	1.5 mV rms
Preamplifier	14 V	8°/V	.2 dB/V	0.08 mV rms
LO Multiplier	14 V	25°/V	.3 dB/V	0.08 mV rms
IF Amplifier	5 V	15°/V	.3 dB/V	0.8 mV rms

Generally, phase sensitivities are the determining factor in setting regulation requirements, although amplitude variations must be considered because of AM to PM conversion in subsequent stages.

Table 3 Module Sensitivity

an air-to-ground system with relatively modest efficiency; 25 watts is indicative of the high average power per transceiver element associated with high PRF air-to-air modes of operation.

To keep everything satisfactorily cophased in the aperture and to satisfy our reliability goals, the RASSR modules are stabilized in the range of 50-70°C. For a military environment, the other end of the problem must also be investigated. When the system is turned on, the modules must be brought to their normal operating temperature quite rapidly. To satisfy requirements for operation in cold environments, there must be some provision for adding heat to the modules. Thus, the thermal control scheme must be a bilateral transfer system. Duty cycle or mode changes must also be considered for the multi-function radar. Orders of magnitude changes in dissipation are experienced when radar modes are varied. The thermal control system must be able to cope with this dynamic range and it also must be capable of handling the transient situation as modes are changed. Only a minimal control lag is acceptable.

Maintenance is a very important factor. When the issue of keeping modules cool and at a constant temperature was first raised, an obvious solution was to immerse them in some sort of oil bath. Maintenance considerations ruled out this approach. On the RASSR system, it seems that the optimum choice was heat pipes. Figure 24 shows what is probably the first and largest application of heat pipes in the radar business. Each of the dual modules has a heat pipe coupling the module to the thermal control system. The back side of the array is the cold plate which doubles as a mechanical support for the modules.

Figure 25 is a cut-away view of the heat transfer system. The cold plate contains liquid flow passages. One end of the heat pipe, the condensor, is embedded in the cold plate. The evaporator end of the pipe is located under the power amplifier assembly. The net change from cold plate temperature to the carrier plate directly under the power amplifier is about 4°C.

Phased array feed techniques are another area of interest (see Figure 26). There are basically two types of feed systems – space-fed arrays and corporate-fed arrays. Of the space-fed techniques, only one has application for solid state radars – the transmission or rear-fed system. This approach is practically limited to solid state radars when operating at the fundamental (i.e., there are no frequency conversions in the module). For the corporate-fed systems using active elements, there are two basic approaches. One deals with the case requiring frequency conversion in the modules, such as multiplication on transmit and down conversion on receive. The other method applies when direct power amplification on transmit and low noise amplification on receive are practical.

On the RASSR system there is a very complex manifolding system because, of all cases, the configuration dealing with the frequency conversion is the most difficult to implement (see Figure 27). An IF manifold collects all the output signals from the 824 dual element modules and provides the amplitude weighting for the antenna patterns. The transmit and the LO excitation are provided through a single S-band manifold. Phase and amplitude errors have very small rms budgets at this level – about 0.3 dB for amplitude errors and 2 degrees for phase. The dc manifold provides the phase control logic, the dc power, and all other control signals to each element. All these distribution systems must be compatible with one another. Note that the IF interface must pass through the RF manifold, the cold plate, and the dc manifold before reaching the module.

Sidelobe Control

Sidelobes require special consideration in active element array radars. In transmit, a uniform taper is the only choice which prevents a significant loss in transmitter efficiency. The result is high peak sidelobes with relatively low average sidelobes on transmit. In receive, amplitude tapering can be implemented following the down conversion which basically sets system noise figure. Consequently, the aperture is heavily weighted on receive. A new perspective on sidelobe performance is thus necessary. For the active array, the product of the transmit and receive patterns establishes antenna performance.

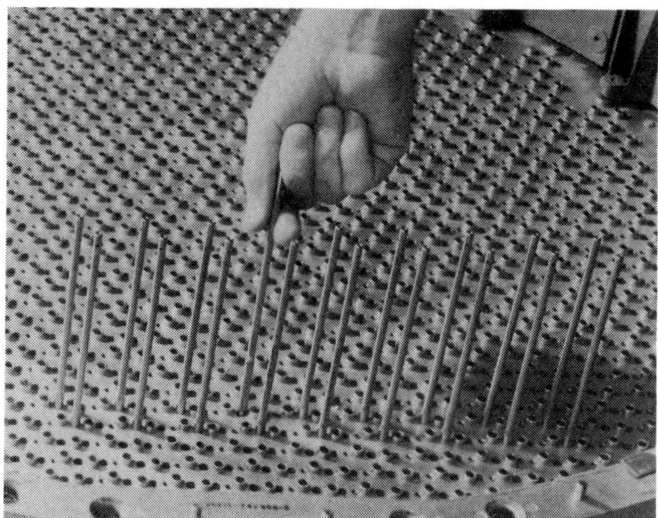

Figure 24 **Heat Pipes Being Installed in the RASSR Cold Plate**

Figure 25 **RASSR Thermal Control System**

Another point is that the first sidelobe on the transmit pattern falls within the main lobe of the receive pattern. From a performance viewpoint, the resulting lobe in the two-way pattern falls within the main lobe of an equivalent passive antenna. Consequently, when discussing sidelobe levels, only sidelobes outside the main lobe of the receive beam are examined. Using this definition, two-

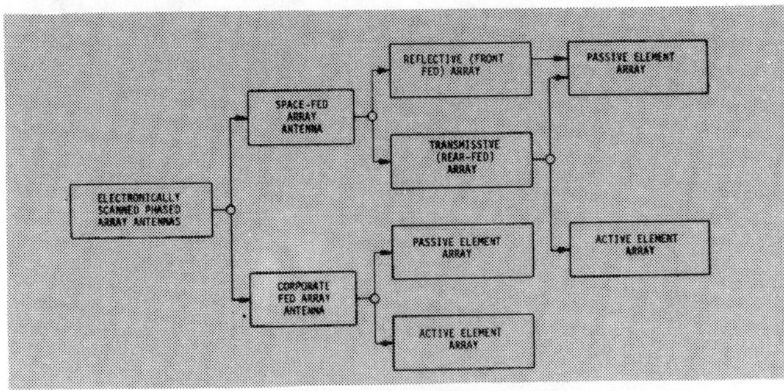

Figure 26 **Phased Array Feed Techniques**

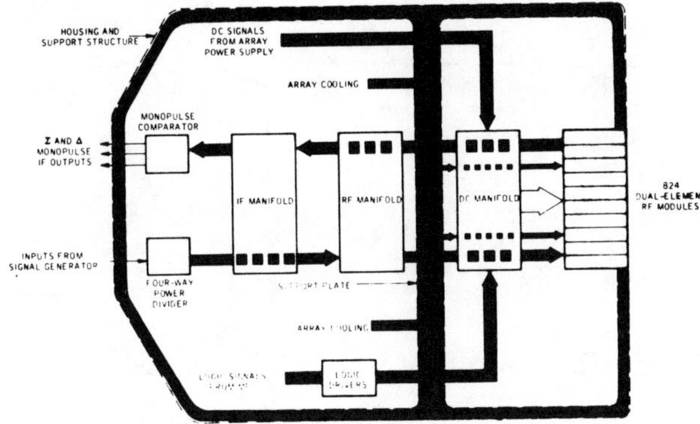

Figure 27 **RASSR Array Signal Flow Diagram**

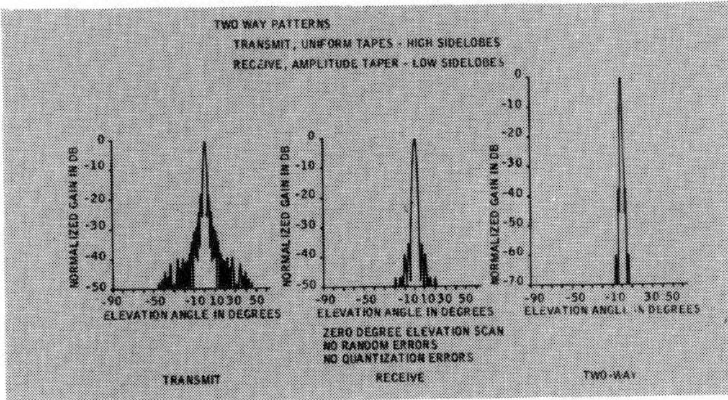

Figure 28 **Active Element Sidelobe Considerations**

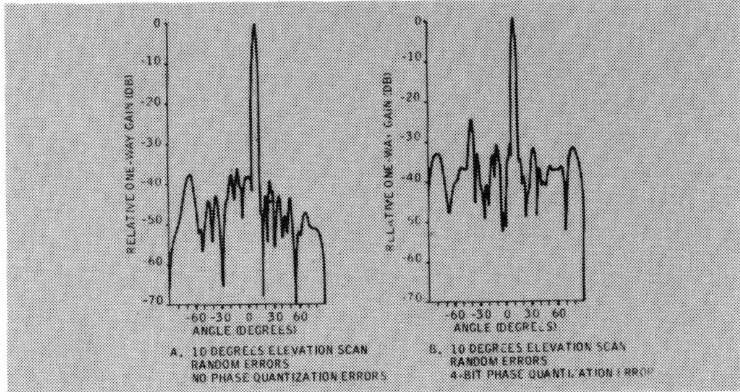

Figure 29 **Quantization Error Effects**

way peak sidelobes on the order of 55 dB and rms sidelobes on the order of 75-80 dB are expected (see Figure 28).

Quantization effects are intrinsic to phased arrays when using digital phase shifters. Figure 29 shows the impact of quantization; (a) is a scanned antenna pattern for the RASSR array neglecting quantization effects. Using a simple phase truncation for a 4-bit phase shifter, quantization errors have degraded sidelobes in excess of 10 dB (b). A rule of thumb is that the peak sidelobe level in dB is about six times the number of bits in the phase shifter – a 4-bit phase shifter yields about 24 dB peak sidelobes.

Figure 30 illustrates one of the techniques to circumvent the full impact of quantization. For randomization of round-off terms, the phase remainder for each element is examined after approximating the required phase to the least significant bit. The round-off technique uses a procedure in which the least phase bits are set randomly by a weighted bit generator. The bit generator is weighted by the value of the remainder term. This action, in effect, permits performance equivalent to that obtained with additional bits in the phase shifter. With RASSR, performance equivalent to that obtained with a 6-bit phasor is expected with only a 4-bit phase phasor in the module. The patterns are an example of the improved performance.

This round-off scheme may sound complicated and seem to be a cause for concern regarding the size of the beamsteering computers. In RASSR, all the steering computations, including those for frequency and beam shape, are executed in the computer, and a single phase control word is transmitted to the module. It takes about 100 microseconds to update the array. Updating at substantially higher rates is possible without adversely affecting the size of the computer simply by properly selecting the way the control signals are distributed. Figure 31 is a photograph of the multi-function computer for the RASSR system; only about 40% of this package is dedicated to the beamsteering computer, the rest providing the TF and TA computations and all the mode control and synchronization functions for the entire radar system.*

Other Radar Applications of Solid State Technology

Figure 32 is a proposed light attack radar, another application of solid state radar technology. Modes of operation are mapping, air-to-ground ranging, terrain avoidance, beacon, and weather avoidance. The system uses RASSR technology with an improved transmitter that is assumed to operate at 15% efficiency. It has 400 elements, 1600 W peak power, 5% duty cycle, 10 dB noise figure, receives 2300 watts of prime power, and yields 29 dB of antenna gain. The aperture is 23 inches by 11 inches. System weight is 380 lbs. The significant thing about this particular system is that it is very close to being competitive with conventional radar technology. The system provides equivalent performance with about a 20% penalty in weight and a 20% greater power consumption. However, greater versatility and reliability offset those penalties.

Figure 33 is an S-band system (which deviates from the standard X-band) that is a very simple, inexpensive air-to-air system. The radar is capable of working 2 square meter targets at a range of about 17 nautical miles with an accuracy of 1° rms in angle and 100 feet in range. Emphasis was on low initial cost, higher reliability, low maintenance, and a minimum impact on the aircraft. The array is about 36 inches in diameter and contains 125 elements. At 27 inches deep, it can easily fit into most airframes. System weight is down to a total of 295 pounds.

The radar that will solve the problem described at the beginning of this chapter must be capable of doing high PRF pulse work with 2500 watts of average power in the long range search and track modes. Also included are low and medium PRF modes. The system must also provide synthetic map modes as well as the conventional modes of the RASSR system.

**Editor's Note* – For further discussion of the effects of roundoff and truncation errors and how to control them, see Brookner, E.: *Practical Phased-Array Systems, Microwave Journal* Intensive Course, Dedham, Massachusetts, 1975.

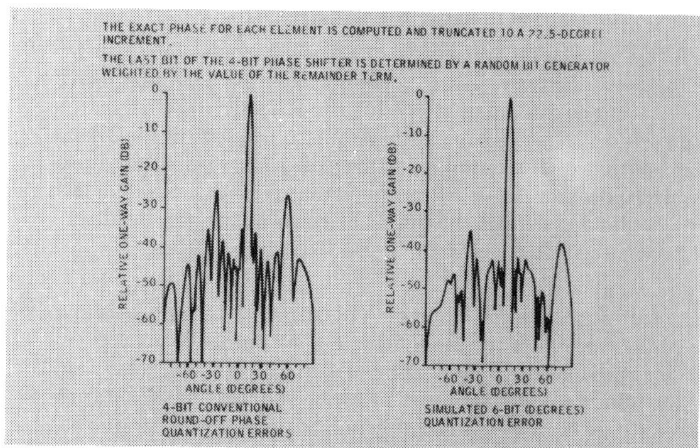

Figure 30 **Randomization of Round-Off Terms**

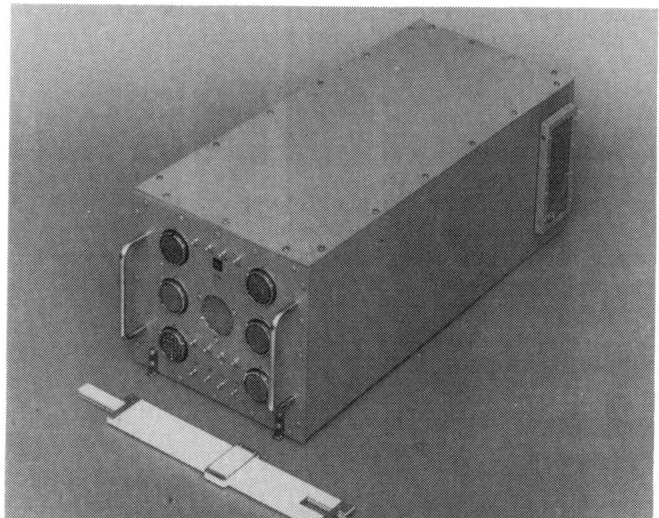

Figure 31 **RASSR System Processor**

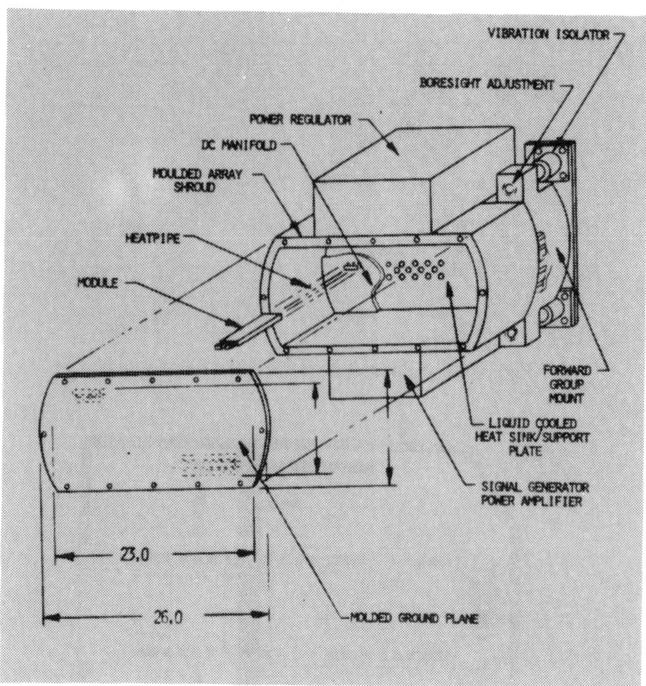

Figure 32 **Solid State Light Attack Radar**

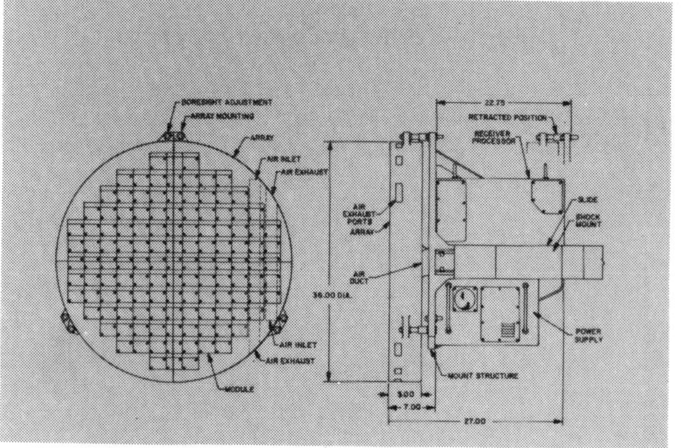

Figure 33 **S-Band Forward Looking Radar**

Out of the configuration study of this system, the most important output was the definition of how net dc power requirements and system weight vary as a function of transmitter efficiency. Figure 34 shows that somewhere in the vicinity of 15% efficiency net power requirements are reached that are competitive with a radar system using tube type transmitters.

Figure 35 shows the weight function to be more demanding. System weight is biased strongly toward average transmitter power. A transmitter efficiency of 20% is required to achieve an acceptable weight of roughly 1000 pounds.

From the data in Table 4, what is needed to build a practical radar system can be discerned. For air-to-ground modes, about 1 watt of peak power from a module operating at 5% duty factor and 20% efficiency is required. A 4 watt peak power level is desired for air-to-air operation (large multi-mode radar), again at 20% efficiency.

Another example of the module described above is illustrated by Figure 36. Westinghouse's MAIR (Molecular Airborne Intercept Radar) effort endeavors to obtain the 20% efficiency, high average power performance.

The block diagram (Figure 37) for such a transmitter differs very little from the low power configuration. The only option available with the power amplifier/multiplier configuration is the choice between parallel output stages (including the multiplication) and a single, large stage. The block diagram highlights one point other

	Air-To-Ground	Air-To-Air
Peak Power	1.0 W	4.0 W
Duty Factor	5%	40%
Pulsewidths	0.2-100 μs	0.2-25 μs
Transmitter Efficiency	20%	20%
Phase Errors, 1σ	20°	8°
Amplitude Error	2 dB	0.7 dB
MTBF	30,000 h	30,000 h

Table 4 **X-Band Module Performance Requirements**

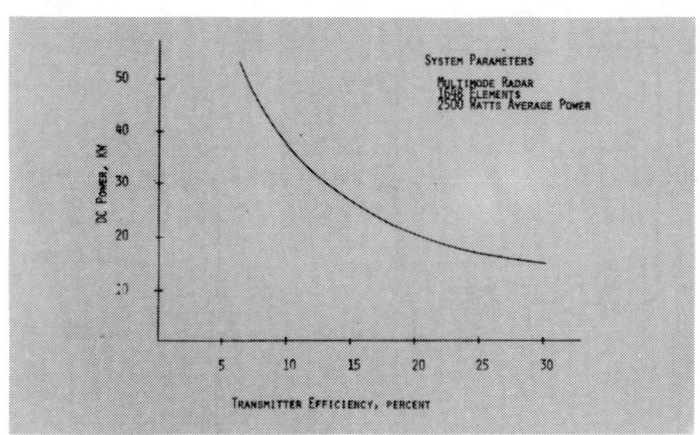

Figure 34 Array Power Requirements

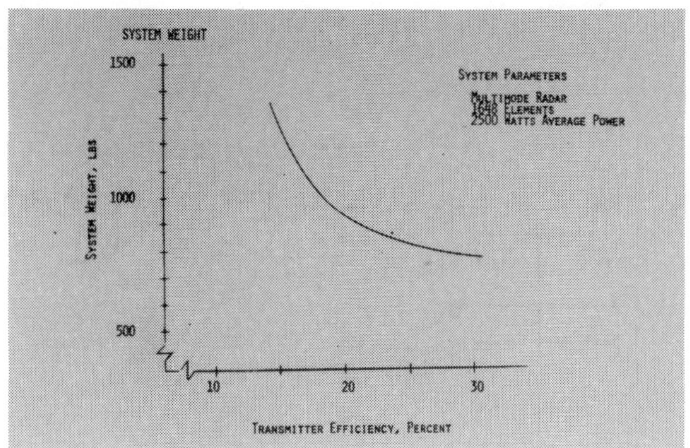

Figure 35 System Weight as a Function of Transmitter Efficiency

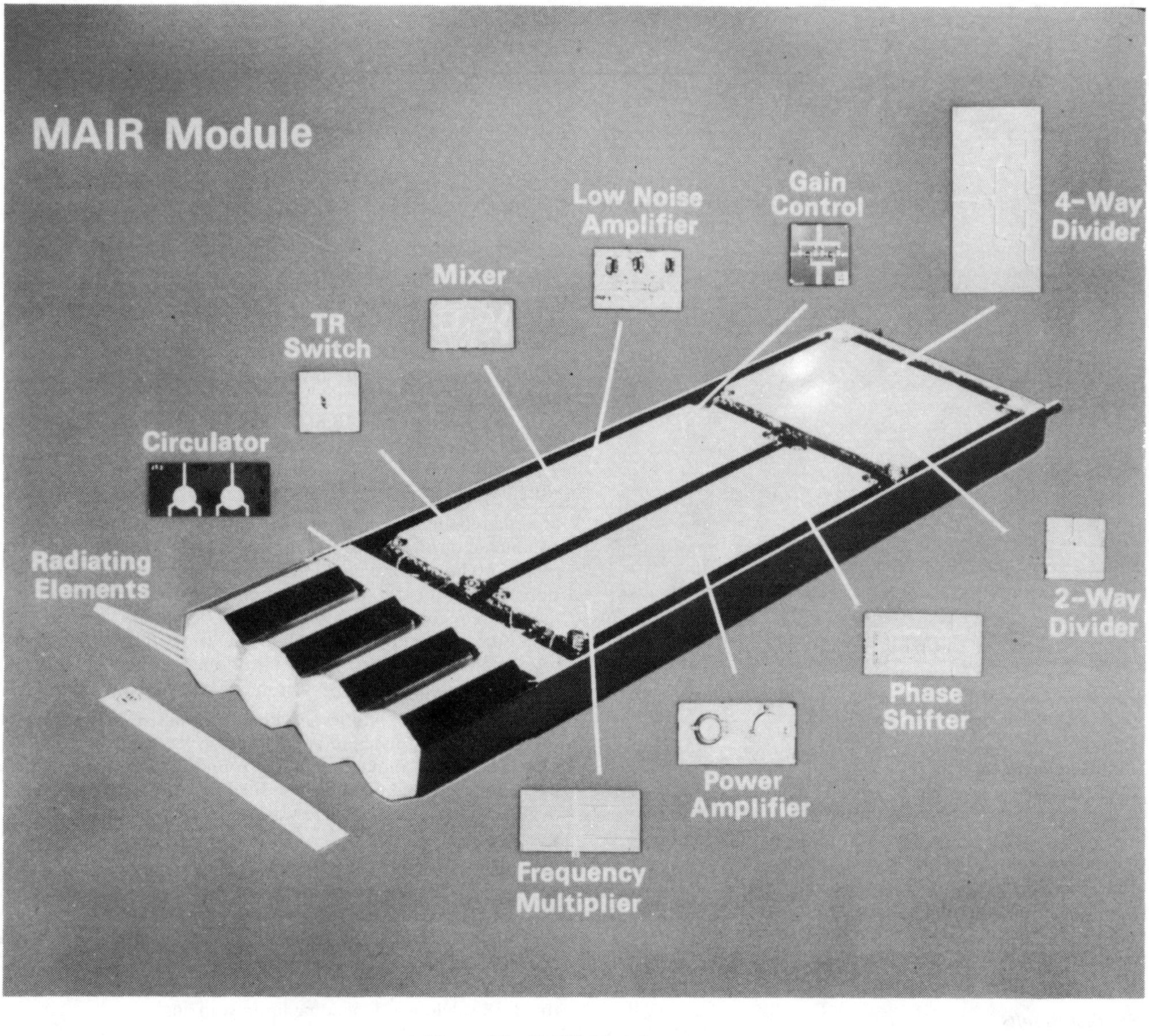

Figure 36 MAIR Module

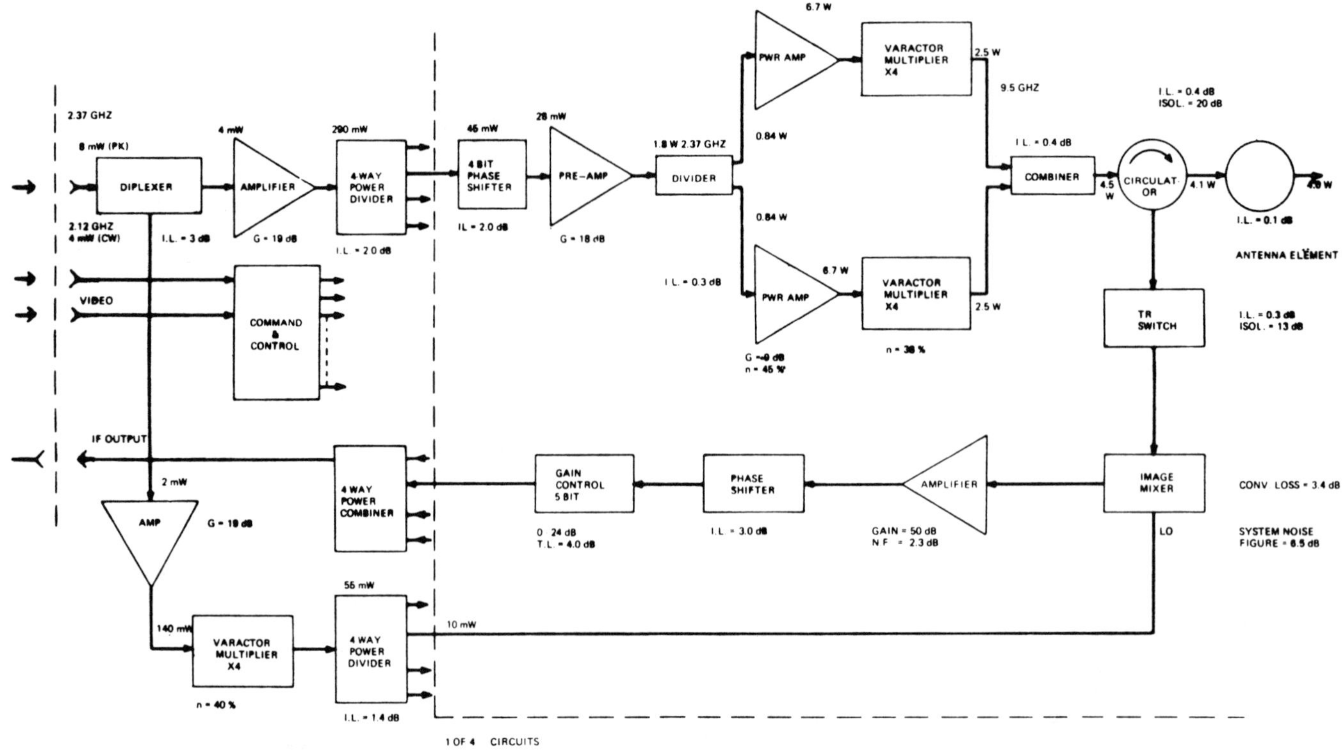

Figure 37 MAIR Module Block Diagram

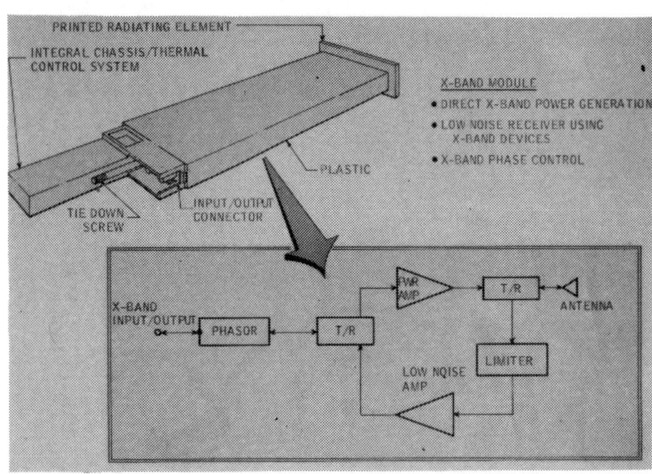

Figure 38 Advanced Solid State Transmit/Receive Module

than outstanding performance – the module is a very complex one. If the long range problem of getting maximum reliability and a minimum cost in the system is to be solved, a move must be made away from frequency conversion in the module, and back to a basic configuration much like those described for the lower frequency band systems, that is, direct X-band power generation and direct X-band low-noise amplification (see Figure 38). (NRL at present (1977) is funding two contracts for such modules at X-band as described in the "Present and Future Trends" section of Part 5.)

RASSR Test Results

The RASSR system was completed in September 1974 and has undergone continuous evaluation since that time. Module performance met or exceeded expectations in almost every category of performance. Nominal performance for the production lot of 825 dual modules was as follows:

Parameter	*Mean*	*Standard Deviation*
Output Power	1.66 W	1.25 dB
Noise Figure	9.1 dB	0.6 dB
Conversion Gain	18.8 dB	1.2 dB
Phase Uniformity, f_o		
Transmit		20.4°
Receive		16.3°

Module reliability data were most impressive. Sample life tests conducted during the course of production indicated a MBTF of 11,740 hours for a dual module (2 transmit and 2 receive functions). Subsequently, during the course of array evaluation, module failure records show a MTBF of 18,000 hours based on more than one million hours of module operation. If functional failure rate is considered (one transmitter or one receiver equals one function), the functional MTBF was in excess of 27,000 hours.

Array performance has been evaluated extensively. In summary, there were no surprises and performance was generally satisfactory. The phase command random round-off technique discussed in the text reduced quantization sidelobes from 5 to 15 dB depending on scan angle. There was no obvious impact on average sidelobe levels.

Sidelobe performance was satisfactory; however, sidelobes far removed from the main beam were somewhat higher than anticipated. Transmit (uniform taper) sidelobes were approximately 17 dB peak and 28 dB average. Receive sidelobe performance was 25 dB peak and 33 dB average. The product of transmit and receive patterns – the measure of two-way radar performance – yielded 55 to 63 dB sidelobes, depending upon scan angle.

X-Band Component Technology – Present and Future Trends

As mentioned previously, to produce an inexpensive and reliable X-band module it is necessary to have direct power generation and low-noise amplification at X-band (see Figure 38). The technology in each of these areas is advancing rapidly.

In the area of microwave power generation, IMPATT, FET, and bipolar technologies have all broken the 20% efficiency barrier at X-band. More than 10 watts of CW power at 20% efficiency is now available with IMPATT technology. FET power transistors have passed through the four watt mark at X-band; bipolar power transistors should follow suit in the near future (having already achieved one watt). Additional data on the state-of-the-art of IMPATTs, GaAs FETs, bipolar transistors, and TRAPATTs are given in Tables 2 through 6 of Chapter 1 and in the "Present and Future Trends" section of Part 5.

The impressive low noise figures presently being obtained and the future potential of low noise amplifier FET devices are indicated in Figure 23 of Chapter 1.

Summary and Conclusions

In summary, the feasibility of solid-state radars is uncontestable, but there are some significant limitations that restrain their application in the area of high frequency airborne systems. Fundamentally, the problems boil down to cost and efficiency. Improved efficiency is key to achieving systems with reasonable power requirements and acceptable weights. Within a very short time, the only remaining barrier to the widespread application of airborne active element array technology will be cost. The cost of modules must be reduced by a factor of 10 to 20 from that experienced during the first half of the 1970 s. Technology advances have simplified module complexity greatly but device cost is still prohibitively high, primarily for lack of volume use. Widespread demand for microwave devices still appears to be the key to solving the cost problem.

Chapter 21 Knittel

Phased Array Antennas - An Overview

Introduction

This chapter is written for those with technical backgrounds who know something about antennas and radiation, but who do not have a detailed knowledge or understanding of phased-array antennas. The emphasis here is on the unique parts of phased arrays – the aperture and radiating elements. Their important aspects are discussed and a perspective is developed which places these components in a proper relationship. Photographs of typical phased-array systems and components are presented.

There are seven major references to phased array antennas which cover in detail the material and concepts presented here; these are recommended for additional study [1-7].* Each of these publications references many papers, which together cover the field comprehensively. Therefore, relatively few additional references are given here.

Figure 1 is a sketch of a typical phased-array designed for transmission. It consists of three main parts – the radiating elements, the phase shifters, and the generators (or feed network, if there is only one generator). For reception, the generators are replaced by receivers (or a feed network and a receiver). The radiating elements are the distinctive part of a phased array; they make the array different from a continuous aperture, such as a parabolic dish. The elements are perhaps the most interesting part of the array because of phenomena like mutual coupling and grating lobes, which are important because of usually-periodic distributions of elements in an aperture.

*See also Reference 18.

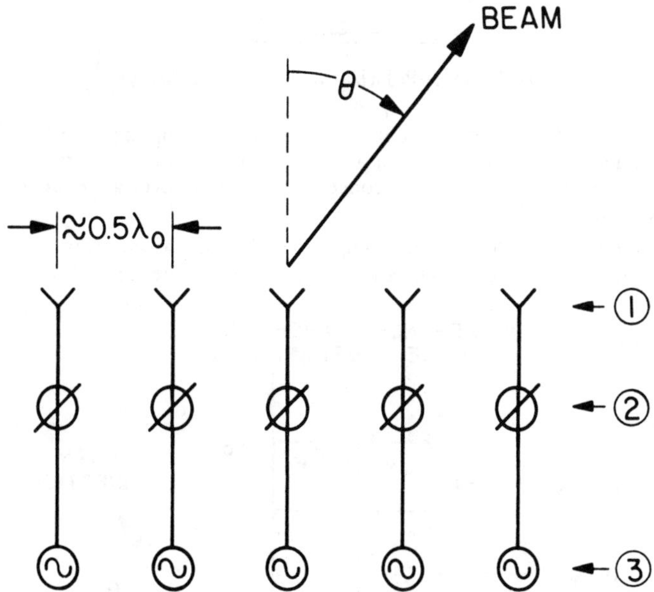

Figure 1 Broadside-Type Array Antenna (Phased Array)

In order to control electronically the phase of the aperture, so that the beam can be steered electronically in space, typically it is necessary to use many (hundreds or thousands) closely spaced individual radiating elements. The phase shifters control the phase of the individual elements and, hence, the aperture phase in a piecewise fashion. It would be desirable to control the aperture phase as a unit, but, since there is no known way to achieve this total control, the phase must be regulated with many phase shifters.* Typically, the elements are more or less a half wavelength apart, for reasons to be discussed later. The material presented here deals primarily with the kind of phased arrays in which there is a separate generator feeding each element – essentially the case for many feed systems for which there is no interaction between the elements in the feed system. Parallel-type feed arrangements are of this kind, having no intercoupling behind the aperture; rather, the coupling is out in front of the aperture between the elements. To first order, the results are applicable to a series network in which there is some intercoupling between the elements behind the aperture.

The principal reasons for using a phased array antenna in an electronic system are listed below.

1) *Electronic Beam Steering (control of aperture phase)*
   a) Multi-function operation
   b) Multi-target track
   c) High-performance targets
2) *Higher Power Capability*
3) *Electronic Control of Aperture Amplitude and/or Polarization*
4) *Multi-Band Operation*
5) *Hybrids Improve Reflector Antennas*
6) *Adaptive Control of Radiation Pattern (adaptive arrays)*

As arrays become less expensive, with lower weight and volume, the number of applications should increase, although arrays may always be "special purpose" antennas.

Aperture and Radiating Elements

Types of Radiating Elements

Figure 2 is a front view of a rectangular waveguide array with a triangular lattice [8]. Typically, either waveguides or dipoles are used for the elements in the array. Most of the arrays that have been built use waveguides because they have been built to operate at L-Band frequencies and above, and the state-of-the-art of waveguide element design is further advanced than is that of dipole element design. So, for a linearly-polarized array, rectangular waveguides probably would be used, arranged in one of the two configurations shown. The dominant mode of the waveguide, the TE_{10} mode, usually is the only propagating mode. It is excited from behind and does most of the radiating. The higher modes of the waveguide are excited at the aperture plane but, in a well-designed array, they should be unimportant and not

*Editor's Note – See page 330 for discussion of continuous-aperture scanning.

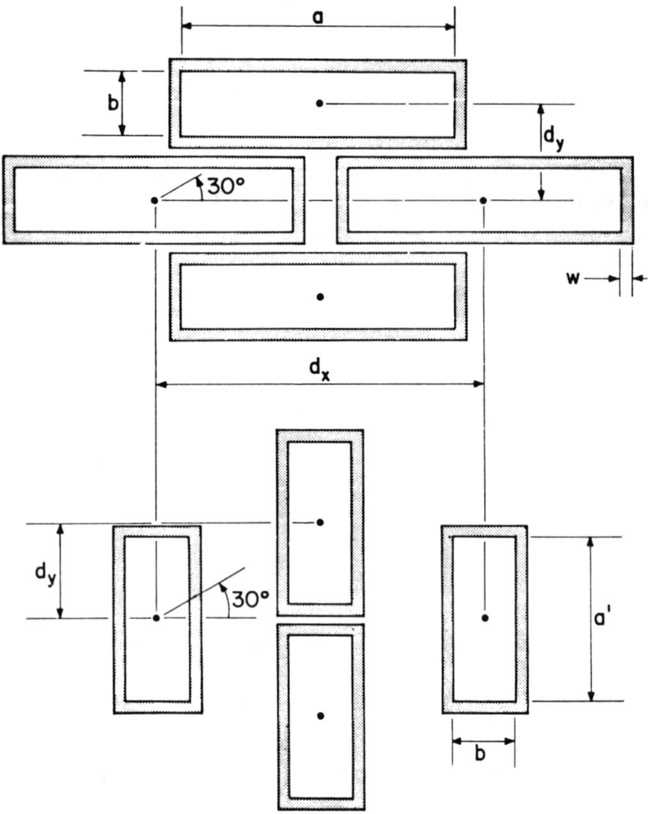

Figure 2 **Rectangular Waveguides in Triangular Lattice, Single Polarization**

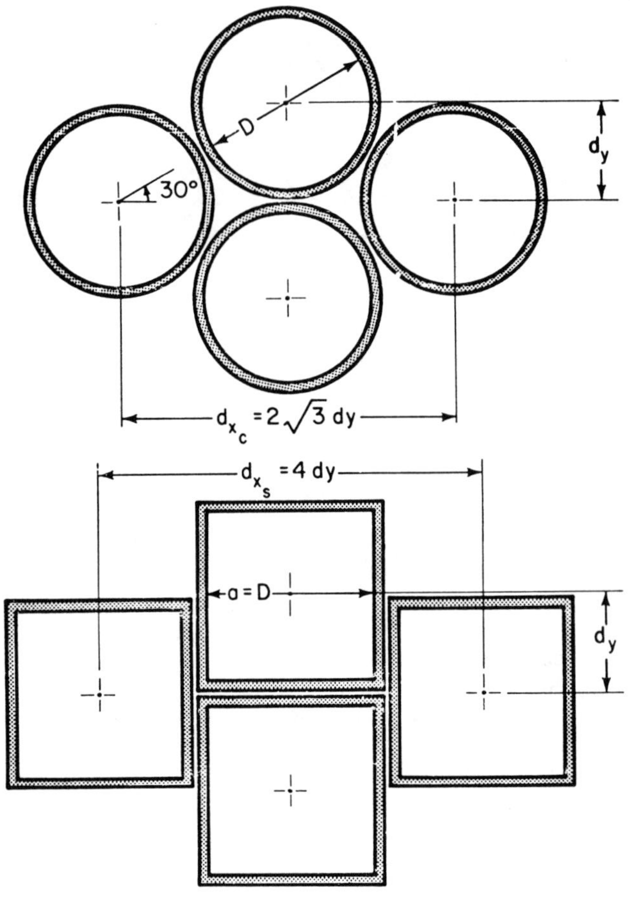

Figure 3 **Dual-Polarization Waveguide Arrays**

contribute significantly to the radiation pattern. If they do, the problem of *blindness effect* or a complete reflection of the incident power arises; these issues are discussed further on.

For dual-polarized or circularly-polarized arrays, either circular or square waveguides, as shown in Figure 3, would be used [8]. Usually, one makes use of the two crossed TE_{10} (or TE_{11}) modes and properly phases them to get the desired polarization in the aperture — circular, linear-vertical, linear-horizontal, or linear in some other plane. The two dominant modes are fed independently from behind. One would not use either of the waveguides in the figure for a linear polarized array because, if only, say, vertical polarization were excited, mutual coupling would excite the cross mode and there would be power coupled back into each element in the cross mode; somehow, that power would have to be dealt with. If the mode is terminated, the power is lost. If one reactively terminates the mode, there is a reflection of that power, changing the radiation pattern and the polarization. In order to avoid that problem, a waveguide in which the cross mode is already cut-off is used for linearly polarized arrays.

Figure 4 is a typical waveguide array viewed from the side. The waves are incident from the left and are radiated toward the right. Usually there is some kind of filling in the aperture, such as a dielectric plug, in order to keep out the weather and insects; sometimes this plug is used as part of the matching network. Because of the proximity of the elements, there is coupling from an element to its neighbors; in fact, all elements couple to all other elements. As the phase of the incident waves is changed to steer the beam, the net reflected signal changes; therefore, the reflection coefficient is a function of the beam steering angle. This relationship poses one of the big problems that occur in phased arrays — a reflection coefficient which is a function of two independent variables, the steering angle and the frequency. The situation is more complicated than the average network in which the main concern is matching over a frequency band.

There are two kinds of matching networks for phased arrays — ordinary matching, which can only center the element impedance

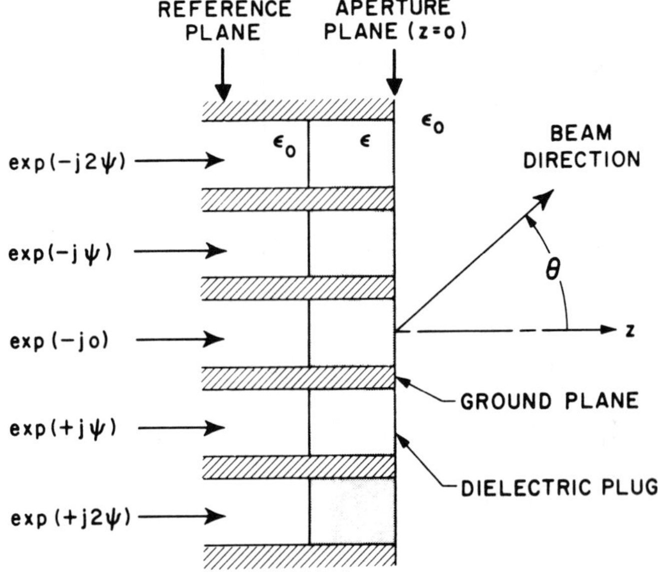

Figure 4 **A Typical Waveguide Array**

on the Reflection or Smith Chart, and wide-angle matching, which actually modifies the element reflection versus scan locus. If impedance matching components are placed inside the element, the variation of reflection versus scan at the aperture plane because of mutual coupling does not change. What does happen is that the impedance variation with scan is centered on the Smith Chart; this process is ordinary, or scan-independent, matching – the usual procedure. Typically, one can match a phased array element to a VSWR of two-to-one or three-to-one for a quarter hemisphere scan and a 10% frequency band. This matching performance is what is to be expected.

The other kind of matching network is scan-dependent. It involves such procedures as putting a dielectric sheet outside the aperture or interconnecting the waveguides behind the aperture so that there is an additional coupling between the elements (which, in turn, is used to partially cancel the coupling that exists at the aperture plane). Such a matching network is scan-dependent and allows one actually to reduce the size of the reflection-versus-scan-and-frequency locus and then, with ordinary matching inside, to center that reduced locus on the Smith Chart. Therefore one can do substantially better with scan-dependent, or wide-angle, matching at the price of additional complexity in the structure. Usually, wide-angle matching is not used because it is considered to be too expensive for the benefit it gives. However, if one has a very wide-angle-scan array or a very wide frequency band and needs good match, it is probably justifiable to use some sort of scan-dependent matching in addition to the ordinary internal matching.

Figure 5 is an example of two ways to feed a dipole element in front of a ground plane; the elements include baluns to go from the unbalanced coax to the balanced dipole. Dipoles generally are not used as frequently for the radiating elements as are waveguides since the performance of a dipole array cannot yet be accurately predicted – general expressions for the driving point impedance of a dipole array which include the effects of the currents on the supporting stubs have not been developed. The problem of a dipole array suspended above a ground plane has been solved and the driving point impedance versus scan can be computed, but the stubs, as yet, have not been included. At least one dipole array has been built in which the stubs have contributed to the blindness effect. This effect exists when the array is scanned to a certain angle and there occurs a very large reflection of the incident power. In an infinite theoretical array structure, the blindness would cause complete reflection; the aperture would actually be "blind" – it could not "see" in the scan direction for which the blindness occurred. The blindness is brought about because all of the mutual coupling coefficients add up and give unity reflection. The blindness effect occurs at large scan angles, but it can take place closer to broadside if there are structures or dielectrics on the array surface which electromagnetically load the surface; metallic stubs supporting dipoles are examples of such structures. Effort is being expended on the problem of solving for the impedance of the dipole array while including the effects of the currents on the stubs. When a solution becomes available, it will be possible to design dipole arrays reliably, to predict performance accurately, and to be sure that no blindness will occur.

Design of Large Linear Arrays

The important considerations when designing an element are:

1) *Electrical*
   a) Impedance matching
   b) Grating-lobe suppression
   c) Polarization control
   d) Power capacity
2) *Non-Electrical*
   a) Environment
   b) Cost

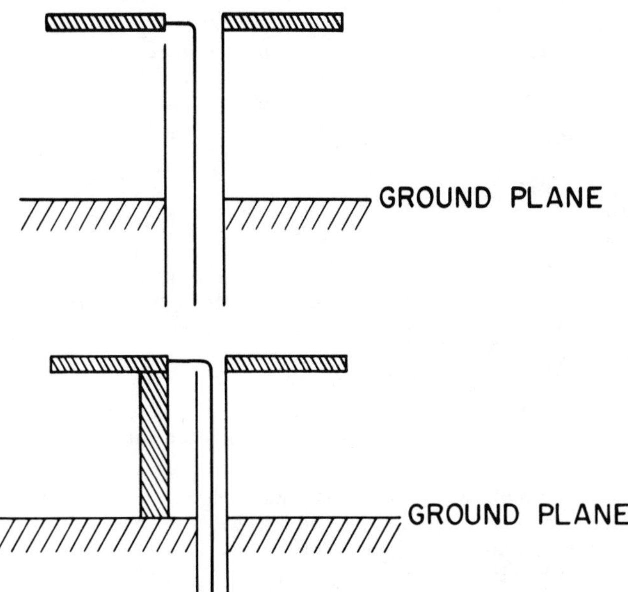

Figure 5 **Dipoles with Baluns**

Today, an element typically costs between \$1 and \$100, depending upon whether it is to be used for a demanding or a non-demanding environment. Of the electrical properties, the one which is considered in some detail here is matching the impedance of the element as a function of scan angle and frequency.

The element factor (or element pattern), $\underline{e}\,(\theta, \phi)$, of an element in an array is defined as the radiation pattern of that element in the presence of all other elements terminated in their generator (or receiver) impedances [9]. Assuming that the patterns of all elements in the array are identical,* the array radiation pattern $\underline{E}(\theta, \phi)$ may be written as [5, 7]

$$\underline{E}(\theta, \phi) = \underline{e}\,(\theta, \phi)\, A(\theta, \phi) \tag{1}$$

The array factor, $A(\theta, \phi)$, is the pattern of an array of isotropic point sources at the element locations, excited with the same amplitudes and phases of fields (or currents) as are the elements. It is clear from Equation (1) that the element pattern determines the polarization state of the array in every direction in space, regardless of how the array is scanned.

Also, under the assumption that the element patterns are identical, two formulas can be derived which relate the element pattern realized gain** in a particular direction, $g(\theta_o, \phi_o)$, to the array performance when all elements are excited with uniform amplitude and the array is scanned to direction θ_o, ϕ_o [5, 7, 9]. One of these formulas concerns the array realized gain, $G(\theta_o, \phi_o)$, and the number of elements, N,

$$G(\theta_o, \phi_o) = N \cdot g(\theta_o, \phi_o). \tag{2}$$

The other formula concerns the power transmitted into the main beam (dominant unit-cell mode) relative to the power incident in the element, $P(\theta_o, \phi_o)$, when all elements are excited with uniform amplitude,

$$g(\theta_o, \phi_o) = \frac{4\pi a}{\lambda^2} P(\theta_o, \phi_o) \cos \theta_o. \tag{3}$$

*This assumption is a reasonable one for a large planar array in which only the relatively few edge elements have substantially different patterns. Pattern is defined here to be the electric field strength of radiation at a specific distance from the array in its far field.

**Realized gain is power gain defined in the usual way except that it is with respect to the power *incident* in the element or array [9].

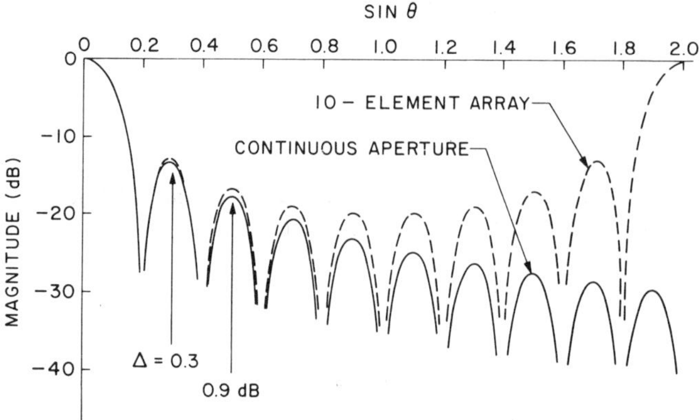

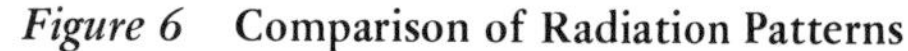

Figure 6 **Comparison of Radiation Patterns**

In this formula, λ is the free-space wavelength, a is the allotted element area (the unit-cell area), and θ_o is the scan angle measured from broadside. From Equation (2), it can be seen that the element-pattern gain determines the array gain.

In view of Equation (1), the problem of designing for a specific array pattern usually is broken into two parts, the design of the radiating element and the design of the array factor. Certain reasonable assumptions are made to simplify the solution for each part. In regard to the array factor, it is usually assumed that the pattern of a *continuous* aperture with the same excitation amplitude is a good approximation. As regards the element, the standard assumption is that the pattern of an element in an infinite-array environment is an accurate estimation. The approximate solutions for each part of the problem then are combined, via Equation (1), to determine array performance.

Figure 6 shows a comparison of array factors from a 10-element linear array of isotropic point sources with λ/2 spacing and from a continuous aperture of length 5λ. Both apertures have uniform amplitude excitation and zero phasing. The patterns are given by

$$A(\theta) = \frac{1}{N}\frac{\sin\left(N\frac{\psi}{2}\right)}{\sin\left(\frac{\psi}{2}\right)} \tag{4}$$

for the array, and by

$$A(\theta) = \frac{\sin\left(N\frac{\psi}{2}\right)}{N\left(\frac{\psi}{2}\right)} \tag{5}$$

for the continuous aperture, where

$$\psi = \frac{2\pi s}{\lambda}\sin\theta$$

In the latter relationship, s is the element spacing and θ is the angle from broadisde; the length of the continuous aperture is Ns. Notice that the patterns are not very different near the main beam. Further toward the edge of real space ($\sin\theta = 1$), they differ somewhat, but this difference is made relatively unimportant in the planar array pattern by the suppression caused by the element pattern in Equation (1)*. Hence, all of the aperture excitation functions studied and used for continuous apertures can be utilized in arrays, as long as grating lobes are not too close to the main beam.

*Also, a tapered excitation amplitude tends to minimize the differences in many practical cases [10].

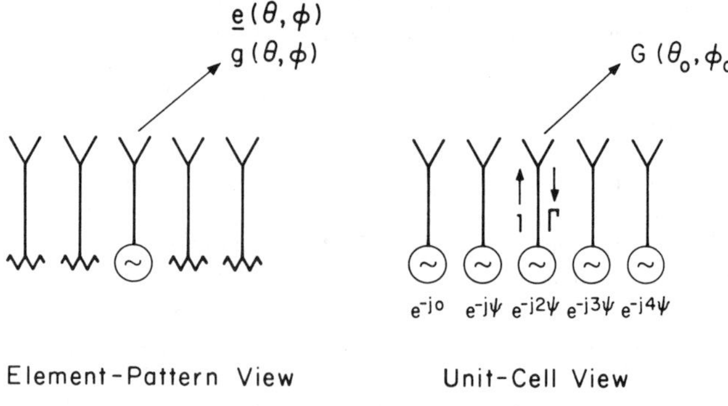

Figure 7 **Two Ways to View Phased-Arrays**

The primary concern in the electrical design of a radiating element is usually good impedance match over the range of scan angles and frequencies of interest. As seen from Equation (3), perfect impedance match for all scan angles of interest means that $P(\theta_o, \phi_o) = 1$ and the element gain is maximized; from Equation (2), this situation maximizes the array gain. A second concern in the electrical design of an element is the polarization of radiation. The infinite-array model for determining the element design yields impedance and polarization data which are excellent approximations of those for an element in a large planar array. The only exception is when the array is scanned to an angle for which a blindness exists; this circumstance could occur at the endfire-grating-lobe scan angle or at some lesser scan angle, depending on the array geometry.

In summary, the usual design and determination of performance for large planar arrays by independently designing $\underline{e}(\theta, \phi)$ and $A(\theta, \phi)$ is a good approximation if:

1) The array is large enough so there are relatively few edge elements.
2) The grating lobes are far enough removed from the main lobe.
3) No blindness or resonance condition exists within the scan sector of interest.

Mutual Coupling and the Blindness Phenomenon

In Figure 7 are shown the two principal ways to view phased-array antennas. On the left is the element pattern view in which one element is excited and all others are terminated; the parameters of interest are the gain of the element pattern ($g(\theta, \phi)$) and the electric field radiated by the element $\underline{e}(\theta, \phi)$). On the right is the all-excited or system view in which all the elements are indeed excited in some fashion, either with generators or a feed network; the parameters of interest are the array gain ($G(\theta_o, \phi_o)$) and the reflection coefficient ($\Gamma(\theta_o, \phi_o)$) or power transmission coefficient ($P(\theta_o, \phi_o) = 1 - |\Gamma(\theta_o, \phi_o)|^2$). For some purposes it is more convenient to use one view; for other purposes it is more convenient to use the other. The sketch on the right is called the unit-cell view because, in an infinite structure, only one unit cell and one period of the array need be analyzed, and the fields solved in that period. The fields in every other cell are uniquely related to those in that period by just a shift in phase, according to the generator phase. As a result, the analysis is simplified tremendously. Note from Equation (2) that the array gain, G, is equal to the number of elements times the element gain. Thus, if the array is thinned by removing elements, the array gain is reduced proportionately – the number of elements determines the array gain. Note from Equation (3) that the element-pattern gain is dependent upon both the allotted area, a, per element and the wavelength squared, and is related to the all-excited condition of the elements via $P(\theta_o, \phi_o)$ and $\cos\theta_o$. If complete power is transmitted for all scan angles (P is unity) the element pattern goes as cosine θ – the ideal element pattern. If there is a complete reflection of the incident power ($|\Gamma|$ is unity), the power transmitted into the unit cell is 0 and the ele-

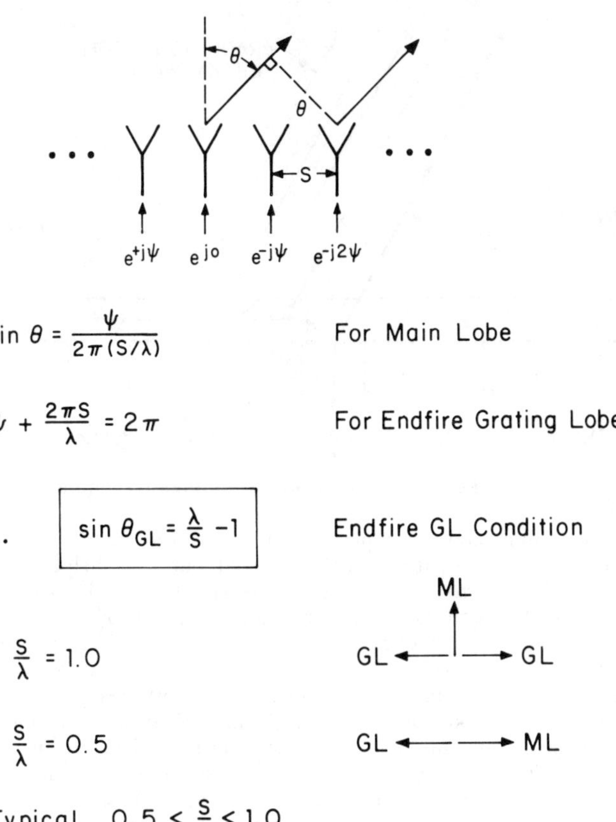

Figure 8 Endfire Grating Lobe

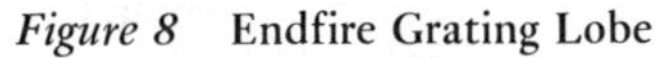

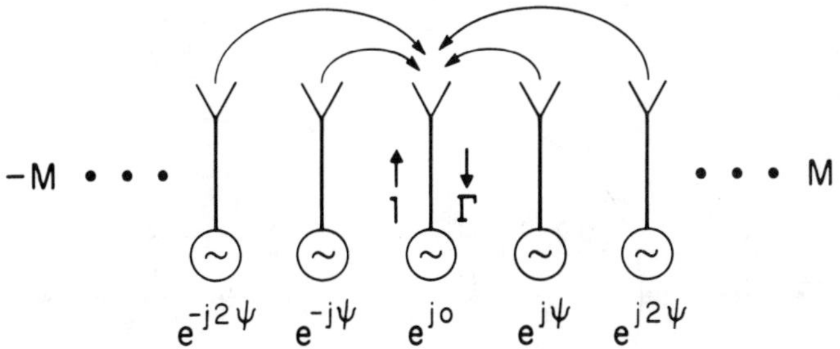

$$\Gamma(\theta) = \sum_{m=-M}^{M} S_{om} V_{im}$$

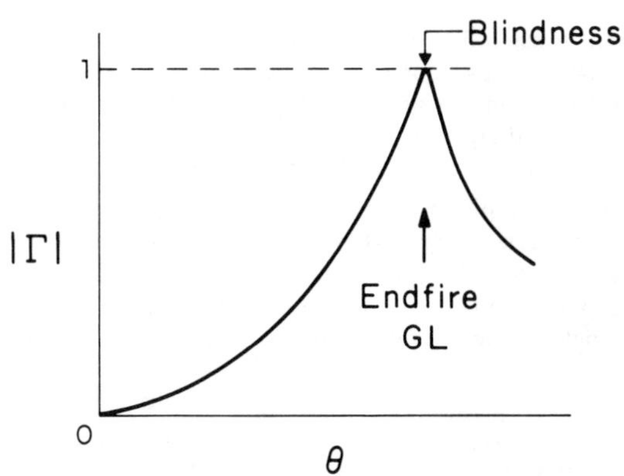

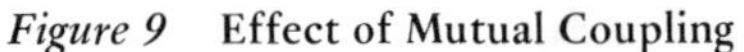

Figure 9 Effect of Mutual Coupling

ment pattern has a null in that particular direction. A complete reflection in the all-excited view corresponds to a null in the element pattern view. Another way to look at this fact is that, if the element pattern has a null in some direction, a complete reflection of the power occurs and the array is blind when the elements are phased to radiate in that direction. Thus, the element pattern null and the complete reflection condition are one and the same and are called a *blindness of the array.* This example is an extreme one of the effects of mutual coupling.

Figure 8 illustrates the formation of a grating lobe in an array. If an array is phased (as shown) to radiate a beam at angle θ, the simple formula in the figure gives the angle, θ_{GL}, for which a second lobe or grating lobe forms at endfire ($\theta = 90°$). If a grating lobe is being formed in the endfire direction, all the elements must add up in both this direction and the direction of the main beam. In the formula, if the element spacing, S/λ, equals one wavelength, the grating lobe occurs when the main beam is scanned to zero degrees. Therefore, when the main beam is at broadside, there are grating lobes in both endfire directions with one wavelength element spacing. If one spaces the elements by a half wavelength, the main beam then must be scanned all the way to sine θ of unity (because λ/S is 2) in order to have an endfire grating lobe. Typically the elements are spaced between a half and one wavelength so that, when the main lobe is scanned to the maximum required scan angle, the grating lobe will not have formed as yet at endfire — it would still be in what is called *imaginary space,* not having appeared in *real* or *radiating space.*

Figure 9 is an illustration of how blindness occurs. The mutual coupling from all the elements couples into, say, the center element. The reflection coefficient in the center element is the sum of all the coupling coefficients from element m to the center element times the incident voltage in element m. Summing all of these coefficients over the total number of elements gives the Γ in the center element. On the lower part of the figure is an example of the nature of |Γ|. If an array is matched at broadisde (the matching network is in the element to match at broadside), the magnitude of Γ typically increases as we scan out, hitting a peak of unity at the endfire grating lobe condition. So, usually, in a well-designed array, a large reflection condition appears simultaneously with the endfire grating lobe condition. In an infinite array (an assumption to simplify the analysis) the reflection is unity; with a large finite array, the reflection peak is about 0.8 or 0.9. This situation is called a blindness because, at this scan angle, all the power is reflected and power cannot be transmitted in either direction through the aperture. If the surface of the array is loaded with dielectric or with metallic stubs (and sometimes even without loading the array surface), blindness can occur substantially closer to broadside than the endfire grating lobe angle. This situation is highly undesirable because it cuts down the scan sector and does not permit scanning out to the maximum angle determined by the endire grating lobe. This example indicates, in the extreme, what can happen due to mutual coupling.

What are some of the reasons that could make the blindness condition occur closer to broadside than the endfire grating lobe condition? To answer this question, a familiarity with grating lobe diagrams for array antennas is needed. Figure 10 is a plot in sin θ space of the grating lobes of a triangular lattice array. The radial coordinate is sine θ where θ is the angle from broadside of the main beam. The circle is $\sin\theta = 1$ and the area inside is called real space. If the beam is inside this circle, it is radiating; if it is outside this circle, it is in imaginary space (it is not radiating). The triangular array lattice has a triangular geometry of grating lobes; the main lobe is shown at broadside scan, and all other dots are grating lobes. If the main lobe is scanned in the horizontal direction, all the grating lobes slide in the same direction on this sine θ plot. Certain grating lobes arrive at real space and begin to radiate after a sufficient scan of the main lobe. This plot is very useful; it is

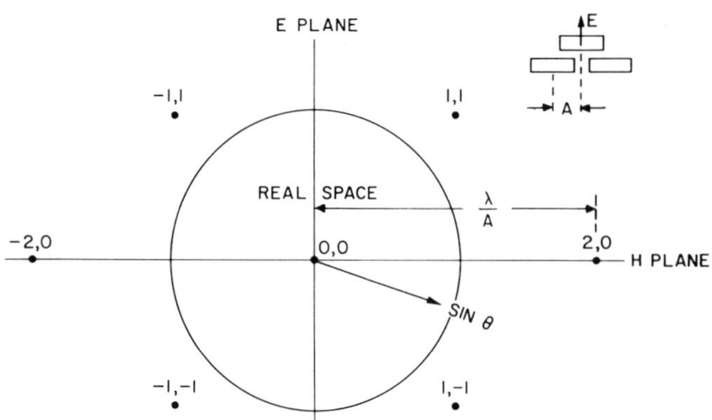

Figure 10 Triangular Lattice Grating Lobes

used frequently when designing phased array apertures. The spacing of the grating lobes in sine θ space is just the reciprocal of spacing of the elements in wavelengths. It turns out that the blindness condition in phased arrays is related to the approach of a grating lobe to the edge of real space, as is demonstrated later.*

Figure 11 is an example of a measured blindness condition [11]. The element pattern shown was measured in a triangular lattice array. Note that the null of the element pattern is before the endfire grating lobe condition; if one were to phase the array to radiate at that angle, complete reflection of the incident power would occur.

*Sine θ space is the same as the T-space and sine α-sine β space described at end of Chapter 18; see also Figure 11 of Chapter 18. Also see Lecture 4 by L.R. Lewis of Reference 18 and Reference 3 and 7.

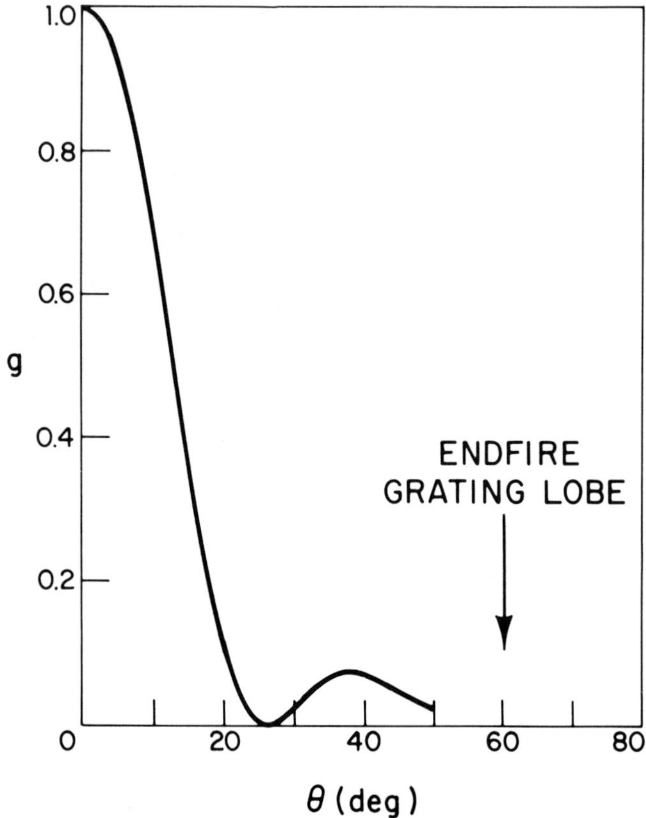

Figure 11 Measured Element Pattern in H Plane of Triangular-Lattice Waveguide Array

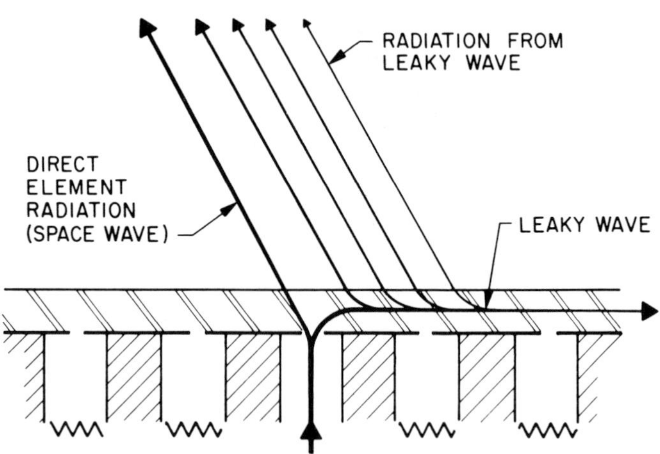

Figure 12 Leaky-Wave Explanation of Blindness Phenomenon

Figure 12 illustrates one explanation that was proposed for the null of the element pattern [12]; it is valid, but is not helpful in determining how to rectify the problem. The structure is an array of waveguides, with the aperture partially closed off by irises; there is a dielectric sheet on top of the array. The incident power in one element is radiated directly and establishes a leaky wave on the surface of the array. The leaky wave draws off energy; the signal which is leaked precisely cancels the radiated signal from the element, thus causing the element pattern null in some direction. This is a correct explanation but it does not describe the conditions under which the leaky wave can be established and supported.

In Figure 13 is shown the most helpful diagram to understand the blindness condition in phased arrays [13, 14] — a k-β diagram. Its vertical axis is proportional to k_o, the free space wave number ($k_o = 2\pi/\lambda$). The horizontal axis is proportional to k_x, the wave number in the x direction. Assume a beam radiating and being scanned in the x direction, which is the H plane of the array. When the beam is at broadside, the wave number in the x direction is zero so that the vertical axis represents a scan angle of 0°. When the beam is at enfire, k_x is equal to k_o; therefore the 45° line is a scan angle of 90°. Scan angles between 0° and 90° are represented by straight lines between these two lines. Also plotted on the figure is a dashed curved line which is the condition for an endfire grating lobe in this array. When the combination of scan angle and frequency lies on this line, there is an endfire grating lobe condition in this array. The dashed horizontal line is the point at which we get a cut-off condition for the TE_{20} mode in the waveguides. The TE_{20} mode is the next higher waveguide mode excited by mutual coupling when the array is scanned. Recall that the incident mode is the TE_{10} mode, and the TE_{20} is only excited by mutual coupling. Notice that the blindness curve for this array

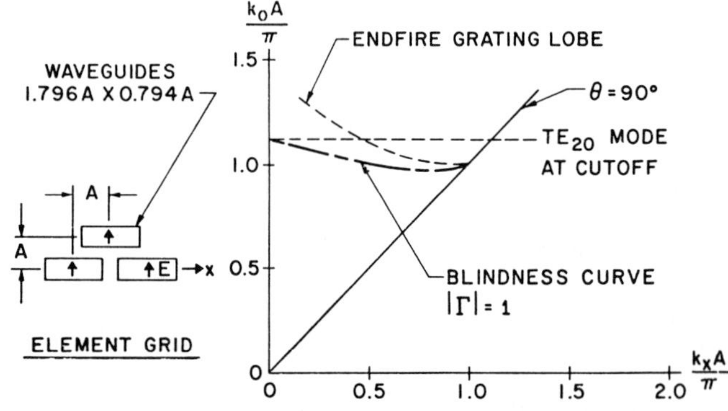

Figure 13 Higher-Mode Explanation of Blindness Phenomenon

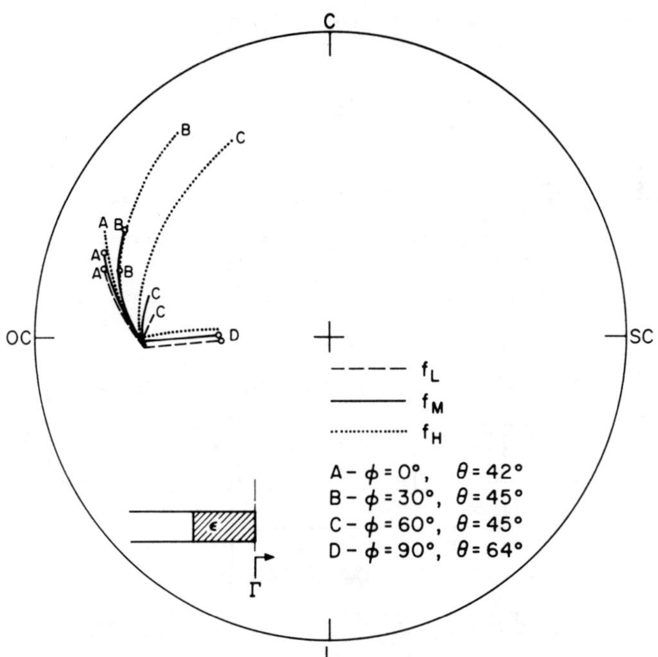

Figure 14 Calculated Γ at Aperture Plane

(|Γ| = 1) is very close to the cut-off conditions for the TE_{20} mode in the waveguide and the nearest grating lobe (or next higher unit-cell mode). This curve clearly shows that the blindness condition in phased arrays is caused by operating the array at a level too close to the cut-off condition of the next higher mode of the waveguide or the next higher mode of the unit cell. If the array is operated sufficiently below these conditions, a blindness condition does not occur. Typically during array operation, the frequency is kept constant and scanning is in angle. In this case, a scan at k_0A/π = 1.0 would hit the blindness condition at about 40° or 30°. If the frequency is lowered during scanning, the blindness condition is avoided, and the scan can proceed through to endfire or 90° without hitting the condition. This explanation of blindness is the most helpful – it tells us what it is related to and how to avoid it.

Impedance Matching of Element

The blindness condition is an extreme example of the effects of mutual coupling. However, even if the blindness condition is avoided, there is still the variation of impedance or reflection coefficient caused by the mutual coupling. Figure 14 shows a typical calculation of the impedance that would be seen at the aperture of an element, versus scan angle and frequency for an element which scans over a quarter hemisphere [4]. The impedance loci are plotted on a reflection chart; the radial scale is |Γ| from 0 to 1, and the circumferential scale is arg Γ from 0° to ±180°. Note the large variation of Γ, primarily with scan angle. The ends of the curves, A, B, C, and D, are the maximum scan angles; the junctions of the curves are broadside scan. The very long extensions at the high frequency, f_h, are caused by an approaching grating lobe; therefore, the elements appear to be too far apart in this design. Notice that there is a substantial variation of impedance with scan angle and a small variation of impedance with frequency (over a 10% band). As matching is added to the array, the finite length of the matching network increases the spread of the reflection with frequency until it becomes comparable with the spread of reflection with scan angle.

Figure 15 shows the performance of an element that was designed to work over a quarter hemisphere scan volume [8]. This plot, on sine θ space, is of the VSWR contours that are seen in the element with all elements excited. This element was designed for a 58%

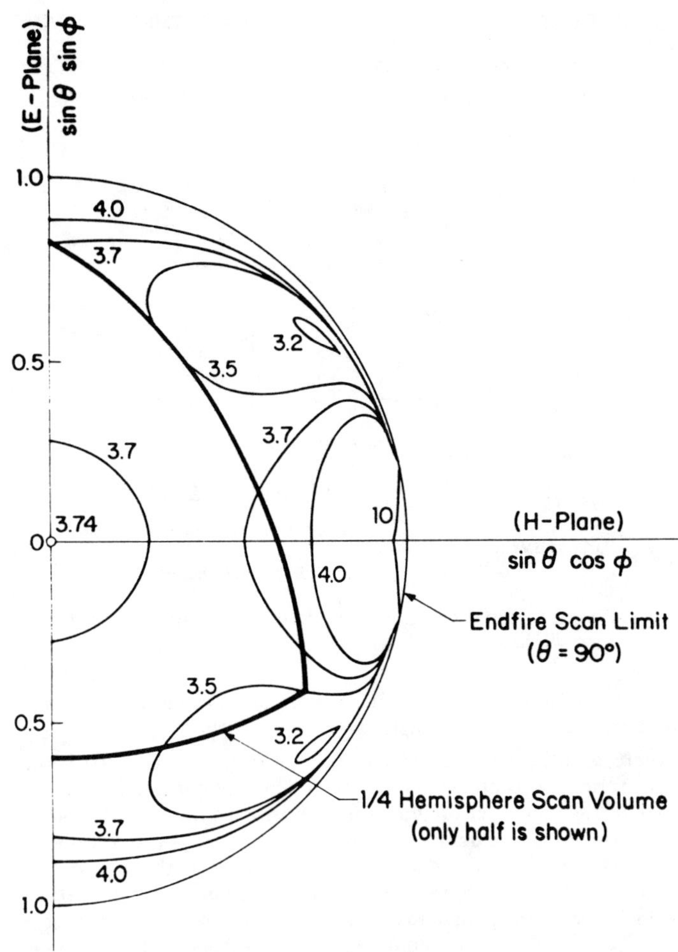

Figure 15 VSWR Contours – Low Frequency, 58% Band Element

band, which is almost an octave bandwidth. (An element operating over a 2:1 frequency range has a 66% bandwidth.) It was possible to match the array to within a VSWR of about 3.7 everywhere within the scan volume over the 58% band. This fact provides an indication of how well a relatively wideband element can be matched over a typical scan volume.

The matching illustration in Figure 15 was done with the aid of a computer program. The computer program was written for the case of an infinite phased array to solve the problem of a waveguide radiating into a unit cell. Figure 16 shows the waveguide and the unit cell. Since the natural modes for each region are known, and since we can write analytically the admittance seen by each mode from the aperture plane, this problem can be solved by the Method of Moments or Galerkin's method; the result is an expression for reflection from and transmission through the aperture. (This computation checks very well against experimental data.) In solving, either TE and TM modes or Longitudinal-Section Electric (LSE) and Longitudinal-Section Magnetic (LSM) modes may be used, according to which method is most convenient. Sufficient conditions for solving this phased array problem via modal analysis are summarized on the figure. At this time, there have been perhaps 8 or 10 different programs written for computing array performance for various kinds of waveguide arrays, including circular waveguides, linear or square waveguides, and planar or cylindrical arrays. Some of the programs include the ability to change the cross-section of the waveguide or put in dielectric plugs to match the elements. The programs used to design the element the performance of which is shown in Figure 15 were of this kind. They included matching networks and the entire design was done by

CONDITIONS SUFFICIENT FOR COMPUTING ELEMENT DESIGN

1. MODAL REPRESENTATION OF FIELDS IN 2 REGIONS
2. Y SEEN BY EACH MODE AT APERTURE PLANE IS KNOWN

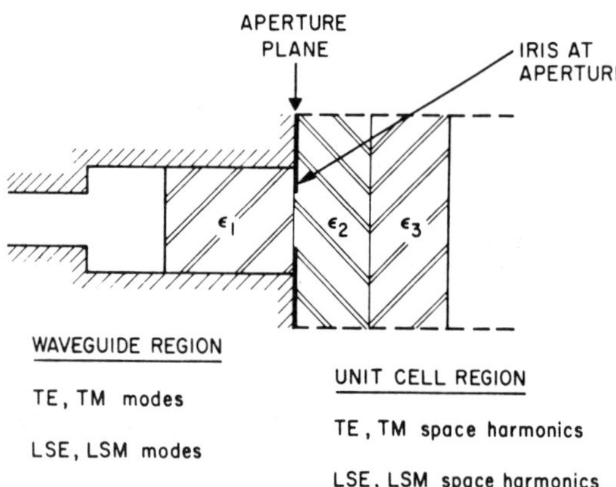

Figure 16 Schematic of a Waveguide Radiating into a Unit Cell (Note: Y = susceptance)

computation, with experimental confirmation of performance at one frequency and scan angle.

There are three current element design procedures. The first is the theoretical or computational method which requires the use of one of the computer programs and a possible experimental confirmation after the design is made. This approach is the recommended one today if the case can be computed. If there is no computable element, an experimental procedure must be employed in which measurements are made in either waveguide simulators or small arrays. Waveguide simulators are simulators which can simulate, with only one or a few elements, what goes on in an infinite phased array. The elements are imaged in waveguide walls to make them appear to be in an infinite array (see Figure 17). An alternate method is to build a small array, make measurements of the coupling coefficients, and then decide what the matching should be on that basis. Such a procedure is laborious; it was used 15 years ago, but it is rarely done today. The third procedure is a combination of the first two; it is not discussed in detail here.

Figure 17 shows a front view of an array for which a waveguide simulator was constructed. The waveguide is partially closed off so that only a slot radiates. The slot is excited from behind by the TE_{10} mode in the waveguide; a rectangular element lattice is used. The heavy solid line outlines a simulator waveguide, which is shown in Figure 18. Notice that only the portion of the array within the simulator waveguide was constructed. The completely-reflecting metal walls of the guide image the 1-1/2 elements into an infinite array. Therefore, measurements made inside the simulator waveguide represent the performance of the infinite phased array at a discrete set of frequencies and scan angles, determined by the dimensions of the simulator waveguide. In the photograph of the simulator in Figure 18 are only a half element and a quarter element — all of the array that was necessary to build. A slot plate is shown which goes over the waveguides to close them up and make them smaller. A dielectric slab is put on the surface of the array to cause blindness at the scan angle of simulation (the simulator was used to study blindness); a desirable feature of this kind of simulator is that there is only one active element. The quarter waveguide does not propagate (it is cut-off at the frequency used) and can be terminated a short distance behind the aperture. The half element is excited from behind and its reflection versus frequency is measured to determine what the infinite array is doing at the frequencies of measurement and at the angles that the TE_{10}

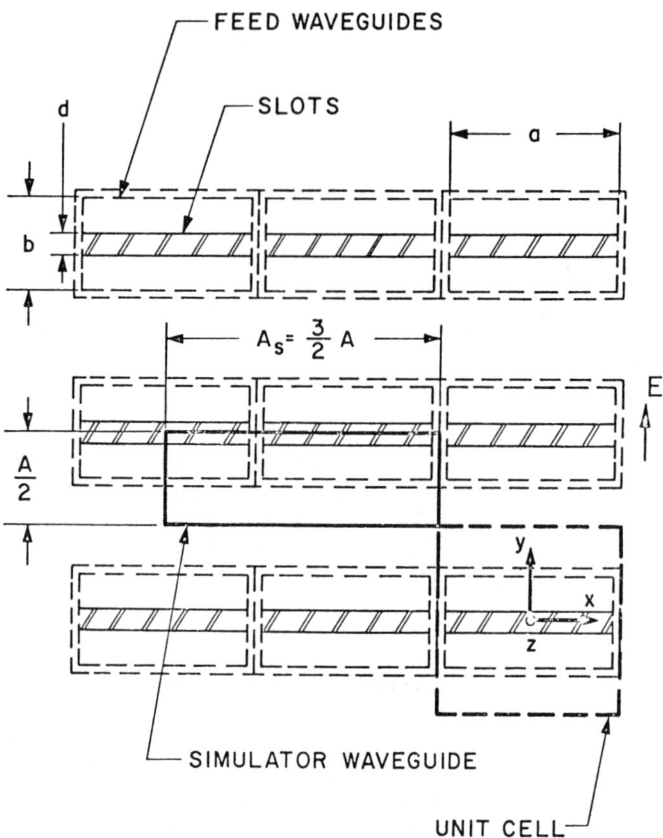

Figure 17 Front View of a Waveguide Array Showing Outline of Simulator Waveguide

mode propagates in the simulator waveguide. This technique of waveguide simulation was developed at Wheeler Laboratories, Inc., over ten years ago; it proved very useful in the early days of element design before computer programs were available [15].

Figure 19 shows the geometry of an array for which a computer program is available. The program is called RWED (Rectangular Waveguide Element Design) and is available from MIT Lincoln Laboratory. It computes the performance of an array element with as many as eight matching sections of cross-section or dielectric change for rectangular waveguides in either rectangular or triangular lattices.

Figure 18 Photo of Waveguide Simulator

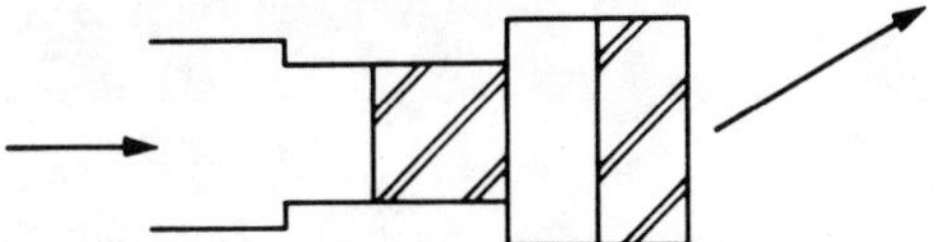

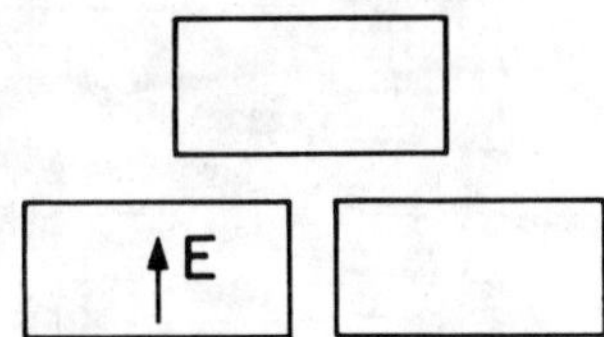

Figure 19 Geometry for RWED (Rectangular Waveguide Element Design) Computer Program

Figure 20 shows the geometry for a similar program which is useful for square waveguides of dual or circularly polarized elements. It calculates the performance of a square waveguide with eight matching sections for polarizations which are either vertical, horizontal, left or right circular, or in or perpendicular to the scan plane. The program first calculates the polarizations of radiation from these values, then finds the cross coupling of these polarizations in space. These two programs developed by Lincoln Laboratory are mentioned, not because they are the best or the only ones, but because they are available.

Figure 21 shows the most useful simulators for phased arrays for triangular and rectangular lattices. Each simulator simulates an infinite phased array at a scan angle of about 40° in the H plane. There is only one active element in each simulator; the others are not active and need not be fed from behind. Today, it is rare that simulators other than these two are built; current design primarily utilizes computer programs and a simple simulator for checking purposes. If there is good agreement between actual and expected performances, the design is regarded as finished from the impedance-matching aspect.

Though this chapter is not appropriate for a discussion of Wide Angle Impedance Matching (WAIM), a brief presentation is made of the most useful technique – use of a thin dielectric sheet with a high dielectric constant. Figure 22 shows a plane wave incident on such a sheet, and the variation of sheet susceptance versus angle of incidence [16, 17]. Notice that for a wave incident in the H plane, the susceptance increases with angle; for a wave incident in the E plane, the susceptance decreases with angle. (The assumption is that the sheet is thin enough for it to be approximated by a susceptance at a plane in space.) The principle behind using such a sheet to wide angle match an array is simply to use the scan-dependent susceptance or angle-dependent susceptance to cancel partially the susceptance of the array aperture.

Figure 23 is an example of a thin sheet in front of an array which has been spaced in an optimum way to cancel partially the susceptance variation at the aperture [4]. In this instance the blindness effect of the sheet is of primary interest; the matching improvement is not shown. It turns out, however, that the match can be improved substantially. If there is a VSWR of 4 to 1 without the sheet structure, it can be reduced to approximately 2 to 1 with an optimum scan-dependent sheet matching structure. Notice, though, that if the reflection versus scan angle peaks out at the endfire grating lobe condition before the sheet is added, the addition of the sheet pulls this condition in a little bit, as shown; thus, the blindness condition moves slightly closer to broadside because of the dielectric loading of the aperture region. This effect is minimal if the sheet is kept very thin and of high dielectric constant.

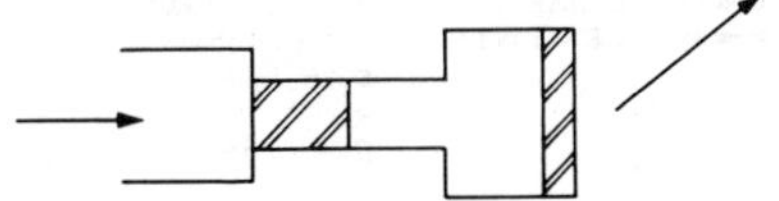

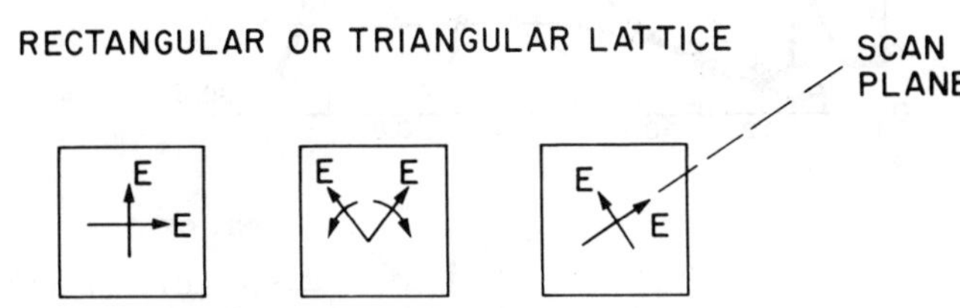

Figure 20 Geometry for SWED (Square Waveguide Element Design) Computer Program

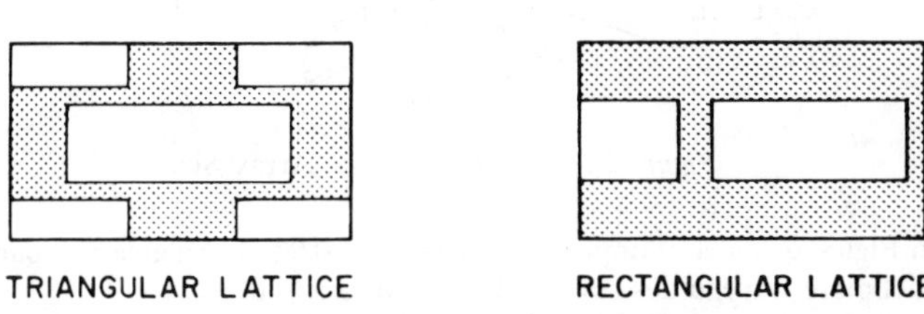

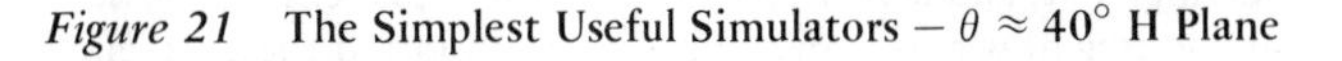

Figure 21 The Simplest Useful Simulators – $\theta \approx 40°$ H Plane

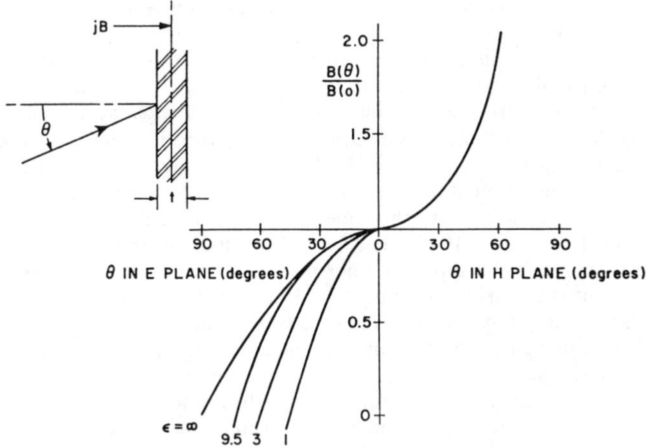

Figure 22 Susceptance of a Thin Dielectric Sheet

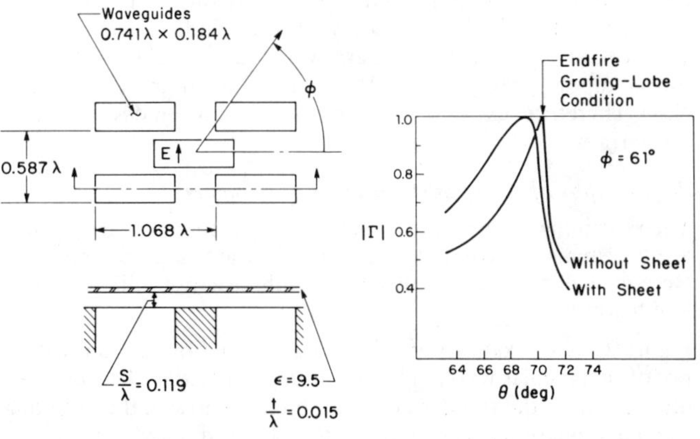

Figure 23 Computation of a Blindness Caused by the Dielectric Sheet WAIM

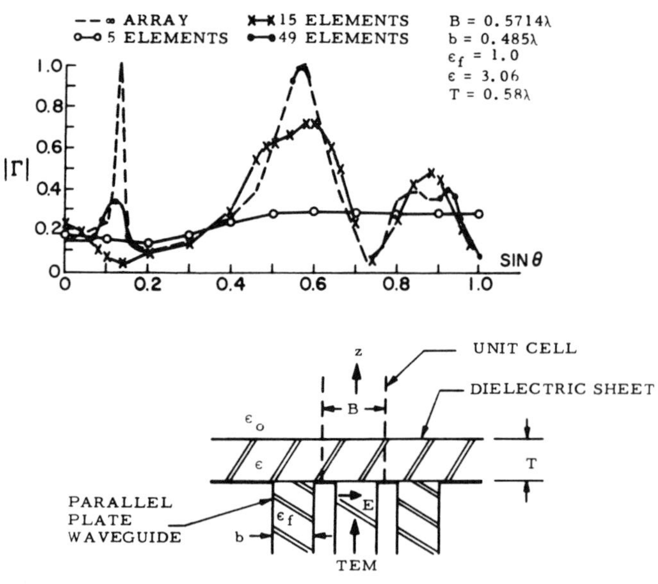

Figure 24 Effect of Finite Array Size

Figure 25 Waveguide Array (6 Sub-Arrays)

In Figure 24, a very important point is illustrated. An array of parallel plate waveguides covered by a dielectric slab has the TEM mode incident, and is phased to scan in the E plane. If the reflection coefficient versus scan angle, sine θ, is plotted for different size arrays, interesting results are obtained. The infinite array (the dashed line) has two blindness conditions present – one very close to broadside, with very high Q, and one further out, with lower Q. The blindness condions are caused, in this case, by the presence of the dielectric slab. If the performance is calculated for the same array with a finite number of elements, say 15 (shown by the x's), the blindness condition near broadside is not observed and that near the endfire grating lobe is seen only partially. The point is that a small finite array does not always illustrate or show blindness conditions to the same extent that a much larger array does. Even with a 49 element array, the blindness near broadside only begins to be apparent, but, out further, the condition is quite evident (the dots indicate that the 49-element array behaves like the infinite array further from broadside). The lesson to be learned from this figure is that, if an array element is designed using a small array (of, say, 15 elements), it is possible that the full size array may have blindness near broadside which was not observed in the small array. Such a phenomenon has been known to occur in at least one recorded case, emphasizing the need for a method of computing the performance of an element in an infinite array. If this calculation can be made and there are no blindness conditions indicated, there are, in fact, no blindness conditions in any small finite array. Hence, the emphasis is on waveguide elements once again because, for them, there are available programs and formulas. For all dipole arrays, though, the computational procedure is not clear. Therefore, extreme care must be taken when designing dipole arrays.

Photographs of Array Systems and Components

The remaining figures in this chapter are photographs of various array systems and components. Since they are all presented in Reference 4, they are discussed here only briefly to illustrate their key features.

Figure 25 is an example of sub-array construction of a phased array of rectangular waveguides in a triangular lattice. This procedure is one of the standard ones for building arrays; the radiating elements, phase shifters, and, in back, the feed networks with connections which sum (or subtract) elements are all obvious.

Figure 26 Reflectarray

Figure 26 is an example of an array which may be familiar – a reflectarray, with a feed out front. The elements are in a planar aperture and the phase shifters are behind the elements, terminated in short circuits. The feed radiates a wave; the wave is captured by the elements, shifted in phase, reflected from the completely reflecting terminations, shifted in phase again, and then radiated off at some angle. This particular feed is quite cumbersome because the array is designed to do many things – it has monopulse capability with dual polarization. The feed is therefore complex and the design is a compromise because of the many necessary functions.

Figure 27 shows an array which was built to scan in only one plane, at 19 GHz, for earth observation from aircraft. This objective greatly simplifies the design because, instead of needing the same number of phase shifters as elements, only a phase shifter per row (or column) is required. Very often, as a compromise, arrays are designed to phase scan in one plane and frequency scan (or scan mechanically) in the other. The array shown uses series lines feeding radiating slots. When frequency is changed, the beam scans in the vertical plane; when the phase shifter settings are changed, the beam scans in the horizontal plane.

Figure 28 is a shot of the FPS-85 antenna at Eglin AFB. It was built as two apertures – a large receive aperture and a smaller transmit one. This array has been performing well for several years now, acquiring space-satellite radar data for the Air Force.

Figure 29 is an example of a feed-through lens in a limited-scan array-reflector system. A feed radiates through a lens onto a reflector. Changing the setting of the phase shifters in the lens scans the beam over a limited, 14° x 20°, sector. This sytem has the benefits of the large gain associated with reflectors and the limited electronic scan associated with arrays.

Figure 28 AN/FPS-85 Radar

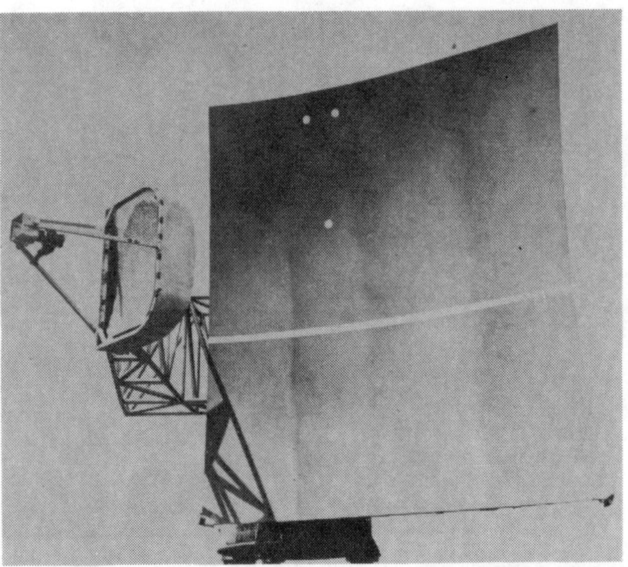

Figure 29 Test Model AN/TPN-19 Antenna – X-Band 14° x 20° Scan and Linearly Polarized Multimode Feed

Figure 27 19 GHz Array for Earth Observation – ±50° Phase Scan

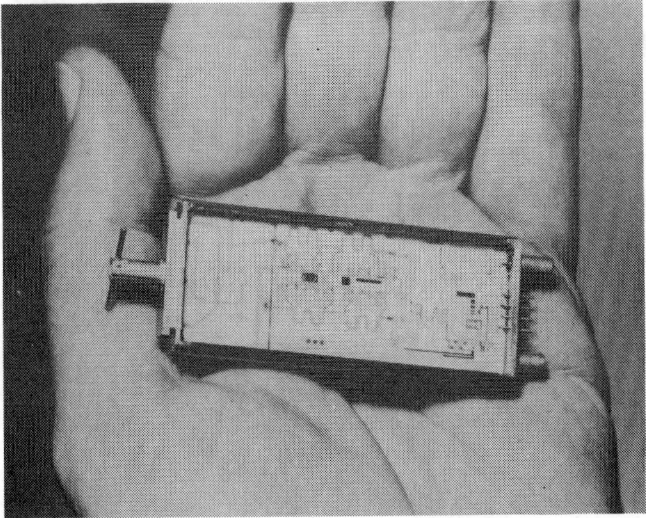

Figure 30 X-Band MERA Module (see page 277 for discussion of MERA system)

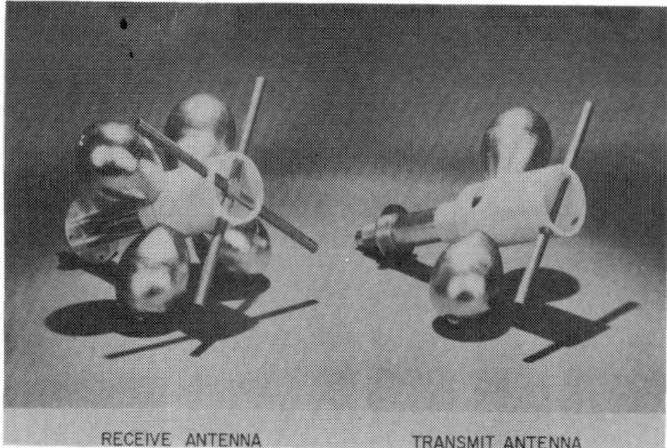

Figure 31 Dipole Radiating Elements for 400 MHz Arrays

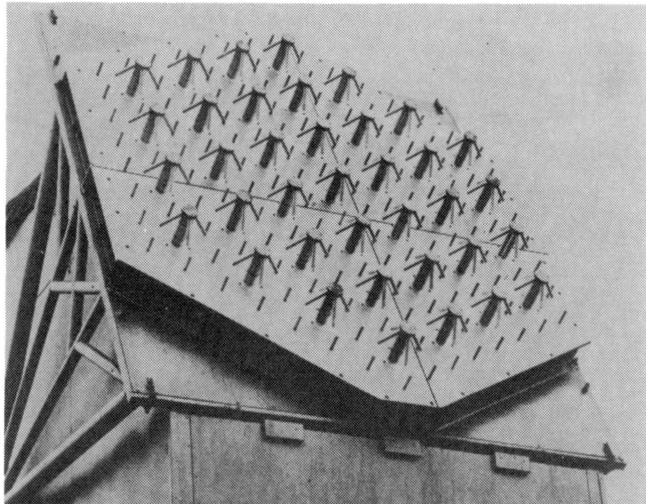

Figure 32 Dual-Polarized Dipole Array

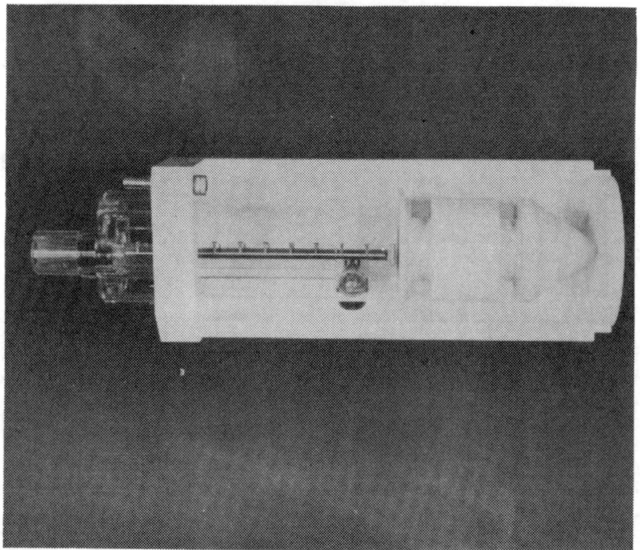

Figure 33 Dual-Polarized Circular-Waveguide Element

Figure 30 is an example of a solid state X-band module (the MERA module) which was built for a demonstration of the capability of solid state arrays. Note its size.

Figure 31 is an example of two elements which were built for a particular phased array – a receive antenna with dual polarization, and a transmit antenna with linear polarization. The elements are dipoles (thickened and rounded for wide bandwidth) with passive director rods used for shaping the element pattern.

In Figure 32 is shown a very interesting array of crossed dipoles, in which blindness was observed during test. When this small array was built and the element pattern measured, a blindness condition was discovered; corrective action was taken by bending the arms of the dipoles down, thereby eliminating the blindness. However, this action ruined the radiated polarization, necessitating insertion of pins between the elements for correction. This example is one illustrating the kind of thing that sometimes has to be done when designing without being able to compute performance.

Figures 33 and 34 are dual-polarized circular-waveguide elements. The former is a design in which the dominant waveguide modes are excited by a coax-waveguide transition, then propagate to the aperture. The latter is a compact element in which the modes are cut-off – the spacing between the dipole and aperture acts as a coupling coefficient between two resonant circuits in the impedance matching design.

The last two figures, Figure 35 and 36 are chosen to make the point that an array feed network can get very complex. Figure 35 is an array which was designed to operate at three bands – L-, S-, and C-bands. The mass of cables is three independent feed networks, one for each band, feeding elements which are interlaced. Figure 36 shows that waveguide feeds can be just as complex, illustrating a waveguide feed for perhaps the most sophisticated array made to date – the HIPSAF test array built at Hughes. The rightmost ends of the waveguides form an array of 8 x 8 elements fed by a Butler matrix with time delay phase shifters on the various ports; the matrix, in turn, is fed by the corporate feed network at the left. The 8 x 8 array feeds a lens with phase shifters in it which steers the beam. This system illustrates that phased arrays can be designed with electronically overlapped sub-arrays. Instead of physical sub-arrays, this configuration generated electronically overlapping sub-arrays with time delay units which permit wideband operation at scan angles substantially off broadside.

Examples of other array systems are given in Chapter 1.

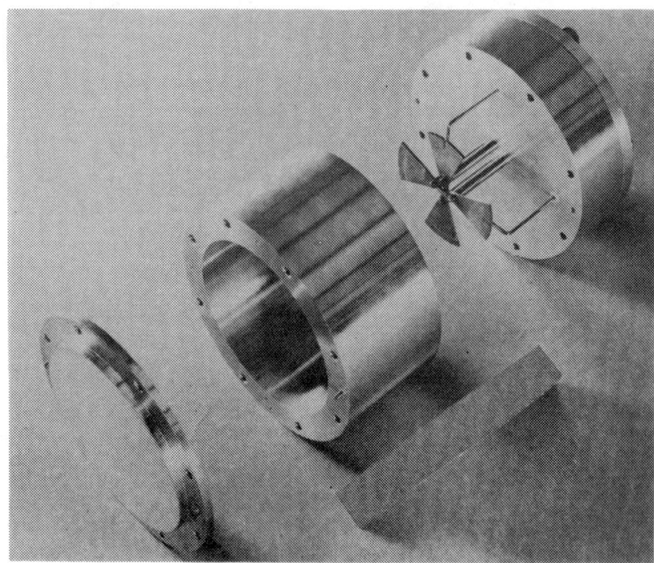

Figure 34 Compact Dual-Polarized Circular-Waveguide Element

Figure 35 Three-Band (L-, S-, and C-) Array Feed

Figure 36 Feed Network for HIPSAF – 8 x 8 Feed Array

References

[1] Steinberg, B.D.: *Principles of Aperture and Array System Design,* John Wiley and Sons, New York, 1976.

[2] Stark, L.: "Microwave Theory of Phased-Array Antennas – A Review," *Proceedings of the IEEE, Vol. 62, No. 12,* pp. 1661-1701, December 1974.

[3] Hansen, R.C. (ed.): *Significant Phased Array Papers,* Artech House, Dedham, Massachusetts, 1973.

[4] Oliner, A.A.; and Knittel, G.H. (eds.): *Phased-Array Antennas,* Artech House, Dedham, Massachusetts, 1972. (Proceedings of the 1970 Phased-Array Antenna Symposium; reference works published in 1970 and earlier.)

[5] Amitay, N.; Galindo, V.; and Wu, C.P.: *Theory and Analysis of Phased Array Antennas,* Wiley-Interscience, New York, 1972. (Reference works published in 1969 and earlier.)

[6] *Proceedings of the IEEE, Vol. 56, No. 11,* November 1968. (Special issue on Electronic Scanning; reference works published in 1968 and earlier.)

[7] Hansen, R.C. (ed.): *Microwave Scanning Antennas, Vol. II* (Array Theory and Practice), Academic Press, New York, 1966. (Reference works published in 1965 and earlier.)

[8] Tsandoulas, G.N.: "Wideband Limitations of Waveguide Arrays," *Microwave Journal,* pp. 52-56, September 1972.

[9] Hannan, P.W.: "The Element Gain Paradox for a Phased-Array Antenna," *IEEE Transactions on Antennas and Propagation, Vol. AP-12,* pp. 423-433, July 1964.

[10] Allen, J.L.; et al.: "Phased-Array Antenna Studies, 1 July 1959 to 1 July 1960," *Technical Report 228,* MIT Lincoln Laboratory, Lexington, Massachusetts, August 1960.

[11] Farrell, G.F. Jr.; and Kuhn, D.H.: "Mutual Coupling in Infinite Planar Arrays of Rectangular Waveguide Horns," *IEEE Transactions on Antennas and Propagation, Vol. AP-16,* pp. 405-414, July 1968.

[12] Knittel, G.H.; Hessel, A.; and Oliner, A.A.: "Element Pattern Nulls in Phased Arrays and their Relation to Guided Waves," *Proceedings of the IEEE, Vol. 56, No. 11,* pp. 1822-1836, November 1968.

[13] Hessel, A.; and Knittel, G.H.: "On the Prediction of Phased-Array Resonances by Extrapolation from Simulator Measurements," *IEEE Transactions on Antennas and Propagation, Vol. AP-18,* pp. 121-123, January 1970.

[14] Knittel, G.H.: "The Relation of Blindness in Phased Arrays to Higher-Mode Cut-off Conditions," *1971 G-AP International Symposium Digest,* pp. 69-72, September 1971.

[15] Hannan, P.W.; and Balfour, M.A.: "Simulation of a Phased Array Antenna in Waveguide," *IEEE Transactions on Antennas and Propagation, Vol. AP-13,* pp. 342-353, May 1965.

[16] Magill, E.G.; and Wheeler, H.A.: "Wide-Angle Impedance Matching of a Planar Array Antenna by a Dielectric Sheet," *IEEE Transactions on Antennas and Propagation, Vol. AP-14,* pp. 49-53, January 1966.

[17] Kelly, A.J.: "Comments on 'Wide-Angle Impedance Matching of a Planar Array Antenna by a Dielectric Sheet,'" *IEEE Transactions on Antennas and Propagation, Vol. AP-14,* pp. 636-637, September 1966. (Authors' reply follows this communication.)

[18] Brookner, E. (ed.): *Practical Phased-Array Systems, Microwave Journal* Intensive Course, Dedham, Massachusetts, 1975.

Chapter 22 Linear Beam Tubes

Staprans

The subject of linear-beam tubes is quite broad; therefore, this presentation is limited to high-power tubes. Oscillators such as the backward wave oscillator or the reflex klystron are not considered at all, nor are helix-type traveling wave tubes discussed to any great extent. The subject matter to be covered consists of high-power cavity-type tubes – klystrons, various coupled-cavity traveling wave tubes, and hybrid tubes.* Helix tubes are touched on in passing only.

A basic characteristic of all these tubes is that they are all coherent amplifiers with fairly high gain (30 to 60 dB). Bandwidths can be very large at low power levels; for example, a helix TWT may have an octave range. For klystrons, however, bandwidths are much narrower (up to a few percent); for high-power coupled-cavity traveling wave tubes, bandwidths up to about 30 percent can be achieved.

Linear-beam tubes are the highest average and peak power generators in the microwave tube family, covering virtually all power and duty cycle ranges. Efficiency of these devices is generally about 30%. All use thermionic cathodes; most of the tubes are liquid-cooled, although air or conduction cooling is used at the lower power levels.

These tubes are most generally used as output tubes in radars (both pulse and CW radars), CW illuminators, and high-power generators for ECM purposes. Often, these tubes are also used as driver tubes for output linear-beam tubes or CFAs. They are used in the largest land-based installations, in transportable and shipboard radars, and in airborne applications where light weight is important.

In this presentation the common elements of all linear-beam tubes are the first to be considered – guns, focusing systems, collectors, and RF windows – to be followed by the particular characteristics of klystrons, traveling wave tubes, and hybrid tubes. This discussion briefly covers the principles of operation, the main characteristics of each type of device, and some of the system interfaces.

Figure 1 shows a brief comparison (schematically) of the linear-beam tubes considered. It should be pointed out that the various types of linear-beam tubes are more like each other than they are different. They all possess an electron gun, a thin well-collimated pencil beam which passes through an interaction structure, a collector to dissipate the spent beam, an input and output coupling (each of which has a microwave window), and a focusing structure (not shown) surrounding the tube.

These tubes differ principally in the interaction structures. For a klystron, the interaction structure represented by the impedance presented to the beam, results in fairly high impedance regions in the cavity gaps separated by drift spaces. The cavities are tunable and generally the Qs are fairly high. Hence, even though high gain can be accomplished, the bandwidth is limited. In a traveling wave tube, the interaction structure (shown as a coupled-cavity type structure, though a helix would be similar) presents a continuous broadband impedance to the beam but of much lower magnitude than in a klystron. However, as a result of many more cavities

*For a more complete discussion of the subject matter of this presentation, see the article by Staprans, A.; McCune, E.; and Ruetz, J.: "High-Power Linear-Beam Tubes," *Proceedings of the IEEE, Vol. 61, No. 3*, pp. 299-330, March 1973.

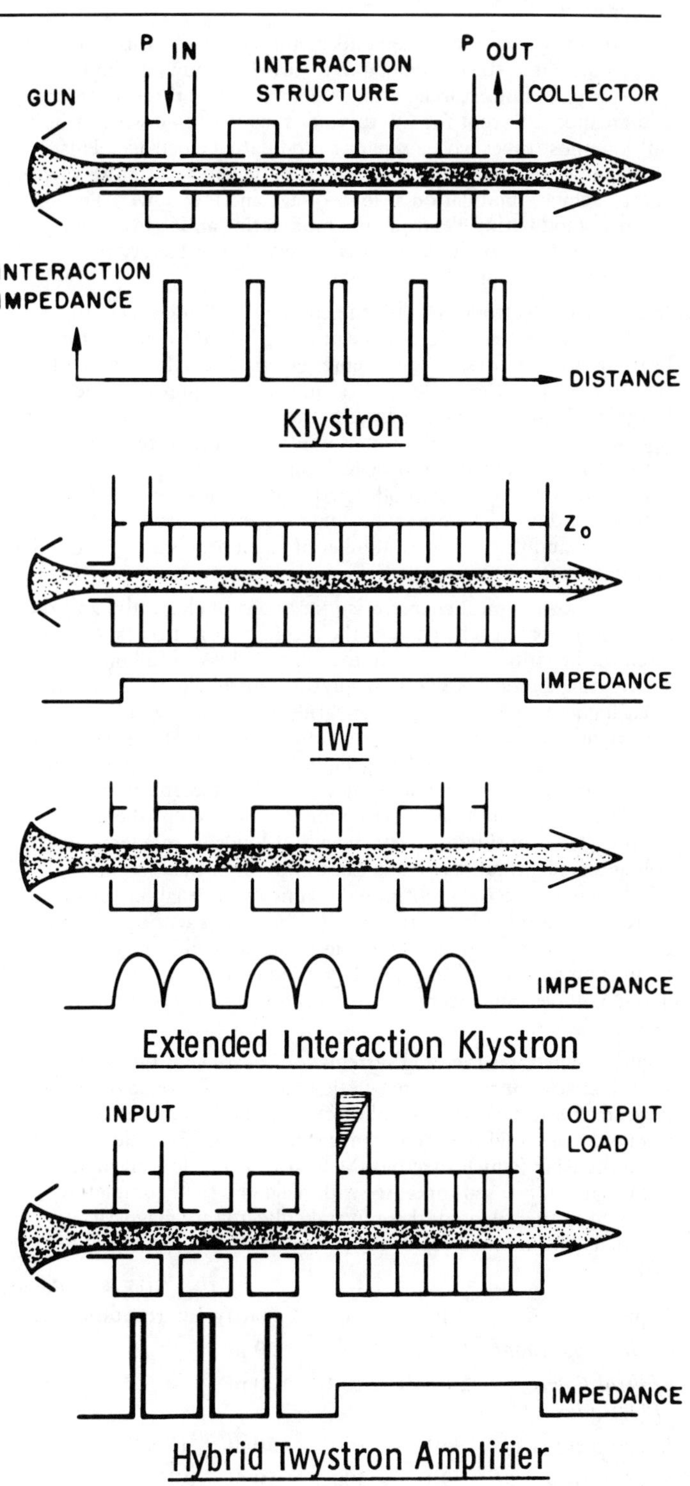

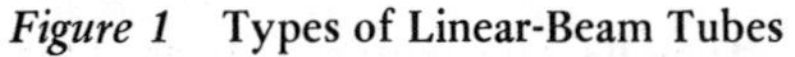

Figure 1 Types of Linear-Beam Tubes

than in a klystron, the same gain – but now together with a broader bandwidth – can be achieved. The extended interaction klystron basically consists of a number of short traveling wave sections which are resonated and coupled to the beam in klystron fashion. It has a somewhat broader bandwidth than a klystron, but not as wide as a TWT. The hybrid Twystron® amplifier is a device used mainly at very high power levels. It combines a klystron input or driver section consisting of several cavities with a traveling wave tube output section; it is the broadest band amplifier at megawatt power levels.

Considering the common elements in all linear-beam tubes, a most important one is the electron gun. Figure 2(a) shows a standard gun – it has a hot cathode, a focus electrode at cathode potential, and an anode. Except for CW tubes or for tubes used with cathode-pulsing, most tubes have a separate modulating electrode. This electrode is necessary to obtain varied pulse widths or pulse shapes and to achieve modulation at low voltage and low power. The simplest modulating electrode is a modulating anode insulated from the tube body; it can be used as a switch or to vary beam current independent of beam voltage.

In a control electrode gun (b), the control electrode is actually an insulated focus electrode capable of being biased to cut off the beam between pulses. Often a center electrode is added which is tied to the focus electrode. This arrangement requires a somewhat lower modulating voltage than does the modulating anode which operates at essentially full beam potential. A still better solution is the shadow-gridded gun (c) which only lately has come into existence. It consists of a cathode, two aligned grids (one of them at cathode potential, the other the control grid), an ordinary focus electrode, and the anode. A number of small beams are generated and then converged into the final single beam.

Table 1 shows typical characteristics of control electrodes and the relative merits of each. The μ is the total amplification factor related to the voltage swing from the negative bias to full on. The modulating anode has a very low μ; the control electrode, a somewhat higher value. In comparison, grids have very high μ's and thus require much lower voltage modulators; the intercepting grid has the highest μ. The shadow grid, however, has the highest cutoff factor, μ_c. Grid or modulating electrode interception is quite low (it is essentially non-existent) for all cases except the intercepting grid. Considering performance at low control electrode voltages (such as pulse rise and fall times), only the modulating anode is seen to have an infinite dynamic range, enabling beam current variation over the full tube voltage range without beam defocusing. For intermediate voltages, the control electron defocuses most severely. In general, gridded guns work best at the design voltage and current; beam transmission suffers at intermediate levels.

Figure 3 illustrates focusing systems for linear-beam tubes; shown are schematics of solenoid magnetic focusing arrangements. Each diagram shows the gun, the collector, tube body, and the focusing solenoid. The solid lines represent magnetic flux. So-called Brillouin focusing (which specifies the lowest theoretical magnetic field that can be used) operates with magnetic field completely enclosed in the solenoid; the gun and collector are magnetically shielded. This method yields the poorest electron optics and high-

Type	μ	μ_c	Capacity	Grid Interception	Focusing at Low Voltage
Modulating Anode	1 to 3		50 pF	0	Good
Control Focus Electrode	2 to 10	2 to 10	100 pF	0	Poor
Intercepting Grid	50	100	50 pF	15%	Fair
Shadow Grid	30	300	50 pF	0.1%	Fair

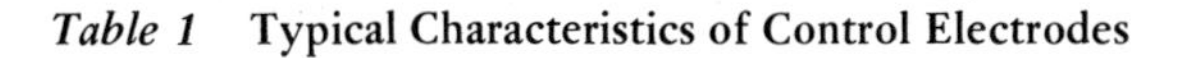

Table 1 **Typical Characteristics of Control Electrodes**

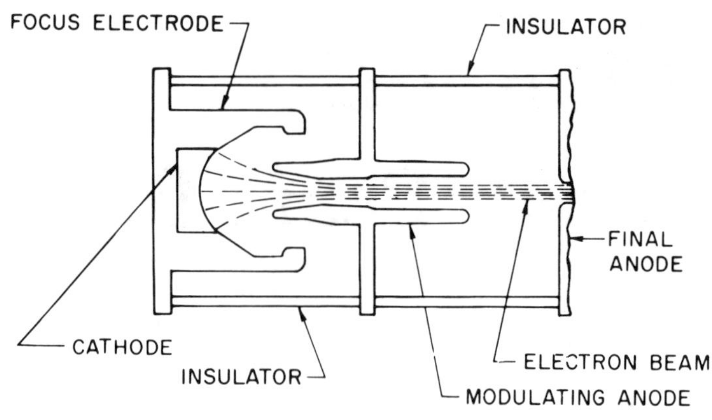

(a) Modulating Anode Gun

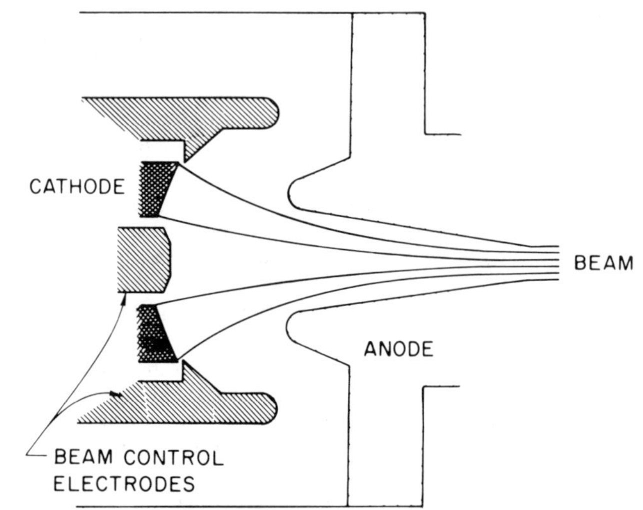

(b) Control Focus Electrode Gun

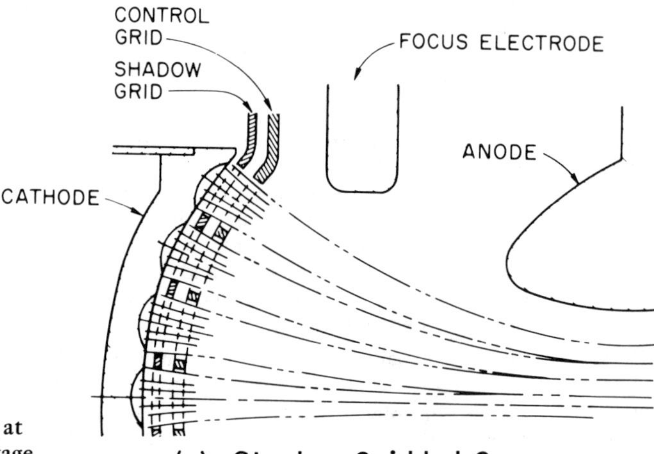

(c) Shadow Gridded Gun

Figure 2 **Electron Gun Control Electrodes**

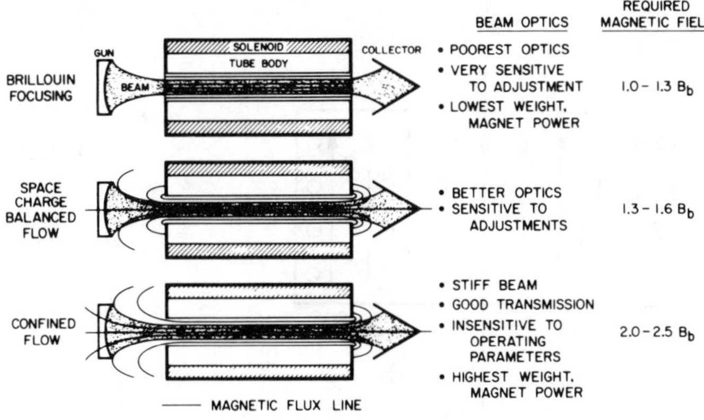

Figure 3 **Comparison of Space-Charge Balanced Flow Methods Used in Uniform Magnetic-Field Focusing of Electron Beams**

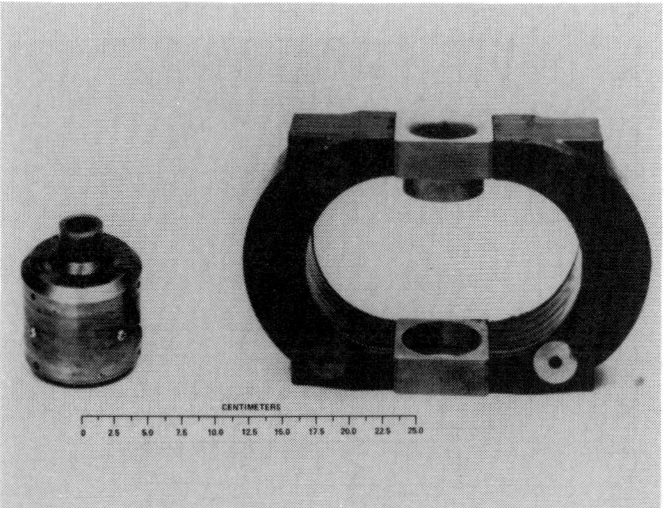

Figure 4 **Comparison of 2 kG Samarium Cobalt (left) and 15 kG Alnico Focusing Magnetic Circuits for a 1.5 kW X-Band Klystron (VKX-7780)**

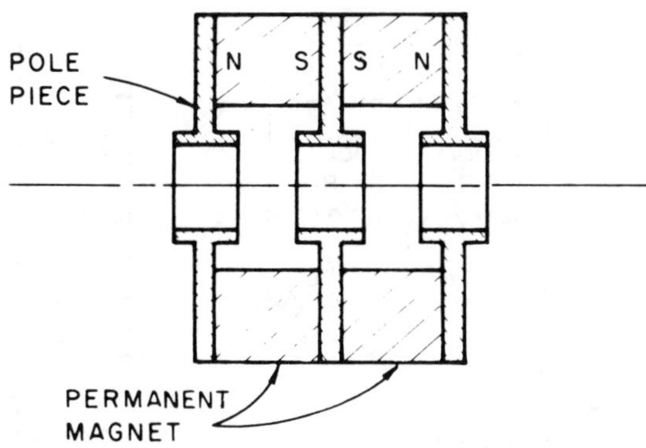

Figure 5 **PPM-Focusing Structure**

est sensitivity to adjustment but it employs the lowest weight device which, in turn, consumes the lowest magnet power. On the other end of the range is confined flow where the magnetic field extends out in the gun region in order for flux to conform to the beam shape. In this case, a higher value of magnetic field must be used – about 2 to 2½ times that of the Brillouin field, B_b. The results are a very stiff beam, good transmission, suitability for high-power operation, relative insensitivity to all operating parameters, and applicability for low-noise operation. Confined flow, however, requires the highest weight and magnet power of all focusing systems. Although tubes are built over the entire range of magnetic fields, most of the high-power tubes in existence use confined flow.

Klystrons, as well as traveling-wave tubes, can be focused by permanent magnets. Figure 4 shows an ordinary double C-type magnet used in an X-band klystron – it weighs 25-30 pounds. On its left is a recent innovation – a samarium cobalt focusing magnet designed for the same tube. The difference in weight is great – the samarium cobalt structure is lighter by about a 7 or 8 to 1 ratio. In the last few years, samarium cobalt has been used extensively both in PPM-focused traveling-wave tubes and in klystrons. Its application is still somewhat restricted because of cost and the limited availability of large pieces, but increasing future use is inevitable.

Many tubes are *Periodic Permanent Magnet* (PPM) focused. Figure 5 shows a short section of such a tube (which could be either a klystron or a TWT). The figure illustrates the cavities, the surrounding longitudinal magnets, and the cavity end-plates (which also serve as polepieces). Many high-power tubes, particularly traveling-wave tubes, are built in this manner.

Figure 6 shows a typical high-power, liquid-cooled collector; in this case, it is used in an L-band tube at a 15-20 kW average power level. It could be a traveling-wave tube or a klystron – the basic construction would not differ. Note the many small water channels. Even though the collector is a rugged, massive device, it's one of the likely failure areas of a linear-beam tube. The channels may clog if the coolant contains particules or impurities, or if it has a high oxygen content. Under such conditions, a boundary layer on the channel surfaces is eventually built up, heat transfer is inhibited, and this most massive part of the tube will, in fact, overheat and melt.

Microwave RF windows for tubes are often of the "pillbox" waveguide type shown in Figure 7, particularly on high-power tubes. The window consists of a disk ceramic in a round waveguide section which is then inserted between pieces of ordinary rectangular waveguide. It is a very broadband window, but it is somewhat limited in average power capability. For the very high power levels used in linear-beam tubes, ceramic block windows are more suitable. Such windows (Figure 8) consist of resonant dielectric blocks inserted in rectangular waveguides. They are extremely rugged and are capable of hundreds of kilowatts of average power (at S-band and X-band frequencies); however, they are narrower band, with a useful bandwidth of around 10%.

The most appropriate way to begin investigating klystrons is with a brief discussion of how they work. Figure 9 shows a klystron cavity, illustrating the electric and magnetic fields, the interaction gap, and the beam. The cavity is derived from the simple pillbox arrangement, and can be represented by a tuned circuit which is loaded by the cavity losses, by the beam loading, and by an externally-coupled load.

In a multi-cavity klystron (Figure 10), there is a succession of cavities from the input to the output cavity. Also shown are the source generator and its impedance, and the useful output load. Coupling between the cavities exists only through the beam. This schematic klystron can be analyzed easily on a computer.

Klystron power output can be varied significantly over a very large range by simply adjusting beam voltage. In Figure 11 are the results of a C-band klystron operated from 5 to 15 kV, yielding from 1 kW to 10 kW of power output. The power obeys the five-halves voltage law. Over some reasonable dynamic range – such as about 5 to 1 in power – the efficiency is not affected very drastically.

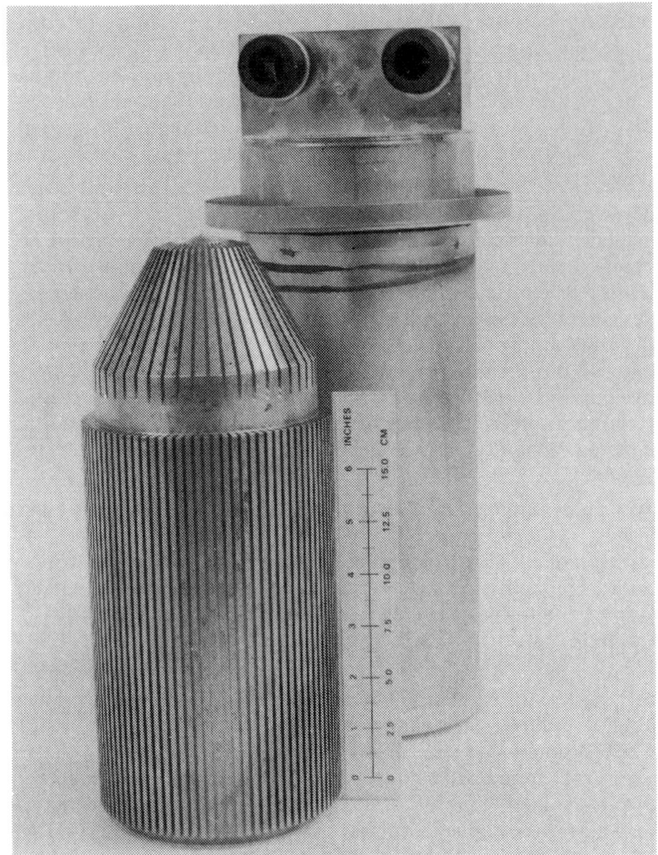

Figure 6 Water-Cooled Collector

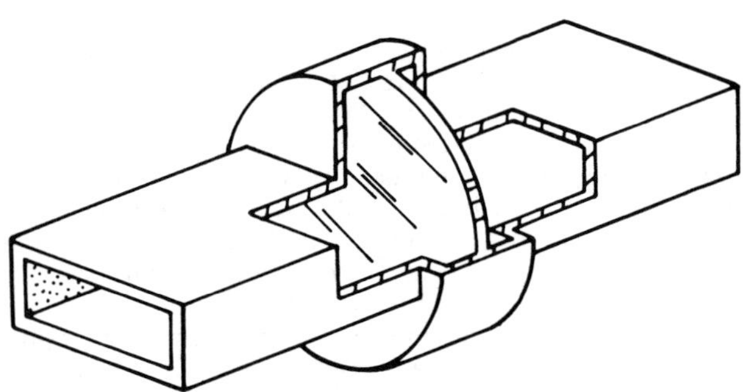

Figure 7 Thin Disk "Pillbox" Window

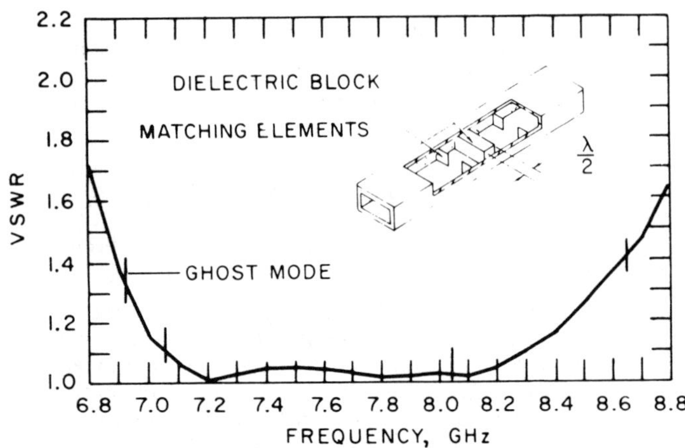

Figure 8 Half-Wave Resonant Block Window

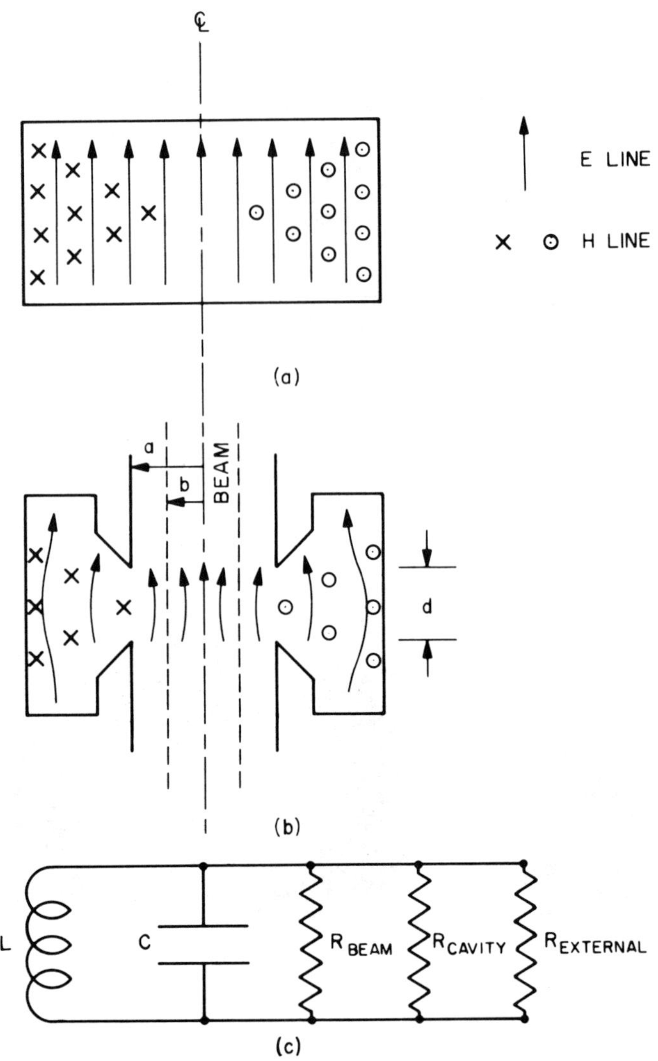

Figure 9 Development of Klystron Cavity

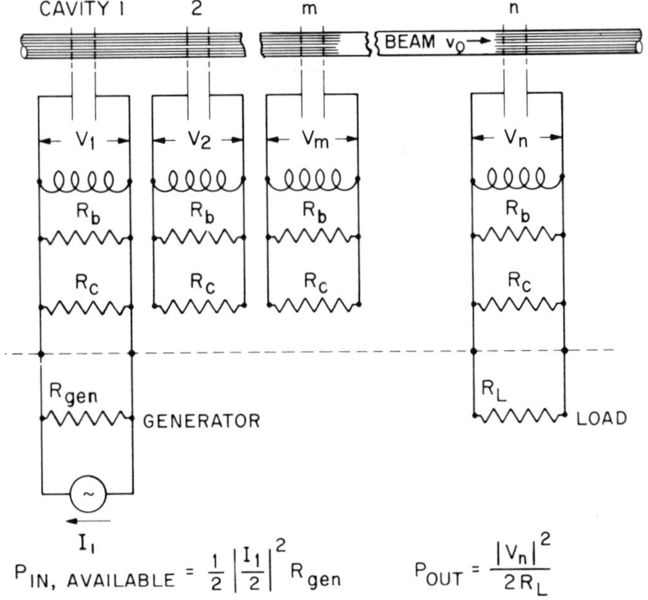

Figure 10 Schematic Diagram of a Multicavity Klystron

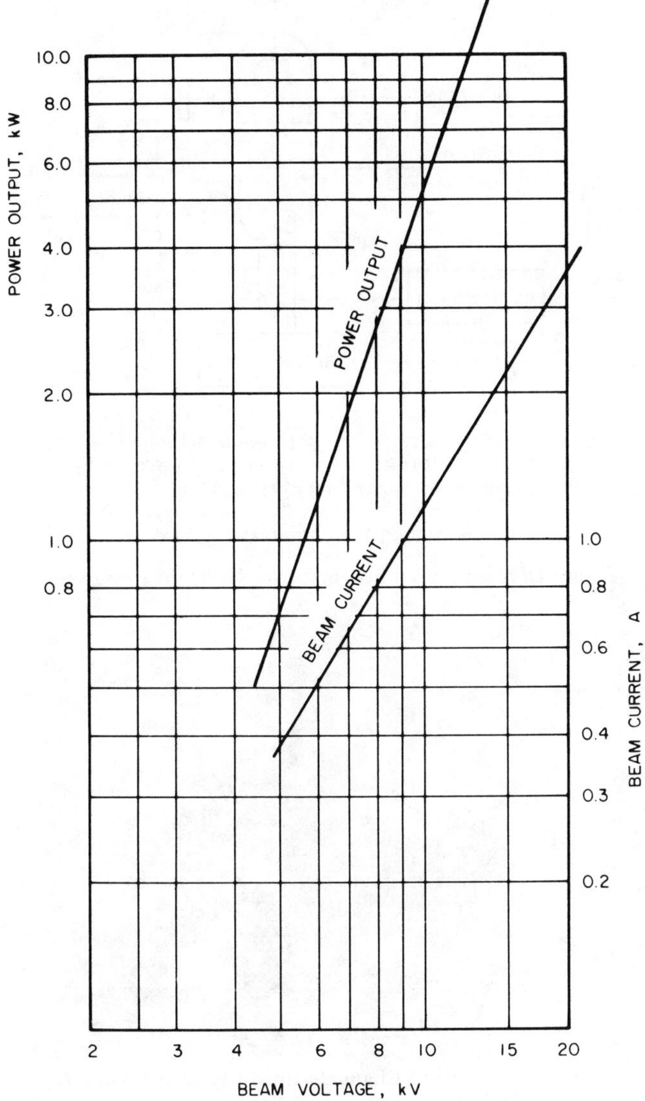

Figure 11 **Beam Current and Saturated Power Output vs. Beam Voltage for the VA-884C and D Klystrons at C-Band**

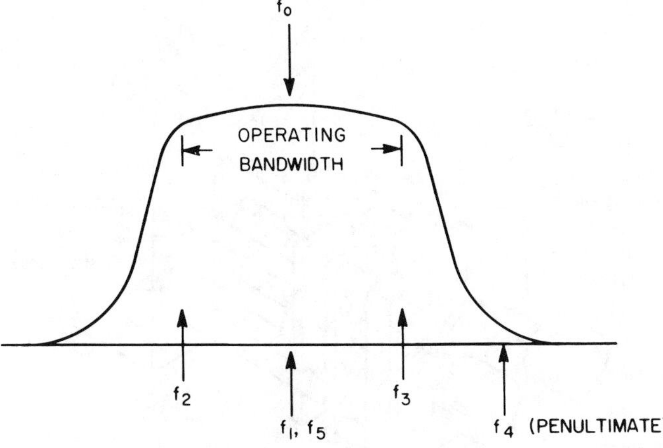

Figure 12 **Typical Cavity Resonant Frequencies for Broadband Tuned Five-Cavity Klystron Amplifier**

The klystron is generally a narrow-band device. Its bandwidth can be significantly improved by so-called staggered tuning – spreading the frequencies of the individual cavities over a frequency range. Figure 12 is a typical broadband klystron, showing the various cavity resonant frequencies. Note that the penultimate cavity is almost always tuned high to achieve good efficiency in the klystron; individually, it does not contribute much to the bandwidth. Depending upon how a klystron is tuned, its operation can differ in gain and efficiency, as seen in Figure 13. If a klystron is tuned for high efficiency, the result is a curve with lower gain but higher power output – a typical narrow-band case. If the cavities are spread out, say, to quadruple the bandwidth of a singly-tuned klystron, the gain may drop 10-20 dB, power and efficiency will drop somewhat. The curves in Figure 13 are from the same klystron with different cavity tunings.

When properly tuned for broadband operation, klystrons tend to preserve the shape of the bandwidth curve reasonably well as drive power is varied. Figure 14 shows the operation of a 10-12 kW klystron fully saturated with an 80 mW drive. Reducing drive to 5 mW, the bandwidth shrinks somewhat because the benefit of the flattening from saturation is lost; overall, though, a fairly flat top is achieved if the tuning has been done correctly.

The relationship showing what factors most significantly affect the bandwidth of a klystron is given by

$$(\text{gain})1/2 \; \frac{\Delta f}{f} \propto \frac{R/Q}{R_0} = \frac{R}{Q} \; P_0^{1/5} K^{4/5}$$

Since the $(\text{gain})^{1/2}$-bandwidth product is proportional to the cavity impedance R/Q, if you double that impedance, bandwidth can be doubled. Bandwidth is relatively insensitive to beam power (P_0), varying only according to 1/5 power. Thus, just increasing and decreasing beam voltage and power level in a given klystron does not change the bandwidth very rapidly. K is the perveance. Bandwidth varies almost linearly with perveance of the beam; thus, one approach to achieving high bandwidth in the klystron is to go to very high perveance. Normally, perveance (current divided by voltage to 3/2 power) tends to be about 10^{-6}; for very broad bandwidth klystrons, it could be three or four times as much.

In Figure 15 is a typical klystron amplifier noise spectrum, taken from actual measured data. Similar results can be obtained for a traveling-wave tube or any other linear-beam amplifier designed for low noise. In the figure, AM and PM noise below carrier in a 1 kHz bandwidth versus frequency from carrier is shown. We see that the PM noise eventually flattens out at about 125 dB below carrier; AM noise, some 10 dB lower. The reason that the AM noise is lower is the compression of the AM noise by saturation effects. Phase response of a klystron or any linear-beam tube does not really saturate; therefore, the difference. The curves shown are close to the best results that can be measured in tubes with well-focused beams with confined flow. For less perfect conditions, certainly the noise will be higher. If PPM-focusing is used, which in itself always generates a rippled beam, the noise generally will be higher. The noise values here approach the theoretical limits of shot noise at the cathode as amplified by the gain of the tube.

Modulation sensitivities of klystrons are also worth mentioning; they are very simple relationships.

AM Modulation $\quad \dfrac{dP}{dV_0} = \dfrac{5}{2}\,\dfrac{P}{V_0}$

PM Modulation $\quad \dfrac{d\theta}{dV_0} = \dfrac{\theta_0}{2V_0}$

The change of power with respect to beam voltage is simply the direct ratio of power to beam voltage multiplied by 5/2; for phase

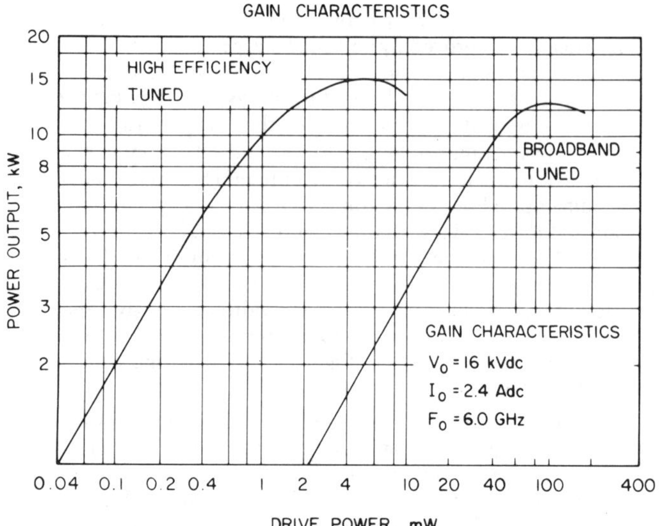

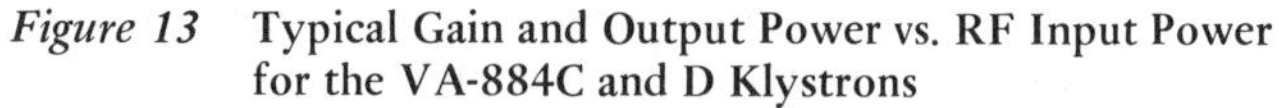

Figure 13 Typical Gain and Output Power vs. RF Input Power for the VA-884C and D Klystrons

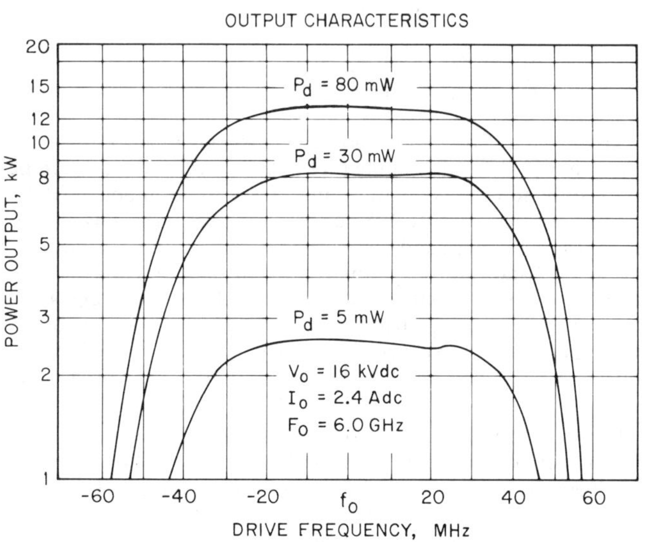

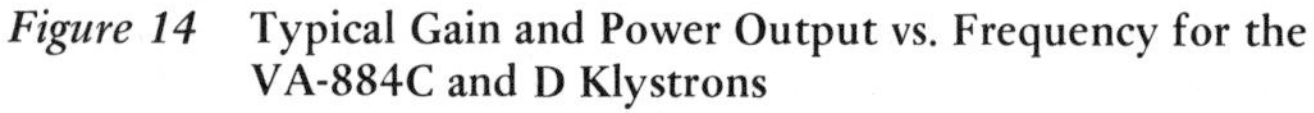

Figure 14 Typical Gain and Power Output vs. Frequency for the VA-884C and D Klystrons

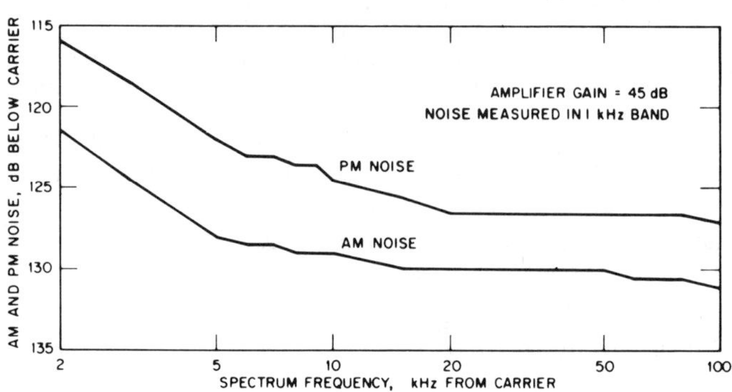

Figure 15 Typical Klystron Amplifier Noise Power

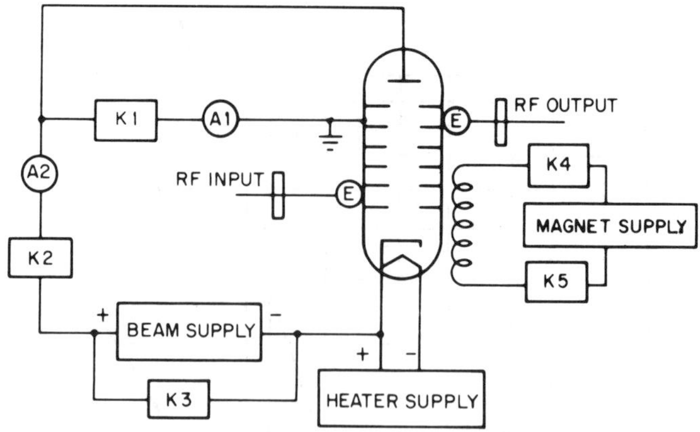

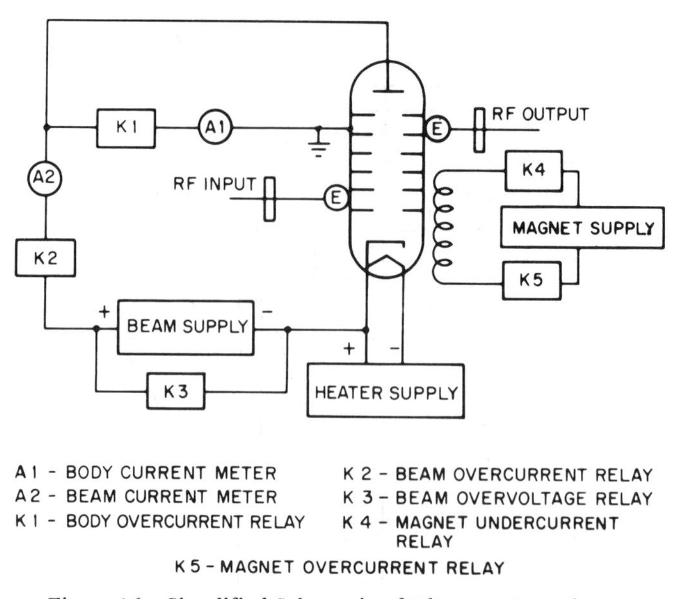

Figure 16 Simplified Schematic of Klystron Operation

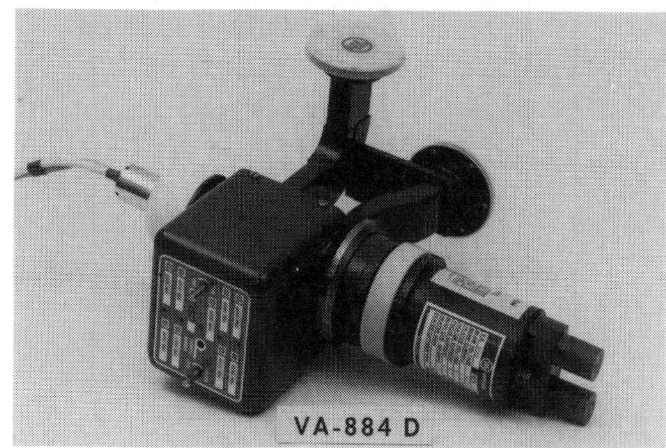

Figure 17 C-Band Channel-Tuned Klystron VA-884D

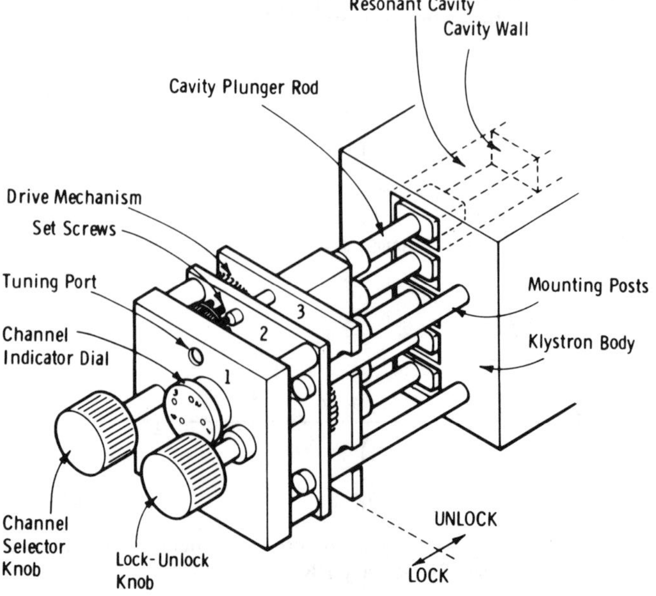

Figure 18 Klystron Channel Tuner Mechanism

modulation, the figure is the total phase length of the tube divided by twice the voltage. The total phase of the multi-cavity klystron may be about 20-30 radians. These sensitivities are both substantially lower than for a traveling-wave tube, which is much more vulnerable to changes in operating voltage since synchronism with the beam is required (unlike in a klystron).

An ordinary system hook-up for a klystron (including protective circuitry), is shown in Figure 16. The klystron tube, its heater supply, a beam supply with a body current measuring meter, a relay or sensor for over-current, beam current meter, and over-current and over-voltage protections are illustrated. The focusing solenoid has its supply and also must have protective circuitry; if magnetic field is lost, most high-power tubes melt in a short time unless beam voltage is removed.

As an example of a typical klystron, Figure 17 is a klystron amplifier at about the 10 kW level at C-band. The body of the tube has an attached outboard box – one of the latest developments in the art of klystrons, the channel tuner mechanism. With this mechanism, the klystron can be cycled through a wide tuning range and reset each time for broadband performance at a different frequency for each tuner position, simply by turning one or two knobs or switches. What goes on inside such a box is shown in Figure 18. There are a number of gears, stops, and plungers; the klystron cavities are tuned by movable plungers in the cavity walls. This sort of tuning can be done manually through the knobs or by remote control. A number of systems simply have push buttons at the console and a servo-motor(s) at the tube; by depressing the button corresponding to a particular channel, its tuning is reproduced within a matter of seconds. Figure 19 shows the data of a channel-tuned klystron – the ten channels cover an entire frequency range with good resettability for as many tuner cycles as desired.

Klystrons are the most efficient of all linear-beam tubes, and recent work attempted to improve that efficiency. While a klystron is normally about 35-40% efficient, some tubes built for specific applications (such as the one in Figure 20) have a much higher efficiency. In the figure is a 100 kW klystron at 805 MHz, which is 70% efficient without a depressed collector. But to achieve this sort of efficiency (at the present time), the tube is limited in other respects; for example, it is narrow-band and usually is not tunable. Some of the characteristics of this high-efficiency tube, tested to about 80 kW, are shown in Figure 21. Efficiency is very close to 70%. Interestingly, as beam voltage is varied over a fairly wide range and output power changes from 25 to 70 kW, efficiency tends to be fairly constant. This fact again illustrates that klystrons have quite a large dynamic range.

Another area where advances in the klystron art are being made is periodic permanent magnet focusing. In Figure 22, there is a comparison of two X-band klystrons which have the same peak power – 100 kW pulse tubes, suitable for 200-400 W average power. The tube on the left uses the old style C-magnet which weighs 35 pounds. The tube on the right uses PPM-focusing, with small magnets alongside tube body, alternating in orientation. This second tube weighs only 9 pounds. In addition, it has a broader bandwidth – a 2% bandwidth versus a 1% figure for the bigger tube. Since bandwidth is achieved by making the tube body longer and adding more cavities, the tube on the left, if made equally long as the one on the right, would have a prohibitively heavy magnet.

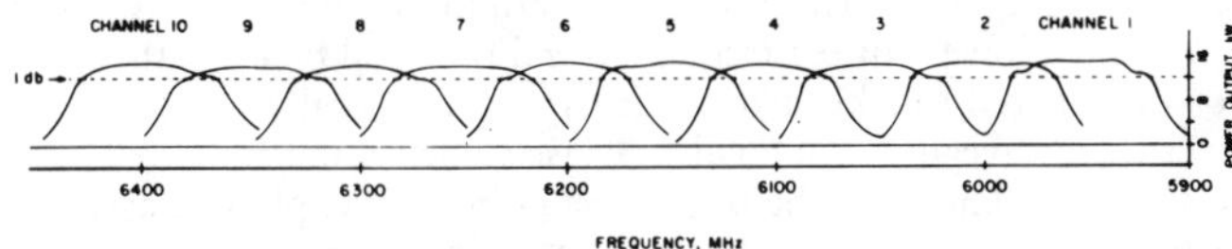

Figure 19 **Passband Characteristics of Channel-Tuned Klystron VA-884D**

Figure 20 High-Efficiency 100 kW Klystron at 805 MHz, Varian X-3074B

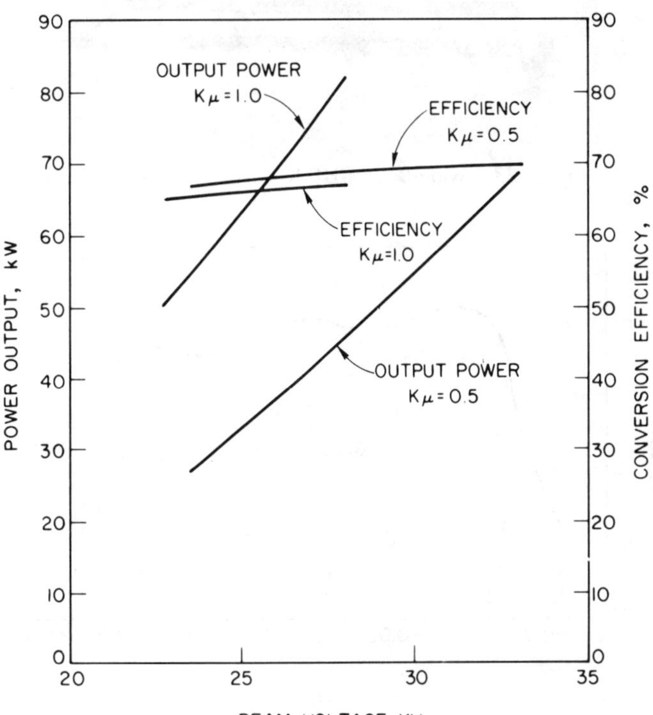

Figure 21 **Performance of X-3074B High-Efficiency Klystron vs. Beam Voltage (Note: $K\mu$ = microperveance)**

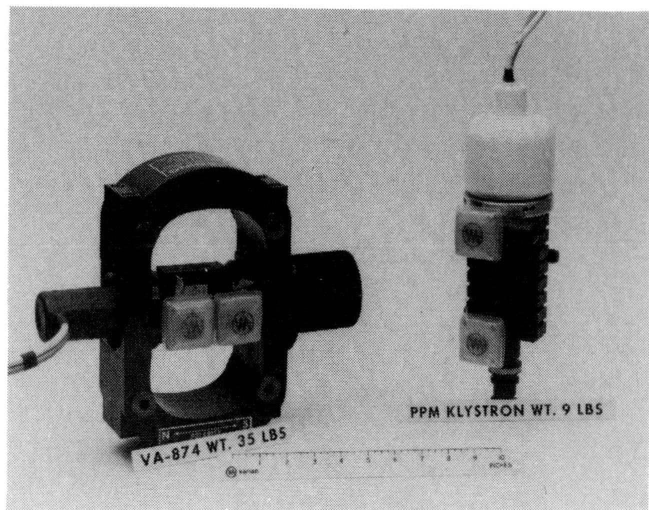

Figure 22 Comparison of VA-874 with VKX-7752

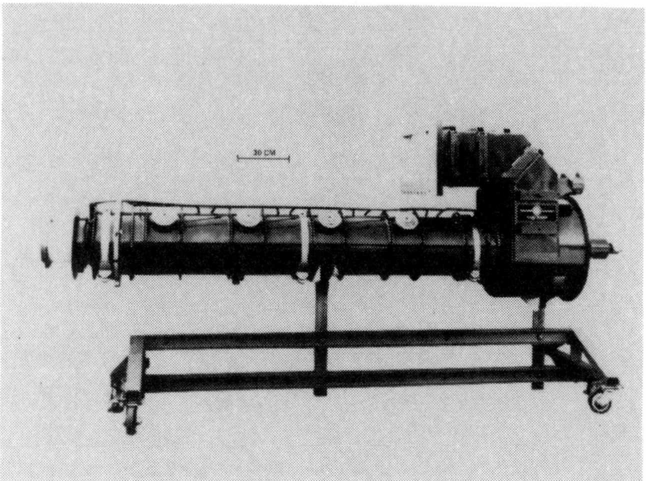

Figure 23 Wide-Band UHF Klystron, VA-812C

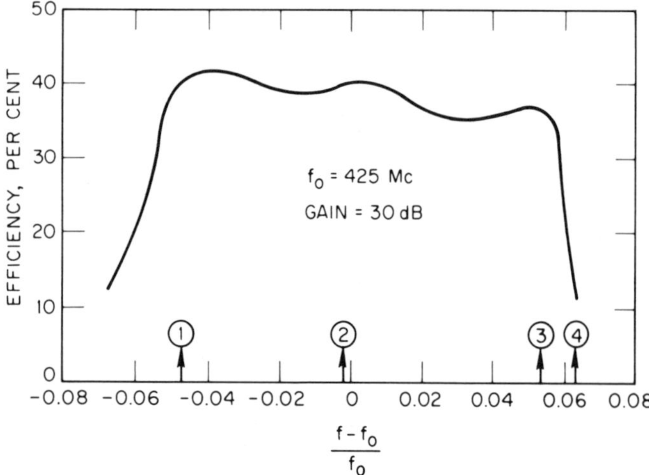

Figure 24 Frequency Response of Broadband Klystron Amplifier, VA-812C

The PPM klystron is a relatively new development. Tubes are now built for only a few applications, but the PPM klystron is likely to become a much more significant device in the future. For the same peak and average powers, the PPM klystron is somewhat more compact than a coupled-cavity TWT and therefore lighter. Of course, though, the klystron does not have the bandwidth of the TWT.

At the other end of the power spectrum are larger tubes of the sort shown in Figure 23. Illustrated is a UHF klystron about 10′ or 11′ long. Tubes of this kind are suitable for tens of megawatts peak power and hundreds of kilowatts average. If a klystron operates at very high peak power (10 megawatts or so), it is capable of broad bandwidth (Figure 24). The bandwidth of this particular tube is about 10%, and it still maintains reasonably good efficiency and gain. This level approaches the limiting bandwidth that can be achieved in a klystron.

Figure 25 gives an example of a yet different klystron. Although most very high-power klystrons tend to be solenoid-focused, they can be permanent magnet focused at almost any power level. The example here is a 30 MW peak power, 30 kW average power klystron used in the Stanford Linear Accelerator; it is an S-band tube. It operates in a barrel-style magnet which is all permanent magnet; hence, no power supplies are required. But this particular magnet weighs 800 pounds – the tube itself weighs only 50-100 pounds – and so may not be quite practical for airborne applications. For more moderate power levels (around 10 kW), however, a permanent magnet focused klystron is smaller and lighter than one focused by an electromagnet.

A number of the comments already made regarding klystrons, such as those on noise behavior, apply to TWTs; of course, these tubes incorporate all of the common elements – the guns, windows, etc. Additionally, klystrons and coupled-cavity TWTs are quite similar in certain other respects. In Figure 26 are two X-band tubes which could be used for communications or as CW illuminators for radar. The tubes operate at the 10 kW level. On top is a klystron without its focusing magnet; on the bottom is a TWT without the focusing solenoid. The two devices can be used in nearly identical applications. The body of the TWT is noticeably longer – it has more cavities and is therefore capable of a greater bandwidth. The collector for approximately the same beam power is much larger for the TWT than for the klystron since it must be depressed in the TWT to achieve a klystron-like efficiency.

Comparing the operating parameters of these two tubes in Table 2, there are some notable differences among the similarities. Both tubes have the same gain, but the klystron has a 0.6% bandwidth, compared with a 5% figure (which could be higher were the device so designed) for the TWT. However, the klystron can make up at least partially for its lack of bandwidth by its channel tuner, if the tuning time is tolerable from the system standpoint. Beam voltage and current of a klystron are somewhat lower because the interaction is more efficient. The TWT, though, has a depressed collector (40% depression is close to the maximum obtainable); therefore, approximately equal efficiencies can be achieved in both the klystron and the TWT. The weights of the two packages (tube and magnet) are quite similar. The klystron, however, is shorter, and its cost is usually one-half to two-thirds that of a similar TWT.

The design of TWTs always involves *Omega-Beta* or *Phase* diagrams (such as those in Figure 27) which illustrate that the phase shift per circuit section in a TWT varies in a certain fashion when plotted versus frequency. The TWT has a definite passband; it has a higher and a lower cut-off frequency and the total bandwidth is broad compared to a klystron cavity, though generally much narrower than waveguide bandwidth. Phase velocity, indicated by the slope of a line from the origin to a point on the phase diagram, has to be in synchronism with the beam velocity (as shown); otherwise useful cumulative interaction does not occur. That require-

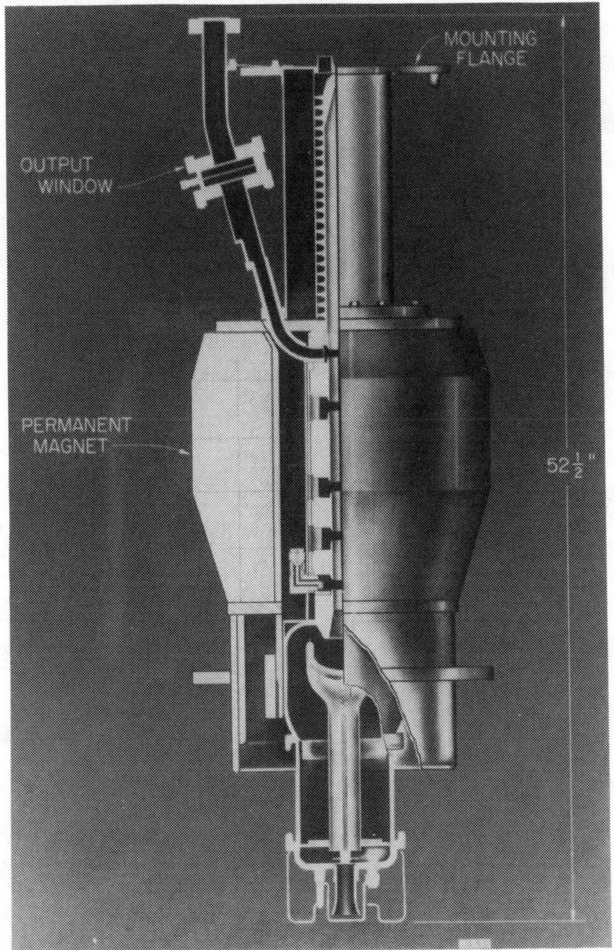

Figure 25 Cutaway of 34 MW S-Band Permanent Magnet Focused Klystron

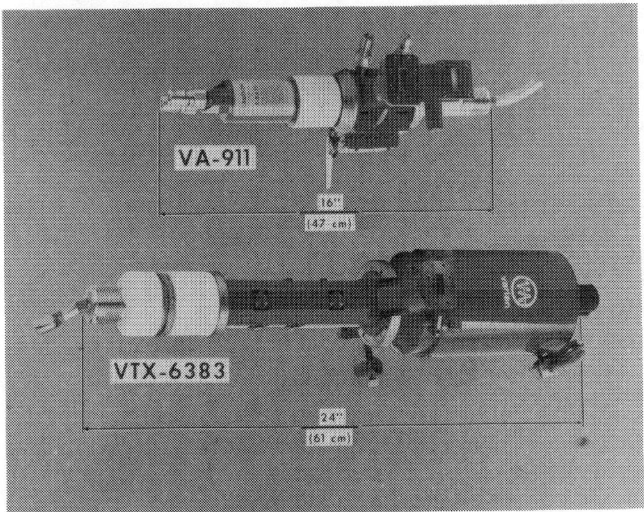

Figure 26 Comparison of VA-911 with VTX-6383

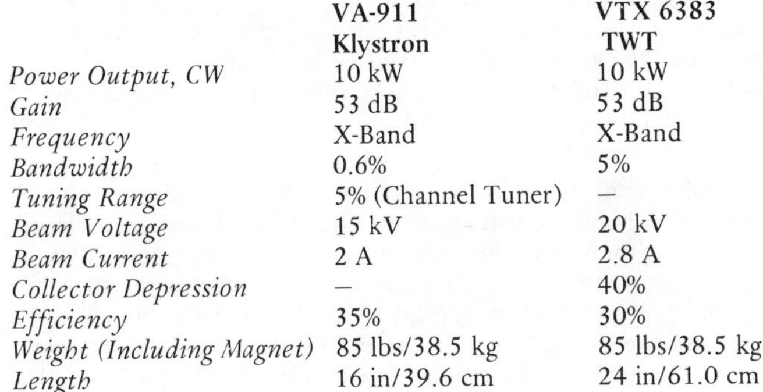

	VA-911 Klystron	VTX 6383 TWT
Power Output, CW	10 kW	10 kW
Gain	53 dB	53 dB
Frequency	X-Band	X-Band
Bandwidth	0.6%	5%
Tuning Range	5% (Channel Tuner)	–
Beam Voltage	15 kV	20 kV
Beam Current	2 A	2.8 A
Collector Depression	–	40%
Efficiency	35%	30%
Weight (Including Magnet)	85 lbs/38.5 kg	85 lbs/38.5 kg
Length	16 in/39.6 cm	24 in/61.0 cm

Table 2 Comparison of Characteristics

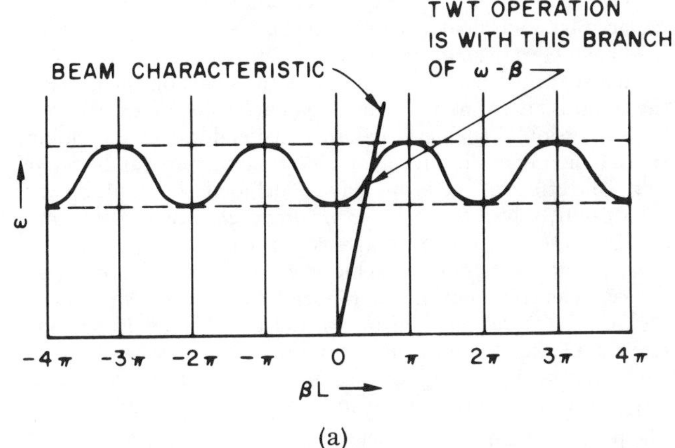

(a)

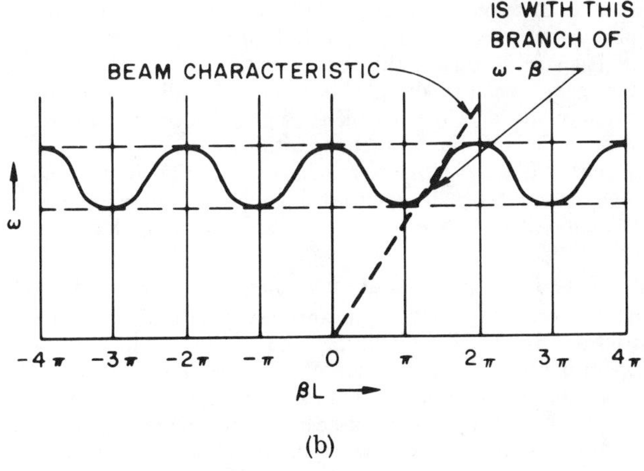

(b)

Figure 27 TWT Phase Diagrams

ment is the major distinction between a TWT and a klystron (for which any voltage works, within limits).

There are two types of TWTs – the so-called fundamental foward-wave tube (shown in (a) in Figure 27) and the space harmonic variety (b). In the forward-wave tube, the fundamental space harmonic (the first branch of the phase diagram) travels in the positive direction – the same direction as that energy flow on the circuit. TWTs of this sort are helix tubes at lower powers, but cloverleaf and centipede (forms of coupled-cavity tubes) at power levels in the megawatt range. Coupled-cavity TWTs use the so-called backward-wave or space harmonic structure, in which the beam interacts with the first space harmonic rather than the fundamental branch. Such a tube has certain characteristics – it is not as broad band as a helix tube, but is more so than is the megawatt cloverleaf-type structure.

The actual hardware – the cavities for the high-power forward-wave coupled-cavity tubes – is shown in Figure 28. From the right are the cloverleaf circuit (single-cavity), showing the shape of the cavity and coupling slots; the centipede, which includes many small coupling loops; and a blank from which the centipede circuit is stamped.

Space harmonic coupled-cavity tubes, which normally are used from a few hundred watts to several hundred kilowatts, are used more commonly than the high-power forward-wave coupled-cavity tubes. Figure 29 shows the circuits for frequencies ranging from Ku to low S-band; all are typical cavity shapes. To obtain an entire slow-wave circuit, it is necessary to stack many of these cavities end-to-end and braze them together; some reasonably simple means exist by which to treat analytically these traveling-wave coupled-cavity structures. In Figure 30 is a schematic single-cavity. If the cavity walls (where most of the current flows) and the interaction gap (where an electric field builds up) and the slots are replaced by equivalent circuit elements, the result is the top left (and center) equivalent circuit. If the center circuit were to be stretched at both pairs of terminals to the shape on the bottom, the result is an ordinary transmission line analog of a single-cavity. If a succession of such cavities are joined end-to-end, a periodic transmission line is produced and TWT parameters can be related to it. The effect of the beam may be included; the resulting equations can be solved through use of computers. Much more computing time is required than for klystrons because of the added complexity, but, again, one can achieve sensible simulation of these devices. The computation for a C-band 8 kW TWT is shown in Figure 31. The calculated gain over the actual bandwidth of interest could be somewhat higher or lower than measured, as the computation leaves something to be desired. But generally, the response shape can be reproduced and predicted fairly well.

The phase and amplitude behavior of TWTs is also worth investigating. In Figure 32 gain and phase variations vs. frequency are shown on an expanded scale over about a 5% frequency range. The amplitude variation is a portion of the 50 or 60 dB total gain; the phase variation is the few degree deviation from linear. It is important to note that, in both a TWT and a klystron, the unsaturated gain variation and the phase variation are strictly related. A linear-beam tube at small signal behaves identically to a minimum-phase network for which a peak of amplitude occurs at the frequency of maximum rate of change of phase, and vice versa. That relationship is a useful one to remember, and it facilitates equalization of these tubes. If amplitude is equalized, phase is equalized at the same time, since equalizers also are minimum phase networks.

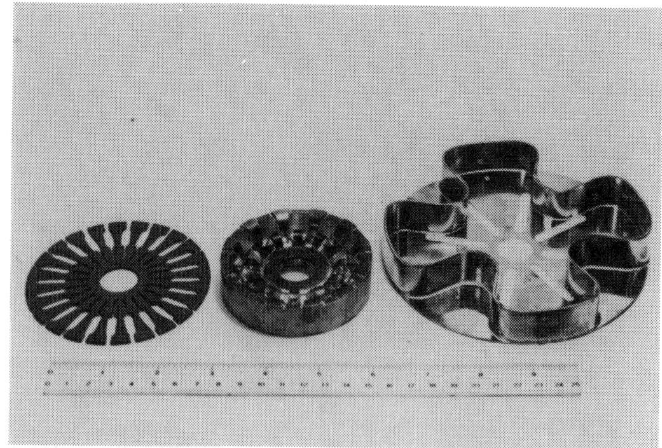

Figure 28 **Centipede and Cloverleaf Coupled-Cavity Circuits. These Circuits Have Negative Mutual Inductive Coupling Between Cavities and Operation is with the Fundamental Space Harmonic.**

Figure 29 **Space Harmonic Coupled-Cavity Circuits. These Circuits Have Positive Mutual Inductive Coupling Between Cavities and Operation is with the First Space Harmonic.**

Table 3 shows a list of typical TWT modulation sensitivities. These parameters are extremely important to systems builders, since they impact the size, weight, quality, and cost of the equipment used in operating the tube. Inspection of the table indicates that a TWT is significantly more sensitive than a klystron, because it relies on a relatively critical adjustment of beam voltage to achieve synchronism of the beam and the wave propagating on the circuit.

TWTs may be operated with several power supply configurations, as shown in Figure 33. The one in (a) is the ordinary grounded collector type – the same arrangement that would be used for a klystron. It includes a beam power supply, a modulator for a grid or mod anode, and a grid bias. However, in order to improve efficiency, TWTs often require depressed collectors so that the beam can be collected at a lower energy (voltage) than the level needed for interaction with the circuit. There are several choices for power supply hook-up. The one in (b) is the series configuration – the collector supply and body supply are in series. In (c) is the parallel hook-up, in which the body is on a supply which is completely separate from that of the collector. Both supply configurations are common. The advantages of the depressed collector are multiple – energy savings; the fact that the collector supply does not need to be very well regulated (collector modulation sensitivity is not high while the body sensitivity is high); and, since it must only provide a small portion of the total beam current, the well-regulated body supply can be fairly small.

The next two figures show some typical TWTs. In Figure 34 is a PPM-focused tube at X-band and about 50 kW peak power output and 150 watts average. It is heat-sinked by conduction cooling to the base plate. Adding air cooling can increase average power capability to a few hundred watts; adding liquid cooling, to a few kilowatts. A much higher power TWT is shown in Figure 35 – a fairly complex device. It operates at about 10 kW peak power at high duty cycle for an airborne application. It has a compact focusing solenoid which, although the entire tube is available as a single package, is removable from the tube body. The tube also incorporates a gridded gun and depressed collector, and it operates with airplane hydraulic fluid, such as Coolanol, as the liquid coolant. Low weight is very important on such a device; this TWT weighs about 50 pounds.

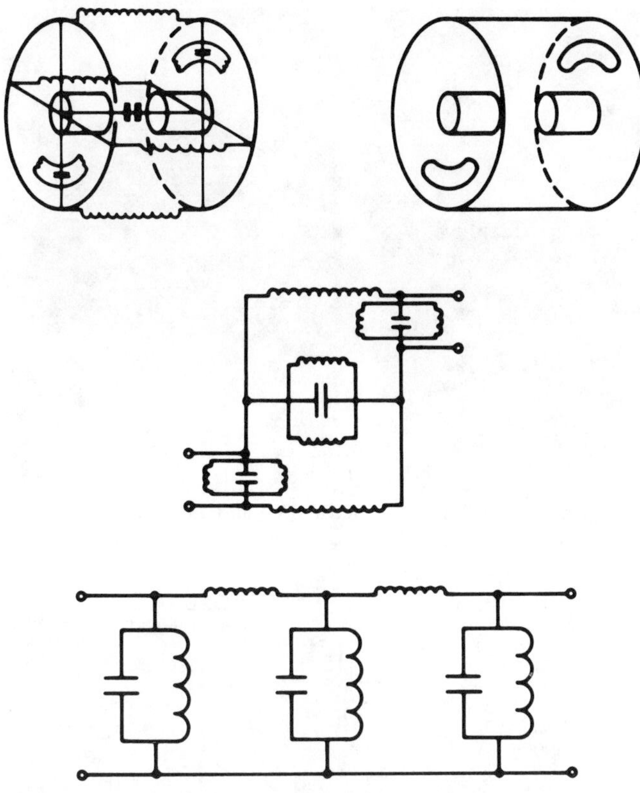

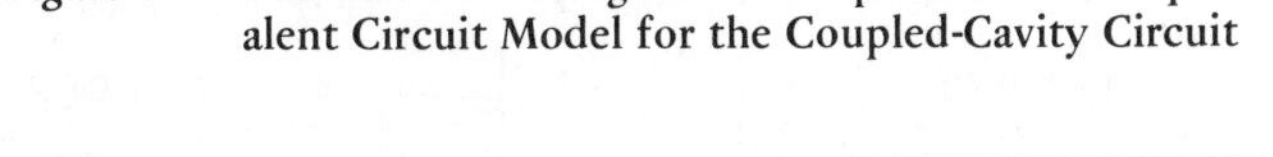

Figure 30 Schematic Showing the Development of the Equivalent Circuit Model for the Coupled-Cavity Circuit

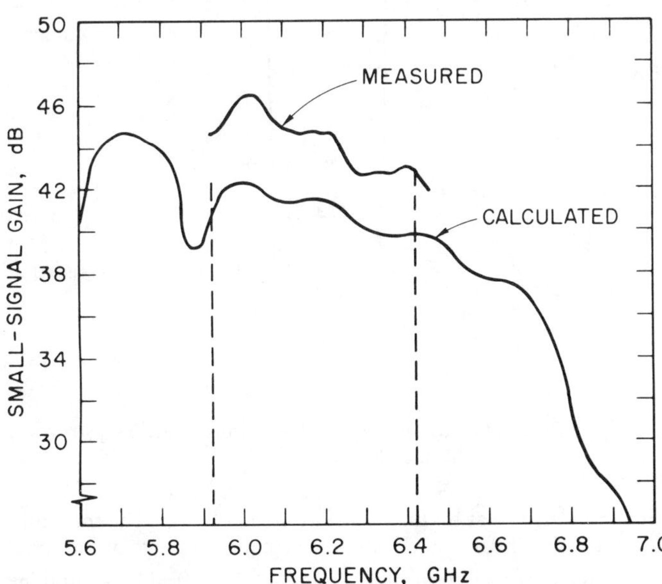

Figure 31 Comparison of Measured and Calculated Gain for an 8 kW CW C-Band TWT

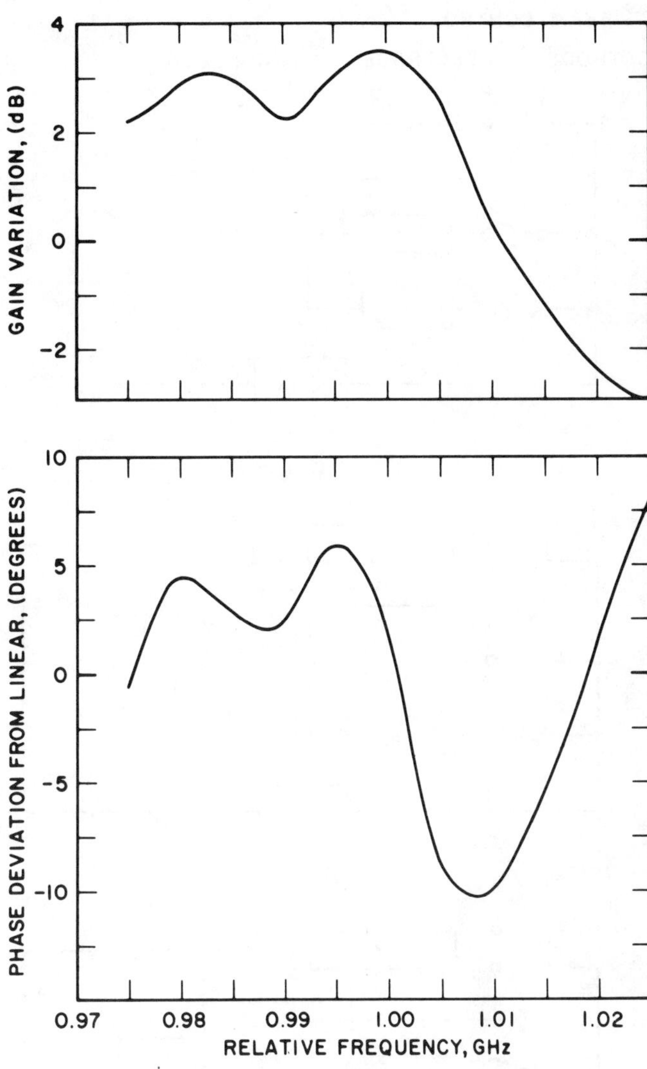

Figure 32 Small-Signal Phase and Amplitude Response for a Coupled-Cavity TWT

Voltage Parameter	AM		PM	
Cathode to Body	0.5	dB/1%	30°	/1%
Anode	0.1	dB/1%	5°	/1%
Cathode to Grid	0.15	dB/1%	7°	/1%
Cathode to Collector	0.02	dB/1%	0.5°	/1%
Heater (Dynamic)	0.00005	dB/1%	0.001°	/1%
Solenoid	0.00001	dB/1%	0.0005°	/1%
Drive Power			2.2°	/1 dB

Table 3 Typical Sensitivites for a 10% Bandwidth 60 dB Gain TWT

In the area of "hybrid tubes," the most common variety is half-klystron/half-TWT, as shown in Figure 36. At low power levels, hybrid tubes have no advantages; but at power levels in the megawatt range, improvements in gain and bandwidth can be afforded over what a klystron or TWT could offer separately. A TWT has reasonably high efficiency over a broad band, but it does not have very uniform gain vs. frequency. A klystron can be stagger-tuned for very broadband operation of its gain or driver section, but the output cavity lacks bandwidth. So by combining a klystron input and a coupled-cavity output section, the advantages of both can be utilized. The tube shown operates at the 5 to 10 MW level at S-band. The gun has a very large insulator for its 150 kV operation. Other parts of the tube are similar to those in standard lower power tubes. The performance of this device is shown by simple diagrams in Figure 37. Gain vs. frequency is plotted over a representative bandwidth. The TWT section has a bell-shaped gain response; the klystron driver section is designed to provide a complementary response. A saddle-like gain vs. frequency characteris-

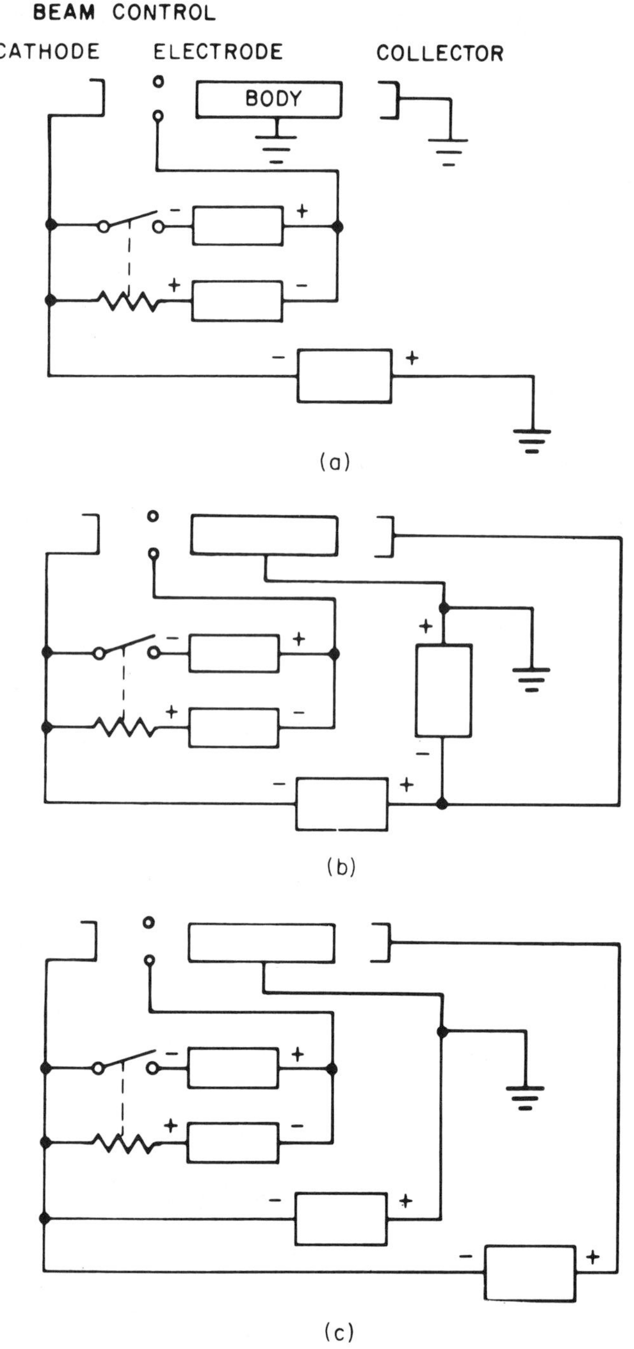

Figure 33 **Typical Power Supply Configurations for High-Power TWT**

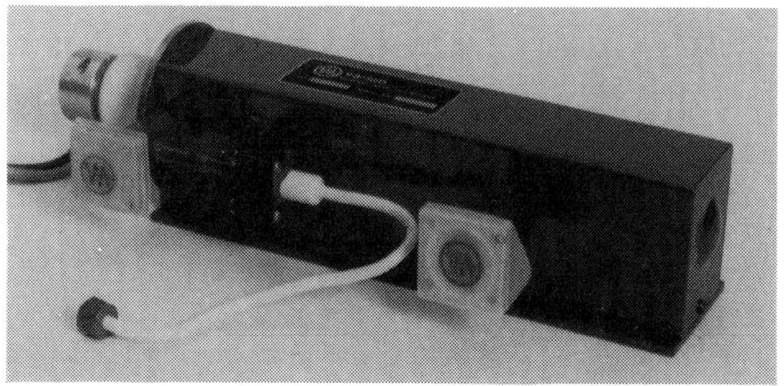

Figure 34 **A PPM-Focused X-Band TWT, VTX-5680**

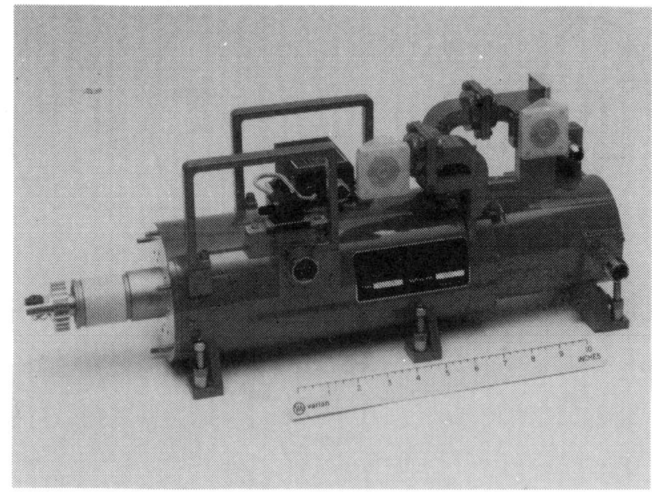

Figure 35 **A Solenoid-Focused X-Band TWT, VTX-5782**

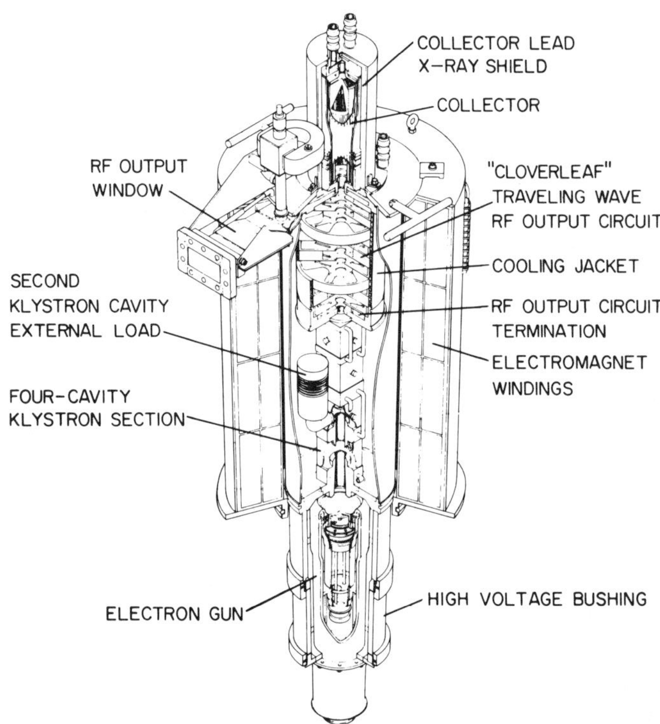

Figure 36 **Cutaway of a Typical Multimegawatt Twystron Amplifier at S-Band**

tic can be arrived at by tuning the klystron cavities. The total tube output is then relatively flat over the frequency range of interest.

The name for this hybrid tube, *Twystron*, is derived by combining the *Tw* from a *TWT* and *ystron* from the kl*ystron*. Table 4 lists several Twystron amplifier types to illustrate typical operating parameters. Peak powers are always in the megawatt range, bandwidth tends to be from 6 to about 15%, efficiency is approximately 30%, and, of course, the voltages and currents are commensurate with megawatt operation.

Table 5 shows where the various high-power, linear-beam tubes fit in an overview. This chart is an extremely coarse one, which lists the basic tube type, peak power, bandwidth and average powers. Klystrons operate at all conceivable power levels; helix TWTs are limited to approximately 10 kW peak, beyond which level synchronous interaction is lost for a reasonable helix struc-

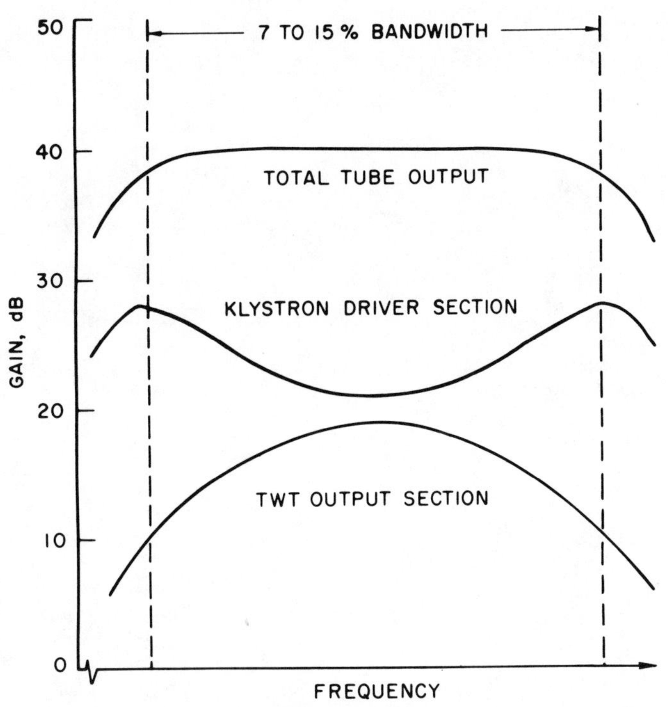

Figure 37 **Gain of a Twystron Amplifier**

powers. Helix TWTs, of course, may have octave bandwidths; coupled-cavity TWTs usually approach 10%, but that specification can be stretched to about 30%. The Twystron amplifier, typically 8 to 10% wide, can, with some effort, be used up to 15%. For average power, klystrons hold all records because of their rugged construction and easy cooling. Megawatt Twystron amplifiers and, to some extent, coupled-cavity TWTs are also capable of large average powers. Helix TWTs are limited to approximately the 1 kW level because of dissipation on the helix circuit. The power values listed are not necessarily the limiting ones, but rather represent the levels that have been achieved in practice.

Tube Type	VA-145	VA-915	VA-146	VA-145LV
Frequency Bandwidth, GHz	2.7-2.9 2.9-3.1 3.0-3.2	3.1-3.6	5.4-5.9	3.1-3.5
Peak Power Output, MW	3.5	7.0	4.0	1.0
Average Power Output kW	7.0	28.0	10.0	1.0
Pulse Width, μs	10.0	40.0	20.0	50.0
Efficiency, %	35.0	30.0	30.0	30.0
Beam Voltage, kV	117.0	180.0	140.0	80.0
Beam Current, A	80.0	150.0	95.0	45.0
Drive Power, kW	0.3	3.0	2.0	1.0

Table 4 **Typical Twystron Amplifier Characteristics**

	Peak Power	Bandwidth	Average Power
Klystrons	All Levels	0.5% (1 kW) 2% (100 kW) 8% (10 MW)	To 500 kW
Helix TWTs	To 10 kW	Octave	To 1 kW
"Coupled-Cavity" TWTs (Space Harmonic)	To 500 kW	To 30%	To 50 kW
Hybrid Twystron® Amplifiers	Above 500 kW	To 15%	To several 100 kW

Table 5 **Typical Tube Usage Summary**

ture. Coupled-cavity TWTs operate up to about 500 kW; beyond that point, bandwidth becomes limited. Hybrid tubes operate from 500 kW up. The bandwidth of klystrons varies from about 1/2% at the 1 kW level to approximately 8-10% at the megawatt level. Klystron bandwidth is a fairly slow function of power level, but klystrons are competitive with TWTs at sufficiently high peak

Chapter 23 Smith

Types of Crossed-Field Microwave Tubes and Their Applications

Crossed-field microwave tubes, the principles of which were first demonstrated by A.W. Hull [1] in 1921, were not used in any practical sense until World War II when the British first introduced the multi-cavity magnetron and deployed it in radar systems. Today, the family of crossed-field tubes includes both rising sun and conventional strapped-vane magnetrons [2], coaxial magnetrons [3], crossed-field amplifiers of both linear and circular format, and both injected electron beam and continuously emitting crossed-field amplifiers.

Early magnetrons were of the unstrapped resonator type; these, though, were soon supplanted due to improved stability afforded by strapped-vane forms. Strapped-vane magnetrons compose about 70% of all magnetrons in use today; the remaining 30% is composed of rising sun configurations (now essentially obsolete) and coaxial magnetrons, (the only form currently being developed in the US). The latter device includes both a strapped-vane system and a high Q resonant cavity which is essentially in cascade within the magnetron, thereby providing greatly enhanced frequency stability to changes of RF load and aberrations of driving voltage. Intriguing potential uses for coaxial magnetrons can be visualized wherein the advantages of coherency afforded by frequency and phase-locking techniques are exploited for low-cost, on-board control of missiles.

All magnetrons and most crossed-field amplifiers (CFAs), except for a specialized device known as the voltage-tuned magnetron, are continuously emitting; i.e., the region of energy exchange, called the *interaction space,* is a cylindrical region bounded by, on the outer side, the microwave circuit (the anode vanes) and, on the inner side, an active surface which emits electrons from any point in its 360° periphery. In this interaction region, electrons emitted from the cathode are accelerated to synchronism with the slowly traveling circuit wave (the *sine qua non* for energy exchange), but the electron stream's total energy is largely potential. As the stream gives up energy to the RF wave, it moves radially outward toward the circuit, thereby losing potential energy. When it finally is collected on the circuit, the remaining energy is only the relatively small amount of kinetic energy originally imparted to the electrons in order for them to attain synchronism. Hence, crossed-field efficiency can be very high; normally, it lies in the range from 50 to 80%.

Virtually all crossed-field amplifiers employ the same cylindrical reentrant interaction space, but the slow wave circuit is matched and nonreentrant. Thus, whereas a magnetron is frequency-controlled by its high Q resonant circuit, the crossed-field amplifier can amplify virtually any drive signal lying within the pass band of its microwave circuit. In contrast to the devices described above with cylindrical reentrant interaction spaces, linear format amplifiers have been developed. The latter's lack of electron feedback (reentrancy) results in lower efficiencies, albeit with simpler power supply requirements.

Electrically nonreentrant amplifiers also are made in both circular and linear forms, utilizing a strip beam generated by a crossed-field electron gun. Such an amplifier form is peak-power-limited by the small cathode area available in the electron gun and, thus, is best suited for quasi-CW applications such as electronic warfare. By combining up-pulses of energy (3 to 10 dB in amplitude) on a CW background through grid pulsing, so-called dual mode capability may be realized, a feature extremely useful in electronic countermeasures.

Magnetrons and crossed-field amplifiers are utilized today not only for military systems but as power sources for linear accelerators and as efficient generators for microwave ovens; they are being seriously investigated for use in the efficient transmission of solar energy from space to earth, as well.

When considering crossed-field tubes, one needs to know why they are used and the conditions attendant to their application. In every situation in radar and ECM, there generally are requirements that move the designer in either the linear beam (e.g., klystrons and traveling-wave tubes; see Chapter 22) or crossed-field direction. For example, if very high power, very high gain, or amplitude linearity were required one probably would lean toward a linear beam tube device. On the other hand, if a combination of high power and low voltage, low voltage and small compact size, or some other such pairing of high power, low voltage, high efficiency, small size and weight, and low cost were called for, one would think primarily of a crossed-field device. In general, when a study of the "nitty-gritty" of a particular system application and its requirements is undertaken, the choice of crossed-field versus linear beam is generally quite clear.

To begin with, one might look at the crossed-field amplifier forms – *continuous cathode* (or *emitting sole)* [4] and *injected beam.* The continuous cathode amplifier is basically a pulsed high peak power, high average power device commonly used in the radar field. The continuous cathode tube develops peak powers up to several megawatts and average powers up to 30 kilowatts. (These figures do not refer to design or theoretical limits, but rather to the tubes which actually exist and which are in use today in field systems.) In terms of pulsewidth, these tubes generally fall in the range of 1 to 100 microseconds.

The other generic form of crossed-field tube is called the injected beam crossed-field amplifier, a device used less frequently than the continuous cathode variety. These tubes are low peak power or CW, capable of octave bandwidth. They are basically ECM devices having the capability of gains up to 20 dB and possessing CW power capability up to and including 2 kilowatts over an octave bandwidth. The injected beam CFA has dual mode pulse-up capability on the order of 3 dB, and it can be grid-controlled with an interceptor type grid.

Applications for crossed-field amplifiers of the continuous cathode type are in all types of radar systems: airborne, mobile, fixed, and transportable. These tubes also are used as power sources in the linear accelerator field and, of course, they are utilized in phased arrays. The continuous cathode tube is capable of a 10% bandwidth and gain up to 16 dB – the latter figure being low in comparison to that obtainable with a linear beam tube. However, because of the continuous cathode tube's high efficiency, it is par-

	Po (kW)	s/n (dB)	Si/Ni (dB/Hz)	T_p (μs)	PRF (kHz)	du (dB)	S/N (dB/Hz)
L-Band	100	75.4	97.2	30	1.0	15.22	112.4
C-Band	500	69.0	92.0	1	5.0	23	115
C-Band	660	77	81.7	20	0.16	25	106.7
S-Band	750	87	104	10	4.0	14	118
S-Band	60	76	86	5.5	1.25	21.6	107.6

s/n = signal-to-noise measured in bandwidth (BW, Column 3)

Si/Ni = signal-to-noise per cycle of bandwidth

S/N = peak signal-to-noise per cycle of bandwidth

Table 1 **Intra-Spectrum Noise Data**

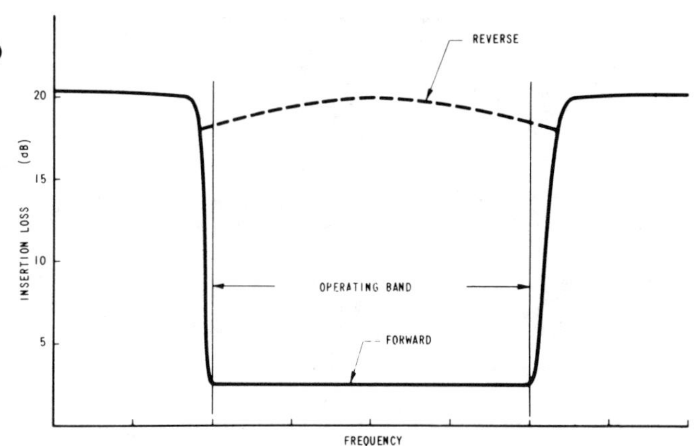

Figure 1 **Injected Beam CFA Insertion Loss Characteristics: Idealized**

ticularly well suited for use as the output stage in amplifier chains. The injected beam tube is limited in application despite its inherent capability for achieving greater gain – up to 20 dB presents no problem. It has octave bandwidth; consequently, its use is almost entirely in dual mode barrage ECM, particularly in airborne applications. However, there is no good reason why the injected beam tube could not be used in radar applications in which relatively low peak power is desired.

Noise Characteristics of Continuous Cathode Crossed-Field Amplifiers

One characteristic of the crossed-field amplifier that generally is not appreciated is its noise performance. Most people think of crossed-field tubes as exceptionally noisy devices which should be avoided, if at all possible, when the situation requires relatively clean operation. Table 1 shows actual measured noise performance data from on-shelf continuous cathode crossed-field amplifiers; a number of points should be made. First of all, note the power levels – these tubes are very high power ones to be thought of in terms of noise performance, as they vary from 60 to 750 kilowatts of power. With this fact in mind, consider first the in-band signal-to-noise ratios. These data are given as carrier spectral line power compared to the noise power between lines in a given bandwidth, as shown in the third column. Such numbers, of themselves, are not compared readily but become more meaningful if one converts from varying bandwidths of measurement to a common bandwidth of 1 Hz, as shown in Column 4.

For pulsed tubes, comparative noise interpretation must be done carefully; for purposes of comparison with the tubes with which most readers are familiar (notably CW tubes) one must convert the measured noise data into peak (equivalent CW) values, as follows:

$$S/N = \frac{S_i}{N_i} \; \frac{1}{du} \quad \text{dB/Hz}$$

where S (peak pulse power) = S_i/du^2 (spectral line power)

N (peak noise power) = N_i/du (average noise power)

du = tube duty cycle T_p (pulse width) x PRF (Pulse Repetition Frequency)

Using that relationship, and returning to the previous data, one derives the final column of data in terms of the peak signal-to-noise performance in a 1 Hz band. The 1 Hz band was chosen because it compares directly with published data. The results show a range from about 107 to 118 dB/Hz. Comparing this data to that from Skolnik [5] for the same frequencies removed from the carrier (that is, 1 to 10 Hz), one finds typical linear beam tube performance on the order of 120 to 140 dB down from the carrier – which is on the order of 20 dB better than for the crossed-field amplifier. The obvious conclusion, therefore, is that the crossed-field tube is certainly not a noisy device; however, it cannot compete with the low noise klystron [5].

New interest in magnetron noise performance has been sparked recently, chiefly by studies of line-conducted Television Interference (TVI) attributable to microwave ovens. It has been found that quiet regions of essentially no noise output can be realized by careful control of electron emission; means for effecting TVI reduction by utilizing this phenomenon are being pursued vigorously.

Use of Ferrites to Reduce Oscillations of Injected Beam and Continous Cathode Crossed-Field Amplifiers

One of the deterrents to the use of the injected beam crossed-field amplifier is the fact that, in applications where one might use it, it is desirable to leave the beam current at its full level while simultaneously backing off the RF drive. However, a tube when operated in this fashion should not exhibit spurious oscillation. Since early types of injected beam crossed-field amplifiers did exhibit self-oscillation when the drive was lowered by 3 dB or more, work was undertaken under the premise that installing ferrites within the tube body itself would produce, in effect, an isolator within the slow wave structure. The objective was a rather obvious one – to get zero or minimum forward insertion loss within the band of interest (as shown in Figure 1) while, at the same time, achieving as high a reverse loss as possible over all frequencies. Of course, this characteristic is an idealized one. The actual results are shown by Figure 2. For an octave bandwidth tube, forward insertion loss was reduced to 3 to 4 dB across the octave. Bear in mind that about half of this 3 to 4 dB represented

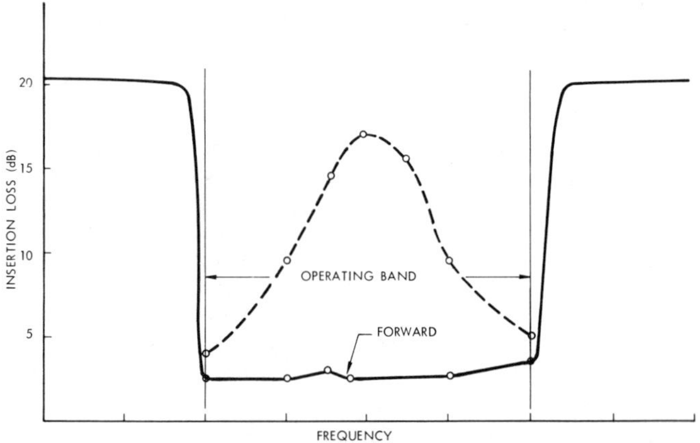

Figure 2 **Injected Beam CFA Insertion Loss Characteristics: Operating Tube**

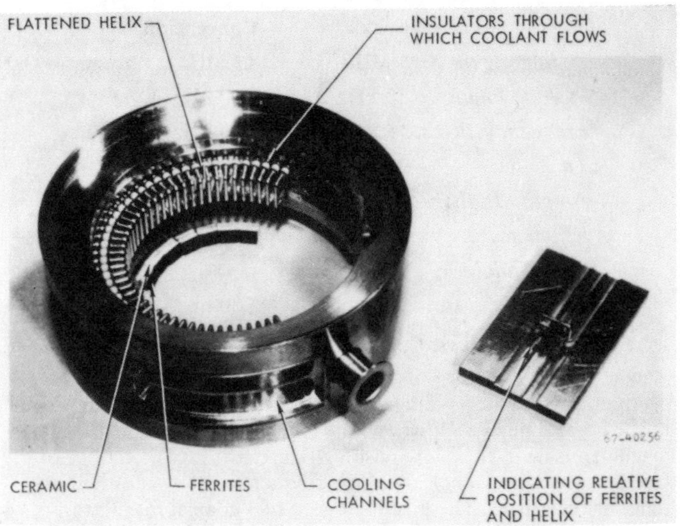

Figure 3 Injected Beam CFA Anode Component

copper losses in the helix circuit itself (the circuit used); only a fraction of it, about 50% was due to the ferrite alone. The reverse losses along the structure peaked out at a 16 dB level at band center, and the bandwidth was limited; improvements though, can be expected. Those readers who are used to working with passive ferrite devices might say that the back-to-front ratio is not really all that it could be. However, things are different in an active device like a tube. The forward traveling wave, for example, grows exponentially as it travels from input to output and, therefore, the 3 dB forward insertion loss has a much smaller effect than it would have in a passive device.

Another area of potential use involves narrow line width ferrites, mostly for continuous cathode tubes. In the continuous cathode tube, cathode modulation usually prevails; that is, on the rise and fall of each pulse, there is a region in which the voltage is synchronous with band edge waves and band edge oscillations are produced. The difficulty is a common problem in microwave amplifiers utilizing a bandpass slow wave circuit. Near the lower cut-off of the iterative structure, the RF match, of necessity, deteriorates and the resulting resonances often interact with the beam in a magnetron-like fashion whenever the electron stream is in synchronism with the associated circuit waves. Therefore, it is desirable to come up with a very narrow line width ferrite which, when inserted in the tube, would produce high bilateral loss at the band edges and, in this way, discourage the onset of spurious power generation during the beginning and end of each pulse. This result has been partially achieved in the injected beam device; it is more difficult to implement with continuous cathode tubes, primarily because of the very high average powers involved in these tubes, on the order of 10 to 30 kilowatts.

Incidentally, ferrites now have been used for some years to limit unwanted radiation from microwave oven magnetrons to levels tolerated by the FCC. Further exploitation of these useful materials is to be expected in controlling TVI, perhaps in conjunction with emission control.

For those unfamiliar with injected beam crossed-field amplifiers, Figure 3 shows a slow wave structure of such a device; the configuration is of interest in that it is different from the usual crossed-field structure. The injected beam variety has a flattened helix structure which has all the properties of the conventional helix except that, in the case of the crossed-field tube, the beam does not go along the axis within the helix but, instead, travels within the inner diameter of this structure between a negative electrode, called the *sole,* and the circuit. The helix itself is flattened so that it fits the crossed-field geometry. The photo shows

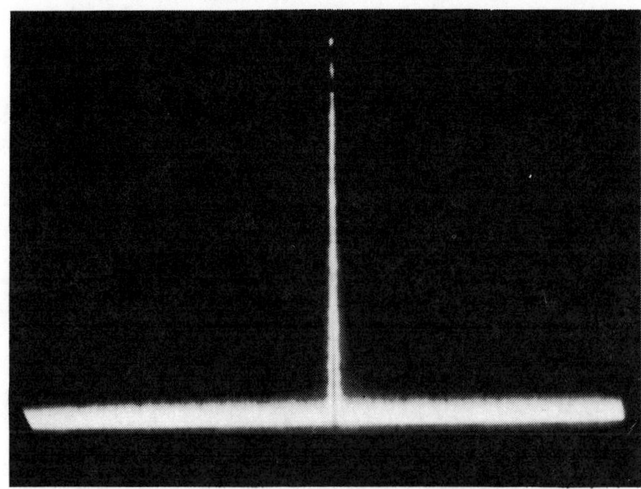

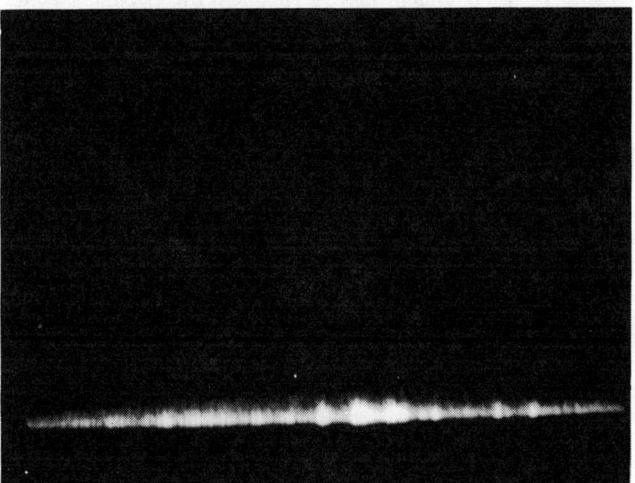

Figure 4 Injected Beam CFA with and without RF Drive; 10 dB/cm

the ferrites and how they are placed in juxtaposition with the circuit. These ferrites (black) are hot pressed to ceramics which match them thermally in expansion; the composites are then brazed directly in back of the helix, between the helix and the wall of the anode, as shown by the inset. The ferrite lies within the fringing field of the helix itself and results in the desired unilateral performance.

Finally, regarding ferrites inside injected beam tubes, Figure 4 shows results that have been achieved in practice. The performance was obtained from an octave bandwidth tube actually using the helix circuit. At the top is the output as seen on an analyzer when the tube was operating at the 1500 W CW power level. The vertical scale is 10 dB per cm; therefore, the noise shelf turns out to be -85 dBm/Hz. The photograph at the bottom shows the same tube at full beam current, but with the RF drive removed. Ordinarily, one would expect the crossed-field amplifier to be oscillating and putting out a great deal of noise power, yet it is obviously relatively clean. The frequency span shown, by the way, is 2.4 to 3.6 GHz – the better part of an octave.

The noise shelf on the right has deteriorated slightly. If one looks closely, some evidence of coherency can be seen in the base line presentation and, by close examination of the peak of these noise regions, it is clear that the noise shelf has deteriorated, in the extreme, by about 10 dB to about -75 dBm/Hz – reasonably good performance for a crossed-field amplifier.

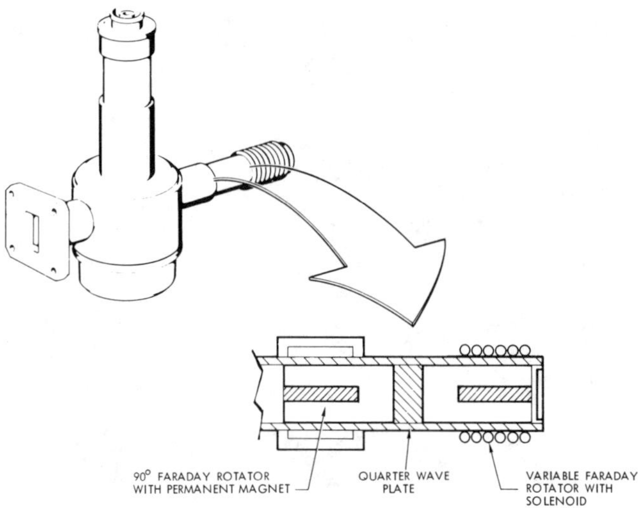

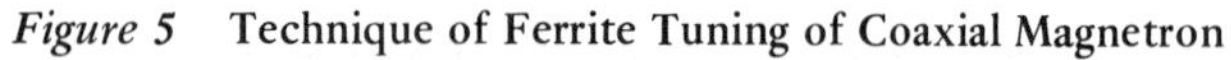

Figure 5 **Technique of Ferrite Tuning of Coaxial Magnetron**

	Coaxial	**Conventional**
Pulling Figure	5 MHz	15 MHz
Pushing Figure	50 kHz/A	200 kHz/A
Frequency Jitter	15 kHz rms	100 kHz rms
Time Jitter	2 ns rms	1 ns rms
Amplitude Jitter	3%	3%
Efficiency	38%	32%
Pulse Stability	0.1%	0.5%
Life	2000 hr	500 hr

Table 2 **Coaxial vs. Conventional Magnetrons at X-Band**

Electronic Tuning of Coaxial Magnetrons by Ferrites

Turning to the use of ferrites for another purpose, we discuss the electronic tuning of coaxial magnetrons, for which there are many possible applications. Many years ago, so-called *double output tuning* was developed [6]. The idea was to see if the mechanical introduction of a change of reactance into a magnetron cavity would result in appreciable tuning. In practice, a kilowatt of CW power over several hundred megahertz was produced in this fashion around 1950. This concept was resurrected recently because it was felt that industry should now know how to effect this result with ferrites. Early prototypes of the ferrite tuned tube developed about 10 kilowatts of peak power (in the X-band region) at very low average power, but tuned 90 MHz electronically. The program was directed toward higher peak power and higher frequencies with about 3% tuning. The components for a ferrite tuner are really quite simple, as shown in Figure 5. A radial transmission line emanates from the cavity of a coaxial cavity magnetron into a Faraday rotation element which introduces 90° of rotation. The waves then pass through a bilateral quarter-wave plate which transforms the field components into circularly-polarized waves. These waves then are phased shifted in an additive manner in a Faraday rotation element so that when the wave arrives back in the cavity it has the same sense as the outgoing wave, but is shifted in phase. In this way, a changing reactance, resulting in electronic tuning, is presented to the coaxial cavity. Another approach to ferrite tuning, not utilizing circularly polarized waves, has been reported on by Bahri [7].

Coaxial Magnetrons

The main applications of coaxial magnetrons are airborne, and they include all kinds of radar systems – terrain avoidance, navigation, air and surface search, weather, terminal homing, and multi-mode attack. A particular one worth noting is terminal homing, a very good application for a frequency-locked magnetron that is small, compact, and lightweight, especially since the coaxial magnetron is relatively inexpensive for missile applications.

Table 2 shows a state-of-the-art assessment of the coaxial magnetron versus the conventional type; a number of interesting comments can be made. For example, the reduced pulling figure of a coaxial tube is attributable to its greater energy storage, which in turn reduces the susceptibility of the tube's frequency to variations of load magnitude and phase. Another item of interest is pulse stability, which is a measure of the percentage of missing pulses due to misfiring, arcing, etc. Efficiency, customarily defined as RF output power minus RF drive power, divided by the product of the tube's voltage and current, is significantly improved in the coaxial device. However, the chief points in favor of coaxial magnetrons lie in their low pushing figure (the rate of change of frequency with current), their low frequency jitter, and their attendant time jitter. The pushing figure of a coaxial magnetron is about four times better than that of the conventional type; this difference provides a great improvement in system performance. Those who have worked with magnetrons are probably familiar with the fact that a magnetron's output spectrum can be very lopsided. Due to pushing, many conventional magnetrons produce a decidedly asymmetrical spectrum. With a four-fold increase in capability due to the reduced pushing of the coaxial magnetron, spectral symmetry is greatly improved and the system designer acquires the ability to control the spectral characteristics (the latter fact because the output spectrum is essentially the Fourier transform of the modulation pulse).

The characteristics of frequency jitter and time jitter are most important when one thinks in terms of MTI performance, and most people using magnetrons today are thinking in terms of MTI. Frequency jitter relates to pulse to pulse frequency changes, while time jitter measures the variable RF delay between the voltage pulse and the subsequent RF pulse. The frequency jitter is approximately six times better, though time jitter, unfortunately, is only about one-half as good as in the conventional magnetron. The time jitter battle is being won, however. The day is close when the time jitter of a coaxial magnetron can equal that of a conventional magnetron. One other point that should be noted is the improvement in life one sees nowadays in the coaxial magnetron, because it utilizes many more vanes, thereby reducing cathode loading. But even more important for increasing life is the existence of low amplitude RF fields in the coaxial type; this characteristic essentially eliminates the occasional arcing associated with high voltage gradients. The improvement of frequency jitter and time jitter has been expressed in Table 3 in terms which may be more meaningful – MTI improvement factors [8].

The improvement from a conventional to a coaxial tube in the area of frequency modulation is 16 dB, accompanied by a 9 dB decrease in the MTI improvement factor concerning jitter. However, the limiting facet of an MTI system is always the worst MTI improvement factor in all the parameters. Viewing these figures in aggregate, it is obvious that, whereas there was a 22.6 dB figure for the conventional magnetron, the worst value for a coaxial magnetron is 38.4 dB – a value very similar to the frequency jitter improvement factor. Therefore, one can expect to get a great improvement in MTI capability using a coaxial tube. Most coaxial magnetron applications run between 100 nanoseconds and a microsecond or two; thus, for comparison of MTI improvement factors (based on the parameters of Table 2), a pulse width of 0.25 microseconds was selected for Table 3.

One other area that should be discussed is a military requirement, MIL-STD-469 – an RFI standard which severely limits the spectral signature of the tube. The conventional magnetron, in most cases, cannot meet this RFI standard, whereas the coaxial magne-

	Coaxial	Conventional
Frequency Modulation		
$20 \log \frac{1}{\pi \cdot T_p \cdot \Delta f}$	38.6 dB	22.6 dB
Time Jitter		
$20 \log \frac{T_p}{T_j}$	38.4 dB	47.9 dB
MIL-STD-469	Yes	No

where:

T_p = pulse width

Δf = change of frequency

T_j = time jitter

Table 3 **Comparison of Spectrum, MTI Improvement Factors**

tron can in almost every case. On the upper photo of Figure 6 is the spectron resulting from the modulating pulse of the bottom photo. The spectrum is well formed, with sidelobes going down regularly. With respect to the MIL-STD-469 requirement, one is allowed an emission bandwidth, the width of which is governed by $20/T_p$ (where T_p is the pulse width). MIL-STD-469 also states that, in the X-band region, for example, the spectral level outside that window should not exceed the specified absolute maximum of -5 dBm/kHz. In S-band, however, the allowable maximum reduces to -16 dBm/kHz. As the frequency is reduced, peak powers tend to get higher; thus it gets more difficult to meet this requirement as the frequency goes lower and lower. Figure 6, however, is an example which shows that the requirement is easily met in the X-band area; the figure in the middle indicates the absence of spurious signals over a wide band. The satisfying of MIL-STD-469, as expressed by the upper photograph, amounts to having a spectrum which clears the cross-hatched bars; in this case, the example does so easily. The data represent a coaxial tube running at 100 kW, typical of X-band peak power levels.

Frequency Agility by Mechanical Means for Coaxial Magnetrons

The technique of frequency agility enables one to achieve clutter reduction and establishes anti-glint capability, etc. Figure 7 shows a presentation from an A-scope showing video output from a rooftop radar. The rooftop radar was fixed to look for low level, small targets over a fairly cluttered ground terrain. Figure 7(a) shows results with the agility on; Figure 7(b) is without agility. There are two hard targets, at ranges of 9 miles and 13 miles. With Agility On, there is a distinct improvement – there is little or no amplitude change of the clutter, but the two hard targets are discerned much more easily. There is little doubt about the target presented; if one were not sure, one could simply flip the Agility switch from On to Off. The agility bandwidth, in this case, was on the order of 75 MHz.

There are a number of ways of achieving this agility using mechanical forms; they all depend, to some extent, upon the use of an electric motor for the mechanical agility. This utilization normally produces a sine wave modulation, although it's possible to get semi-triangular waveforms. There are several different techniques used to effect this agility within the microwave tube, however, and Figure 8 shows one form. Referring to the tuner plunger, most coaxial tubes have what is called broadband tuning which is accomplished by moving the end wall of the coaxial cavity up and down in a slow fashion. In this particular adaptation, a moving platform, in essence, superimposes a sinusoidal perturbation over the broadband tuned frequency curve.

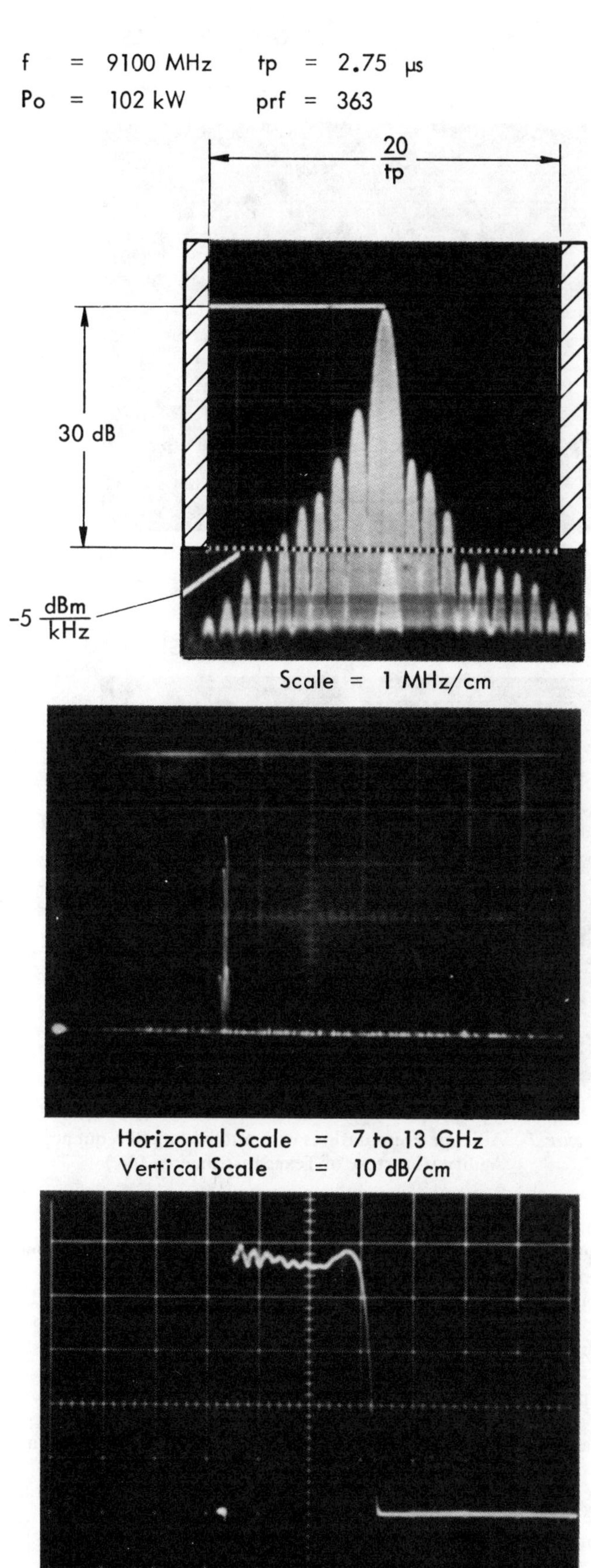

Figure 6 Coaxial Magnetron (Raytheon QK1664) Performance Relative to MIL-STD-469

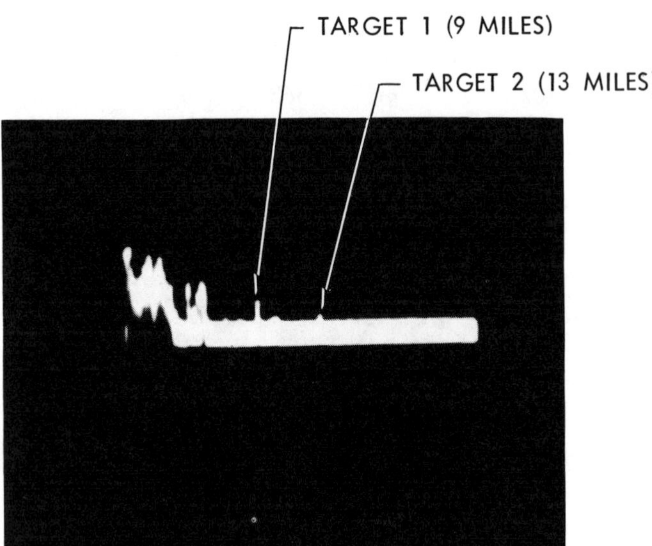

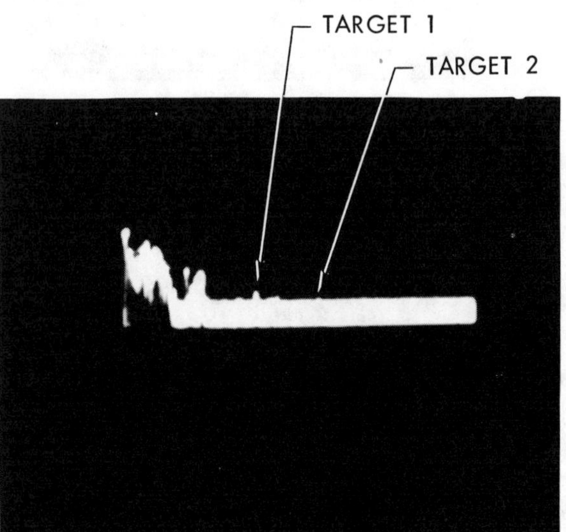

Figure 7 **A-Scope Presentations with and without Frequency Agility (Courtesy of Texas Instruments Inc.)**

An alternative means of achieving frequency agility is shown in Figure 9. Here, broadband tuning is achieved in the normal manner by moving, in effect, the end wall of the cavity in an axial direction (shown by the arrow). Frequency agility, on the other hand, is produced by an independent means. A ring secured at the two sides of the coupling slot is contained within a channel on the outer periphery of the coaxial cavity. The two ends of the ring opposite the output coupling slot are brought outside the vacuum envelope in such a way that the ring's diameter can be varied mechanically by means of a motor drive. This scheme has attraction in that the ring has low inertia. Furthermore, the ring in normal use is always close to the axial center of the cavity and, therefore, the tuning sensitivity with the so-called *ring tuner* is not a function of the broadband tuner position. For this reason, *slope correction* (that is, the variation of agility tuning bandwidth with broadband tuned frequency) is not a factor which must be considered. Typical characteristics of the ring tuner (on the X-band tube) are tuning speeds ranging from 80 to 200 Hz and tuning bandwidth up to 100 MHz at the present state-of-the-art. A photograph of the ring-tuned agility magnetron is shown in Figure 10. At the very top of the tube is a housing containing the tuning motor, the shaft of which actuates the cam driven plunger (shown enclosed in a spring at the rear of the tube). This plunger actuates the lever arm which, in turn, alternately withdraws and inserts the ring into the cavity of the magnetron itself. Control of the position of the tuner (position readout) is given by a resolver readout, the resolver actually being connected to the motor at the front end of the magnetron tube as shown. Tubes of the type in the picture are being utilized in operational systems today which are in volume production and have given very satisfactory service.

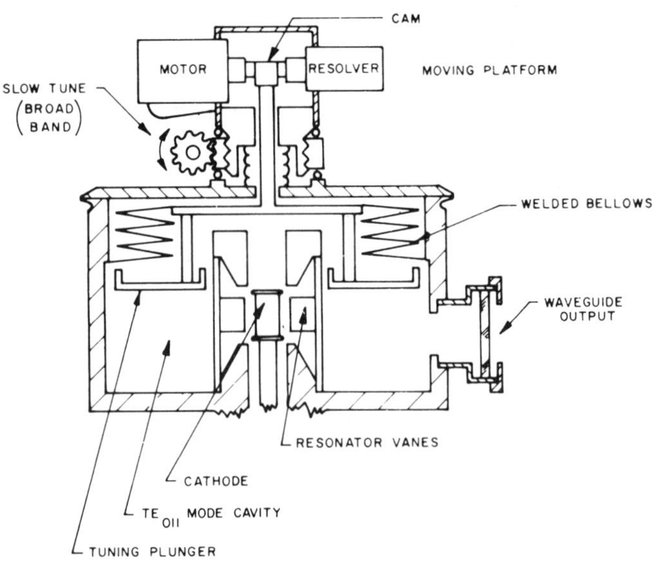

Figure 8 **Mechanical Broadband Tuning of Coaxial Magnetron**

Priming and Locking of Coaxial Magnetrons

Final subjects of interest are the priming and locking of coaxial magnetrons. Frequency locking of magnetrons has been of interest for many years, but has not become very popular because of the relatively low locking gains (i.e., the ratio of controlled output power to the required locking power) which were available. More recently, the subject of magnetron priming has gained attention. Priming does not result in total phase control of the output signal, whereas locking does, but the lower input power requirement of priming brings about a considerable improvement in magnetron output spectral quality and noise reduction. The pertinent characteristics are shown in Table 4. Both priming and locking produce a significant reduction in time jitter, a vitally important end for MTI in magnetrons. Priming permits introduction of a fast rise time on the modulating pulse, suppresses spurious oscillations, and reduces noise. Priming power relative to the output power is termed as giving 30-60 dB gain and, in this range, it can effect spectral improvements. At the same time, the frequency of the modulating, or priming signal, is not critical with respect to the natural oscillation frequency of the magnetron. A variation of 20-40 MHz between the priming signal frequency and the natural oscillation frequency of the magnetron is allowable to procure interesting results. A typical priming gain is on the order of 45 dB which, in many cases, can be high enough to permit the use of solid state locking or priming sources. Locking, on the other hand, provides complete phase control of the output signal; in effect, it makes a coherent system out of a normally incoherent device. But, as might be expected, the locking bandwidth is quite limited. However, 2-6 MHz of bandwidth is still available in the X-band region. Locking gain limits have not been thoroughly explored but, at the time of this writing, a locking gain of 23 dB is typical of the state-of-the-art.

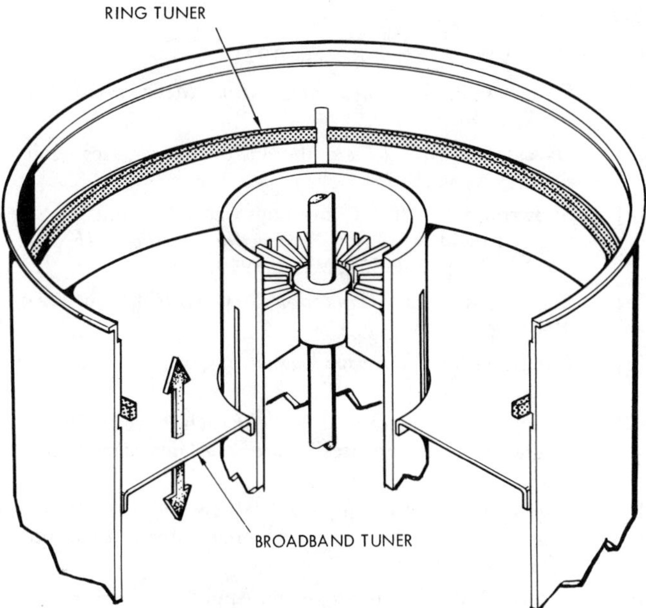

Figure 9 **Combined Broadband Tuning with Rapid Ring Tuning in Coaxial Magnetron**

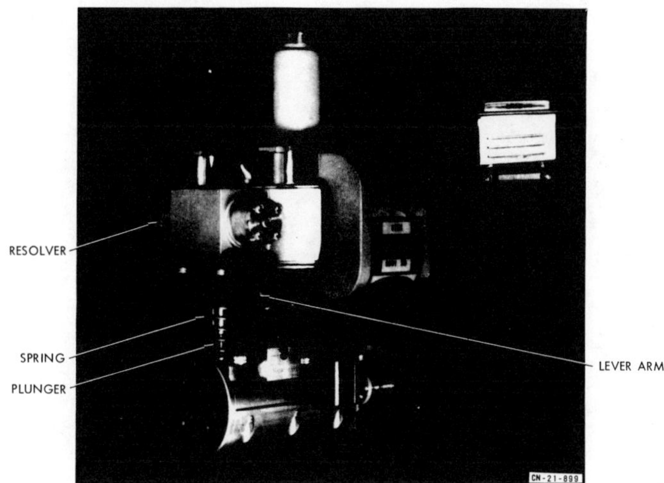

Figure 10 **Photograph of Ring-Tuned Agility Magnetron**

Returning to priming, some interesting data was published a number of years ago by Microwave Associates. Figure 11 shows time jitter as a function of the difference between the priming frequency and the natural resonant frequency of the magnetron. Time jitter can be reduced markedly over quite useful bandwidths by the introduction of relatively low peak power priming signals. It is by introduction of techniques of this type that one could make the coaxial magnetron a highly useful device for MTI applications.

Summary and Conclusions

In conclusion, crossed-field devices represent both a class of microwave elements, the state-of-the-art of which is improving continually, and one of the most interesting low cost approaches to microwave power available today. In the crossed-field amplifier area, of the two available types, continuous cathode tubes continue to be employed largely in field systems in applications where combinations of peak and average power and high efficiency (efficiencies of 70-80% in very high power S-band crossed-field amplifiers are common) are not realizable in competing devices. Injected beam devices are still in the laboratory stage but are expected to be a factor of significance in future ECM, and possibly also in radar, systems. Of late, the combination of tube and ferrite technologies has introduced ferrites into the crossed-field amplifier for stability and electronic tuning purposes. In both cases, success has been realized. The coaxial magnetron, a device rapidly supplanting conventional magnetrons, is of particular interest because of its relatively low cost, small size, and improved specifications (particularly in the area of frequency jitter and spectral characteristics). A particularly interesting application of the coaxial magnetron is in modern day frequency agile radar systems. This device has improved the capability of many radars and permits the use of techniques which are likely to be explored more in ensuing years. Frequency agility in coaxial magnetrons can be achieved in a number of ways (currently all mechanical); several approaches are peculiar to the particular company involved.

A new and very interesting area is the priming and locking of coaxial magnetrons. Priming can be of particular interest for MTI systems in the sense that it is very beneficial in jitter reduction. Locking, on the other hand, promises to be an important factor in relatively inexpensive coherent systems of the future, particularly missile applications.

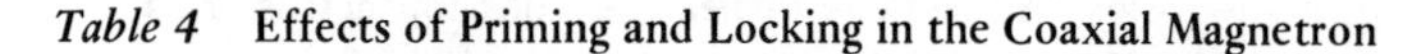

Priming	**Locking**
30 to 60 dB Gain	10 to 30 dB Gain
Jitter Reduction *Permits Fast Rise* *Suppresses Spurious* *Reduces Noise*	
20 to 40 MHz Bandwidth @ 45 dB Gain	2 to 6 MHz Bandwidth @ 23 dB Gain

Table 4 **Effects of Priming and Locking in the Coaxial Magnetron**

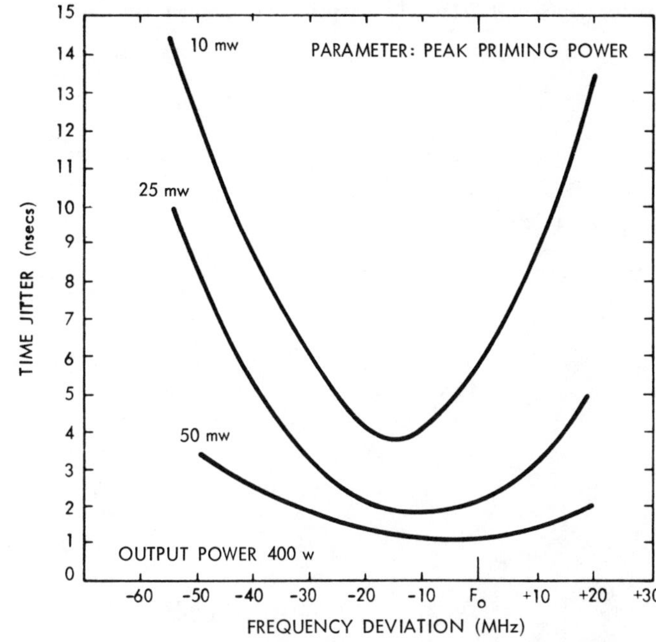

Figure 11 **Magnetron Time Jitter vs. Frequency Deviation Between the Magnetron Natural Frequency and the Priming Frequency**

Finally, post-Vietnam trends have resulted in a growing understanding of the control of conventional magnetron noise and TVI properties as related to microwave cooking, while the continuous cathode amplifier is being actively developed as a solar power transmitter. The latter fascinating application [9] involves the transmission by a pencil RF beam of up to 10,000 megawatts from synchronous orbit (where it is generated by solar cells) to a receiving station on earth capable of supplying, for example, the entire electric energy needs of all New England. Considering the advantages of crossed-field devices as a group, one must conclude that there will always be a place for the crossed-field device in each of its many forms. In these days, when cost consciousness is of paramount importance, crossed-field devices offer the equipment designer several possible ways of achieving state-of-the-art system performance with relatively inexpensive components.

References

[1] Hull, A.W.: *Physical Review,* 1921.

[2] Collins, G.B.: *Microwave Magnetrons,* McGraw-Hill, New York, 1948.

[3] Okress, E.: *Crossed-Field Microwave Devices,* Academic Press, New York, 1961.

[4] Skowron, J.F.: "The Continuous Cathode (Emitting Sole) Crossed-Field Amplifier," *Proceedings of the IEEE, Vol. 61, No. 3,* pp. 330-356, March 1973.

[5] Skolnik, M.: *Radar Handbook,* McGraw-Hill, New York, 1970, pp. 14-16.

[6] Collins, G.B.: *Microwave Magnetrons,* McGraw-Hill, New York, 1948.

[7] Bahri, J.L.: "Electronic Tuning of High Power X-Band Magnetrons with Ferrites," *European Microwave Conference,* 1974.

[8] Hayes, D.D.; and Logan, S.V.: "Microwave Tube Requirements for Radar Applications," *Microwave Journal,* pp. 37-44, April 1973.

[9] Brown, W.C.: "Technology and Application of Free Space Power Transmission by Microwave Beams," *Proceedings of the IEEE, Vol. 62, No. 1,* pp. 11-25, January 1974.

Present and Future Trends in Radar Systems and Components

Solid State Technology

L-Band Solid State Radars

Solid state L-band system technology has advanced rapidly over the last seven years. This fact is illustrated by Table 1, which lists key parameters of four L-band solid state modules developed by Raytheon Company between 1970 and 1974 [1]. Device CW power has gone from 6W to well over 50W, achieved over a fixed tuned bandwidth of up to 250 MHz. 1976 performance and future potential are indicated in columns 5 and 6 of Table 1.

The advances in L-band solid state systems is illustrated dramatically by the recent development and testing of the AN/TPS-59 (see Figure 90 and Table 1e of Chapter 1), the development of the Transportable Surveillance Radar (TSR) (see Figures 1, 2, and 3 and Tables 2 and 3) [1], and the current development of the L-band transceiver module for an airborne phased array system (see Figures 4 and 5) [2]. The reliability of transistor devices is improving through improved manufacturing, screening, and circuit techniques. During 546 hours of round-the-clock operation of the AN/TPS-59, only 18 out of 925 installed operating modules failed – fewer than 2% [3].

Extensive CW testing of 1975 state-of-the-art L-band transistors indicated that, at an operating junction temperature of 200°C, a gold metalized device has extremely long (virtually indefinite) life [4]. Somewhat shorter lifetimes were exhibited for pulsed operation; however, indications are that the difficulties experienced for short pulsed operation have been solved by the use of proper manufacturing techniques and screening of the wafers and devices.

The newly developed L-band bipolar microwave power transistors show high reliability while operating at high power levels. The Power Hybrids Inc. (PHI) PH-8444 (having diffused ballast resistors and a gold metalization system), the Microwave Semiconductor Corp. (MSC) 1214-125P (which is also gold metalized), and the Communications Transistor Corp. (CTC) 1214-70P (which is aluminum metalized) show short pulse powers of 100 watts peak per device; see Figures 6 and 7 and Table 4. Measurements made on the CTC device indicate operation with a junction temperature below 80°C (for all cells) for a 250 μs, 50W peak pulse output at a 400 pps PRF over the band from 1200-1300 MHz.

Figure 1 **All Solid State, Lightweight, Highly Reliable, L-Band Transportable Surveillance Radar (TSR) (From Hoft [1] with permission)**

	I 1970	II 1972	III 1973	IV 1974	V 1976 Potential	VI 1977-8 Projection
Power Output (Nom)	50 ± 10	50 ± 10 W	50 ± 10 W	150 W min	300 W min	750 W min
Number of Parallel Output Stages	8	4	2	4	4	4
Bandwidth	135 MHz	135 MHz	250 MHz	200 MHz	200 MHz	200 MHz
Pulsewidth (Max)	1500 μs	1500 μs	1500 μs	45 μs	50 μs	50 μs
DF	30%	30%	30%	14%	14%	14%
Device Internally Matched/Ballasted	No	No	Yes	Yes	Yes	Yes
Typical Junction Temp, T_i	140°C	120°C	90°C	90°C	120°C	120°C
MBTF, Design for		$> 10^5$ hr	$> 10^5$ hr	$> 10^5$ hr	$> 10^5$ hr	$> 10^5$ hr
Module Efficiency	> 30%	>30%	> 30%	> 30%	> 30%	> 30%
Cooling	Water Cooled Cold Plate			Forced Air	Forced Air	Forced Air

Table 1 **Parameters of Solid State L-Band Modules (From Hoft [1] with permission)**

Frequency	1200-1400 MHz
Power Output	8 kW
Average Power Output	1.25 kW
Bandwidth	200 MHz
Pulsewidth	46 μs (45+1) μs
MTI Limitation	-70 dB
T_R *and* T_F	100 ns max
MTBF	10^4 hr Graceful Degradation
Load VSWR	1.2:1 Operating ∞:1 Without Damage

Table 2 **All Solid State L-Band TSR Transmitter Characteristics (From Hoft [1] with permission)**

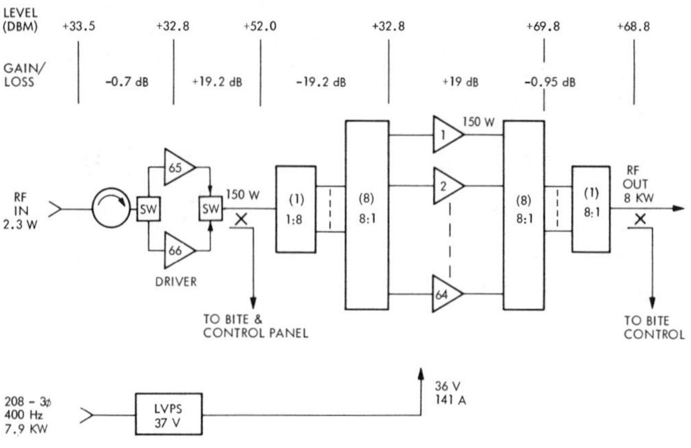

Figure 2 **TSR Solid State Transmitter. 64 150W peak power (see Table 3) solid state modules are combined out of one port to provide a 8kW peak power transmitter output. The modules are combined first in groups of 8 with an 8:1 combiner (see Figure 3); these groups are then combined with a 8:1 combiner. (From Hoft [1] with permission)**

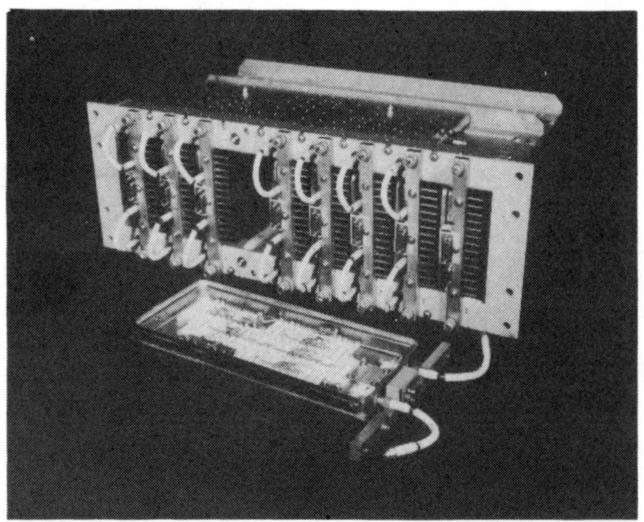

Figure 3 **L-Band Module Group Consisting of 8 Modules Combined with an 8:1 Combiner. (From Hoft [1] with permission)**

Power Output	150 W (peak) min
Frequency	1200-1400 MHz
Gain	20 dB
Efficiency	33%
Bandwidth	200 MHz
Pulsewidth	46 μs nom (45+1) μs
Voltages	36 V @ 14 A
Phase Tracking	10° rms
Amplitude Tracking	0.3 dB rms
Interpulse Noise	-104 dBm/MHz
Intrapulse Noise	-52 dB/MHz
MTBF	10^5 hr

Table 3 **TSR Module Characteristics (From Hoft [1] with permission)**

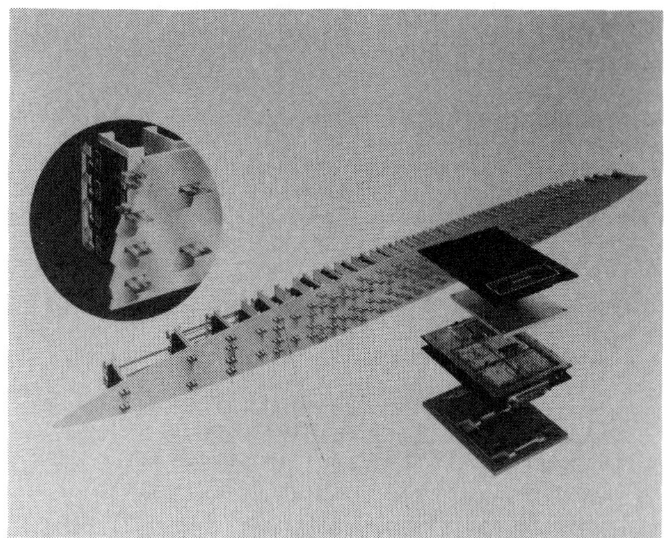

Figure 4 **Phased Array Antenna for Airborne L-Band Solid State Radar System. Antenna sidelobe shaping and reduction is achieved by a combination of the physical elliptical shaping of antenna, spatial density tapering through thinning and amplitude tapering on receive. Note that no dummy elements are used. (From Davis, Smith, and Grove [2] with permission)**

UHF Solid State Radars

The rapid advance in UHF modules is indicated in Table 5, which lists the parameters of two modules developed prior to 1976 [1]. Also shown are the parameters for a special purpose application [1]. The most important development at UHF is the on-going building of the solid state PAVE PAWS radar; see Figure 17 of Chapter 1 and Table 6.

S-Band Technology

Table 7 shows the parameters of an S-band module developed by MSC for the US Army Advanced Ballistic Missile Defense Agency (ABMDA) [5]. Also shown are the parameters of a module under development by MSC for ECOM. The parameters of the bipolar transistor being developed for this module are indicated in Table 8.

X-Band Technology

NRL is presently providing technical program management for the transmitter portion of an X-band solid state module which

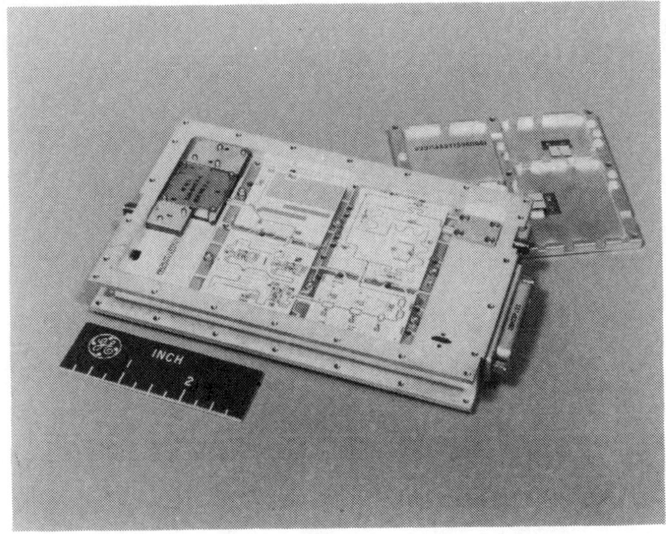

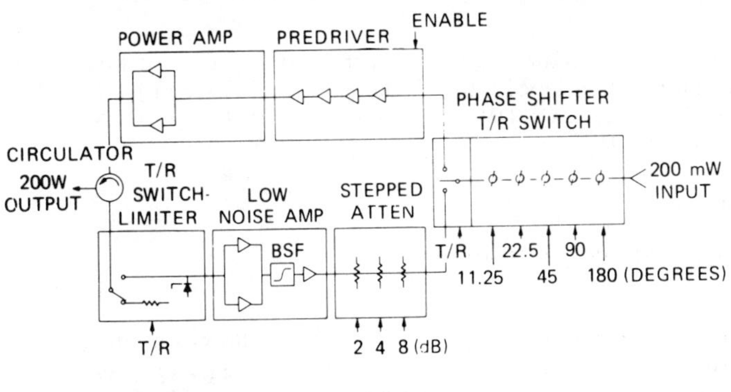

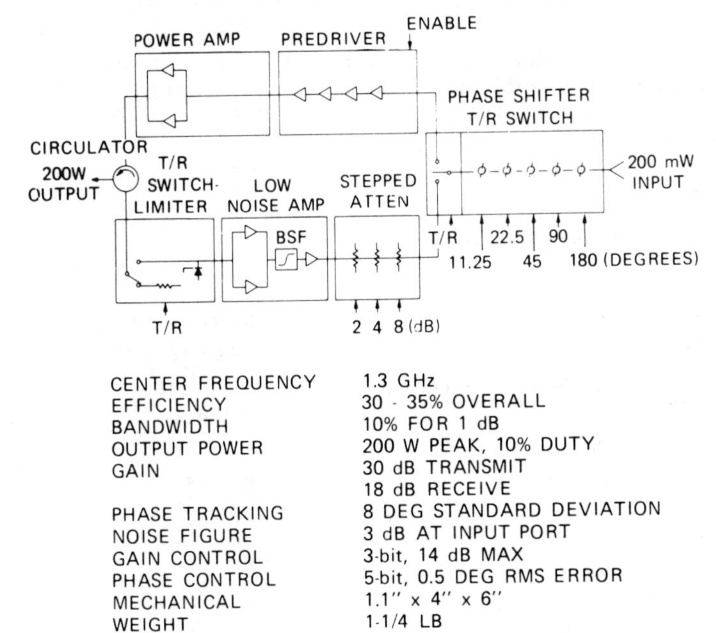

CENTER FREQUENCY	1.3 GHz
EFFICIENCY	30 - 35% OVERALL
BANDWIDTH	10% FOR 1 dB
OUTPUT POWER	200 W PEAK, 10% DUTY
GAIN	30 dB TRANSMIT
	18 dB RECEIVE
PHASE TRACKING	8 DEG STANDARD DEVIATION
NOISE FIGURE	3 dB AT INPUT PORT
GAIN CONTROL	3-bit, 14 dB MAX
PHASE CONTROL	5-bit, 0.5 DEG RMS ERROR
MECHANICAL	1.1" x 4" x 6"
WEIGHT	1-1/4 LB

Figure 5 **Module for Solid State Airborne Phased Array (From Davis, Smith, and Grove [2] with permission)**

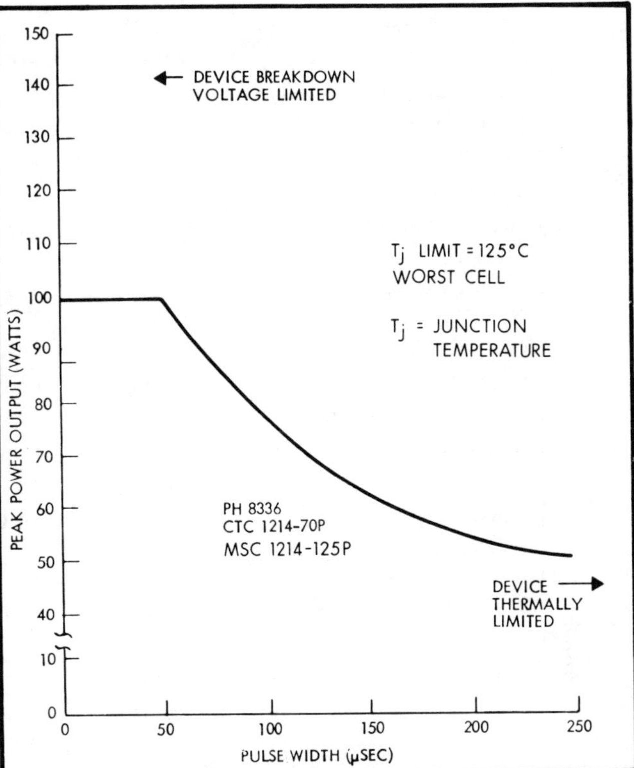

Figure 6 **Power Output vs. Pulse Width Limitations for State-of-the-Art L-Band Devices (Compliments of David Laighton, Raytheon)**

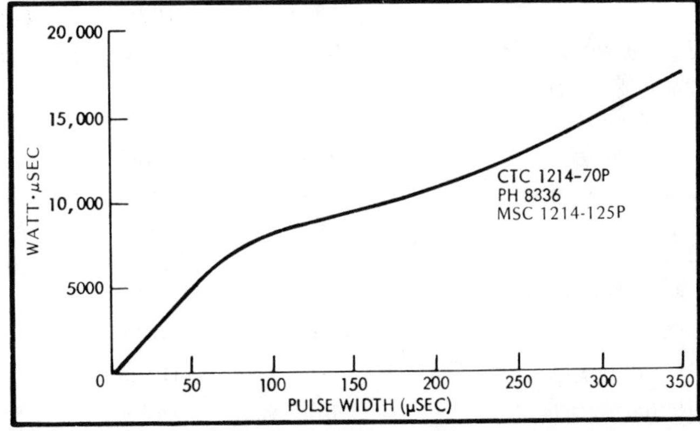

Figure 7 **Power-Time Product vs. Pulse Width (Compliments of David Laighton, Raytheon)**

uses direct power amplification at X-band (9.5 GHz) to generate the radiated power rather than the inefficient frequency conversion technique which was used for the MAIR (Molecular Airborne Intercept Radar), MERA, and RASSR; see Chapter 20. Eliot Cohen is program manager for the first phase of Naval Air System Command-funded contracts to RCA and Texas Instruments for the development of the devices and power amplifier for this module. Module power amplifier specifications are 1 dB bandwidth of 10% and a 5W peak power output for a 50% duty cycle, with pulses ranging from 2 to 20 μs wide [6]. Efficiencies approaching 30% are expected. Texas Instruments has obtained 5.1W CW power at 8 GHz using a 4-cell power FET device (6400 μm gate width) operating with a 13V bias [6]. The Air Force Avionics Laboratory also has a power GaAs FET device development program with RCA and Texas Instruments.

Distributed- and Lumped-Element and Monolithic Module Technology

For some applications, a low power, low weight transceiver solid state module is required. One such case is for a space-based phased array radar in which modules having a peak power of less than 1 W are required. Three approaches are available for the construction of such low power, low weight modules – distributed-element and lumped-element circuits and a monolithic construction [7,8].

Distributed-element circuits are what are used today in solid state modules such as the PAVE PAWS and the AFAR (CAMEL) module; see Figures 16 and 17 of Chapter 19 and Figure 5 of this section. For this technology, the inductors and capacitors are made using photoetched microstrip circuitry. (A capacitor is made of an open-circuited microstrip shunt stub the length of which

Company	PHI	PHI	CTC	MSC	TRW
Part Number	PH8336	PH8444	1214—70P	1214—125P	1214—60H
Power Output/Device (100 μs Pulse Width) (watts)	80	80	70	85	70
Gain (dB)	7	7	7	5	7
Efficiency (%)	50	55	50	45	45
Junction Temperatures (°C) 1.5:1 VSWR	120	115	125	125	–
Hermetic Package (Inorganic)	Yes	Yes	Yes	No	Yes
Metalization	Gold	Gold	Aluminum	Gold	Gold
Emitter Ballast Resistor	Tantalum	Diffused	Nichrome	Tungsten	Tungsten
Input/Output Matching	Yes	Yes	Yes	Yes	Yes
Device Geometry	Fishbone	Fishbone	Herringbone	Matrix	Interdigitized
Number of Cells	28 Cell R11	24 Cell R11A	12 Cells E3	12	10
Breakdown Voltage Between Collector and Emitter (with Emitter to Base Shorted)	65 V	70 V	70 V	70 V	65 V
Device History	R11 Geometry Originally Used for TACAN, DME, IFF, Telemetry	R11A Developed for Advanced Solid State Radar, J-TIDs Application	Structure Developed into TACAN, DME, IFF	Cell Structure Used in TACAN, DME, IFF, J-TID	Cell Structure Used in J-TID, IFF, TACAN

Table 4 **L-Band Transistor Performance Comparison**

	I 1971	II 1975	III 1976 Potential
Power Output (nom)	400 W	350 W	600 W
Number of Parallel Output Stages	32	4	4
Bandwidth	175 MHz	30 MHz	120 MHz
Pulsewidth (max)	60 μs	16 ms	2 ms
DF	25%	25%	3%
Device Internally Matched/Ballasted	No	Yes	Yes
Typical Junction Temp, T_J	140°C	100°C	120°C
MTBF, Design for		$> 10^5$ hr	$> 10^5$ hr
Module Efficiency	40%	42%	40%
Cooling	Forced Air Cooled	Water Cooled Cold Plate	Forced Air Cooled

Table 5 **Solid State UHF Module Parameters (From Hoft [1] with permission)**

Transmit	**Specifications Limits**
Frequency	433 ± 13 MHz
RF Peak Power Output	284 - 440 W 350 W avg
Pulse Width	0.250 to 16 ms
Duty Cycle	0 to 25%
Efficiency	42% avg 35% min
Antenna Port Tracking (Circularity)	
Amplitude	0.25 dB
Phase	3°
Phase Tracking Error	14° rms
Phase Settling	25° peak
Pulse Droop	0.7 dB max
Rise Time	OTP req 3 μs nominal
Harmonics	-90 dBc
Receive	
Gain	27 dB min ± 1 dB
Noise Figure	2.9 dB rms max
Limiter Power Handling	440 W, 16 ms, 25% DF
Phase Tracking	10° rms
Dynamic Range	KTBF to -28 dBm (1 dB CPRSN)
No. Phase Shifter Bits	4
Phase Shifter Error	4.6° rms

Table 6 **PAVE PAWS Transceiver Module Electrical Specifications (From Hoft [1] with permission)**

	ABMBA/ECOM Module*	ECOM Module**
Peak Power (W)	27-32	200
Duty Cycle	30% long term	15%
Average Power (W)	8.1-9.6 long term	30
Module Efficiency	19-25%	25%
Life of Module (hr)	8.8 x 10^5	300,000
Module Gain (dB)	24-25	20
Device	Silicon Bipolar Transistors	Silicon Bipolar Transistors
Metalization	Aluminum	Gold
Collector Efficiency (%)	50***	50
Module Voltage (V)	28	40-45
Junction Temperature (°C)	125 (typical all)	120
Pulse Width (μs)	10-2000	100
Date Completed	1974	Expected Late 1977

*Funded by US Army Advanced Ballistic Missile Defense Agency (ABMDA); monitored by ECOM; built by Microwave Semiconductor Corp. (MSC).

**Under development by MSC; planned for late 1977.

***Output device since modified to take ∞:1 mismatch; originally capable only of handling 10:1 mismatch. As a result, collector efficiency has been reduced to 40%. Also, gold metalization used instead of aluminum.

Table 7 **S-Band (3.1-3.5 GHz) Solid State Modules**

Frequency	3.1-3.5 GHz
Peak Power	65 W*
Duty Cycle	15%
Pulse Width	100 μs
Collector Efficiency	35%
Junction Temperature	120°C (for 30°C heat sink)
Internal Circuitry	Ballasted and matched

*65 W presently (January 1977) could be achieved in a single device if a larger package than the standard AMPAC (Internally-Matched Microwave Power Transistor) were available. A 40 W device could be packaged into the AMPAC for short pulse operation; it would provide 27 W for CW or long pulse operation.

Table 8 **ECOM High Power S-Band Bipolar Transistor Development (Developer: MSC)**

(ℓ) is less than a quarter wavelength [$\ell < \lambda/4$] ; a shunt indicator is made by shorting the stub; and a series inductor is made by using a series line section.) Photoetched lumped-element circuits use discrete inductors (made from spirals, for example) and capacitors (possibly using metal-oxide-metal (MOM)). A lumped inductor or capacitor must be much smaller than a wavelength. Consequently, lumped-element circuits are much smaller in size than distributed-element circuits at frequencies like UHF or L- or S-band. For example, at L-band, a simple LC circuit consisting of a series inductor shunted by a parallel capacitor (which occupies about 1″ by 2″ of substrate area using distributed-circuitry) takes about 0.1″ by 0.2″ using lumped-circuitry. Lumped-element circuits also weigh less and, in large quantities (like 10,000 or more), cost less. As you go higher in frequency, like to X-band,

Figure 8 **Lumped-Element 1W, 2 GHz Transistor Amplifier Circuit (150 x 120 mil sapphire chip) (From Caulton [8] with permission)**

the much-less-than-a-wavelength criterion for lumped-elements is more difficult to meet, the tolerances required for small element values being difficult to realize. As a result, there is no clear cut advantage for lumped-elements at the higher frequencies. Lumped-element circuits are more complex to fabricate (consisting of dielectric sandwiches) than are microstrip circuits (consisting of one layer).

Figure 8 shows a 1 W, 2 GHz lumped-element amplifier having a 40% collector efficiency. Complete transceiver modules for space applications using lumped-element circuits are expected to weigh less than 0.1 pound.

Monolithic-technology involves growing the active devices (transistors, diodes) along with passive components on a common semiconductor (silicon, GaAs) substrate. RF amplifiers (power and low noise), phase shifters, logic, and transmit/receive (TR) switches, together with interconnecting circuitry, are fabricated monolithically on a common substrate; hence, die attach and wire bonding operations are eliminated in assembly. Monolithic-technology promises to be the lowest cost technology in very large quantities (one million units or more), and it would be the ultimate in small size and weight.

Other Results

Table 9 [9] summarizes the state of the art of CW bipolar transistors from L- to X-band. The state-of-the-art of other devices is given in Tables 2 through 6 in Chapter 1 and in Table 1 of Chapter 19.

Antenna Arrays and Their Components

Table 10 lists some of the major phase-phase steered arrays built since the early work of John Allen's [10] group at Lincoln Laboratory.

Dome Antennas [12, 13]

An interesting new technology in phased arrays is the development of the dome antenna; see Figure 84 of Chapter 1 for the first dome antenna (1974, receive only) and Table 10 for the Sperry Dome Radar (1976) [12, 13]. This type of antenna incorporates a lens over a planar array in order to achieve hemispherical coverage. Figure 9 gives the basic principle behind the dome antenna system. For the situation depicted, the dome lens causes a beam generated by the horizontal planar array to point slightly below the horizon. In general, the lens increases the off-boresite angle of the beam radiated from the planar array. For the case illustrated, the planar

Freq. (GHz)	P_o (W)	Power Gain (dB)	Eff.* (%)	V_{cc} (volts)	Status
1.0	35-40	8-9	50-65	28	Prod
1.1	250-300**	6-7	40-45	50	Prod
2.0	24	7	40-45	24	Prod
2.0	40	8	48-52	24	Dev (TRW)
3.0	8	6-7	40-45	28	Prod
4.0	5	5	30-35	28	Prod
5.0	5	6	22-25	24	Dev (TRW)
6.0	1.5	4	36	28	Dev (HP)
8.0	1.0	6	25	12	Dev (TI)
10.0	1.0	4.5	20	12	Dev (TI)

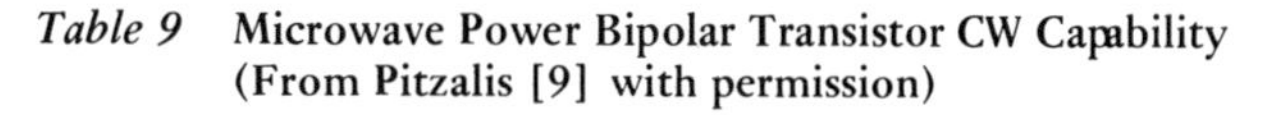

*Collector Efficiency, $\eta_C = \frac{P_C}{P_{DC}}$

**Pulsed (10 μs Pulse Width, 1.0% Duty Factor)

Table 9 **Microwave Power Bipolar Transistor CW Capability (From Pitzalis [9] with permission)**

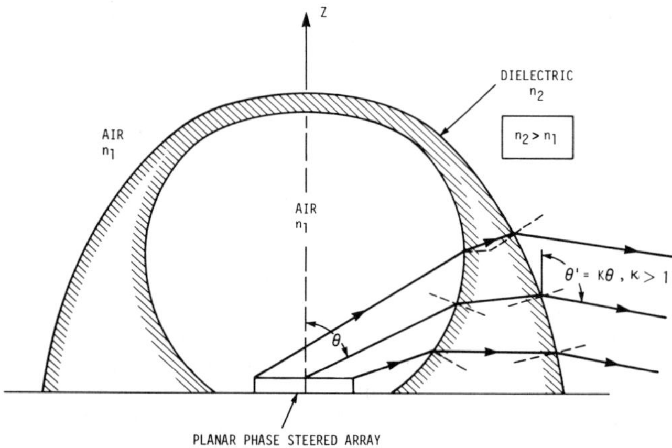

Figure 9 **Principle by which dome lens provides greater than hemispherical coverage for a planar phased array; n_1 and n_2 represent indices of refraction of air and dielectric, respectively.**

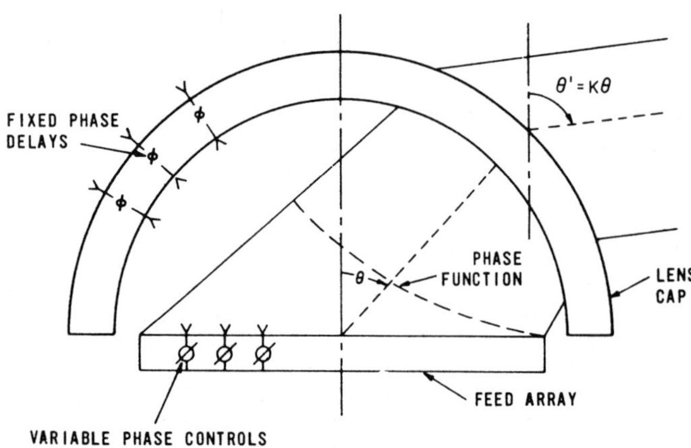

Figure 10 **Use of dome consisting of fixed phase shifters to simulate dielectric lens of Figure 9 (From Liebman, Schwartzman, and Hylas [12] with permission)**

array is scanned to generate a beam which is 60° off boresite. The lens increases this off-boresite angle to produce a beam which is 110° off-boresite. Only for zero off-boresite angle is there no magnification of the off-boresite angle of the beam produced by the planar array. Beam steering control is achieved by the phase shifters in the planar array.

The lens, instead of being made of a dielectric of varying thickness as shown in Figure 9, actually is made of a dome of fixed phased shifters, as shown in Figure 10. The values of the phase shifters over the dome are selected to simulate the varying thickness of the dielectric lens of Figure 9.

It is instructive to view the operation of the dome antenna from another viewpoint. The dome antenna can be thought of as a lens which alters the element pattern of the radiating elements of the planar array; see Figure 11. The area under the element gain pattern remains fixed. Hence, the dome has the effect of redistributing the planar array gain versus the off-boresite angle. For an amplification constant, K, (see Figures 9 and 10) equal to 1.5, the loss in antenna gain at boresite relative to the ideal gain obtained without the dome (i.e., relative to that obtained with the ideal cosθ element pattern fall-off) is 4.2 dB; see Figure 12. Note that for K = 1.5, hemispherical coverage is achieved for off-boresite scans to 60° with the planar array. The amplification constant can actually vary with θ to produce gain patterns tailored to the application; see Figure 12. The applications for which the dome antenna is cost effective remain to be seen.

A phase shifter using a titanium dioxide resistive gate has the potential of reducing the cost of future phased arrays [14]. It has the potential of switching 100 W of S-band RF power with less than 1 μW of control power — 1/50,000 the amount needed with conventional PIN diode phase shifters.

Continuous-Aperture Scanning

All phased arrays that have been built or are under construction to date divide the antenna into many individual radiating elements which are roughly a half-wavelength apart with each requiring a phase shifter. Therefore, a large number of radiating elements and phase shifters are required. It would be desirable to use a continuous aperture (i.e., continuous-aperture scanning) instead.

One technique that looked promising is the Ferroscan technique [15]. With this ferrite scanner, it was hoped to be able to replace ten or more radiating elements and their phase shifters with a single radiating horn and phase shifter; however, a competitive technique did not result. A second technique which appears more promising for special applications is the series ferrite scan technique [16]. It may be useful for applications requiring small arrays and not very large instantaneous bandwidths. The series ferrite scan technique uses a waveguide having one long series-connected phase shifter inserted lengthwise, to which the radiating elements of the guide are connected; see Figure 13. Electronic scanning is achieved by controlling the propagation velocity or phase shift per unit length of the ferrite loaded waveguide. The phase shift is obtained by applying a current to the solenoid, which induces a longitudinal magnetic field in the ferrite. This technique requires one driver circuit per column or row of a phased array, with each row or column being made up of the ferrite loaded waveguide. Two-dimensional scanning is achieved by driving the feeds to the rows or columns with another ferrite loaded waveguide.

Using the series ferrite scan technique, a one-dimensional X-band phased array has been developed for the Naval Electronics Laboratory in San Diego by Rockwell International. This array is capable of scanning ±45°.

One way to reduce the cost of thinned arrays (such as that of the Cobra Dane array; see Figure 18 of Chapter 1) is to not use dummy elements. However, the design of thinned arrays without dummy elements is not well enough understood at present. Investigation of thinning without dummy elements may be fruitful.

Radar	Design	Date Completed
ESAR (Bendix)	One tetrode per radiating element 746 radiating elements IF phase shifting	1960
AN/FPS-85 (Bendix)	High-power-multiple transmitters Separate transmit and receive arrays Confined feed Thinned receive array Diode phase shifters	1968
HAPDAR (Sperry)	Monpulse space feed Thinned Diode phase shifters	1965
PAR (GE)	High power-multiple transmitters Monopulse confined feed Subarrays Diode phase shifters	1974
MSR (Raytheon)	High power Monopulse space feed Fully filled Diode phase shifters	1969
AN/TPN-19 PAR (Raytheon)	Offset monopulse space feed Optical magnification reflect array Limited scan Ferrite phase shifters	1971
PATRIOT (Raytheon)	Monpulse space feed Fully filled Ferrite phase shifters	1975
AEGIS (RCA)	Multiple transmitters Monopulse confined feed Varying size subarrays Ferrite phase shifters	1974
Sperry Dome (Feasibility) Radar (Sperry)	360° in azimuth; zenith to 30° below horizon C-band, 1 MW peak, 5 kW average, 50 ft range resolution 2 s volume search frame time, 427 pps Dome-cylinder lens, 6 ft diameter; confined feed	
COBRA DANE (Raytheon)	High power-multiple transmitters Very wide bandwidth Monopulse confined feed Thinned Subarrays	1976
PAVE PAWS (Raytheon)	Solid State Thinned	Under construction

Table 10 **Phase-Phase Steered Arrays (After Kahrilas [11])**

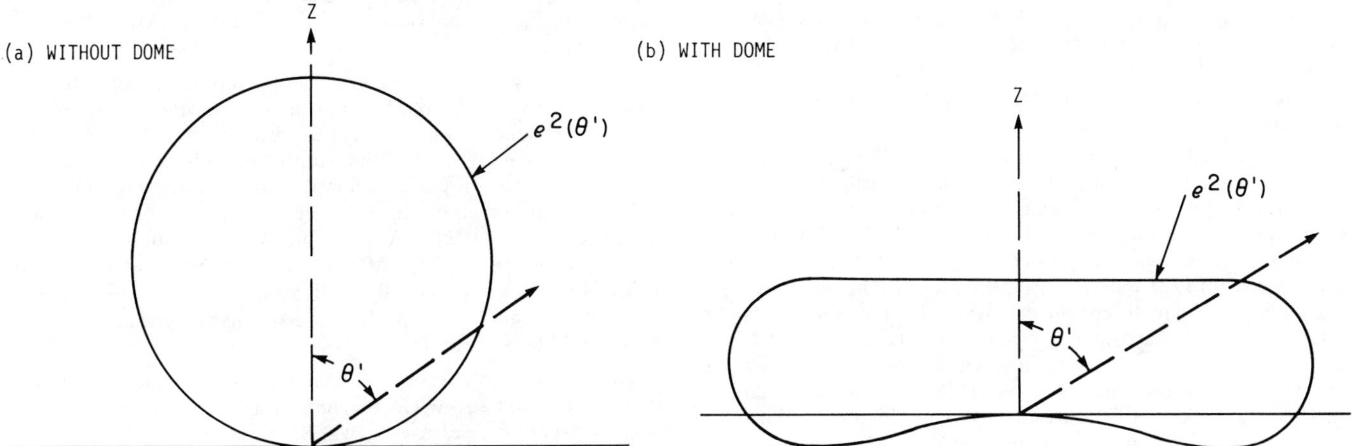

Figure 11 Effect of dome on planar phased array element pattern: (a) element pattern without dome lens and (b) with dome lens

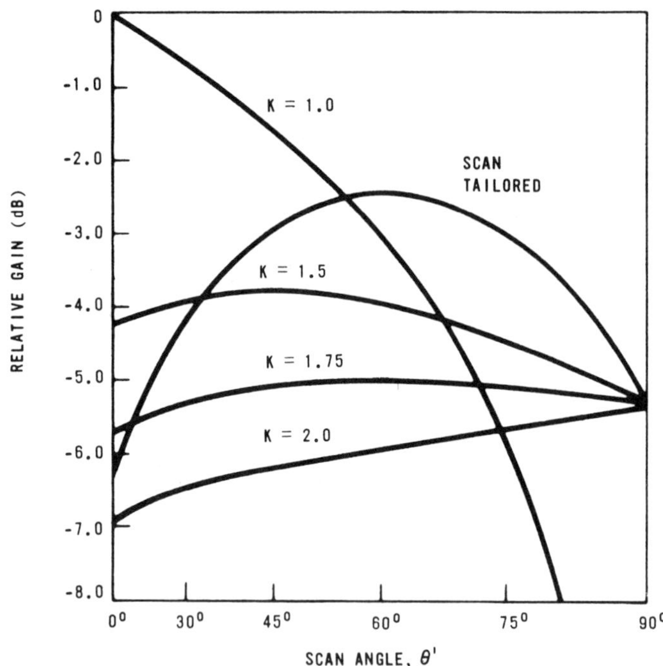

Figure 12 **Dome Antenna Gain (or, equivalently, Element Pattern Gain) vs. Scan Angle, θ', for Different Amplification Constants, K. K = 1 corresponds to the antenna gain obtained without the dome lens, the ideal $\cos\theta'$ antenna pattern. For the scan-tailored pattern, K varies with scan angle, θ, of Figure 10. (From Liebman, Schwartzman, and Hylas [12] with permission)**

Adaptive Arrays [17]

There are three types of adaptive array systems; in order of decreasing complexity, they are fully adaptive arrays, partially adaptive arrays (the subarray approach), and sidelobe canceller systems. The fully adaptive array is one in which each element of the array is controlled adaptively. This technique is useful for small arrays having up to approximately 100 radiating elements. The technique is not attractive for larger arrays because the number of arithmetic steps is proportional to the cube of the number of radiating elements and high numerical accuracy is required [17]. The fully adaptive array technique has the desirable feature of being able to compensate for the antenna element-to-element errors which produce poor sidelobes. This ability was demonstrated by Chapman [17] for a linear array consisting of 32 elements and for which the element-level errors produced a sidelobe level 15 dB below isotropic; see Figure 14a. Making this array adaptive and subjecting it to severe jammer interference spread throughout most of the sidelobe region resulted in a sidelobe level which ranged from 35 to 48 dB below isotropic in the region of the interference; see Figure 14b.

Sidelobe canceller systems use adaptive control for a small number of array elements (or groups of array elements) called auxiliary antennas. These systems place nulls in the sidelobes of the antenna in the direction the jammers (as do the fully adaptive arrays). The maximum number of jammers that can be handled is approximately equal to the number of independently controllable auxiliary antenna elements for the case where the receiver processing bandwidth, B_S, is narrow (i.e., when $B_S \ll 1/\tau_p$, where τ_p is the propagation time across the main antenna or, more exactly, across the projection of the antenna along the line-of-sight to the jammer having the largest off boresite angle); see Figure 15. When wideband jammers are used, the number of auxiliary antennas that must be controlled adaptively is larger by a factor approximately proportional to $B_S\tau_p$. The sidelobe cancellation for a

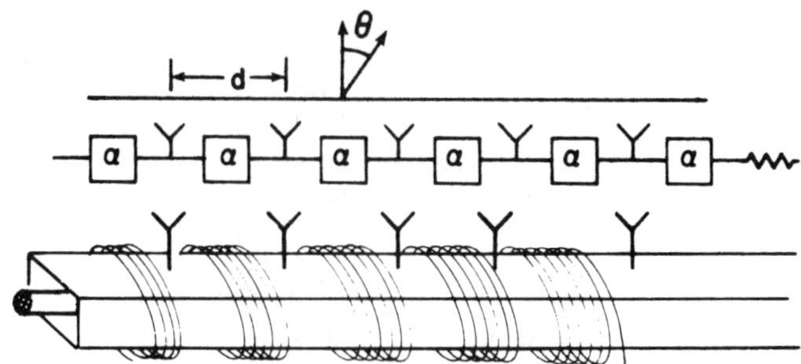

Figure 13 **Series Ferrite Scan Technique. Waveguide-enclosed long series connected ferrite phase shifter to which the radiating elements are connected. Solenoid coils, placed around the nonmagnetic waveguide, control the phase shift per unit length [16].**

single loop and single jammer is given (in dB) approximately by $10 \log_{10}(1/3)(\pi B_S\tau_p)^2$.

The partially adaptive array is a compromise between the fully adaptive array and sidelobe canceller system. For the partially adaptive array groups (subarrays) of radiating elements are controlled instead of the individual radiating elements. The number of degrees of freedom thereby is reduced from n (the number of elements in the array) to somehwere on the order of n/10 to n/50 [17]. For some applications, this type of array performs as well as a fully adaptive array, provided the element errors are sufficiently low [17]. However, for cases where the errors are high, the partially adaptive array behaves like a sidelobe canceller system.

Tubes

Efficiency

In the last decade, the dc to RF power conversion efficiency of TWTs has gone from 20% to over 50% (saturated operation). Litton has built a 200 W CW, 12 GHz PPM TWT having an overall efficiency of 53% (on the ground, 50% in space) [18]. This tube is being used on the Communication Technology Satellite (CTS) built for the NASA Lewis Research Center. The tube uses eight depressed collector stages to achieve its high efficiency, a technique proposed and funded by Dr. H.G. Kosmahl of the NASA Research Center. Dr. Kosmahl indicates that an efficiency of 75% is achievable at 4 GHz with a PPM TWT.

An octave bandwidth (4.8-9.6 GHz) 330-550 W CW electronic counter measures (ECM) TWT is under development by Dr. Kosmahl which should provide in excess of 40% overall power efficiency [19]. Multistage depressed collectors and spent-beam refocusing techniques are being used in this joint USAF–NASA funded program.

High efficiency crossed field amplifiers (Amplitrons) and klystron tubes are being considered for transmission of solar generated power from synchronous-altitude space stations. William C. Brown (Raytheon), the inventor of the Amplitron, indicates that 90% overall efficiency should be achievable by 1990 for a 6.5 kW CW, 2.45 GHz Amplitron having 7 dB gain; efficiencies of 80% already have been achieved [20]. A.D. LaRue (Varian Associates) indicates that the VKS-7773, a 50 kW 2.45 GHz, 50 dB klystron, provides 74% efficiency [21]. By adding a depressed collector and making other modifications, he indicates that an efficiency of 87% would be achievable.

A 92% dc to RF conversion efficiency has been achieved at L-band with a Raytheon 25 kW CW magnetron. Due to output circuit losses, the overall tube efficiency was 85%.

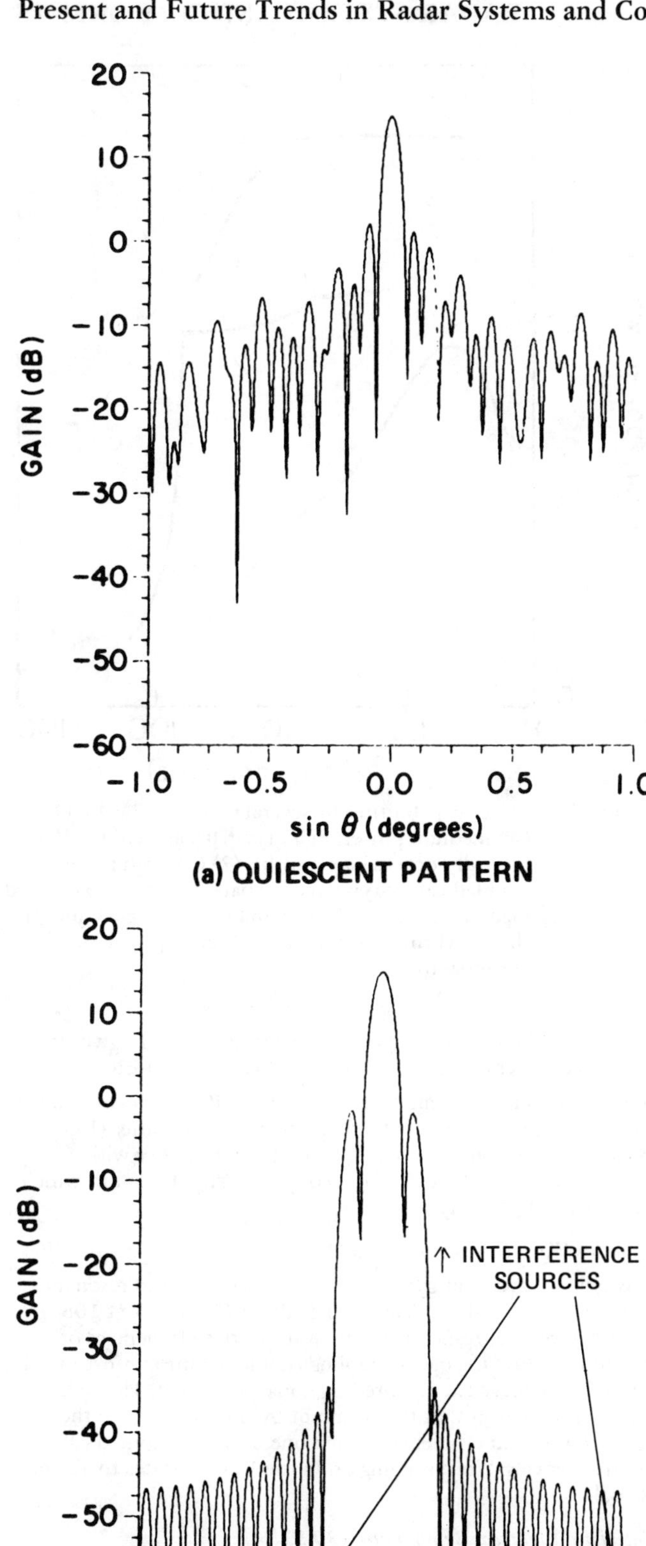

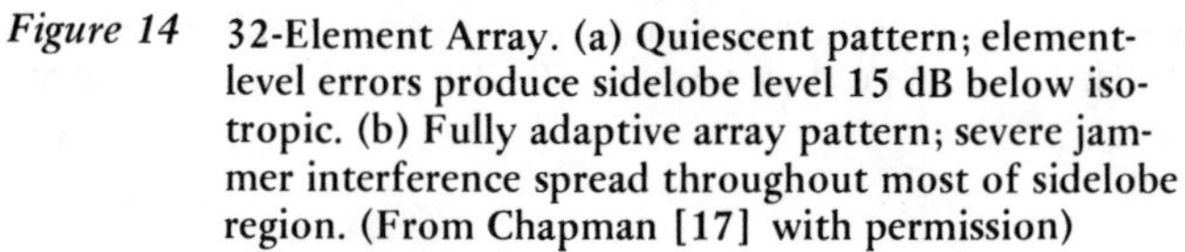

Figure 14 32-Element Array. (a) Quiescent pattern; element-level errors produce sidelobe level 15 dB below isotropic. (b) Fully adaptive array pattern; severe jammer interference spread throughout most of sidelobe region. (From Chapman [17] with permission)

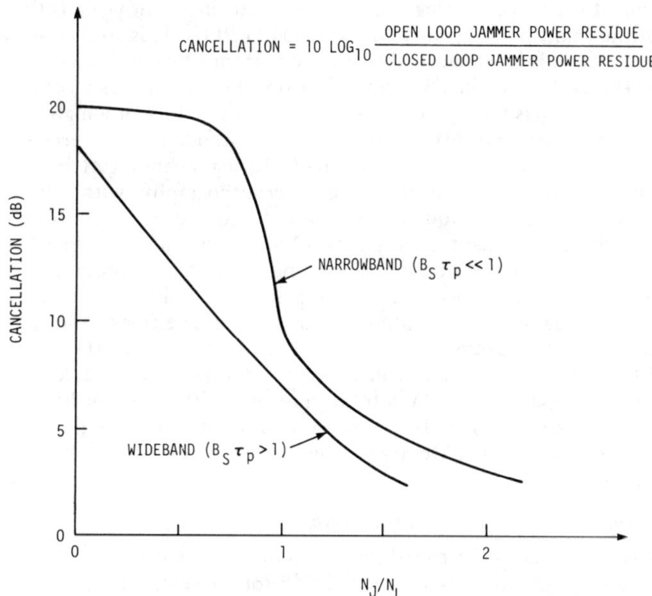

Figure 15 Typical Jamming Signal Rejection vs. N_J/N_L, where N_J and N_L equal number of jammers and loops, respectively.

Life

Non-gridded TWTs and klystrons have achieved very long life. The non-gridded klystron tubes used on the BMEWS have achieved an average life of 83,000 hours, with some tubes having achieved over 100,000 hours [21]. The 10 W CW TWT put on the Pioneer V is still in operation (1977) after twelve years of use. The OGO TWT has been in operation for ten years (1977) without failure. The Hughes 294H TWT (100 W CW, 12 GHz) is designed to operate in excess of 30,000 hours, it uses three depressed collectors to achieve a minimum overall efficiency of 52% (between 11.95 and 12.13 GHz). It is a variation of a second tube funded by the NASA Lewis Research Center for the Communication Technology Satellite.

Newly-developed cathode materials have aided in increasing the life of tubes. The coated-particle cathode (CPC) developed by Bell Laboratories [22] allows for cathode operation at low temperatures, 785°C versus 1,000°C used for conventional dispenser cathodes (also called tungsten matrix, barium impregnated, and B-type cathodes). A CPC is used by Raytheon in the Cobra Dane 1.175-1.375 GHz QKW-1723 TWT having an average power of 10.5 kW and peak power of 175 kW. Accelerated life tests on the entire cathode-gridded gun structure of this tube indicate an emitter life of 100,000 hours for a cathode loading of 0.7 A/cm^2.

Dr. Kosmahl [18] indicates that, for cathode loadings of 0.5 A/cm^2, extrapolation based on emission rates obtained with the new Phillip's M-type cathode (which operates at the low temperature of 850°C) predict lifetimes of 20 to 40 years. However, he also points out that the extrapolation may not be valid.

The life expectancy of gridded tubes usually is not as good as that of non-gridded tubes. However, appreciable progress has been made in recent years. At UHF and L-band, gridded ring-bar helix TWTs having average powers of 10-15 kW, peak powers of 200 kW, and expected operating lifetimes of 50,000 to 100,000 hours are in service [23]. A CPC and shadow grid (nonintercepting grid) are used. Lifetimes of over 20,000 hours have already been achieved in the field for these tubes. The MTBF realized thus far in high power phased array service exceeds 30,000 hours. At S-band, it should be possible to build, using techniques now available, a pulsed TWT having an average power of 10 kW and peak power of 250 kW with a life of between 15,000 and 50,000 hours.

One of the most exciting tube developments in recent years is that of the thin-film field emission cathode (TFFEC). This cold cathode, which is being worked on by the Stanford Research Institute [24], can be used in TWTs and klystrons. Because it uses a cold cathode, it has the potential of virtually infinite life for a high current emission of 50 A/cm^2 [25]. This cathode is the interesting marriage of semiconductor manufacturing technology (thin-film-technology and electron beam microlithography) with tube technology. The cathode is composed of a square array of molybdenum cones which typically have 12 μm center-to-center spacings. Each cone has a base diameter and height on the order of 0.5 and 1.5 μm, respectively, and is placed in a hole in the cathode having a diameter and depth of about 1.5 μm. The tips of the cones are very sharp (500Å radius) so that only a low voltage (100 to 300 V) is needed for electron emission from the cathode. Each cone emits about 100 μA; a total emission of 20A/cm^2 has been demonstrated [25]. A 100 cone array providing 2 mA total emission (equivalent to 3A/cm^2) has operated in excess of 12,000 hours.

Cathode-Driven Crossed-Field Amplifier

Raytheon has a CFA called the "cathode-driven CFA" which operates at high gain (30 dB versus 10 dB for conventional CFAs) while still realizing the CFA's high efficiency (greater than 50%), low noise characteristics, and wide bandwidth (> 10%). The tube uses a cold cathode which makes it possible to incorporate in the cathode surface a slow wave structure into which the drive signal is fed. The RF synchronized space charge spokes therefore are formed where the electrons are emitted at the cathode surface rather than by fringing RF fields from the anode circuit (as in the conventional Amplitron). An added advantage of this approach is the natural isolation one achieves between output (anode) and input (cathode) RF circuits, preventing oscillations at high gain levels.

Electronically Tuned Pulse Magnetrons [26]

English Electric Valve (EEV) Company Ltd. is now marketing two magnetrons which can be tuned electronically to one of a preselected number of frequencies (the number being dependent on the number of auxiliary resonant cavities). The tuning is achieved in a few nanoseconds by a control voltage of 1 to 3 kV applied to the multipacitor-cavity electrode. The auxiliary resonant cavity is designed so that when an appropriate RF voltage is applied between its electrodes, a controlled electron multipactor discharge occurs, causing a resonance shift and changing the magnetron frequency. Two auxiliary cavities permit a 100 MHz tuning range at X-band with 50 kW output, and 30 MHz at S-band with 200 kW. The auxiliary cavities can be built into most any existing magnetron to achieve electronic tuning. EEV is exploring obtaining intrapulse FM modulation with this technique.

EEV also uses a piezoelectric crystal to control a ring tuner (see Figure 9 of Chapter 23) for fast magnetron tuning.

Millimeter Wave TWTs

TWTs are capable of high power in the millimeter region. For example, the Hughes 819H TWT can supply 6 kW of CW power at 54.5-55.5 GHz with an efficiency of 16%. Using depressed collectors, better efficiency could be achieved. Figure 16 shows the capabilities of CW and pulsed millimeter TWT tubes as well as low frequency TWTs [27].

Extended Interaction Oscillator (EIO) Tubes [28]

A recent development in millimeter tubes is the EIO which, like a magnetron, is an oscillator. Its region of application is from 30-300 GHz. The tube's efficiency ranges from 1-20% CW; its electronically tunable bandwidth ranges from 0.2-0.4%; its mechanically tunable bandwidth is up to 7%; and its weight is 5-40 lbs., 90% of which is in the permanent magnet. Figure 17 gives estimates of the CW and pulsed power achievable. Table 11 gives typical characteristics of a medium power 94 GHz pulsed EIO tube.

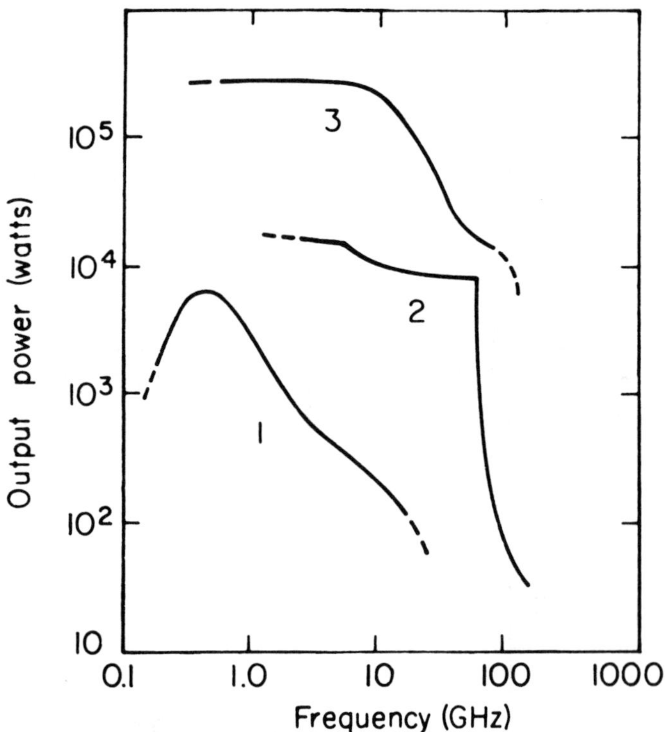

Figure 16 **Power capabilitites of several classes of TWTs. (1) CW medium power, helix type; broadband (> 40%, typically octave bandwidth). (2) CW high power, coupled-cavity type; narrowband (< 15%). (3) Pulsed high power type; narrowband (⩽ 10%), peak power shown. (From Osepchuk and Bierig [27] with permission)**

Under development is a modification of the EIO which could be used as an amplifier – the extended interaction amplifier (EIA). It is expected to result in a small, lightweight amplifier with a bandwidth about 10% of that of a coupled cavity TWT, but which would cost considerably less.

Millimeter Wave Gyrotron [29, 33]

The Russians have built gyrotron oscillators (cyclotron resonance masers) which provide 12 kW and 1.5 kW of CW power at 108 GHz and 326 GHz, respectively, with respective efficiencies of 30% and 6% [29]. Cryogenic cooling (liquid helium) unfortunately is needed to achieve the required high magnetic field. Theoretically, there is no reason that tubes cannot by built to provide these output powers and efficiencies when operating as amplifiers at these frequencies. Work is going on in the United States to achieve this goal.

Relativistic-Electron Beam Tube [30-33]

NRL has developed a technique for generating 1 MW pulses at the sub-millimeter wavelength of 0.4 mm (750 GHz). The technique involves stimulated magneto-Raman scattering of a pump electromagnetic wave from the electrons in a relativistic-beam traveling at 0.9 ft/ns (close to the speed of light). A 15 GHz pump of hundreds of megawatts is used. The 15 GHz pulsed power for the pump is generated using a gyrotron (cyclotron resonance maser).

By scattering from the beam-vacuum interface, 300 kW at about 40 GHz has been produced using a 10 GHz, 150 kW pump source.

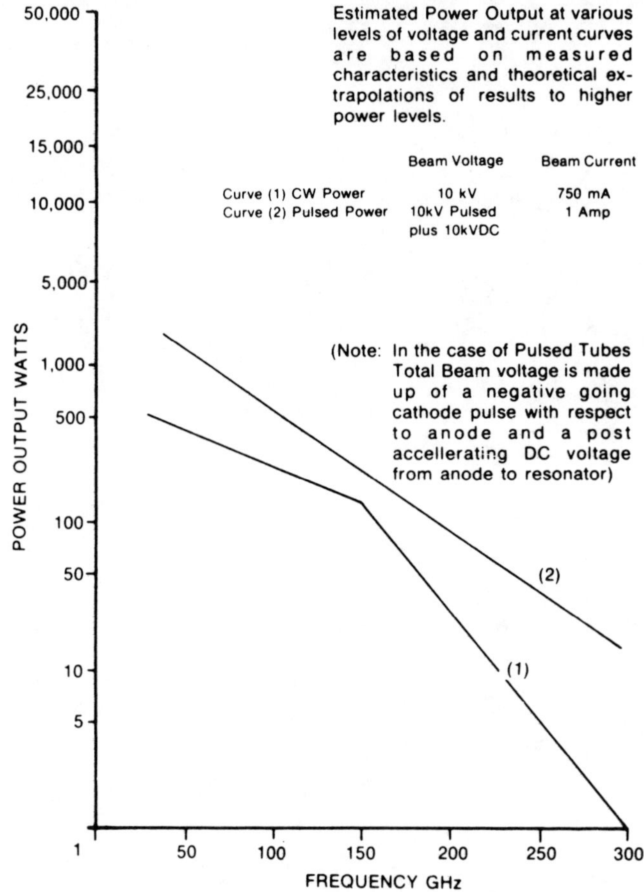

Figure 17 **RF Power Output Capabilitites for Extended Interaction Oscillator (From Varian Canada [28] with permission)**

Frequency	94 GHz
Power Output (Pulsed Peak Power)	> 1000 W
Electronic Tuning Range	150 MHz
Temperature Coefficient	-2.0 MHz/C°
Cathode Modulation	
Sensitivity	0.15 MHz/V
Cathode to Anode Voltage (Pulsed)	10-12 kV peak
Anode to Body Voltage (CW)	8-10 kV
Cathode Current	500 mA peak
Pulse Length	100 μs
Duty Cycle (Water cooled or Forced Air)	.025
Form Factor	≈ 3″ x 3″ diameter
Weight (Rare Earth Magnets)	< 6 lb

Table 11 **Typical Medium Powered EIO Parameters (From Varian Canada with permission)**

References

[1] Hoft, D.J.: "Devices and Techniques for All Solid State Radars," Paper No. 80, 1976 IEEE EASCON, Washington, DC, 26-29 September 1976.

[2] Davis, M.E.; Smith, J.K.; and Grove, C.E.: "L-Band T/R Module for Airborne Phased Array," *Microwave Journal, Vol. 20,* pp. 54-60, February 1977.

[3] Klass, P.J.: "Marines to Test New Surveillance Radar," *Aviation Week and Space Technology, Vol. 105,* pp. 56-58, 6 December 1976.

[4] LaCombe, D.J.; and Naster, R.J.: "Reliability Prediction for Transistors," General Electric Co., *Report R-76ELS 022,* funded by Rome Air Development Center, April 1976.

[5] Lazar, S.: "Solid State Power Amplifiers for S-band Phased Array Radars," *MSN, Vol. 5,* pp. 77-84, February/March 1975.

[6] Cohen, E.D.: *Private communication,* Naval Research Laboratory; see also Davis, R.: "Improved Power FETs and Impatts Spark Interest at Solid-State Conference," *MSN, Vol. 7,* pp. 13-20, April 1977.

[7] Caulton, M.; Hershenov, B.; Knight, S.P.; and DeBrecht, R.E.: "Status of Lumped Elements in Microwave Integrated Circuits – Present and Future," *IEEE Transactions on Microwave Theory and Techniques, Vol. MTT-19, No. 7,* pp. 588-599, July 1971.

[8] Young, L.; and Sobol, H. (eds.): *Advances in Microwaves, Volume 8,* Academic Press, 1974; see sections by M. Caulton.

[9] Pitzalis, O.: "Bipolar Transistor and FET Devices: Present and Future," Boston IEEE Lecture Series *Modern Radar Techniques, Components, and Systems,* 28 October 1976; see also Pitzalis, O.: "Status of Power Transistors – Bipolar and Field Effect," *Microwave Journal, Vol. 20,* pp. 30-34 and 61, February 1977.

[10] Allen, J.L.; et al.: "Phased Array Radar Studies," *MIT Lincoln Lab Technical Report 228,* 12 August 1960; see also Allen, J.L., et al.: "Phased Array Radar Studies," *MIT Lincoln Lab Technical Report 236,* July 1960-July 1961.

[11] Kahrilas, P.J.: "Phased Array Trends," Paper No. 22-2, IEEE ELECTRO '76, Boston, Massachusetts, 11-14 May 1976.

[12] Liebman, P.M.; Schwartzman, L.; and Hylas, A.E.: "Dome Radar – A New Phased Array System," *The Record of the IEEE 1975 International Radar Conference,* Washington, DC, 21-23 April 1975, pp. 349-353.

[13] Stangel, J.J.; and Valentine, P.A.: "Phased Array Fed Lens Antenna," *US Patent No. 3,755,815,* 28 August 1973.

[14] *Microwaves, Vol. 15,* p. 16, January 1976.

[15] Stern, E.; and Tsandoulas, G.N.: "Ferroscan: Toward Continuous-Aperture Scanning," *IEEE Transactions on Antennas and Propagation, Vol. AP-23, No. 1,* pp. 15-20, January 1975.

[16] *Microwaves, Vol. 14,* pp. 14 and 16, August 1975.

[17] Chapman, D.J.: "Partial Adaptivity for the Large Array," *IEEE Transactions on Antennas and Propagation, Vol. AP-24, No. 5,* pp. 685-695, September 1976.

[18] Kosmahl, H.: "Microwave Power Tubes for Space Applications," IEEE International Electron Devices Meeting, Washington, DC, December 1976.

[19] Kosmahl, H.G.; and Ramins, P.: "Small-Size 81- to 83.5-Percent Efficient 2- and 4-Stage Depressed Collectors for Octave-Bandwidth High Performance TWTs," *IEEE Transactions on Electron Devices, Vol. ED-24, No. 1,* pp. 36-44, January 1977.

[20] Brown, W.C.: "Considerations in the Design of a CFA for the Power from Space Application," IEEE International Electron Devices Meeting, Washington, DC, December 1976.

[21] LaRue, A.D.: "High Efficiency Klystron CW Amplifier for Space Applications," IEEE International Electron Devices Meeting, Washington, DC, December 1976.

[22] *Bell System Technical Journal,* pp. 2375-2404, December 1967.

[23] Handy, R.A.: *Private Communication,* Raytheon Company.

[24] Spindt, C.A.; Brodie, I.; Humphrey, L.; and Westerberg, E.R.: "Physical Properties of Thin-Film Field Emission Cathodes with Molybdenum Cones," *Journal of Applied Physics, Vol. 47, No. 12,* pp. 5248-5263, December 1976.

[25] Kosmahl, H.G.: "Microwave Tubes in Space Applications," Boston IEEE Aerospace and Electronic Systems Chapter Meeting, 5 April 1977.

[26] Pickering, A.H.; Lewis, P.F.; and Brady, M.: "Electronically Tuned Pulse Magnetron," IEEE International Electron Devices Meeting, Washington, DC, pp. 145-148, December 1975.

[27] Osepchuk, J.M.; and Bierig, R.W.: "Future Trends in Microwave Power Sources," IEEE ELECTRO '76, Paper No. 1, Boston, Massachusetts, pp. 1-8, 11-14 May 1976.

[28] Cunningham, M.: "Millimeter Wave Power Tube Developments," *Countermeasures, Vol. 2,* pp. 34-36 and 64, June 1976.

[29] Zaytsev, N.I.; Pankratova, T.B.; Petelin, M.I.; and Flyagin, V.A.: "Millimeter- and Submillimeter-Wave Gyrotrons," *Radio Engineering and Electronic Physics, Vol. 19,* pp. 103-106, March 1975.

[30] Davis, G.R.: "Navy Researchers Develop New Sub-Millimeter-Wave Power Sources," *Microwaves, Vol. 15,* pp. 12-13, December 1976.

[31] Granatstein, V.L.; Schlesinger, S.P.; Herndon, M.; Parker, R.K.; and Pasour, J.A.: "Production of Megawatt Submillimeter Pulses by Stimulated Magneto-Raman Scattering," *Applied Physics Letters, Vol. 30,* pp. 384-386, 15 April 1977.

[32] Granatstein, V.L.; Sprangle, P.; Parker, R.K.; Pasour, J.; Herndon, M.; Schlesinger, S.P.; and Settor, J.L.: "Realization of a Relativistic Mirror: Electromagnetic Backscattering from the Front of a Magnetized Relativistic Electron Beam," *Physical Review A, Vol. 14,* pp. 1194-1201, September 1976.

[33] *IEEE Transactions on Microwave Theory and Techniques,* June 1977; see *Microwaves, Vol. 16,* p. 14, July 1977, for brief summary of some of the articles in this issue.

Part 6
Special Topics

The laser (invented in 1958 by Townes and Schawlow [1]) has become a mature technology. In Chapter 24, Jelalian covers the fundamentals of laser theory and technology for radar application. Covered are the laser radar equation (with its similarities and differences to the microwave radar equation), laser noise terms, laser radar detection and estimation, laser transmitters and detectors, and several laser example systems.

In Chapter 25, Morrison introduces the complicated subject of tracking and smoothing in simple terms, also providing interesting insight into the development of this field which began with the work of Gauss. Introduced are the process equation, observation relation, transition matrix and equation, the concept of state variables, and polynomial smoothing. Least squares smoothing, minimum variance smoothing, and the Kalman filter (for which Kalman received the IEEE's highest award, its Medal of Honor in 1975) are introduced and related. In Chapter 26, Sheats relates the Kalman filter to the classical Weiner filter.

In Chapter 27, Weil discusses the important subject of spectrum control, a topic that must be dealt with in all radars.

Finally, in the "Present and Future Trends" section at the end of this part, the latest developments in the use of lasers for precise range measurements, the state-of-the-art of fiber-optic communication systems, and some recent results on pulsed radar out-of-band radiation are summarized.

References

[1] Schawlow, A.L.; and Townes, C.H.: "Infrared and Optical Masers," *Physical Review, Vol. 112,* pp. 1940-1949, 15 December 1958.

Chapter 24 Jelalian

Laser Radar Theory and Technology

Laser radars are a subject that can be viewed from different perspectives. Every individual looks at the electromagnetic spectrum a little differently, selecting those wavebands of interest and ignoring the remaining ones. Because the subjects basically are the same, there are similarities that run through all observations; however, because of the different wavebands, there are also significant differences.

The discussion here centers on comparisons between radars at microwave and laser wavelengths. The signal-to-noise equations associated with both incoherent and coherent detection systems at laser wavelengths are covered, along with a description of the more popular lasers, atmospheric transmission factors, detectors, detection statistics, and the various laser radars that have been built.

The electromagnetic spectrum is divided into a variety of wavebands, as shown in Figure 1. Those involved with the initial nomenclature selection for the microwave wavelengths used all the imaginable superlatives – high frequencies, ultra-high frequencies, super-high frequencies, etc. As a result of cryptic World War II efforts, letter nomenclature evolved for subdivisions of the microwave band, utilizing such designations as S-, C-, L-, X-, and K-band moving to the shorter wavelengths, the millimeter waveband, J, eventually is reached. Up to the latter region exists the conventional radar transmission band. The other end of the electromagnetic spectrum, with no superlatives left, is named for where the human eye had response (the visible band – from 0.5 to 0.7 μm); as the wavelengths became longer compared to 0.7 μm (red), the wavebands were successively called the near-infrared, the mid-infrared, and the far-infrared bands, as shown in Figure 1.

In this chapter, the discussion is not in terms of centimeters (10^{-2} m), but rather in terms of micrometers (μm) (10^{-6} m); therefore, any microwave considerations that are wavelength-dependent should be scaled by about 3 or 4 orders of magnitude. This scaling changes the context to one of typical laser operation. Useful lasers operate at wavelengths in the ultraviolet, visible, near-infrared, mid-infrared, and far-infrared regions where atmospheric windows occur. Figure 2 illustrates the transmission of over a thousand feet of the atmosphere at sea level [1]. Approaching the electromagnetic spectrum from the radar wavebands, the first atmospheric window which allows propagation to occur is the 8-14 μm band. Continuing past that point, the atmosphere is highly non-transmissive until the mid-infrared waveband between 3-5 μm. The absorption that occurs in the atmosphere, preventing transmission at the far- and mid-infrared bands, is caused basically by the absorption of atmospheric constituents such as ozone, water vapor, and carbon dioxide. Proceeding along, the next atmospheric windows occur in the near-IR region from 0.7-2.5 μm and, subsequently, the visible region. When utilizing these smaller wavebands, haze and scattering attenuation caused by particulates in the atmosphere are of prime concern.

Table 1 illustrates the amount of attenuation as a function of wavelength for both laser and microwave wavebands. The chart is compiled from broadband data and does not include the fine-grain atmospheric absorption values into which laser lines may drift occasionally (and which can be prevented by suitable transmitter design); see examples in Chapter 15.

In the visible waveband where the ruby laser operates (0.7 μm) the atmospheric "seeing" conditions have a very dramatic effect upon

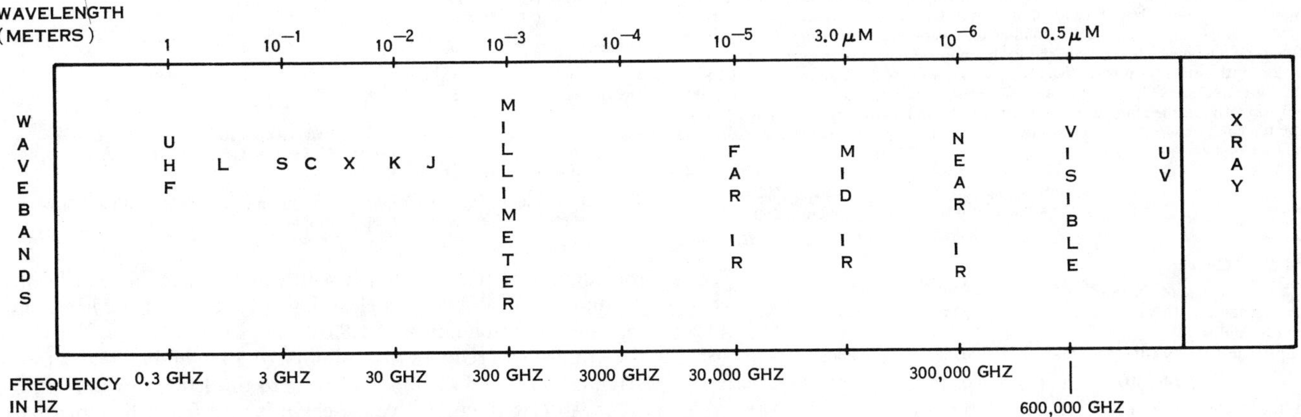

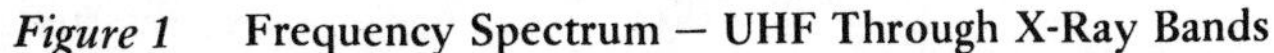

Figure 1 Frequency Spectrum – UHF Through X-Ray Bands

dB/nmi	0.7 µm	1.06 µm	3.8 µm	10.6 µm	0.3 cm [6,7]	1 cm [6,7]	3cm [6,7]
0.4	Extremely Clear	Extremely Clear	Clear (V=23km)	Sub-Arctic Winter			12.5 mm/hr
1.0	Standard Clear (V=25km)*	Standard Clear	Haze (V=5km)	Clear		2.5 mm/hr	
1.4		Clear			Rain 2.5 mm/hr	Fog (V=100 ft)	
1.8	Clear (V=15km)			Haze; Sub-Arctic Summer; Light Fog**[3] (V=2km)	Fog (V=400 ft)		50 mm/hr
2.8	Light Haze						
3.2	Light Haze (V=8km)			Mid-Latitude Summer; Light Rain [4] 2.5 mm/hr			
4.3	Medium Haze			Fog**[3] (V=0.5km)			
5.2	Medium Haze (V=5km)			Tropical [2]		12.5 mm/hr	
7.1		Haze					
9.5	Haze (V=3km)					2.5 mm/hr	
11.0				Medium Rain [4,5] 12.5 mm/hr	Rain 12.5 mm/hr		
16.5				25 mm/hr	Fog (V=100 ft)	30 mm/hr	

*V = Visibility

Table 1 **Attenuation (one-way) vs. Wavelength**

***Editor's Note* – Fog usually is characterized by its optical visibility. However, this description is inadequate in that the same visibility conditions can arise for fog of different liquid water content and/or different water droplet size distribution (see Figure 28 of Chapter 15). It is the liquid water content alone which determines fog's attenuation at microwave frequencies, with the attenuation directly proportional to the liquid water content. Hence, fogs with the same visibility but with different particle size distributions and liquid contents can produce different attenuations at microwave frequencies. An advection fog can have five times the liquid water content of a radiation fog with the same visibility conditions [a]. This fact adds complexity to the problem of comparing optical and microwave frequency fog attenuation.

For example, the Goodwin and Nussmeier [3] fog and light-fog conditions have respective visibilities of 500m (1640 ft) and 2000m (6560 ft) and liquid water contents of 2.06 mg/m^3 and 0.58 mg/m^3 – corresponding to radiation fog conditions. An advection fog (the type fog on which the Ryde and Ryde empirical relationship in the caption of Figure 28 in Chapter 15 was based) yields liquid water contents of 42 mg/m^3 and 5.8 mg/m^3 for the same visibility conditions. The table below gives the attenuation at both 10 and 100 GHz for the liquid water contents corresponding to these radiation and advection fog conditions, and it compares the results to the measurements obtained at 10 µm by Goodwin and Nussmeier.

Visibility		Fog 500m (1640 ft)		Light Fog 2000m (6560 ft)	
Type Fog		*Advection*	*Radiation*	*Advection*	*Radiation*
Liquid Water Content (mg/m^3)		42	2.06	5.8	0.58
One-Way Attenuation (due only to fog particles) (dB/nmi)	10.6µm	–	2.00	–	0.75
	10 GHz	0.0037	0.00018	0.00052	0.000052
	100 GHz	0.34	0.017	0.047	0.0047
Total One-Way Attenuation (dB/nmi)	10.6µm	–	3.5	–	1.7
	10 GHz	0.037	0.033	0.034	0.033
	100 GHz	1.12	0.80	0.83	0.78

The second through fourth rows give the attenuation due to the fog alone, as does Figure 28 of Chapter 15. The total attenuation includes the CO_2 absorption and the H_2O vapor absorption for the 10.6µm range; for the microwave frequencies, it includes the H_2O vapor and O_2 absorptions [b].

For 10.6µm, the attenuation due to the fog alone was obtained from the total measured attenuation by subtracting the estimated CO_2 absorption of 0.29 dB/km and the estimated H_2O vapor absorption (found using Equations (21) and (21a) of Chapter 15). The meteorological conditions for the 10.6µm measurements and the assumed microwave propagation are:

	10.6µm [3] *Light Fog*	10.6µm [3] *Fog*	Microwave [b]
Temperature (°F)	60	58	68
(°C)	15.6	14.4	20
Humidity (%)	40	75	43
Water Vapor Density (g/m^3)	5.3	9.3	7.5

[a] Richard, V.W.: "Millimeter Wave Radar Applications to Weapon Systems," *Memorandum Report No. 2631,* US Army Ballistic Research Laboratory, Aberdeen Proving Ground, Maryland.

[b] Rosenblum, E.S.: "Atmospheric Absorption of 10 to 400 kMCPS Radiation," *Microwave Journal, Vol. 4,* pp. 91-96, March 1961.

the amount of attenuation per nautical mile that the laser beam encounters. Visibility on a clear day is 15 km, with a corresponding one-way attenuation of 1.8 dB/nmi. Introduction of haze into the atmosphere reduces the visibility to 3 km and increases the attenuation to 9.5 dB/nmi.* Moving to the 1.06 µm wavelength (that associated with a YAG laser), it is seen from the table that these conditions produce less attenuation than they did in the visible region; further movement to the mid-infrared at 3.84 µm (the wavelength of a strong emission line of the DF chemical laser) results in only 1 dB/nmi of attenuation in a haze – close to the lowest value for the visible/IR region. Progression into the far-

*See Equation (22) and caption of Figure 25 in Chapter 15 for the approximate relationship between visibility and attenuation.

infrared at 10.6 μm increases the haze attenuation to 1.8 dB/nmi; the predominant attenuation mechanism in that region is absorption by H_2O. The second column under the 10.6 μm wavelength addresses the problems of humidity in sub-arctic, mid-latitude, and tropical environments. McClatchey's [2] data for the atmospheric attenuation under tropical environments result in a factor of 5.2 dB/nmi; correspondingly, a chief concern is system operation in fog and rain. The third column under λ = 10.6 μm illustrates specific data accumulated by Goodwin [3], Rensch [4], and Chu [5]. From this listing it is clear that light fog, having a visibility (in the visible waveband) of 2 km, would yield an attenuation of approximately 1.8 dB/nmi. A radiation fog having visibility of 0.5 km would increase the attenuation to about 3.5 dB/nmi. These figures may be contrasted with those commonly found in microwave radar textbooks (Barton [6] and Skolnik [7]) for attenuation at microwave wavelengths from 3 cm to 0.3 cm. These last three columns indicate the ability of longer wavelength electromagnetic systems to penetrate fog with less attenuation.

The data with regard to rain are not necessarily so apparent. The table indicates that a rainfall rate of 12.5 mm/hr has equivalent attenuation at 10.6 μm and at 0.3 cm. Subsequent columns in this chart show the attenuation of 1 cm (5.2 dB/nmi) and 3 cm (0.4 dB/nmi) systems in 12.5 mm/h of rain. (See Chapter 15 by Brookner for further data comparing laser and microwave atmospheric attenuation.)

Because the subject is electromagnetic propagation, the microwave radar range equation still applies.*

$$P_R = \frac{P_T G_T}{4\pi R^2} \times \frac{\sigma}{4\pi R^2} \times \frac{\pi D^2}{4} \times \eta_{ATM}\ \eta_{SYSTEM}$$

where:

P_R = received signal power

P_T = transmitter power

G_T = transmitter antenna gain $(4\pi/\theta_T{}^2)$

σ = effective target cross section

R = system range to target

D = receiving aperture diameter

η_{ATM} = atmospheric transmission factor

η_{SYSTEM} = system transmission factor

This equation shows that the received power is a function of transmitter power, the gain of the antenna, the effective scattering target cross section (σ), the target range, antenna collection area, and the atmospheric and system transmission factors. The transmitter antenna gain may be expressed by the approximate relation $4\pi/\theta_T^2$. Typical laser beamwidths are on the order of 1 mrad and, as a result, the typical antenna gain at laser wavelengths is about 70 dB.

This standard range equation, however, applies only in the far field of the antenna. At typical microwave bands of λ = 1 to 10^{-3} m, the far field distances are quite short, as shown in Figure 3. The far field (Fraunhofer) region of an antenna is typically concerned with the distance from $2D^2/\lambda$ to infinity; in this vicinity, the generalized range equation applies. In some cases, the far field distance occurs within the feed horn assembly of a microwave antenna. As illustrated by the figures, at a 1 μm wavelength a 10 cm aperture has a far field distance of approximately 20 km. As a result, it is not unusual to operate in the near field of the antenna or optical system; thus, modification of the range equation to account for near field operation is required.

The effective target cross section is found as shown below:

$$\sigma = \frac{4\pi}{\Omega}\ \rho_T\, dA$$

*See Glossary at the end of this chapter for a list of symbols

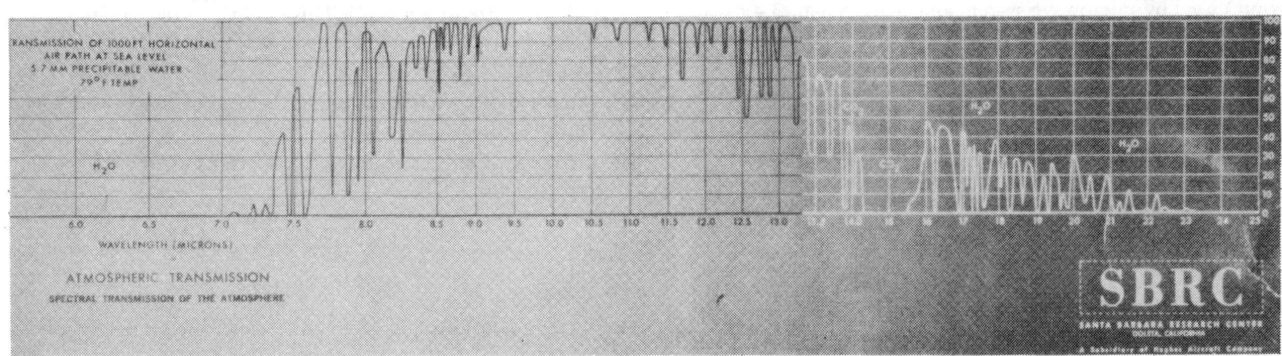

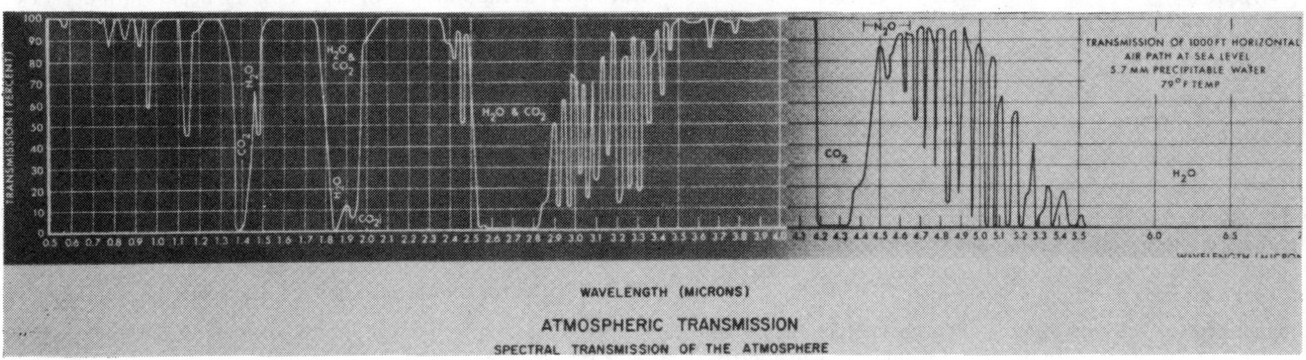

Figure 2 **Atmospheric Transmission for 1000 ft (0.3 km) Path at Sea Level for 0.5 μm to 25 μm (see Chapter 15, Figures 15-24 for additional data)**

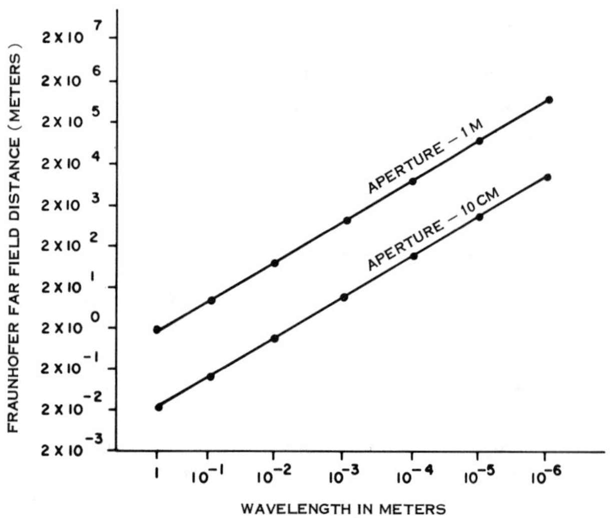

Figure 3 **Far Field Distance vs. Wavelength for 1 m and 0.1 m Apertures**

where:

Ω = scattering steradian solid angle of target

ρ_T = target reflectivity

dA = target area

The target gain is defined as $4\pi/\Omega$. Physicists or "opticers" in the optical sciences tend to replace Ω with the value associated with the standard scattering diffuse target (Lambertian target) having a solid angle of π steradians, thereby reducing the equation to

$$\sigma = 4\rho_T \, dA$$

This situation may be compared to the standard diffuse target (isotropic) used by microwave system engineers, having a 4π scattering steradian angle. For an optical target, Ω is considered to be equal to π; for a diffuse radar target, Ω is considered to be equal to 4π. As a result, conversion between radar and optics must include appropriate corrections.

At laser wavelengths, the cross section in the far field for an extended target (σ_{EXT}) (i.e., one in which the footprint is smaller than the target itself) results in the following relationships:

$$dA = \frac{\pi R^2 \theta_T^2}{4}$$

Thus:

$$\sigma_{EXT} = \tau \rho_T R^2 \theta_T^2$$

resulting in

$$P_R = \frac{\pi P_T \rho_T D^2}{(4R)^2} \eta_{ATM} \eta_{SYS}$$

With narrow laser beams, standard targets for microwave systems may become inverse R^2 extended targets at the optical waveband. Other received power inverse range-dependent functions ($1/R^4$, $1/R^3$, etc.), developed similarly in the microwave region, can also occur in the optical (see Chapters 1 through 3).

If a mirror-type target is used instead of a π steradian diffuse target, a transmitting beamwidth of 1 mrad would yield an approximate 65 dB increase in target cross section. This gain would occur when a mirror target was aligned properly to the receiver to allow the specular reflection from the target to be redirected toward the receiver. Because typical targets have a probabilistic distribution associated with their specular facets, the likelihood that the signal would return to the receiver is very low. As a result, wide signal fluctuation is seen to exist from such a target.

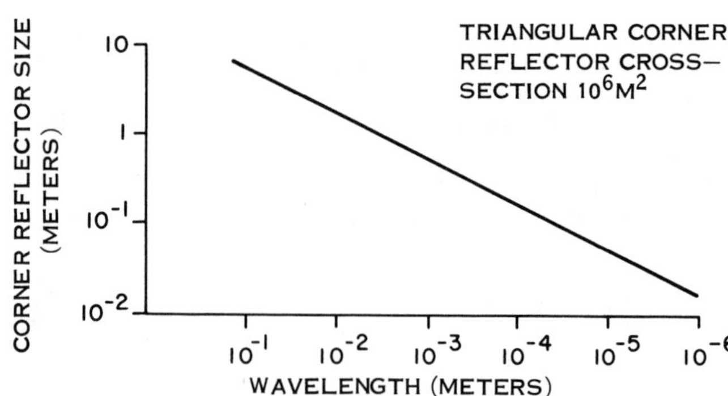

Figure 4 **Triangular Corner Reflector Size vs. Wavelength for $10^6 m^2$ Cross Section**

Figure 4 graphs corner reflector-size as a function of wavelength, according to the following equation:

$$\sigma = \frac{4\pi D^4}{3\lambda^2}$$

where:

σ = target cross section

D = corner reflector size

λ = wavelength

Corner reflectors at microwave wavebands are usually very large; a cross section of 10^6 m^2 would require a 2.2 m corner reflector at 1.0 cm wavelength. Obtaining this same cross section at a 1 μm wavelength requires a corner reflector having a size of 2.2 cm (basically the size of a nickel). This effect is often utilized at optical wavebands to enhance cooperative target cross sections; very small corner reflector sizes can yield very high target gains.

The impact that wavelength has on the beamwidth of the laser transmitter is illustrated below.

The diffraction limited beamwidth of a transmitting source may be expressed as

$$\theta = C \frac{\lambda}{D}$$

where C is a constant and, if λ and D are in the same units, θ is in radians.

Obviously, as the wavelength decreases, the beamwidth decreases. For a 1 cm waveband and a 30.5 cm optical system, the beamwidth is approximately 33 mrad. Similarly, if the wavelength is decreased to 10^{-3} cm (10 μm), the beamwidth becomes 33 μrad for the same antenna size. A constant, C, is utilized in these expressions to illustrate the difference between radar and various optical beamwidth definitions. Optical beamwidths are often specified as the 1/e points; the standard radar notation of beamwidth is typically measured at the half-power points. One should, therefore, be aware of the differences in beamwidth definition at different points in the electromagnetic spectrum.

The equation for the Doppler shifted backscatter signal may be expressed as:

$$f_D = \frac{2v}{\lambda} \cos \gamma$$

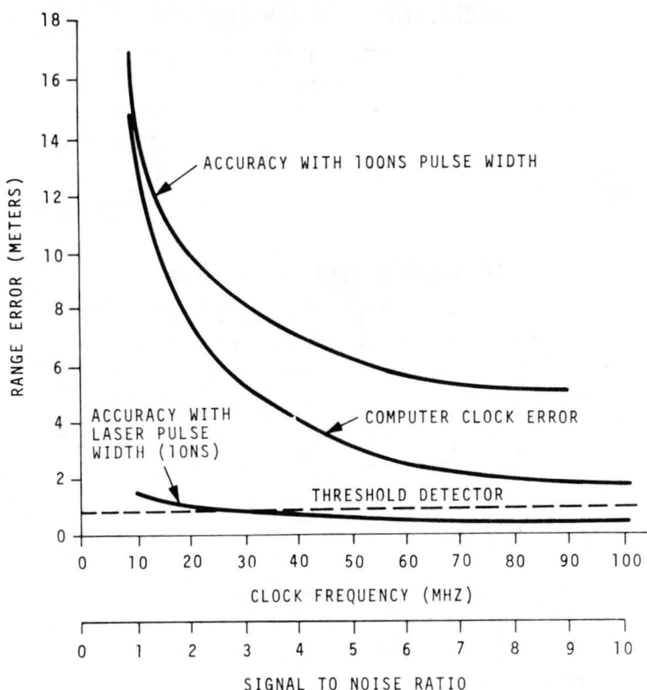

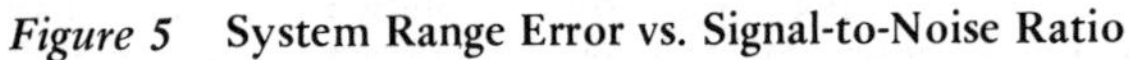

Figure 5 **System Range Error vs. Signal-to-Noise Ratio**

where:

v = target velocity

γ = angle between target velocity vector and line-of-sight

Because of the shorter operating wavelength associated with lasers, there is greater Doppler shift per meter per second of velocity.

A 3 cm wavelength radar would have a Doppler shift on the order of 67 Hz per meter per second of target motion. Correspondingly, operation at 1 μm would yield 2000 kHz of Doppler shift for every meter per second of target motion. The best sources of coherent laser radiation for Doppler detection work occur at approximately 10 μm – at this wavelength, Doppler shift is approximately 200,000 Hz per meter per second of target velocity. A logical question might be – what does this all mean? In microwaves, if it is desired to extract a Doppler shift of a target, that Doppler shift typically is masked within the (sin x)/x response of the transmitted pulsewidth and, as a result, PRF sampling is utilized to sample the small shift. The by-product of this situation is that the usual microwave system is of the high repetition rate pulsed Doppler variety. At optical wavebands, the Doppler shifts are sufficiently large for instantaneous measurement to be performed on a single pulse; therefore, low PRF systems can be configured.

The expressions for the rms measurement errors associated with radars operating at the microwave bands are listed below.

Range Error

$$\sigma_R \doteq \frac{c}{2}\tau\sqrt{\frac{1}{2\ S/N}}$$

Velocity Error

$$\sigma_V \doteq \frac{\lambda}{2\pi\tau}\sqrt{\frac{3}{2\ S/N}}$$

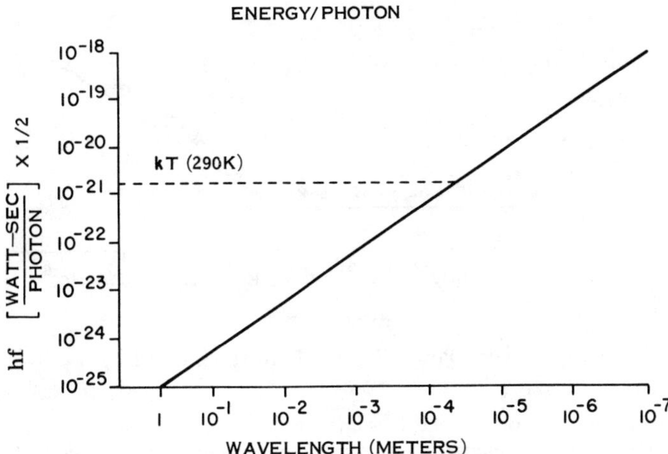

Figure 6 **Photon Energy vs. Wavelength**

Angular Error

$$\sigma_\theta \doteq \frac{K\,\theta}{\sqrt{2\ S/N}}$$

Because of the very narrow pulsewidths (typically on the order of 10 ns) which are obtained easily with solid-state lasers, the range error is decreased significantly (as indicated in Figure 5). Correspondingly, the velocity errors are decreased by the shorter wavelengths involved; similarly, because the beamwidths of laser systems are narrower, the angular error sources are smaller.

These advantages of laser systems indicate that configurations utilizing microwave radars and optical adjuncts may be possible. However, one of the problems associated with lasers regards their ability to scan out a given solid angle. Because their beams are very narrow and their repetition rates are typically very small, lasers suffer a distinct disadvantage as search system devices. A situation in which such an adjunct system might be useful involves a radar footprint too large to resolve multiple targets. A laser system could be used to scan out the radar footprint both to resolve the number of targets within that footprint and to determine the range, velocity, and angular position of each target. Additionally, because the beamwidth is so narrow, the problems of multipath which are prevalent with radar systems are, to all intents and purposes, eliminated.

Detection Considerations

In basic physics books, the dual nature of electromagnetic propagation is discussed frequently, in the sense that electromagnetic fields travel sometimes in a wave train and sometimes as a discrete packet of energy. The packets of energy are detected at laser wavelengths whereas in the microwave region, the context is wave trains. The energy of a quantum of light is measured in units of joules (watt-seconds). The energy associated with one photon (this basic particle of light) is equal to hf, where h is Planck's constant (6.626 x 10^{-34} J-s) and f is the frequency.

It may be difficult to envision, but transmitter frequency can be related to energy. The number of photons arriving in an optical receiver in a second is the received power divided by hf. In order to keep the measurement units related to those of microwave terminology, the discussion is in terms of the number of photons arriving per second, as a result the notation "watt" is used.

Figure 6 illustrates the energy associated with the photon as a function of wavelength. As the wavelength becomes shorter, the energy of a particle increases. Indicated with a dotted line is the stand-

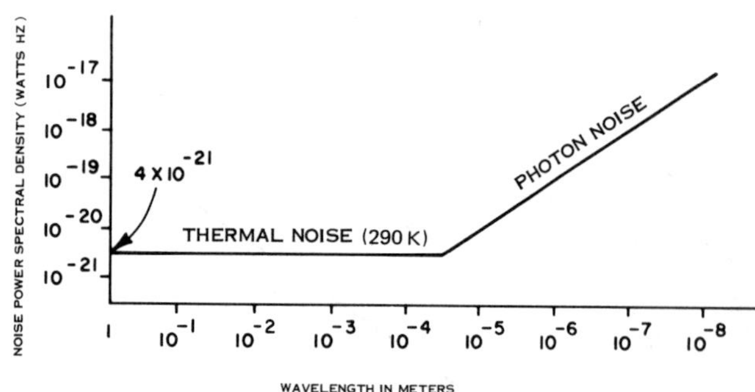

Figure 7 Noise Power Spectral Density vs. Wavelength

ard thermal noise effect. It is impossible to detect a photon at microwave wavelengths — it is the minimum discernable signal in the electromagnetic spectrum, but thermal noise obscures the individual photon in that region. Thermal noise (kT) is derived from a term in physics called *blackbody radiation.* The blackbody radiation law relates the power emitted from a blackbody radiator to the fourth power of the temperature of the radiator. The total blackbody radiation, P_{BBT}, may be expressed as the product of the radiant emittance, I, and the surface area within the field-of-view, A_S. Radiant emittance of a blackbody over all wavelengths is expressed by the Stefan-Boltzmann law as:

$$I = \sigma_\tau T^4$$

Therefore:

$$P_{BBT} = IA_S = \sigma_\tau T^4 A_S$$

where:

σ_τ = Stefan-Boltzmann constant = 5.67×10^{-12} W cm^{-2} K^{-4}

T = absolute temperature (in K)

Utilizing a selective optical filter allows the receiving system to pass only a portion of P_{BBT}, thereby reducing the blackbody background noise contribution.

The spectral radiant emittance of a blackbody into a hemisphere in the wavelength from λ to $\lambda + d\lambda$ may be expressed as:

$$I(\lambda)\,d\lambda = \frac{2\pi c^2 h}{\lambda^5} \frac{1}{e^{(hc/\lambda kT)} - 1} d\lambda$$

where:

$2\pi c^2 h = 3.74 \times 10^{+4}$ W μm^4 cm^{-2}

$$\frac{ch}{k} = 1.438 \times 10^4\ \mu\text{mK}$$

$I(\lambda)\,d\lambda$ is the flux radiation into a hemisphere per unit area of the source, measured in watts per square centimeter.

$$I(\lambda)\,d\lambda = \frac{3.74 \times 10^4}{\lambda^5 \left[\exp\left(\frac{1.438 \times 10^4}{\lambda T}\right) - 1\right]} d\lambda$$

The noise power spectral density versus wavelength is shown in Figure 7.

a) INCOHERENT DETECTION RECEIVER

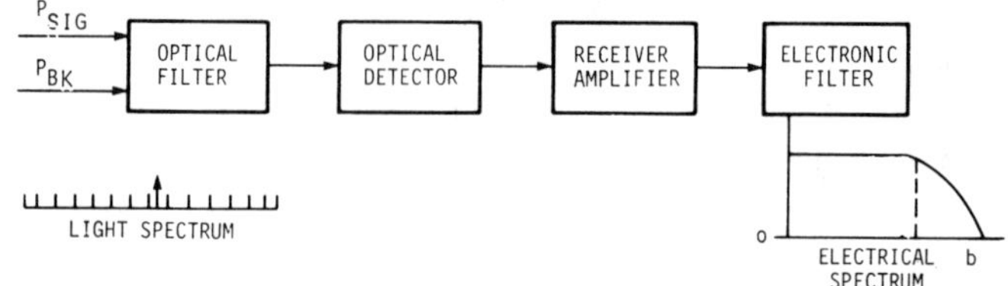

b) COHERENT DETECTION RECEIVER

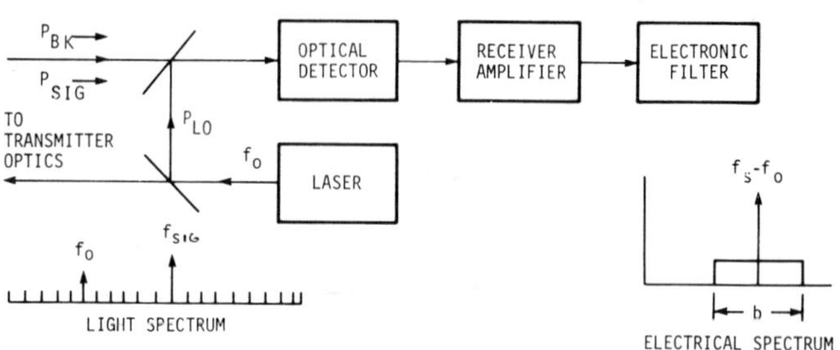

Figure 8 Receiver Systems

Noise Power Spectral Density

Quantum mechanical analyses [8] have shown that an ideal amplifier has a noise power spectral density referred to the input (Ψ) [watts/hertz] as follows:

$$\Psi(f) = \frac{hf}{\exp\left(\frac{hf}{kT}\right) - 1} + hf$$

where:

T = absolute temperature of source

h = Planck's constant

f = transmitter frequency

k = Boltzmann's constant

Expanding exp(hf/kT) in the denominator of this expression in terms of an exponential series results in $1 + (hf/kT) + (hf/kT)^2 + \ldots$. When operating in the microwave band, kT is very large compared to hf, and the exponential series tends to $(1 + hf/kT)$. As a result, the hf terms in the denominator and numerator cancel, leaving kT — thermal noise. Thermal noise then is derived from Planck's blackbody radiation equation.

Plotting this relationship as a function of wavelength indicates that, at wavelengths longer than 10^{-4} μm, thermal noise is the minimum discernable noise source; in the region ranging from that point toward shorter wavelengths, the minimum detectable signal is actually the photon noise.

Having discussed the minimum discernable signal, attention naturally is turned to the noise figure of an optical receiver. As demonstrated thus far, the minimum discernable signals at microwaves and at optical regions are different and comparisons are difficult. However, if the "noise figure" of a coherent optical receiver were to be compared to the thermal noise in the microwave receiver, the following equations could be utilized:

Noise Figure

The SNR power of a microwave receiver may be expressed as:

$$SNR_M = \frac{P_{SIG}}{kT_s \Delta f}$$

where:

P_{SIG} = received signal power

k = Boltzmann's constant (1.38×10^{-23} J/K)

T_s = system noise temperature

Δf = receiver bandwidth

The SNR power for a quantum noise limited coherent optical receiver may be expressed as:

$$SNR_O = \frac{\eta P_{SIG}}{hf\,\Delta f}$$

where:

h = Planck's constant

f = transmission frequency

η = quantum efficiency

The effective noise figure of the coherent optical receiver compared to that of a 290 K microwave system is

$$(NF_{EFF})_O = \frac{hf}{\eta k T_s}$$

at 10.6 μm, hf = 2×10^{-20}, $kT_s = 4 \times 10^{-21}$, and $\eta = 0.5$. Therefore

$$(NF_{EFF})_O = 10$$

These equations compare the signal-to-noise ratio of the microwave receiver in terms of its minimum discernable signal (kT_s) to that of a quantum noise limited optical receiver having a minimum discernable signal of hf. If a 50% quantum efficiency detector were used, the noise figure of the optical system compared to a microwave system having a 290 K system temperature would be about 10 dB poorer at 10.6 μm; at shorter wavelengths, the noise level would be even higher.

The term "quantum efficiency" is basically one which relates the efficiency by which photons of light are converted to electrical signals by the detector. If one photon were to arrive on a detector surface and one photoelectron were to be emitted, the device would have 100% quantum efficiency. For the analysis utilized (that is, a quantum efficiency of 50%), two photons must arrive in order to generate one photoelectron in the receiving system. In effect, this receiver would have an optical noise figure of 3 dB relative to the minimum discernable optical noise.

Receiver Detection Techniques

In Figure 8, diagrams are shown for incoherent and coherent detection receivers. The incoherent detection receiver at optical wavelengths is similar to a video radiometer receiver (i.e., an envelope detector at microwave wavelengths). However, an optical receiver has an additional term besides the signal term, P_{SIG} – the optical background power, P_{BK}, which is due to undesired signals such as sunlight, cloud reflections, and flares. The received signal competes with these external noise sources at the receiver. The received optical power, after suitable filtering, is applied to the optical detector; square-law detection then occurs, producing a video bandwidth electrical signal.

The coherent detection receiver is similar to the incoherent; however, a portion of the laser signal, f_O, is coupled to the optical detector via beamsplitters. As a result, the optical detector has the local oscillator power (P_{LO}) in addition to the received signal power, P_{SIG}, and the competing background terms, P_{BK}.

Comparison of Incoherent and Coherent Detection Receivers

The SNR equation for *incoherent* detection may be expressed as:

$$SNR = \frac{\eta P^2_{SIG}}{2hfB[P_{SIG} + P_{BK} + K_1 P_{DK} + K_2 P_{TH}]}$$

The SNR equation for *coherent* detection may be expressed as:

$$SNR = \frac{\eta P_{SIG} P_{LO}}{hfB(P_{LO} + P_{SIG} + P_{BK} + K_1 P_{DK} + K_2 P_{TH})}$$

where:

η = quantum efficiency

f = transmission frequency

B = electronic bandwidth

P_{SIG} = received signal power

P_{BK} = background power

P_{DK} = equivalent dark current power

P_{TH} = equivalent receiver thermal noise power

P_{LO} = reference local oscillator power

$$K_1 = \frac{1}{\rho_i}$$

$$K_2 = \frac{1}{2\rho_i q}$$

where:

ρ_i = the detector current responsivity

q = the electron charge

The signal-to-noise ratio for the incoherent system has the received signal power squared in its numerator and has a summation of noise terms associated with the return signal, the background signal, the dark current, and the thermal noise of the receiver. The returned signal power and the background power are included as noise sources in the detection process because of the random photon arrival rate. In the coherent detection system, the local oscillator power is an additional source of noise (compared to the incoherent system) and the numerator is related to the product of the received signal power and the local oscillator power. The local oscillator power is very important in the detection process: here, it may be increased so that it overwhelms all of the noise sources. As a result, the local oscillator power in the denominator cancels out the local oscillator power in the numerator; the SNR equation becomes related directly to the received signal power, rather than to the received signal power squared (as with the incoherent system). Additionally, because the local oscillator power becomes the predominant noise source, the coherent detection system typically is background-immune.*

Noise Terms

Noise terms in an optical receiver are not the typical ones considered in the microwave field. As a result, major differences exist between optical receivers and microwave receivers. Background noise in optical receivers includes reflections of signals from the earth, the sun, the atmosphere, clouds, or any other scatterer which contributes an undesired signal to the receiver. An analogy to this background effect is the driving of a car on a dark, foggy evening with the headlights on – a lot of light energy is scattered

**Editor's Note* – For further discussions on incoherent and coherent detection systems performance limits, including atmospheric turbulence effects and multiple adaptive dish systems, see Brookner, E.; Kolker, M; and Wilmotte, R.M.: "Deep Space Optical Communications," *IEEE Spectrum, Vol. 4*, pp. 75-82, January 1967, and Brookner, E.: "Performance of Single- and Multiple-Dish Laser Communication Systems," *Fifth Space Congress*, Cocoa Beach, Florida, 12 March 1968.

from the fog particles, producing a significant amount of backscattered fog energy as well as backscattered road light. As a result, the visibility of the road is limited. In this case, the backscatter from the fog represents an undesired signal because it masks the radiation returned from the ground, the desired signal. This same problem can occur for ungated laser systems propagating radiation into clear air, haze, and clouds.

Signal-induced noise refers to the shot noise caused by the received signal itself coming into a detector. The received signal is not only a source of desired radiation, but it also causes a noise to be generated. This noise is called quantum noise because it is induced by the signal when the signal exists.

Additionally, the detector has a source of noise, commonly referred to as dark current. Even though the detector surface may be blocked from having any radiation applied to it, its internal physics cause a leakage current to exist, typically referred to as dark current. Thermal-receiver noise, 1/f noise, and generation recombination noise can occur in both microwave and optical receivers.

The following equations are those associated with calculating the amount of background radiation which may be incident upon a receiver.

Blackbody Radiation, P_{BB}

$$P_{BB} = \frac{\epsilon \sigma_\tau T^4 \Delta\lambda\, \Omega\, A_R}{\pi} \eta_{SYS}$$

Solar Backscatter, P_{SB}

$$P_{SB} = kS_{IRR} \times \Delta\lambda\, \Omega\, \rho\, \eta_{SYS}\, A_R$$

Atmospheric Solar Scatter, P_{NS}

$$P_{NS} = kS_{IRR} \times \Delta\lambda\, \Omega\, I_S\, \eta_{SYS}\, A_R$$

where:

ϵ = target emissivity
ρ = target reflectivity
T = temperature (K)
$\Delta\lambda$ = optical bandwidth
A_R = receiver area
k = fraction of solar radiation penetrating Earth's atmosphere
S_{IRR} = solar irradiance
I_S = atmospheric scatter coefficient
η_{SYS} = system optical efficiency
Ω = solid angle over which energy radiates from radiating body
σ_τ = Stefan Boltzmann constant = 5.67×10^{-12} W cm^{-2} K^{-4}

The blackbody radiation equation as a product of the radiant emittance, I, and the surface area within the field-of-view, A_S, has been discussed earlier under *Detection Considerations.* The radiant emittance, I, is described by the Stefan-Boltzmann law as $\sigma_\tau T^4$; it includes the blackbody radiation over all wavelengths. The background noise impinging upon a detector may be filtered optically to reduce the background energy competing with desired signal radiation. The equations associated with the calculation of the spectral radiant emittance into a particular waveband over a hemisphere are also covered in that earlier section. Figure 9 illustrates the radiant emittance as a function of wavelength for a variety of blackbody temperatures. As the temperature increases, the peak of the radiant emittance curve also increases. Figure 10 [9] shows the spectral irradiance of the sun (6000 K) as a function of wavelength; these values can be utilized in the equations previously noted. Inasmuch as the energy from the sun may be either reflected or emitted from terrains, clouds, or targets, the reflectivities and emissivities of various terrain materials as a function of wavelength are given in Table 2; more details may be found in Reference 10.

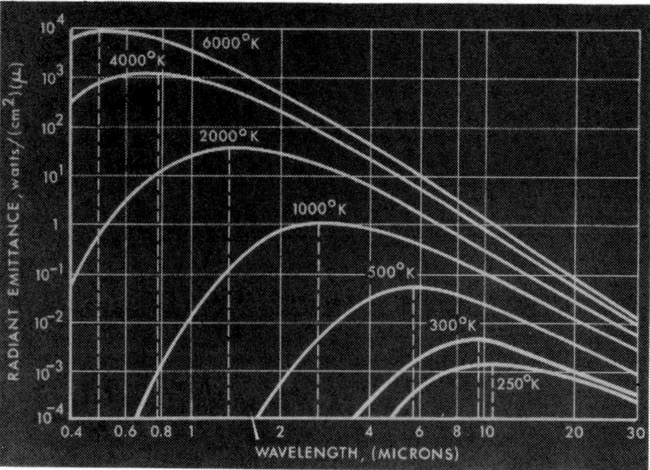

Figure 9 **Blackbody Radiant Emittance vs. Wavelength**

SNR Expression Development

Having discussed the methods by which the background power can be calculated, let us now turn to the derivation of the SNR expressions.

$$SNR = \frac{i_{SIG}^2}{i_{SN}^2 + i_{TH}^2 + i_{BK}^2 + i_{DK}^2 + i_{LO}^2}$$

where:

i_{SIG}^2 = mean square signal noise current
i_{SN}^2 = mean square shot noise current
i_{TH}^2 = mean square thermal noise current
i_{BK}^2 = mean square background noise current
i_{DK}^2 = mean square dark noise current
i_{LO}^2 = mean square local oscillator noise current

The signal-to-noise expression above is shown to be related to the signal current squared divided by the summation of noise current terms squared. As indicated above, the summation of noise terms involves shot noise, thermal noise, background noise, dark current, and (with a coherent detection system) a local oscillator noise. The photons or energy collected from the background result in a fluctuation of the carrier or electron densities in the detector, thereby contributing shot noise.

In the absence of photons at the detector, there is a current flowing, termed the detector dark current. The thermal mean square noise current, which is conventionally referred to as the receiver Johnson noise, is expressed in terms of 4KTBNF/R, where B is the electronic bandwidth, NF is the noise factor of the amplifier following the detector, and R is the detector load resistance. If a local oscillator signal is utilized for a coherent detection system, the local oscillator signal itself generates a shot noise similar to the received signal and background radiation. Expressions for these noise current terms are now given.

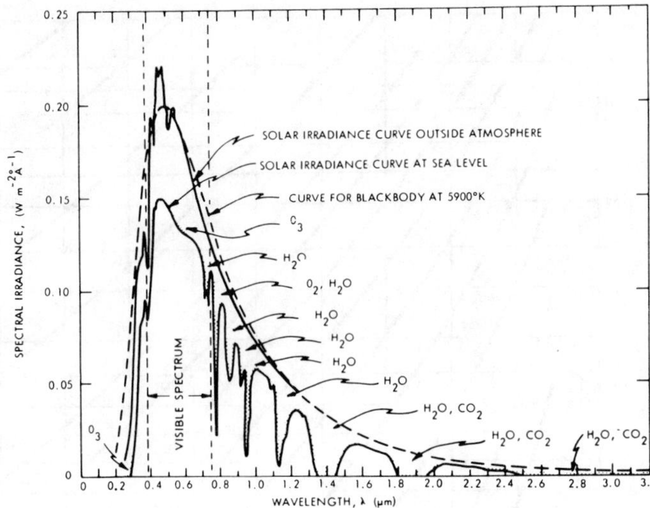

Figure 10 **Spectral Irradiance of the Sun at Mean Earth-Sun Separation (From *RCA Electro-Optics Handbook*)**

Shot Noise Current

Photons or energy collected from backgrounds result in a fluctuation of carrier or electron densities in the detector and thereby contribute shot noise.

$$i_{BK}^2 = 2q\, P_{BK}\, \rho_i B$$

where:

q = electron charge - 1.602×10^{-19} coulombs
P_{BK} = background power
ρ_i = current responsivity
B = electronic bandwidth

Similarly there is a fluctuation in the detector output caused by the random arrival of signal photons.

$$i_{SN}^2 = 2q\, P_{SIG}\, \rho_i B$$

Detector Dark Current

$$i_{DK}^2 = 2q\, I_{DK} B$$

Thermal Noise Current

$$i_{TH}^2 = \frac{4kTBNF}{R}$$

where:

NF = receiver noise factor
R = detector load resistance

Local Oscillator Induced Noise (assuming coherent detection)

$$i_{LO}^2 = 2q\, P_{LO}\, \rho_i B$$

where:

P_{LO} = local oscillator power

For a photo conductor detector, the following noise term can arise:

Generation-Recombination Noise (Photo Conductors)

$$i_{GR}^2 = 4q\, \rho_i (P_{LO} + P_{SIG}) B$$

The signal current is determined as:

$$i_{SIG} = \frac{\eta\, q\, P_{SIG}}{hf} \text{ (incoherent)}; \quad i_{SIG} = \frac{\eta q \sqrt{2\, P_{SIG} P_{LO}}}{hf} \text{ (coherent)}$$

where: η = detector quantum efficiency

	0.7-1.0 μm	1.8-2.7 μm	3-5 μm	8-13 μm
Green Mountain Laurel	ρ = 0.44	ϵ = 0.84	ϵ = 0.90	ϵ = 0.92
Young Willow Leaf (dry, top)	0.46	0.82	0.94	0.96
Holly Leaf (dry, top)	0.44	0.72	0.90	0.90
Holly Leaf (dry, bottom)	0.42	0.64	0.86	0.94
Pressed Dormant Maple Leaf (dry, top)	0.53	0.58	0.87	0.92
Green Leaf Winter Color – Oak Leaf (dry, top)	0.43	0.67	0.90	0.92
Green Coniferous Twigs (Jack Pine)	0.30	0.86	0.96	0.97
Grass – Meadow Fescue (dry)	0.41	0.62	0.82	0.88
Sand – Hainamanu Silt Loam – Hawaii	0.15	0.82	0.84	0.94
Sand – Barnes Fine Silt Loam – South Dakota	0.21	0.58	0.78	0.93
Sand – Gooah Fine Silt Loam – Oregon	0.39	0.54	0.80	0.98
Sand – Vereiniging – Africa	0.43	0.56	0.82	0.94
Sand – Maury Silt Loam – Tennessee	0.43	0.56	0.74	0.95
Sand – Dublin Clay Loam – California	0.42	0.54	0.88	0.97
Sand – Pullman Loam – New Mexico	0.37	0.62	0.78	0.93
Sand – Grady Silt Loam – Georgia	0.11	0.58	0.85	0.94
Sand – Colts Neck Loam – New Jersey	0.28	0.67	0.90	0.94
Sand – Mesita Negra – lower test site	0.38	0.70	0.75	0.92
Bark – Northern Red Oak	0.23	0.78	0.90	0.96
Bark – Northern American Jack Pine	0.18	0.69	0.88	0.97
Bark – Colorado Spruce	0.22	0.75	0.87	0.94

*Estimated average values of reflectance ρ, or emissivity $\epsilon = 1 - \rho$, in the indicated wavelength bands, read from spectral reflectance curves.

Table 2 **Reflectance (ρ) and Emissivity (ϵ) of Common Terrain Features* [10]**

The detector responsivity, $\eta q/hf$, is concerned with the conversion of optical power to receiver current and, as such, is represented by a term ρ_i, the current responsivity, in units of amperes per watt.

Substituting these current expressions into the generalized SNR equation gives the expressions at the beginning of the section on Receiver Detection Techniques.

Detection Probabilities and False Alarm Rates

At high signal-to-noise ratio or operating far from the photon detection limit (i.e., there are many photons arriving), the central limit theorem results in the Gaussian detection statistic used in microwave radar system calculations. Swerling class targets utilized in radar detection theory thus can be employed for laser systems (see Chapters 2 and 3). Because of the shorter wavelength of laser systems, targets which are slow fading at microwave frequencies can be fast fading at laser wavelengths; see Chapter 2.

At low photon or photoelectron levels, where the photon arrivals in one interval are independent of those in any other, Poisson statistics are used for detection. This condition applies to the energy detection or incoherent detection receiver. Here the system designer must compute the average number of the noise photoelectrons emitted ($\overline{n}_N$) in the time interval of interest as well as the average number of signal photoelectrons in the time interval of interest ($\overline{n}_S$) in order to set the decision threshold detecting device at a level of photoelectrons (n_t) corresponding to a distinct probability of detection (P_D) and probability of false alarms (P_{FA}). For the Poisson statistics case, the probability of detection may be expressed as [6, 11, 12]:

$$P_D = \sum_{r=n_t}^{\infty} P(S + N)r = \sum_{r=n_t}^{\infty} \frac{(n_S + n_N)^r}{r!} e^{-(\overline{n}_S + \overline{n}_N)}$$

where: $P(S + N)$ = the probability that r signal plus noise photoelectrons are emitted in time interval, τ.

and the probability of false alarm as:

$$P_{FA} \geqslant \sum_{r=n_t}^{\infty} \frac{(\overline{n}_N)^r}{r!} e^{(-\overline{n}_N)}$$

Figures 11 and 12 permit the computation of P_D and P_{FA}, respectively.

The following example illustrates the use of these curves.

Laser Detection Statistics

Given: signal photoelectrons = 6

Noise photoelectrons in the absence of signal = 1

Pulse width, τ, = 20 ns (video detection)

Average False Alarm Rate = 1 in 10^3

Problem: What is probability of detection?

For Figure 12, enter $P_{FA} = 10^{-3}$, with $\overline{n} = 1$ PE; obtain threshold setting $n_t = 5.8$ PE ≈ 6 PE

For Figure 11, enter $n_t = 5.8$ PE and $\overline{n} = 7$ photoelectrons (6 signal + 1 noise); obtain $P_D \approx 0.7$.

Coherent detection systems, because of the strong local oscillator signals, are shot noise limited, resulting in Gaussian detection statistics similar to those developed for microwave radars. Probability of detection and false alarm statistics may be found in Chapter 2 and the Appendix.

Lasers

Laser action has been observed at a wide number of wavelengths, covering the optical spectrum from the UV through the far IR; there even are masers that operate in the microwave region. Neon lasers have operating wavelengths in the UV, argon lasers have wavelengths in the blue or green region of light, and helium neon lasers operate in the red region of light; such a laser can be purchased for approximately \$100 per milliwatt of power. Additionally, the ruby (chromium aluminum oxide) laser, operates in the red region of the visible spectrum. Semiconductor lasers such as gallium arsenide and gallium arsenide phosphide emit radiation in the near-IR and visible spectra, respectively. Neodymium-type lasers operate at 1.06 μm; such units have been utilized for accurate range measurements and as designators for precision ordinance. Progressing to the mid- and far-IR portions of the electromagnetic spectrum, the lasers are chemical, operating with hydrogen fluorine (2.8 μm), duterium fluorine (3.8 to 4.3 μm), carbon monoxide (5 μm), or carbon dioxide (10.6 μm). At the 10.6 μm wavelength, almost any aspect of microwave system operation may be attained because of the high power and good coherence capabilities of these transmitters.

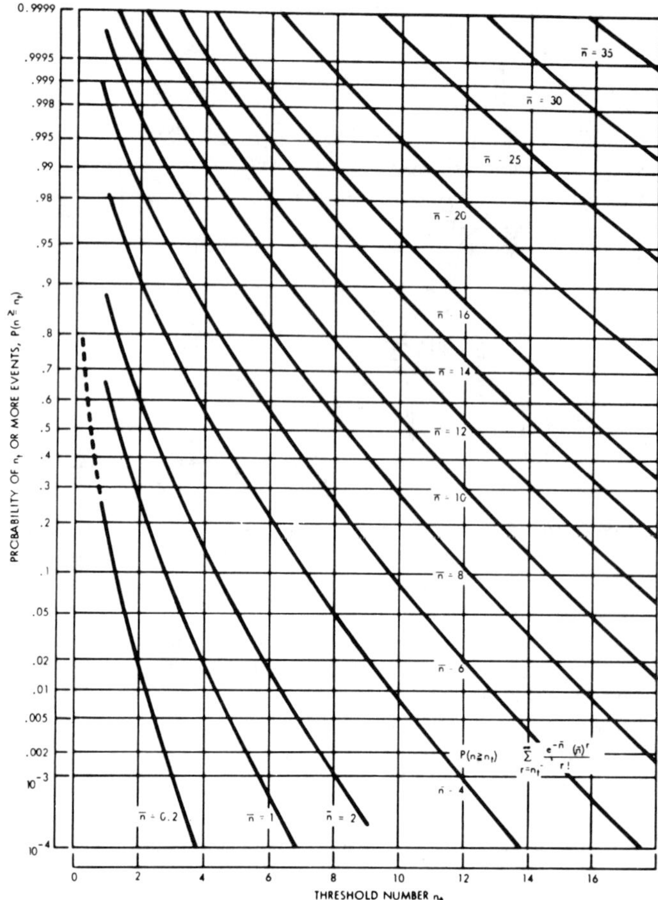

Figure 11 **Probability of n_t or More Random Events with Poisson Distribution When the Expected (or Mean) Number of Events is $\overline{n}$ as a Function of the Threshold Number n_t (From *RCA Electro-Optics Handbook*)**

The laser shown in Figure 13 has a cavity which looks similar to a fluorescent bulb. On one end of the cavity is a fully-reflecting mirror; on the other is a partially-reflecting mirror. Coupled between these two mirrors is an active medium – a ruby rod, as in a ruby laser; a gas, such as CO_2; or a semi-conductive material, such as gallium arsenide. Basically, this material is one which absorbs energy. Energy is coupled into this active medium by means of a pumping medium. The energy could be coupled into the active medium absorption band via electrical, RF, light, or chemical energy. Assume, for example, an active medium that absorbs 0.53 μm energy (green light). Because green light is at a higher frequency than red light, the green region of the optical spectrum has

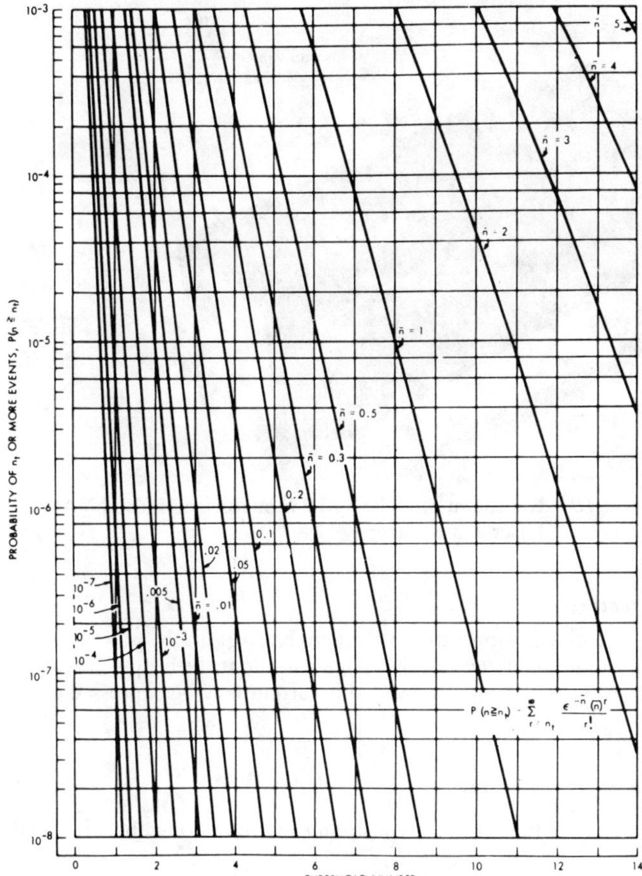

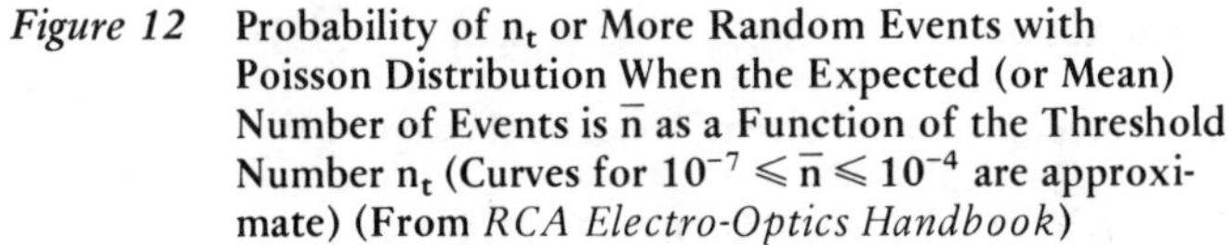

Figure 12 **Probability of n_t or More Random Events with Poisson Distribution When the Expected (or Mean) Number of Events is $\bar{n}$ as a Function of the Threshold Number n_t (Curves for $10^{-7} \leq \bar{n} \leq 10^{-4}$ are approximate) (From *RCA Electro-Optics Handbook*)**

more energy. The pumping energy applied to the active medium results in excitation of the molecules; if enough energy is coupled into this rod from the light source, lasing action occurs — all photons within the active laser medium move at the same rate, forming a narrow coherent beam of energy. In the case of the ruby chromium aluminum laser, the chromium absorbs the green. That green light may be obtained from an electrically excited flash lamp which elevates the energy of the atoms in the ground state of the rod, making them "climb up an atom hill". When the atoms get to the top of this "hill" (inverted population), some drop back to the "ground." Returning to the ground indicates a change of states — the molecules have changed energy states from one of high energy to a region of low. As a result, an energy difference exists. If the top of the "atom hill" is thought of as the region of green light and the "ground" state is conceptualized as the level to which the atoms would like to drop, the difference between the height of the "atom hill" and the "ground" represents an energy increment. As these atoms drop to the "ground" they emit radiation proportional to the difference between the two energy states. For the case of the chromium aluminum laser, the radiation emitted is at 0.6943 μm — red light. Descriptions of this process are shown in Figures 14 and 15. The excited energy causes the atoms to climb the "atom hill" to position 3. Some of the energy drops down to energy state 2; it subsequently falls to the ground state (1), resulting in the emission of red light. Basically, what has happened is that the normal population distribution of the material has been inverted so that there are more atoms in an elevated state than there are in a ground state.

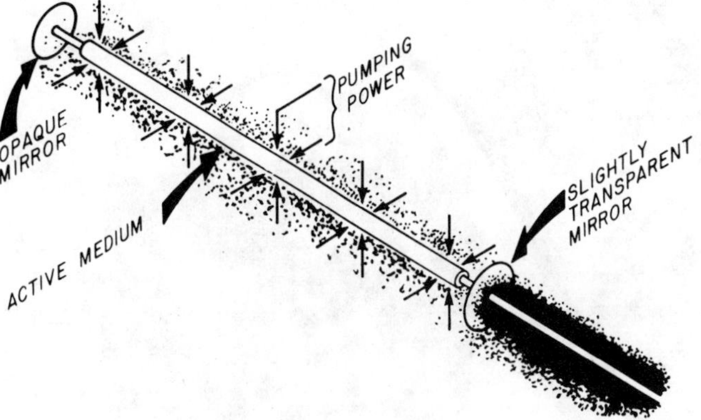

Figure 13 **Basic Laser Elements**

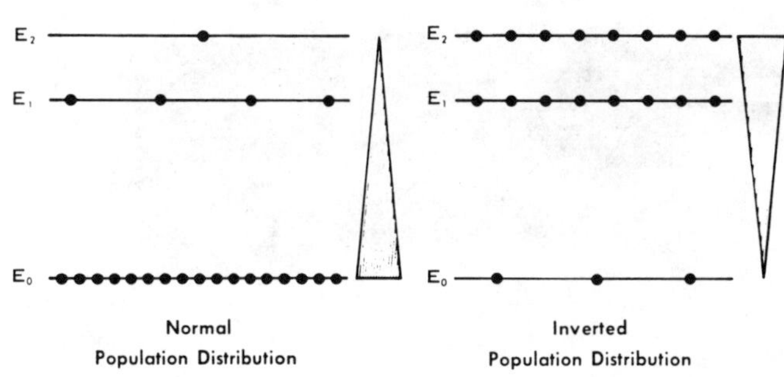

Figure 14 **Illustration of Population Inversion**

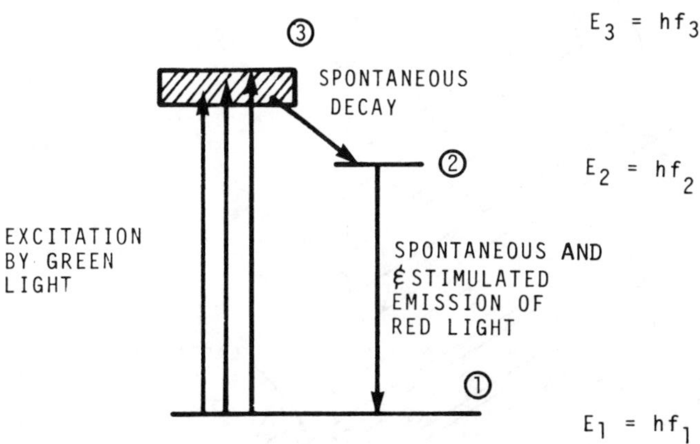

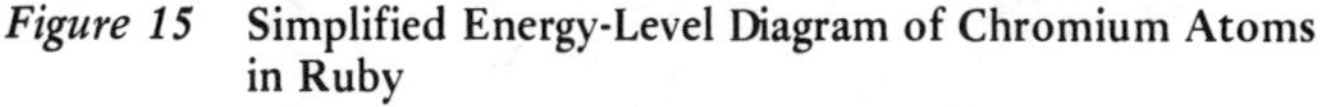

Figure 15 **Simplified Energy-Level Diagram of Chromium Atoms in Ruby**

Figure 16 illustrates a neodymium YAG laser. It is approximately 8" long, 3" wide, and 3" high. It has a heat exchanger which utilizes a fan to blow air through radiating fins, thereby cooling the liquid surrounding the flash lamp and laser rod (see Figure 17). This type of laser and power supply typically is used in rangefinder applications. Typical specifications for such a device are shown in Figure 18 and Table 3.

Table 4 illustrates some of the more popular laser sources.

Figure 16 **Neodymium YAG Laser (Compliments of Raytheon Laser Advanced Development Center)**

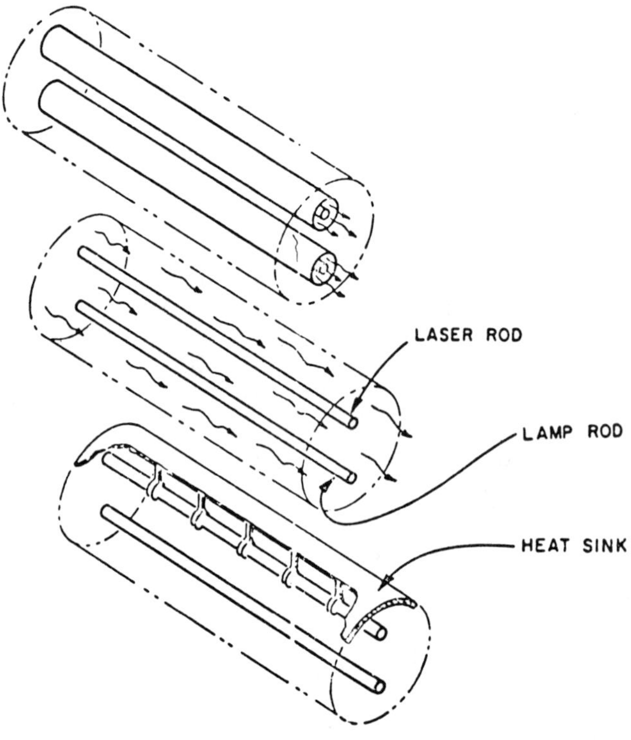

Figure 17 **Laser Rod Cooling Techniques**

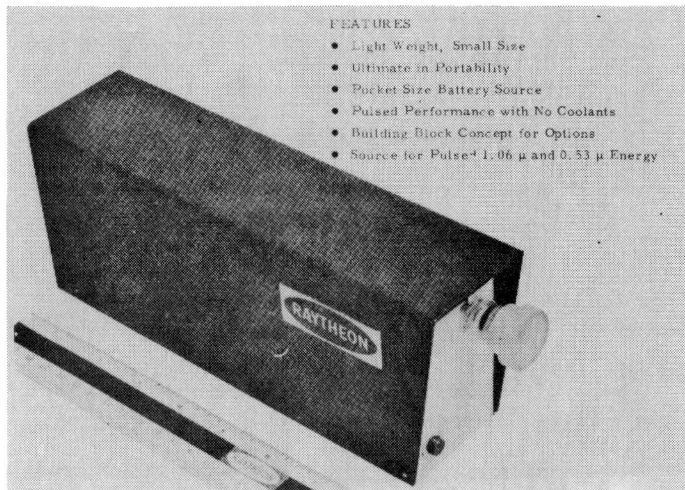

Figure 18 **Raytheon Model SS-219 Miniaturized Pulsed YAG Laser Transmitter**

Detectors

When a detector operates in a square law region, the power incident upon the detector results in a current at the output directly related to the incident power; this current may be expressed as:

$$i = \rho_i E^2$$

where: ρ_i = constant

The electric field, E, incident upon the detector may be expressed as:

$$E = A_1 \cos\omega_s t$$

The detector output current, i, can be shown to be:

$$i = \rho_i A_1^2 (1/2 + 1/2 \cos 2\, \omega_s t)$$

For an optical detector, ρ_i relates to the conversion of optical energy to electrical energy; as such, it may be termed the conversion factor.

Detector Responsivity

For ideal optical detection, one photon of light releases one electron from the detector. The number of photons, M_p, required to emit these electrons for the non-ideal detector then becomes a measure of how efficiently the detector operates and is typically termed the device quantum efficiency, η.

$$M_p = \frac{1}{\eta}$$

The current produced by M photons per second may be expressed as:

$$i = \eta\, M\, q$$

where: q = electron charge

As indicated before, the energy level associated with each photon is expressed as the product of Planck's constant, h, and transmitting frequency, f.

Correspondingly, the number of photons per second may be expressed as:

$$M = \frac{P_{SIG}}{hf}$$

Mechanical Specifications
Total Transmitter (less accessories) Size . . . . 2" x 3.25" x 7"
Weight (24 volts dc input). . . . 3 lb
(115 V ac 50/60 Hz input) . . . . 5 lb
Enclosure . . . . Dustproof
Laser Head (less accessories) Size . . . . 1.125" x 1.25"; 6.75"
Weight . . . . 10 oz

Performance Specifications
Laser Output (Normal Model) . . . . Threshold to 100 mJ (adjustable) at 1.06 μm
Pulse Duration . . . . = 100 μs
Laser Output (Q-Switched) . . . . 50 mJ (nominal) at 1.06 μm
Q-Switched Pulse Duration . . . . 10-20 ns
Pulse Repetition Rate (Q-Switched and Normal Mode) . . . . 1 pps (max.) continuous
Trigger . . . . Manual Auto Rep. (1 pps) External (+15 V, 10 μs)
Laser Crystal . . . . Nd^{+3}: YAG 4 x 50 mm
Polarizer . . . . Calcite Glan Prism
E/O Q-Switch . . . . $LiNbO_3$ Pockels Cell
E.M.I. . . . . Unit shielded to minimize electro-magnetic interference

Electrical Input
DC Option:
Voltage . . . . 24 Vdc (nominal) 18-32 V
Current . . . . 2.7 A (nominal) 2.2-3.3 A
Charge Time . . . . 0.7 s (nominal) 0.35-0.9 s
Battery Pak . . . . 250 laser shots between recharges
AC Option:
Voltage . . . . 115 Vac 50/60 Hz
Power . . . . 25 Watts (nominal)

Available Accessories
115 Vac Adapter for 24 Vdc Unit
Pockels Cell Q-Switch
Rechargeable Battery Pak
Frequency Doubler
Boresighting Telescope
Beam Shaping Telescope

Table 3 **Model SS-219 Miniaturized Pulsed YAG Transmitter**

Lasers	Type	Wavelength of Operation	Energy/Pulse or Power Energy or Power	Pulse Width	PRF
He-Cd	Gas	0.325-0.44 μm	10 mW	–	CW
Argon	Gas	0.4880 μm 0.5145 μm	5 W	–	CW
HeNe	Gas	0.6328 μm 1.1523 μm 3.3913 μm	150 mW 25 mW 10 mW		CW
Ruby	Crystal	.6943 μm	1 joule +	30 ns	6 ppm
YAG	Crystal	1.06 μm	150 mJ 1 W	20 ns	1-10 pps CW
CO_2	Gas	10.6 μm	40 W + 2 kW peak 10 kW peak	 300 ns 1-10 μs	CW 50-400 Hz 200 pps-10 kHz
CO	Gas	5.1-5.3 μm	40 W	–	CW
DF	Gas	3.8 μm	40 mW	100 ns- 25 pps	Pulse
HF	Gas	2.8-3 μm	1 joule	1-2 pps- 0.5 μs	Pulse
GaAs	Semi-conductor	0.9 μm	10 W peak @ 5 kHz	200 ns	Pulse
GaAs	Semi-conductor	0.82 μm	6 mW	–	CW

Table 4 **Popular Laser Sources**

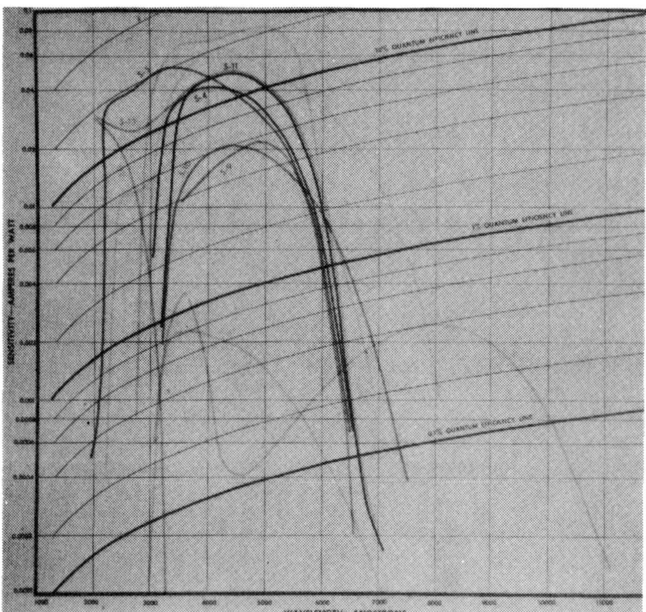

Figure 19 **Typical Absolute Spectral Response Characteristics of Photoemissive Devices**

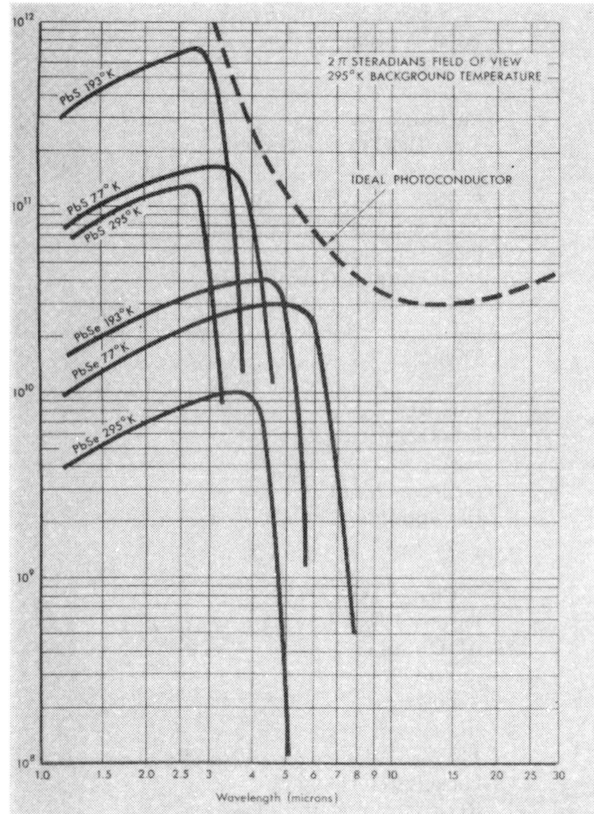

Therefore the detector current is:

$$i = \frac{\eta \, q \, P_{SIG}}{hf}$$

Thus ρ_i, the detector current responsivity, is equal to $\eta q/hf$. The detector current responsivity, ρ_i, physically represents the ratio of the rms current out of the detector to the rms power incident onto the detector; it has units of amperes/watt. Correspondingly, the voltage responsivity, ρ_v, has units of volts/watt.

Noise Equivalent Power

The noise equivalent power, NEP, may be expressed as:

$$\text{NEP} = \frac{\text{Detector Noise Current}}{\text{Responsivity}}$$

and represents the amount of rms modulated power applied to the detector (an amount equal to the rms detector noise voltage), usually specified as:

NEP (500 K,	1000,	1)
Radiation Source Temperature	Measurement Frequency (Hz)	Detected Bandwidth (Hz)

Because NEP is dependent upon the area of the detector, the term D* was developed to standardize detector comparisons; this term references detector measurements to a 1 cm^2 area, in a bandwidth of 1 Hz.

$$D^* = \frac{(A \, \Delta f)^{1/2}}{\text{NEP}} \qquad \frac{\text{cm - (Hz)}^{1/2}}{\text{W}}$$

Often D* is specified as D* (λ, f_m) where f_m is the modulating frequency in Hz referred to a 1 Hz receiver bandwidth.

Note that D* is also dependent on the signal wavelength. Figures 19 [13] and 20 [1] and Table 5 indicate typical detector sensitivities and characteristics throughout other visible, near-, mid-, and far-infrared wavebands [10].

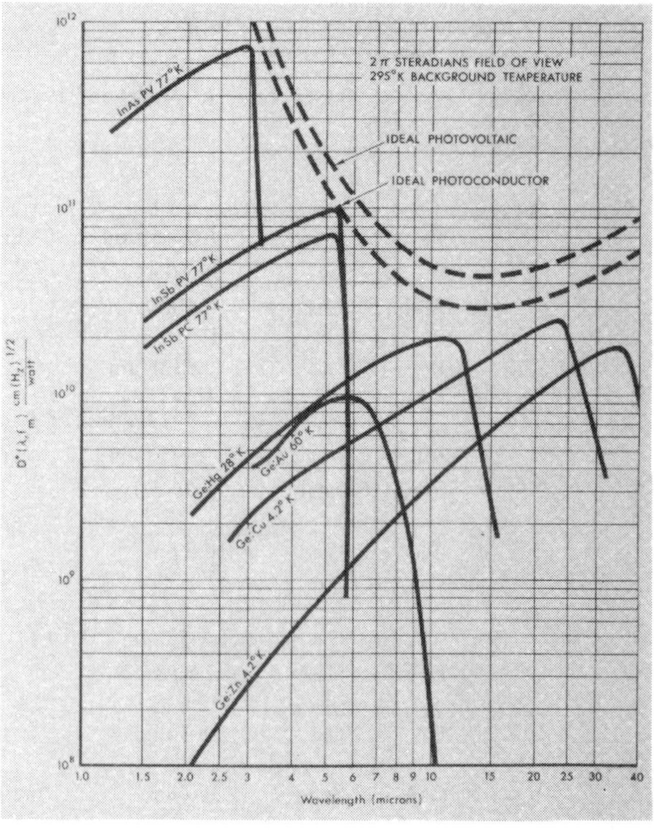

Figure 20 **Spectral Detectivities for Above Average Detectors Fabricated by SBRC (Santa Barbara Research Center of Hughes Aircraft Company); a Reduction in Background Photon Flux Produces Higher Detectivities [1]**

Photoemissive Detectors				
Spectral Range (μm)	0.145 – 0.87	0.145 – 1.1	0.4 – 1.2	
Responsivity (μA/μW @ λ)	128 @ 0.4	18 @ 1.06	0.0019 @ 0.8	
Dark Current (a)	5×10^{-10}	3×10^{-10}	3×10^{-7}	
Quantum Efficiency (%)	15	2	0.2 @ 0.9	
Rise and Fall Times (ns)	0.32	0.12	1.5	
Photocathode Type	S-20	InGaAs P	$AgOC_s$	
Pyroelectric Detectors				
Spectral Range (μm)	0.001 – 1000	0.2 – 1000	2.5 – 30	0.2 – 500
Peak Detectivity (cm-Hz½/W)	2×10^6	10^7	10^8	10^8
Responsivity (μV/μW @ λ)	200 @ 10	0.3×10^{-6} μA/μW)	10^{-6} A/W	10^{-5} - 100 V/W
Rise Times (s)	10^{-7}	10^{-8}	6×10^{-8}	5×10^{-4} - 5×10^3
Type	$LiTaO_6$	Ferroelectric	PVF_2	SBN
Dimensions	1 – 5 mm	0.5 – 4 mm	10 mm	3 – 20 mm
Semiconductor Detectors (Visible – Near IR)				
Spectral Range (μm)	0.3 – 1.8	0.4 – 1.1	0.35 – 1.13	0.4 – 1.1
NEP (W) or Detectivity (cm-Hz½/W)	10^{-16}	4.5×10^{-13}	6×10^{11}	5×10^{-15}
Responsivity μA or μV / μW @ λ	0.7 μA/μW	0.6 @ 0.9	0.5 @ 0.9	110 μA/μW
Peak Wavelength (μm)	1.5	0.9	0.95	0.9
Rise Time (μs)	10^{-2}	0.12	0.01	0.02
Type	Ge	Silicon	Silicon PIN	Silicon Avalanche
Dimensions	0.8 mm²	11 mm diameter	11 mm diameter	3.2 mm diameter
Semiconductor Detectors (Mid IR)				
Detector Type	Indium Antimonide (PV)	Lead Sulfide (PC)	Lead Selenide (PC)	Lead Selenide (PC)
Temperature (K)	77 or 243	77	195	300
Spectral Response	0.5 – 6 μm	0.5 – 5 μm	0.5 – 6 μm	0.5 – 4.7 μm
D* (cm-Hz½/W)	$1\text{-}3 \times 10^{11}$	$1.5\text{-}2.5 \times 10^{11}$	2×10^{10}	1.5×10^9
Peak Wavelength	5 μm	2.8 μm	4.8 μm	3.8 μm
Responsivity	1.5 – 2 A/W	0.1 – 1 A/W	0.01-0.05 A/W	$4\text{-}30 \times 10^{-3}$ A/W
Rise Time (μs)	0.02 – 0.2	$2 – 5 \times 10^3$	10 – 40	1 – 3
Active Area (mm)²	0.008 – 7	0.01 – 100	0.01 – 100	0.01 – 100
Semiconductor Detectors (Far IR)				
Detector Type	Lead Tin Telluride (PV)	Mercury Cadmium Telluride (PC)	Mercury Cadmium Telluride (PV)	
Temperature (K)	77	77	77 or 243	
Spectral Response	0.5 – 14	0.5 – 13	2 – 12	
D* (cm-Hz½/W)	2×10^{10}	$1 – 4 \times 10^{10}$	$2 – 5 \times 10^9$	
Peak Wavelength	10	10.6	10.6	
Responsivity	3 A/W	3 – 30 A/W	1.3 – 4.3 A/W	
Rise Time (μs)	1 – 2	0.2 – 0.8	$1.6 – 16 \times 10^{-4}$	
Active Area (mm)²	0.008 – 3	$2.5 \times 10^{-3} – 9$	0.008 – 0.8	

Table 5 **Detector Characteristics**

In order to optimize system SNR, the system engineer must choose wisely from a wide variety of variables. The equation below illustrates some of the considerations involved with detector selection and illustrates that, if the optical background energy is the limiting noise source after suitable filtering, any choice should be predicted upon the device having high quantum efficiency (in which case the signal to noise ratio is increased).

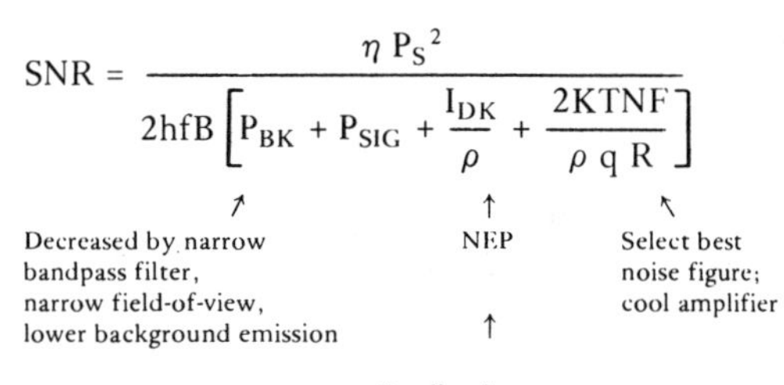

$$\mathrm{SNR} = \frac{\eta P_S^2}{2hfB\left[P_{BK} + P_{SIG} + \frac{I_{DK}}{\rho} + \frac{2KTNF}{\rho q R}\right]}$$

Decreased by narrow bandpass filter, narrow field-of-view, lower background emission (↑ P_{BK})

NEP (↑ I_{DK}/ρ) ↑ Smaller detector – smaller dark current; higher responsivity – lower NEP

Select best noise figure; cool amplifier (↑ $2KTNF/\rho q R$)

where:

ρ = responsivity

I_{DK} = detector dark current

Correspondingly, if the detector dark current is the limiting noise term, the detector having the smallest NEP should be chosen. Detector NEP may be reduced by choosing small area or high responsivity detectors. Photomultiplier detectors have significant amplification that can reduce the effect of detector dark current. Similarly, choice of a receiver noise figure consistent with other receiver noise expressions is required to insure proper system operation.*

Laser Radars (Incoherent Detection)

Figure 21 illustrates a functional block diagram of a laser rangefinder. A source of electrical energy is utilized to charge up an energy storage network which, upon application of a triggering network, transfers this energy to a flashlamp located within an elliptical cavity along with a solid state material (in this case, a ruby crystal rod). Less than 1% of the light energy emitted from the flashlamp couples into the absorption band (previously noted to be in the green light waveband) of the ruby rod. Other energy emitted from the flashlamp is not effective and results in internal heating. This heat may be removed by a number of cooling system configurations. The energy coupling effectively into the ruby rod results in the ruby rod providing a source of gain to the photons reflecting back and forth between the reflectors at the ends of the cavity. At an appropriate time after the flashlamp is energized, a polarization switch, such as a Kerr cell, may be energized to change the effective coupling of the output mirror. As a result, a pulse of light energy at the wavelength of 6943 Å, typically having a pulse width of 30 ns, propagates through the optical system. An optical system provides the desired transmitter beam divergence.

Mounted within the optical cavity is a high bandwidth optical diode which detects the output ruby laser energy and provides a reference timing signal to the receiver electronics and elapsed time counter. Energy backscattered from the target then passes through suitable collection optics to a photomultiplier. Standard threshold detection circuitry, range gating, and elapsed time counting are used to provide accurate range and range rate information. Because of the short pulse width and narrow beamwidth associated with this type of optical system, precision target location information can be obtained.

*Editor's Note – For the practical design of an IR-surveillance receiver for maximum detectability, see Berger, T.; and Brookner, E.: "Practical Design of Infrared Detector Circuits," *Applied Optics, Vol.6,* pp. 1189-1193, July 1967.

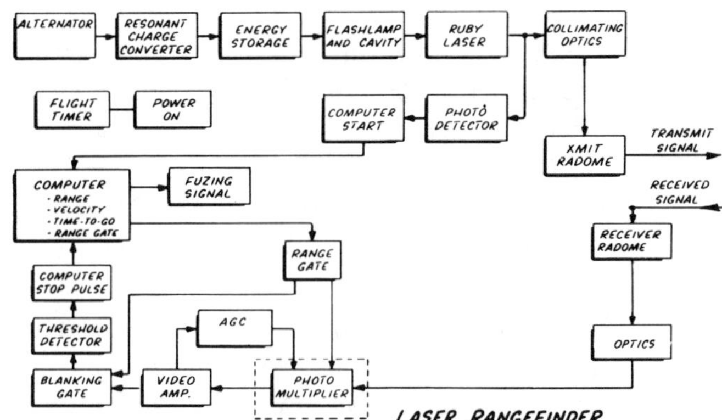

Figure 21 **Functional Block Diagram of Laser Rangefinder**

Figure 22 **Apollo Laser Altimeter, manufactured by RCA Automated Systems for NASA, Lyndon B. Johnson Space Center, under Contract NAS-9-10600 (Compliments of Woodward, RCA)**

Figure 22 illustrates an artist's conception of the laser altimeter system utilized on Apollo 15, 16, and 17 to provide accurate ranging information for the camera system flown on these flights. As a result of having accurate ranging information to a known point within the camera field, accurate scale factoring and significant improvement in map accuracy were obtained. Equipment specifications for this system may be noted in Table 6; Figure 23 shows the hardware.

Laser
Q-Switched Ruby – Mechanical Q-Switch
Energy Output > 250 mJ/pulse
PRF: 1 pulse per 15 seconds maximum
"Solo" Mode 1 per 20 seconds
"Camera" Mode – Slaved to 3″ metric mapping camera: 1 per 15 seconds or slower typical at 60 Nautical Mile Altitude 1 per 25 seconds
Beam Divergence: 80% of output energy in 0.3 milliradians (Collimating Telescope 16 x 4″ Clear Aperture, Catadioptric)

Receiver
Extended red S-20 Photomultiplier detector
25 Å bandwidth ca 6943 Å
Field of View 0.2 milliradians

Altitude Measurement
Min altitude 40 nautical miles
Max altitude 80 nautical miles
Resolution: 1 meter
Accuracy = ± 2 meters

Altitude Output
18 bit binary to telemetry
Different format for mapping camera – recorded on each film frame.
Also analog and digital status outputs to telemetry.

Table 6 **Laser Altimeter Flown on Apollo 15, 16, 17; Manufacturer – RCA Corp., Burlington, Mass.**

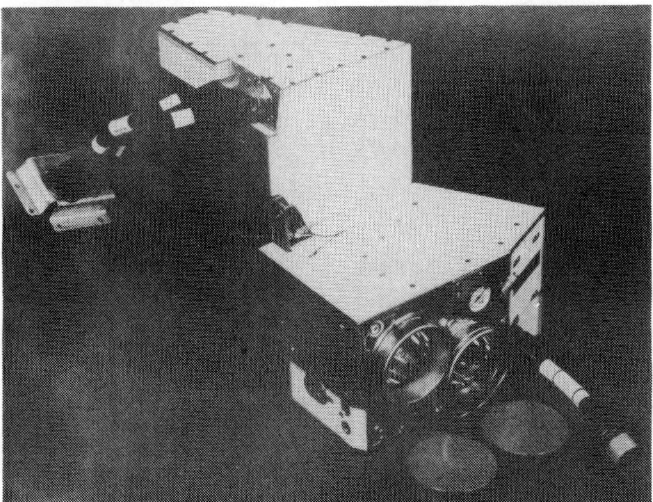

Figure 23 **Apollo Laser Altimeter Hardware, manufactured by RCA Automated Systems for NASA, Lyndon B. Johnson Space Center, under contract NAS-9-10600 (Compliments of Woodward, RCA)**

Table 7 [9] compares the relative performance of ruby and neodymium YAG lasers for a range of 5 km. Here the ruby system is evaluated using an S-20 photo cathode surface and a silicon semiconductor detector. Similarly the neodymium system is evaluated for an S-1 photo surface photomultiplier and a silicon detector. The sun irradiance and the optical filter bandpass is twice as large for the ruby system than it is for the YAG. As a result, the background power, P_{BK}, is four times larger.

Evaluation of the noise terms indicates that the silicon detector systems are receiver thermally noise limited; correspondingly, the photomultiplier systems are signal shot noise limited, with the background noise power approximately an order of magnitude or more below the signal shot noise level. The incoherent detection SNR equation above indicates where further effort should be concentrated in order to improve the system's performance; Table 8 illustrates the transmitter power requirements for each system. Tables 9 and 10 and Figure 24 illustrate the configuration and specifications of an incoherent detection YAG laser radar system developed by Sylvania for cooperative target tracking [14].

Figure 24 **Sylvania Incoherent Detection YAG (1.06 μm) Laser Radar Used for Cooperative Target Tracking**

Coherent Detection System Requirements

In order to utilize a coherent detection system, temporal and spatial coherence requirements must be maintained. Spatial photomixing requirements necessitate that the local oscillator and received signal optical phase front be aligned over the active detector surface area. Figure 25 utilizes a laser transmitter, emitting a spatial and temporal coherent signal which is coupled through an amplifier chain and transmitted to the atmosphere as E(t) = B cos ω_0t. This signal, backscattered and Doppler shifted, is then collected by the receiver optics in the form $E_s = C \cos (\omega_0 \pm \omega_d)t$; after suitable receiver optical transmission, it impinges on the receiver detection area along with a local oscillator signal, $E_{LO} = A \cos \omega_0 t$, derived from the reference transmitting laser.

The signal power derived from the cross-product term is determined by standard range equation relationships as well as the degree to which temporal and spatial coherences have been maintained. The requirements for photomixing spatial coherent detection are given in Figure 26 [11]. Phase front distortion due to atmospheric turbulence perturbation of the refractive index can result in limitations on the receiver aperture area of a coherent detection system; see next section and Chapter 15.

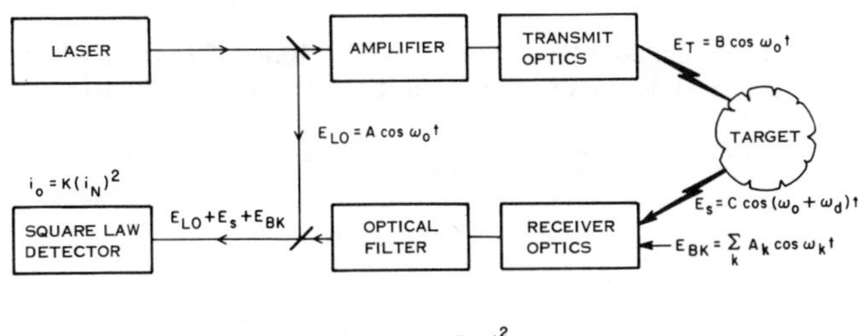

Figure 25 **Coherent Detection System**

	Units	Ruby S-20	Ruby Si	Neodymium S-1	Neodymium Si
Range R	km	5 →			
Wavelength, λ	μm	0.694 →		1.06 →	
Attenuation Coefficient, σ	km^{-1}	0.139 →		0.114 →	
Atmospheric Transmittance, T_a	–	0.449 →		0.565 →	
Sun Spectral Irradiance, $H_{\lambda s}$	$W\ m^{-2}\ Å^{-1}$	0.12 →		0.06 →	
Filter Leakage Transmittance, X	–	0 →			
Backscatter Coefficient, σ_s	km^{-1}	0 →			
Background Power, P_{BK}	W	1.66×10^{-10} →		4.6×10^{-11} →	
Pulsewidth, τ	ns	20 →			
Bandwidth, B	MHz	25 →			
Optical Filter Bandpass, B_o	Å	40 →		20 →	
Receiver Lens Diameter, d_r	in	2.8 →			
Geometry Factor, M	–	1 →			
Receiver Beamwidth, α_r	mrad	1 →			
Transmitter Beamwidth, α_t	mrad	1 →			
Target Reflectance, ρ	–	0.1 →			
Receiver Noise Factor, F	–	1.5 →			
Detector Load Resistor, R_L	Ω	1000 →			
Transmittance, receiver optics, T_r	–	0.7 →			
Transmittance, transmitter optics, T_t	–	0.7 →			
Target Incidence Angle, θ	deg	0 →			
Signal-to-Noise Ratio, SNR	–	53 →			
Detector Gain, G	–	5×10^4	1	1.5×10^5	1
Cathode Dark Current, I_D	pA	0.030	10^5	12.9	10^5
Detector Responsivity, β	$A\ W^{-1}$	0.028	0.517	3.5×10^{-4}	0.152
Peak Received Signal Power, P_s	W	1.23×10^{-9}	2.5×10^{-7}	1.64×10^{-7}	8.48×10^{-7}
Single Pulse Range Accuracy, δR	m	0.82	0.82	0.82	0.82

Table 7 **Summary of Parameters and Performance Calculations for Four Laser Rangefinders**

Compliments: *RCA Electro-Oprtics Handbook,* by permission.

Laser Material	Detector	Peak Laser Power Required, P_t (kW)
(1) Ruby	S-20 Photomultiplier	2.6
(2) Ruby	Silicon Photodiode	405
(3) Neodymium YAG	S-1 Photomultiplier	282
(4) Neodymium YAG	Silicon Photodiode	1,120

Table 8 **Calculated Peak Laser Power to Range 5 km in a Clear Standard Atmosphere for Four Laser Rangefinders.**

Compliments: *RCA Electro-Optics Handbook,* by permission.

Transmitter	Q-Spoiled Nd:YAG
Peak Power	10^6 Watts
PRF	100 pps
Beamwidth	10 mr
Pulsewidth	25 ns
Range Measurement Accuracy	±½ foot
Detector Quantum Efficiency	> 50%
Receiver	Quadrant Photodiode
Angle Tracking (Az & El) Accuracy	±0.1 mrad
Target	3″ Retro Reflector
Range	10 nmi
Initial Acquisition	TV

Table 9 **Sylvania 1.06 μm Laser Radar**

Compliments: R. Cooke, Sylvania, by permission.

1. Absolute Accuracy*
   Azimuth: ±0.01 percent of range (0.1 mrad) (for target ranges of 500 to 65,000 feet)
   Elevation: ±0.01 percent of range (0.1 mrad) (for target ranges of 500 to 65,000 feet)
   Range: ±1 foot for target ranges of 700 to 30,000 feet
   ±2 feet for target ranges of 30,000 to 65,000 feet
2. Maximum Range 100,000 feet
   Accuracy for target at 100,000 feet
   a. Azimuth: ±0.3 mrad b. Elevation: ±0.3 mrad c. Range: ±5 feet
3. Data Sample Rate: 100, 50, 20, or 10 sample sets per second, selectable
4. Angular Coverage
   Azimuth: ±170° (multiple turn capability available)
   Elevation: -5° to +85° (dynamic specifications apply for elevation angles between -5° to +45°)
5. Acquisition Dynamics (manual – using joystick and TV monitor)
   Maximum angular rate (azimuth and elevation): 100 mrad/sec
   Maximum angular acceleration (azimuth and elevation): 80 mrad/sec^2
6. Acquisition Dynamics (automatic-aided with 100 sample per second coordinate data)
   Maximum angular rate (azimuth and elevation): 500 mrad/sec
   Maximum angular acceleration (azimuth and elevation): 80 mrad/sec^2
7. Autotrack Dynamics
   Maximum angular rate (azimuth and elevation): 500 mrad/sec
   Maximum angular acceleration (azimuth and elevation): 80 mrad/sec^2
8. Operator Displays
   Range: Digital in 1-foot increments Azimuth: Digital in 1° increments
   Elevation: Digital in 1° increments
9. Acquisition Field of View: 5° to 20° with zoom control
10. Viewfinder Field of View: 3 mrad
11. Power Requirements: 208-volt, 3-phase, per electrical performance specifications of MIL-STD-633
12. Environmental Conditions (Operating):
    Ambient temperature: -20° to 120°F Wind: 0 to 50 knotts
13. Set-Up Time: < 1 hour

Table 10 **PATS Incoherent Detection YAG Laser Radar, based on use of 3-inch diameter 4-arc-second retroreflector**

*After computer smoothing, a weighted averaging technique is used with a sliding window of 0.1 second and a sample rate of 100 per second.

Compliments: Sylvania

Effective Aperture Diameter for Coherent Receivers

Perturbations in the refractive index of a medium caused by turbulence results in limitations on the receiver aperture area of a coherent detection system [15-19]; the effective aperture (D_{EFF}) represents the actual aperture diameter at which the heterodyne signal is reduced 3.0 dB below that level expected in the absence of turbulence:

$$D_{EFF} = (0.0588 \frac{\lambda^2}{C_N{}^2 L})^{3/5}$$

where:

C_N = atmospheric structure function for index of refraction variation (see Chapter 15) ($cm^{-1/3}$)

λ = wavelength (cm)

L = path length (cm)

D_{EFF} = effective optics diameter (cm)

The structure function coefficient function, C_N, as a function of height may be observed in Figure 27.

Laser Radars (Coherent Detection)

Figures 28 and 29 and Table 11 illustrate the ground based Lincoln Laboratory Firepond 10.6 Micron Laser Doppler Radar Facility and its block diagram and system specifications [20,21].

This equipment was utilized recently at a reduced power level to track the GEOS-III satellite at an approximate range of 1100 km [22], with the coherent optical monopulse receiver illustrated in Figure 30. Initial target acquisition was accomplished with the Millstone Hill Radar, with the aid of a low light level TV tracker. Elevation angle tracking statistics are shown in Figure 31.

Vernier mirror tracking data portraying the difference between the tracking algorithm (Kalman Filter*) pointing data and the monopulse line of sight indicated that the coarse pointing errors of approximately 100 microradians were tracked out to an rms value of one microradian with a tracker bandwidth of about 1 Hz.

Figures 32 and 33 give the block diagram and photograph of an airborne pulse Doppler laser radar developed by Raytheon for NASA. Figure 34 illustrates the equipment mounted in a NASA Convair 990 aircraft, utilizing a pod configuration to direct the laser beam forward of the aircraft. Table 12 illustrates some of the equipment parameters. Figure 35 shows a photograph of a superimposed picture of an A-scope display (lower photograph) and a Velocity Range Display of a non-cooperative target 13.2 nautical miles in front of the aircraft [23].

In conclusion, both microwave and optical systems designers might find increased system capability by blending the techniques and capabilities of the microwave and optical communities.

*See Chapters 24 and 25.

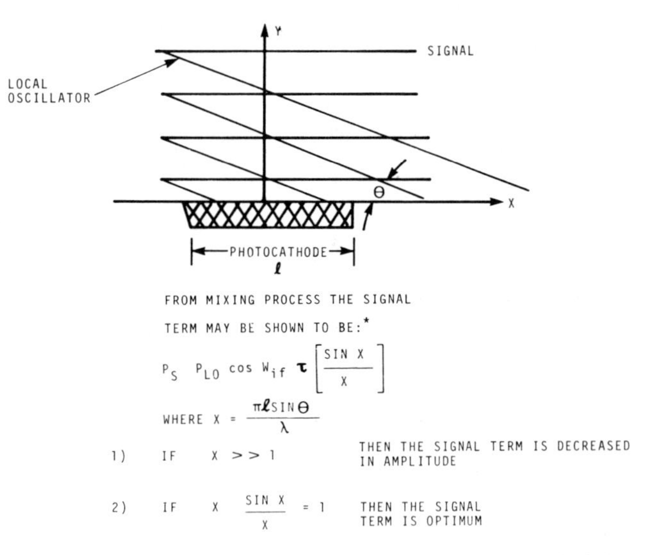

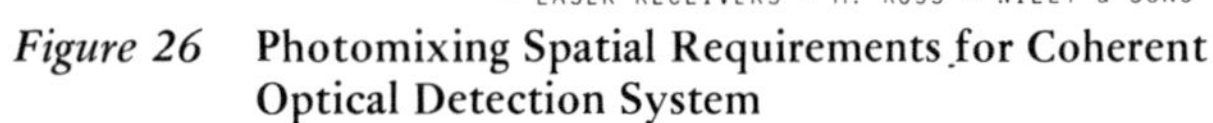

Figure 26 **Photomixing Spatial Requirements for Coherent Optical Detection System**

Wavelength	10.6 microns
Configuration	MOPA
Transmitter Power	15 kW Peak, 1.4 kW Average
PRF	10 kHz
Laser Frequency	3 x 10^{13} Hz
Laser Stability	20 Hz over 50 ms
Telescope	48 cm; f/7 Cassegrain
Offset Laser Frequency	5 MHz
Beat Frequency Difference Between Two Lasers	< 1 kHz over seconds
Visual Track	Wide Angle TV
IR Angle Track	Conical Scan Amplitude – Monopulse
Detector	Cu:Ge
Detector Bandwidth	1.2 GHz (15,000 mph)

Compliments: R. Kingston, T. Gilmartin; Lincoln Laboratory

Table 11 **Lincoln Laboratory Firepond 10.6 μm Laser Doppler Radar**

System Parameters	
Operating Wavelength	10.6 μm
Pulse Width	Variable 1-10 μs Selectable Pulses of 2 μs, 4 μs, 8 μs
Repetition Rate	200 pps
Optics Size	12″ Cassegrain
Output Polarization	Circular or Vertical
Display	Range-Velocity
Recording Single Range Cell	Amplitude-Velocity
Recording System	Analog or Digital

Compliments: Raytheon Company, Sudbury, Massachusetts; and NASA Marshall Space Flight Center, Huntsville, Alabama.

Table 12 **Raytheon Pulse Doppler Laser Radar**

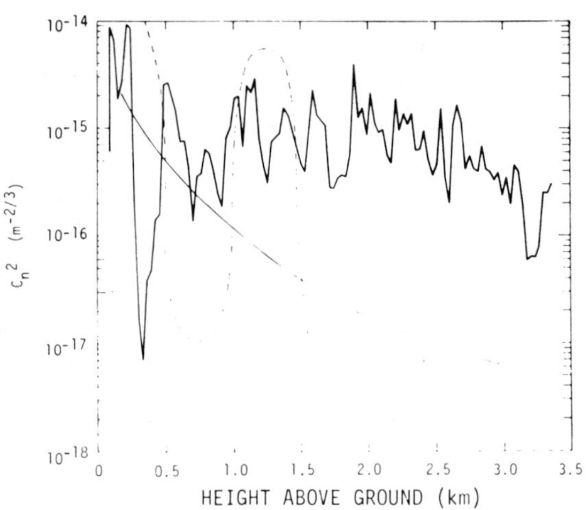

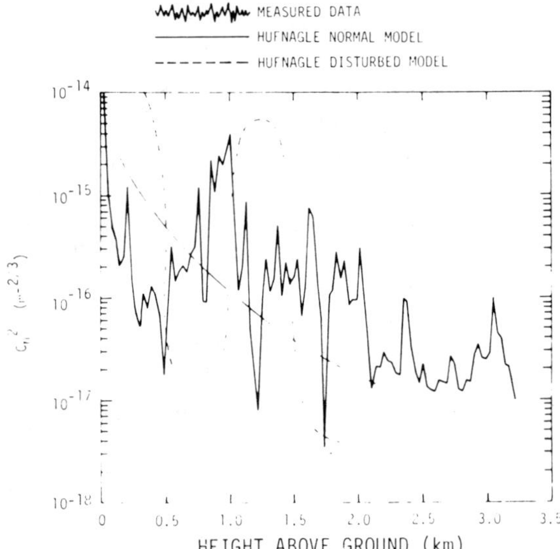

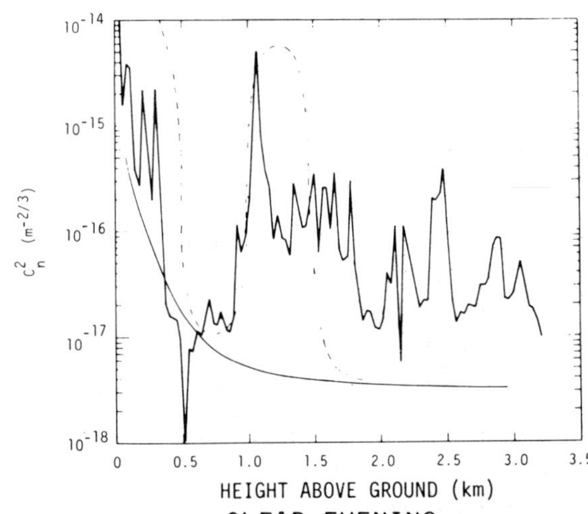

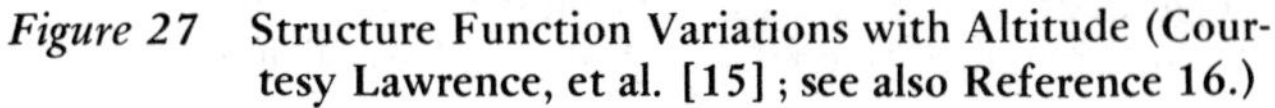

Figure 27 **Structure Function Variations with Altitude (Courtesy Lawrence, et al. [15]; see also Reference 16.)**

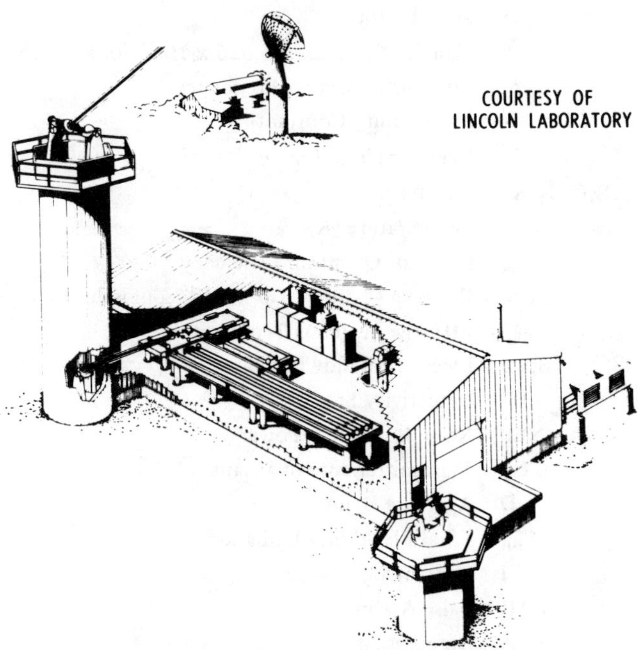

Figure 28 Lincoln Laboratory Firepond 10.6 μm Laser Doppler Radar Facility (Courtesy of Lincoln Laboratory)

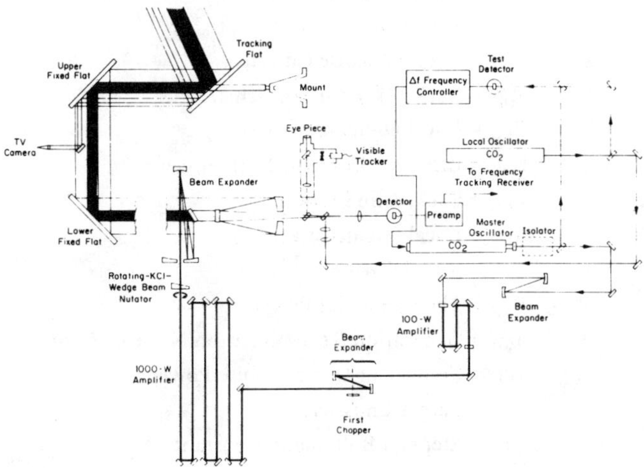

Figure 29 Lincoln Laboratory Firepond CO_2 Laser Radar Schematic (Courtesy of Lincoln Laboratory)

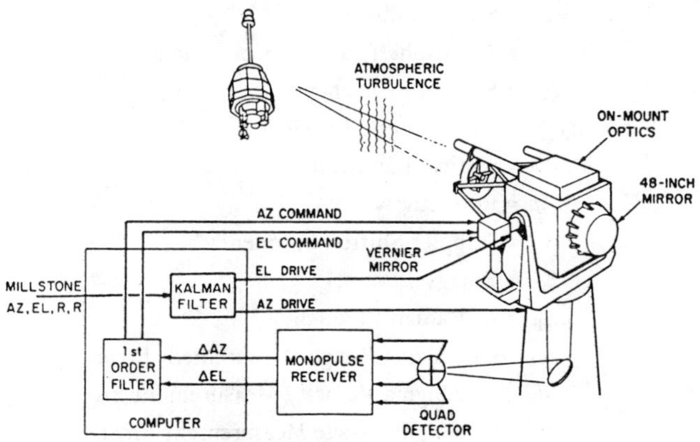

Figure 30 Satellite Tracking Mode (Long Range Mode) of Lincoln Laboratory Firepond 10.6 μm Laser Radar

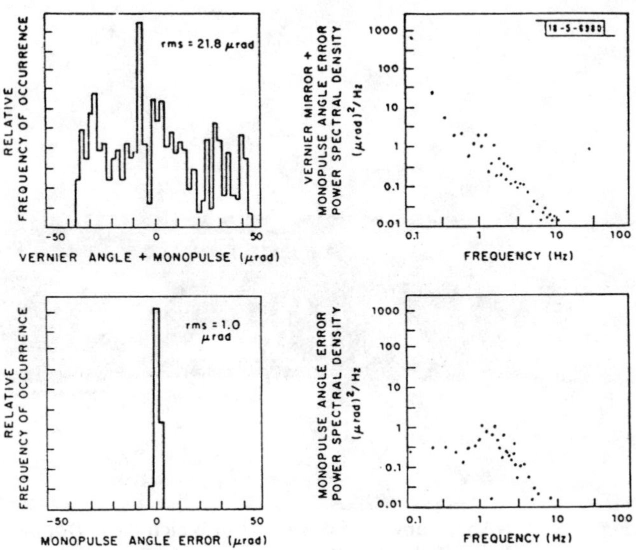

Figure 31 Tracking Data Analysis in Elevation Only of Monopulse Errors from GEOS-III Retroreflector at Range of About 1100 km

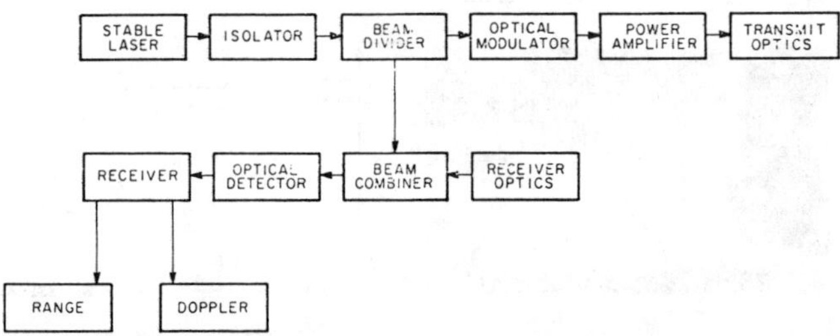

Figure 32 Block Diagram of Airborne Pulse Doppler Laser Radar Developed for NASA by the Raytheon Company

Figure 33 Airborne Pulse Doppler Laser Radar Equipment Built for NASA by Raytheon Company

Figure 34 **NASA Convair 990 Aircraft in Which Pulse Doppler Laser Radar was Mounted**

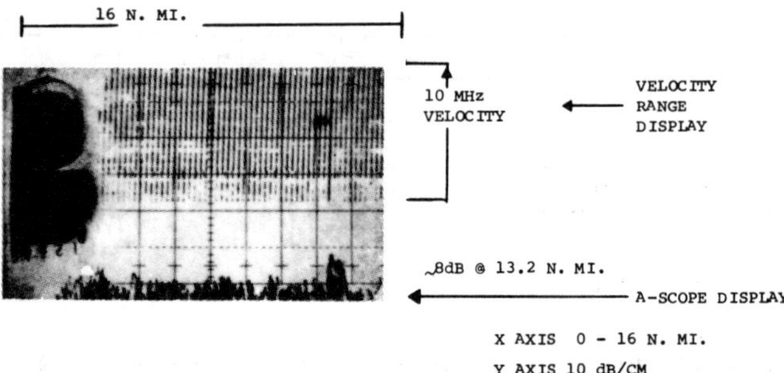

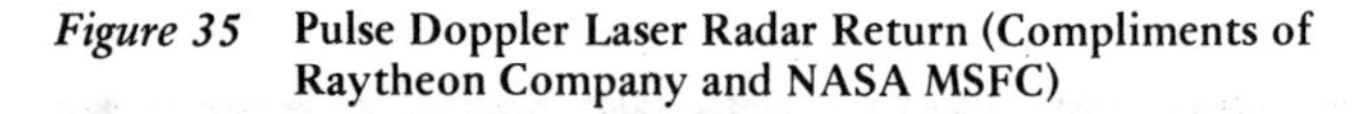

Figure 35 **Pulse Doppler Laser Radar Return (Compliments of Raytheon Company and NASA MSFC)**

Glossary

Å = Angstroms
P_{FA} = Probability of False Alarm
P_D = Probability of Detection
E = Electric Field
σ = Target Cross Section
Ω = Solid Angle
ρ_T = Target Reflection
dA = Target Area
P_{SIG}, P_R = Received Signal Power
R = Range
θ = Beamwidth
λ = Wavelength (meters)
v = Velocity
γ = Angle Between Target Velocity Vector and Line of Sight
c = Speed of Light
SNR, S/N = Signal-to-Noise Ratio
τ = Pulse Width
h = Planck's Constant = 6.626×10^{-34} joule-second
f_o = Transmitter Center Frequency
k = Boltzmann's Constant
NF = Receiver Noise Figure
A, B, C, K_1, K_2 = Constant
T = Temperature (K)
η = Detector Quantum Efficiency
q = Electron Charge = 1.6×10^{-19} coulombs
T_{EFF} = Effective Antenna Temperature
B, Δf = Electrical Bandwidth
D* = Detectivity Star
NEP = Noise Equivalent Power
G_T = Transmitter Antenna Gain
D = Aperture Diameter
P_{BB} = Total Blackbody Radiation
I = Radiant Emittance
FAR = False Alarm Rate
A = Detector Area
I_{DK} = Dark Noise Current
G = Detector Current Gain
ρ_i = Detector Current Responsivity
I_{SN} = Shot Noise Current
i_T = Thermal Noise Current
i_{DK} = Dark Noise Current (rms)
P_{LO} = Local Oscillator Power
P_{DK} = Effective Detector Dark Power
P_{BK} = Background Power
i_{LO} = Local Oscillator Current
i_{SIG} = Signal Current
P_{AMP}, P_{TH} = Thermal Noise Power
i_{GR} = Generation-Recombination Noise Current
P_N = Equivalent Optical Noise Power
ϵ = Target Emissivity
σ_τ = Stephan Boltzmann Constant

$$= 5.67 \times 10^{-12} \frac{W}{cm^2 K^4}$$

S_{IRR} = Solar Irradiance
I_s = Atmospheric Scatter Coefficient
η_{SYS} = System Efficiency
η_{ATM} = Atmospheric Efficiency
dλ, $\Delta\lambda$ = Optical Bandwidth
ρ = Reflectivity
f_D = Doppler Shifted Frequency
E_N = Energy Level, N
ω_{IF} = IF Radian Frequency
σ_R = One Sigma Range Measurement Error
σ_v = One Sigma Velocity Measurement Error
σ_θ = One Sigma Angle Measurement Error
A_S = Surface Area
R_L = Detector Load Resistance

References

[1] *Santa Barbara Research Center Wallchart*

[2] McClatchey, R.; Selby, J.: *AFCRL-72-0611, Environmental Research Paper No. 419,* AFCRL, Bedford, Massachusetts, 12 October 1972.

[3] Goodwin, F.; and Nussmeier, T.: *IEEE Journal of Quantum Electronics, Vol. QE-4, No. 10,* p. 616, October 1968.

[4] Rensch and Long: *Applied Optics, Vol. 9, No. 7,* pp. 1563-1573, July 1970.

[5] Chu, T.S.: *Bell System Technical Journal,* pp. 723-759, May/June 1968.

[6] Barton, D.K.: *Radar System Analysis,* Artech House, Dedham, Massachusetts, 1976.

[7] Skolnik, M.I.: *Introduction to Radar Systems,* McGraw-Hill, New York, 1962.

[8] Standberg, M.W.P.: "Inherent Noise of Quantum Mechanical Amplifiers," *Physics Review, Vol. 106,* pp. 610-617, 15 May 1957.

[9] *Electro-Optics Handbook,* RCA, Commercial Engineering, Harrison, New Jersey, May 1968.

[10] Wolfe, W.F.: *Handbook of Military IR Technology,* US Government Printing Office, Washington, DC, 1965.

[11] Ross, M.: *Laser Receivers,* John Wiley and Sons, Inc., New York, 1966.

[12] Goodman, J.: *IEEE Transactions on Aerospace and Electronic Systems, Vol. AES-2, No. 5,* September 1962.

[13] *ITT Wall Chart*

[14] *Laser Focus Buyer's Guide,* 1977.

[15] Lawrence, R.S.; Ochs, G.R.; and Clifford, S.F.: "Measurements of Atmospheric Turbulence Relevant to Optical Propagation," *Journal of the Optical Society of America, Vol. 66,* pp. 826-830, June 1970.

[16] Brookner, E.: "Atmospheric Propagation and Communication Channel Model for Laser Wavelengths – An Update," *IEEE Transactions on Communication Technology, Vol. COM-22,* pp. 265-270, February 1974; and Brookner, E.: "Atmospheric Propagation and Communication Channel Model for Laser Wavelengths," *IEEE Transactions on Communication Technology, Vol. COM-18,* pp. 396-416, August 1970. Both articles are reprinted in Goldberg, B. (ed.): *Communication Channels: Characterization and Behavior,* IEEE Press, 1976.

[17] Huffnagel, R.E.: "An Improved Model Turbulent Atmosphere in Restoration of Atmospherically Degraded Images," Volume 2 of *Woods Hole Summer Study* (Appendix 3), Woods Hole, Massachusetts, pp. 15-18, July 1966.

[18] Fried, D.L.: "Optical Heterodyne Detection of an Atmospherically Distorted Signal Wave Front, *Proceedings of the IEEE, Vol. 55,* pp. 57-66, January 1967.

[19] Goldstein, I.; Miles, P.A.; and Chabot, A.: "Heterodyne Measurements of Light Propagation through Atmospheric Turbulence," *Proceedings of the IEEE, Vol. 53,* pp. 1172-1180, September 1965.

[20] Gilmartin, T.; Bostick, H.; and Sullivan, L.: *NEREM Record,* Massachusetts Institute of Technology, Lincoln Laboratory, May 1973.

[21] Kingston, R.; and Sullivan, L.: "Optical Design Problems in Laser Systems," *SPIE, Vol. 69,* 1975.

[22] Teoste, R.; and Scouler, W.J.: Massachusetts Institute of Technology *Report 1976-19a* (10.6 μm Coherent Monopulse Tracking), Lincoln Laboratory, 7 May 1976.

[23] Jelalian, A.; Kawachi, D.; and Miller, C.: "Radar Electronic Counter-Countermeasures," Electro '76, Boston, Massachusetts, 1976.

Chapter 25 Morrison

Tracking and Smoothing*

The intent of this chapter is to provide an overall picture of the techniques of smoothing and prediction. Therefore, esoteric discussions of individual methods and dwelling unnecessarily on minor details are avoided.

Upon my entry into the field in the mid-sixties, I found a highly fragmented and diverse set of approaches. Amidst the testimonies, deprecations, and general confusion, there seemed little or no central theory. But after a number of years of doing extensive reading and research, I realized that there indeed was an underlying concept, discovered by Gauss in the late 18th and early 19th centuries. In fact, the only progress in the 20th century compared to the work by Gauss was in matrix notation, which is little more than symbolic shorthand. Had Gauss wanted to work out multidimensional problems, chances are he would have invented matrices – he had already invented probability theory and the normal density function.

When Gauss reached the age of eighteen, it was announced that a planetoid or asteroid had been observed. A gap had been known to exist in the solar system – there was Mercury, Venus, Earth and Mars, then a perplexing void. A rule predicted where the next planet should have been, yet there was a blank. Someone came up with a theory that there had been an instability at that location during the formation of the solar system, and the missing body had fragmented into thousands of pieces – but no one had ever seen one. Then, in 1795, an Italian astronomer accidently sighted a planetoid. But an immediate concern arose amidst the excitement and celebration. Since it had taken over a hundred and fifty years to find a planetoid by *accident,* would it take that long to find this object again once it disappeared in the next few days?

Gauss (then nineteen) had some ideas on the subject. He gathered all the observations that astronomers had made and quickly formulated a mature body of thought. The theory, of course, was that of smoothing, prediction, and orbital determination. After fourteen months of frantic tabulation and calculation of all the necessary transcendental functions, he made the announcement that the planetoid would reappear at a certain time, azimuth, and elevation.

Skeptically, telescopes were trained to that location at the specified time – and there was the object! What Gauss had accomplished in that period was a revolutionizing of the whole procedure of making observations and drawing conclusions. He had laid the foundation in that short time for the theories of smoothing and prediction, regression, and probability. Thus, there is nothing in this chapter that Gauss did not know already, except for a few refinements in technique owing to the advent of the digital computer. There are things we do now that he could not have done computationally or that he was not interested in doing – we touch on those areas.

A typical problem is to make observations on a comet so as to be able to predict its future position. A telescope is located on the earth, and a comet is moving in a parabola around the sun. The telescope provides only two parameters – azimuth and elevation – while six orbital parameters are desired. By making successive observations in time, understanding the underlying processes, and using mathematical techniques, the six parameters can be obtained with arbitrary precision.

Another problem which is perhaps more applicable in the 20th century occurs when making observations on an artificial satellite in order to be able to direct at it a very narrow beam antenna and to provide the option of making predictions. In this case (using a radar), we have three observables – azimuth, elevation, and range; again, we want six parameters.

A more critical problem occurs when observations are made on an intercontinental ballistic missile. Time is of the essence, since one has only about 15 minutes from the moment the missile comes over the horizon, and only eight or nine minutes to launch an interceptor. We need to know where the ICBM came from, in order to decide against what country to retaliate. We want to predict the impact point quickly and to determine a mid-course rendezvous at which to (hopefully) destroy the missile. Again there are three observables – range, elevation, and azimuth; the list of needed parameters has increased to seven items with one which relates to drag ratio added to the six orbital ones. That new parameter aids in deciding whether the object is authentic or a decoy.

Moving away from these extraterrestrial problems, there is an application to the printing industry made at a place where I worked some time back. When a man uses a printing press, he makes an adjustment on that press and the ink starts to come down what is called the ink train onto the paper. Each successive sheet of paper represents a sample of a first order lag process. What the man really would like to know when he makes an adjustment is "What is the steady state going to be?" What we did was to add to the printing press a densitometer that samples density on each sheet, feeds those figures into a computer, uses an appropriate algorithm, and, within 10 or 15 sheets of the adjustment, indicates the eventual steady state value. Long before that value is achieved, further corrections can be made, thereby bringing that press up much faster (see Figure 1).

There was another intriguing situation in a company which is one of the major suppliers for the broadcast industry. One of the big problems they faced involved the transition to unattended equip-

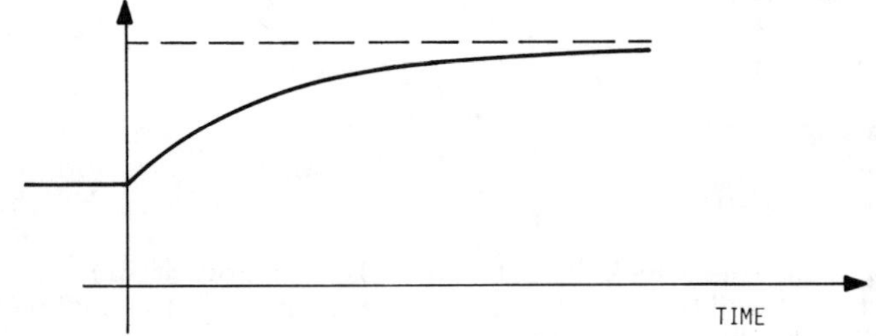

Figure 1 **Printing Press Example. Observations made using a color densitometer to predict steady-state values of color density. (First-order lag system with unknown time constant.)**

**Editor's Note* – First draft of this chapter was edited with the help of the late Dr. Paul Buxbaum, formerly with the Advanced Development Laboratory of the Raytheon Company. The editor has since made extensive modifications to portions of the text and takes responsibility for errors that may have arisen.

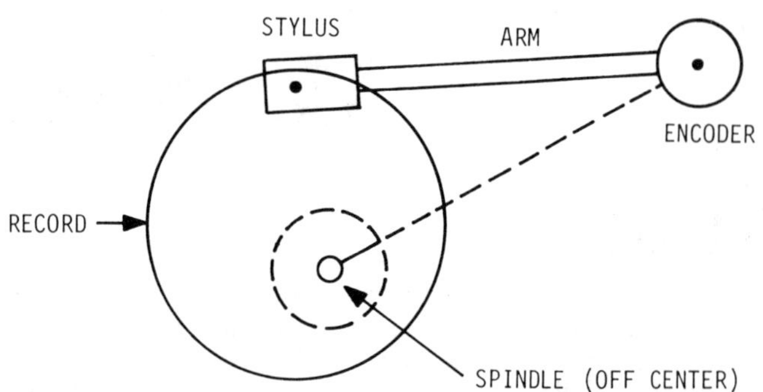

Figure 2 **Unattended Broadcasting Example. Observations made on record player to decide if groove jump has occurred.**

ment, in an effort to save labor and money. A difficulty with unattended broadcasting is that when you are playing a disc with a bad groove, the phonograph arm will jump repeatedly; if access is not fast enough, listeners can tell that this station is unattended. In Figure 2, the dot at the center is the center of the grooves. When a disc is pressed, the spindle hole is not always dead at the center; thus, the spindle hole is displaced as shown; typically on the order of 1/8 of an inch. As a result you have an oscillatory motion caused by that 1/8 of an inch. An encoder or code wheel was attached to the arm; the reading from the encoder went into a computer. With the appropriate algorithm, we were able to determine if the tracking was smooth. When a groove jump occurred, there was a discontinuity in the behavior of the system. The groove edges are spaced on the order of a few mils apart; thus, you are looking for a few mils discontinuity superimposed on the 1/8 inch of motion in this complicated system. It turned out to be a task for a fairly sophisticated and clever smoothing and prediction scheme. Once it was instrumented, the plan worked like a charm; unfortunately, it turned out to be too expensive. In those days, the chip computer was not available; now with the five dollar chip, such a system could be built at a reasonable price.

In general, we want to make observations on a process for two reasons. We want to obtain knowledge of the present state, and we want to predict the future state. The observations are cluttered by noise, errors, and inaccuracies. Whether it is the stylus, the ballistic missile, the satellite, or the density value on the printing press, we want to know where it is right now and, equally important (if not more so), we want to know where it will be at some point in the future (so that some anticipatory action can be taken).

Things now become a trifle more mathematical. What we are observing is called a *process* – it could be a chemical plant, it could be a printing press, or it could be a ballistic missile. The process is governed, in general, by either an algebraic or, equivalently, differential equation. The equations for the printing press example are, respectively:

$$x(t) = C + D\, e^{-\alpha t} \tag{1a}$$

and

$$\ddot{x}(t) = -\alpha\, \dot{x}(t) \tag{1b}$$

The ink density looks like a first-order lag* with two arbitrary constants, C and D. It can start at any level and it can end at any level. When working with an algebraic equation, the procedure is called regression analysis; at some time or another, we have all fitted a quadratic function to a set of data. When working on a differential equation, the technique is known as smoothing and prediction.

*For those who have forgotten, a first-order lag system having two arbitrary constants is a system whose response to a step function is of the form given by Equation (1a). Equivalently, it is a circuit whose response is governed by the differential equation, Equation (1b). A simple example of such a system is the current or voltage of an RC circuit. For our example, it is the density of the ink, x(t).

It is convenient to write Equation (1b) in an equivalent form, as two equations.

$$\frac{d}{dt}\, x(t) = \dot{x}(t) \tag{2a}$$

$$\frac{d}{dt}\, \dot{x}(t) = -\alpha\, \dot{x}(t) \tag{2b}$$

Equations (2a) and (2b) are equivalent to the differential equation, Equation (1b). This form of the differential equation which describes the first-order lag process is the *state variable* description of the process. The parameters x(t) and $\dot{x}(t)$ represent the states of the system. The first-order lag system with two arbitrary constants is thus a two-state process. Notice that the differential equation (Equation (1b)) is written in Equations (2a) and (2b) with the use of only *first* derivatives; there are no second derivatives, just the first derivatives of the states x(t) and $\dot{x}(t)$ – the ink density and its rate of change.

These two differential equations (Equations (2a) and (2b)) govern the dynamics of the process. If we knew, or could estimate, C and D, we could determine the ink density, x(t), at the present time, t, and could predict its value at some point in the future. If there were no measurement errors, we could estimate C and D *perfectly* based solely on observations of x(t) at two different times. However, errors are present in our measurement of x(t), and this fact precludes perfect smoothing and prediction. It is because of these errors that tracking and smoothing theory is needed.

Another source of error is in the model used to describe the process. Specifically, the form of the differential equation given by Equation (1b), or its restatement given by Equations (2a) and (2b), may not represent the process dynamics correctly. A third order differential equation may actually be needed to describe x(t). Tracking and smoothing theory also handles this type of model error. We discuss this aspect of the theory only briefly. (See Reference 1 for further details.)

In smoothing and prediction, two sets of equations govern the whole technique. If you understand those two sets of equations, you understand the whole concept. The first set of equations is the differential equations which describe the process. The second set relates the parameters being measured to those to be estimated. In our printing press example, this second set of equations indicates how measurement noise corrupts the quantity that we desire to measure – x(t). The first set is called the *process equations,* the second is called the *observation relations.* Let us elaborate further on the process equations and the various forms they take in practice.

Gather the terms on the left of Equations (2a) and (2b) and put them in vector form. The differential equation then can be expressed as the derivative of a vector. The derivative equals a constant matrix times the same vector that appears on the left.

$$\frac{d}{dt}\begin{pmatrix} x(t) \\ \dot{x}(t) \end{pmatrix} = \begin{pmatrix} 0 & 1 \\ 0 & -\alpha \end{pmatrix}\begin{pmatrix} x(t) \\ \dot{x}(t) \end{pmatrix} \tag{3}$$

This form is perfectly general; it governs any differential equation which does not include products of the derivatives. Products of derivatives are undesirable; fortunately, we can avoid them.

There are three forms of processes that are encountered in practice; they are described by the differential equations below.

$$\frac{d}{dt}\, X(t) = A \cdot X(t) \tag{4a}$$

$$\frac{d}{dt}X(t) = A(t) \cdot X(t) \tag{4b}$$

$$\frac{d}{dt}X(t) = F(X(t)) \tag{4c}$$

where X(t) is a column matrix such as that of Equation (3). (Capital letters are used for matrices.) Equation (4a) is a constant coefficient linear differential equation, as is Equation (3). Equation (4b) arises when A is no longer a constant but is permitted to vary with time — a time-varying linear differential equation. Equation (4c) is a nonlinear function of the vector; it indicates the maximum complexity we encounter. It turns out that the last case is the one that occurs most frequently in practice. Ballistic missiles, satellites, and comets are examples of the third form. All of the theory derived relates to the first two. However, although most practical applications occur in the third form, that form can be reduced to the linear one shown in Equation (4b) by the technique of *local linearization.* For details, see Reference 1.

An example of the form of a nonlinear differential is:

$$\begin{aligned} \frac{d}{dt}(x_0(t)) &= f_0(x_0, x_1, \ldots x_5) \\ \frac{d}{dt}(x_1(t)) &= f_1(x_0, x_1, \ldots x_5) \\ &\cdot \\ &\cdot \\ &\cdot \\ \frac{d}{dt}(x_5(t)) &= f_5(x_0, x_1, \ldots x_5) \end{aligned} \tag{5}$$

The nonlinear form is given for the assumption of six coordinates — let us say three Cartesian coordinates of position and three Cartesian coordinates of velocity. This definition is complete for a moving body in Cartesian space. The six orbital parameters of a planet or a comet in orbit around the sun might form the state variables. The first derivative of each is some function of all six; this sort of situation is what is meant by a nonlinear differential equation.

Whether the process differential equation is nonlinear or the observation scheme is nonlinear is immaterial — either forces you into the nonlinear domain, and recourse to linearization techniques is necessary. Figure 3 shows that the simple set of observation coordinates — range, azimuth, and elevation — are related very nonlinearly to Cartesian coordinates.

We have discussed the differential equations that define the model; mathematically, a model is represented by a differential equation. But that is only half the story; the other half is how we are going to *observe* the process.

Figure 3 illustrates an observation scheme in which we are going to *observe* an object with a radar. Given are the polar radar coordinates, the Cartesian coordinates, and the relationships between them. Assume the object is moving at a constant speed in some arbitrary direction. Let x_0, x_1, and x_2 indicate its position at time t = 0. At time t, range, azimuth, and elevation are functions of all six state variables (x_0, x_1, and x_2, and the three Cartesian velocity coordinates); i.e.,

$$Y_t = \begin{pmatrix} r \\ \psi \\ \theta \end{pmatrix} = \begin{pmatrix} y_0 \\ y_1 \\ y_2 \end{pmatrix}_t \tag{6a}$$

or (see Figure 3)

$$r = \sqrt{x_0^2 + x_1^2 + x_2^2}$$

$$\psi = \text{ARCTAN}\left(\frac{x_1}{x_0}\right)$$

$$\theta = \text{ARCSIN}\left(\frac{x_2}{\sqrt{x_0^2 + x_1^2}}\right)$$

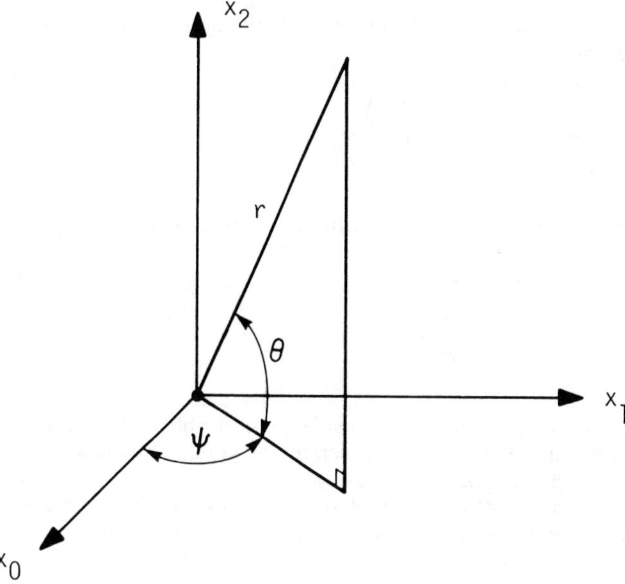

Figure 3 Nonlinear Observation Scheme, r, θ, ψ observed; x_0, x_1, x_2 desired.

$$\begin{pmatrix} y_0 \\ y_1 \\ y_2 \end{pmatrix}_t = \begin{pmatrix} g_0\,(x_0, x_1, \ldots x_5) \\ g_1\,(x_0, x_1, \ldots x_5) \\ g_2\,(x_0, x_1, \ldots x_5) \end{pmatrix}_t \tag{6b}$$

which we write in condensed form as

$$Y(t) = G(X(t)) \tag{6c}$$

where $x_3 = \dot{x}_0$, $x_4 = \dot{x}_1$, and $x_5 = \dot{x}_2$. The subscript indicates dependence of the matrix component variables on time t. (Equation 6 is horrendously nonlinear! Nevertheless we are forced to deal with it. If anybody has any ideas of how to build a Cartesian radar, there are many persons who would like to hear about it.)

In the observation scheme above, the vector Y defines the vector of coordinates that are being observed at any one time. At some time t, we observe range, azimuth, and elevation, and see that y_0 (range) is some nonlinear function of our six state variables, as are azimuth and elevation. We use vector form so that the vector on the left is some nonlinear function of the state variables. This expression is the beginning of an observation relation. It is not complete because, in practice, we do not get uncorrupted measurements of the object's coordinates. As indicated before, we have measurement errors such as noise in the system.

What really happens is that there is a vector, N(t), which is a vector of noise sources.

$$Y(t) = G(X(t)) + N(t) \tag{7}$$

The situation is now complete. A model is given as a differential equation in vector notation. We then add an observation relation which includes the noise sources, N(t). It was at this point that Gauss realized that he would have to develop a body of theory related to how to reduce the effects of that N(t) vector. What he had at his disposal was Y(t), and he was being asked to come up with an X(t); in Y(t) there were two observables – azimuth and elevation – from the telescope. He realized that he would have to come up with the six numbers in X; from two numbers on the left he had to calculate six numbers in the presence of noise.

Consider a simpler problem for which the process differential equation is:

$$\frac{d}{dt}\begin{pmatrix} x \\ \dot{x} \\ \ddot{x} \end{pmatrix} = \begin{pmatrix} 0 & 1 & 0 \\ 0 & 0 & 1 \\ 0 & 0 & 0 \end{pmatrix}\begin{pmatrix} x \\ \dot{x} \\ \ddot{x} \end{pmatrix} \tag{8a}$$

and for which the following observation relation holds:

$$y_n = (1\ \ 0\ \ 0)\begin{pmatrix} x \\ \dot{x} \\ \ddot{x} \end{pmatrix}_n + \nu_n \tag{8b}$$

The subscript n indicates dependence on the observation instant n. Differential Equation (8a) generates the family of quadratics, i.e., second degree polynomials using a single derivative of three state variables. The observation relation indicates that the observation at time n is simply the zero derivative at time n, plus noise.

We now have touched on the fundamentals of the whole problem. The process is defined by a differential equation to which we add an observation relation. The tremendous power of using vector and matrix notation has become obvious. We are able to reduce things to very simple algebraic quantities though, in fact, the matrices carry with them great power and ability.

Here is Equation (8a) again, rewritten as a quadratic:

$$x(t) = a + bt + ct^2 \tag{9a}$$

It is an arbitrary quadratic so it has three arbitrary constants. If we replace a by x(0), b by $\dot{x}(0)$, and c by ½$\ddot{x}(0)$, we see another way to write it:

$$x(t) = x(0) + t\,\dot{x}(0) + \frac{t^2}{2}\,\ddot{x}(0) \tag{9b}$$

which can be written, using matrix notation, as:

$$\begin{pmatrix} x \\ \dot{x} \\ \ddot{x} \end{pmatrix}_t = \begin{pmatrix} 1 & t & t^2/2 \\ 0 & 1 & t \\ 0 & 0 & 1 \end{pmatrix}\begin{pmatrix} x \\ \dot{x} \\ \ddot{x} \end{pmatrix}_0 \tag{9c}$$

or, defining the vectors and matrices appropriately:

$$X(t) = \Phi(t)\,X(0) \tag{9d}$$

Thus, starting with a differential equation, we reach a form in which the state vector is some matrix times an initial vector. In control theory this matrix is called the *transition matrix.* This method is another and, as it turns out, better way of writing the process differential equations.

A differential equation cannot be put into a digital computer – it must first be quantized and made discrete. Hence the differential equation is always replaced by what turns out to be the *difference* equation – such as Equation (9d). Instead of taking X(0), as in Equation (9d), take X(t) and multiply it by Φ(h), thereby moving X(t) forward by amount h and giving X(t + h). Thus:

$$X(t+h) = \Phi(h)\,X(t) \tag{10a}$$

or, if the interval τ between observations is unity:

$$X_n = \Phi(1)X_{n-1} \tag{10b}$$

Thus:

$$X_{n+h} = \Phi(h)X_n \tag{10c}$$

Equation (10c) is a discrete numerical way of computing the values down a trajectory. Given the values at any one point, t_n, on that trajectory, they are multiplied by the transition matrix to move the sampling point to any other point on the trajectory. If we have the vector value of the process at any point in time, we can move forward or backward simply by multiplying by the transition matrix. This method is how we do all predictions. The problem, therefore, is to get the state vector at some instant of time.

Returning to the simple example of Equation (8b), we are observing the zero derivative and have not yet obtained a trajectory. The row matrix (100) is the observation matrix to be designated as M; it multiplies the state vector. The measurement error is ν_n. Equation (8b) can be rewritten as:

$$y_n = MX_n + \nu_n \tag{11a}$$

Let us assume that τ seconds prior to time n we made an observation (which we call y_{n-1}) where:

$$y_{n-1} = MX_{n-1} + \nu_{n-1} \tag{11b}$$

This equation has the same matrix M, but a different state vector – the state vector that prevailed τ seconds earlier (i.e., one observation instant before) plus whatever noise prevailed at that time. Let us use the transition matrix. To move X back one observation instant, multiply it by the transition matrix with negative τ:

$$X_{n-1} = \Phi(-\tau)X_n \tag{11c}$$

This equation indicates that the previous state vector is related to the present one by the differential equation. We next take Equation (11b) and rewrite it as:

$$y_{n-1} = M\,\Phi(-\tau)X_n + \nu_{n-1} \tag{11d}$$

Combining (11a) and (11d) yields:

$$\begin{pmatrix} y_n \\ y_{n-1} \end{pmatrix} = \begin{pmatrix} M \\ M\,\Phi(-\tau) \end{pmatrix} X_n + \begin{pmatrix} \nu_n \\ \nu_{n-1} \end{pmatrix} \tag{11e}$$

Although our observations have been taken at *different* points in time, Equation (11e) has related them to the state vector at *one* instant in time. This final equation is written with a matrix T which can be evaluated prior to the measurements and be built into the computer program:

$$Y_{(n)} = TX_n + N_{(n)} \tag{12}$$

where the subscript n in parentheses is used to indicate a matrix composed of variates obtained from more than one observation instant, $Y_{(n)}$ and $N_{(n)}$ consisting of the observations and measurement noise for all the time instants to be used in the estimation process, the measurements at times other than just n and n−1 being included if available [1]. X_n is our state vector at *one* instant in time, even though the observations occurred at *several* instants. Equation (12) is the most general form of an observation relation and is one of the two fundamental elements that govern smoothing and prediction.

The problem, then, restated, is given the set of numbers, $Y_{(n)}$, we want to obtain estimates of X_n. We know what the T matrix is because we are constructing the observation scheme. We do not know the elements of $N_{(n)}$ but we know that they exist. We assume that we know their statistics. Thus we assume that their means are 0 – in boresiting our instrument, we have taken out any biases. We will also assume that we are able to estimate or measure the variances or the standard deviations of these error sources; we do not know the instantaneous values, but we can estimate or measure their second order statistics.

Let X_n^* be an estimate of X_n. ($X_{m,n}^*$ represents the estimate of X at time m based on the measurements made up until time n. Sometimes the second or both subscripts are omitted, in which case the estimate is given at time n based on the measurements up until time n.) Predicting X_{n+1} from the X_n^* is done by moving along the trajectory one observation instant forward, based on the observation that prevailed up to time n. We move forward simply by integrating the differential equation or by multiplying by the transition matrix. Thus, if we obtain an estimate of the state vector somewhere in the process, the problem of prediction is trivial – it is just a matter of integrating the differential equation forward or backward. This integration forward or backward is made simple through the use of the transition matrix. Thus:

$$X_{n+1,n}^* = \Phi(\tau)\, X_{n,n}^* \qquad (13a)$$

$$X_{n+h,n}^* = \Phi(h\tau)\, X_{n,n}^* \qquad (13b)$$

The problem then boils down to obtaining the "best" estimate of X_n, which we call $X_{n,n}^*$. This is the smoothing problem. The prediction problem is that of obtaining the "best" estimation of $X_{n+h,n}^*$. But just what "best" means is a good question, and we now direct our attention to that subject.

Two solutions of how to extract the "best" smoothing estimate are:

1. Least Squares

$$\hat{X}_{n,n}^* = (T_n{}^T T_n)^{-1}\, T_n{}^T Y_{(n)} \qquad (14a)$$

$$\text{i.e., } \hat{X}_{n,n}^* = \hat{W}_n Y_{(n)} \qquad (14b)$$

where:

$$\hat{W}_n = (T_n{}^T T_n)^{-1} T_n{}^T \qquad (14c)$$

2. Minimum Variance

$$\mathring{X}_{n,n}^* = (T_n{}^T R_{(n)}{}^{-1} T_n)^{-1} T_n{}^T R_{(n)}{}^{-1} Y_{(n)} \qquad (15a)$$

$$\text{i.e., } \mathring{X}_{n,n}^* = \mathring{W}_n Y_{(n)} \qquad (15b)$$

where:

$$\mathring{W}_n = (T_n{}^T R_{(n)}{}^{-1} T_n)^{-1} T_n{}^T R_{(n)}{}^{-1} \qquad (15c)$$

$$Y_{(n)} = T_n X_n + N_{(n)} \qquad (16a)$$

$$R_{(n)} = \text{cov}\,(N_{(n)}) \qquad (16b)$$

$\hat{X}_{n,n}^*$ and $\mathring{X}_{n,n}^*$ are the least squares and minimum variance estimates, respectively. Both use the matrix T of Equation (12). The subscript n on T is used to allow for the more general case where T and, in turn, M are dependent on time.

(The oldest solution is least squares. Legendre, a French mathematician of the 19th century, published his paper in 1803 on the method of least squares, thinking he had earned his immortality after 40 or 50 years in mathematics. In 1809 Gauss told the world how he had solved the orbital prediction problem – and how he had used the method of least squares in 1799!)

The matrix X_n is possibly six state variables, and there must be at least six numbers in $Y_{(n)}$ because, to satisfy the original goal of getting six parameters from two observations, we need more than one set of measurements. If we observe azimuth and elevation at three discrete instances, we have six numbers in $Y_{(n)}$. From six numbers in $Y_{(n)}$ we can get six state variables by *interpolation.* The trajectory in some hyperspace is made to pass exactly through the observations – nothing is being done to smooth out the measurement noise.

Suppose we have six observations and then take another six. From these twelve observations we are going to attempt to get six estimated quantities. We have more degrees of freedom and more observations than are needed for a minimal solution, and we use that freedom of the other six observations to reduce the effects of $N_{(n)}$, the noise. An overabundance of observations which we reduce to a small number (say, going from a hundred to three) is the essence of least squares. We use the other 97 degrees of freedom in this example to work on the random vector, $N_{(n)}$. By the least squares method, the matrix T_n of Equation (16a) is inverted, ignoring $N_{(n)}$. T_n may be a hundred rows by three columns but we invert it by using what is known as the pseudo-inverse, which is simply of the form of Equation (14c), where $\hat{W}_n$ can be computed ahead of time. We can take our observations, $Y_{(n)}$, multiply by the matrix, $\hat{W}_n$, and produce the estimated state vector, $\hat{X}^*$, prevailing at time n and based on observations up to n; see Equation (14b). $\hat{W}_n$ is a matrix that can be computed once and for all; if it is computed from T_n in this way, we are using the method of least squares. What least squares does is to take our fitting function and fit it to the data in such a way as to calculate the residuals (the differences between the observations and our fitting function), square them, add them, and produce a single number. Least squares then minimizes that number, in the most general form for any arbitrary number of dimensions. The catch, of course, is to calculate $\hat{W}_n$. In many cases, this task is difficult because we must invert a matrix. That inversion, though, can be done in practice on a computer.

The minimum variance method (Equations (15a) through (15c)) is similar to least squares, but it is more generalized. It turns out that Gauss was one step beyond Legendre in that he went past least squares to generalized least squares. (He went further than Legendre had ever dreamed which, in the long run, is the measure of the two men.) We take the vector of errors, $N_{(n)}$, which has known second order statistics, then assemble the second order statistics into a matrix known as the *covariance matrix.* This matrix, $R_{(n)}$, is the covariance matrix of $N_{(n)}$ – a matrix of numbers that we know. We then calculate $\mathring{X}^*$ from $Y_{(n)}$ again, in the form $\mathring{X}^* = \mathring{W}_n Y_{(n)}$. This time $\mathring{W}_n$ is even more inscrutable – it is similar to the one before, except that an inverse has been placed between the T_n's; see Equation (15c). If $R_{(n)}$ is the identity matrix, it simply drops back to least squares; however, $R_{(n)}$ is generally not an identity matrix, so minimum variance is a more sophisticated technique. It also reverts to least squares when $R_{(n)}$ is a diagonal matrix with every element on the diagonal having the same value. The latter situation occurs very frequently; thus, minimum variance often is least squares. Minimum variance and least squares are related but, in general, not identical. Minimum variance takes every observation in $Y_{(n)}$ and weights each precisely according to its merits. For instance, observations from better radars deserve more attention than those from poorer ones.

Let us say that some of the errors have a very large variance. Are they good or bad observations? Obviously, they are very bad. Other errors have a very small variance; they would be classed as good observations. If the variance could be put into the denominator, poor observations would get weighted by the inverse of their variance; good observations would be boosted relatively. Minimum variance weights every observation according to its value. Each value, as it turns out in a one-dimensional case, is the inverse of the variance.

Minimum variance is one way to obtain the Kalman filter. The relation between minimum variance and the Kalman filter is significant. Moreover, the *Wiener theory* can be obtained from least squares.*

We now ask the question – how good are our estimates? We need to know the covariance matrix of $\hat{X}^*$. For the least squares and minimum variance estimates, the variance of the estimates $\hat{X}^*$ and $\mathring{X}^*$ are given respectively by:

Least Squares

$$\hat{S}^*_{n,n} = \hat{W}_n R_{(n)} \hat{W}_n^T \tag{17a}$$

Minimum Variance

$$\mathring{S}^*_{n,n} = \mathring{W}_n R_{(n)} \mathring{W}_n^T \tag{17b}$$

We start with a vector of observations and the covariance matrix of their errors, then process them through a computer (in an appropriate algorithm). The output is a state vector (in this case, an estimate) and its covariance matrix obtained using Equations (14b) and (17a) or Equations (15b) and (17b). This process is on-going. For example, X* may be an ingredient of another estimate. Two $Y_{(n)}$'s and $R_{(n)}$'s could have been pre-digested, producing two X*'s and S*'s. Houston's X* and the satellite's X*, and Houston's S* and the satellite's S*, can be treated respectively as a new $Y_{(n)}$ and a new $R_{(n)}$. The covariance matrix weights an observation precisely according to its value. Information is carried along according to its merit; the procedure is coherent. This processing is rooted in fundamental physics; when Gauss invented it, his genius was in that he invented something that was so essential and basic.

Figure 4 illustrates a batch of observations in one dimension; time is plotted horizontally. The observations bounce around; the wavy lines are the possible trajectories of our differential equation. The differential equation runs out the two-dimensional plot of trajectories, the family of all possible curves that it can generate. What these algorithms do is to scan through all of those curves and all possible trajectories that the differential equation carries with it, and find the one solution that best fits the observations in the sense of least squares (or minimum variance). If, of course, our trajectory is one type of shape because the differential equation generates a family of, say, circles, and the observations look like those in Figure 4, we are using the wrong model for the observations. We are going to get a very poor fit though the algorithm does its best. Given the universe of curves, it picks out the one that fits best. If a differential equation in the family of trajectories is in fact a good representation of what is going on, the algorithm seeks it out and there is a good fit. Thus, the results depend on the match of the process and the observations.

Do the methods we have discussed so far extend to more complex results like elections, the stockmarket, the economy, and the gross national product? Yes, as long as a differential equation can be used as a model.

What we are discussing in this chapter are the basic techniques of smoothing and prediction. In practice, you have to combine these techniques, shuffle them around, use your own ingenuity, and create the structure of the eventual scheme which may resemble its basic components only slightly.

*See Chapter 26 for a discussion of some of the fundamental differences between the Wiener and Kalman theories; also see Editor's Note at end of chapter.

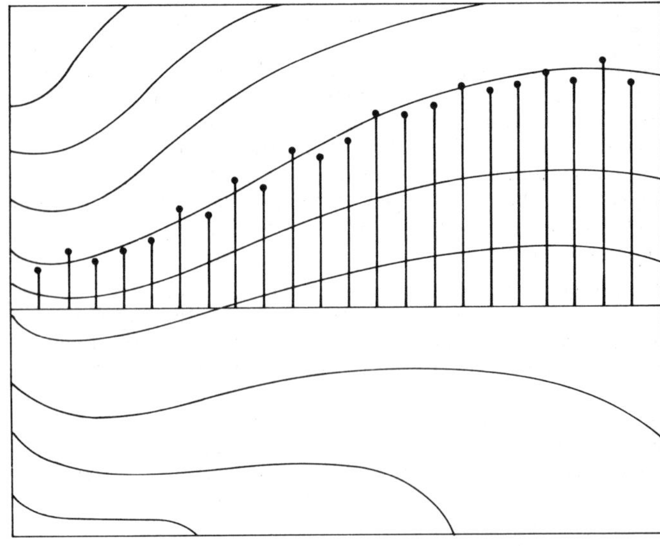

Figure 4 **Fitting Data Points to One of a Family of Possible Trajectories Obtained for Different Constants of the Process Differential Equation.**

As shown in Figure 5, there are three basic families of filters – the *fixed* memory, the *expanding* memory, and the *fading* memory.

The *fixed* memory family takes a flat window and moves it in time; the distance between the front and back of the window is fixed. Any observations that are on the time axis and which land in this window are put into the algorithm to produce an estimate based on whatever is "seen" with all observations weighted equally.

In the *expanding* memory, every observation adds to the total. The observation window, again flat, expands with time and, theoretically, any estimate gets increasingly better due to the expanded body of information.

Fading memory is based on the exponential function which is at the root of all linear systems. The observation window moves forward but is not flat. The emphasis has an exponential decay; theoretically, the tail goes to infinity but the weighting diminishes to zero. It is in this manner that observations are weighted – they are "forgotten" at an exponential rate.

There are two cases in each of the families above – the *polynomial* case and the *generalized* model. As it turns out, polynomials are excellent for use in filters because the resulting filters are so simple and so computationally compact.

A practical problem is illustrated in Figure 6. The radar is providing range, azimuth, and elevation; we want six orbital parameters. If we make enough range observations and put them into a polynomial filter, we can fit a second or third degree polynomial to these observations. We can differentiate that polynomial to give the range rate, producing R* and $\dot{R}$* estimates based on the observations. The same things can be done for azimuth – we can get an estimate of azimuth and its first derivative. Similarly, we can get an estimate of elevation and its first derivative. From three numbers at various instants in time, processed in three polynomial filters, we have produced six good numbers. Putting those six numbers into a coordinate transformation (an algebraic process) produces six orbital parameters.

The three filters shown in the figure are extremely simple and very compact. The big rectangular box is a well known set of algebraic transformations; the whole thing can be put together quite rapidly.

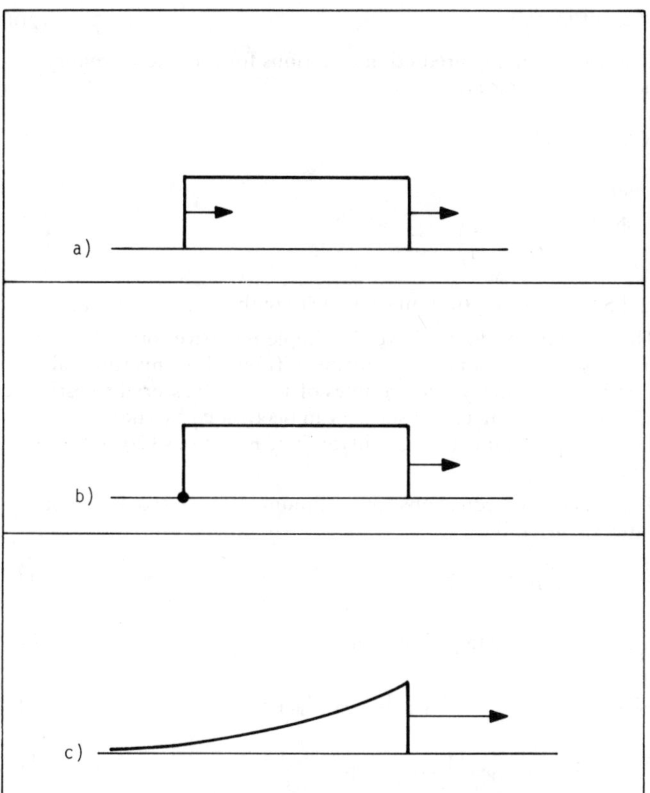

Figure 5 **Three Basic Families of Filters: a) fixed memory, b) expanding memory, c) fading memory.**

The prediction equations for each of the three filters for first degree expanding memory polynomial filtering (Figure 5b) are*

$$\dot{x}^*_{n+1,n} = \dot{x}^*_{n,n-1} + \frac{6}{(n+2)(n+1)}(y_n - x^*_{n,n-1}) \tag{18a}$$

$$x^*_{n+1,n} = x^*_{n,n-1} + \dot{x}^*_{n+1,n} + \frac{2(2n+1)}{(n+2)(n+1)}(y_n - x^*_{n,n-1}) \tag{18b}$$

The variance of the estimate of the state vector for each filter is

$$S^*_{n+1,n} = \frac{\sigma^2}{(n+1)n}\begin{pmatrix} 4n+6 & 6 \\ 6 & \dfrac{12}{n+2} \end{pmatrix} \tag{18c}$$

where σ is the single sample rms measurement error of R, ψ, or θ. With the expanding memory we want to compute the first derivative and a zero derivative of the state vector. The process is recursive as an observation adds to a previous estimate to produce a refined estimate. The previous prediction is $x^*_{n,n-1}$; the new observation is y_n. They are subtracted and the very simple function of n in Equation (18a) is computed, starting at 0, 1, 2, 3 for each successive observation. We have $x^*_{n,n-1}$ in the computer memory, so we take the observation less $x^*_{n,n-1}$ multiplied by the very simple function of n, and add to that the old estimate of the first derivative, producing the new prediction of the first derivative. We calculate this same difference for the second equation (i.e., the new observation minus the old estimate), multiply this difference by another very simple function of n, add to that the new estimate of the first derivative, and then add the old estimate, $x^*_{n,n-1}$. The result is the new prediction (which is based on Legendre's polynomials). [1] In the computer we carry only two numbers — $x^*_{n+1,n}$ and $\dot{x}^*_{n+1,n}$ — because we use $x^*_{n,n-1}$ and $\dot{x}^*_{n,n-1}$ to compute a new $x^*_{n+1,n}$ and $\dot{x}^*_{n+1,n}$. The process is very compact. You can run hundreds of these procedures simultaneously in a computer, as you sometimes must do. All that is needed are two memory slots per filter.

*This filter obtains the least squares estimate by minimizing:

$$e_n = \sum_{r=0}^{n} (y_{n-r} - [p^*(r)])^2$$

where $[p^*(r)]$ is a polynomial of degree m that is being fit to the data. For Equations (18a) and (18b), m = 1; i.e., $[p^*(r)]$ is a first degree polynomial.

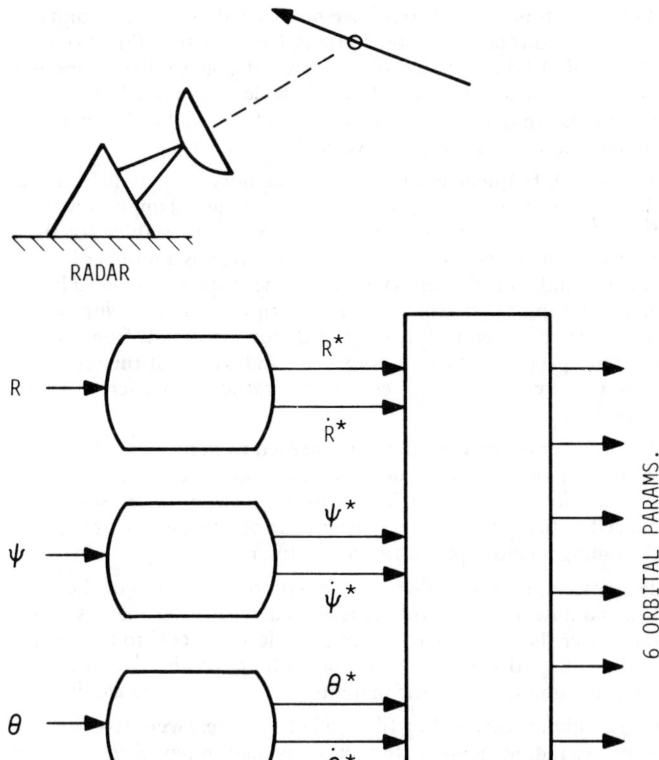

Figure 6 **Example Illustrating the Use of Three Separate Sets of Simple Polynomial Filters: one for R and Ṙ, one for ψ and $\dot{\psi}$, and one for θ and $\dot{\theta}$ to estimate 6 orbital parameters.**

The prediction equations corresponding to those of Equations (18a) to (18c) for first degree fading memory polynomial filtering of Figure 5c are:

$$\dot{x}^*_{n+1,n} = \dot{x}^*_{n,n-1} + (1-\theta)^2(y_n - x^*_{n,n-1}) \tag{19a}$$

$$x^*_{n+1,n} = x^*_{n,n-1} + \dot{x}^*_{n+1,n} + (1-\theta^2)(y_n - x^*_{n,n-1}) \tag{19b}$$

$$S^*_{n+1,n} = \frac{\sigma^2(1-\theta)}{(1+\theta)^3}\begin{pmatrix} (5+4\theta+\theta^2) & (1-\theta)(3+\theta) \\ (1-\theta)(3+\theta) & 2(1-\theta)^2 \end{pmatrix} \tag{19c}$$

For the above fading memory polynomial filter, also called a critically damped α-β or G-H filter [2], the form is the same. Subtract $x^*_{n,n-1}$ from the observation, multiply the result by the constant $(1-\theta)^2$ (theta is the exponential decay constant), and add $\dot{x}^*_{n,n-1}$ (which is now in the memory), giving $\dot{x}^*_{n+1,n}$.* A similar calculation yields $x^*_{n+1,n}$. This process can be generalized easily to any degree using the orthogonal polynomial expressions (Legendre and Laguerre for expanding and fading memory filters, respectively [1]).

*This filter gives the least squares estimate by minimizing:

$$e_n = \sum_{r=0}^{\infty} (y_{n-r} - [p^*(r)])^2\theta^r$$

where $[p^*(r)]$ is a polynomial of degree m that is being fit to the data. For Equations (19a) and (19b) m = 1.

As regards time, these filters have normalized implicity to unity the separations between observations. On the left in Equations (18a) and (19a) is not $\dot{x}_n$ but t times $\dot{x}_n$. If you want real time out of the normalized time, you have to divide $\dot{x}_n$ obtained using the Equations (18a) and (18b) or Equations (19a) and (19b) by the actual number of seconds between observations.

The natural frequencies of the fading memory polynomial filter in the z-space are all at the point θ. It is a critically damped system. If all the poles are in the unit circle, the system is stable. For these filters, all of the poles lie at the point θ, which is a number between 0 and 1 on the real axis; hence, the system is stable. They all lie on top of each other and are multiple, thereby giving two natural frequencies at θ. If so desired, you can return from the discrete space to the continuous space and work out the set of differential equations with continuous, rather than discrete, functions [1].

The number of memory locations needed to implement the filter in the computer, even though there appears to be an infinitely long window, is only two. Therefore, only two memory slots are needed in the computer. The magnitude of theta is what provides the fading memory properties of the filter.

The optimum value of theta depends on the dynamics of the problem and how much smoothing is desired. The closer theta is to 1, the longer the smoothing interval; the closer theta is to 0, the faster the fading. If you want heavy smoothing, you have to accept poor dynamics, that is, the inability to follow fast changes in x.

Gauss did not use a fading filter; all of his filters were fixed memory or expanding memory. In fact, Gauss had so few observations that they were all thrown together — a fixed memory. Later on, when he obtained another observation, he learned how to combine that with what he already had — an expanding filter. Some time later when the planetoid returned, new observations were used to refine the orbit; he did not recalculate the whole thing based on all the data. This refining was the beginning of Swerling's work. Swerling took that scheme and produced the recursive filter (the Bayes filter); after Swerling's work came Kalman's work. Peter Swerling, at the Rand Corporation in 1959 when the first satellites went up, worked out the mechanism of implementing the minimum variance filter effectively and compactly in a computer. Swerling demonstrated this method of recursion by combining new observations with the old estimate. The generalized fading memory filter was first worked out by Dr. Paul Buxbaum at the Bell Telephone Labs.

The least squares smoothing equations for the fixed memory polynomial filter are:

$$X^*_{n,n} = (T^T T)^{-1} T^T Y_{(n)} \tag{20a}$$

$$= W Y_{(n)} \tag{20b}$$

$$S^*_{n,n} = W R_{(n)} W^T \tag{20c}$$

where:

$$T = \begin{pmatrix} 1 & 0 \\ 1 & -1 \\ 1 & -2 \\ \cdot & \cdot \\ \cdot & \cdot \\ \cdot & \cdot \\ 1 & -L \end{pmatrix} \tag{20d}$$

$$T^T T = \begin{pmatrix} L+1 & -\frac{L}{2}(L+1) \\ -\frac{L}{2}(L+1) & \frac{L}{6}(L+1)(2L+1) \end{pmatrix} \tag{20e}$$

$$W = (T^T T)^{-1} T^T \tag{20f}$$

The corresponding prediction equations for the fixed-memory polynomial filter are:

$$X^*_{n+1,n} = \Phi(1)\, X^*_{n,n} \tag{20g}$$

where:

$$\Phi(t) = \begin{pmatrix} 1 & t \\ 0 & 1 \end{pmatrix} \tag{20h}$$

and $S^*_{n+1,n}$ is given by Equation (18c) with n replaced by L.

These equations do not have the simple recursive form of the expanding memory and fading memory filters. For this filter, all the past L (the memory size) samples of y_n must be stored to estimate x^*_n. The matrix inversion shown can be avoided by the use of orthogonal polynomials to calculate W of Equations (20b), (20c), and (20f) [1].

The generalized equations for minimum variance fixed memory filter are given by:

$$\mathring{X}^*_{n+1,n} = \Phi(n+1,n)\, \mathring{X}^*_{n,n} \tag{21a}$$

$$\mathring{S}^*_{n+1,n} = \Phi(n+1,n)\, \mathring{S}^*_{n,n}\, \Phi(n+1,n)^T \tag{21b}$$

$$\mathring{S}^*_{n+1,n+1} = (\mathring{S}^{*\,-1}_{n+1,n} + T_{n+1}{}^T R_{(n+1)}{}^{-1} T_{n+1})^{-1} \tag{21c}$$

$$\mathring{H}_{n+1} = \mathring{S}^*_{n+1,n+1} T_{n+1}{}^T R_{(n+1)}{}^{-1} \tag{21d}$$

$$\mathring{X}^*_{n+1,n+1} = \mathring{X}^*_{n+1,n} + \mathring{H}_{n+1}(Y_{(n+1)} - T_{n+1}\, \mathring{X}^*_{n+1,n}) \tag{21e}$$

where $\Phi(n+1,n)$ is the transition matrix for a time-varying linear differential equation such as Equation (4b). It permits moving the state variable X_n forward (or backward) to time X_{n+h}, as in Equation (10a).

Equations (21a) through (21e) illustrate the manner by which to deal with the most complex problems. The complete differential equation is used, and the computational procedure is expensive. This method was used in the Mercury and subsequent space programs. The cost was considered acceptable for estimates of the required precision.

In Equations (21a) through (21e) we have T_n, the observation scheme; Φ, the transition matrix; and $R_{(n)}$, the covariance matrix of the observations. The equations provide the minimum variance filter. Swerling noted that one begins with an updated estimate (X at time n based on observations up to n) and its covariance matrix. An observation vector, $Y_{(n)}$, is taken which is related to X_n, the state vector, by matrix T plus errors (Equation (12)); along with $Y_{(n)}$, its covariance matrix is needed. The desired result is an updated version of X_n and its covariance matrix, based on new observations.

The first thing we do is a simple prediction (Equation (21a)) moving X_n forward in time, but adding nothing to it. The covariance matrix is updated by Equation (21b). Equation (21c) gives us a new matrix, the covariance matrix of the new state vector — via three matrix inversions. We must invert R; that inversion generally can be done once and for all because the statistics are usually stationary. $\mathring{S}$ cannot be inverted just once. If $\mathring{S}$ is a six vector, two 6 by 6's must be inverted — it is a non-trivial amount of work for the computer to maintain stability in the matrix inversion. But using that new covariance matrix, we compute $\mathring{H}$. To our prediction, we add $\mathring{H}$ times the difference between the observations and what we call their *simulated* or predicted values. The result is an updated state vector.

This algorithm is known as the Bayes algorithm (though perhaps it should be called the Swerling algorithm) – it is at the foundations of the powerful smoothing techniques that are used in all satellite work, much of the orbital work, much of the Apollo program, and so forth. It is very stable; it maintains precision. As time passes, the estimates get better and better and the covariance matrix S gets smaller and smaller. But it is a costly filter because of these two matrix inversions.

The Kalman filter without driving noise can be arrived at from the Swerling filter.* In the Swerling filter, there is the equation $\mathring{S}^*_{n+1,n+1} = (\mathring{S}^{*\,-1}_{n+1,n} + T_{n+1}{}^T R_{(n+1)}{}^{-1} T_{n+1})^{-1}$. Schur, a mathematician in the 19th century, developed the rather improbable looking relationship shown below:

$$(S^{-1} + T^T R^{-1} T)^{-1} = S - ST^T(R + TST^T)^{-1} TS \tag{22}$$

where, for simplicity, the subscripts have been dropped.

The term on the left has three inversions, but the term on the right has just one. This right-hand term is an inversion of a single matrix of the same order as R, the covariance matrix of our observation. Suppose we are observing azimuth and elevation with a telescope. Our observation vector is then a two vector and R is a 2 by 2 matrix. To calculate the six orbital parameters, S is the covariance matrix of that estimate and is 6 by 6. The T's knock the 6 by 6 down to a 2 by 2, so all we have to do is invert a single 2 by 2 matrix. By using this so-called *Matrix Inversion Lemma* (Equation (22)), we can produce the special case of the Kalman Filter without driving noise. The Kalman filter equations, derived by the method indicated, become, for no driving noise**:

$$\mathring{X}^*_{n+1,n} = \Phi(n+1,n)\,\mathring{X}^*_{n,n} \tag{23a}$$

$$\mathring{S}^*_{n+1,n} = \Phi(n+1,n)\,\mathring{S}^*_{n,n}\,\Phi(n+1,n)^T \tag{23b}$$

$$\mathring{H}_{n+1} = \mathring{S}^*_{n+1,n} M^T_{n+1} (R_{n+1} + M_{n+1} \mathring{S}^*_{n+1,n} M_{n+1}{}^T)^{-1} \tag{23c}$$

$$\mathring{S}^*_{n+1,n+1} = (I - \mathring{H}_{n+1} M_{n+1})\,\mathring{S}^*_{n+1,n} \tag{23d}$$

$$\mathring{X}^*_{n+1,n+1} = \mathring{X}^*_{n+1,n} + \mathring{H}_{n+1} (Y_{n+1} - M_{n+1} \mathring{X}^*_{n+1,n}) \tag{23e}$$

where the state vector is updated based on the observations of one time, Y_{n+1}, rather than those of several observation instants, $Y_{(n+1)}$, as done in Equations (21a) through (21e). Again we make a prediction, update the covariance matrix, and then apply the inversion lemma. We get the weight matrix (Equation (23c)), then compute the updated covariance matrix of the estimate (Equation (23d)). Equation (23e) gives an updated estimate.

A logical question to ask is why anyone would use the Swerling or Bayes algorithm when the Kalman algorithm (without driving noise) appears to be easier to use in computation?

Usually we observe fewer parameters than we have in the state vector. We may make two or even only one observation. If we make a single observation, the inversion of a scalar is trivial. In the Bayes algorithm we always have to invert two matrices of the order of the estimate vector; for any sort of powerful estimating, six is a minimum order and it can often rise to nine. But the numerical properties of the Bayes filter are very solid; those of the Kalman filter (without driving noise) tend to become obscured very quickly.

The term on the left in Equation (22) is the Bayes-Swerling method of updating the covariance matrix – the most important factor in these filters. (The method of updating the covariance matrix is the key to the stability of these filters.) The right side of Equation (22) is the way the Kalman filter (without driving noise) updates its covariance matrix.

The left side is of the form 1/a + 1/b; the inverse is smaller than a. A series of *divisions* drive the covariance matrix down. (Divisions in a computer are stable and precision is not lost.) Using the Swerling approach, you don't lose numerical precision.

In the implementation of the Kalman approach, we make reductions by successive *subtractions.* Repeatedly subtracting (either floating point or fixed point) causes a loss in precision. When the matrix is sufficiently small, it is low in precision. Starting with ten decimal places, using the Kalman filter could leave you with one, but the Bayes-Swerling approach always maintains ten decimal places. The choice is clear – you pay for the extra precision.

Matrix inversion does involve subtraction, but matrix inversion routines can be implemented so that precision is maintained. However, you cannot maintain precision over successive subtractions in the Kalman filter; the loss is inherent in the formula.

To avoid inversions, one observation can be taken at a time. However, allow for a faster computation of the estimate. The same estimate is produced if you combine one observation at a time or all six observations. On the ballistic missile defense problem discussed earlier, the Kalman filter was used. In the satellite work done at the Bell Telephone Labs on the Telstar satellites, the Bayes algorithm was used. In one case we have to work fast – there is a ballistic missile coming at you, so money is no object. In such an instance, implement the Kalman filter! In satellite work, you use a much smaller computer, and you have much more time. Time is not of the essence, so the Bayes algorithm is used.

Driving noise adds a single matrix to the Kalman filter. When implemented with driving noise, a term is added as below [1].

$$\frac{d}{dt} X(t) = A(t)\,X(t) + D(t)\,U(t) \tag{24a}$$

where U(t) is a vector of white noise sources.

The above differential equation can be discretized to yield the difference equation:

$$X_{n+1} = \Phi(n+1,n) X_n + V_{n+1,n} \tag{24b}$$

where $V_{n+1,n}$ results from U(t) acting over the time-span $t_n \leqslant t \leqslant t_{n+1}$. The equations for the Kalman filter with driving noise are identical to those of Equations (23a) through (23e) with Equation (23b) replaced by:

$$S^*_{n+1,n} = \Phi(n+1,n)\,S^*_{n,n}\,\Phi(n+1,n)^T + Q_{n+1,n} \tag{23b}$$

where $Q_{n+1,n}$ is the covariance matrix of $V_{n+1,n}$.

The resulting filter has the desirable property of having a fading memory, which the Swerling filter does not.

Editor's Note – Whether the Kalman filter is, or is not, a simple extension of Gauss's method of least squares is somewhat controversial. There are those who feel that the modern theory is fundamentally different from the older work, since Gauss dealt, at best, with the estimation of a set of random parameters, while Wiener, Swerling, and Kalman have dealt with the estimation of the state of a random process. It is the continuous introduction of new driving noise random variables (discussed at the end of the chapter) into the process, in addition to the random observation noise, that distinguishes modern estimation theory from regression theory.

*See Editor's Note at the end of chapter.

**The driving noise, in contrast to the measurement noise, is a noise added on to the process noise to account for the inexactness of the differential or polynomial equations used to describe the process and to allow for other sources of errors (other than measurement) that arise.

The addition of the driving noise has the desirable feature of compensating for inaccuracies in the model used. It permits handling a changing model, such as caused by a maneuver of the object being tracked. Also it insures the stability of the filter.

References

[1] Morrison, N.: *Introduction to Sequential Smoothing and Prediction,* McGraw-Hill, New York, 1969.

[2] Brookner, E.: "Tracking, Prediction and Smoothing," Boston IEEE *Modern Radar Techniques, Components, and Systems* Lecture Series, Fall 1976.

Chapter 26

Sheats

The Kalman Filter

Since the introduction by Kalman and Bucy in 1960 of a new technique for treating estimation problems, considerable amounts of literature have been published on the original work and significant advances have been made. Unfortunately, most of this literature has been written by experts working actively with estimation theory and, consequently, make considerable assumptions regarding the reader's knowledge. Even if one can follow the mathematical manipulations, which frequently appear by the second paragraph, one usually is left somewhat confused as to exactly what problem is being solved and why the Kalman-Bucy technique is advantageous.

The questions, then, that we hope to answer are — what is the Kalman-Bucy filter and in what situations is it useful? Let us begin by considering the problem of estimation; as has been indicated already, this concern is what motivated a new approach.

Consider first a very simple problem in "classical" estimation. Suppose you wished to determine the average height of male adults. If you are a statistician, you might begin by measuring ten men at random. Having no prior knowledge of height distribution, you weight all measurements the same and find an average height. The answer is a random variable because of the inherent randomness of the parameter, and possibly because the measurements were inaccurate. A typical result might be — average height = $\bar{X}$ = 68 in; variance = σ^2 = 10 in^2. The classical 95% or 2σ confidence interval is —

$$\text{estimated height} = \theta = \bar{X} \pm 1.96 \frac{\sigma}{\sqrt{N}} = 68 \pm 1.96 \text{ in}$$

That is, 19 out of 20 times the estimate, extended in either direction by the error shown, brackets the "true" answer. Note that the sample average, $\bar{X}$, has been used as the estimator, either for intuitive or theoretical reasons.

The way in which the estimate is formed is unimportant. Instead of waiting for all ten measurements, a "running" average can be computed. Thus, when the tenth measurement is available, the estimate is the average of the first nine measurements plus 1/10 the "residual," i.e., the deviation of the last measurement from the previous estimate

$$\bar{X}_{10} = \bar{X}_9 + 1/10\,(X_{10} - \bar{X}_9)$$

This observation and, in particular, the form of equation that results, may seem trivial but are, in fact, central to the Kalman Filter.

Suppose we now receive some new information — someone else has investigated this same problem and assures us the answer is — mean height = θ_0 = 72 inches with variance $\sigma_0{}^2$ = 4. Are we in any better position? The answer is yes — we can use a result known as Bayes' Theorem (which originated some 200 years ago) and write a general result; derived by statisticians without resorting to calculus*

$$P(\theta/\bar{X}) = N\left(\frac{W_1\bar{X} + W_2\theta_0}{W_1 + W_2},\ \frac{1}{W_1 + W_2}\right)$$

*P(A/B) is the probability of A given that B was observed

This expression is the joint probability density function of our own measurements and the prior information given to us. The notation, $N(M,\sigma^2)$, denotes a normal distribution with mean M and variance σ^2. The factors W_1 and W_2 are "weights" to be applied respectively to the mean of our observations, $\bar{X}$, and the mean of the "known" or *a priori* distribution. The mean of the above joint probability density function is the new estimate of the height and is designated as $\bar{X}'$ where

$$\bar{X}' = \frac{W_1 \cdot \bar{X} + W_2 \cdot \theta_0}{W_1 + W_2}$$

W_1 and W_2 are given by

$$W_1 = \frac{1}{\sigma^2/n} \qquad W_2 = \frac{1}{\sigma_0{}^2}$$

Note that the more reliable the information (the smaller the variance), the more heavily it is weighted in forming the new estimate. Applying this result to our example

$$W_1 = (1)/(10/10) = 1, \qquad W_2 = 1/4$$

$$\text{mean} = \bar{X}' = [1\,(68) + (1/4)72]\,/(1 + 1/4) = 68.8 \text{ in}$$

$$\text{variance} = 1/(1 + 1/4) = .8$$

$$\text{estimated height} = 68.8 \pm 1.96 \times .8 = 68.8 \pm 1.76 \text{ in}$$

The important points to notice are the following:

1) We choose as our new estimate the conditional mean of the *a posteriori** distribution.
2) Our estimate is "better" (has a smaller variance) if we use the *a priori* information.
3) The credence given the measurements and prior knowledge is inversely related to the variance of each.

Note again that we may write our new estimate explicitly in the form

$$\bar{X}' = \theta_0 + \frac{W_1}{W_1 + W_2}\,(\bar{X} - \theta_0)$$

Thus we have formed the new estimate $\bar{X}'$ from the *a priori* mean, θ_0, and a "suitably weighted" residual. It is, of course, not clear at present that this procedure is optimum, but we see later that it is.

This simple example from classical estimation theory — some of which is more than two centuries old — illustrates the fundamental strategy of the Kalman Filter.

Before abandoning the role of statistician we must first consider the meaning of covariance. We choose a simple example to illustrate.

**A posteriori* knowledge is "after the fact" knowledge that we obtain by combining *a priori* (prior) knowledge with observed data.

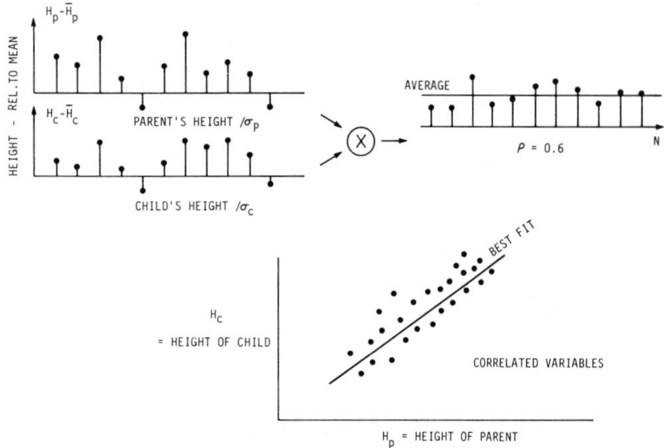

Figure 1 **Deviation Plot**

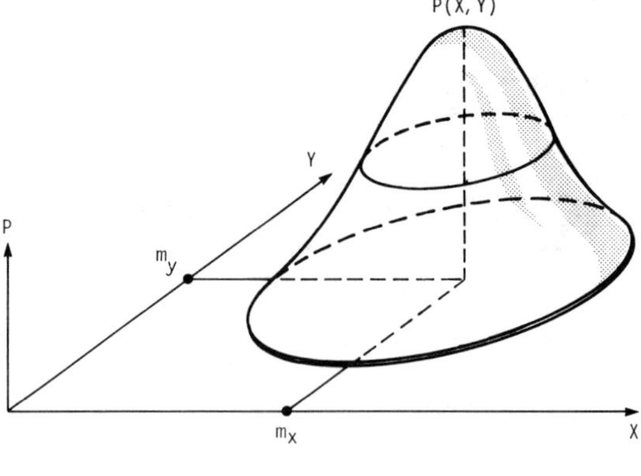

(a) BIVARIATE NORMAL DENSITY

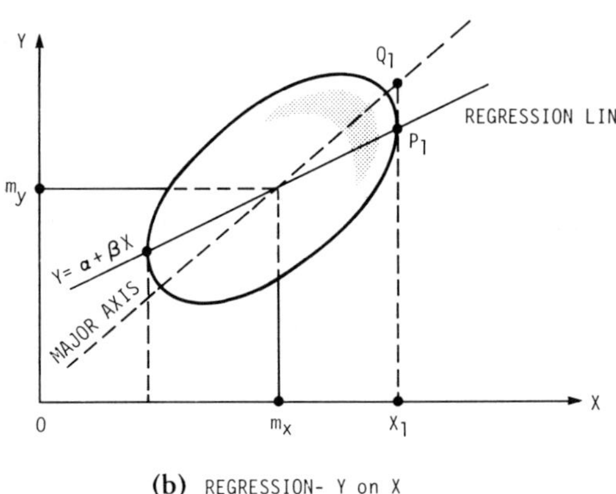

(b) REGRESSION- Y on X

Figure 2 **Ellipses of Constant Likelihood**

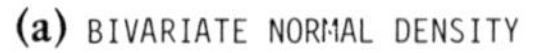

Suppose in addition to measuring ten men you also measure each man's son – assume that all the men were fathers and that the sons were magically all the same age. We would discover that height of the father and his son – each a random variable – are correlated; this result is not surprising. But what is meant by "correlated"?

Proceeding in the same manner as the statistician, we plot the deviation of each measurement from the mean, multiply father and son deviations, and average as illustrated in Figure 1. We are careful to express each in proper units so that the results do not depend on the particular units chosen – each variable is expressed in units of standard deviation. The result is a single number between 0 and 1, called the *correlation,* which tells how closely the variables move together, with the zero result indicating no tendency for the variables to act alike. If we find correlation, we assume a linear dependence of one variable on the other

$$H_C = \alpha + \beta H_P$$

We can now estimate α and β from our data by choosing the "best" straight line that goes through a plot of the data.

The variables H_C and H_P have a joint (bivariate) distribution that appears as shown in Figure 2a where H_C and H_P have been replaced by X and Y, respectively. We slice this volume and view the resulting two-dimensional figures – ellipses of constant "likelihood" (illustrated in Figure 2b). If we draw vertical tangents to an ellipse and join the two points of tangency, the result is the line we have defined

$$Y = \alpha + \beta X$$

called the line of linear *regression* of Y on X. If X varies over some increment, Y undergoes a variation. Part of this variation is due to the stochastic nature of Y and part is due to the dependence on the variable X as illustrated in Figure 3. The latter variation is, as the statistician says, "explained by the regression" and the former "unexplained by the regression."* The total uncertainty in Y due to uncertainty in X is thus seen to be made up of two parts – the variance of Y itself and the covariance of X and Y (which is due to the regression). Thus in the covariance matrix

$$\begin{bmatrix} \sigma_X{}^2 & \rho\sigma_{YX} \\ \rho\sigma_{XY} & \sigma_Y{}^2 \end{bmatrix}, \quad \rho = \frac{\sigma_{XY}}{\sigma_X \sigma_Y}$$

the diagonal terms are the variances; the off-diagonal terms are the covariances. Referring to Figure 3, it is clear that the latter terms can be eliminated by a suitable rotation of coordinates.

*It can be noted here that the regression line, $Y = \alpha + \beta X$, defines the conditional mean of the *a posteriori* distribution, which in turn yields the optimum estimate of the variable Y.

Refer to Figure 2 and note that, if we know that the father's height is X_1, for example, we predict the son's height, not at Q_1, but rather at P_1. That is, our estimate "regresses" toward the average (horizontal) line through center of ellipse); this movement accounts for the origin of the term *regression theory.* Thus, if you are 74 inches in height, don't predict your son's height to be 74 inches – "regress" it toward the population mean. Tall persons sometimes marry short persons, thereby causing statisticians to invent regression theory.

Now consider another example that illustrates a situation somewhat more familiar to engineers (perhaps less familiar to statisticians) – we want to determine the magnitude, A, of a voltage (see Figure 4). Our meter is inaccurate, and, in an inconsistent way, introduces errors in measurements. Furthermore, we know nothing about A except that it (hopefully) is bounded by the limits ($\pm$V) of our meter scale. We wish to take our readings (corrupted by the random inaccuracies of the meter), put them into an estimator, and find a value of A, $\hat{a}$, that is in some sense "best." For mathematical simplicity, we assume that the meter inaccuracies are modeled by an additive noise with zero mean.

How we solve this problem depends on our assumptions concerning A. If we assume that A is simply a deterministic but unknown quantity, we use the maximum likelihood estimate. To obtain the maximum likelihood estimate of A, we choose that value of the measurement noise that is the "most likely" value – in this case, zero – and we subtract it from our reading, R_i; then we average the differences. The estimate of A is

$$\hat{a} = \frac{1}{N} \sum_{i=1}^{N} R_i$$

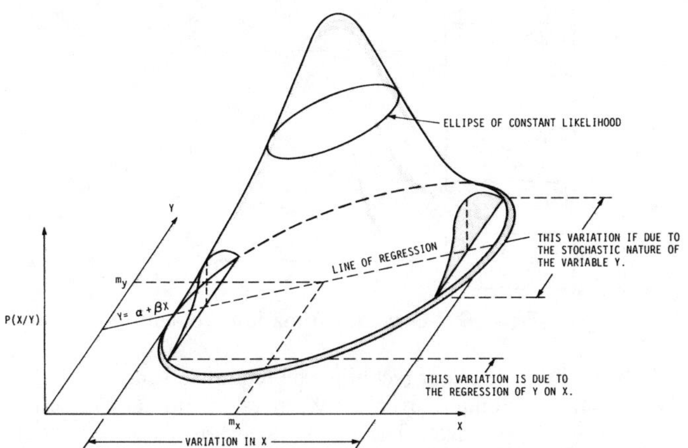

Figure 3 **Bivariate Gaussian Density Function**

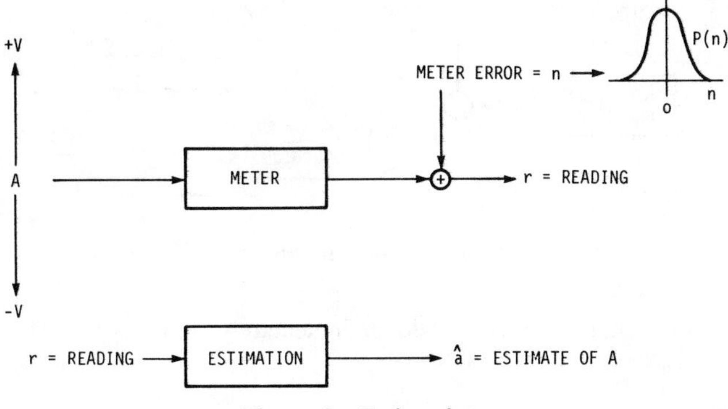

Figure 4 **Estimation**

As intuition would suggest, we simply average a number of readings to obtain the estimate of A.

If we assume that A is not a steady parameter but is rather a random variable of "known" statistics, our strategy is changed somewhat. Instead of a maximum likelihood estimate, we form a "maximum *a posteriori*" (MAP) estimate. For the best estimate of A we take the value that corresponds to the maximum of the joint-density function, p(A/R) (see Figure 5).

Since this function is in the form of an exponential with a negative exponent, we minimize the magnitude of the exponent and thus choose the maximum of the joint density function. The result is

$$\hat{a}(R) = \frac{\sigma_A^2}{\sigma_A^2 + \sigma_n^2/N} \frac{1}{N} \sum_{i=1}^{N} R_i$$

where

σ_n = standard deviation of measurement "noise"

σ_A = standard deviation of random variable A (voltage)

Note now that if the fluctuation of A is large compared to the measurement noise (or if we make a large number of measurements)

$$\hat{a}(R) \cong \frac{1}{N} \sum_{i=1}^{N} R_i$$

since $\sigma_n^2/N\sigma_A^2 \rightarrow 0$.

This expression means that our *a priori* knowledge of the statistics of A was of little value – we use the observed data to form the estimate. If, on the other hand, the fluctuation due to n is large and the variation of A is small, the best choice for the estimate is simply the *a priori* knowledge – the known mean value of A (assumed to be equal to zero in this case). This problem is the same as the one of height estimation just considered, only cast in more familiar engineering terms.

So far we have considered two simple estimation problems, leaving out the mathematics. The point we wish to emphasize is that the choice of a formula for the best estimate is different depending on what we know about the quantity we are trying to estimate. If the voltage A is assumed to be non-fluctuating but unknown, we use the maximum likelihood procedure. If A is a random variable (statistics known), we use the MAP estimate.

Unfortunately, the foregoing discussion has not defined the notion of "best" estimate – the MAP estimate, the ML (Maximum Likelihood), or some other. Fortunately, the Gauss-Markov Theorem assures us that the proper one to use is that which minimizes the mean square error (MMSE). For the examples we consider, any strategy we wish to adopt leads to the same equation for the estimate – based on a model of Leondes' Postulate which asserts that the world is discrete, linear, and Gaussian.

At this point we pause to make the problem more general. Suppose that instead of using a meter to read voltage A we use a strip chart recorder. Each day we insert a new roll of paper and obtain a new record, as illustrated in Figure 6. The readings, R, recorded on the chart paper, no longer are a random variable but a random process, and we need a new way to characterize this phenomenon. Questions arise as to whether today's readings are statistically different from yesterday's, etc. We must know the joint-density function of the daily readings, such as the 9 am readings, the 10 am readings, etc. The simple concepts of mean and variance are generalized into an expected value and covariance for the random process. Furthermore, we can generalize the problem and try to estimate not only the amplitude but also the phase of the voltage; the data, R, can be a vector as well as a scalar random variable.

The problem of estimating voltage A now requires little further generalization to be identical to the problem solved by Weiner and Kolomogorov in the early forties. Refer to Figure 7, which illustrates the general estimation problem we wish to consider. The following additional refinements have been added:

1) Voltage A can be passed through any linear system, such as an amplifier or filter.
2) We may not wish to estimate A directly; rather, some linear function of A, such as its derivative, or A + τ (a predicted future value). The solution thus embraces the filtering as well as the prediction problem.

We still have the random process R – our observations – to work from, and we wish to estimate d(t), which is some "desired" function of A (and which could be just A itself). Note that we are now considering estimation of a continuous waveform.

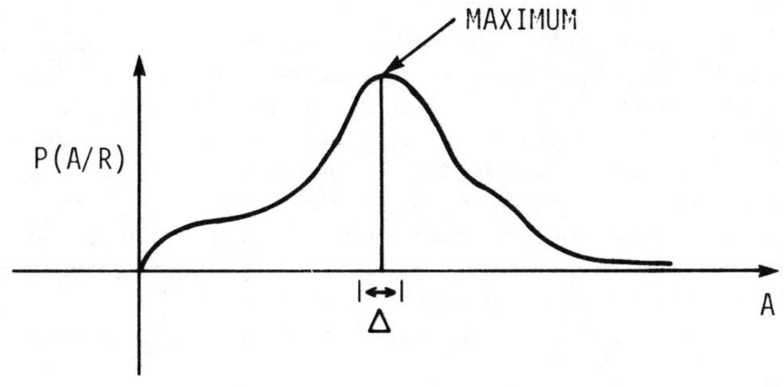

Figure 5 ***A Posteriori* PDF**

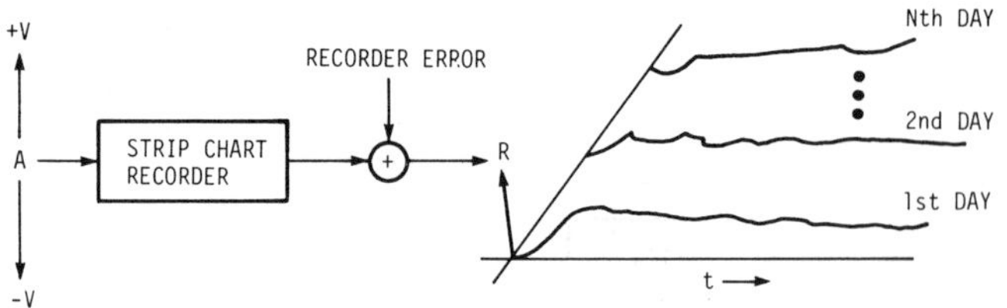

Figure 6 **Voltage Recording**

The solution can be written down immediately, and it is the celebrated Wiener-Hopf Equation:

$$\int_0^{T_1} h_0(t,u)K_R(u,t)d\tau = K_{DR}(t,u)$$

where T_1 is the observation time. We can skip over the mathematics at this time and simply explain what the terms of the equation are:

1) $h_0(t,u)$ is the impulse response at time t (due to an impulse applied at time u) of the "optimum" filter that produces the estimate of d(t).
2) $K_R(u,t)$ is the covariance of our readings R obtained from our recorder.
3) $K_{DR}(t,u)$ is the cross-covariance of our readings R and the desired output d.

What may not be obvious here is that the answer to our problem is the quantity $h_0(t,u)$. It does not stand alone by itself on the left hand side of an equation; thus, the solution is implicit rather than explicit. For the sake of simplicity, stationary processes may be assumed. That is, all statistics of random processes are assumed to be independent of t. For example, in the previous example, the 9:00 readings and the 10:00 readings have the same mean, variance, and distribution. Then $h_0(t,u)$ becomes simply $h_0(u)$.

To form the estimate of d(t) we simply apply r(t) as input to our optimum estimator, as illustrated in Figure 7. The equation for $h_0(t)$ is an integral equation, but it can be solved, for mathematicians have devised methods for solving integral equations.*

By imposing some restrictions, we can even write down the solution in closed form:

$$H_0(j\omega) = \frac{1}{G^+(j\omega)} \frac{S_{DR}(j\omega)}{G^+(j\omega)^*}$$

where the asterisk indicates complex conjugate.

This expression assumes that R has a rational (hence factorable) spectrum.

$$S_R(\omega) = G^+(j\omega)G^-(-j\omega)$$

where $S_R(\omega)$ is factored above into parts having poles only in the right- (+) and left- (−) hand planes.

S_{DR} is the cross spectrum of d(t) and r(t).**

This discussion seems to have drifted far afield of a consideration of the Kalman-Bucy filter, but this excursion into Wiener filters is not without a purpose. To summarize all this discussion, we note that we have gone from a very simple estimation problem — that of estimating a voltage from a series of meter readings — to an entirely general problem, namely, that of estimating some linear function of a continuous waveform which may have undergone yet another linear transformation (in addition to corruption by measurement noise) before it can be observed. Yet, we have a solution (in the Wiener-Hopf equation) of the general problem —

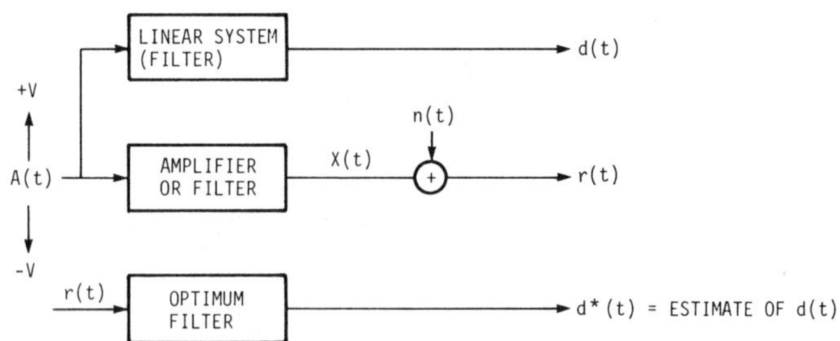

Figure 7 **General Estimation Problem**

why would any different technique be required? Is not the Weiner-Hopf technique sufficiently general to solve any reasonable estimation problem? The answer is obviously not an unequivocal yes since the Kalman-Bucy technique has dominated in practical situations.

The advantages of the latter approach can be noted by considering the disadvantages of the Wiener procedure. First we note that random processes are always characterized by their covariance function. Furthermore, in order to obtain a solution in a general case, we first have to solve an integral equation and then perform a convolution. Neither process is well adapted to implementation in either hardware or software for a practical application such as estimating target range, given a radar signal. The Kalman-Bucy approach to estimation makes use of differential (rather than integral) equations and introduces a new way of characterizing a random process. Both features simplify the actual realization of the estimator.

We return to the question of estimation. First, the strategy that we adopt is that of sequential estimation which we have already discussed. Returning to the example of height estimation, we first measure ten men, obtaining an estimate in the form of a random variable, characterized by a mean and variance. We combine this estimate with the *a priori* information given and generate a new estimate, $\bar{x}$. We then repeat the process measuring ten more men. The result — again a random variable — is called an "observation." That which we just called a "new estimate" now becomes the *a priori* information and we use Bayes' strategy. We apply the simple formula already given.

$$\overline{x''} = \bar{x} + \frac{w_1}{w_1 + w_2} (\overline{x'} - \bar{x})$$

where

$\overline{x'}$ is the new observation

$\bar{x}$ is the old estimate

$\overline{x''}$ is the updated estimate

Thus we obtain a new estimate, formed by adding to the old estimate a suitably weighted "residual" (also called an "innovation"*) which is just the deviation of the observation from the old estimate. This procedure is repeated indefinitely. The observations are made and used in a time ordered sequence, and the estimate is updated recursively.

Note that we are using the MAP estimate, for, as has been pointed out, the parameter that has been chosen as the estimate is the conditional mean of the *a posteriori* density function. This situation raises the question of what do we assume about the quantity we are estimating? Is it deterministic? Stochastic?

Actually, when we consider the Kalman filter we make a very definite assumption regarding the nature of what we are estimating. We assume that the variable (such as the voltage, A) follows some natural law, and we can write a differential equation for it.

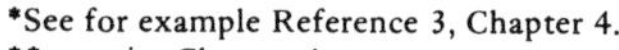

*See for example Reference 3, Chapter 4.
** *op. cit.*, Chapter 6

*This terminology due to Kailath [4].

A, for example, is now called a "process," in using the Kalman-Bucy approach, we write a differential equation for A. The differential equation leads naturally into the sequential method of using the observations and a recursive updating of the estimate.

We now have all the tools we need.

1) The optimum estimate is the conditional mean of the *a posteriori* distribution.
2) Statisticians have already derived the formula for this distribution.

We then are ready to proceed to consider an example of an application of the Kalman filter.

We must not forget the nature of the parameter we are going to estimate – it is a "process," a somewhat unfortunate case for, before we can proceed, we must pause briefly to consider how this "process" is to be described.

We first note that we may think of the process as either a differential equation or description of a linear system. Most engineers think of a linear system in terms of its poles and zeros. The equation

$$H(s) = \frac{b_{N-1}s^{N-1} + \ldots + b_0}{s^N + \rho_{N-1}s^{N-1} + \ldots + \rho_0}$$

and the "simulation diagram" shown in Figure 8 describes a linear system.

The linear system can be described in terms of the equations

$$x_1(t) = y(t)$$
$$x_2(t) = \dot{x}_1 - \rho_{N-1}y + b_{N-1}u(t)$$
$$x_3(t) = \dot{x}_2 - \rho_{N-2}y + b_{N-2}u(t)\ldots$$

The variables $x_1(t)$, $x_2(t)$, $x_3(t)$, . . . , $x_N(t)$ are called the state variables defined by the equations given above or the simulation diagram of Figure 8. It can be noted that these variables are just the outputs of the integrators.

The equations for the above linear system (which is completely general) can be written in matrix form in three ways

$$\dot{x} = Fx + Gu$$

where either

$$1)\quad F = \begin{bmatrix} -\rho_{N-1} & 1 & 0 & & \\ -\rho_{N-2} & 0 & 1 & & \\ \cdot & & & 1 & 0 \\ \cdot & & & & 1 \\ \cdot & & & & 0 \\ -\rho_0 & 0 & & & \end{bmatrix} \qquad G = \begin{bmatrix} b_{N-1} \\ b_{N-2} \\ \cdot \\ \cdot \\ \cdot \\ b_0 \end{bmatrix}$$

$$2)\quad F = \begin{bmatrix} 0 & 1 & 0 & & 0 \\ 0 & 0 & 1 & & \\ \cdot & & & & \\ \cdot & & & & 0 \\ \cdot & & & & 1 \\ -\rho_0 & -\rho_1 & -\rho_2 & \ldots & -\rho_{N-1} \end{bmatrix} \qquad G = \begin{bmatrix} b_{N-1} \\ b_{N-2} \\ \cdot \\ \cdot \\ \cdot \\ b_0 \end{bmatrix}$$

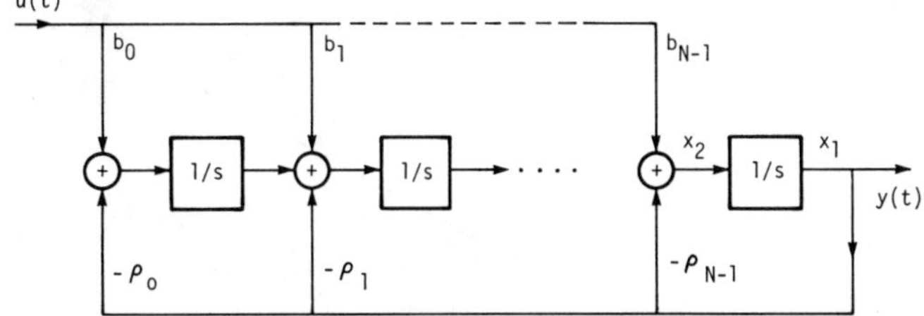

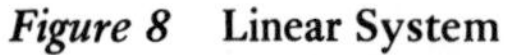

Figure 8 **Linear System**

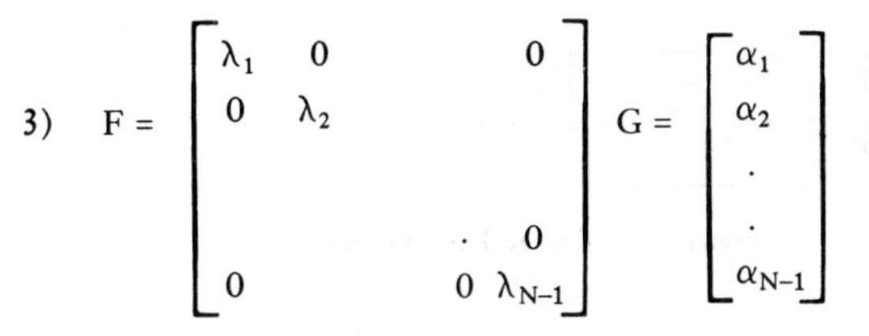

$$3)\quad F = \begin{bmatrix} \lambda_1 & 0 & & 0 \\ 0 & \lambda_2 & & \\ & & \cdot & 0 \\ 0 & & 0 & \lambda_{N-1} \end{bmatrix} \qquad G = \begin{bmatrix} \alpha_1 \\ \alpha_2 \\ \cdot \\ \cdot \\ \alpha_{N-1} \end{bmatrix}$$

In this last form, the (λ)'s are the eigenvalues and the (α)'s are the residues. Their identities are explicit if H(s) is expressed in a partial fraction expansion of the form

$$H(s) = \sum_{i=1}^{N} \frac{\alpha_i}{s_i - \lambda_i}$$

The first two F-matrices are shown in column and row companion form, respectively. The last form shown is referred to as the Jordan form. Three techniques are available for solving the above state equations.

1) Cayley Hamilton Method
2) Modal Matrix Method
3) Laplace Transform Method

Either approach allows us to find an expression for a matrix exponential, $e^{F(t)}$. The solution of the state equations then is

$$x(t) = e^{F(t)}x(0) + \int_0^t e^{F(t-\tau)}G\,u(t)d\tau$$

The matrix exponential $e^{F(t)} = \Phi(t)$ is called the state transition matrix and has an interesting characteristic

$$x(t) = \Phi(t, t_0)x(t_0)$$

Thus, Φ propagates the state of the system from time t_0 to time t in the manner shown in the last equation. An excellent treatment of the state variable approach to linear systems is given in Reference 5.*

With this brief background in state-variable techniques, we may again consider the question of the Kalman-Bucy filter. The example used thus far, however, becomes somewhat far-fetched, so another example is chosen to illustrate the concept. Consider the configuration illustrated in Figure 9. An ionospheric probe, which is inertially guided, is to be initialized during its early flight using radar measurements of its position. Assume the radar data are uplinked to the probe control system. To make the example simple assume that the probe moves only radially outward along the radar line of sight so that only the single vector x (the range to the target) is considered. Assume that the process (that is, the motion x(t)) is described by a second order polynomial

$$x(t) = x_1 + x_2\,t + x_3\,t^2/2$$

Thus the motion has a constant acceleration. The state variables are:

*See also p. 366.

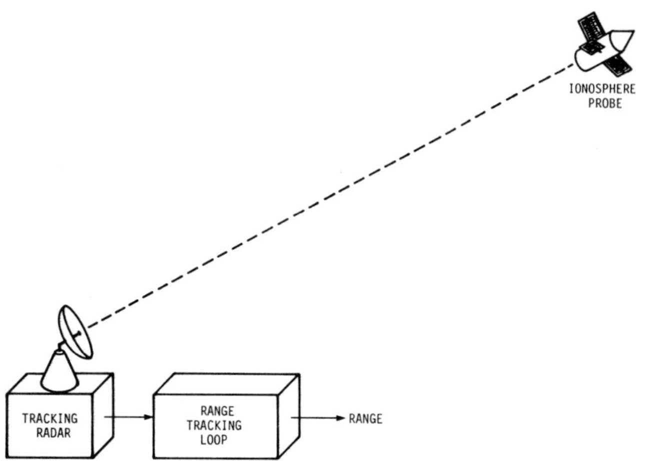

Figure 9 **Range Estimation**

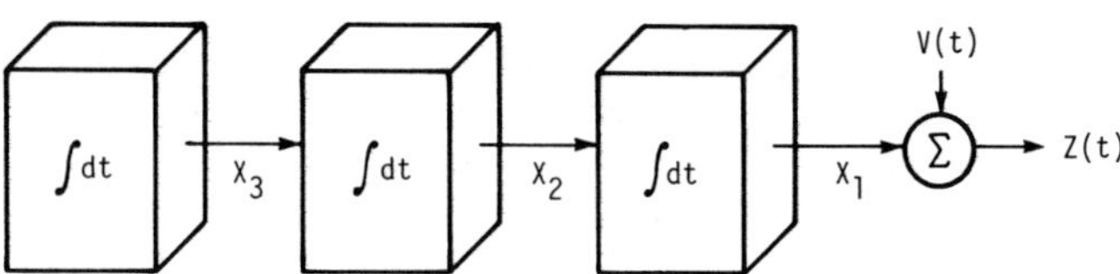

Figure 10 **System and Measurements**

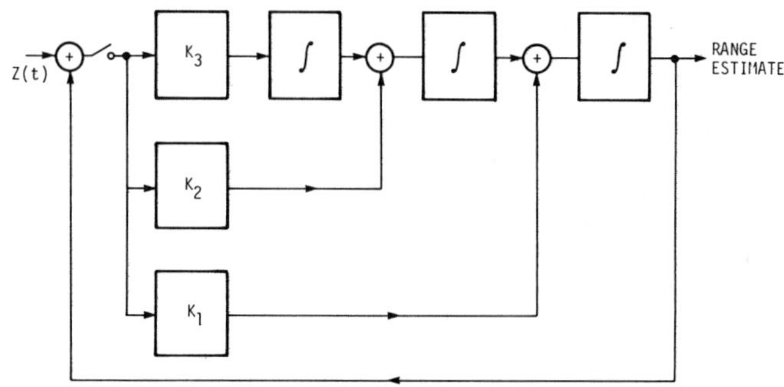

Figure 11 **Optimum Filter**

x_1 = position

x_2 = velocity

x_3 = acceleration

and $x_2 = \dot{x}_1$, $x_3 = \dot{x}_2$, $\dot{x}_3 = 0$

or $\dot{x}(t) = F\, x(t)$

$$F = \begin{bmatrix} 0 & 1 & 0 \\ 0 & 0 & 1 \\ 0 & 0 & 0 \end{bmatrix}$$

The transition matrix $\Phi(t)$ is given by

$$\Phi(t) = \begin{bmatrix} 1 & t & t^2/2 \\ 0 & 1 & t \\ 0 & 0 & 1 \end{bmatrix}$$

The equation for the vector x above is the "process" we wish to estimate, as illustrated in Figure 10. We do not observe x, however. Instead, we observe a linear transformation of x corrupted by the radar measurement error, v.

$$z = H\, x + v$$

The Kalman filter optimum estimate of x at time N + 1 is

$$x_{N+1} = \Phi_N x_N + K_N (z_N - H_N \Phi_N x_N)$$

Thus the estimate of the vector x at time N + 1 is the previous estimate (extrapolated forward in time) plus a "suitably weighted" residual, just as in the simple example shown earlier. The weighting factor, K, is

$$K_N = P_N H'(HP_N H' + R)^{-1}$$

where:

P_N = covariance matrix of the residuals at time N.

R = covariance matrix of v

and the ′ stands for transpose.

The expression above for K_N is analogous to the term $w_1/(w_1 + w_2)$ encountered in the simple example. The estimator configuration is shown in Figure 11.

Just as we had to compute a new w_2 after each "measurement" in the height estimation example, we must compute a new P matrix after each radar measurement, as given by

$$P_{N+1} = (P_N^{-1} + H^T R^{-1} H)^{-1}$$

The actual calculations involved in this simple example are quite tedious, which probably explains why it took so long for this method to gain acceptance. However, currently available computers are sufficiently powerful that even the three matrix inversions shown in the last equation are no problem.

The next logical step at this point is a higher level treatment of the fundamental strategy of the Kalman filter which has already been established. Some suggestions are given in the references. Reference 1 is an excellent treatment of the modern-day theory against a fascinating historical background; must of this historical background is given by Morrison in Chapter 25.

References

[1] Morrison, N.: *Introduction to Sequential Smoothing and Prediction,* McGraw-Hill, 1969.

[2] Bryson, A.E.; and Ho, Y.-C.: *Applied Optimal Control,* Ginn and Co., 1969.

[3] Van Trees, H.L.: *Detection, Estimation and Modulation Theory (Part 1),* John Wiley & Sons, 1968.

[4] Kailath, T.: "An Innovations Approach to Least Squares Estimation Part I: Linear Filtering in Additive White Noise," *IEEE Transactions on Automatic Control, Vol. AC-13, No. 6,* December 1968.

[5] DeRusso, P.M.; Roy, R.J.; and Close, C.M.: *State Variables For Engineers,* John Wiley & Sons, 1965.

Efficient Spectrum Control for Pulsed Radar Transmitters

A pulsed radar transmitter that generates a wider RF spectrum than necessary is generating a kind of pollution. Like so many of our other resources, the radio frequency spectrum has been wasted by careless use. And, just as the old term "conservation" has been replaced by the more popular terms "ecological action" and "pollution control," the old term RFI (Radio Frequency Interference) has been replaced by the more positive-thinking term EMC (Electromagnetic Compatibility); EMC requirements are now commonly specified on radar equipments. MIL-STD-469 sets limits on unnecessary RF radiation; because the limits are in absolute power rather than relative, the problem of controlling excess radiation is more severe for higher-power systems.

Part of the problem is the harmonic power output of radar systems, plus possible out-of-band spurious signals generated by undesired operating modes in RF tubes. Although these problems may be real ones, out-of-band spurious signals can be attacked with reasonable high-power microwave filters. Therefore, the main emphasis of this paper is on the more difficult problem of using the transmitted RF pulse to minimize the width of the RF spectrum generated in the desired operating band. Since this spectrum is already within the desired operating band, filtering is more difficult, requiring sharply tuned narrowband filters that tend to be costly and have temperature stability problems. If we are to avoid a narrowband filter for in-band spectrum control, the transmitter must *generate* an adequately narrow RF spectrum to meet the requirements.

An ideal rectangular transmitted pulse (Figure 1) has the familiar (sinx)/x RF spectrum, with x defined as shown. f_0 is the operating frequency and τ is the pulse width. If we choose to call $1/\tau$ the nominal bandwidth of the signal, the envelope of the peaks of the (sin x)/x spectrum falls off at the rate of 6 dB per bandwidth octave. This reduction continues until the envelope reaches the inherent output noise level of the transmitter [5]; i.e., each time we double how far we are from the carrier, the spectrum envelope falls another 6 dB. This fall-off of power is a fairly slow one, and MIL-STD-469 requires a more rapid decrease for high-power radars.

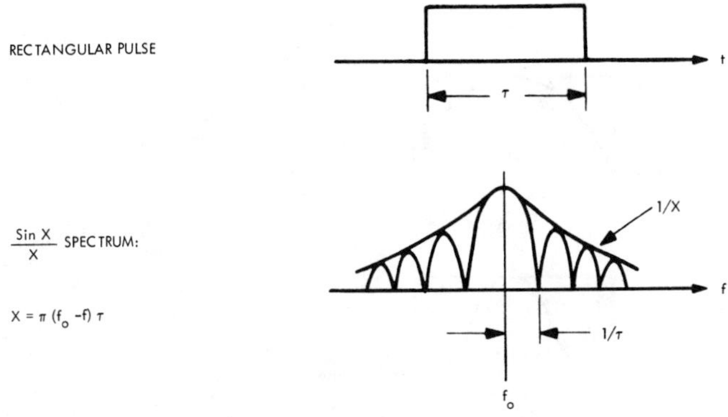

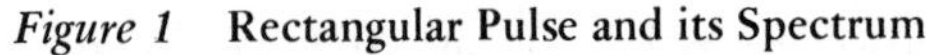

Figure 1 Rectangular Pulse and its Spectrum

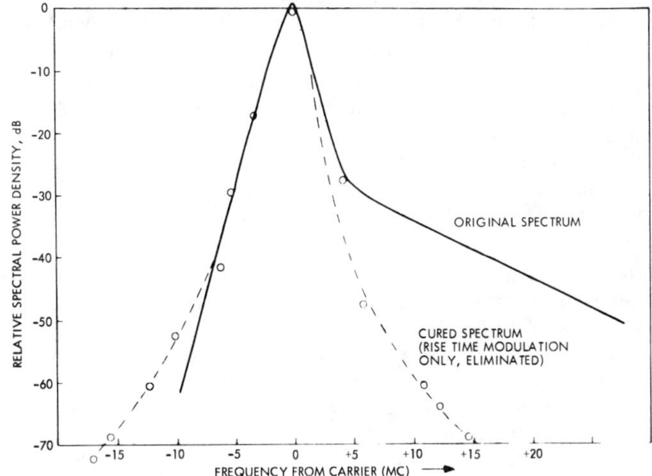

Figure 2 Original and "Cured" Transmitter Spectra (After Miller [2])

For an example, we discuss the ARSR-2 Air Route Surveillance Radar, operating at 4 MW peak, 4 kW average RF power at L-band. If this radar transmits an ideal rectangular pulse, it will *fail* to meet the MIL-STD-469 requirement, which is specified at a frequency $10/\tau$ from the carrier, by about 30 dB. Thus, high-power radars must do appreciably better than (sin x)/x.

In some cases, radars have actually been *worse* than (sin x)/x. In one case (Figure 2), interference was occurring between similar systems operating on nearby frequencies because of the high level of spectrum energy far from the carrier [2]. It turned out that the high sideband energy on one side of the carrier was being caused by accidental phase modulation on the leading and trailing edges of the pulse during the finite rise and fall time of the beam-pulse modulator in this system, which uses a cathode-pulsed klystron. The solution to this problem (called "CURE," for Control of Unwanted Radiated Energy) was simply to withhold the RF drive during the rise and fall of the modulator pulse, and the spectrum improved significantly (about 30 dB), as shown in Figure 2. It actually became a little *better* than (sin x)/x because of the inherent filtering action of the tunable klystron, at least at frequencies far from the carrier. Figure 3 illustrates how the RF drive to the klystron was gated to avoid the severe phase modulation during the modulator rise and fall times. Although the total RF power output clearly is reduced by this gating, it should be noted that the RF spectral energy beyond approximately $1/\tau$ from the carrier, generated by the phase modulation during the rise and fall time with RF drive present, is not used by the receiver anyway [5].

Since the energy beyond $1/\tau$ is not useful to the system, and since MIL-STD-469 strictly limits the amount of energy that can be transmitted beyond $10/\tau$, transmitter designers must consider ways of avoiding transmitting energy beyond those limits [5].

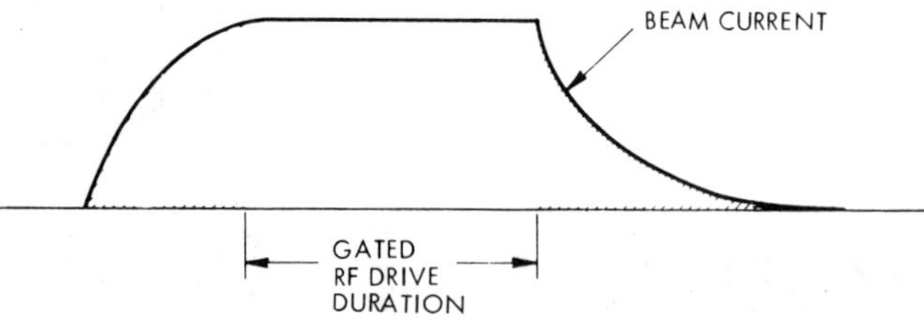

Figure 3 **RF Drive Gating to Avoid Operation During Klystron Beam Current Rise and Fall**

These objectives may be approached by using a pulse shape which is different from the usual and convenient rectangular pulse. The pulse shape having the greatest rate of attenuation of sideband energy for a given effective pulse width is the Gaussian pulse [5].

Figure 4 shows the spectra of a rectangular pulse and a Gaussian pulse of the same half-voltage width (1 μs, in this case). The rectangular pulse spectrum has high sidelobes, the first peak being only 13 dB down, with successive peaks falling off rather slowly. The Gaussian pulse spectrum not only falls off rapidly, but also continues to fall at a rapidly increasing rate.

Figure 5 shows the pulse shapes for rectangular and Gaussian pulses. The Gaussian pulse shape is extremely inconvenient – it extends in time from minus infinity to plus infinity. At the plus and minus 1.5 μs points from the center of the pulse (which are at three times the half-voltage width points), the amplitude appears very small on the linear scale used here, but what is happening mathematically is that, gradually, more and more higher-order derivatives of the voltage (which determine the spectrum amplitude far from the carrier) are approaching zero. However, we can truncate the pulse at the points shown (plus and minus 1.5 μs), and the spectrum does not differ significantly from Gaussian until sidelobes are more than 70 dB down.

Besides a truncated Gaussian pulse, cosine-to-the-nth-power pulse shapes are suitable. In fact, cosine-to-the-nth-power approaches Gaussian as n becomes large. Figure 6 shows the first alternatives in this method of approximation, together with the resulting rate of attenuation of sideband energy that occurs with these shapes. Notice, as shown by the shaded areas, that the higher the order of the approximation, the more of the available transmitter power that is wasted by shaping the pulse. To compare different pulse shapes properly, we should keep the transmitted pulse energy constant so that radar target detection performance remains constant; energy is essentially constant for these shapes if the half-power pulse width is kept constant. Scaling these pulses this way increases the duration of the higher-order waveforms, which in turn makes their spectra even narrower when compared with the rectangular pulse, but also makes the amount of wasted energy even greater.

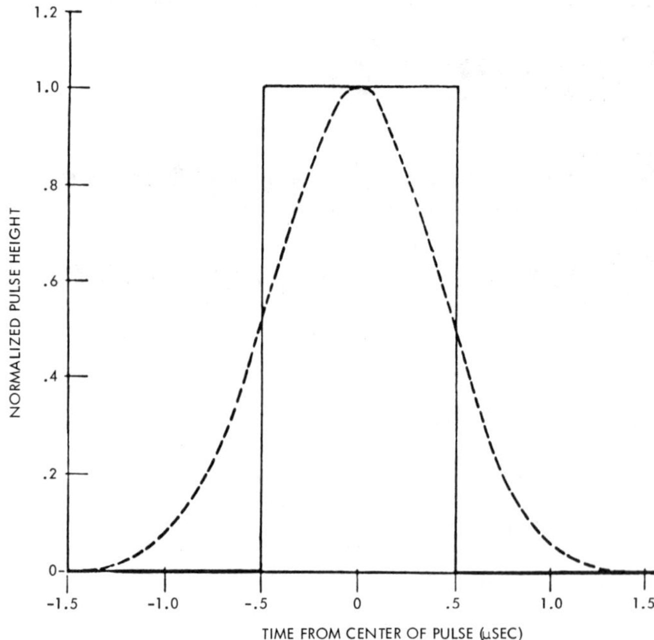

Figure 5 **Comparison of Rectangular and Gaussian Shaped Pulse Envelopes (After Ashely & Sutherland [4])**

Figure 7 shows spectra for several cosine-to-the-nth waveforms which were matched at the 6 dB (half-voltage) points; if they had been matched at the half-power points, the spectra of the higher-order waveforms would be even narrower. Note that the spectrum for a cosine-to-the-6th power pulse envelope matches the Gaussian spectrum to nearly 80 dB down. Higher-order waveforms are seldom used, because the shaping usually cannot be carried out perfectly enough to achieve even that good a spectrum.

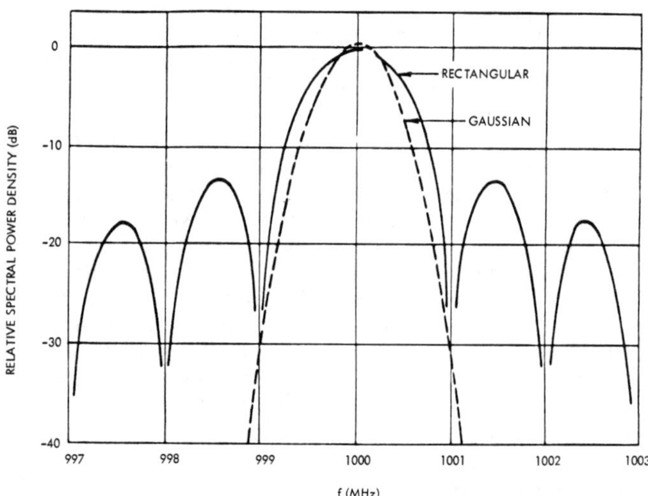

Figure 4 **Comparison of Rectangular and Gaussian Shaped Pulse Spectra (After Ashley & Sutherland [4])**

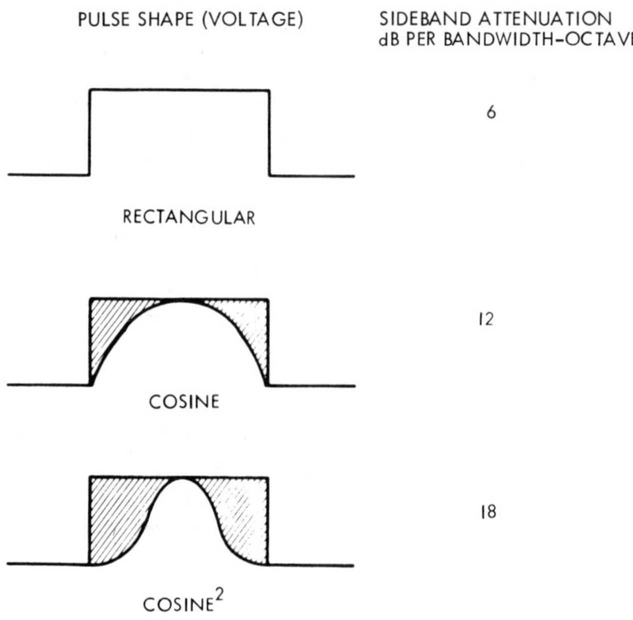

Figure 6 **First Few Cosine-to-the-nth Approximations to Gaussian Pulse Waveform**

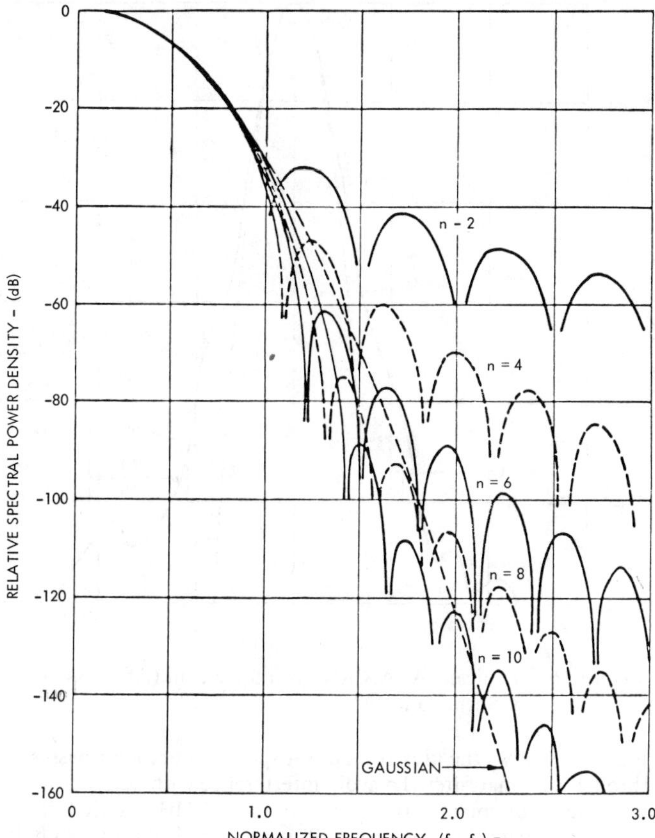
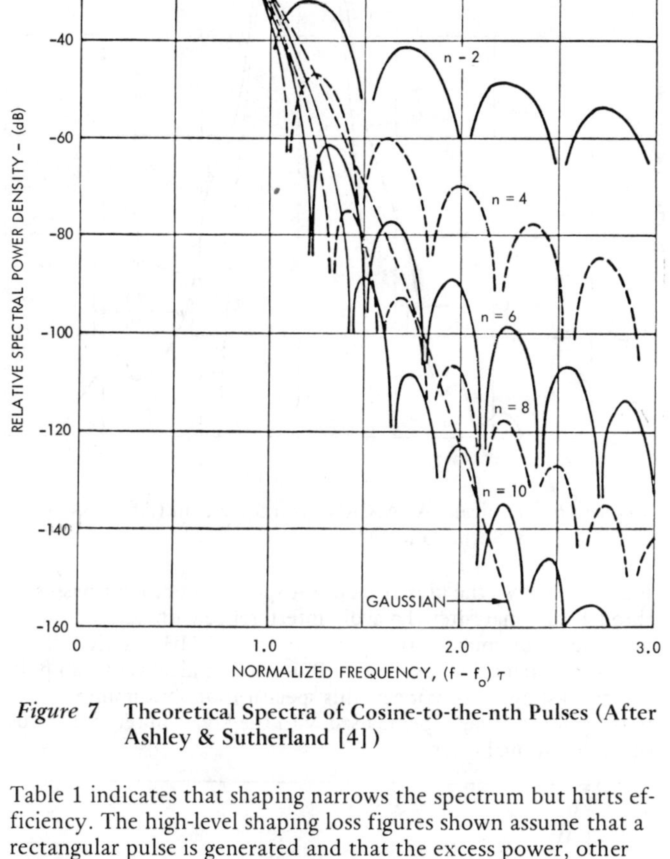

Figure 7 Theoretical Spectra of Cosine-to-the-nth Pulses (After Ashley & Sutherland [4])

Table 1 indicates that shaping narrows the spectrum but hurts efficiency. The high-level shaping loss figures shown assume that a rectangular pulse is generated and that the excess power, other than the desired shape, is thrown away (into a dummy load, for example). The linear-amplifier shaping loss applies to amplifiers, such as an ideal class-B linear RF amplifier, in which the current drawn by the stage is proportional to the RF output voltage as the amplitude is varied. The losses shown are degradations in efficiency below the inherent amplifier stage efficiency that would be obtained with a rectangular pulse at full rated power output. Note that the high-level shaping losses are much more severe than the linear-amplifier shaping losses, and that the linear-amplifier shaping losses climb very little as the order (exponent) of the waveform shaping increases.

Figure 8 shows three methods of high-level shaping. The first requires coherent amplifier chains. The chains are operated at full peak power with a rectangular RF pulse, and the combined power may be delivered to the output arm of the combiner, to the dummy load, or partially to each of them by properly phasing the two

	Sideband Attenuation (dB/bandwidth-octave)	High-Level Shaping Loss (dB)	Linear Amplifier Shaping Loss (dB)
Rectangular	6	0	0
Cosine	12	3.0	1.05
Cosine^2	18	4.3	1.25
Cosine^3	24	5.1	1.33
Cosine^4	30	5.6	1.37

Table 1 Shaped Pulse Waveform Characteristics

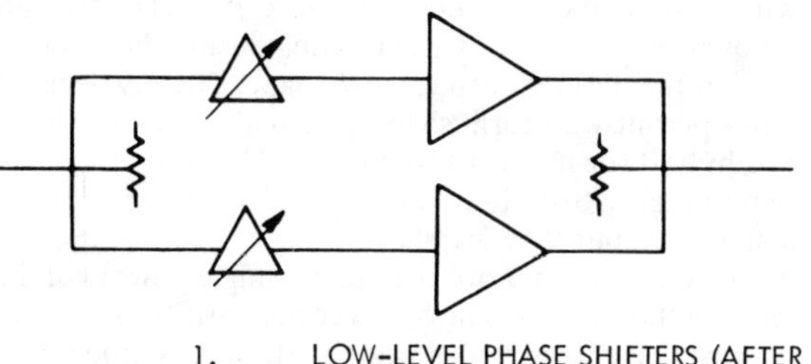

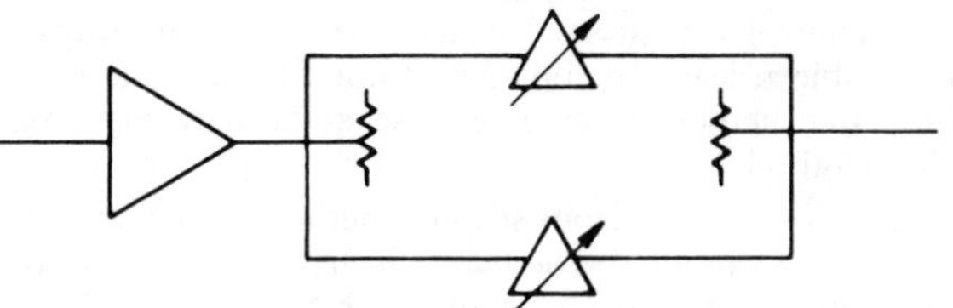
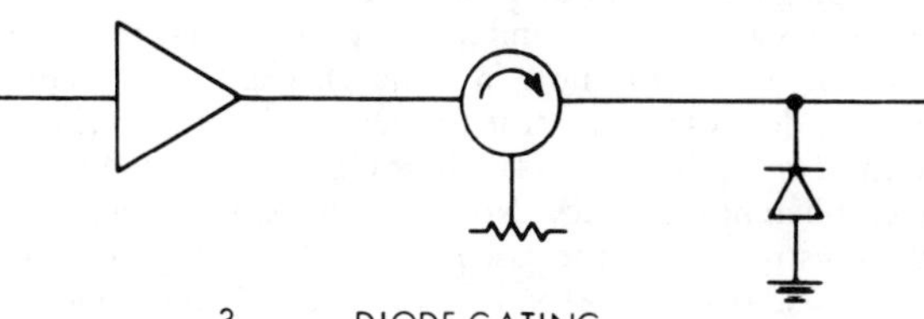

Figure 8 High-Level Shaping Methods

	Low-Level Shaping*		Plate Modulation	High-Level Shaping	
	Grid Modulation	*RF Drive Shaping*		*Hybrid Shaping*	*Diode Shaping*
Triode, Tetrode	OK	OK	OK	LLPS	OK
Magnetron	X	X	X	HLPS	OK
CFA	X	X	X	LLPS	OK
Klystron	OK	OK	X	LLPS	OK
TWT	OK	OK	X	LLPS	OK
Transistors	**	**	**	LLPS	OK

*Requires linear amplifier or predistortion to cancel nonlinearity
**Not demonstrated; may be feasible
LLPS = Low-level phase shifters useable
HLPS = High-level phase shifters required

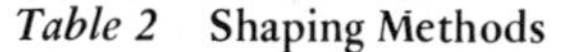

Table 2 Shaping Methods

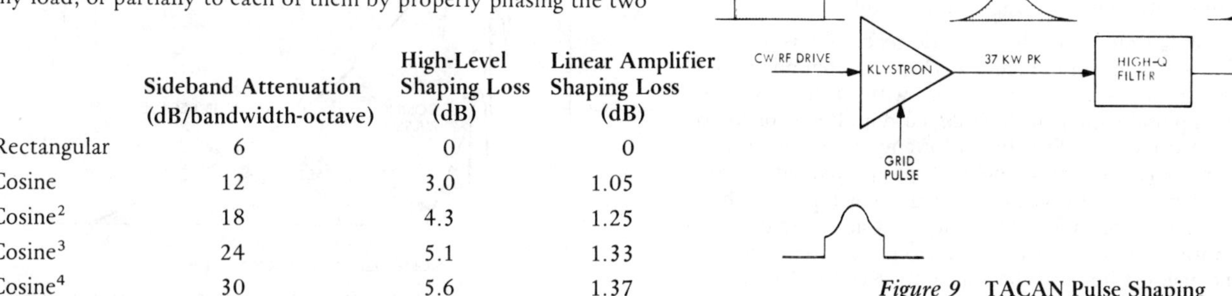

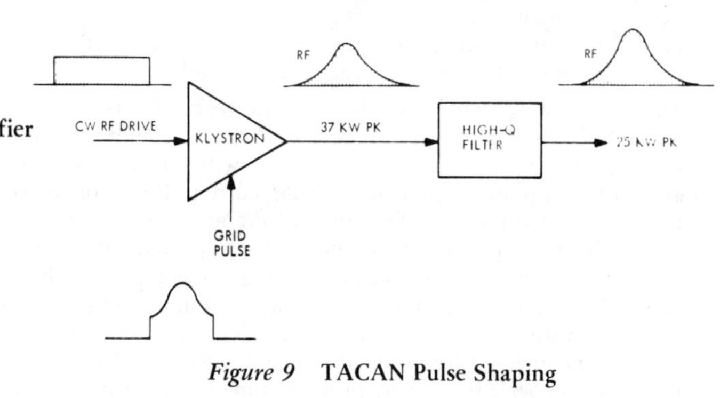

Figure 9 TACAN Pulse Shaping

chains with respect to each other [5 (Figure 29)]. The two phase shifters shown are used to vary the phasing during the pulse duration to obtain the desired output pulse shape. The low-level phase shifters thus permit waveform shaping by making use of the characteristics of hybrid (or magic-T) combiners. The other two methods [3] shown in Figure 8 apply to any type of RF power source, even a magnetron, but they are slightly lossier because the full power must pass through more than just a single power combiner. The second method requires high-power phase shifters (or a Faraday-rotator equivalent switch), while the third uses diodes to reflect all, some, or none of the power. The diode method is more difficult to control accurately at intermediate values of output, but it is suitable at least for clipping off noisy leading and trailing edges from an otherwise acceptable pulse, and it may cost less than other methods.

Table 2 summarizes the various shaping methods. An X in the table means "not suitable." The two right-hand columns are the high-level methods of Figure 8, all of which have the high losses associated with high-level shaping as noted in Table 1. The linear-amplifier shaping loss figures of Table 1 apply to the first three OKs after triodes and tetrodes and possibly to the first three OKs after transistors (see ** footnote in Table 2). The first two columns for these devices require true (or nearly) linear operation, while the third is applicable to class B or class C operation with their higher inherent efficiency. Note that the X in the third column in the klystron row is the case noted earlier in Figure 2 as resulting in *worse* than (sin x)/x; this way is clearly *not* good to shape a pulse for improved spectrum. For TWTs and Klystrons, the low-level methods noted as OK result in poor pulse-shaping efficiency, similar to high-level shaping losses. These difficulties occur because RF-drive shaping in linear-beam tubes requires that the beam be on at full current during the entire pulse; even with grid-pulse shaping [5], the gain and efficiency drop so rapidly as beam current is reduced below the full value that a much greater-than-proportional fraction of the beam current must be used during shaping.

To improve TWT or Klystron efficiency on shaped pulses, various techniques have been considered that vary the voltage on a depressed collector as the RF drive level is varied. This approach works because it is harmless to depress the collector to a very low voltage when there is little or no RF drive, and yet the collector voltage is high enough to collect the beam current during the middle of the pulse when there is strong RF drive. Normal efficiency thus is obtained in the middle of the pulse, and efficiency is not degraded too severely at the ends of the pulse because greater collector depression reduces the power input to the tube.

In linear-beam tubes with a multiplicity of depressed collectors operated at graduated voltages it may be possible to obtain the same effect automatically. As the RF drive is varied, the beam is collected at the lowest possible voltage by the appropriate section of the collector.

Using these techniques, the reduction in basic tube efficiency to permit handling shaped pulses is no worse than the efficiency loss in the linear amplifier approaches; the cost, however, is that of increased complexity in power supplies for the tube.

The best existing example of shaped-pulse generation is the TACAN system, as shown in Figure 9. The most common version uses a 37 kW peak power L-band klystron with constant RF drive and with grid-pulse shaping of the beam current. Predistortion of the grid-pulse shape is used to correct for amplitude nonlinearity in the klystron characteristics. The resulting approximately $\cos^2$ RF pulse is then passed through a very-high-Q high-power RF filter [5]. The filter has about 1.5 dB loss but improves the spectrum by another 10 dB in the critical region about $4/\tau$ from the carrier. The high-Q filter must be temperature controlled or temperature compensated to stay properly tuned. If the filter is tuned to the wrong frequency, the klystron can be damaged by the high VSWR that results [5].

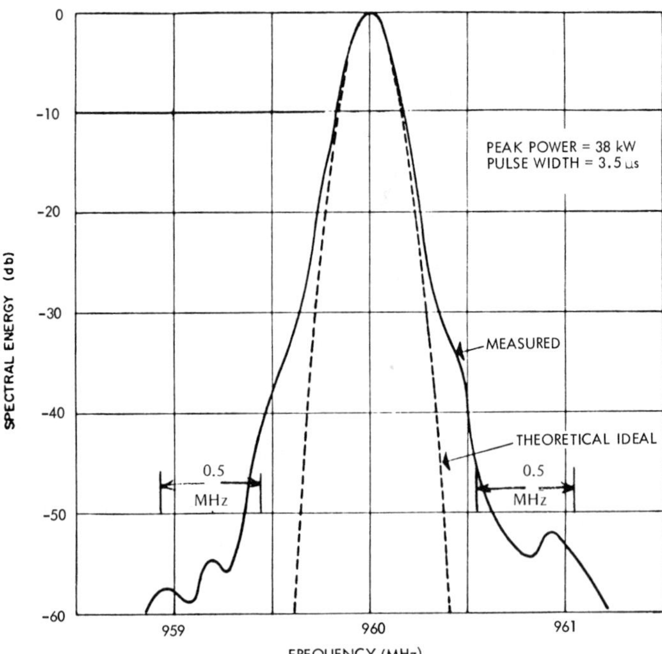

Figure 10 **Typical TACAN Klystron Spectrum (After Ashley & Sutherland [4])**

Figure 10 shows the klystron output spectrum before it passes through the final filter. To avoid interference with adjacent channels, the requirement is that power in the 0.5 MHz bandwidth shown on each side must be 60 dB down; the ideal Gaussian pulse spectrum shown easily meets this specification, but nonlinearities in the pulse shaping scheme used with the klystron require the use of the additional filter.

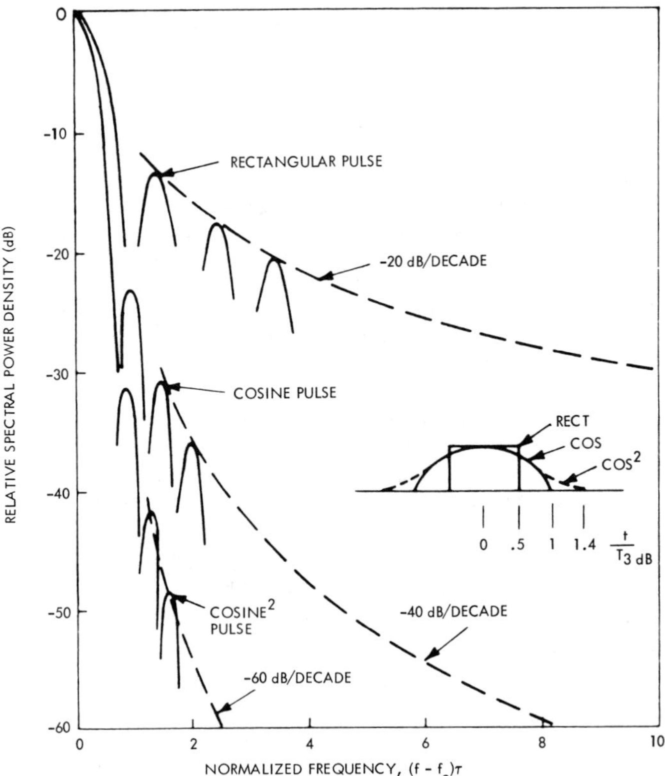

Figure 11 **Spectra for Shaped Pulses of Equal Half-Power Duration (After Ward [1])**

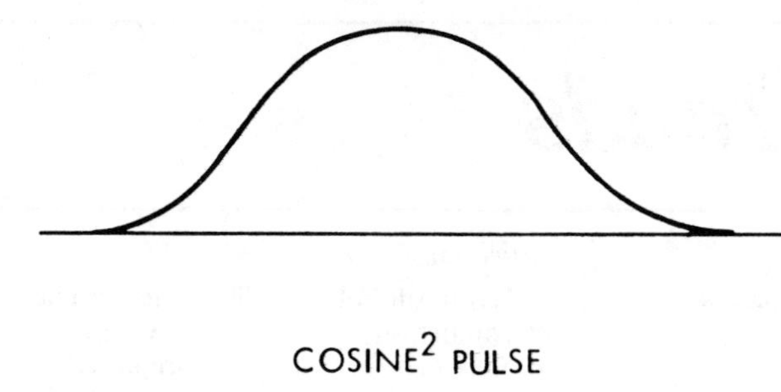

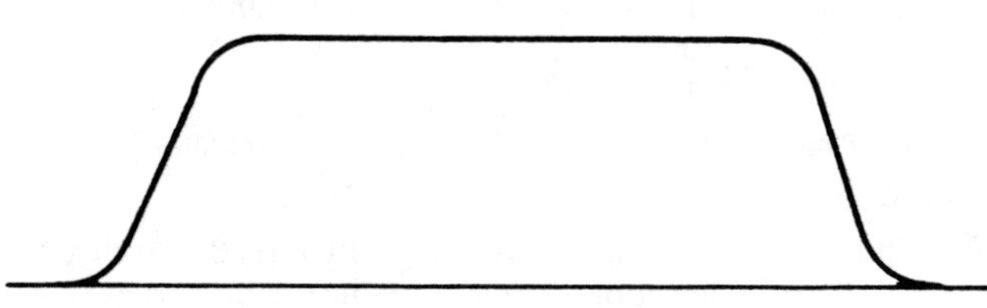

RECTANGULAR PULSE WITH COSINE^2 EDGES

Figure 12 **Comparison of Fully-Shaped Pulse and Shaped Edges**

The TACAN approach is not attractive, generally, because of the many adjustments required when frequency is changed and because of the very narrowband filter required. On the other hand, MIL-STD-469 requirements in general are not nearly as tight as the TACAN requirement.

Figure 11 shows three spectra for pulses correctly compared on the basis of equal duration at their half-power points [1]. The MIL-STD-469 requirements are specified at $10/\tau$ from the carrier, and we noted earlier that an ARSR-2 radar with a (sin x)/x spectrum would fail to meet MIL-STD-469 by about 30 dB; thus, the lower-right hand corner of Figure 11 is approximately the requirement in this case, since it is about 30 dB below the rectangular pulse spectrum. Note that an ideal cosine-squared pulse spectrum is about three times better than required, in terms of spectrum width at the -60 dB point. Therefore, if we could generate the cosine-squared shape accurately, we could meet the spec with a pulse 1/3 as wide as the full actual pulse duration. Figure 12 shows this comparison. Clearly, the pulse with only its edges shaped suffers less shaping loss than does the fully shaped pulse. The higher the order of the shaping, the faster the spectrum falls off away from the carrier; thus, the higher the order of shaping used, the smaller the fraction of the pulse that must be shaped and the smaller the shaping loss that must be incurred. The practical limit on the order of shaping that can be used is the distortion and accidental phase modulation that occur in the shaping process [6], which limit how accurately the desired shaping can be achieved. There may also be other limitations, such as turn-on and turn-off transients, depending on the type of shaping used (see Reference 1). If we use cosine-squared shaping of 1/3 of the pulse, as suggested in Figure 12, the shaping losses are significantly reduced – the high-level shaping loss drops from 4.3 dB (see Table 1) to 1.1 dB, and the linear-amplifier shaping loss drops from 1.25 dB to 0.4 dB, averaged over the full pulse duration.

Table 3 notes that shaping edges, rather than full-pulse shaping, is always desirable to minimize efficiency loss. Suitable methods of pulse shaping are shown for various RF amplifier types. It is also possible that in some cases the coaxial magnetron may approach or meet MIL-STD-469 without special shaping because it, in effect, has a high-level RF filter built in that automatically tracks the operating frequency. On the other hand, many magnetrons produce spurious outputs, especially during starting and stopping, and may require at least the use of a bandpass filter.

Shaped Edges to Preserve Efficiency	
Magnetrons CFAs	*High Level Shaping*
Klystrons TWTs	*RF Drive Shaping*
Triodes Tetrodes Transistors*	*Linear with RF Drive Shaping, or Class C with Plate Modulation*

*See note in Table 2; otherwise, high-level shaping may be used.

Table 3 **Most Promising Methods of Pulse Shaping**

In conclusion, four points are apparent:

1) Pulse shaping is becoming increasingly important in meeting EMC and MIL-STD-469 requirements.
2) Shaping is feasible, at a price in complexity and efficiency.
3) More efficient shaping methods than high-level shaping can be considered in some cases.
4) The efficiency loss can be minimized by shaping only enough of the pulse duration to get the needed spectrum improvement.

References

[1] Ward, H.R.: "Radar Spectrum Control," *EASCON 1969 Convention Record.*

[2] Miller, S.N.: "The Source of Spectrum Asymmetry in High Power RF Klystrons," *MIL-E-CON-9,* 22 September 1965.

[3] Bernstein, A.; and Miller, S.N.: "Radar Frequency Spectrum Control Circuit," *US Patent 3,603,991.*

[4] Ashley, J.R.; and Sutherland, A.D.: "Microwave Spectrum Conservation by Means of Shaped Pulse Transmitters," *Sperry Tube Division Report NJ-2761-0168,* July 1964.

[5] Skolnik, M.I. (ed.): "Transmitters," *Radar Handbook* (Chapter 7), McGraw-Hill, 1970.

[6] Brookner, E. and Bonneau, R.J.: "Spectra of Rounded-Trapezoidal Pulses having AM/PM Modulation and Its Application to Out-of-Band Radiation," *Microwave Journal, Vol. 16, No. 12,* pp. 49-51, 80, December 1973.

Present and Future Trends

Laser Cost

Helium neon (HeNe) lasers (~0.5 mW, 0.6328μm) presently cost less than $100 (1977); in the near future, it is projected that they will be available for $10.

Laser Technology and Systems

Laser Measurement Systems

Lasers are being used for miniscule range measurements of the Earth's crust along faults. One such system, developed by the University of Washington, permits distance measurements to within 1 part in 10^7 over paths of 1 to 10 km [1]. Operating along two fault lines in Hollister, California, this system detected fault motions after a month of operation. Conventional systems would have required years to detect the same motion. The system transmits on three carrier frequencies over the path – a helium-cadmium laser at 0.4416μm, a helium-neon laser at 0.6328μm, and a microwave signal at 9.6 GHz.

NASA plans to include a laser on the space shuttle in 1981 in order to measure movements of the Earth's crust to within a few centimeters. Corner reflectors will be placed along Earth fault lines. Measurements of the sort described above are believed by geologists to signal impending earthquakes.

A laser was used to align the face of the Cobra Dane phased array (see Figure 18 of Chapter 1) to 25/1000 of an inch.

Fiber Optics

Increasing attention is being given to the use of fiber optics in radar systems for the transmission of information between subsystems and for signal processing. Fiber optic links have the advantage of providing a method of communication between subsystems which is immune to electromagnetic interference (EMI). A fiber optics link consists of an optical source (a light emitting diode [LED] or injection laser), a fiber optic cable, and a silicon photodiode (PIN or avalanche). LED sources are more efficient than incandescent bulb sources because they rely on the jumping of electrons from high to low energy states at the semiconductor's junction for the generation of light rather than on the superheating of a filament. Table 1 compares LED with injection laser optical fiber systems. The attenuation per kilometer for high quality fibers is given in Table 2. High quality fibers can cost about $2,000/km. For systems requiring runs of only 50 feet, high-loss inexpensive 1000-2000 dB/km lines can be used. This latter type of fiber is what is being proposed for auto control and monitoring systems [4].

Wideband optical fiber lines are being considered for the construction of tap delay line processors [6] and pulse code generators [7]. Taps can be placed as close as 0.01 ns. The frequency dependent dispersion in single-mode fiber optic guide, due to both material and waveguide properties and which is on the order of 1 ps/μs/GHz, can be used to generate linear chirp waveforms at optical frequencies for laser radars [6].

Efficient Spectrum Control for Pulsed Radar Transmitters

Weil, in Chapter 27, points out that it is desirable to shape a small fraction of the pulse (the leading and trailing edges) and that a reasonable type of shaping is the cosine-squared shaping illustrated in Figure 12 of Chapter 27 and in Figure 1 of this section. He also indicates that one limit to the control obtainable is the accidental phase modulation occurring during the rise and fall of the pulse.

	Injection Laser	*LED*
Frequency	0.8-1.5μm (double-heterojunction structure)	Ultraviolet, visible, or infrared (most efficient types are in red region of spectrum)
Typical Output Power	10 mW	1 mW
Device Life	15,000 h, now 100,000 h, before 1980	> 100,000 h
Divergence Angle	20°	2π steradians
Spectral Width	2 nm	36 nm
How modulated	By directly varying injection current	By directly varying injection current
Type modulation	Digital or Analog	Digital or Analog
Response Time	< ns (after laser thresholding)	1-200 ns
Maximum Data Rate	300 Mbps	1 Gbps
Optical Fiber time Dispersions	1 ns/km (multimode delay dispersion for parabolically graded index of refraction across the fiber)	3.6 ns/km (material dispersion)
Optical fiber bandwidth limit	1000 Mb-km/s (for multimode fiber optic line having parabolically graded index)	278 Mb-km/s (material dispersion)

Table 1 **Comparison of Injection Laser and LED Fiber Optic Communication Systems [2, 3]**

Brookner and Bonneau [8] have analyzed the limit due to AM/PM phase modulation. The results were obtained for the pulse shape of Figure 1 for a worse case type of AM/PM phase modulation, specifically

$$\phi(E) = \Delta\phi E^3 \qquad (1)$$

where $\phi(E)$ is the phase modulation as a function of the applied input voltage, E, which is normalized to E = 1 at saturation so that the phase shift is $\Delta\phi$ at that point. The results are summarized in Figure 2.

Fiber	Attenuation Research 0.85 μm	1.06 μm	Production 0.85 μm	1.06 μm	Bit-Rate-Length Product Research	Production
Graded Index	3 dB/km	< 2 dB/km	10 dB/km	7 dB/km	1000 Mb-km/s	200 Mb-km/s
Plastic-Clad	6 dB/km	5 dB/km	20 dB/km	20 dB/km	30 Mb-km/s	30 Mb-km/s

Historical Note – The reduction in attenuation has been rapid. Prior to 1970, the best available was 1000 dB/km; in 1970, Corning Glass Works achieved 20 dB/km, presently (1977), < 2 dB/km has been obtained by a number of industrial and university groups [5].

Table 2 **Fiber Attenuation Status and Multimode Group Delay Spread (After Campbell [3])**

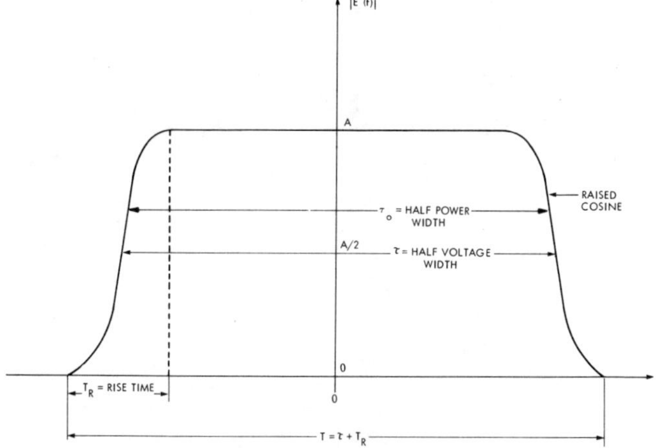

Figure 1 **Rounded-Trapezoid Pulse (From Brookner and Bonneau [8])**

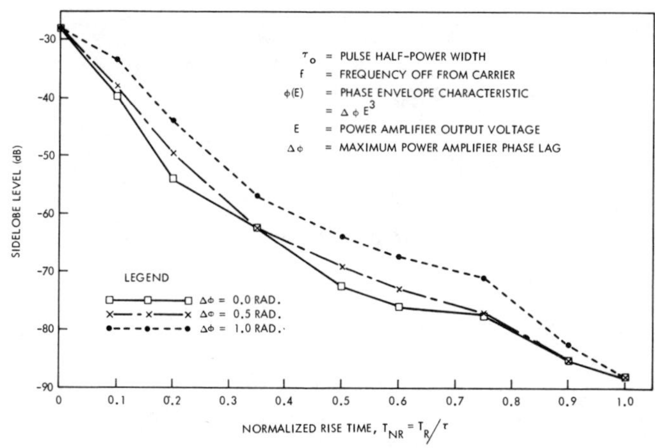

Figure 2 **Sidelobe Level at f = $10/\tau_o$ for Rounded-Trapezoidal Pulse (From Brookner and Bonneau [8])**

References

[1] "Distances are Measured to 1 Part in 10^7 With Two Lasers and a Microwave Beam," *Laser Focus, Vol. 12,* pp. 25-26, July 1976.

[2] Stigliani, D.: "Optical Fibers for Data Transmission," *Digital Design,* pp 48-52, May 1976.

[3] Campbell, L.L.: "Status of Fiber Optic Research in the USA," *Optical Engineering, Vol. 15, No. 5,* pp. 472-476, September-October 1976.

[4] Brennesholtz, A.H.; and Balkenhol, T.C.: "Fiber Optics Cuts Weight in Car Systems," *Digital Design,* pp. 70-75, August 1976.

[5] Miller, S.E.: "Photons in Fibers for Telecommunication," *Science, Vol. 195,* pp. 1211-1216.

[6] Kalman, W.; and van den Heuvel, A.P.: "Fiber-Optic Delay Lines for Microwave Signal Processing," *Proceedings of the IEEE, Vol. 64, No. 5,* pp. 805-807, May 1976.

[7] Ohlhaber, R.L.; and Wilner, K.: "Fiber Optic Delay Lines for Pulse Coding," *Electro-Optical Systems Design, Vol. 9,* pp. 33-35, February 1977.

[8] Brookner, E. and Bonneau, R.J.: "Spectra of Rounded-Trapezoidal Pulses having AM/PM Modulation and Its Application to Out-of-Band Radiation," *Microwave Journal, Vol. 16, No. 12,* pp 49-51, 80, December 1973.

Appendix

How to Look Like a Genius in Detection Without Really Trying*

Simple cookbook procedures for determining the detection performance of a radar system are presented here. Single-scan detection procedures are given for both the problem of solving for the signal-to-noise ratio (SNR) per pulse, given the desired detection probability, P_d, and the reverse problem of solving for P_d given the SNR/pulse. Approximate and more precise methods are presented. The procedures are given in the form of simple flow diagrams and easy-to-use worksheets which are illustrated extensively with example calculations. The approximate procedure of Table 1 is a modification, formalization, and extension of the method outlined by Barton in Chapter 2; see Table A.

Cookbook Procedures

Tables 1 through 15 and Figures 1 through 13 give the procedures and are self explanatory. Table A is provided for guidance in using Tables 1 through 15. Illustrative examples are given in Tables 8, 13, 14, and 15. These tables also indicate the decibel differences between the results obtained by the approximate and more precise procedures given. Table 10 indicates the accuracies of the different procedures. Tables 14 and 15 provide worksheets which give step-by-step instructions on the approximate methods; they can be used immediately without one having studied the other tables or figures which outline the procedures. In fact, to gain familiarity quickly with the approximate procedure methods, it is urged that the reader first carry out some of the example cases of Tables 8, 13, 14, and 15 using the worksheets before studying the other tables and figures. The values worked out in detail can be used as an in-progress check. It is further recommended that the reader start with the simple cases (i.e., no collapsing loss [$\rho = 1$], no frequency diversity [F = 1; see Table 2], and standard Swerling target models [Cases 0 through 4; see Tables B and 2]) using the approximate procedure of Tables 1, 9, and 12. To become acquainted with the more precise procedures of Tables 7 and 11 it is highly recommended that the reader regenerate the curves of Figures 12 and 13 independently (for the simple target models first), then obtain SNR/pulse and P_d and check the results with those given in Tables 8 and 13 for the more precise procedure. Definitions of all the terms used in the tables and figures are given in the glossary.

The approximate methods given apply for a target the cross section distribution of which is chi-square with 2K degrees of freedom (K duo-degrees of freedom). Table B gives the *exact* relationship between the standard Marcum (nonfluctuating), Swerling, and Weinstock target models with the chi-square distribution. Also, it shows how targets with cross sections with the Rice and log-normal distributions can be approximated by the chi-square distribution so that the performance of these target models also can be handled for many cases of interest.

The single-scan probability of detection results apply for a system consisting of a matched filter IF amplifier, followed by a square-law envelope detector, which in turn is followed by a video pulse-to-pulse integrator; see Figure 7 of Chapter 2. For a mismatched filter, the matching loss, L_m, of Figure 14 must be included. The applicability of the results to the case of a linear envelope detector is discussed at the end of this section.

The matched filter can be matched to a burst of coherently integrated pulses. For this case, the signal-to-noise ratios per pulse ($D_o(1)$, $D(\rho N, K)$, etc.) at the IF output are defined for the coherently integrated pulses (i.e., after the matched filter) and N refers to the number of matched filtered burst waveforms that are video integrated.

The procedures apply independently of whether noise-alone video samples are added in with the signal-plus-noise samples; i.e., independently of whether or not a collapsing loss exists. The single-scan detection procedures take into account possible loss in detection sensitivity due to beam scanning, constant false alarm rate (CFAR) processing, and binary or quaternary integration (as opposed to analog integration).

The applicability of the results to the case of a linear envelope detector is discussed in the next paragraph. Those not familiar with detection theory would be best advised to skip this paragraph on first reading.

Consider the special case where no collapsing loss exists (i.e., $\rho = 1$) and the amplitudes of the signal components of the pulses in a scan have unity correlation (i.e., F = 1, as is the situation for the Swerling Cases 1 and 3 target models and the nonfluctuating target model). For this case the single-scan detection performance SNR results obtained for a square-law envelope detector apply for when a linear envelope detector is used, to within an accuracy of 0.2 dB [29, 32]. Trunk [14], however, has shown that the performance results obtained for a square-law detector do not apply for a linear envelope detector when collapsing loss exists (i.e., when $\rho \neq 1$).

It is held generally in the literature that when no collapsing loss exists ($\rho = 1$) the performance results obtained with a square law envelope detector apply as well when a linear envelope detector is used for the case of signal amplitude pulse-to-pulse decorrelation (i.e., for the Swerling Cases 2 and 4 target models and, in turn, for the chi-square target model with arbitrary F) [5]. However, for this case, no published data is available indicating the accuracy to which the square-law envelope detector results apply for the linear envelope detector system. Such an analysis is needed.

*Condensed and revised version of Brookner, E.: *How to Look Like a Genius in Detection Without Really Trying*, 1974 IEEE NEREM, Boston, Massachusetts, pp. 37-64. Part removed is covered in Chapter 3.

Glossary

b = bandwidth of narrow-band-amplifier or matched-filter following limiter for Dicke-fix CFAR; see Figure 8.

B = bandwidth of broad-band amplifier preceding limiter for Dicke-fix CFAR; see Figure 8.

CFAR = Constant False Alarm Rate; receiver processing involving the adaptive variation of the detection threshold in direct proportion to any variations in the receiver noise power (or clutter or jamming) so as to keep the false alarm rate constant in the presence of such variations. An estimate of the receiver noise power is obtained by averaging a specified number of noise alone receiver output reference samples [$M(\rho N)$ per scan]. The receiver detection threshold is made proportional to this estimate.

$C_x(\rho N)$ = detector loss; specifically, the increased signal energy required per scan because analog video integration is used instead of having all the signal energy and receiver noise in one pulse with known signal phase and ideal IF coherent receiver processing and thresholding (see Table 6). (*Note* – Consider a case where $\rho = 1$ for which N_1 signal-plus-noise samples are integrated; also assume $N_1 = \rho N$. Then $C_x(N_1) = C_x(\rho N)$, in contrast to the situation for $D_o(\rho N)$ [and $D(\rho N, K)$] where $D_o(N_1) \neq D_o(\rho N)$; see definition for $D_o(\rho N)$.

$D(\rho N, K)$* = definition same as for $D_o(\rho N)$ except for fluctuating target, specifically for chi-square target model having K duo-degrees of freedom (see text of Appendix for further discussion as to type of envelope detector assumed).

$D_o(\rho N)$* = IF output power SNR/pulse required for case of nonfluctuating target to achieve P_d and P_{fa} with analog video integration of N signal-plus-noise video samples together with $m(= \rho N\text{-}N)$ noise alone video samples. Additional assumptions are ideal matched filter IF receiver, rectangular antenna beam pattern, fixed threshold (no CFAR), and for $\rho = 1$ linear or square-law envelope detector or $\rho \neq 1$ square-law envelope detector [14]. (It is important to note that $D(\rho N) \neq D(N_1)$ for $N_1 = \rho N$.* Without the parameter ρ, no collapsing loss is assumed; instead, it is assumed that N_1 signal-plus-noise video samples are integrated.) The SNR/pulse is measured at the time at which the signal peaks at the output of the matced filter. However, the ratio of average signal power over an IF cycle to average noise power is specified instead of the peak power (occurring at the IF cycle peak) to average noise power. This definition is the one used generally [1, 5, 18-20, 30-32, 36, 37]. It is, though, 3 dB lower than the SNR/pulse used for the curves in Reference 22.

F = number of independently fading signal pulse groups in the N signal-plus-noise single-scan samples.

j = number of pulses between 3 dB points of beam per scan.

K = number of duo-degrees of freedom for equivalent chi-square target model for analog pulse-to-pulse integrator output (N pulses integrated); see Tables B and 2; = Fk.

k = number of duo-degrees of freedom for equivalent chi-square target for single pulse; see Table 2.

*For $N_1 = \rho N, \frac{1}{\rho} D_O(\rho N) = D_O(N_1)$

L_B = loss of detection sensitivity due to the use of quantized video samples rather than analog samples for integration process.

L_{CFAR} = CFAR loss; i.e., the additional SNR/pulse needed when CFAR processing is used over that required when fixed thresholding is used, the same P_d and P_{fa} being specified for both cases.

$L_c(\rho, N)$ = collapsing loss; specifically, the loss in sensitivity resulting from the integration of $m = \rho N\text{-}N$ noise alone video samples together with the N signal-plus-noise video samples; see Table 6 and Equation (6).

L_{fK} = fluctuation loss for chi-square target having K duo-degrees of freedom.

$= D(\rho N, K)_{dB} \text{-} D_o(\rho N)_{dB}$ (9)

L_{f1} = L_{fK} for K = 1, i.e., for Swerling Case 1 target.

$L_i(\rho N)$ = video integration loss; specifically, increased signal energy required per scan because analog video integration (of N signal-plus-noise video pulses and $m[= \rho N\text{-}N]$ noise alone video pulses) is used instead of having all the signal energy and receiver noise in one pulse; see Table 6. (*Note* – As for $C_x(\rho N)$, $L_i(N_1) = L_i(\rho N)$ for $N_1 = \rho N$.)

L_p = beam shape factor; specifically, the additional IF SNR/pulse needed when a non-rectangular beam pattern is used instead of a rectangular beam to achieve the specified P_d and P_{fa} where the SNR/pulse for nonrectangular beam pattern is measured on the beam axis. The width of the rectangular beam pattern is assumed equal to the one-way 3 dB width of the nonrectangular beam pattern.

M = number of reference noise cells averaged per pulse integrated per scan for CFAR processing. (For $\rho \neq 1$, the number of pulses integrated is ρN; hence, $M\rho N$ reference noise cells are averaged per scan.)

N = number of signal-plus-noise video samples integrated per scan. (n used in Chapter 2 for N.)

n = false alarm number for ρN=1; specifically, average number of video samples in absence of signal observed before one or more false alarms occur with probability 50% [29]. (Designated as n_{fa} in Chapter 2.)

$= 0.693/P_{fa}$ (14) [5, 22, 29, 30]

n′ = false alarm number for ρN video integrated signal-plus-noise samples; specifically, average number of video integrator output samples observed in absence of signal before one or more false alarms occur with probability 50% [5, 22, 29, 30].

$= 0.693/P_{fa}$

$= n/\rho N$ (15) [29]

P_d = probability of single-scan detection. (In general, based on integration of N signal-plus-noise video samples and m $[= \rho N\text{-}N]$ noise alone video samples.)

P_{fa} = false alarm probability at output of receiver after integration (analog, binary, or quaternary) and CFAR processing (if used).

$= 0.693/n'$ (16)

P_{fa1} = P_{fa} at output of first threshold for binary integrator.

P_{fa2} = P_{fa}

P_n = P_{fa}

p(x) = probability density function for x.

s = ratio of cross section of specular component to that of Rayleigh component for target having Rice cross section statistics.

SNR/pulse = same as $D_o(\rho N)$ and $D(\rho N,K)$, except that increased SNR/pulse resulting from presence of L_p, L_{CFAR}, and L_B losses included, if present, for Tables 1, 8, and 12-15.

T = expanded pulse width for dispersive or pulse-compression CFAR; see Figure 8.

X = see Table B.

X_o = IF output power SNR/pulse required for case of nonfluctuating target to achieve specified P_d and P_{fa} for $\rho N = 1$ when ideal matched IF receiver is used for assumption that received signal phase is known [22 (pp. 291-298)]. SNR/pulse is measured at the time at which the signal peaks at output of the IF matched filter. Furthermore, ratio of the peak power occurring at the peak of IF cycle to the average noise power is used in specifying X_o. Thus, by definition, X_o is 3 dB higher than $D_o(\rho N)$ for the same signal and noise power at the IF output – which is as it should be because, when the received phase is known, one can and does sample at the peak of the signal IF cycle [1, 22].

x_i = see Table B.

Y_b = IF output detector threshold voltage.

ΔD_o = correction needed for obtaining SNR/pulse for log-normal target from equivalent chi-square target model; see Tables 1 and 12 and Figure 5.

$\Delta\theta$ = angle moved between pulses for scanning antenna.

ρ = collapsing ratio.

$= (N+m)/N \quad (17)$

where m = number of noise alone video samples integrated together with N signal-plus-noise video samples [5, 22, 29].

$\underline{\rho}$ = ratio of mean to median for log-normal variate.

σ = standard deviation of the natural logarithm of log-normal variate.

$= \sqrt{2 \ln \underline{\rho}}$

τ = compressed pulse width for dispersive or pulse-compression CFAR; see Figure 8.

TYPE OF DETECTION	PROBLEM SOLVED	TYPE ACCURACY	TABLE	SOURCE
SINGLE-SCAN DETECTION	GIVEN P_d FIND SNR/PULSE	APPROXIMATE	1	AN EXTENSION OF BARTON [CHAP. 2, 1] AND CANN [2] PROCEDURES: SEE [3], [4], [5] FOR BASIS FOR EXTENSION: FLOWGRAPH PATTERNED AFTER [39].
		MORE PRECISE	7	REFINEMENT OF MEYER AND MAYER [5] PROCEDURES: SEE ALSO [3].
		APPROXIMATE	9	NEUVY'S [6] EQUATIONS.
	GIVEN SNR/PULSE FIND P_d (REVERSE PROBLEM)	APPROXIMATE	11 AND 12	URKOWITZ [7] AND CANN [8]; NEUVY [6].
		MORE PRECISE	11	REFINEMENT OF MEYER AND MAYER [5] PROCEDURES: SEE ALSO [3].

Table A **Brief Table of Contents**

PROBLEM SOLVED:

GIVEN: P_d, P_{fa}, N, ρ, TARGET MODEL, TYPE OF PROCESSING (CFAR; ANALOG, BINARY, OR QUATERNARY INTEGRATION), AMOUNT OF FREQUENCY DIVERSITY, TYPE OF SCANNING.

FIND: SNR/PULSE REQUIRED.

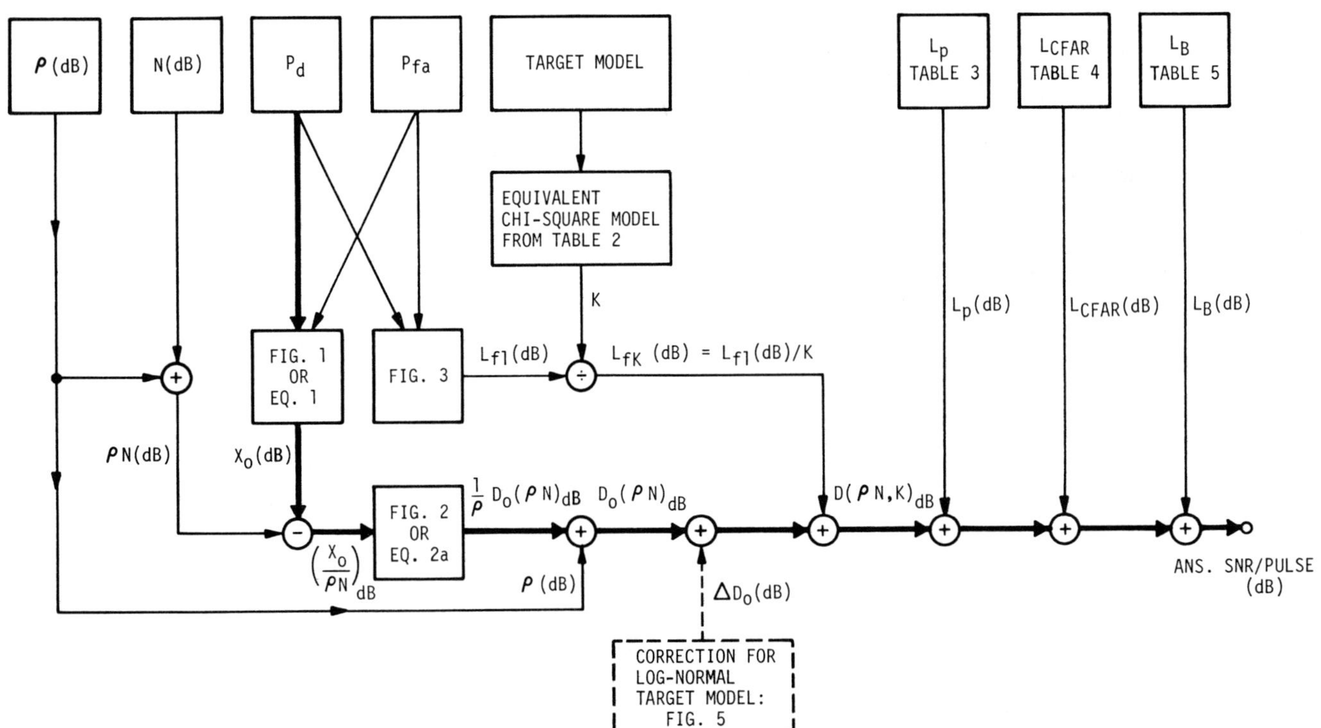

NOTES:

(1) $L_i(\rho N)$, $L_c(\rho, N)$, AND $C_x(N)$: SEE TABLE 6 FOR DETERMINATION OF THESE PARAMETERS.

(2) MORE PRECISE PROCEDURE FOR DETERMINATION OF $D(\rho N,K)$, AS WELL AS L_{fK}, ETC., GIVEN IN TABLE 7.

(3) RESULTS APPLY WITHOUT RESTRICTIONS FOR ASSUMPTIONS OF SQUARE-LAW ENVELOPE DETECTOR AND WITH SOME RESTRICTIONS FOR ASSUMPTION OF LINEAR ENVELOPE DETECTOR; SEE TEXT.

Table 1 **Find Required SNR/Pulse to Achieve a Specified Single-Scan Probability of Detection; Approximate Procedure**

NOTATION: x_i=POWER SNR FOR i^{TH} RECEIVED PULSE AND HENCE PROPORTIONAL TO CROSS SECTION OF TARGET AS SEEN BY i^{TH} PULSE.

$$X = \sum_{i=1}^{N} x_i$$ = SUM OF THE SNRs FOR THE N PULSES

TARGET MODELS GIVEN IN TERMS OF DISTRIBUTIONS FOR x_i AND X

TARGET MODEL	ALTERNATE DESCRIPTIONS	EXAMPLE
NONFLUCTUATING (ALSO CALLED: MARCUM OR SWERLING CASE 0 or 5)	1) x_i = CONSTANT* 2) CONSTANT CROSS SECTION TARGET	SPHERE
SWERLING CASE 1	1) x_i HAS EXPONENTIAL DISTRIBUTION** , $\sqrt{x_i}$ HAS RAYLEIGH DISTRIBUTION** , $x_i = x_j$ FOR ALL i, j SUCH THAT $1 \leq i$, $j \leq N$.* 2) X HAS CHI-SQUARE DISTRIBUTION WITH 2 DEGREES OF FREEDOM, HENCE 1 DUO-DEGREES OF FREEDOM, i.e., K=1. 3) $p(X) = \frac{1}{\bar{X}} \text{EXP} - \frac{X}{\bar{X}}$, $X \geq 0$	JET [15]
SWERLING CASE 2	SAME AS FOR SWERLING CASE 1 EXCEPT THAT x_i INDEPENDENT OF x_j FOR $i \neq j$.	PROPELLER AIRCRAFT [15]
SWERLING CASE 3	1) x_i HAS CHI-SQUARE DISTRIBUTION WITH 4 DEGREES OF FREEDOM; $x_i = x_j$ FOR ALL i, j SUCH THAT $1 \leq i, j \leq N$.* 2) X HAS CHI-SQUARE DISTRIBUTION WITH 4 DEGREES OF FREEDOM, i.e., K=2. 3) $p(X) = \frac{4X}{\bar{X}^2} \exp - \frac{2X}{\bar{X}}$, $X \geq 0$	STABILIZED MISSILE TANKAGE; RANDOMLY ORIENTED PROLATE SPHEROID [5], [16].
SWERLING CASE 4	SAME AS FOR SWERLING CASE 3 EXCEPT THAT x_i INDEPENDENT OF x_j FOR $i \neq j$.	PROPELLER AIRCRAFT [15]
WEINSTOCK	X HAS CHI-SQUARE DISTRIBUTION WITH K DUO-DEGREES OF FREEDOM WHERE $K < 1$.	RANDOMLY ORIENTED CYLINDERS [3], [5], [16].
CHI-SQUARE	1) X HAS CHI-SQUARE DISTRIBUTION WITH K DUO-DEGREES OF FREEDOM 2) $p(X) = \frac{1}{(K-1)!} \frac{K}{\bar{X}} \left(\frac{KX}{\bar{X}}\right)^{K-1} \text{EXP}\left(-\frac{KX}{\bar{X}}\right)$, $X > 0$	AIRCRAFT [15], [17]
RICE	1) $X = X_S + X_R$ WHERE X_S IS A CONSTANT AND $\sqrt{X_R}$ HAS RAYLEIGH DISTRIBUTION. 2) TARGET CROSS SECTION COMPOSED OF SPECULAR PLUS RAYLEIGH SIGNAL ECHO AMPLITUDE	[3], [4], [5], [17]
LOG-NORMAL	LOG X HAS NORMAL DISTRIBUTION	RANDOMLY ORIENTED FLAT PLATE; BATTLESHIP: SATELLITES; MISSILES; [17], [18].

* NOTE THAT FOR SQUARE-LAW ENVELOPE DETECTOR ONE CAN HAVE UNEQUAL $x_i = a_i x_0$ WHERE a_i, i=1, . . ., N ARE CONSTANTS AND x_0 HAS CHI-SQUARE DISTRIBUTION WITH k=1 DUO-DEGREES OF FREEDOM FOR THE SWERLING CASE 1 TARGET MODEL AND OBTAIN IDENTICAL PERFORMANCE AS WITH EQUAL x_i AS LONG AS $\bar{X}$ IS THE SAME FOR BOTH CASES [38]. SAME STATEMENTS APPLY FOR SWERLING CASES 0 AMD 3 TARGET MODELS.

** x_i HAS CHI-SQUARE DISTRIBUTION WITH 2-DEGREES OF FREEDOM, i.e., K = 1 FOR TARGET HAVING RAYLEIGH DISTRIBUTED ECHO AMPLITUDE, IN OTHER WORDS FOR $\sqrt{x_i}$ RAYLEIGH DISTRIBUTED.

Table B **Target Models**

TARGET MODEL			NUMBER OF DUO-DEGREES OF FREEDOM, FOR EQUIVALENT CHI-SQUARE MODEL K, **
DISTRIBUTION	CASE	NUMBER OF INDEPENDENTLY FADING PULSE GROUPS #	
NONFLUCTUATING TARGET (MARCUM)	0≠	-	∞
SWERLING	1##	F	F
SWERLING	2	N	N
SWERLING	3##	F	2F
SWERLING	4	N	2N
WEINSTOCK	-*	F	kF**
CHI-SQUARE	-*	F	kF
RICE (S)	-*	F	kF WHERE $k = 1 + \frac{s^2}{1+2s}$ (3)
LOG-NORMAL ($\underline{\rho}$)	-*	F	kF WHERE k GIVEN BY FIG. 4
ARBITRARY MODEL	-	F	kF WHERE k OBTAINED FOR EQUIVALENT CHI-SQUARE MODEL USING FIGS. 4-8 AND 4-9 OF [5]; SEE ALSO [3].

\# COULD ARISE FROM FREQUENCY DIVERSITY CONSISTING OF F DIFFERENT FREQUENCY TRANSMISSIONS DURING THE INTEGRATION TIME OR FROM TARGET INTERNAL FLUCTUATIONS ALONE OR A COMBINATION OF THE TWO; SEE CHAPTER 2, EQUATIONS (27) TO (30).

≠SOMETIMES DESIGNATED AS SWERLING CASE 5 INSTEAD OF SWERLING CASE 0.

\##FOR THE DEFINITION ORIGINALLY GIVEN BY SWERLING [15] AND USED STANDARDLY IN THE LITERATURE F=1. SPECIFICALLY, THE SWERLING 1 AND 3 MODELS HAVE CHI-SQUARE DISTRIBUTIONS FOR THEIR ECHO POWERS WITH, ONE AND TWO DUO-DEGREES OF FREEDOM, RESPECTIVELY, (i.e., TWO AND FOUR DEGREES OF FREEDOM RESPECTIVELY). FURTHERMORE, THEIR ECHO SIGNAL POWERS ARE CONSTANT OVER THE N RECEIVED PULSES. FOR THESE TWO TARGET MODELS F INDEPENDENTLY FADING GROUPS CAN BE OBTAINED BY USING FREQUENCY DIVERSITY. F FREQUENCIES ARE TRANSMITTED HAVING SUFFICIENT SEPARATION SO AS TO CAUSE THE AMPLITUDES OF THE ECHOS FOR THE DIFFERENT FREQUENCIES TO BE INDEPENDENT OF EACH OTHER: SEE CHAPTER 2, EQUATIONS (29) AND (30). IF F IS NOT SPECIFIED THE DEFINITION STANDARDLY USED IN THE LITERATURE FOR THE SWERLING CASE 1 AND 3 MODELS WITH F=1 IS IMPLIED IN THE TABLES AND FIGURES TO FOLLOW.

* SOMETIMES REFERRED TO IN THE LITERATURE [18] AS CASE 1 WHEN F=1 FOR NO FREQUENCY DIVERSITY AND AS CASE 2 WHEN F=N. THIS CONVENTION USED IN TABLES 8 AND 11 AND IN FIG. 13.

**k = NUMBER OF DUO-DEGREES OF FREEDOM OF EQUIVALENT CHI-SQUARE MODEL FOR SINGLE PULSE. IN CONTRAST K IS THE NUMBER OF DUO-DEGREES OF FREEDOM FOR EQUIVALENT CHI-SQUARE MODEL FOR THE OUTPUT OF THE PULSE-TO-PULSE ANALOG INTEGRATOR.

Table 2 **Equivalent Chi-Square Target Model**

L_p(dB)		
FIXED BEAM	SCANNING GAUSSIAN ANTENNA	
	N > 3	N ≤ 3
0	~1.6*[23]	FIG. 6 OR 7 AND [23].

*EXACT FOR N=30. LOSS 0.2 dB LESS FOR LARGE N (≥1000) AND 0.2 dB LARGER FOR SMALL N (≐4). ASSUMPTIONS: GAUSSIAN ANTENNA BEAM PATTERN; ONE-DIMENSIONAL SCANNING; SQUARE-LAW ENVELOPE DETECTOR; GAUSSIAN WEIGHTING FOR VIDEO INTEGRATED PULSES. IF UNIFORM WEIGHTING OF VIDEO INTEGRATED PULSES USED THEN LOSS INCREASES BY ABOUT 0.2 dB. FOR TWO-DIMENSIONAL SCANNING ABOVE LOSS DOUBLED.

Table 3 **Beam Shape Factor, L_P**

L_{CFAR}(dB)		
CONSTANT THRESHOLD	CFAR ADAPTIVE THRESHOLDING	
0	ρN = 1 FIG. 8	ρN ≥ 1 FIG. 9

Table 4 **CFAR Loss L_{CFAR}**

TYPE INTEGRATION	L_B (dB)
ANALOG	0
BINARY OR QUATERNARY	FIGS. 10 OR 11 DEPENDING ON TARGET MODEL AND P_{fa}.

Table 5 **Quantization Integration Loss, L_B**

1. $L_i(\rho N)$:

Alternate Methods*

1) $L_i(\rho N)_{dB} = [ND_o(\rho N)]_{dB} - D_o(1)_{dB}$ (4)

where $D_o(\rho N)$ and $D_o(1)$ are determined using Table 1, or 14 or Meyer Plots [5].

2) For $\rho = 1$:

$$L_i(N) = \frac{4.6N}{D_o(1)}\left[\sqrt{1 + \frac{9.2N[D_o(1) + 2.3]}{D_o^2(1)}} - 1\right]^{-1} \quad (5)\ [7]$$

$$L_i(N) = \frac{ND_o(N)}{4.6}\left[\sqrt{1 + \frac{9.2[D_o(N) + 2.3]}{ND_o^2(N)}} - 1\right] \quad (5a)$$

$$= \frac{2[D_o(N) + 2.3]}{D_o(N)}\left[\sqrt{1 + \frac{9.2[D_o(N) + 2.3]}{ND_o^2(N)}} + 1\right]^{-1}$$

(5b) [7]

For $\rho \neq 1$, in Equations (5a) and (5b), replace N by ρN and $D_o(N)$ by $(1/\rho)D_o(\rho N)$.

Equation (5) is used when $D_o(1)$ is specified (or, equivalently, P_d and P_n are given) and $L_i(N)$ (and, in turn, possibly $D_o(N)$) is to be determined. Equation (5a) and (5b) used when $D_o(N)$ given and $L_i(N)$ (and possibly $D_o(1)$ and P_d, in turn) is to be determined.

3) Figure 17.

Equations (1), (2), (5), (5a), and (5b) are useful if a computer program computation of $L_i(N)$ is to be made; otherwise Figure 8 of Chapter 2 (or Figures 1 and 2 in conjunction with Equation (4)) is sufficiently accurate.

2. $L_c(\rho, N)$:

$L_c(\rho,N)_{dB} = L_i(\rho N)_{dB} - L_i(N)_{dB}$ (6) [1]

$L_c(\rho,N)_{dB} = C_x(\rho N)_{dB} - C_x(N)_{dB}$ (6a) [1]

3. $C_x(N)$:

$C_x(N) = [D_o(N) + 2.3]/D_o(N)$ (7) [1]

$C_x(N) = [ND_o(N)]_{dB} - [(X_o)/2]_{dB}$ (7a)

*$L_i(N)$ also given by

$L_i(N)_{dB} = C_x(N)_{dB} - C_x(1)_{dB}$ (8) [1]

For $\rho \neq 1$ in Equations (7), (7a), and (8) replace N by ρN and $D_o(N)$ by $(1/\rho)D_o(\rho N)$.

Table 6 **Determination of Integration Loss, $L_i(\rho N)$; Detector Loss, $C_x(N)$; and Collapsing Loss, $L_c(\rho, N)$.**

Problem to be Solved: Same as for Tables 1 and 14.

Procedure: For one of standard Swerling Cases 0-4, SNR/Pulse obtained directly from Meyer Plots* [5]. (Procedure described below in detail for $\rho \neq 1$.)

For other target models of Table 2 procedure identical to that of Tables 1 and 14 except in determination of $D_o(\rho N)$ and L_{fK} as now described.

1. $D_o(\rho N)$: Obtain from Meyer Plots [5]:* for Swerling Case 0, SNR/pulse required in dB for specified P_d and P_{fa} when integrating ρN signal plus noise pulses;** increase value obtained by ρ in dB to obtain $D_o(\rho N)_{dB}$.# See Figure 15 for example Meyer Plot.
2. L_{fK}: 1) First obtain L_{fK} for Swerling Cases 1 to 4 using

$$L_{fK}(dB) = D(\rho N, K)_{dB} - D(\rho N, \infty)_{dB} = D(\rho N, K)_{dB} - D_o(\rho N)_{dB} \qquad (9)$$

where K = 1, 2, N and 2N for Swerling Cases 1, 3, 2, and 4, respectively. $D(\rho N, 1)_{dB}$ for K = 1 obtained from Meyer Plots* for Swerling 1 model by finding SNR/pulse required in dB for specified P_d and P_{fa} when integrating ρN signal plus noise pulses, then adding ρ in dB.

$D(\rho N, K)$ for K = 2 is obtained in a similar manner from the Meyer Plots* for Swerling Case 3.

$D(\rho N, K)$ for K = N and 2N and $\rho = 1$ obtained directly from Meyer Plots* for Swerling Cases 2 and 4, respectively. When $\rho \neq 1$, L_{fK} evaluated using Equation (9) with $\rho = 1$ for K = N and 2N.

3. Using $L_{fK}(dB)$ for K = 1, 2, N, and 2N, plot L_{fK} vs. K. (Figure 13 shows example plot).
4. From this plot obtain $L_{fK}(dB)$ for K of equivalent chi-square model; see examples of Figure 12.

*In place of Meyer Plots one could use the curves of DiFranco and Rubin [22], Marcum [29], or Whalen [32], the Felner curves [30], tables [31], Robertson curves [42], or a computer computation of SNR/pulse for Swerling Case 0-4; see References 5, 27, 31, and 33.

**The value of the SNR/pulse so derived physically represents the SNR/pulse required if the N signal plus noise pulses had their signal energy equally distributed among ρN pulses instead of N. Symbolically, the SNR/pulse so derived is equal to $(1/\rho) D_o(\rho N)$. Because the signal in actuality is distributed among only N pulses, the SNR/pulse obtained from the Meyer Plots for the integration of ρN signal plus noise pulses has to be increased by the factor ρ. In general, the Meyer Plots give the SNR/pulse required for the integration of any arbitrary number of signals plus noise pulses, let us say N_n, with the signal energy and noise energy equally distributed among these N_n pulses. However, for a quare-law envelope detector, the performance results really only depend on the total integrated signal energy to total integrated noise energy, and do not depend on how the signal energy or noise energy is distributed among the pulses as long as the number of degrees of freedom for the signal model and for the noïse model are unchanged; see References 5 and 34 and footnote of Table B.

#Plot of P_d vs. $D_o(1)$ given in Figure 6 of Chapter 2 by Barton. A more detailed plot (Robertson curve) is given in Figure 16.

Table 7 **Find SNR/Pulse Given P_d; More Precise Procedure**

METHOD OF CALCULATION	PARAMETER	WHERE PARAMETER DERIVED FROM	NONFLUCTUATING TARGETS (MARCUM)		SWERLING FLUCTUATION TARGETS									RICE TARGET s = 1	WEINSTOCK k = 0.6	LOGNORMAL TARGET		
																ρ = 1(8)	ρ = 2	ρ = 2
	EXAMPLE NO.		1	2	3(1)	4(2)	5(3)	6	7(1)	8(2)	9	10(3)	11(7)	12	13	14	15	16
	CASE		0	0	1	3	2	4	1	1	1	2	2	1	1	1	1	1
	F		—	—	1	1	30	30	1	2	5	30	30	1	1	1	1	3
	SCANNING ?		NO	NO	NO	NO	NO	NO	NO	NO	NO	YES	YES	NO	NO	NO	NO	NO
GIVEN	CFAR ? M ?	GIVEN	NO	NO	NO	NO	NO	NO	NO	NO	NO	YES, 10	YES, 10	NO	NO	NO	NO	NO
	TYPE INTEGRATION		—	ANALOG	ANALOG	ANALOG	ANALOG	ANALOG	ANALOG	ANALOG	ANALOG	ANALOG	BINARY	ANALOG	ANALOG	ANALOG	ANALOG	ANALOG
	N		1	30	30	30	30	30	15	30	30	30	30	30	30	30	30	30
	ρ		1	1	1	1	1	1	2	1	1	1	1	1	1	1	1	1
	K	Fk; TABLE 2	∞	∞	1	2	30	60	1	2	5	30	30	1.33	0.6	∞	1.17	3.51
	X_0(dB)	FIG. 1 OR EQ. 1	16.85	16.85	16.85	16.85	16.85	16.85	16.85	16.85	16.85	16.85	16.85	16.85	16.85	16.85	16.85	16.85
	$N\rho$ (dB)	$N_{dB} + \rho_{dB}$	- 0.0	-14.77	-14.77	-14.77	-14.77	-14.77	-14.77	-14.77	-14.77	-14.77	-14.77	-14.77	-14.77	-14.77	-14.77	-14.77
	$X_0/\rho N$ (dB)	$X_0(dB) - (\rho N)$ dB	16.85	2.08	2.08	2.08	2.08	2.08	2.08	2.08	2.08	2.08	2.08	2.08	2.08	2.08	2.08	2.08
	$\frac{1}{\rho} D_0(\rho N)$(dB)	FIG. 2 OR EQ. 2a	14.2	2.60	2.60	2.60	2.60	2.60	2.60	2.60	2.60	2.60	2.60	2.60	2.60	2.60	2.60	2.60
	$D_0(\rho N)$(dB)(9)	$\frac{1}{\rho} D_0(\rho N)_{dB} + \rho_{dB}$	14.2	2.60	2.60	2.60	2.60	2.60	5.60	2.60	2.60	2.60	2.60	2.60	2.60	2.60	2.60	2.60
TABLE 1 APPROXIMATE PROCEDURE	L_{fi}(dB)/K	L_{f1} FROM FIG. 3	+ 0	+ 0	+ 8.2	+ 4.1	+ 0.27	+ 0.14	+ 8.2	+ 4.1	+ 1.64	+ 0.27	+ 0.27	+ 6.15	+13.7	+ 0	+ 7.0	+ 2.3
	$D(\rho N,K)$(dB)	$D_0(\rho N)_{dB} + (L_{f1}/K)_{dB}$	14.2	2.60	10.8	6.7	2.87	2.74	13.8	6.7	4.24	2.87	2.87	8.75	16.3	2.60	9.1(5)	4.4(5)
	L_p (dB)	TABLE 3	+ —	+ —	+ —	+ —	+ —	+ —	+ —	+ —	+ —	+ 1.60	+ 1.60	+ —	+ —	+ —	+ —	+ —
	L_{CFAR} (dB)	TABLE 4	+ —	+ —	+ —	+ —	+ —	+ —	+ —	+ —	+ —	+ 0.40	+ 0.8(6)	+ —	+ —	+ —	+ —	+ —
	L_B (dB)	TABLE 5	+ —	+ —	+ —	+ —	+ —	+ —	+ —	+ —	+ —	+ —	+ 1.7	+ —	+ —	+ —	+ —	+ —
	SNR/PULSE (dB)	$D(\rho N,K) + L_p + L_{CFAR} + L_B$	14.2	2.60	10.8	6.7	2.87	2.74	13.8	6.7	4.24	4.9	7.0	8.75	16.3	2.60	9.1	4.4
TABLE 7 MORE PRECISE PROCEDURE	SNR/PULSE (dB)	TABLE 7 MORE PRECISE PROCEDURE	14.3	2.78	11.2	7.3	3.16	3.0	14.2	7.3	4.75	5.2	7.3	9.2	16.0	2.78	11.0(7)	5.0
	ERROR (dB)	(TABLE 1)-(TABLE 7) VALUES FOR SNR/PULSE	- 0.1	- 0.2	- 0.4	- 0.6	- 0.3	- 0.3	- 0.4	- 0.6	- 0.5	- 0.3	- 0.3	- 0.5	+ 0.3	- 0.2	- 1.9(8)	- 0.6
NEUVY'S EQS. IN TABLE 9	SNR/PULSE (dB)	TABLE 9 EQ. (10)	15.1	2.15	11.6	7.6	2.15	2.15	14.6	—	—	4.2	6.3	—	—	—	—	—
	ERROR (dB)	(TABLE 9)-(TABLE 7) VALUES FOR SNR/PULSE	0.8	- 0.6	+ 0.4	+ 0.3	- 1.0	- 0.85	+ 0.4	+0.3	—	-1.0	-1.0	—	—	—	—	—

(1) FOR BOTH EXAMPLES 3 AND 7, 30 PULSES ARE INTEGRATED EXCEPT THAT FOR EXAMPLE 7 THE SIGNAL IS ONLY PRESENT IN 15 OF THESE PULSES ($N = 15$ AND $\rho = 2$). OTHERWISE, THESE TWO EXAMPLES ARE IDENTICAL. AS A RESULT, THE SNR/PULSE FOR EXAMPLE 7 IS 3 dB LARGER THAN FOR EXAMPLE 3.

(2) NOTE THAT EXAMPLES 4 AND 8 HAVE IDENTICAL PERFORMANCE, THIS FOLLOWS FROM THE FACT THAT THE TARGET MODELS ARE EXACTLY THE SAME, BOTH HAVING TARGET MODELS CORRESPONDING TO A CHI-SQUARE DISTRIBUTION WITH K = 2 DUO-DEGREES OF FREEDOM; FOR EXAMPLE 4 THE TARGET IS THE SWERLING CASE 3 MODEL; FOR EXAMPLE 8 THE TARGET IS A SWERLING CASE 1 BUT WITH A FREQUENCY DIVERSITY OF 2 (F = 2) TO GIVE THE SAME DEGREES OF FREEDOM.

(3) EXAMPLES 10 AND 11 ARE IDENTICAL TO EXAMPLE 5 EXCEPT THAT CFAR PROCESSING AND BEAM SCANNING ARE INCORPORATED IN EXAMPLE 10 AND IN ADDITION BINARY INTEGRATION IS INCORPORATED IN EXAMPLE 11. WEIGHTED VIDEO INTEGRATION IS ASSUMED SO AS TO YIELD A 1.6 dB BEAM SHAPE LOSS FOR EXAMPLES 10 AND 11; SEE TABLE 3.

(4) THE LOG-NORMAL TARGET MODEL FOR $\rho = 1$ IS IDENTICAL TO THE NONFLUCTUATING TARGET MODEL HENCE EXAMPLES 14 AND 2 ARE IDENTICAL.

(5) CORRECTION FACTOR FOR LOG-NORMAL VARIATE ΔD_0 OF FIG. 5 (EQUALS $-0.48 \doteq -0.5$ dB FOR $k = 1.17$ FOR EXAMPLES 15 AND 16) ADDED IN CALCULATION OF D(N,K).

(6) CFAR APPLED TO THE SINGLE PULSE DETECTION, I.E., TO THE FIRST THRESHOLD. THUS N = 1 CURVE OF FIG. 8 MUST BE USED WITH FALSE ALARM VALUE FOR FIRST THRESHOLD USED FOR P_{fa} WHICH IS $0.273 = 10^{-1.56}$ SO THAT x_1 OF FIG. 8 EQUALS 1.56. (TO CALCULATE P_{fa1} USE IS MADE OF APPROXIMATION $P_{fa_2} \doteq C_T^N (P_{fa_1})^T$ WITH T = 10 FOR SECOND THRESHOLD, A DETECTION OF T OUT OF N PULSES NEEDED FOR SINGLE-SCAN DETECTION WITH THE BINARY INTEGRATOR. C_T^N IS THE BINOMIAL COEFFICIENT N!/N-T)!T!)

(7) SNR/PULSE VALUE OBTAINED FROM INTERPOLATION OF HEIDBREDER AND MITCHELL DETECTION PROBABILITY CURVES FOR LOG-NORMAL DISTRIBUTED SIGNALS [18].

(8) ERROR BECOMES -1.4dB IF MORE PRECISE PROCEDURE OF TABLE 7 USED TO DETERMINE SNR/PULSE.

(9) AN ALTERNATE METHOD FOR CALCULATING $D_0(\rho N)$ IS TO USE FIGURE 16 TO CALCULATE $D_0(1)$, FIGURE 17 TO CALCULATE $L_i(\rho N)$ AND THEN EQUATION (4) OF TABLE 6.

Table 8 **Example of Calculations – Find SNR/Pulse Given Single-Scan Probability of Detection, P_d; P_d = 90% and P_{fa} = 0.693×10^{-8} ($n' = 10^8$)**

NEUVY'S EQUATION FOR DETERMINATION OF SNR/PULSE GIVEN P_d:

$$D(N,K) = \frac{\alpha}{N^{2/3}} \frac{LOG_{10} n}{(LOG_{10} 1/P_d)^{\beta}} \qquad (10)^{*\#} \quad [6]$$

WHERE α AND β GIVEN BY FOLLOWING TABLE FOR DIFFERENT SWERLING AND CHI-SQUARE MODELS:

SWERLING CASE	EQUIVALENT CHI-SQUARE MODEL K	α	β
1**	1	$\frac{2}{3}(1+\frac{2}{3} e^{-N/3})$	1
2≠	N	1	$\frac{1}{6}+e^{-N/3}$
3**	2	$\frac{3}{4}(1+\frac{2}{3} e^{-N/3})$	$\frac{2}{3}$
4	2N	1	$\frac{1}{6}+\frac{2}{3}e^{-N/3}$
0	∞	$1+2e^{-N/3}$	$\frac{1}{6}$

INVERSE OF NEUVY'S EQUATION FOR REVERSE PROBLEM : DETERMINATION OF P_d GIVEN SNR/PULSE:

$$P_d = 10^{-\left[\frac{\alpha}{N^{2/3}} \frac{LOG_{10} n}{D(N,K)}\right]^{1/\beta}} \qquad (11)^{*,\#}$$

---

* IF THERE IS A COLLAPSING LOSS PRESENT, I.E., $\rho \neq 1$, THEN FOR SWERLING CASES 0,1, AND 3, N IS REPLACED BY ρN AND D(N,K) BY $\frac{1}{\rho}$ D(ρN,K) IN EQS. (10) AND (11).

\# $n' = n/N$.

** F = 1, I.E., THE CLASSIC DEFINITION FOR THE SWERLING CASE 1 AND 3 TARGETS MODELS IS BEING USED; SEE TABLE 2.

≠ FOR N = 1 USE SWERLING CASE 1 PARAMETERS FOR α AND β FOR BETTER ACCURACY.

NOTES: (1) EQ. (10) CAN BE USED TO DERIVE D(N,K) FOR CHI-SQUARE TARGET MODEL WITH ARBITRARY DUO-DEGREES OF FREEDOM K. THIS IS DONE BY USING EQUATION (10) TO FIND $D(N,\infty) = D_0(N)$, D(N,1) AND L_{f1} [EQUATION (9) OF FIG. 12 USED TO OBTAIN L_{f1}]; THEN $D(N,K)_{dB} = D_0(N)_{dB} + L_{fK}(dB)$ WHERE $L_{fK}(dB) = L_{f1}(dB)/K$.

(2) EQ. (10) USEFUL FOR OBTAINING INITIAL ESTIMATE OF D(N,K) IN MORE PRECISE COMPUTOR COMPUTATION PROCEDURE OF [33].

Table 9 Neuvy's Equation

TARGET MODEL	METHOD	ERROR (dB)
NONFLUCTUATING	APPROXIMATE (TABLE 1)	$\leq \pm 0.5$ FOR $0.05 \leq P_d \leq 0.99$; SEE [1].
	MORE PRECISE (TABLE 7)	$\leq \pm 0.1$*
	NEUVY'S EQS. (TABLE 9)	$\lesssim +1.4$ FOR $0.5 \leq P_d \leq 0.99$; SEE[6].
SWERLING CASES 1,2,3 AND 4	APPROXIMATE (TABLE 1)	$\leq \pm 1$ FOR $0.05 \leq P_d \leq 0.95$ SEE [1].
	MORE PRECISE (TABLE 7)	$\leq \pm 0.1$*
	NEUVY'S EQS. (TABLE 9)	$\lesssim \pm 1$ FOR $0.1 \leq P_d \leq 0.9$#, $1 \leq N \leq 1000$; SEE [6].
SWERLING CASE 1 AND 2 FOR 1<F<N, i.e., CHI-SQUARE FOR $1 < k < 2N$	APPROXIMATE (TABLE 1)	$\leq \pm 1$ FOR $0.05 \leq P_d \leq 0.95$
	MORE PRECISE (TABLE 7)	$\lesssim \pm 0.1$ FOR $0.05 \leq P_d \leq 0.99$
RICE	APPROXIMATE (TABLE 1)	$\leq \pm 2$
	MORE PRECISE (TABLE 7)	$\leq \pm 1$; SEE [3] AND [5].
LOG-NORMAL FOR $1 \leq \rho \leq 2$	APPROXIMATE (TABLE 1)	$\lesssim 3.3 \ln \rho + 1$ — EX. 1) ρ=1.2, ERROR=1.6dB; EX. 2) ρ=2, ERROR=3.3dB
	MORE PRECISE (TABLE 7)	$\lesssim 3.3 \ln \rho + 0.1$≠ — EX. 1) ρ=1.2, ERROR=0.7 dB; EX. 2) ρ=2.0, ERROR=2.4dB

*ERROR DUE TO READING CURVES, e.g., MEYER PLOTS.
#FOR SWERLING CASE 1 TARGET MODEL ACCURACY STATED ACHIEVED FOR $0.01 \leq P_d \leq .99$.
≠FOR F≠1 THE ERROR IS A FUNCTION OF P_d. IT OSCILLATES FROM ABOUT $+3.3 \ln \rho$ dB AT P_d=99% TO ABOUT $-3.3 \ln \rho$ AT P_d=60% AND BACK TO ABOUT $+3.3 \ln \rho$ dB AT P_d =1%. AT ABOUT P_d =95% AND 12% THE ERROR GOES THROUGH ZERO. FOR F = 1 THE PEAK ERROR IS GENERALLY SMALLER.

Table 10 Accuracy of $D(\rho N, k)$ Determination

Problem to be Solved:

Given: SNR/pulse, P_{fa}, N, ρ, Target Model, Type of Processing (CFAR; Analog, Binary, or Quaternary Integration), Amount of Frequency Diversity, Type of Scanning.

Find: P_d

1. Nonfluctuating Target

Approximate Procedures

1. The procedure outlined in Table 1 can be used in reverse with the + and - signs reversed and the direction of flow reversed for the bold line path; see Table 12. (This procedure cannot be used directly for the fluctuating target models because one needs to know P_d to specify L_{fK} of Table 1. For the same reason there are some restrictions on the situations that can be handled with L_p and L_B losses. For example, it is necessary that $N > 3$ in order to handle scanning antennas with beam shape losses. L_B given by Figures 10 and 11 in general can be considered independent of P_d (to within ±0.3 dB accuracy worst case) for $0.1 \leqslant P_d \leqslant 0.9$. Similarly L_{CFAR} given by Figures 8 and 9 generally can be considered independent of P_d for $0.1 \leqslant P_d \leqslant 0.99$).
2. Neuvy's Equation 11 of Table 9. (Not as accurate as above procedure but simpler.)

More Precise Procedure

(i) Determine $(1/\rho)D_o(\rho N)$ using procedure of Table 12.
(ii) Direct lookup of P_d using Meyer Plots or any equivalent curves (Marcum [29], DiFranco and Rubin [22], Felner [30], or Whalen [32]) with $(1/\rho)D_o(\rho N)$ used for SNR/pulse and ρN used for number of pulses integrated.

2. Fluctuating Targets

More Precise Procedures

$\rho = 1$

(i) Find D(N,K) using procedure of Table 12 to find $D_o(N)$*.
(ii) Find equivalent chi-square model (Table 2). If it corresponds to Swerling Case 1, 2, 3, or 4 (i.e., if K = 1, 2, N, or 2N, respectively), use Meyer Plots directly; if not, use steps (iii) and (iv).
(iii) For SNR/pulse = D(N,K) and given P_{fa}, using Meyer Plots or equivalent curves, find P_d for Swerling Cases 1-4 and plot P_d vs. K on probability paper having two or three log cycles for the abscissa, as shown in Figure 13.
(iv) Determine P_d for equivalent chi-square model from curve, as illustrated in Figure 13 for Case 1** Rice target model, Swerling Case 1 target model for F = 5, Case 1** log normal target model, and Case 1** Weinstock target model. Tabulated results given in Table 13.

$\rho \neq 1$

(i) Find $(1/\rho)D(\rho N,K)$ using procedure of Table 12 to find $(1/\rho)D_o(\rho N)$*.
(ii) Find equivalent chi-square model (Table 2). If it corresponds to Swerling Case 1, 2, 3, or 4 (i.e., if K = 1, 2, ρN or 2ρN, respectively), use Meyer Plots directly. If not, use precise method of Table 7 to obtain plots such as those of Figure 12 for several P_d's for the ρ and P_{fa} specified. Use these plots to obtain curve of P_d versus $D(\rho N,K)$ for K.

Approximate Procedures

1. Neuvy's Equation (11) of Table 9. (Limited to Swerling Cases 1, 2, 3, and 4).
2. Could use Table 1 to plot P_d vs. SNR/pulse.
3. Table 12 can be modified to calculate P_d for fluctuating target by iterative procedure. This approach works best for small L_{fK} (i.e., K is large).

*For log-normal target model, ΔD_o correction must be made in determining D(N, K) or $(1/\rho)D(\rho N,K)$; see Table 12.
**See Table 2 footnote "*".

Table 11 The Reverse Problem – Find P_d Given SNR/Pulse

PROBLEM STATEMENT:

GIVEN: SNR/PULSE, P_{fa}, N, ρ, TYPE AT PROCESSING (CFAR; ANALOG, BINARY, OR QUATERNARY INTEGRATION), TYPE OF SCANNING.

FIND: P_d

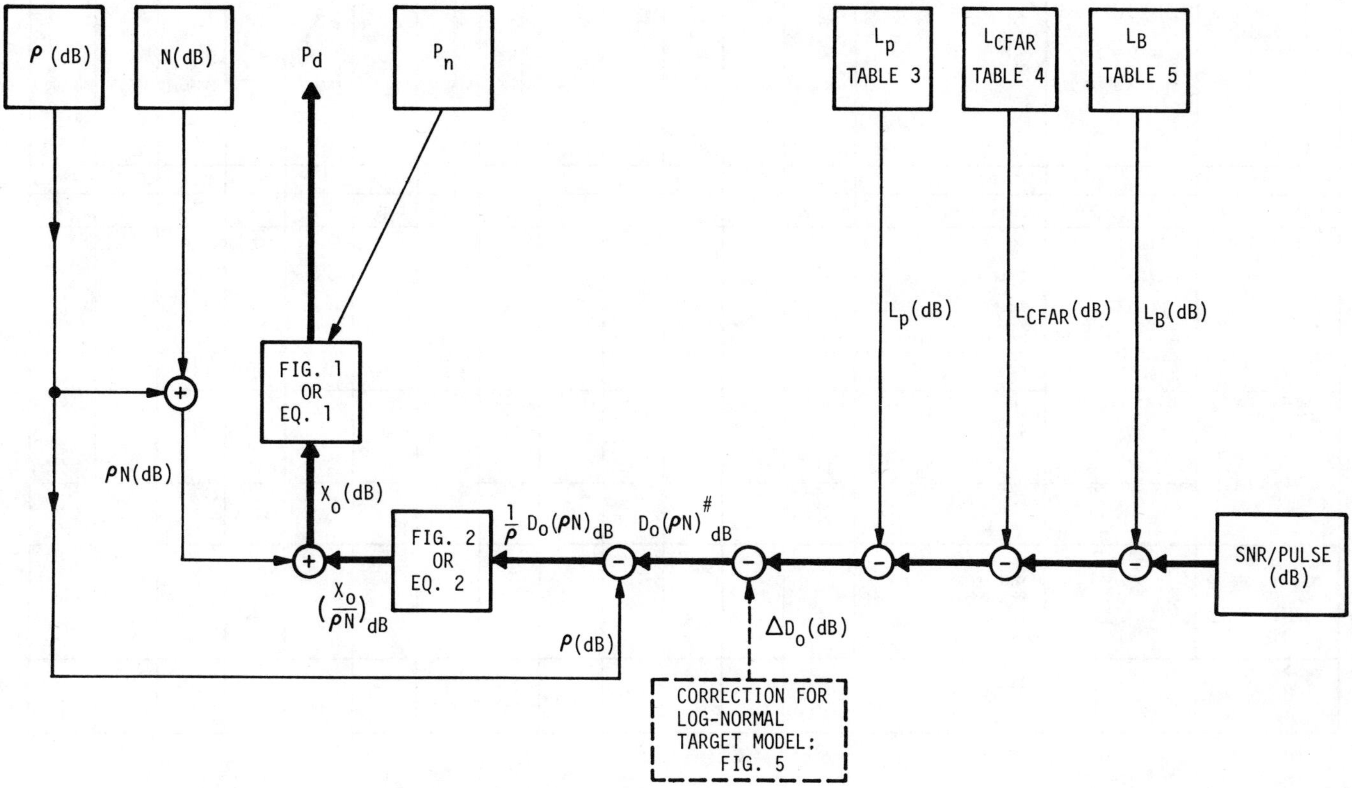

\# BECOMES $D(\rho N,K)_{dB}$ FOR FLUCTUATING TARGET MODEL.

Table 12 **Flow Diagram for Reverse Problem – Find P_d Given SNR/Pulse; Approximate Procedure for Nonfluctuating Target Model**

METHOD OF CALCULATION	PARAMETER	WHERE PARAMETER DERIVED FROM	NONFLUCTUATING TARGETS (MARCUM)		SWERLING FLUCTUATING TARGETS								RICE TARGET s = 1	WEIN-STOCK TARGET k = 0.6	LOG-NORMAL TARGET ρ = 2
	EXAMPLE NO. (1)		1	2	3	4	5	6	7	9	10	11	12'	13'	16'
↑	CASE	↑	0	0	1	3	2	4	1	1	2	2	1	1	1
	F		1	1	1	1	30	30	1	5	30	30	1	1	3
	SCANNING ?		NO	NO	NO	NO	NO	NO	NO	NO	YES	YES	NO	NO	NO
GIVEN	CFAR ?, M ?	GIVEN	NO	NO	NO	NO	NO	NO	NO	NO	YES, 10	YES, 10	NO	NO	NO
	TYPE INTEGRATION		ANALOG	ANALOG	ANALOG	ANALOG	ANALOG	ANALOG	ANALOG	ANALOG	ANALOG	BINARY	ANALOG	ANALOG	ANALOG
	N		1	30	30	30	30	30	15	30	30	30	30	30	30
↓	ρ	↓	1	1	1	1	1	1	2	1	1	1	1	1	1
	K	TABLE 2	∞	∞	1	2	30	60	1	30	30	30	1.33	0.6	3.51
GIVEN	SNR/PULSE (dB)	GIVEN	14.2	2.78	11.2	7.3	3.2	3.0	14.2	4.75	5.2	7.3	4.75	4.75	4.27
TABLE 12 CALCULATION OF $D_o(\rho N)$	L_B (dB)	TABLE 5										-1.7			
	L_{CFAR}(dB)	TABLE 4									-0.40	-0.8			
	L_p (dB)	TABLE 3									-1.60	-1.6			
	$D_o(\rho N)$ (2) (dB)	SNR/PULSE-L_B-L_{CF}-L_p (3)	14.2	2.78	11.2	7.3	3.2	3.0	14.2	4.75	3.2	3.2	4.75	4.75	4.75 (5)
	$\frac{1}{\rho} D_o(\rho N)$(dB)	$D_o(\rho N)$(dB)- ρ(dB)	14.2	2.78	11.2	7.3	3.2	3.0	11.2	4.75	3.2	3.2	4.75	4.75	4.75
TABLE 11 AND 12 APPROXIMATE PROCEDURE	$X_o/\rho N$ (dB)	FIG. 2 OR EQ. 2	16.8	2.34											
	$N\rho$ (dB)	$N_{dB}+\rho_{dB}$	0	+14.77											
	X_o (dB)	$(X_o/\rho N)_{dB}+(\rho N)_{dB}$	16.8	17.1											
	P_d (%)	FIG. 1 OR EQ. (1)	89%	93%											
TABLE 11: MORE PRECISE PROCEDURE	P_d (%)	TABLE 11: MORE PRECISE PROCEDURE	90%	90%	90%	90%	90%	90%	90%	90% (4)	90	90	67 (4)	52 (4)	85 (4)
INVERSE OF NEUVY'S EQ. (TABLE 9)	P_d(%)	EQUATION (11) OF TABLE 9.	69%	95.7%	89.2%	89.0%	97.6%	96.8%	89.2%		97.6	97.6			

(1) EXAMPLES WHICH ARE EQUIVALENT TO THOSE OF TABLE 8 ARE GIVEN THE SAME NUMBER. THOSE WHICH ARE SIMILAR BUT NOT EXACTLY THE SAME ARE INDICATED BY THE SAME NUMBER PRIMED.

(2) $D_o(\rho N)$ BECOMES $D(\rho N,K)$ FOR FLUCTUATING TARGET MODELS; SEE TABLE 12 FOOTNOTE.

(3) ALL PARAMETERS ARE IN dB.

(4) SEE FIG. 13.

(5) $D(\rho N,K)$ 0.48dB LARGER THAN SNR/PULSE BECAUSE OF ΔD_o = -0.48dB CORRECTION FOR LOG-NORMAL VARIATE GIVEN IN FIG. 5; SEE TABLE 12.

Table 13 **Example Calculations — Find P_d Given SNR/Pulse; $P_{fa} = 0.693 \times 10^{-8}$ ($n' = 10^8$)**

GIVEN: P_d = ______ ; P_{fa} = ______ (n' = ______)

SNR/PULSE = ?

METHOD OF CALCULATION	PARAMETER	WHERE PARAMETER DERIVED FROM	LOGNORMAL TARGET ρ=2					
	EXAMPLE NO.		Ill. EX.*					
GIVEN	CASE	GIVEN	1					
	F		3					
	SCANNING ?		YES					
	CFAR ? M ?		YES, 10					
	TYPE INTEGRATION		ANALOG					
	N		30					
	ρ		1					
	k	TABLE 2	1.17					
	K	Fk	3.51					
TABLE 1 APPROXIMATE PROCEDURE	L_{f1}(dB)	FIG. 3	8.2					
	X_o(dB)	FIG. 1 OR EQ. 1	16.85					
	-Nρ (dB)		-14.77	-	-	-	-	-
	$X_o/\rho N$ (dB)	X_o(dB)-(ρN) dB	2.08					
	$\frac{1}{\rho}D_o(\rho N)$(dB)	FIG. 2 OR EQ. 2a	2.60					
	$D_o(\rho N)$(dB) **	$\frac{1}{\rho}D_o(\rho N)_{dB}+\rho_{dB}$	2.60					
	ΔD_o(dB)	FIG. 5	+- 0.48	+	+	+	+	+
	L_{fK} (dB)	L_{f1}(dB)/K	+ 2.34	+	+	+	+	+
	D(ρN,K)(dB)	$D_o(\rho N)_{dB}+(L_{f1}/K)_{dB}$ $+\Delta D_o$(dB)	4.5					
	L_p (dB)	TABLE 3	+ 1.6	+	+	+	+	+
	L_{CFAR} (dB)	TABLE 4	+ 0.5	+	+	+	+	+
	L_B (dB)	TABLE 5	+ 0.0	+	+	+	+	+
	SNR/PULSE (dB)	D(ρN,K)+L_p+L_{CFAR}+L_B	6.6					
TABLE 7 MORE PRECISE PROCEDURE	SNR/PULSE (dB)	TABLE 7 MORE PRECISE PROCEDURE	7.1					
	ERROR (dB)	(TABLE 1)-(TABLE 7) VALUES FOR SNR/PULSE	- 0.5					
NEUVY'S EQS. IN TABLE 9	SNR/PULSE (dB)	TABLE 9 EQ. (10)	-					
	ERROR (dB)	(TABLE 9)-(TABLE 7) VALUES FOR SNR/PULSE	-					

* P_d = 90% AND P_{fa} = 0.693x10^{-8} (n'=10^8)

SIMILAR TO EXAMPLE NO. 16 OF TABLE 8.

**AN ALTERNATE METHOD FOR CALCULATING $D_o(\rho N)$ IS TO USE FIGURE 16 TO CALCULATE D_o(1), FIGURE 17 TO CALCULATE $L_i(\rho N)$ AND THEN EQUATION (4) OF TABLE 6.

Table 14 **SNR/Pulse Calculation Worksheet**

GIVEN: SNR/PULSE = ______ ; P_{fa} = ______ (n' = ______)

P_d = ?

METHOD OF CALCULATION	PARAMETER	WHERE PARAMETER DERIVED FROM	NONFLUCTUATING TARGET					
	EXAMPLE NO.		Ill. EX.*					
GIVEN	CASE	GIVEN	0					
	F		1					
	SCANNING ?		YES					
	CFAR ?, M ?		YES, 10					
	TYPE INTEGRATION		ANALOG					
	N		30					
	ρ		1					
	k	TABLE 2	∞					
	K	Fk	∞					
GIVEN	SNR/PULSE (dB)	GIVEN	4.8					
TABLE 12 CALCULATION OF $D_0(\rho N)$	$-L_B$ (dB)	TABLE 5	- 0.0	-	-	-	-	-
	$-L_{CFAR}$(dB)	TABLE 4	- 0.40	-	-	-	-	-
	$-L_p$ (dB)	TABLE 3	- 1.60	-	-	-	-	-
	$-\Delta D_0$(dB)	TABLE 5	- 0.0	-	-	-	-	-
	$D_0(\rho N)$(dB)	SNR/PULSE-L_B-L_{CF}-L_p-ΔD_0	2.8					
	$\frac{1}{\rho}D_0(\rho N)$(dB)	$D_0(\rho N)$(dB)- ρ(dB)	2.8					
TABLE 11 AND 12 APPROXIMATE PROCEDURE	$X_0/\rho N$ (dB)	FIG. 2 OR EQ. 2	2.37					
	ρN(dB)	$N_{dB}+\rho_{dB}$	14.77					
	X_0 (dB)	$(X_0/\rho N)_{dB}+(\rho N)_{dB}$	17.14					
	P_d (%)	FIG. 1 OR EQ. (1)	94%					
TABLE 11: MORE PRECISE PROCEDURE	P_d (%)	TABLE 11: MORE PRECISE PROCEDURE	90%					
INVERSE OF NEUVY'S EQ. (TABLE 9)	P_d (%)	EQUATION (11) OF TABLE 9.	95.7%					

* $P_{fa} = 0.693 \times 10^{-8}$ ($n' = 10^8$)

SIMILAR TO EXAMPLE NO. 2 OF TABLE 13.

Table 15 **Probability of Detection Worksheet**

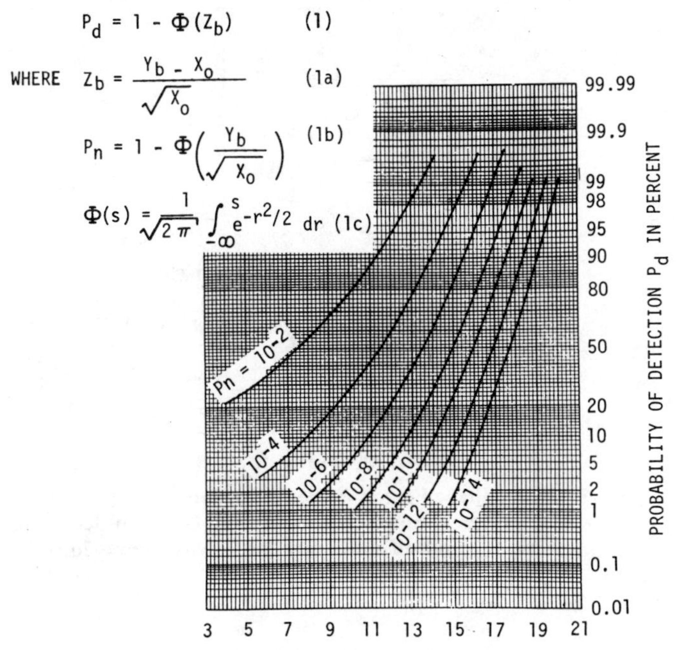

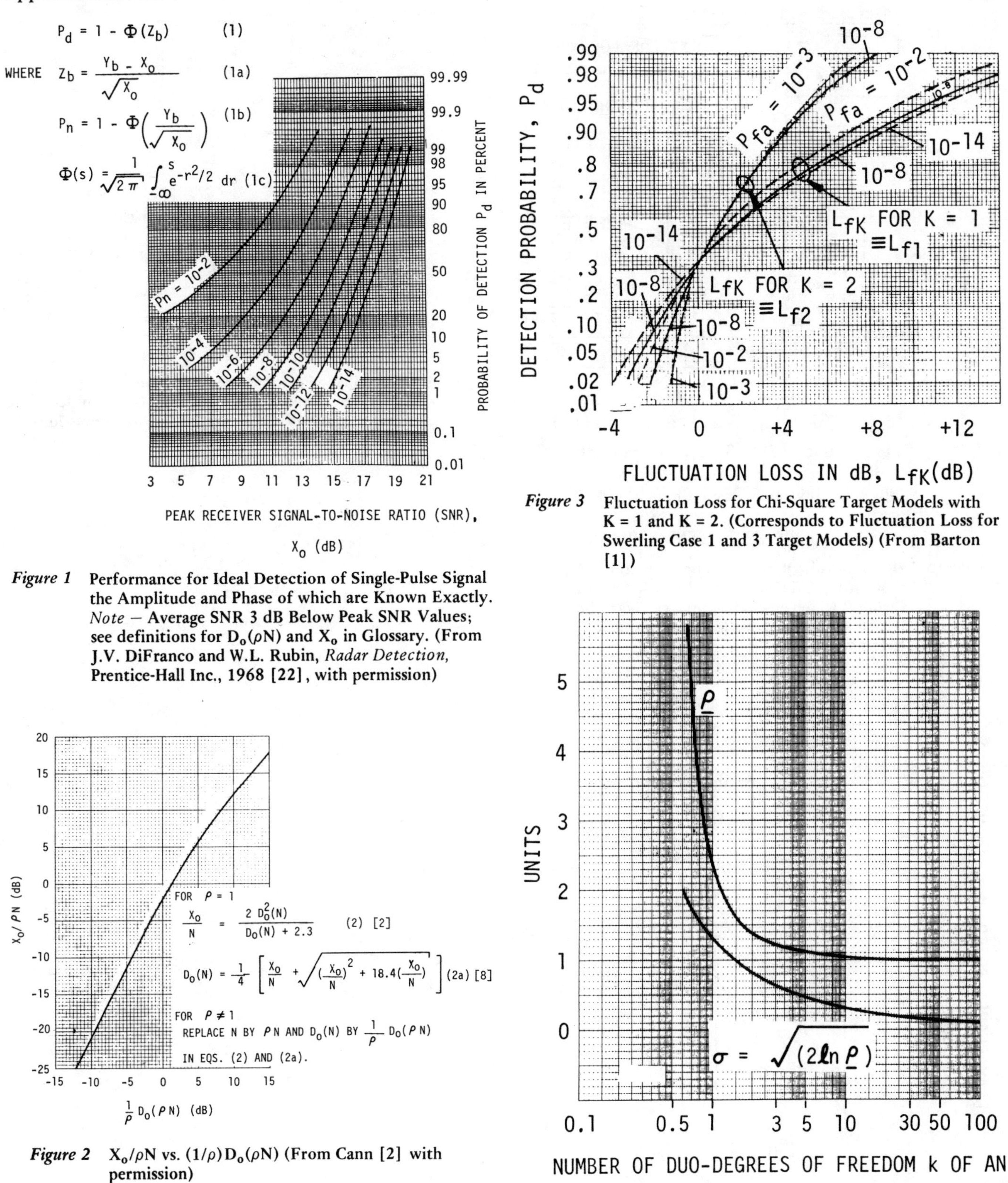

Figure 1 Performance for Ideal Detection of Single-Pulse Signal the Amplitude and Phase of which are Known Exactly. *Note* – Average SNR 3 dB Below Peak SNR Values; see definitions for $D_o(\rho N)$ and X_o in Glossary. (From J.V. DiFranco and W.L. Rubin, *Radar Detection,* Prentice-Hall Inc., 1968 [22], with permission)

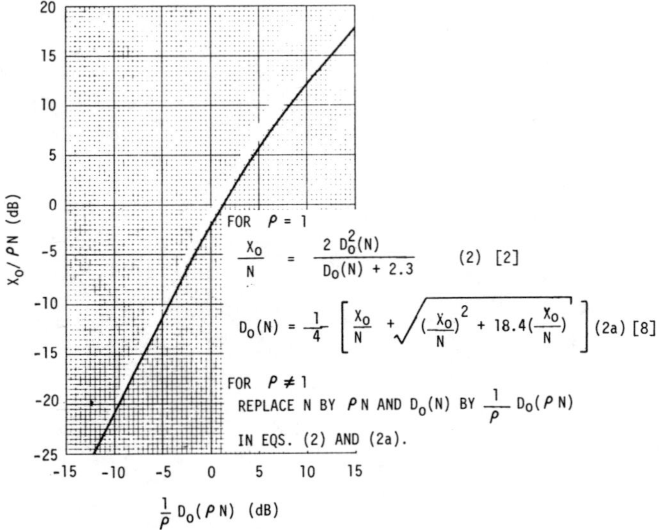

Figure 2 $X_o/\rho N$ vs. $(1/\rho) D_o(\rho N)$ (From Cann [2] with permission)

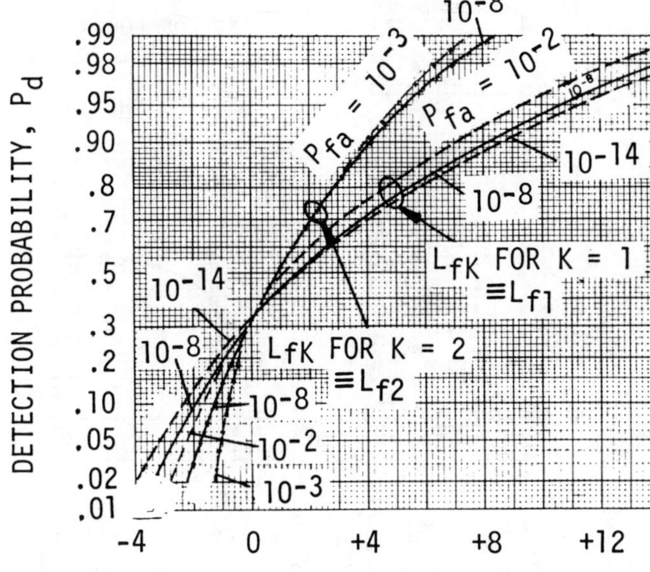

Figure 3 Fluctuation Loss for Chi-Square Target Models with K = 1 and K = 2. (Corresponds to Fluctuation Loss for Swerling Case 1 and 3 Target Models) (From Barton [1])

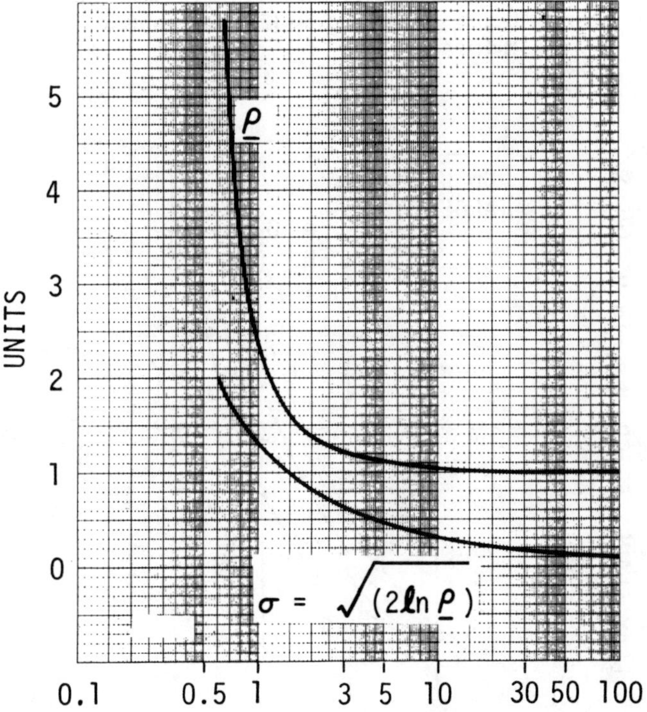

Figure 4 Value of k for Best Fitting Chi-Square Target Model to Log-Normal Model Having Mean to Median Ratio, $\underline{\rho}$. (From D.P. Meyer and H.A. Mayer, *Radar Target Detection,* Academic Press [5]; see also Reference 3)

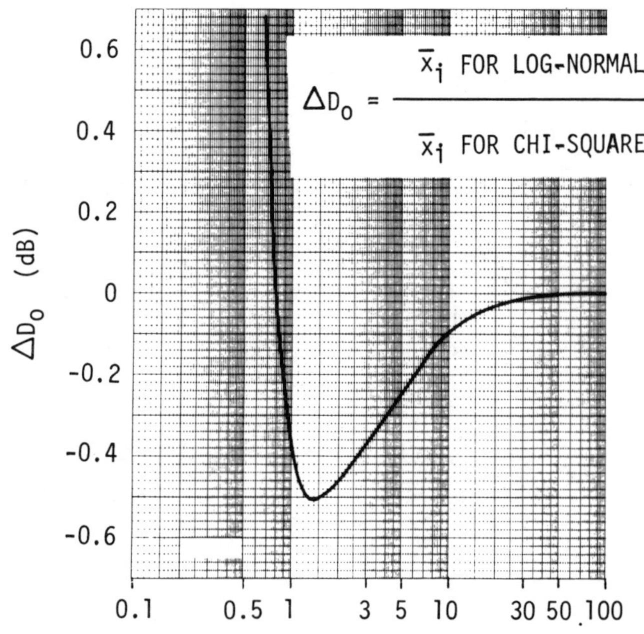

Figure 5 Signal-to-Noise Ratio per Pulse Correction ΔD_0 for Best Fitting Chi-Square Model to Log-Normal Model (From D.P. Meyer and H.A. Mayer, *Radar Target Detection,* Academic Press [5] with permission; see also Reference 3)

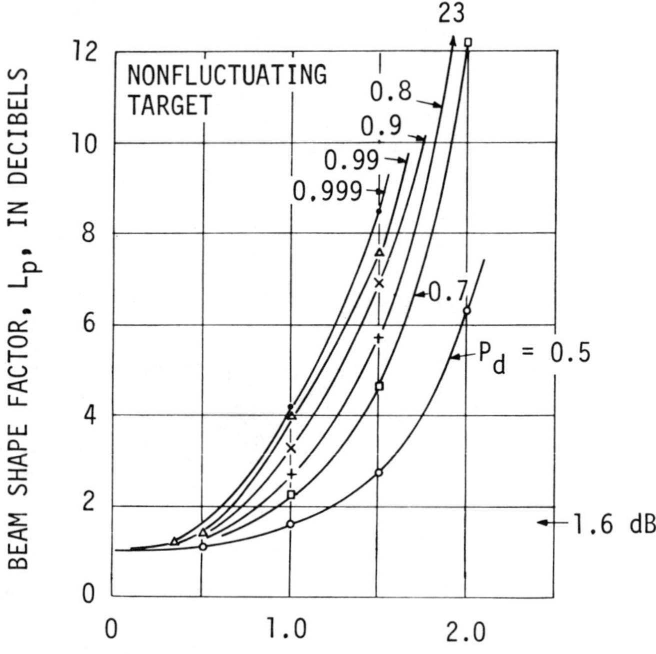

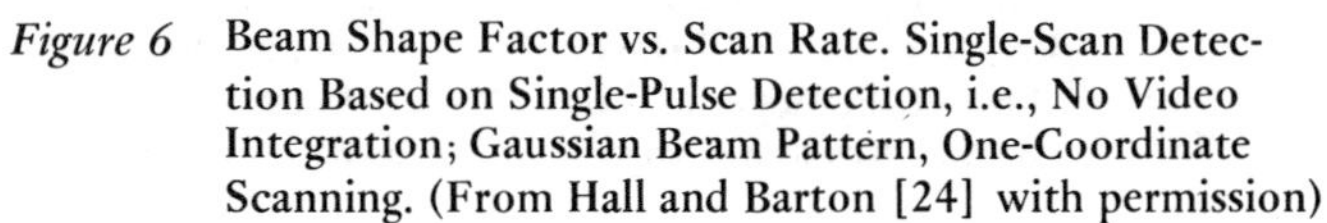

Figure 6 Beam Shape Factor vs. Scan Rate. Single-Scan Detection Based on Single-Pulse Detection, i.e., No Video Integration; Gaussian Beam Pattern, One-Coordinate Scanning. (From Hall and Barton [24] with permission)

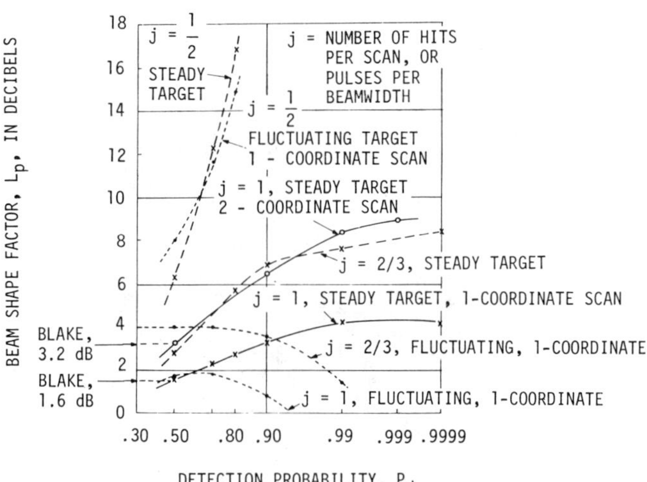

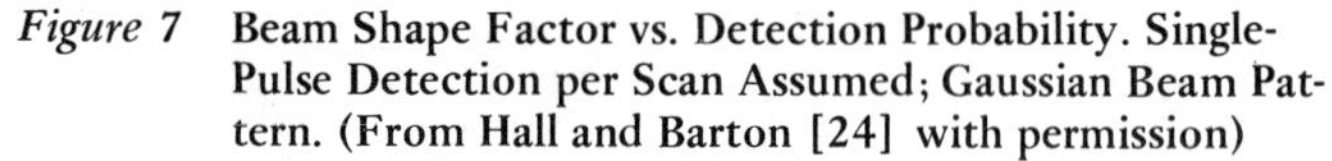

Figure 7 Beam Shape Factor vs. Detection Probability. Single-Pulse Detection per Scan Assumed; Gaussian Beam Pattern. (From Hall and Barton [24] with permission)

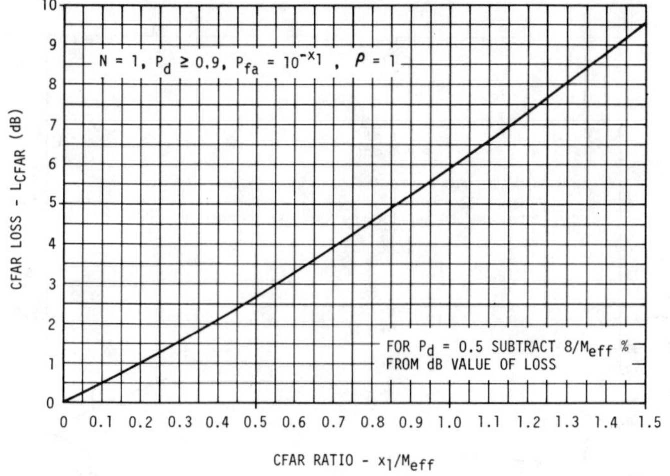

ASSUMPTIONS:

SINGLE-HIT DETECTION; I.E., N = 1 AND $\rho = 1$

NONFLUCTUATING AND SWERLING CASE 1 OR 2 TARGET MODEL

M_{eff} IS THE EFFECTIVE NUMBER OF REFERENCE OBSERVATIONS, i.e.;

M_{eff} = M FOR CELL-AVERAGING CFAR USING SQUARE-LAW DETECTOR

$M_{eff} = \frac{M+h}{1+h}$ WHERE h = 0.09 FOR CELL-AVERAGING CFAR USING LINEAR ENVELOPE DETECTOR

h = 0.65 FOR LOG CELL-AVERAGING CFAR

$M_{eff} \doteq \frac{B}{b} - 1$ FOR DICKE-FIX BUT ADD 1 dB LIMITING LOSS TO RESULT

$M_{eff} \doteq \frac{T}{\tau} - 1$ FOR DISPERSIVE CFAR OR PULSE-COMPRESSION WITH HARD-LIMITING BUT ADD 1 dB LIMITING LOSS TO RESULT

$M_{eff} = \frac{M+h}{1+h}$ FOR CELL-AVERAGING CFAR USING 'GREATEST-OF' SELECTION WITH

h = 0.37 FOR SQUARE-LAW DETECTOR

h = 0.5 FOR LINEAR ENVELOPE DETECTOR

h = 1.26 FOR LOG DETECTOR

FOR 'GREATEST-OF' SELECTION, NOISE IN M/2 RANGE CELLS ON BOTH SIDES OF SIGNAL RANGE CELL BEING EXAMINED FOR PRESENCE OF TARGET ARE AVERAGED SEPARATELY AFTER ENVELOPE DETECTION AND THE GREATER OF THE TWO AVERAGES USED FOR SETTING THE DETECTION THRESHOLD.

Figure 8 **Universal Curve of CFAR Loss vs. CFAR Ratio. (From Hansen [25, 26] with permission)**

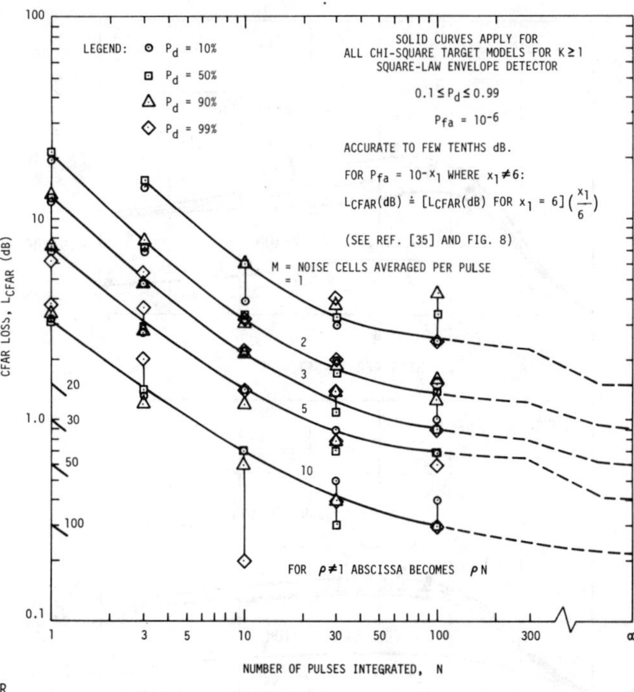

Figure 9 **CFAR Loss for Video Integration of N Pulses. (Data points plotted for Swerling Case 1 target model from results of Reference 27. Results for Swerling Case 2 and nonfluctuating target generally within few tenths of a decibel of those for Swerling Case 1 target model. Solid curve can be used to give CFAR loss for all chi-square target models of Table 2 for K ⩾ 1 when $0.1 \leqslant P_d \leqslant 0.99$. Results accurate to few tenths of a decibel. See also Reference 25.)**

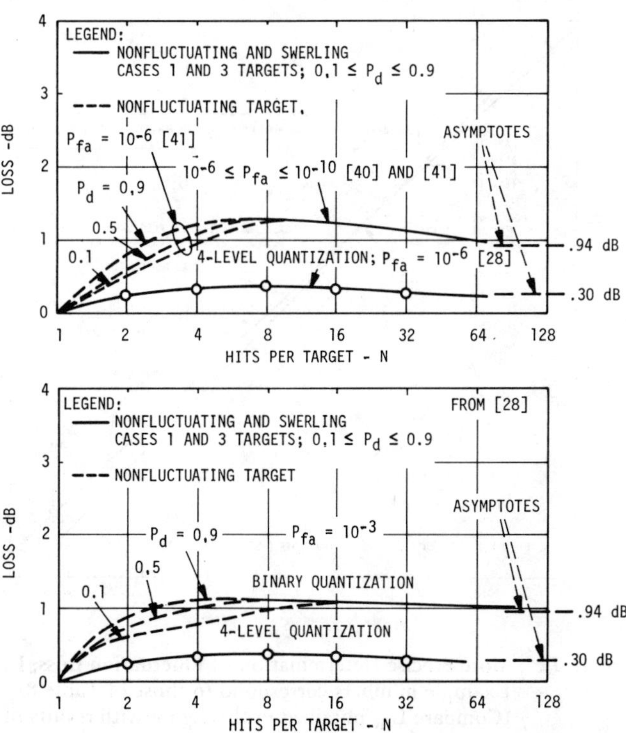

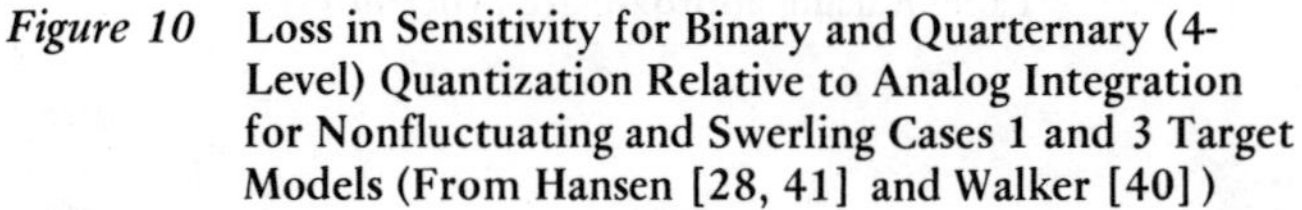

Figure 10 **Loss in Sensitivity for Binary and Quarternary (4-Level) Quantization Relative to Analog Integration for Nonfluctuating and Swerling Cases 1 and 3 Target Models (From Hansen [28, 41] and Walker [40])**

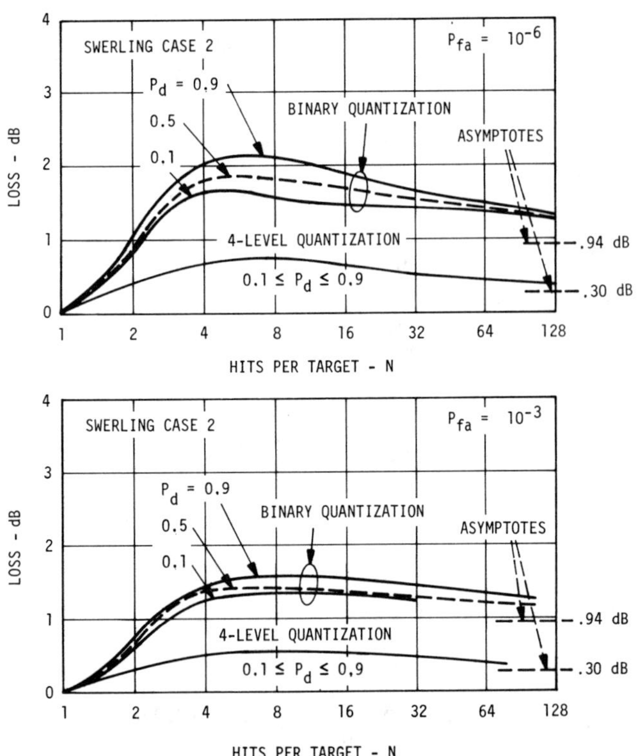

Figure 11 **Loss in Sensitivity for Binary and Quarternary (4-Level) Quantization Relative to Analog Integration for Swerling Case 2 Target Model. (From Hansen [28, 41] with permission)**

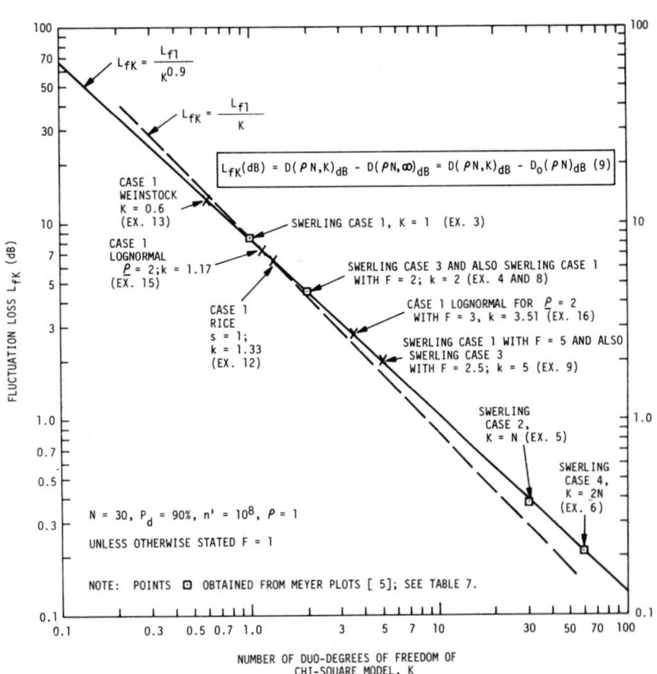

Figure 12 **More Precise Determination of Fluctuation Loss, L_{fK}. Example numbers correspond to those of Table 8. (Compare L_{fK} obtained in this figure with results of Table 8 using approximate procedure.)**

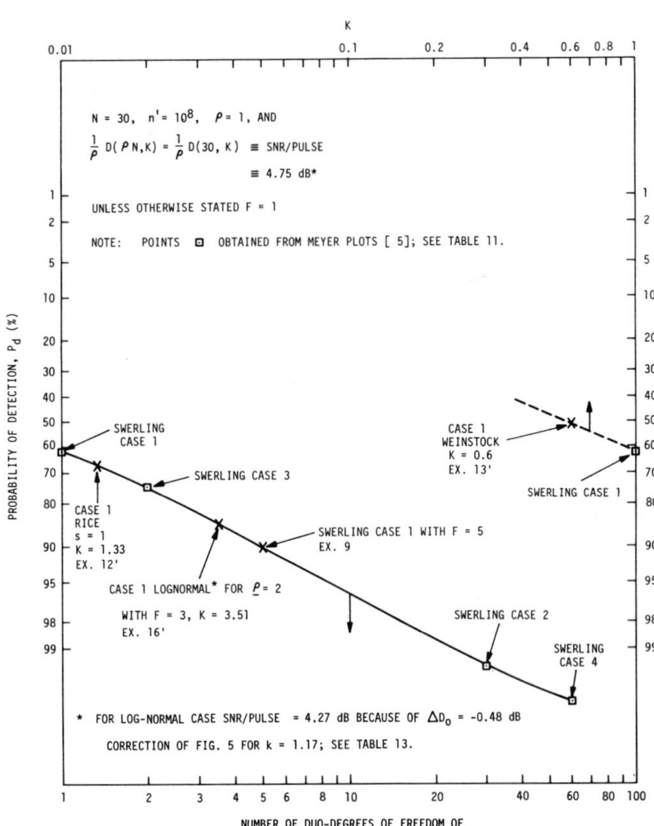

Figure 13 **Probability of Detection P_d vs. Number of Duo-Degrees K of Freedom for Chi-Square Model. Example Numbers Correspond to Those of Table 13.**

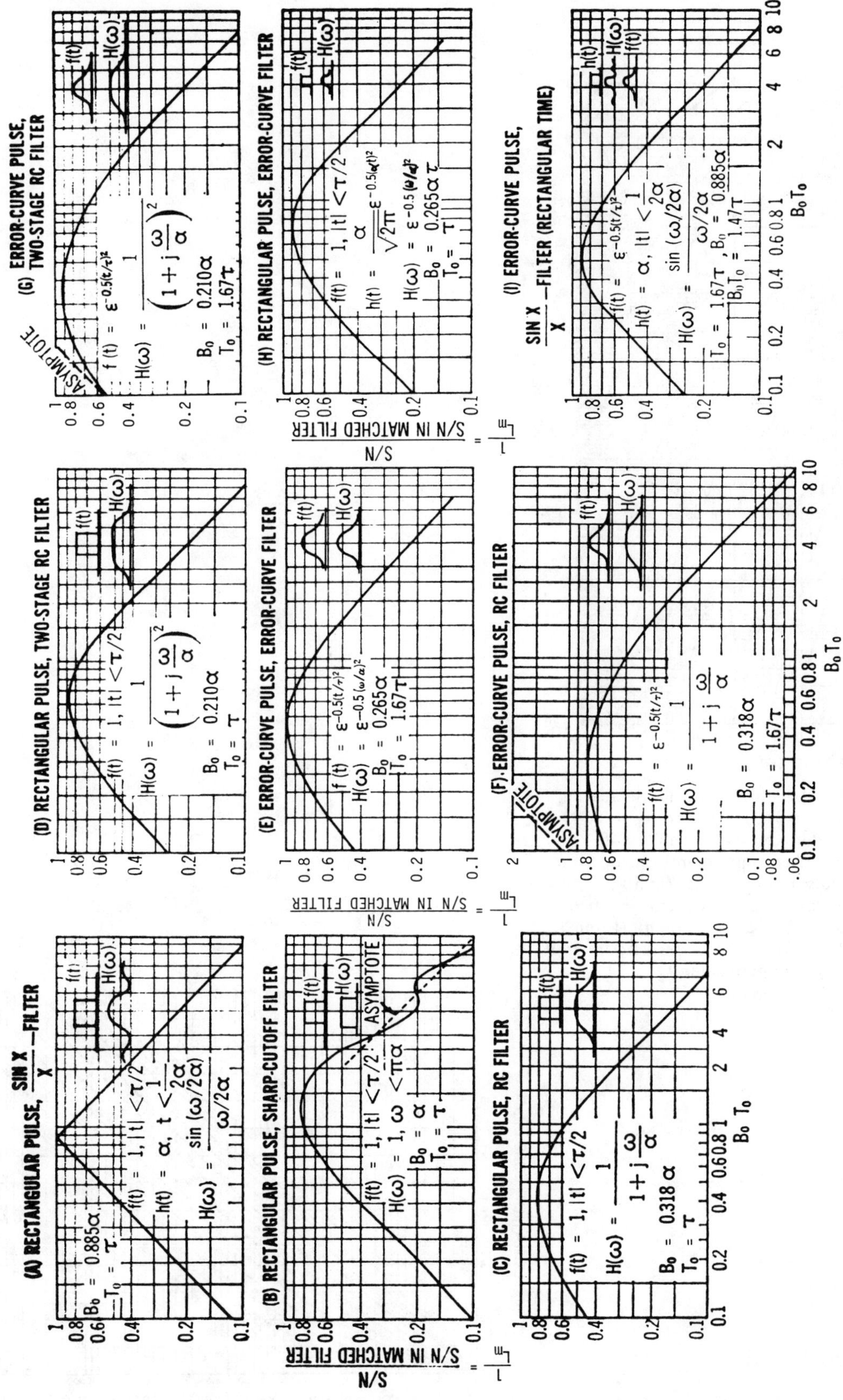

Figure 14 IF filter mismatch loss, L_m (actually $1/L_m$ as power ratio), for various IF filter transfer characteristics and various signal pulse shapes. S/N equals maximum IF output power signal-to-noise ratio; B_o, the effective bandwidth between the half-power points (in hertz); ω, radian frequency measured relative to IF center frequency; and T_o, the pulse duration between the half-power points (in seconds). (From Hall [38] with permission)

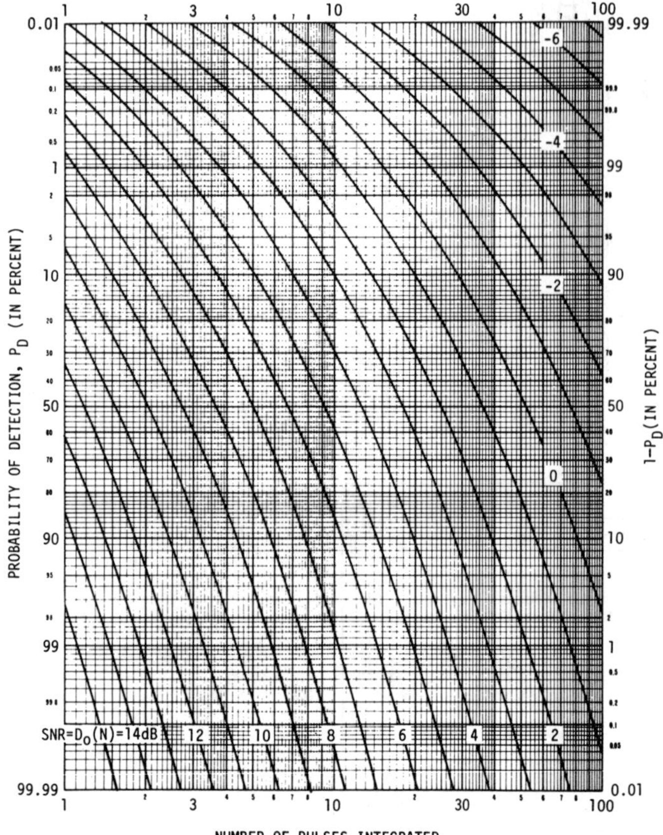

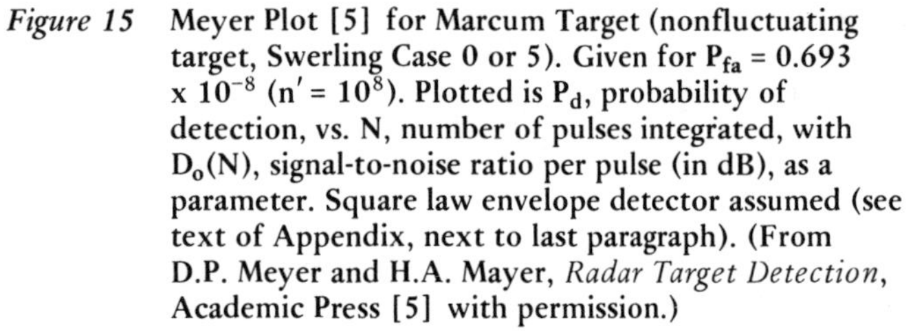

Figure 15 Meyer Plot [5] for Marcum Target (nonfluctuating target, Swerling Case 0 or 5). Given for $P_{fa} = 0.693 \times 10^{-8}$ ($n' = 10^8$). Plotted is P_d, probability of detection, vs. N, number of pulses integrated, with $D_o(N)$, signal-to-noise ratio per pulse (in dB), as a parameter. Square law envelope detector assumed (see text of Appendix, next to last paragraph). (From D.P. Meyer and H.A. Mayer, *Radar Target Detection*, Academic Press [5] with permission.)

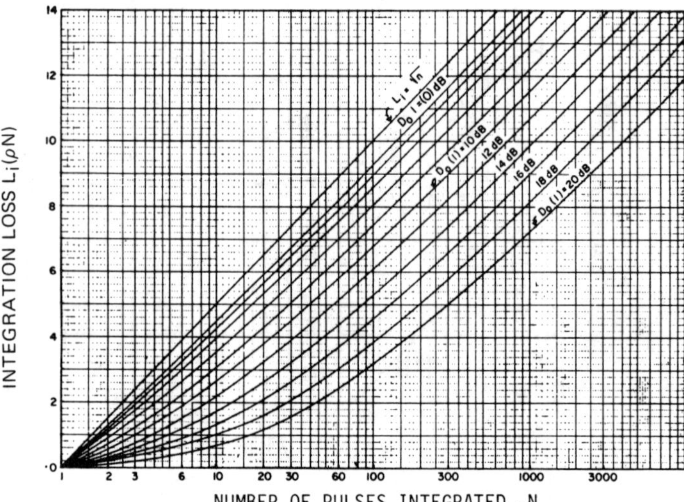

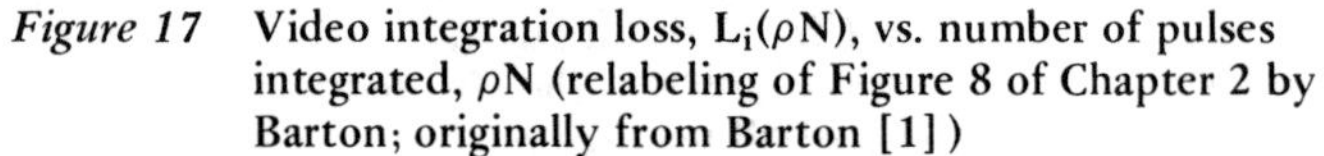

Figure 17 Video integration loss, $L_i(\rho N)$, vs. number of pulses integrated, ρN (relabeling of Figure 8 of Chapter 2 by Barton; originally from Barton [1])

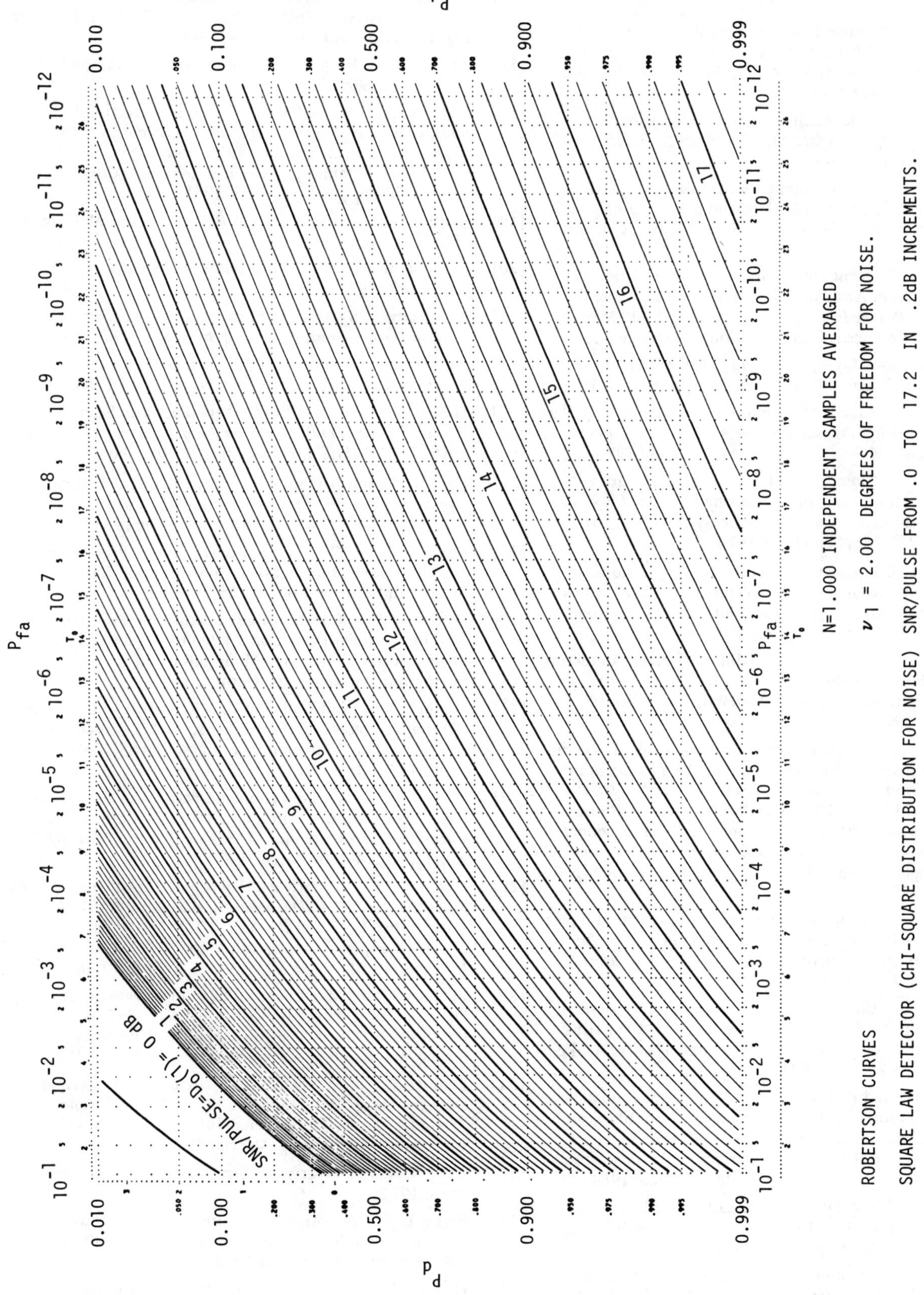

Figure 16 Robertson Curves [42] for Marcum Target (nonfluctuating target, Swerling Case 0 or 5) for Case of Single Pulse Detection (no video integration). P_d, probability of detection, vs. P_{fa}, probability of false alarm, with $D_o(1)$, single pulse signal-to-noise ratio (in dB), as parameter. Abscissa T_o is related to the normalized threshold voltage, Y_b (of Marcum [29], [29], Fehlner [30], and Meyer and Mayer [5]) at output of square-law envelope detector; $Y_b = R_b{}^2/2\sigma^2$ where R_b = envelope of IF voltage and σ is IF rms noise voltage. $Y_b = T_o + 1$; hence $P_{fa} = \exp - Y_b = \exp - (T_o + 1)$. (From Robertson [42] with permission)

References

[1] Barton, D.K.: "Simple Procedures for Radar Detection Calculation," *IEEE Transactions on Aerospace and Electronic Systems, Vol. AES-5, No. 5,* pp. 837-846, September 1969; see also Chapter 2.

[2] Cann, A.J.: "Simple Radar Detection Calculation," *IEEE Transactions on Aerospace and Electronic Systems, Vol. AES-8, No. 1,* pp. 73-74, January 1972.

[3] Swerling, P.: "Lecture Notes on Radar Target Signatures: Measurements, Statistical Models, and Systems Analysis," Technology Service Corporation, Santa Monica, California, August 1968.

[4] Swerling, P.: "Recent Developments in Target Models for Radar Detection Analysis," AGARD *Avionics Technical Symposium Proceedings,* Istanbul, Turkey, 25-29 May 1970; this reference is a condensation of Reference 3.

[5] Meyer, D.P.; and Mayer, H.A.: *Radar Target Detection, Handbook of Theory and Practice,* Academic Press, 1973.

[6] Neuvy, J.: "An Aspect of Determining the Range of Radar Detection," *IEEE Transactions on Aerospace and Electronic Systems, Vol. AES-6, No. 4,* pp. 514-521, July 1970.

[7] Urkowitz, H.: "Closed-Form Expressions for Noncoherent Radar Integration Gain and Collapsing Loss," *IEEE Transactions on Aerospace and Electronic Systems, Vol. AES-9, No. 5,* pp. 781-783, September 1973.

[8] Cann, A.J.: "Comment on Closed-Form Expressions for Noncoherent Radar Integration Gain and Collapsing Loss," *IEEE Transactions on Aerospace and Electronic Systems, Vol. AES-10, No. 2,* p. 295, March 1974.

[9] Brookner, E.: "Cumulative Probability of Target Detection Relationships for Pulse Surveillance Radars," *Raytheon Report,* Raytheon Company, Wayland, Massachusetts, November 1972.

[10] Brookner, E.: "Cumulative Probability of Target Detection Relationships for Pulse Surveillance Radars," *Journal of the Institution of Engineers* (India), *Vol. 51, No. 10, Part 3,* pp. 125-134, May 1971; also "Symposium on Radar Techniques and Systems," sponsored by Radar Engineering Group, Telecommunication Engineers Division, Institution of Engineers (India), New Delhi, India, *Abstract,* p. 7, 1-3 May 1970.

[11] Brookner, E.: "Cumulative Probability of Detection," part of Lectures Numbers 1 and 3, IEEE Boston Section Series on "Modern Radar Theory," Lexington, Massachusetts, October 1972.

[12] Brookner, E.: "Cumulative Probability of Target Detection Relationships for Pulse Surveillance Radars," 1973 IEEE International Symposium on Information Theory, Ashkelon, Israel, 25-29 June 1973.

[13] Mallett, J.D.; and Brennan, L.E.: "Cumulative Probability of Detection for Targets Approaching a Uniformly Scanning Search Radar," *Proceedings of the IEEE, Vol. 51, No. 4,* pp. 596-601, April 1963; and "Correction," *Proceedings of the IEEE, Vol. 52, No. 6,* pp. 708-709, June 1964.

[14] Trunk, G.V.: "Comparison of the Collapsing Losses in Linear and Square-Law Detectors," *Proceedings of the IEEE, Vol. 60, No. 6,* pp. 743-744, June 1972.

[15] Swerling, P.: "Probability of Detection for Fluctuating Targets," *RAND Corporation Report RM-1217,* March 1954; also reprinted as part of a special monograph in *IRE Transactions on Information Theory, Vol. IT-6, No. 2,* April 1960.

[16] Weinstock, W.: "Target Cross Section Models for Radar Systems Analysis," PhD Dissertation in Electrical Engineering, University of Pennsylvania, 1964.

[17] Nathanson, F.E.: *Radar Design Principles,* McGraw-Hill, New York, 1969.

[18] Heidbreder, G.R.; and Mitchell, R.L.: "Detection Probabilities for Log-Normally Distributed Signals," *IEEE Transactions on Aerospace and Electronic Systems, Vol. AES-3, No. 1,* January 1967; also *Aerospace Corporation Report TR-669(9990)-6,* April 1966.

[19] Skolnik, M.I.: *Introduction to Radar Systems,* McGraw-Hill, New York, 1962.

[20] Skolnik, M.I. (ed.): *Radar Handbook,* McGraw-Hill, New York, 1970.

[21] Berkowitz, R.S.: *Modern Radar Analysis, Evaluation and System Design,* John Wiley and Sons, New York, 1965.

[22] DiFranco, J.V.; and Rubin, W.L.: *Radar Detection,* Prentice-Hall, Englewood Cliffs, New Jersey, 1968.

[23] Hall, W.M.: "Antenna Beam-Shape Factor in Scanning Radars," *IEEE Transactions on Aerospace and Electronic Systems, Vol. AES-4, No. 3,* pp. 402-409, May 1968.

[24] Hall, W.M.; and Barton, D.K.: "Antenna Pattern Loss Factor For Scanning Radars," *Proceedings of the IEEE, Vol. 53, No. 9,* pp. 1257-1258, September 1965.

[25] Hansen, V.G.: "Constant False Alarm Rate Processing in Search Radars," International Conference on Radar – Present and Future, London, England, 23-25 October 1973.

[26] Hansen, V.G.: "Constant False Alarm Rate Processing and Automatic Radar Detection," Lecture Number 2 of IEEE Boston Section Series on "Modern Radar Technology," Lexington, Massachusetts, October 1973.

[27] Mitchell, R.L.; and Walker, J.F.: "Recursive Methods for Computing Detection Probabilities," *IEEE Transactions on Aerospace and Electronic Systems, Vol. AES-7, No. 4,* pp. 671-676, July 1971.

[28] Hansen, V.G.: "Optimization and Performance of Multilevel Quantization in Automatic Detectors," *IEEE Transactions on Aerospace and Electronic Systems, Vol. AES-10, No. 2,* pp. 274-280, March 1974.

[29] Marcum, J.I.: "A Statistical Theory of Target Detection by Pulsed Radar," *RAND Corporation Reports RM-754,* 1 December 1947 and *RM-753,* 1 July 1948; also reprinted as part of a special monograph in the *IRE Transactions on Information Theory, Vol. IT-6, No. 2,* April 1960.

[30] Fehlner, L.F.: "Marcum's and Swerling's Data on Target Detection by a Pulsed Radar," *Report TG-451,* Applied Physics Laboratory, The Johns Hopkins University, Silver Spring, Maryland; also *Report AD-602-121,* July 1962.

[31] Fehlner, L.F.: "Supplement to Marcum's and Swerling's Data On Target Detection by a Pulsed Radar," *Report TG-451A,* Applied Physics Laboratory, The Johns Hopkins University, Silver Spring, Maryland, September 1964.

[32] Whalen, A.D.: *Detection of Signals in Noise,* Academic Press, 1971.

[33] Blake, L.V.: "A Fortran Computer Program to Calculate the Range of a Pulse Radar," Naval Research Laboratory, NRL Report 7448, August 1972.

[34] Hall, W.M.: "Prediction of Pulse Radar Performance," *Proceedings of the IRE, Vol. 44, No. 2,* pp. 224-231, February 1956.

[35] Kaye, M.: "CFAR Loss With Integration in The Signal Channel," *RS-69-320,* Raytheon Company, Bedford, Massachusetts, October 1969.

[36] Barton, D.K.: *Radar System Analysis,* Artech House, Dedham, Massachusetts, 1976.

[37] Hovanessian, S.A.: *Radar Detection and Tracking Systems,"* Artech House, Dedham, Massachusetts, 1973.

[38] Hall, W.M.: "General Radar Equation," *Space/Aeronautics R and D Handbook,* 1962-1963.

[39] Barton, D.K.; and Hansen, V.G.: "Radar Signal Detectability Calculator," *Technical Memo DKB-70-44/VGH-70-15,* Raytheon Company, 24 September 1970; see also Barton, D.K. (ed.): *The Radar Equation,* Artech House, Dedham, Massachusetts, 1974.

[40] Walker, J.F.: "Performance Data for a Double-Threshold Detection Radar," *IEEE Transactions on Aerospace and Electronic Systems, Vol. AES-7, No. 1,* pp. 142-146, January 1971.

[41] Hansen, V.G.: *unpublished data.*

[42] Robertson, G.H.: *NEMAC (Normal Error Model Analysis Charts),* May 1977 (available from the author at PO Box 445, Summit, New Jersey, 07901).

Index